PACIFIC SYMPOSIUM ON
BIOCOMPUTING 2016

PACIFIC SYMPOSIUM ON BIOCOMPUTING 2016

Kohala Coast, Hawaii, USA,
4 – 8 January 2016

Edited by

Russ B. Altman
Stanford University, USA

A. Keith Dunker
Indiana University, USA

Lawrence Hunter
University of Colorado Health Sciences Center, USA

Marylyn D. Ritchie
The Pennsylvania State University, USA

Tiffany Murray
Stanford University, USA

Teri E. Klein
Stanford University, USA

World Scientific

NEW JERSEY • LONDON • SINGAPORE • BEIJING • SHANGHAI • HONG KONG • TAIPEI • CHENNAI • TOKYO

Published by

World Scientific Publishing Co. Pte. Ltd.
5 Toh Tuck Link, Singapore 596224
USA office: 27 Warren Street, Suite 401-402, Hackensack, NJ 07601
UK office: 57 Shelton Street, Covent Garden, London WC2H 9HE

ISSN: 2335-6936

British Library Cataloguing-in-Publication Data
A catalogue record for this book is available from the British Library.

BIOCOMPUTING 2016
Proceedings of the Pacific Symposium

ISBN 978-981-4749-41-1 (ebk)

Print-on-Demand in Singapore

WORKSHOPS

PACIFIC SYMPOSIUM ON BIOCOMPUTING 2016

2016 marks the 21st Pacific Symposium on Biocomputing (PSB)! This is an exciting time for Biocomputing—there are now a diversity of terms associated with the use of computer science, statistics and informatics to solve problems in biology and medicine. Most recently, we have seen a rise in the use of the term "biomedical data science." In the US, the National Institutes of Health (NIH) has recognized the importance of biomedical data science with the appointment in 2014 with Phil Bourne as Associate Director for Data Science. Don Lindberg recently stepped down as the Director of the National Library of Medicine (NLM). The NLM is a long-time supporter of biomedical data science (including biomedical informatics, clinical informatics, and bioinformatics). We are grateful that NLM regularly has supported PSB as well. This summer, an NIH working group wrote a vision for the future of NLM that calls for it to be the "epicenter of data science at NIH." All these developments are good for the field—which has evolved from a niche activity among (visionary) investigators to a mission-critical effort in support of biological and medical science. While some are worried about shifting labels (biocomputing, biomedical informatics, biomedical data science, etc.), the core challenges to the field remain the same, and the new labels simply reflect the influx of new talent and the need periodically for some rebranding. We are comfortable with the large umbrella and the legacy of the term "biocomputing." There are no plans to rename PSB!

The mission of PSB is to provide a forum for the best *emerging* science in Biocomputing, providing both formal and informal mechanisms for scientific communication—with an emphasis on work in the pacific rim. PSB depends on the community to define emerging areas in biomedical computation. Its sessions are usually conceived at the previous PSB meeting as people discuss trends and opportunities for new science. The typical program includes sessions that evolve over two to three years as well as entirely new sessions. This year we revisit new dimensions of precision medicine (which continues to advance at dazzling speed since the announcement of the US Precision Medicine Initiative by President Obama at the State of the Union address), and add new sessions on reproducibility and social media.

In addition to being published by World Scientific and indexed in PubMED, the proceedings from all PSB meetings are available online at http://psb.stanford.edu/psb-online/. PSB has published more than 800 papers. These papers are often cited in journal articles and represent early contributions in emerging subfields—many times before there is an established literature in more traditional journals; for this reason, many papers have garnered hundreds of citations. The Twitter handle PSB 2016 is @PacSymBiocomp and the hashtag this year will be #psb16.

The efforts of a dedicated group of session organizers have produced an outstanding program, including introductory tutorials. The sessions of PSB 2016 and their hard- working organizers are as follows:

Discovery of Molecularly Targeted Therapies
Philip R.O. Payne, Kun Huang, Nigam Shah

Innovative Approaches to Combining Genotype, Phenotype, Epigenetic, and Exposure Data for Precision Diagnostics
Melissa Haendel, Nicole Washington, Maricel Kann

Methods to Enhance the Reproducibility of Precision Medicine
Arjun Manrai, Chirag Patel, Nils Gehlenborg, Nicholas Tatonetti, John Ioannidis, Isaac Kohane

Precision Medicine: Data and Discovery for Improved Health and Therapy
Bruce Aronow, Steven Brenner, Sean Mooney, Alexander Morgan

Regulatory RNA
Drena Dobbs, Steven Brenner, Robert Jernigan, Alain Laederach, Vasant Honavar, Quaid Morris

Social Media Mining for Public Health Monitoring and Surveillance
Graciela Gonzalez, Matthew Scotch, Karen Smith, John Brownstein, Abeed Sarker, Michael Paul, Azadeh Nikfarjam

We are also pleased to present five workshops in which investigators with a common interest come together to exchange results and new ideas in a format that is more informal than the peer-reviewed sessions. For this year, the workshops and their organizers are:

Computational Approaches to Study Microbes and Microbiomes
Casey Greene, James Foster, Bruce Stanton, Deborah Hogan, Yana Bromberg

Biomedical Applications of Topology and Abstract Algebras
Eric Neumann, Svetlana Lockwood, David Spivak, Bala Krishnamoorthy

Use of Genome Data in Newborns as a Starting Point for Life-Long Precision Medicine
Steven E. Brenner and Sean D. Mooney

Translational Bioinformatics 101
Jessica D. Tenenbaum, Subha Madhavan, Robert Freimuth, Josh Denny, Lewis Frey

Social Media Mining Shared Task Workshop
Graciela Gonzalez, Abeed Sarker, and Azadeh Nikfarjam

We thank our keynote speakers Nancy Cox (Science keynote) and Winter Mason (Ethical, Legal and Social Implications keynote).

Tiffany Murray has managed the peer review process and assembly of the proceedings since 2003, and also plays a key role in many other aspects of the meeting. We are grateful for the support of the Institute for Computational Biology, a collaborative effort of Case Western Reserve University, the Cleveland Clinic Foundation, and University Hospitals their support of PSB 2016. We also thank the National Institutes of Health, the National Science Foundation, and the International Society for Computational Biology (ISCB) for travel grant support. We are particularly grateful to the onsite PSB staff Al Conde, Brant Hansen, Georgia Hansen, BJ Morrison-McKay, Jackson Miller, Kasey Miller, and Paul Murray for their assistance. We also acknowledge the many busy researchers who reviewed the submitted manuscripts on a very tight schedule. The partial list following this preface does not include many who wished to remain anonymous, and of course we apologize to any who may have been left out by mistake.

We look forward to a great meeting once again. Aloha!

Pacific Symposium on Biocomputing Co-Chairs,
October 15, 2015

Russ B. Altman
Departments of Bioengineering, Genetics & Medicine, Stanford University

A. Keith Dunker
Department of Biochemistry and Molecular Biology, Indiana University School of Medicine

Lawrence Hunter
Department of Pharmacology, University of Colorado Health Sciences Center

Teri E. Klein
Department of Genetics, Stanford University

Marylyn D. Ritchie
Department of Biochemistry and Molecular Biology, Pennsylvania State University

Thanks to the reviewers...

Finally, we wish to thank the scores of reviewers. PSB aims for every paper in this volume to be reviewed by three independent referees. Since there is a large volume of submitted papers, paper reviews require a great deal of work from many people. We are grateful to all of you listed below and to anyone whose name we may have accidentally omitted or who wished to remain anonymous.

Zach Abrams
Denis Agniel
Shandar Ahmad
Frank Albert
Dvir Aran
Paul Avillach
Chloe-Agathe Azencott
Andrew Beam
Steve Bedrick
Adrian Benton
Elmer Bernstam
Gaurav Bhatia
Olivier Bodenreider
Paul Bryar
Janusz Bujnicki
Binghuang Cai
José Caldas
Fabien Campagne
Emidio Capriotti
Peter Carbonetto
Gregory Carter
Jake Chen
James Chen
Jonathan Chen
Rong Chen
Shi-Jie Chen
Elissa Chessler
Kevin Cohen
Brian Connolly
Adrien Coulet
David Crosslin
Christian Darabos
Dan Day
Phillip Dexheimer
Drena Dobbs
Joel Dudley
Jesse Eickholt
Olivier Elemento
Ehsan Emazadeh
Alal Eran
Shahim Essaid
Samuel Flores
Robert Freimuth
Nils Gehlenborg
Anthony Gitter
Ben Glass
Rachel Goldfeder
Bryan Gorman
Peyton Greenside
Tudor Groza
Melissa Haendel
Hao Han
Xu Han
Zhi Han
Sam Handelman
Steven Hart
Winn Haynes
Elena Helman
Aron Henrikssson
Daniel Himmelstein
Lynnette Hirschman
Gabriel Hoffman
Vasant Honavar
Kun Huang
Lei Huang
Youngsook Jung
Konrad Karczewski
Rebekah Karns
Jonathan Karr
Purvesh Khatri
Warren Kibbe
Brian Kidd
Daniel Kim
Keiichi Kodoma
Sek Won Kong
Ioannis Korkontzelos
Diane Korngiebel
Jens Roat Kultima
Alain Laederach
Chirag Lakhani
Nicholas Larson
Allison Lazard
Ross Lazarus
Robert Leaman
Nathan Lewis
DingCheng Li
Fuhai Li
Haiquan Li
Li Li
Yue Li
Jianminmg Liang
Olivier Lichtarge
Xiangtao Liu
Luliana Lonita-Laza
Bo Luo
Yuan Luo
Yves Lussier
Yael Mandel-Gutfreund
Arjun Manrai
Dan Masys
Sergio Matos
Shannon McWeeney
Tom Metz
Cristian Micheletti
Julie Mitchell
Alex Morgan
Jonathan Mortensen
Sara Mostafavi
Ruchit Nagar
Eric Neumman
Azadeh Nikfarjam
Tim O'Connor
Wilma Olson
Gustavo Parisi
Leopold Parts
Jyoti Pathak
Philip Payne
Peggy Peissig
Sarah Pendergrass
Thomas Peterson
Nico Pfeifer
Snehit Prabhu
Abhishek Pratap
John Quackenbush
Gerald Quon
Predrag Radivojac
Satyajeet Raje
Luke Rasmussen
Kelly Regan
Marylyn Ritchie
Peter Robinson
Karin Rodland
Andrey Rzhetsky
Janice Sabin
Nathan Salomonis
Mayur Sarangdhar
H. Andrew Schwartz
David Searls
Arash Shaban-Nejad
Nigam Shah
Brian Shirts
Franco Simonetti
Marina Sirota
Damian Smedley
Karen Smith
Thomas Smith
Tim Sterne-Weiler

Shanhkar Subramaniam
Tim Sweeney
Tasnia Tahsin
Nima Tajbakhsh
Suzanne Tamang
Peter Tarczy-Hornoch
Gian Gaetano Tartaglia
Jessie Tennenbaum
Paul Thompson
Ming-Hsiang Tsou
Lyle Ungar
Rohit Vashisht
Shankar Vembu
Jean-Phillippe Vert
Mauno Vihinen
Bjarni Vilhjalmsson
Byron Wallace
Yuhong Wang
Griffin Weber
Davy Weissenbacher
Chunchua Weng
Scott Williams
Rainer Winnenburg
Chunlei Wu
Rong Xu
Yun Xu
Kun-Hsing Yu
Judith Zaugg
Jie Zhang
Xiaobo Zhou

DISCOVERY OF MOLECULARLY TARGETED THERAPIES

KELLY REGAN
Department of Biomedical Informatics
The Ohio State University
Columbus, OH 43210
Kelly.Regan@osumc.edu

ZACHARY ABRAMS
Department of Biomedical Informatics
The Ohio State University
Columbus, OH 43210
Zachary.Abrams@osumc.edu

MICHAEL SHARPNACK
Department of Biomedical Informatics
The Ohio State University
Columbus, OH 43210
Michael.Sharpnack@osumc.edu

ARUNIMA SRIVASTAVA
Department of Biomedical Informatics
The Ohio State University
Columbus, OH 43210
srivatava.1@osu.edu

KUN HUANG
Department of Biomedical Informatics
The Ohio State University
Columbus, OH 43210
Kun.Huang@osumc.edu

NIGAM SHAH
Center for Biomedical Informatics Research
Stanford University
Stanford, CA 94305
nigam@stanford.edu

PHILIP R.O. PAYNE
Department of Biomedical Informatics
The Ohio State University
Columbus, OH 43210
Philip.Payne@osumc.edu

1. Introduction

The delivery of personalized healthcare is predicated on the application of the best available scientific knowledge to the practice of medicine in order to promote health, improve outcomes and enhance patient safety [1-3]. Unfortunately, current approaches to basic science research and clinical care are poorly integrated, yielding clinical decision-making processes that do not take advantage of up-to-date scientific knowledge [2-4]. Basic scientists investigating the biological basis for a given disease may regularly encounter synergistic effects spanning two or more bio-molecular entities or processes that can contribute to our understanding of the mechanisms underlying phenomena such as the etiologic basis of the targeted disease state or potential response to therapeutic agents [5]. However, systematic approaches to the use of that knowledge in order to directly inform the selection of targeted molecular therapies for "real world" patients are extremely limited [1, 3, 6-9]. There are an increasing number of multi-modelling and in-silico knowledge synthesis techniques that can provide investigators with the tools to quickly generate hypotheses concerning the relationships between entities found in heterogeneous collections of scientific data — for example, exploring potential linkages among genes, phenotypes and molecularly targeted therapeutic agents, thus enabling the "forward engineering" of treatment strategies based on knowledge generated via basic science studies [1, 4, 6, 10, 11]. Ultimately, the goal of such methodologies is to accelerate the identification of actionable research questions that can make direct contributions to clinical practice. Given increasing concerns over the barriers to the timely translation of discoveries from the laboratory to the clinic or broader population settings, such high-throughput hypothesis generation and testing is highly desirable [1, 4, 6, 8, 12]. These needs are particularly critical in numerous disease areas where the availability of new therapeutic agents is constrained, thus calling for the re-use and repositioning of existing treatments [13, 14].

In response to the challenges and opportunities enumerated above, there exits an emerging body of research and development focusing on multi-modeling approaches to the discovery of molecularly targeted therapies, including experimental paradigms spanning a spectrum from the identification of molecular targets for drugs, to the repurposing or repositioning of existing agents that utilize such targets, to the systematic identification of novel combination therapy regimens that amplify or enhance the effectiveness of their constituent components. This focus is motivated by recent and significant advances in the state of systems biology and medicine that have demonstrated that the ability to generate and reason across complex and scalar models is essential to the discovery of high-impact biologically and clinically actionable knowledge [1, 4, 12]. Such approaches are designed to overcome the limitations of reductionist approaches to scientific discovery, replacing decomposition-focused problem-solving with integrative network-based modeling and analysis techniques [4, 8]. Systems-level analysis of complex problem domains ultimately enables the study of critical interactions that influence health and wellness across a scale from molecules to populations, and are not observable when such systems are broken down into constituent components. The use of systems-level analysis methodologies is well supported by the foundational theory of vertical reasoning first proposed by Blois [15]. This theory holds that effective decision-making in the biomedical domain is predicated on the vertical integration of multiple, scalar levels of reasoning. This fundamental premise is the basis for a correlative framework put forth by Tsafnat and colleagues, which states that the ability to replicate expert reasoning relative to complex biomedical problems using computational agents (e.g., in-silico knowledge synthesis) requires the replication of such multi-scalar and integrative decision-making [16]. In order to achieve such an outcome, Tsafnat posits that multi-scalar decision-making in an in-silico context requires both: 1) the generation of component decision-making models at multiple scales; and 2) the similar generation of interchange layers that define important pair-wise connections between entities situated in two or more component models, often referred to as vertical linkages [16]. When such component models and interchange layers are combined in a computationally actionable format, they yield what can be referred to as a multi-model for a given domain that is able to satisfy the premises of Blois' vertical reasoning axiom, and therefore facilitate the replication of expert performance in a high-throughput manner [16]. Of note, this type of approach is extremely reliant upon graph-theoretic reasoning and representational models, using a network paradigm that allows for the application of logical reasoning operations spanning the entities and relationships that make up a multi-model [8]. Network paradigms have been regularly shown to be the ideal representational model for naturally occurring systems, such as the 'scale-free' networks encountered in biological and clinical phenomena [8]. At the most basic level, network-based multi-modeling across scales presents an elegant and computationally tractable approach to understanding and evaluating complex biological and clinical systems in order to discover the knowledge incumbent to such constructs. This type of approach benefits from a robust set of foundational theories and frameworks that can inform and shape the application of multi-modeling techniques to a variety of knowledge discovery use cases. As such, there is a growing body of evidence concerning the application of network-based approaches to multi-modeling with an emphasis on therapeutic agent discovery, re-positioning and molecular targeting. Examples of such evidence include reports and perspectives published by Hood and Perlmutter [1], Butcher and colleagues [12], and Lussier and Chen [13].

2. Overview of Session Contributions

The utility and impact of multi-modeling approaches to integrative biological and clinical analyses, including hypothesis discovery operations such as those related to the identification of molecularly targeted therapies as noted above, have been explored in a number of instances by the biology, computer science and translational bioinformatics communities. At a high level, the exemplary efforts made by authors contributing to this session of PSB 2016 provide a broad cross-section of such novel methods, and focus on: 1) the development of factorization-based models to traverse multiple large-scale database comprising types of drug-disease and drug-target relationships (**Zitnik *et al*** and **Regenbogen *et al***); 2) network-theoretic approaches in a variety of applications including: linking environmental risk factors for disease via systematic analysis of biological pathways (**Darabos *et al***), the prioritization of gene mutations causing drug resistance (**Verkhivker**), and the facilitation of viable community detection (**Yu *et al***); and 3) the incorporation of prior knowledge into in silico methods in order to optimize large-scale regression-based association studies (**Verma** *et al*) and to discover dependencies between genes differ across disease conditions (**Speyer** *et al*). Brief synopses of these reports are provided below:

2.1 Factorization-based Models for Traversing Databases

Zitnik *et al* describe a novel collective pairwise classification (COPACAR) model for analysis of multi-relational data, including clinical manifestations of diseases, molecular interactions of diseases, drug-drug and drug-target interactions and drug-drug similarities. Their model combines factorization models that are optimized for large relational data with classification pairwise ranking loss for classification. Importantly, their model incorporates prior knowledge that is also scalable to highly complex, large-scale data. The authors address the issue of ranking in their predictions, where relationship are ranked according to their relevance, which is ideal for prioritizing large-scale, diverse relationships. They distinguish their approach to other widely recognized collective relational learning approaches optimized to minimize error rate are not well-suited to rank high-confidence relationships integral to applications of precision medicine and drug repurposing. The COPACAR method optimizes a ranking metric using pairwise classification in order to estimate latent factors of entities, which are use to parameterize the model's predictions about pairwise entity relationships. Another particularly significant contribution is the implementation of an application the authors term "category-jumping," which permits the generation of novel hypotheses relating heterogeneous biomedical entities that may be unrecognized by other models that rely on data of a single relation type. The authors demonstrated a widely observed phenomenon that shared clinical manifestations of disease, in particular high-level symptom characteristics, indicate shared molecular interactions (e.g. genetic associations and protein interactions). Finally, hierarchical clustering of the disease matrix demonstrated that diseases with sparse molecular information could be grouped to disease with molecular-rich relations based on clinical manifestations, thus resulting in novel hypotheses for molecular basis of these diseases.

Regenbogen *et al* address an important problem of extrapolating knowledge across diverse, large-scale sources for small-scale, high-resolution problems in personalized medicine, including individual patient drug prediction and drug repositioning. The authors employed a technique called collaborative filtering (CF), which is extensively used in online recommendation systems. Specifically, non-negative matrix factorization (NMF) was used to analyze knowledge of connections, rather than entity features, in order predict interactions among chemicals, genes, and diseases contained within the Comparative Toxicogenomics Database (CTD). Although NMF has been widely used in the analysis of genomics data and for predicting protein-protein and drug-target interactions, a particular novel contribution of this work is the authors' integration across multiple entity types. One benefit of this framework is that it can be easily extended to new entity classes without extensive pre-processing or abstraction, unlike other methods highly specific to entity attributes; however, it is limited to predict interactions among entities without details regarding how entities interact (e.g. directionality, causality, etc.). Their method was able to accurately predict protein-protein interactions in an independent database and successfully predict CTD entity relationships between successive versions of the database. Furthermore, integrating data across these two independent databases increased the performance of the CF method. Importantly and similar to Zitnik *et al*, the authors confirmed a high degree of precision in their results in addition to a high sensitivity, which is crucial to precision medicine and drug repurposing initiatives that focus on pursuing a small number of hypotheses relative to the total interaction space.

2.2 Network-theoretic Analyses

Darabos *et al* presents a methodology for determining the effect of environmental factors in complex diseases. This is an important problem to address since it is often difficult to distinguish environmental causality in disease development. The authors utilize a tripartite network linking diseases, environmental chemicals and biological pathways in order to identify potential biological effects of environmental chemicals relating to disease. This tripartite network allows for the connecting environmental factors with disease through shared biological processes. The utility of this model is demonstrated in one instance through the linkage of arsenic to multiple diseases through its role in disrupting signal transduction pathways. Overall, this work supports the use of multi-modeling network approaches to elucidate the effects of environmental exposure related to disease states. The authors also show how linking disparate datasets together can help answer large-scale questions through creation of a hypothesis generating system that can help fuel future research areas such as population health and epigenetics.

Verkhivker investigates mechanisms of resistance to lapatinib caused by EGFR mutations. Using genetic and structural data, they are able to prioritize mutations by their ability to affect a residue interaction network, computed using molecular dynamics simulations. The centrality of the residue in the network predicts its ability to disturb the effect of EGFR inhibition. Their results provide a framework for understanding the spectrum of resistance causing mutations, with the added benefit of implying causality of the associated mutations. They suggest that a wide range of mutations within the EGFR protein could cause resistance to lapatinib therapy. Their simulations also recover known resistance mutations, further validating the success of their method.

Yu *et al* propose innovative extensions on the Markov clustering methodology for community detection in networks. While viable community detection has implications in a variety of fields, the authors propose an integrative methodology that is especially apt for garnering a holistic picture in biological networks. They propose two subsequent extensions to the well-known Markov Clustering and regularized Markov Clustering algorithms in order to, firstly, focus on information or influence flow in a non-exclusive manner (inverse regularized Markov Clustering – irMCL) and subsequently integrating network structure with node attributes of biological significance such as phenotype, gene expression or demographic information (attribute inverse regularized Markov Clustering-airMCL). The authors have ideated a method which allows for node attributes to be incorporated in the community detection paradigm, utilizing and weighing attributes with respect to their effect on inter and intra community information flow. They have modeled the connections between node attributes and network structure in way that is malleable with statistical classification approaches. They prove the validity and robustness of this method by employing it on a simulated as well as real world dataset, utilizing the requisite statistical models and measures for rigor. Their results showcase that the methodology was immune to weak attributes whereas attribute similarity that predicted the structure was highlighted. This eliminates the need for a user-based selection of attribute importance. In the real world Breast Cancer dataset, the algorithm was able to isolate a variety of pathways, including, but not limited to the cell cycle pathway, signal transduction pathway and ribosome biogenesis. Also, the modules isolated showed significant association with time to survival. The authors have aimed to examine and stratify attribute impact by its connection to network structure. This is a novel ideology that promotes multi-modal data integration without succumbing to formation of overly complex models. Finally, with the inclusion and use of classification methodologies in community detection, the authors plan to utilize the inherent classification properties to better select models and features for future work.

2.3 Incorporation of Prior Knowledge into in silico Methods

Verma ***et al*** describe a system of discovering associations utilizing a novel method called Phenome-wide interaction study (PheWIS), which builds on the authors' previous work with phenome-wide association studies (PheWAS). This work seeks to address the problem of discovering associations between single nucleotide polymorphisms (SNPs) and phenotypes on a large scale. The authors approach this problem of large-scale association assessment by modeling the variance of the SNPs. They identified genetic variants that are associated with multiple phenotypes by prioritizing previously published results from both genome-wide and phenome-wide association studies using the AIDS Clinical Trials Group (ACTG) and the Roadmap Epigenome project. They discovered that by filtering out variance from low functional regions of the genome they could conduct a pair-wise search using linear regression analysis to identify associations. With their system the authors were able to identify 50,798 statistically significant associations related to 26 different phenotypes. This work helps to demonstrate not only the importance of modeling genotypic and phenotypic information together but also shows the strength of utilizing previously published information to help inform novel hypothesis driven systems.

Speyer ***et al*** investigate the effect of injecting biological knowledge into a previously developed method, Evaluation of Differential Dependency (EDDY). Their method seeks to answer the question, how do dependencies between genes differ across conditions? They apply their method to the TCGA glioblastoma multiforme data, to find differential dependencies between proneural and nonproneural, and mesenchymal and non-mesenchymal tumors. The result is a list of gene sets whose dependencies most differ between two cancer subgroups. Specifically, they find that the mesenchymal subset is defined by changes to metabolic processes and the proneural subset is defined by changes to AKT-ERK signaling. These pathways are strongly implicated in cancer, which shows the power of this method to find cancer-related results. They compare their results to knowledge-fused differential dependency network (KDDN) and find that the EDDY method appears to be more sensitive to differential dependencies, although there is substantial overlap for a subset of pathways.

3. Discussion and Conclusions

The goal of PSB 2016 is to demonstrate advances relative to *"work in databases, algorithms, interfaces, natural language processing, modeling and other computational methods, as applied to biological problems, with emphasis on applications in data-rich areas of molecular biology."* Further *"a major goal of PSB is to create productive interaction among the rather different research cultures of computer science and biology."* The body of work represented by this session, focusing on the development and application of methods for the discovery of molecularly targeted therapies, is emblematic of the vigorous and highly productive exchange of knowledge and ideas surrounding the aforementioned foci. Further, the work summarized herein serves to emphasize:

1) *The state-of-the-art in terms of in-silico knowledge synthesis methods that can be used to identify, aggregate and instantiate component-level models and that can be used to construct application-specific multi-models for therapeutic targeting (e.g., having a specified disease or biological context);*
2) *Ongoing challenges and opportunities surrounding the creation of "interchange layers" and the execution of "vertical reasoning" tasks across and between scalar multi-models in order to generate hypotheses linking synergistic bio-molecular entities or processes of interest and correlative molecularly targeted therapeutic agents; and*
3) *Exemplary instances where the preceding theories and methods have been applied to create an "end to end solution" in which multi-modeling approaches have been used to generate scalar multi-models, identify hypotheses concerning molecularly targeted therapeutics informed by such multi-models, and ultimately evaluate those hypotheses using some combination of in-silico, laboratory, animal or human study paradigms.*

As such, these report amplify the highly promising future for the molecular targeting of therapeutics in a variety of disease states, all in support of what are ultimately envisioned as precision medicine paradigms with the ensuing benefits relative to the quality, safety, outcomes, and costs of such data-driven and adaptive healthcare.

.

References

1. Hood L, Perlmutter RM. The impact of systems approaches on biological problems in drug discovery. Nature Biotechnology. 2004;22(10):1215-7.
2. Auffray C, Chen Z, Hood L. Systems medicine: the future of medical genomics and healthcare. Genome Med. 2009;1(1):2.1-2.11.
3. Hood L, Friend SH. Predictive, personalized, preventive, participatory (P4) cancer medicine. Nat Rev Clin Onc. 2010(8):184-7.
4. Ahn AC, Tewari M, Poon CS, Phillips RS. The Limits of Reductionism in Medicine: Could Systems Biology Offer an Alternative? PLOS Medicine. 2006;3(6):709-13.
5. Jones PA, Baylin SB. The Epigenomics of Cancer. Cell. 2007(128):683-92.
6. Payne PR, Embi PJ, Sen CK. Translational Informatics: Enabling High Throughput Research Paradigms. Physiological Genomics. 2009.
7. Fitzgerald JB, Schoeberl B, Nielsen UB, Sorger PK. Systems biology and combination therapy in the quest for clinical efficacy. Nature Chemical Biology. 2006;2(9):458-66.
8. Barabasi AL, Oltvai ZN. Network Biology: Understanding The Cell's Functional Organization. Nature Reviews Genetics. 2004;5(February):101-13.
9. Anastassiou D. Computation analysis of the synergy among multiple interacting genes. Molecular Systems Biology. 2007;3(83):1-8.
10. Schadt EE, Bjorkegren JL. Network-enabled wisdom in biology, edicine, and health care. Science Translational Medicine. 2012;4(115):115rv1
11. Payne PR, Johnson SB, Starren JB, Tilson HH, Dowdy D. Breaking the translational barriers: the value of integrating biomedical informatics and translational research. J Investig Med. 2005 May;53(4):192-200.
12. Butcher EC, Berg EL, Kunkel EJ. Systems biology in drug discovery. Nature Biotechnology. 2004;22(10):1253-9.
13. Lussier YL, Chen JL. The Emergence of Genome-Based Drug Repositioning. Science Translational Medicine. 2011;3(96):1-3.
14. Ainsworth C. Networking for new drugs. Nature Medicine. 2011(17):1166-8.
15. Blois M. Medicine and the nature of vertical reasoning. N Engl J Med. 1988;381(13):847-51.
16. Tsafnat G, Coiera EW. Computational Reasoning across Multiple Models. J Am Med Inform Assoc. 2009;16(6):768-74.

AN INTEGRATED NETWORK APPROACH TO IDENTIFYING BIOLOGICAL PATHWAYS AND ENVIRONMENTAL EXPOSURE INTERACTIONS IN COMPLEX DISEASES

CHRISTIAN DARABOS[1,2], JINGYA QIU[1], JASON H. MOORE[2]

[1] *The Geisel School of Medicine, Dartmouth College*
Lebanon, NH 03756, U.S.A.
[2] *The Perelman School of Medicine, University of Pennsylvania*
Philadelphia, PA 19104, U.S.A
E-mail: jhmoore@upenn.edu

Complex diseases are the result of intricate interactions between genetic, epigenetic and environmental factors. In previous studies, we used epidemiological and genetic data linking environmental exposure or genetic variants to phenotypic disease to construct Human Phenotype Networks and separately analyze the effects of both environment and genetic factors on disease interactions. To better capture the intricacies of the interactions between environmental exposure and the biological pathways in complex disorders, we integrate both aspects into a single "tripartite" network. Despite extensive research, the mechanisms by which chemical agents disrupt biological pathways are still poorly understood. In this study, we use our integrated network model to identify specific biological pathway candidates possibly disrupted by environmental agents. We conjecture that a higher number of co-occurrences between an environmental substance and biological pathway pair can be associated with a higher likelihood that the substance is involved in disrupting that pathway. We validate our model by demonstrating its ability to detect known arsenic and signal transduction pathway interactions and speculate on candidate cell-cell junction organization pathways disrupted by cadmium. The validation was supported by distinct publications of cell biology and genetic studies that associated environmental exposure to pathway disruption. The integrated network approach is a novel method for detecting the biological effects of environmental exposures. A better understanding of the molecular processes associated with specific environmental exposures will help in developing targeted molecular therapies for patients who have been exposed to the toxicity of environmental chemicals.

Keywords: Exposure; Complex Diseases; Chemical Agents; Biological Pathways; Human Phenotype Network.

1. Introduction

Complex diseases are believed to be the result of non-linear genetic, epigenetic and environmental interactions. Epistatic and pleiotropic genetic interactions, though ubiquitous in nature, only explain a fraction of disease occurrences.[1] Both acute and prolonged exposure to environmental factors such as chemical agents present in water, soil, or air also contribute to human disease.[2] GWAS partially reveal causal genetic interactions in complex diseases. Well-established studies of specific chemical agents link tobacco smoke to cardiovascular and respiratory diseases, and asbestos dust to several types of cancer. Integrating data from multiple sources helps to gain a better understanding of the way genetic risk factors, environmental exposures, as well as lifestyle choices all contribute to causing complex diseases.

Human phenotypes, including physical traits, diseases and behaviors, have been successfully linked through their shared biology to form networks of diseases. These networks and the interactions they reveal have been thoroughly studied using mathematical and statistical analyses.[3,4] Indeed, networks offer a comprehensive array of analytical tools while at the same time providing an intuitive representation of interactions within otherwise inextricably complex data.[5] Additionally, the concept of *exposome*[6] encompasses all human environmental exposures and complements the genome in predicting complex disease.

At a systems biology level, the interaction between genetic predisposition and environmental factors is poorly understood. The discovery of novel personalized molecular drugs that target specific pathways rely on the development of methods to study the intricate interactions between our environment and the biological pathways that govern human complex disease. The focus of this work is to provide a novel tool to identify candidates for potential environmental chemical agent and biological pathway interactions.

The sheer combinatorial complexity of chemical agents and pathways drives us to explore new approaches that prioritize the interaction of potential interest. Therefore, we propose to build an extension of the Human Phenotype Network (HPN)[7] based on biological pathway interactions, and overlay it with the HPN based on environmental exposure data.[8] The resulting model is a tripartite network constituted of three different types of vertices: human phenotypes, biological pathways, and environmental chemical agents. By projecting the network onto the space of human phenotypical traits, we are able to identify disorders that share only biological pathways, those that share environmental factors, and those that interact both at the environmental and the genetic level. We speculate that by integrating pathways and environmental exposure data in a single network, we are able to generate plausible hypotheses about the disruptive nature of chemical agents on certain biological pathways.

We analyze the resulting integrated networks both in quantitative and qualitative terms. We show how focusing on the double-edges of disorders that share both environmental and genetic origins can help identify potential candidates for environmental chemical agent and biological pathway interactions.

2. Methods

The expansion of systems biology has given rise to a trend towards studying disease from a global perspective, reaching beyond the silos of traditional medicine. Graphs, or network, are commonly used to study the interactions between phenotype and genotype. In the Human Disease Network (HDN),[3] or its extension, the Human Phenotype Network,[7] nodes representing diseases and phenotypes are linked by edges that represent various connections between disorders. These connections can be established by identifying shared causal genes,[3] genetic variants (SNPs),[9] linkage-disequilibrium SNP clusters,[7] biological pathways,[4] or clinical symptoms.[10] The underlying connections of these networks contribute to the understanding of the molecular basis of disorders, which in turn lead to a better understanding of human disease.

In previous works, we presented the concept of Human Phenotype Networks (HPNs), which represent the interactions between human traits and diseases based on their shared biological background, such as SNPs, genes, or pathways.[4,7] This approach has proven useful in analyzing epistatic and pleiotropic effects at the systems level.[11] Additionally, we have proposed an extension to the HPN based on shared environmental chemical agents.[8] When considered separately, both environmental and genetic HPNs are bipartite networks[5] composed of two distinct sets of vertices. Edges can only connect members of the opposite set. Bipartite networks can be projected in the space of either vertex set. Projecting the network increases the readability and interpretability of the data represented, but results in information loss. Figure 1 shows a schematic representation of a bipartite network in the center panel (b) and the resulting projection in either the space of *circle* vertices (a) and the space of *rectangle* vertices (c). In the case of the genetic HPN presented below, the vertex sets are composed of diseases and biological pathways. In the environmental HPN, the vertex sets are composed of diseases and chemical substances.

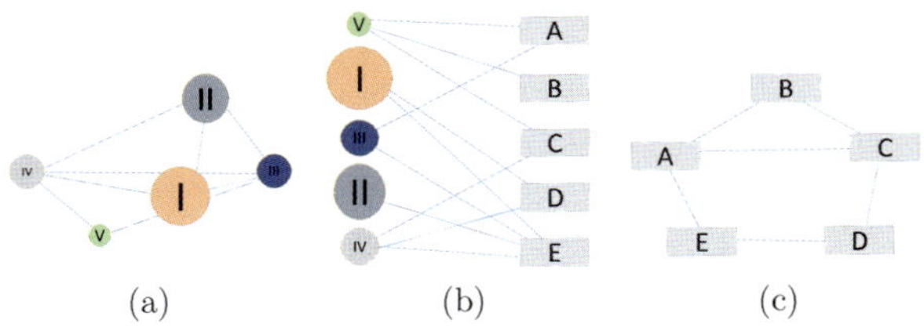

Fig. 1. Schematic representation of a Bipartite Network (b) and its projection in the space of either vertex set (a) and (c).

Because both HPNs share the disease vertex set, we can combine the two HPNs into a single "tripartite" network composed of three distinct vertex sets: traits, biological pathways, and chemical agents. Figure 2 represents a tripartite network (a) and its projection onto the *rectangle* vertex set (b). In tripartite networks, the edges are also divided into two categories. In our example, the blue edges only connect *circle* and *rectangle* vertices, whereas the red edges connect *rectangles* to *diamonds*. The resulting projection network has vertices linked by blue edges, red edges or both red and blue edges. Naturally, a tripartite network can be projected onto the space of either vertex set.

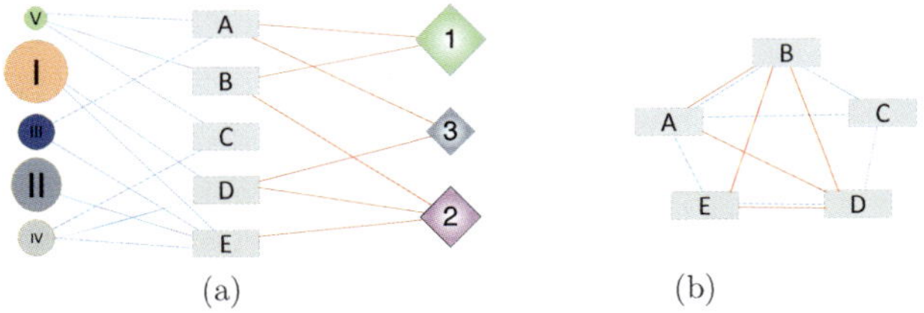

Fig. 2. Schematic representation of a Tripartite Network (a) and its projection in the space of the "rectangle" vertex set (b).

In the following sections, we discuss the two bipartite HPNs and the third novel tripartite HPN that combines exposure and genetic data.

2.1. *Human Phenotype Network based on Exposure Data*

In our previous study, we proposed a novel approach to bridging the gap between environmental exposure data and information on the diseases they may cause.[8] To the best of our knowledge, the exposure-to-disease data has not been aggregated in publicly accessible sources. To establish possible causal effects at a global level, we integrated data from the CDC's Fourth National Report on Human Exposure to Environmental Chemicals (`http://www.cdc.gov/exposurereport/`), including its subsequent updated tables, and the data of the NHGRI GWAS Catalog, accessed on 05/06/2014. Through a meticulous *PubMed* and *Google Scholar* literature survey, we compile a list of the diseases and traits that have been associated with any 60 environmental chemicals of the CDC's report. The CDC has identified these chemical agents as potentially harmful to human health and categorized them into 11 groups such as tobacco smoke, heavy metals, pesticides, etc. Figure 8 (X-axis) recapitulates all the chemical agents and their group in square brackets. Causal association between a chemical substance and a disease is based on compelling evidence found in the literature and confirmed in multiple studies, limiting uncertain associations to a minimum. We subsequently use the phenotype list from the GWAS catalog and the International Classification of Diseases Ninth Revision (ICD-9) codes to classify all traits and eliminate redundancies. Our survey inventories 548 well-established causal effects between these 60 substances and 151 human phenotypic traits and disorders. We note, however, that the data collected might contain a bias towards phenotypes and exposures that are more heavily studied.

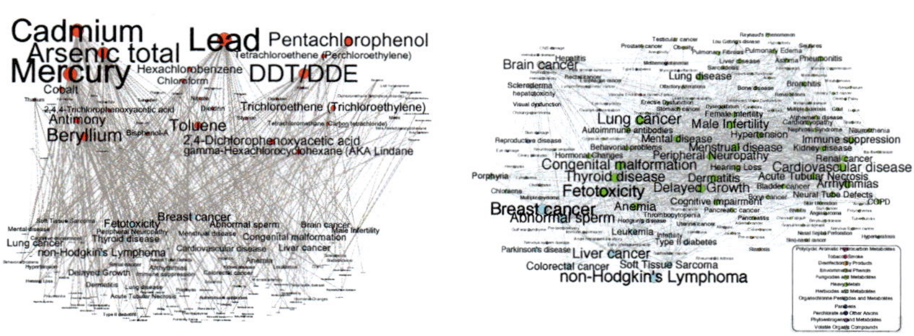

(a) (b)

Fig. 3. Phenotype-Substances Network. (a) The Bipartite Network. Top row, red vertices: environmental chemical substances. Bottom row, blue vertices: human phenotypes and diseases. Vertex size is proportional to the degree. (b) Projections onto the Phenotype Space. Nodes are colored according to their (majority) substance group according to the legend. Node sizes are proportional to the number of substances associated. Edge weights and width represent the number of shared substances.

The data aggregated in the survey is arranged in a bipartite network of diseases and environmental chemical compounds linked by "probable causality" edges. The resulting graph is depicted in Figure 3(a). This bipartite network shows the 548 relationships between the 60 chemical substances (top row, red vertices) and the 151 human disorders (bottom row, light blue vertices). The node sizes are proportional to vertex degree, i.e. the number of connections to the opposite set of vertices. The resulting projection onto the disease space is presented in Figure 3(b), where edges display common chemical factors associated with disorders. Furthermore, each node in the network is annotated with the substance classification group(s) to which it belongs. In the case of chemicals, the annotation is straightforward, as each substance belongs to exactly one class. For diseases, we identify all groups that contain at least one causal substance. A detailed description of the environmental HPN and our findings is available in our previous study.[8] The projection onto the chemical substance space is not shown in this study to save space, but it can be found in our previous study.[8] Nodes are color coded according to their (majority) substance class. The phenotype network (b) has 151 nodes and is densely connected (average degree of 40+), where each edge signifies that the two diseases they connect are associated with one or more common chemical agents.

2.2. *Human Phenotype Network based on Biological Pathways*

In their seminal work, Goh *et al.*,[3] explored the Human Disease Network, limiting their analysis to the genes shared by different diseases. Another study by Li *et al.*[9] traced the genetic variants connecting disease traits. In 2009, Silpa Suthram *et al.*[12] analyzed diseases by their related messenger RNA in combination with the human protein interaction network. They found significant genetic similarities between certain diseases, some of which shared drug treatments. Also in 2009, Barrenas *et al.*[13] further studied the genetic architecture of complex diseases by doing a GWAS, and found that complex disease genes contribute less and are less represented than the single-gene diseases in the human interactome. In 2014, Zhou *et al.*[14] have presented yet another way of finding overlap in disease commonalities, that is, they link disorders that share symptoms.

In the present work, we expand on the biological SNP-based HPN presented first in our previous studies.[11,15] We update the data to the most recent versions of the GWAS catalog (05/15/2015), NIH database of Genotypes and Phenotypes (dbGaP), and the Kyoto Encyclopedia of Genes and Genomes (KEGG).[16] We integrate the 1,252 phenotype information from both sources, the over 37,000 SNPs annotation, and 16,000 gene/loci association, as well as the biological pathways data to build the present pathway-based HPN. By aggregating these associations, we were able to link phenotypes with shared pathways, i.e. with genes involved in the same pathways. Furthermore, we have used the International Classification of Diseases Ninth Revision (ICD-9) codes to classify all traits and identify redundancies. The HPN encompasses all phenotypes listed in the GWAS catalog and dbGaP, provided that they are connected to at least one other trait. It is comprised of 986 phenotypic traits, 1,424 biological pathways, and over 260,000 edges, with an average connectivity of 500+.

2.3. *Combining the Human Phenotype Networks: Tripartite and Projection*

The main focus of this work is to help identify potential candidates for environmental chemical agent and biological pathway interactions. These interactions can in turn guide the development of novel targeted and personalized therapies. To help tease out potentially relevant pathway-environment interactions out of all the possible combinations, we build the tripartite HPN by combining the pathway-based HPN and the environmental HPN into one graph. The resulting model is comprised of 2,529 vertices of three different types: 1,045 diseases (142 overlapping between the two HPNs), 1,424 biological pathways, and 60 environmental chemical substances. Moreover, the tripartite HPN includes two different types of interaction edges: about 80,000 disease-to-pathway and over 1,500 disease-to-substance links.

Because of the sheer size, density and complexity of the tripartite network, we choose to present only its projection onto the disease space in Figure 4. Red edges represent biological pathway interactions, green edges represent environmental chemical agent interactions, and blue edges are double edges that share both biological and environmental interactions.

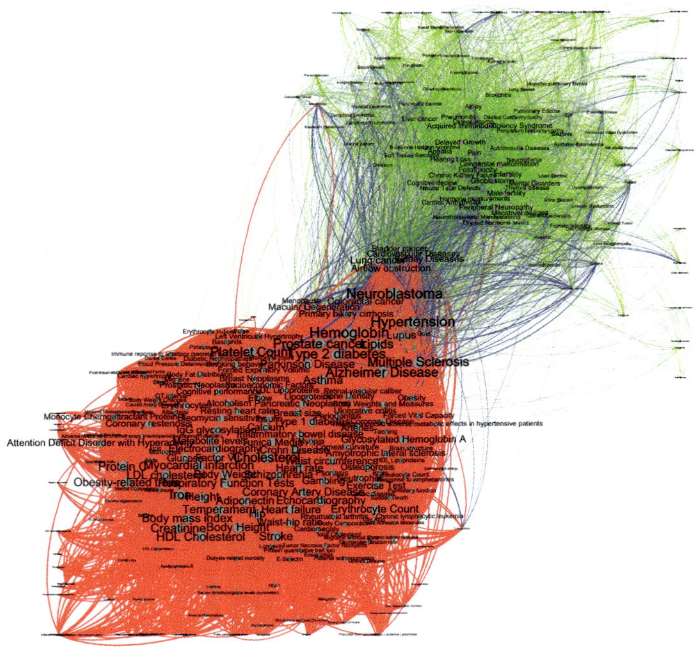

Fig. 4. Combined Human Phenotype Network based on Biological Pathways and Chemical Substances Exposure: Projection of the substance-phenotype-pathway tripartite network onto the phenotype space. Red edges represent pathway interactions only. Green edges show identified substance interactions. Finally, blue edges show pairs of traits that share both biological pathways and chemical substance exposure. The vertex and label size are proportional to its degree (i.e. the number of impinging edges).

To further facilitate the interpretation of the results and focus our analysis on shared genetic and environmental candidates, we filter the combined HPN to retain only the traits and diseases that have at least one edge of each kind impinging on them. In other words, we extract the subnetwork made of only blue edges and the vertices that are connected by those blue edges. The resulting HPN is presented in Section 3 below.

3. Results

In this section, we present the results of the quantitative analysis of the projected tripartite HPN. Quantitative network and graph analysis relies on strict statistical and mathematical tools and can be applied to networks of arbitrary size and complexity.[5] In this study, we focus on a subnetwork that shows the shared interactions between traits associated with both environmental and genetic factors. Therefore, we reduce the size and the complexity of the projected HPN to a manageable number of diseases and interactions in order to allow both quantitative, qualitative, and visual interpretation of the results. The final HPN integrating only vertices that share both genetic and environmental background, pictured in Figure 5, is composed of 74 phenotypes and 1,000 edges. The node (and label) size is proportional to the total number of associated environmental chemical agents; the color hue represents the number of biological pathways associated (green for fewer and red for more). The edge weight (i.e. its width) is proportional to the number of pathways shared between the disease endpoints whereas the color hue represents the number of shared environmental factors (green for fewer, red for more).

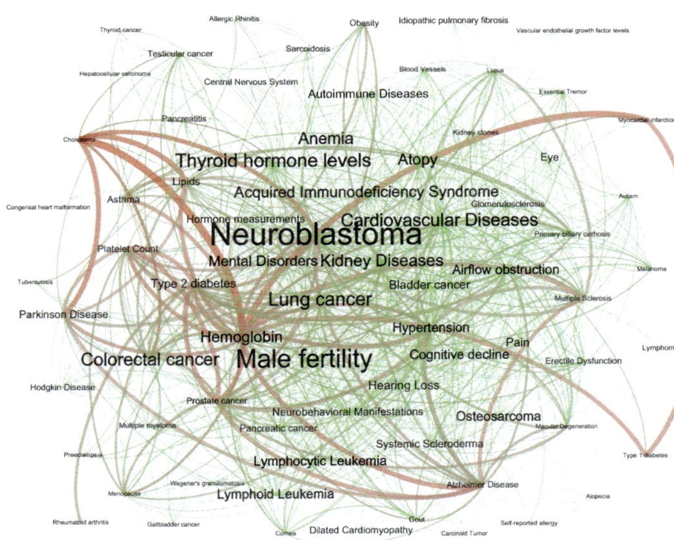

Fig. 5. Filtered Substance-Exposure HPN: Projection of the substance-phenotype-pathway tripartite networks onto the phenotype space filtered to retain only edges and vertices that share both substance and genetic interactions. Vertex size is proportional to the total number of associated substances; vertex color is proportional to the number of biological pathways associated (green for fewer and red for more). Edge width is proportional to the number of shared pathways; edge color is proportional to the number of shared substances (green for fewer, 1 pathway, and red for most: 10 pathways).

Further study of the individual edges and their distribution reveals that the vast majority of disease pairs are connected by heavy metals (59%), pesticides (20%) or organic compounds (7%). Mercury is potentially a common cause for almost 300 pairs of disorders, closely followed by lead, cadmium, DDT and arsenic. Figure 6 provides the detailed distribution of each substance of amongst the edges of the HPN. In the inset of Figure 6, the pie chart shows the same distribution by chemical agents classification groups, no by individual compound.

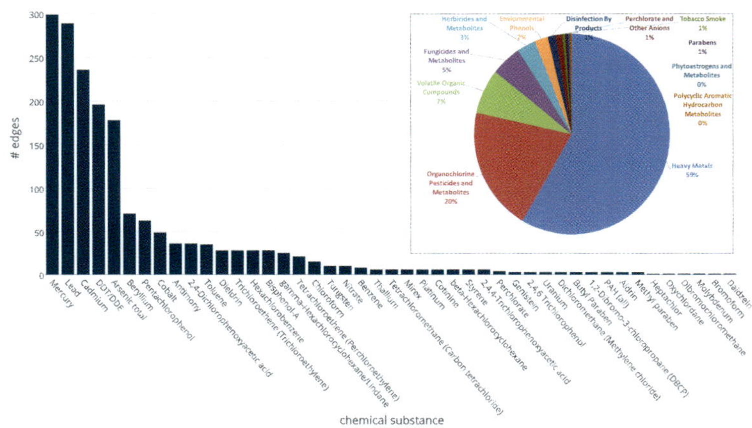

Fig. 6. Distribution of the Chemical Agents among the Edges within the Combined Network. The number of edges connecting two traits for each substance. Inset: the distribution of the substance classification groups among all edges in the combined network.

A similar study of the biological pathway edges reveals that the signal transduction, the immune system and metabolism pathways are the most represented. This comes as no surprise because these are "generic" pathways involving hundreds or even thousands of genes, therefore statistically highly probable to be represented more within the network. The complete breakdown of the 25 most represented biological pathways is shown in Figure 7.

Finally, we studied the distribution of biological pathway and chemical agent interactions within the projected network. The heatmap in Figure 8 shows the frequency of co-occurrence (double edge) for chemical substances and pathways between pairs of diseases.

In Figure 8, the biological pathways are approximately sorted by ascending frequency along the Y-axis. The chemical substances are arranged in groups along the X-axis. Heavy metals, in particular lead, cadmium, arsenic and mercury, and a pesticide (DDT) appear to interact with the most biological pathways. To assess the significance of these co-occurrences, we test the statistical probability of each existing pair in a null-model by running a 10,000-fold permutation test on all the edges of the tripartite network. For lack of space, we cannot present these data in detail. The results of the permutation test show that the most represented

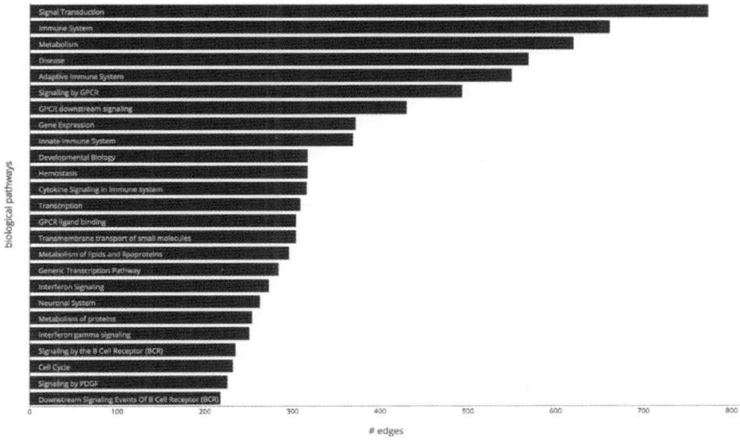

Fig. 7. Distribution of the biological pathways amongst the edges within the combined network. The number of edges connecting two traits for each the top 50 most represented pathways.

chemical agent-pathway pairs of have less than a 3% probability of occurring by chance.

4. Qualitative Observations, Biomedical Implications & Discussion

In this study, we integrated genetic and environmental exposure data in a tripartite network to identify interactions between environmental agents and biological pathways. Although the effects of environmental agents on disease have been studied extensively, the mechanisms of exposure are still poorly understood. Using the network approach, we aim to identify specific biological pathways disrupted by environmental agents. Identifying these pathways would not only help to establish more effective, more precise treatment therapies for patients who have been exposed, but can also provide insight to the mechanisms behind complex diseases.

To identify potential exposure-pathway interactions, we analyze overlapping edges in the final integrated HPN. Each edge distinguishes two phenotypes that are associated with the same environmental and genetic risk factors. We conjecture that the number of co-occurrences between pairs of environmental substances and biological pathways is correlated with a higher likelihood of an interaction between substance and pathway involved. Our permutation test shows that due to the combinatorial complexity of our HPN, the statistical probability of having our identified pathway-environment interaction occur by chance is generally below 3%.

The integrated network approach is a novel method for detecting the biological effects of environmental exposures. A better understanding of the molecular processes associated with specific environmental exposures will help in developing targeted molecular therapies for patients who have been exposed to the toxicity of environmental chemicals. We qualitatively analyze the HPN and propose possible biomedical applications. To establish the validity of

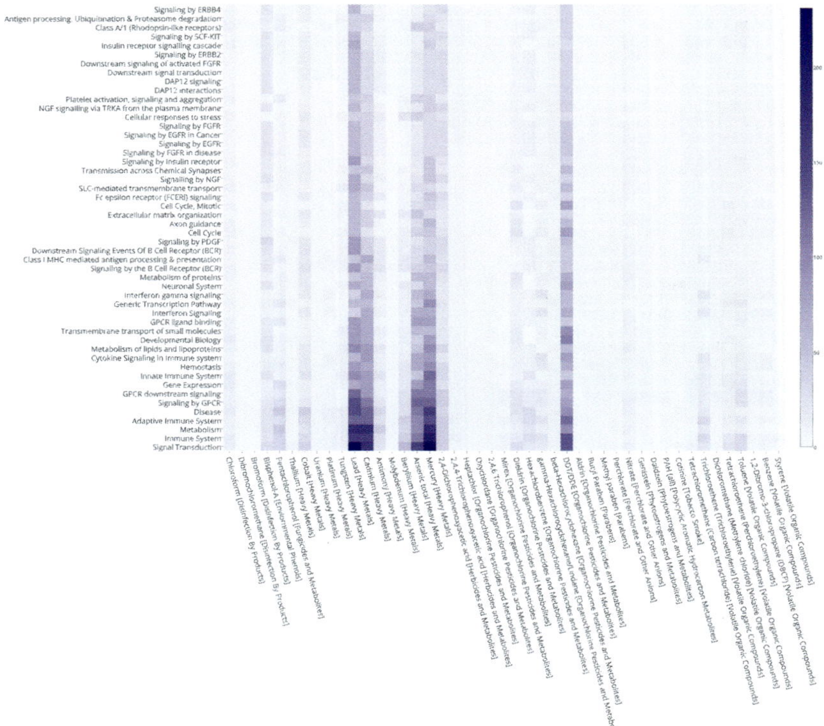

Fig. 8. Pathway-Substance Interaction Heatmap.

the HPN, we assess its ability to detect known environmental substance-pathway interactions. To construct the tripartite network, we used epidemiological data that associated environmental exposure and genetic data to phenotypic disease. In order to validate our network and generate new hypotheses, we used distinct publications of cell biology and genetic studies that associate environmental exposure to pathway disruption. There is no overlap between the publications we used to build the network and the publications we used to validate it and generate hypotheses.

Arsenic and signal transduction: Arsenic is a heavy metal toxin found naturally in the soil, minerals, and groundwater. Because of the many health risks it poses for humans, arsenic and its associated molecular mechanisms have been investigated extensively. By now, it is well known that arsenic severely disrupts signal transduction pathways.[17,18] Thus, we expect to see arsenic exposure overlap with signal transduction pathways at a high frequency in the HPN. Indeed, arsenic occurred most frequently in conjunction with "signaling by GPCR" and "GPCR downstream signaling", with a combined 225 co-occurrences and an approximate

1.4% chance co-occurrence, and less than 2.3% chance respectively. Arsenic exposure also had a high number of co-occurrences with more specific signaling transduction pathways such as T-cell receptor signaling (28 co-occurrences, 1% probability)[19] and B-cell receptor signaling (36 co-occurrences, 1.2% probability),[20] both of which have been supported by scientific literature.

4.1. *Generating hypotheses for substance-pathway interactions*

Beyond the HPN's capability of replicating recognized environment-pathway interactions, we can further use it to search for undiscovered exposure-pathway interactions and identify possible molecular targets candidates for environmental exposure treatments. We generate hypotheses by first looking for high frequency co-occurrences that are less established in the literature. Pathways that are highly incident on a particular disease are evaluated to establish a possible link between the exposure, biological pathway, and disease using biological knowledge and scientific literature. Using this approach, our integrated HPN can narrow in on plausible exposure-pathway interactions that are worth studying further in order to elucidate the molecular mechanisms involved in environmental toxicity.

Cadmium and cell-cell junction organization: Occupational studies from the 1980s and 1990s suggest that kidney stones, a highly recurrent and hard calcium deposit in the kidneys, are more common among workers exposed to cadmium.[21–23] A subsequent study analyzed NHANES data from 1999-2006 and concluded that low levels of exposure to cadmium increase the risk of chronic kidney disease.[24] There has been little elucidation of how cadmium contributes to kidney disease, however. We use the combined HPN to generate a hypothesis about which biological pathways are disrupted by cadmium exposure and how they might contribute to kidney disease. We observe on the network that cell-cell junction organization pathway is highly incident on the kidney stones phenotype node. We also observe the cell-cell junction organization pathway occurs most often with cadmium exposure (21 co-occurrences, 0.9% probability). From this, we can hypothesize that cadmium increases risk of kidney stones by obstructing tight junction functionality. Recent studies have provided preliminary support for this hypothesis. A recent study provided evidence that claudin-14, a gene associated with tight junction function, is responsible for a genetic predisposition to kidney stones.[25] The study suggested that claudin-14 mutations blocks calcium from entering tight junctions of the kidneys and causes excess calcium to go into urine, leading to kidney stones. Additionally, a literature survey indicates preliminary evidence that cadmium affects the distribution of tight junction proteins.[26] These studies suggest that both claudin-14 and cadmium confer risk for developing kidney stones. Using the combined HPN, we identified cell-cell junction pathway disruption as one way cadmium exposure might confer this risk.

Most complex diseases are synergistic outcomes of genetic and environmental effects. In order to develop effective therapies, we must understand the molecular processes modulated by both genetic variants and environmental exposures. The combined HPN provides a method to detect pathways that are disrupted by environmental exposures and proposes potential molecular targets for therapies.

Acknowledgments

This work was supported by the National Institute of Health (NIH) grants R01 EY022300, LM009012, LM010098, AI59694.

References

1. J. H. Moore, *Hum Hered* **56**, 73 (2003).
2. A. A. Rooney, A. L. Boyles, M. S. Wolfe, J. R. Bucher and K. A. Thayer, *Environ Health Perspect* **122**, 711 (Jul 2014).
3. K.-I. Goh, M. E. Cusick, D. Valle, B. Childs, M. Vidal and A.-L. Barabasi, *Proceedings of the National Academy of Sciences* **104**, 8685 (2007).
4. C. Darabos, M. J. White, B. E. Graham, D. N. Leung, S. Williams and J. H. Moore, *BioData Min* **7**, p. 1 (Jan 2014).
5. M. Newman, *Networks: An Introduction* (Oxford University Press, Inc., New York, NY, USA, 2010).
6. C. P. Wild, *Cancer Epidemiol Biomarkers Prev* **14**, 1847 (Aug 2005).
7. C. Darabos, K. Desai, R. Cowper-Sallari, M. Giacobini, B. Graham, M. Lupien and J. Moore, Inferring human phenotype networks from genome-wide genetic associations, in *Evolutionary Computation, Machine Learning and Data Mining in Bioinformatics*, eds. L. Vanneschi, W. Bush and M. Giacobini, Lecture Notes in Computer Science, Vol. 7833 (Springer Berlin Heidelberg, 2013) pp. 23–34.
8. C. Darabos, E. D. Grussing, M. E. Cricco, K. A. Clark and J. H. Moore, *Pac Symp Biocomput* , 171 (2015).
9. H. Li, Y. Lee, J. L. Chen, E. Rebman, J. Li and Y. A. Lussier, *Journal of the American Medical Informatics Association : JAMIA* **19**, 295 (January 2012).
10. X. Zhou, J. Menche, A.-L. Barabasi and A. Sharma, *Nat Commun* **5** (06 2014).
11. C. Darabos, S. H. Harmon and J. H. Moore, *Pac Symp Biocomput* , 188 (2014).
12. S. Suthram, J. T. Dudley, A. P. Chiang, R. Chen, T. J. Hastie and A. J. Butte, *PLoS Comput Biol* **6**, p. e1000662 (02 2010).
13. F. Barrenas, S. Chavali, P. Holme, R. Mobini and M. Benson, *PLoS ONE* **4**, p. e8090 (11 2009).
14. X. Zhou, J. Menche, A.-L. Barabasi and A. Sharma, *Nat Commun* **5**, p. 4212 (2014).
15. J. Qiu, C. Darabos and J. H. Moore, Studying the genetics of complex diseases with ethnicity-specific human phenotype networks: The case of type 2 diabetes in east asian populations, in *5th Translational Bioinformatics Conference*, 2014.
16. M. Kanehisa and S. Goto, *Nucleic Acids Res* **28**, 27 (Jan 2000).
17. I. L. Druwe and R. R. Vaillancourt, *Archives of toxicology* **84**, 585 (08 2010).
18. Y. Kumagai and D. Sumi, *Annu Rev Pharmacol Toxicol* **47**, 243 (2007).
19. A. S. Andrew, D. A. Jewell, R. A. Mason, M. L. Whitfield, J. H. Moore and M. R. Karagas, *Environ Health Perspect* **116**, 524 (Apr 2008).
20. E. J. Cordova, A. Martinez-Hernandez, L. Uribe-Figueroa, F. Centeno, M. Morales-Marin, H. Koneru, M. A. Coleman and L. Orozco, *PLoS One* **9**, p. e88069 (2014).
21. C. G. Elinder, C. Edling, E. Lindberg, B. Kågedal and O. Vesterberg, *British Journal of Industrial Medicine* **42**, 754 (11 1985).
22. R. Scott, C. Cunningham, A. McLelland, G. S. Fell, O. P. Fitzgerald-Finch and N. McKellar, *Br J Urol* **54**, 584 (Dec 1982).
23. L. Järup and C. G. Elinder, *British Journal of Industrial Medicine* **50**, 598 (07 1993).
24. P. M. Ferraro, S. Costanzi, A. Naticchia, A. Sturniolo and G. Gambaro, *BMC Public Health* **10**, p. 304 (2010).
25. Y. Gong, V. Renigunta, N. Himmerkus, J. Zhang, A. Renigunta, M. Bleich and J. Hou, *EMBO J* **31**, 1999 (Apr 2012).
26. G. Jacquillet, O. Barbier, I. Rubera, M. Tauc, A. Borderie, M. C. Namorado, D. Martin, G. Sierra, J. L. Reyes, P. Poujeol and M. Cougnon, *Am J Physiol Renal Physiol* **293**, F1450 (Nov 2007).

COMPUTING THERAPY FOR PRECISION MEDICINE: COLLABORATIVE FILTERING INTEGRATES AND PREDICTS MULTI-ENTITY INTERACTIONS

SAM REGENBOGEN
Department of Pharmacology, Baylor College of Medicine
Houston, TX 77030, USA
Email: regenbog@bcm.edu

ANGELA D. WILKINS
Department of Molecular and Human Genetics, Baylor College of Medicine
Houston, TX 77030, USA
Email: aw11@bcm.edu

OLIVIER LICHTARGE
Department of Molecular and Human Genetics, Baylor College of Medicine
Houston, TX 77030, USA
Email: lichtarg@bcm.edu

Biomedicine produces copious information it cannot fully exploit. Specifically, there is considerable need to integrate knowledge from disparate studies to discover connections across domains. Here, we used a Collaborative Filtering approach, inspired by online recommendation algorithms, in which non-negative matrix factorization (NMF) predicts interactions among chemicals, genes, and diseases only from pairwise information about their interactions. Our approach, applied to matrices derived from the Comparative Toxicogenomics Database, successfully recovered Chemical-Disease, Chemical-Gene, and Disease-Gene networks in 10-fold cross-validation experiments. Additionally, we could predict each of these interaction matrices from the other two. Integrating all three CTD interaction matrices with NMF led to good predictions of STRING, an independent, external network of protein-protein interactions. Finally, this approach could integrate the CTD and STRING interaction data to improve Chemical-Gene cross-validation performance significantly, and, in a time-stamped study, it predicted information added to CTD after a given date, using only data prior to that date. We conclude that collaborative filtering can integrate information across multiple types of biological entities, and that as a first step towards precision medicine it can compute drug repurposing hypotheses.

1. Introduction

At the same time as advances in biomedical research have enabled humanity's knowledge to grow far beyond the limits of any one person, that knowledge is being applied on ever-smaller scales. Specialized therapies are benefiting smaller subsets of the population, using all available knowledge to design a therapy for a specific case or to repurpose an existing drug for a novel use.

Online databases that compile this knowledge have become invaluable resources for researchers. Massive interaction networks can be powerful sources for hypothesizing novel relationships between biological entities. However, most of these networks are either focused on one particular type of entity (STRING[1] – genes/proteins) or interaction (DrugBank[2], ChEMBL[3] – drug-gene interactions). A full representation of biomedical knowledge would integrate the interactions among these physical entities and associate them with more abstract entities, such as pathways (KEGG[4], REACTOME[5,6]) and diseases (CTD[7]).

Several approaches to data integration have been explored. One approach is to predict how two classes of entity interact (e.g., drugs and targets) by integrating multiple types of feature data about the entities[8–10], or taking this a step farther, propagating this information to a third entity type[11]. These methods utilize information about the entities themselves, so they are specific to certain classes of entity. We will show an alternative approach, which can predict interactions among chemicals, genes, and diseases utilizing only information about how they connect to one another, and which benefits from the integration of disparate forms of information.

Collaborative filtering (CF) is a computational approach used in online recommendation systems, in which large-scale knowledge of how entities interact is used to predict likely connections[12,13]. Non-negative matrix factorization (NMF) is a popular tool for CF that compresses a matrix into two smaller factors whose product approximates the original[14,15]. NMF has long been used in biomedical science for clustering and classifying microarray data[16], but recent works have used NMF, or related algorithms, in CF strategies to predict drug-target[17,18] or protein-protein[19] interactions. We hypothesized that this basic approach could be pushed farther, to incorporate more than two types of biological entity, improving prediction of novel interactions among them.

Testing this hypothesis required multiple interaction networks, comprising connections between at least three entity types, so we turned to the Comparative Toxicogenomics Database (CTD). CTD is a publicly available resource that employs a team of human "biocurators" to comb the literature, extracting and annotating Chemical-Gene, Chemical-Disease, and Disease-Gene relationships[7]. In this paper, we will demonstrate that NMF can be used to recover hidden interactions in each of these networks individually and that NMF over any two of these networks can predict back the third. To show that this is not an artifact of the data source (CTD), we will demonstrate that NMF over the combined CTD networks recapitulates experimental protein-protein interactions in the STRING database. We will focus in on the CTD Chemical-Gene interaction network, and show that our ability to predict missing connections improves when we perform NMF over a network incorporating Chemical-Gene, Chemical-Disease, and Disease-Gene interactions from CTD and also Protein-Protein interactions from STRING.

2. Methods:

2.1. *Construction of datasets:*

Tables of interactions from CTD were obtained and processed as follows. Unless otherwise noted, all data processing and manipulation was performed in Matlab. Chemical-Gene and Chemical-Disease interactions were downloaded on April 2, 2014[a], each as a single tab-delimited text file. The full Chemical-Gene interactions file was imported into Matlab as a table containing 878,594 rows, each representing one unique curated relationship between one chemical and one gene, or between other relationships. This initial table comprised 10,520 unique chemicals and 32,248 unique genes. Relationships containing nested relationships were removed, as were any relationships whose "Gene Form" was not given as "protein" ("mRNA," for example.) The result of this filtering was a table of direct relationships involving 8,653 unique chemicals and 8,288 unique genes. A binary adjacency matrix was built in which each row and column corresponded to one chemical or gene, respectively, with interacting pairs assigned a value of 1, and all other pairings 0. The resulting sparse 8,653-by-8,288 matrix contains 82,168 unique, binary Chemical-Gene interactions.

The Chemical-Disease interactions file was similarly imported into Matlab, but was filtered to remove all CTD-inferred relationships by deleting any row for which the "Direct Evidence" column was blank. The filtered table was used to build a binary adjacency matrix as described above, which in this case comprised 8,226 chemicals, 3,031 diseases, and 80,433 unique, curated interactions.

The full Disease-Gene interactions file was too large to process in the same way, so CTD's Batch Query tool[b] was used to retrieve only the curated interactions. On April 18, 2014, the CTD Disease Vocabulary file was downloaded, and the Disease IDs were input to the Batch Query tool, which was set to export all Curated Gene Associations for each disease. The output tab-delimited interactions were then imported into Matlab and, as before, used to build a sparse, binary adjacency matrix of 4,907 Diseases by 7,362 Genes, with 23,133 unique interactions.

For construction of a combined Chemical-Gene-Disease (CGD) interaction matrix, the interaction tables used to build the individual matrices were used. A single list of 30,102 unique entities was obtained from the union of the three individual matrices' unique entity lists, comprising 12,119 Chemicals, 6,333 Diseases, and 11,650 Genes. Each of the three interaction tables was then used to populate a matrix in which each of the 30,102 entities was represented as both a row and a column. Thus, for each row in the three tables, the interacting entities' positions in the combined entity list defined two symmetrical pairs of indices in the 30,102-by-30,102 matrix at which to represent the interaction.

For later experiments, we used the STRING network of human protein-protein interactions[c], which we mapped to the CTD CGD matrix. When comparing our predictions to STRING, we focused on 7,604 genes whose IDs we could map between databases, and used the confidence scores

---

[a] from http://ctdbase.org/downloads - dates noted because CTD updates monthly; previous versions are unavailable

[b] http://ctdbase.org/tools/batchQuery.go

[c] STRING v9.1, now archived at http://string91.embl.de/newstring_cgi/show_download_page.pl

assigned by STRING (ranging from 0 to 999) to define the positive class at various thresholds. To construct a CTD+STRING CGD matrix, we added protein-protein interactions from this STRING[d] network to the Gene-Gene diagonal block of the CTD CGD matrix. Interactions among the 7,604 genes also in CTD were dropped directly into the corresponding cells in the CGD matrix symmetrically. The matrix was extended by 6,699 rows and columns, corresponding to the genes that were not matched to CTD. The final matrix contains 254,929 nonzero Gene-Gene interactions, 66,685 with values of 0.5 or greater, and 1,405 with the maximum value of 0.999.

2.2. *Non-negative Matrix Factorization (NMF):*

NMF describes several closely related algorithms that, given a non-negative matrix **A** with size $m \times n$ and a positive integer $k \ll \min(m,n)$, attempt to find $m \times k$ matrix **W** and $k \times n$ matrix **H** such that **W** and **H** are non-negative, and such that $\mathbf{A} \approx \mathbf{WH}$. This is done by solving the optimization problem

$$\min_{W,H \geq 0} f(\mathbf{W},\mathbf{H}) = \frac{1}{2} \|\mathbf{A} - \mathbf{WH}\|_F^2 \tag{1}$$

Throughout this work, NMF was run using the *nnmf*() function of Matlab's Statistics Toolbox with all input arguments (other than **A** and *k*) left at default settings. Consequently, the optimization method used was Alternating Least Squares (ALS), in which initial **W** and **H** matrices are randomly generated, and then alternatingly solved for in the following matrix equations, until the minimization function converges or until the maximum number of iterations has been reached:

$$\text{Solve for } \mathbf{H}\text{: } \mathbf{W}^T\mathbf{W}\mathbf{H} = \mathbf{W}^T\mathbf{A} \tag{2.1}$$

$$\text{Solve for } \mathbf{W}\text{: } \mathbf{H}\mathbf{H}^T\mathbf{W}^T = \mathbf{H}\mathbf{A}^T \tag{2.2}$$

In our applications of NMF to datasets of various sizes, we tested multiple *k* values for each, to find a value that would give optimal performance without overfitting.

NMF is known to converge at solutions that are local, rather than global, minima of the optimization problem, meaning the product **WH** is not unique. We found that calculating the average of **WH** across multiple replicate factorizations increased performance in our experiments; all results we discuss below were obtained by averaging the output of 4 NMF replicates[e].

2.3. *10-fold Cross-validation Experiments:*

In *N*-fold cross-validation experiments, each point in a dataset is randomly assigned to one of *N* subsets. Then, one at a time, every subset is removed, and the remaining *N*-1 subsets are used as training data for the algorithm to be tested. In the end, the algorithm's predicted values for each dropped subset form a test set covering all of the original data. An algorithm's ability to successfully recover data in cross-validation depends not only on the algorithm itself, but also on the internal consistency of the dataset. Entities with only 1 known interaction were not considered, because NMF would have no way to recover that interaction.

---

[d] inserted as the confidence score divided by 1000 to match the range of the rest of the CGD matrix, which is binary.

[e] Data not shown. We chose 4 replicates to balance diminishing returns in improvements v. computational cost.

2.4. *Performance evaluation for NMF predictions*

The performance of NMF in each experiment was evaluated by calculating the Receiver Operator Characteristic (ROC) curve, comparing predicted scores to an input positive class, and computing the number of correct predictions at varying score thresholds. An ROC curve can be understood as sorting the list of predictions by score and, beginning at the origin, moving up on the y-axis for each true prediction and moving right on the x-axis for each false prediction. The area under an ROC curve (AUC) can serve as a broad measure of performance, representing the probability that a randomly chosen positive (known) interaction will have been assigned a higher score by NMF than a randomly chosen negative (not known) interaction.

3. Results and Discussion

3.1. *10-fold cross-validation for NMF of individual CTD matrices.*

In order to determine whether a CF approach can integrate interactions between multiple classes of biological entity, we first made certain that NMF can be used to recover unknown pairwise interactions among Chemicals, Diseases, and Genes from incomplete interaction data. 10-fold cross-validation was performed on three adjacency matrices constructed from CTD's Chemical-Disease (CD), Chemical-Gene (CG), and Disease-Gene (DG) networks, respectively.

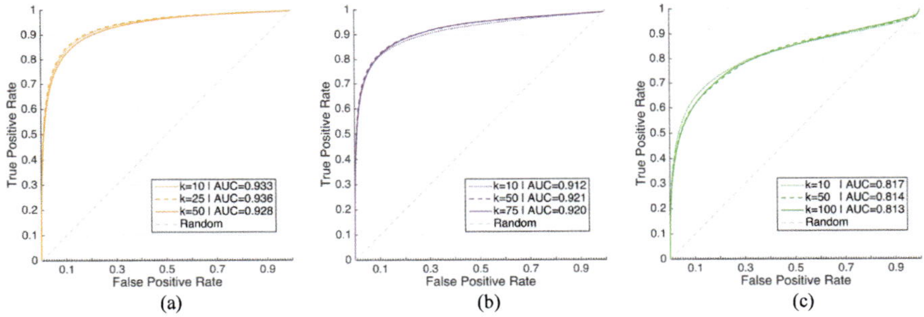

Fig. 1. Receiver Operator Characteristic (ROC) curves of NMF at varying k values in 10-fold cross-validation experiments over individual CTD interaction networks. (a) Chemical-Disease, (b) Chemical-Gene, (c) Disease-Gene

Figure 1 shows NMF performs much better than random guessing in 10-fold cross-validation for the three CTD networks, with performance plotted as Receiver Operator Characteristic (ROC) curves, with k varying over a range to find values that optimize AUC. The best results were AUC of 0.94 (CD), 0.92 (CG), and 0.82 (DG). The results in Figure 1 show these three networks are internally consistent enough to recover missing interactions using NMF, and that the interactions involving Chemicals (CD and CG) are particularly well-suited to prediction by NMF.

3.2. *CTD Chemical-Gene-Disease matrix and leave-one-matrix-out experiments*

Once we verified that the three networks from CTD were, individually, amenable to prediction of missing interactions via NMF, we considered how to utilize this multifaceted data more effectively. The data encompassed three classes – Chemicals, Diseases, and Genes – of biological entity, with information about each category spread across two matrices. When it factorizes an interaction matrix, NMF represents each entity (row/column vector) as a compressed vector that approximates all available information. Therefore, we reasoned that simply combining the asymmetric CD, CG, and DG matrices into one symmetric "all-vs-all" Chemical-Gene-Disease (CGD) matrix would allow NMF access to more information about the relationships between Chemicals, Diseases, and Genes, and thus improve our ability to predict missing ones.

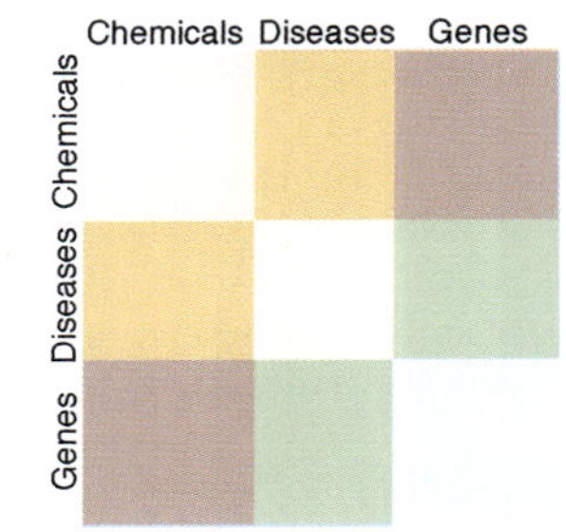

Fig. 2. Illustration of the combined, symmetric CGD matrix. The CTD CD, CG, and DG matrices are orange, purple, and green, respectively. The diagonal blocks are empty before factorization.

In order to test the ability of our CF approach to integrate different types of interaction, we devised a "leave-one-matrix-out" experiment (Fig. 3a-c). From the combined CGD matrix in Figure 2, we removed all interactions of one class (CD, CG, or DG) at a time, and attempted to predict them from only the other two interaction classes. We performed this test, using NMF with various k values, for each of the three interaction types and calculated ROC curves. Fig. 3d shows the AUC for each *k* value used to predict the missing matrices.

Table 1. Amount of data dropped and re-predicted in Leave-One-Matrix-Out Experiments, followed by AUC when NMF was performed over the remaining two interaction matrices. Column headings indicate which interaction matrix was left out.

Dropped Matrix	Chemical-Disease	Chemical-Gene	Gene-Disease
Size	4760x1605	4760x3940	3940x1605
# Interactions	59,766	60,831	15,522
AUC k=100	0.801	0.833	0.802
AUC k=200	0.810	0.840	0.801
AUC k=300	0.813	0.837	0.802
AUC k=500	0.817	0.832	0.795

These results show that NMF is able to predict the interactions contained in each of the matrices created from CTD's datasets, given only information contained in the other two matrices, despite the distinctly different biological connections they represent. Put another way, this demonstrates that combining these binary interaction matrices can unlock new layers of information that was not accessible from the individual matrices. Because all three networks share an origin in CTD's manual curation process, however, we need to determine that the latent information tapped by NMF for these predictions provides a meaningful insight to the workings of biology, and not just an insight into the CTD curation pipeline.

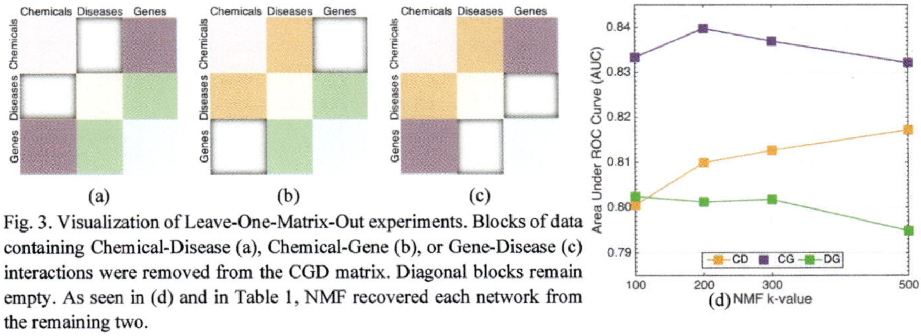

Fig. 3. Visualization of Leave-One-Matrix-Out experiments. Blocks of data containing Chemical-Disease (a), Chemical-Gene (b), or Gene-Disease (c) interactions were removed from the CGD matrix. Diagonal blocks remain empty. As seen in (d) and in Table 1, NMF recovered each network from the remaining two.

3.3. *Prediction of Gene-Gene associations from CTD Chemical-Gene-Disease matrix*

The diagonal blocks of the combined matrix, which would correspond to Chemical-Chemical, Disease-Disease, and Gene-Gene associations, contain no data initially from CTD, but are also filled in when we use NMF. We sought to compare predictions in these regions to an external data source, in order to find out if the values predicted by NMF represent real biological relationships.

Although it is unclear what the disease-disease network might represent, comparing the gene-gene block to existing protein interaction databases was a natural next step. We compared the values from NMF to known protein-protein interactions from the STRING database. 7,604 genes were present in both the combined CTD matrix and the STRING experimental network. Among these 7,604 genes, STRING contained 67,763 experimentally supported protein-protein interactions, of which 38,424 have been assigned confidence scores by STRING of at least 500, and 902 have been assigned the highest confidence score of 999.

As shown in Fig. 4., the values produced by NMF over the CTD CGD matrix predicted these interactions with an ROC AUC of 0.69, which increased to AUC=0.73 for interactions ≥500 confidence score, and to AUC=0.75 when only the highest-confidence STRING interactions (999) were considered.

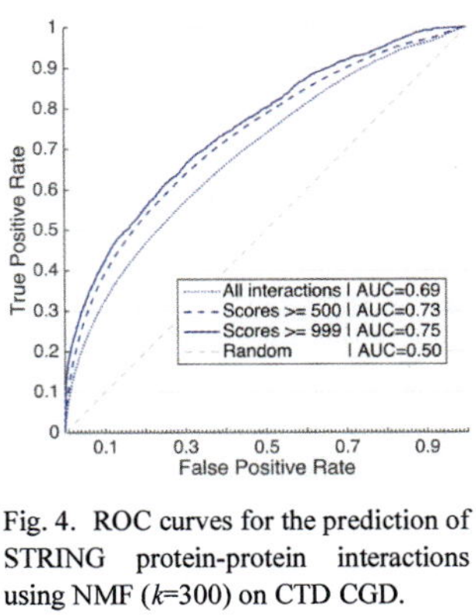

Fig. 4. ROC curves for the prediction of STRING protein-protein interactions using NMF (k=300) on CTD CGD.

These results show that the Gene-Gene associations filled into the Chemical-Gene-Disease matrix by NMF correspond to real, experimentally known protein-protein interactions. This result is important because, unlike the Leave-One-Matrix-Out experiments, these predicted edges were never part of CTD, reducing the chance that positive results are due to some inherent bias in the CTD curation process. This also suggests that the predictions in the Chemical-Chemical and Disease-Disease blocks may be biologically meaningful, potentially representing drug interactions and disease co-morbidity, for example. At the same time, these results suggest that some of the information contained within the STRING network was not found by NMF in the combined CGD matrix. We created a second Chemical-Gene-Disease matrix containing all the same interactions from CTD, but with protein-protein interactions from STRING added to the Gene-Gene block of the diagonal.

3.4. *10-fold cross-validation of Chemical-Gene edges within combined CGD matrix*

In order to determine if additional data can improve upon the prediction of Chemical-Gene interactions observed in Fig. 1b, we performed an experiment similar to 10-fold cross-validation, which only removed Chemical-Gene edges from the larger matrix. We performed this experiment using the CTD CGD matrix, and also using the CTD+STRING CGD matrix, both with k=200. We also repeated the 10-fold cross-validation using the CG matrix alone, using the best-performing k value, k=50. For this comparison, a single set of randomized cross-validation classes was generated first, and then was used for all three input matrices, to ensure that the only differences in available information were those we were testing.

Table 2. Comparison of NMF performance in 10-fold cross-validation of Chemical-Gene edges without added data, with the addition of CD and DG information from CTD, or with that plus GG information from STRING

Matrix	k	ROC AUC[f]	p-value[f] vs CTD_{CG}	p-value[f] vs CTD_{CGD}
CTD_{CG}	50	0.920	–	–
CTD_{CGD}	200	0.927	$4.6x10^{-109}$	–
$CTD_{CGD}+S_{GG}$	200	0.932	$4.9x10^{-244}$	$5.8x10^{-117}$

As shown in Table 2 and Figure 5a, the Chemical-Gene cross-validation performance after the addition of Chemical-Disease and Disease-Gene interactions yielded AUC=0.927, an increase over the highest-performing k value with Chemical-Gene interactions only (AUC=0.920 at k=50). Moreover, when Gene-Gene interactions from STRING were added to the CGD matrix, performance further improved to AUC=0.932. To measure this improved performance, we used the StAR method[20], which implements an approach based on Mann-Whitney U-statistics[21], to determine if the ROC curves were significantly different. Although the increases in AUC appear small, so many data points were used to calculate the ROC curves that they were found to be highly significant.

AUC of the ROC curve provides an overall indicator of how well a method recovers true interactions. However, practical applications (e.g., drug repurposing,) are likely to focus on relatively few predictions compared to the total interaction space. For this reason, it is often more important that the top predictions have high precision (i.e., few false positives). To be sure the CGD matrices were not only increasing AUC by improving recall of the low-confidence interactions, we calculated precision-recall curves for the cross-validation (Figure 5b). As the inset shows, the

[f] Output by StAR tool, standalone version[20], rounded for table

precision of the highest-scoring 10% of interactions[g] is high for all three test cases, with the CTD+STRING Chemical-Gene-Disease matrix showing precision improvements across the range.

These results show that we can improve the ability of NMF to predict missing Chemical-Gene relationships by incorporating information about how those Chemicals and Genes interact with Diseases, and, further, how those Genes interact with one another.

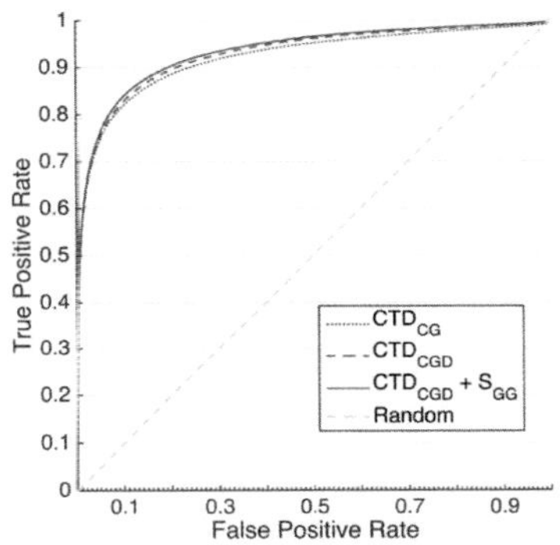

Fig. 6a. ROC curves showing NMF performance for 10-fold cross-validation of Chemical-Gene interactions, improving with more data. AUCs with statistical comparison are in Table 2 above.

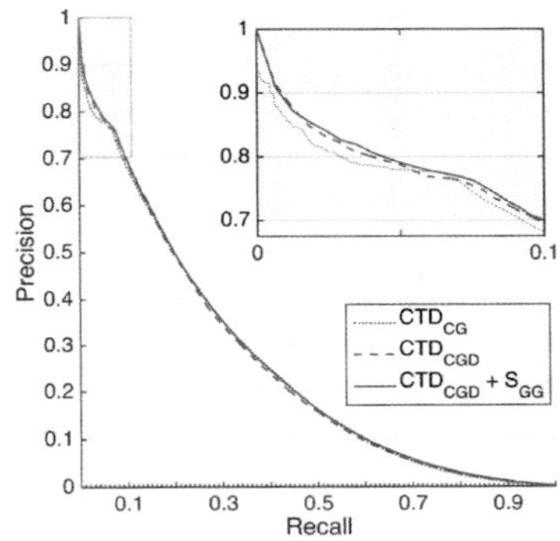

Fig. 6b. Plot of Precision vs. Recall for the same experiments shows precision approaches 1.0 for the top recovered interactions. Focusing on top 10% (inset) shows improvement with more data.

3.5. *Retrospective prediction of new Chemical-Gene interactions*

Finally, to corroborate these results in a more realistic context, we retrospectively predicted Chemical-Gene interactions that had been added to CTD over one year. Following the same process described in Section 2.1, we downloaded the CTD Chemical-Gene network on April 5, 2015, and again built a binary matrix of direct interactions. We mapped this to the 2014 CTD CGD matrix, removing entities that were not present in both versions, resulting in a 2015 matrix of 8,706 Chemicals by 8,304 Genes with 5,879 new interactions.

We calculated an ROC curve (shown in Figure 6) comparing these new interactions to the predictions for the same 8,706 Chemicals and 8,304 Genes that were obtained from NMF (k=200) on our CTD+STRING CGD matrix. The

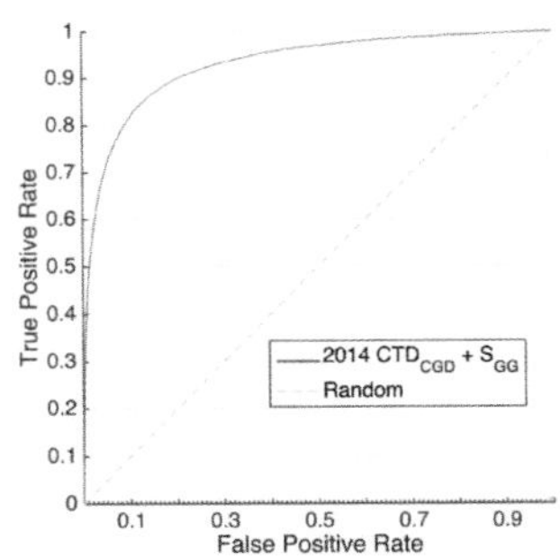

Fig. 6. Retrospective prediction of new CTD Chemical-Gene interactions (added between 4/2014 and 4/2015), using NMF (k=200) on CTD+STRING CGD matrix AUC=0.930.

[g] The positive class comprised 75,804 interactions, so the inset shows precision for over 7500.

resulting curve, with AUC=0.93, indicates that our approach was able to correctly anticipate missing or undiscovered interactions.

3.6. *Example: prediction of Chemical-Disease interactions for Pancreatic Neoplasms*

One potential application of our approach is to identify unknown or overlooked drugs with connections to a particular disease. In Table 3, we present an example of this involving pancreatic cancer, a disease with high lethality and few effective treatments[22]. Following NMF (*k*=200) over the CTD+STRING CGD matrix, we inspected the highest[h] values corresponding to new interactions (that is, interactions that have not been curated by CTD at this time) between Chemicals and the disease entity "Pancreatic Neoplasms." Examples were chosen in which the Chemical is a drug[i]; as the primary focus of CTD is toxicology, much of the information therein concerns environmental toxins and disease-causing interactions. As Table 3 shows, literature searches found evidence supporting a connection to pancreatic cancer for 14 of the top 15 drug predictions, over half of which were studied in clinical trials. This shows that, at minimum, our approach generated hypotheses worth testing clinically.

Table 3. Top[h] 15 drugs[i] predicted to interact with Pancreatic Neoplasms by NMF using the CTD+STRING CGD matrix. These interactions were not present in the CTD CD matrix, but 14 are supported by papers or clinical trials in associated PubMed ID (PMID).

Drug Name	Support for Connection to Pancreatic Cancer	Reference
Indomethacin	Pre-clinical cell line study	PMID: 1890839
Carboplatin	Phase II clinical trial	PMID: 15802284
Mitoxantrone	Phase II clinical trial	PMID: 16334117
Simvastatin	Phase II clinical trial	PMID: 24162380
Cytarabine	Phase III clinical trial	PMID: 1833042
Topotecan	Phase II clinical trial	PMID: 11218186
Sorafenib	Phase II clinical trial	PMID: 24574334
Rosiglitazone	Pre-clinical mouse study	PMID: 22864396
Melphalan	Pre-clinical rat study	PMID: 4075299
Methamphetamine	-	-
Thiotepa	Use in other cancers	PMID: 4183076
Thalidomide	Phase I clinical trial	PMID: 15753541
Caffeine	Phase III clinical trial	PMID: 1833042
Sirolimus	Patient Case Report	PMID: 19581741
Gefitinib	Phase II clinical trial	PMID: 19258727

[h] Values above a threshold of 0.425. To provide context for this choice of threshold, the inset in Figure 5b shows cross-validation performance as precision versus recall at varying thresholds; 0.1 Recall in that graph corresponds to a threshold value of 0.425. Thus, we chose predictions whose precision should be at least 0.7.

[i] Approved by the FDA, according to http://www.accessdata.fda.gov/scripts/cder/ob/default.cfm

At the same time, this example highlights key pitfalls. Because we created binary interaction matrices from CTD, we can not say these drugs are predicted to treat pancreatic cancer or to cause it, only that they interact in some way. Indeed, the clinical trial we reference for simvastatin found no significant effect, but suggested further study in specific circumstances that could benefit from it[23]. Incorporating more detail from the interactions in CTD into our CGD matrix will, we believe, help resolve some of the ambiguity in our current predictions. For truly personalized treatments, we foresee a use case in which therapy suggestions are derived from a subset of predicted drug-gene interactions. That subset would be determined by a patient's unique situation; for example, the somatic mutations driving a tumor, or the germ line mutations linked to a disease phenotype (the latter being a possible application for our approach's gene-disease predictions).

4. Conclusions

Taken as a whole, our results show that Collaborative Filtering can integrate biological interaction networks in order to reveal missing connections between diverse entities. This approach depends only on knowledge of connections, so it can be extended to new classes of entity with minimal customization, unlike more specialized methods. Consequentially, our approach is limited to predicting *that* entities interact, rather than *how*. Matrix tri-factorization, which has been used to classify entities by fusing interaction networks with entity feature data[24,25], may enable more detailed predictions. Ultimately, however, we see this as an initial component in a pipeline that will harness the ever-expanding universe of knowledge and focus it on a small point, illuminating a patient's unique situation or highlighting a new use for a drug. This will need to be done rapidly, affordably, and accessibly. Importantly, implementations of NMF have been developed that can efficiently handle matrices with millions of times more entities than we have so far attempted[13,26]. Ultimately, this work may offer a step towards computing therapy.

5. Acknowledgements

This work was funded through DARPA SIMPLEX N66001-15-C-4042, NSF DBI1356569, and NIH NIDCR 1U01DE025181-01.

6. References

1. Franceschini, A, *et al.* STRING v9.1: protein-protein interaction networks, with increased coverage and integration. *Nucleic Acids Res.* 41, D808–15 (2013).
2. Law, V, *et al.* DrugBank 4.0: shedding new light on drug metabolism. *Nucleic Acids Res.* 42, D1091–7 (2014).
3. Gaulton, A, *et al.* ChEMBL: a large-scale bioactivity database for drug discovery. *Nucleic Acids Res.* 40, D1100–7 (2012).
4. Kanehisa, M, *et al.* Data, information, knowledge and principle: Back to metabolism in KEGG. *Nucleic Acids Res.* 42, 199–205 (2014).
5. Croft, D, *et al.* The Reactome pathway knowledgebase. *Nucleic Acids Res.* 42, 472–7 (2014).

6. Milacic, M, *et al.* Annotating cancer variants and anti-cancer therapeutics in Reactome. *Cancers* 4, 1180–211 (2012).
7. Davis, AP, *et al.* The Comparative Toxicogenomics Database's 10th year anniversary: update 2015. *Nucleic Acids Res.* 9880, 1–7 (2014).
8. Xu, T, *et al.* Quantitatively integrating molecular structure and bioactivity profile evidence into drug-target relationship analysis. *BMC Bioinf.* 13, 75 (2012).
9. Huang, L-C, *et al.* A weighted and integrated drug-target interactome: drug repurposing for schizophrenia as a use case. *BMC Syst. Biol.* 9, S2 (2015).
10. Yang, F, *et al.* Drug-target interaction prediction by integrating chemical, genomic, functional and pharmacological data. *Pacific Symp. Biocomp.* 148–59 (2014).
11. Huang, Y-F, *et al.* Inferring drug-disease associations from integration of chemical, genomic and phenotype data using network propagation. *BMC Med. Geno.* 6 Suppl 3, S4 (2013).
12. Zhang, S, *et al.* Learning from Incomplete Ratings Using Non-negative Matrix Factorization. *SDM* 549–53 (2006).
13. Zhou, Y, *et al.* Large-scale parallel collaborative filtering for the netflix prize. *Algo. Asp.* (2008).
14. Paatero, P & Tapper, U. Positive matrix factorization: A non-negative factor model with optimal utilization of error estimates of data values. *Environmetrics* 5, 111–26 (1994).
15. Lee, DD & Seung, HS. Learning the parts of objects by non-negative matrix factorization. *Nature* 401, 788–91 (1999).
16. Kim, H & Park, H. Sparse non-negative matrix factorizations via alternating non-negativity-constrained least squares for microarray data analysis. *Bioinformatics* 23, 1495–502 (2007).
17. Cobanoglu, MC, *et al.* Predicting drug-target interactions using probabilistic matrix factorization. *J. Chem. Inf. Model.* 53, 3399–409 (2013).
18. Zheng, X, *et al.* Collaborative matrix factorization with multiple similarities for predicting drug-target interactions. *Proc. 19th ACM SIGKDD* 1025 (2013).
19. Wang, H, *et al.* Predicting protein-protein interactions from multimodal biological data sources via nonnegative matrix tri-factorization. *J. Comput. Biol.* 20, 344–58 (2013).
20. Vergara, I a, *et al.* StAR: a simple tool for the statistical comparison of ROC curves. *BMC Bioinf.* 9, 265 (2008).
21. DeLong, ER, *et al.* Comparing the areas under two or more correlated receiver operating characteristic curves: a nonparametric approach. *Biometrics* 44, 837–45 (1988).
22. Ryan, DP, *et al.* Pancreatic Adenocarcinoma. *N. Engl. J. Med.* 371, 1039–49 (2014).
23. Hong, JY, *et al.* Randomized double-blinded, placebo-controlled phase II trial of simvastatin and gemcitabine in advanced pancreatic cancer patients. *Cancer Chemother. Pharmacol.* 73, 125–30 (2014).
24. Zitnik, M & Zupan, B. Matrix factorization-based data fusion for gene function prediction in baker's yeast and slime mold. *Pacific Symp. Biocomp.* (2014).
25. Žitnik, M & Zupan, B. Data fusion by matrix factorization. *IEEE Trans. Pattern Anal. Mach. Intell.* 37, 41–53 (2015).
26. Liu, C, *et al.* Distributed nonnegative matrix factorization for web-scale dyadic data analysis on mapreduce. *WWW 2010* (2010).

KNOWLEDGE-ASSISTED APPROACH TO IDENTIFY PATHWAYS WITH DIFFERENTIAL DEPENDENCIES*

GIL SPEYER, JEFF KIEFER

Integrated Cancer Genomics Division, The Translational Genomics Research Institute
Phoenix, AZ 85004, U.S.A.
Email: gspeyer@tgen.org, jkiefer@tgen.org

HARSHIL DHRUV, MICHAEL BERENS

Cancer Cell Biology Division, The Translational Genomics Research Institute
Phoenix, AZ 85004, U.S.A.
Email: hdhruv@tgen.org, mberens@tgen.org

SEUNGCHAN KIM

Integrated Cancer Genomics Division, The Translational Genomics Research Institute
Phoenix, AZ 85004, U.S.A.
Email: skim@tgen.org

We have previously developed a statistical method to identify gene sets enriched with condition-specific genetic dependencies. The method constructs gene dependency networks from bootstrapped samples in one condition and computes the divergence between distributions of network likelihood scores from different conditions. It was shown to be capable of sensitive and specific identification of pathways with phenotype-specific dysregulation, i.e., rewiring of dependencies between genes in different conditions. We now present an extension of the method by incorporating prior knowledge into the inference of networks. The degree of prior knowledge incorporation has substantial effect on the sensitivity of the method, as the data is the source of condition specificity while prior knowledge incorporation can provide additional support for dependencies that are only partially supported by the data. Use of prior knowledge also significantly improved the interpretability of the results. Further analysis of topological characteristics of gene differential dependency networks provides a new approach to identify genes that could play important roles in biological signaling in a specific condition, hence, promising targets customized to a specific condition. Through analysis of TCGA glioblastoma multiforme data, we demonstrate the method can identify not only potentially promising targets but also underlying biology for new targets.

* This work was supported in part by the National Cancer Institute, National Institutes of Health [1U01CA168397].

1. Introduction

1.1. *Gene set analysis, DDN and EDDY*

Identification of biological features underlying disease phenotypes or conditions (e.g. differentially expressed or mutated genes) is critical in identifying therapeutic targets. As specific pathways are capable of complex rewiring between conditions, methods such as Gene Set Enrichment Analysis (GSEA) (1) and network-based analyses (2-4) have become increasingly attractive for extraction of such biological features from genomic data. One can use known genetic interactions as a ground truth network and overlay genomic data from different conditions to statistically evaluate regions with differential activities (5) or condition-specific sub-networks (6-8). Differential Dependency[†] Network (DDN) approaches are able to identify individual differential dependencies (9-13) or condition-specific sub-networks from genome-wide dependency networks such as a protein-protein interaction networks. Differential co-expression analysis methods (14), such as Gene Set Co-expression Analysis (GSCA), test gene sets for differential dependencies, but they are often overly sensitive to minor correlation changes and produce biased results with respect to the size of gene sets (15).

In our previous work, we have developed a novel, network-based computational method that overcomes the limitations of other network-based approaches (15). This novel computational approach – *EDDY: Evaluation of Differential DependencY* – combines GSEA's gene-set-assisted advantages with the robustness of assessment of differential network dependency. It interrogates gene sets (pathways) in a database to test if dependencies across genes are significantly rewired between conditions (see Fig. 1). It was shown to be capable of sensitive and specific identification of pathways with phenotype-specific dysregulation, i.e. *rewiring of dependencies between genes in different conditions*, with its robust network inference and low false discovery rate (15).

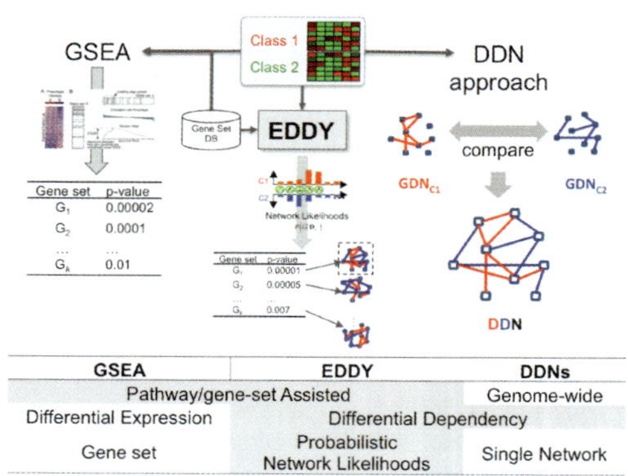

Figure 1. Advantages of EDDY compared to other tools

In this paper, we present a method to integrate known biological interactions to improve the performance of network inference and to enable better interpretation of inferred DDNs. The effect of the degree of prior knowledge integration on inferred DDNs is also analyzed. Finally, we describe the application of prior-knowledge assisted EDDY to glioblastoma (GB) gene expression downloaded from the Cancer Genome Atlas (TCGA).

† In this manuscript, we use 'dependency' to denote statistical dependencies derived from data such as co-expression, or conditional dependencies, and 'interaction' to denote *known* direct or indirect relationships between genes.

2. Methods

From two sets of samples representing different conditions, EDDY computes the discrepancy of gene dependency in a specific gene set by contrasting the two resulting probability distributions of candidate network structures (based on a likelihood of each network), constructed via a resampling approach, and evaluates its statistical significance to determine if the network structures are rewired between the conditions.

2.1. *EDDY: Evaluation of differential dependency*

Let a set of variables $\boldsymbol{G} = \{g_1, g_2, \ldots\}$ (each variable corresponds to a gene) denote the activity levels of the genes. For $\boldsymbol{G}$, there are N possible gene dependency network (GDN) structures $d_1, d_2, \ldots, d_N$ for the variables. Let a discrete random variable D take on $d_1, d_2, \ldots, d_N$ as its discrete values, then the posterior probability distribution $\Pr(D|\boldsymbol{S}_C)$ for a data $\boldsymbol{S}_C$ of a given condition C can represent the probability distribution of dependency network structures for $\boldsymbol{G}$ in the condition C. When two data sets, $\boldsymbol{S}_{C_1}$ and $\boldsymbol{S}_{C_2}$, are given for two different conditions C_1 and C_2, the divergence between the two corresponding probability distributions $\Pr(D|\boldsymbol{S}_{C_1})$ and $\Pr(D|\boldsymbol{S}_{C_2})$ is computed as a measure of difference between the conditions. The divergence between the conditions C_1 and C_2 is measured using the Jensen-Shannon (JS) divergence, an information-based metric to measure the similarity between two probability distributions (16) and the statistical significance of the divergence is computed using a permutation approach. This approach is a generalization of comparing the best networks from different conditions by considering many possible networks and their likelihoods instead of comparing the single best networks. The benefit of this generalization is a more reliable measure of discrepancy (15), especially when data is limited. Thus, there is a high chance of finding many local optima for the best network. By considering many probable dependency networks instead of one local optimal network, our approach can represent a more complete picture of dependencies at the cost of additional computation. EDDY then iterates through all gene sets in a database, for example, MSigDB (http://www.broadinstitute.org/gsea/msigdb/) to identify the dysregulated pathways.

2.2. *Inference of gene dependency network supported by known interactions*

To reduce computational complexity, EDDY uses a heuristic method that proposes probable dependency structures by independently evaluating each dependency between two variables. Specifically, χ^2-test is applied to test the independence between every pair of two variables g_i and g_j ($\in \boldsymbol{G}$), obtaining the resultant p-value p_{ij} ($=p_{ji}$). An edge e_{ij} between g_i and g_j is included when

$$\Pr(i;j|\boldsymbol{S}_C) = \left(1 - p_{ij}\right)^{\lambda} > \theta \qquad (1)$$

where $\lambda \geq 1$ and a user-specified parameter θ together control sensitivity of dependency discovery. We integrate known interactions retrieved from pathway databases to support dependency discovery. Formally, let $w_P \in [0,1]$ denote a prior weight to control the level of prior knowledge to be incorporated into the inference of GDN and $E_P(i;j)$ be a binary-valued variable indicating the existence of known interaction between g_i and g_j. Known interactions can be retrieved from a pathway database such as Pathway Commons 2. Edge-specific threshold is given,

$$\theta_P(i;j) \leftarrow \theta \cdot [1 - w_P \cdot E_P(i;j)]. \qquad (2)$$

Prior weight (w_P) can be varied between 0 and 1, where $w_P = 0$ specifies no influence of the known gene interactions in GDN inference and all edges in inferred GDN requires full support from the data $\theta_P(i;j) = \theta$, and $w_P = 1$ makes inferred GDN include all the known interactions unconditionally, $\theta_P(i;j) = 0$. When $w_P = 0.5$, edges with half the support from the data will be included in the network. Edges are included in a network if they satisfy:

$$\Pr(i;j|\boldsymbol{S}_C) > \theta_P(i;j). \quad (3)$$

Since information on the condition-specificity of known interaction is generally not available, incorporating known interactions into GDN inference could potentially decrease the divergence between GDNs, hence, the sensitivity of the EDDY algorithm to detect pathways with condition-specificity. The specific effect of prior weight (w_P) on the sensitivity of EDDY will be discussed in the Results section.

Considerations: As opposed to data-derived edges, prior edges can have a direction, indicating, for example, the influence of one gene on another. While it is straightforward to incorporate the direction of an edge into EDDY, this may conflict with the acyclic requirement of Bayesian networks. For the computations in this work, directionality was determined not to create cycles. In addition, prior edge encompasses many types of interactions such as catalysis or phosphorylation. It also may describe various degrees of influence from explicitly controlling a state change to simply being a neighbor gene. For the work described here, we excluded these so-called "neighbor" interactions. In future work, we may examine a nuanced means of weighting other types of interactions.

2.3. *Estimating divergence between two conditions-specific probability distributions of GDNs*

The empirical estimate of the probability distribution, $\Pr(D|\boldsymbol{S}_C)$, is yielded from bootstrapping samples and the construction of GDNs as described above. Once the probability distribution of dependency network structures $\Pr(D|\boldsymbol{S}_{C_1})$ and $\Pr(D|\boldsymbol{S}_{C_2})$ are computed, the divergence between the conditions C_1 and C_2 is measured using the Jensen-Shannon (JS) divergence and the statistical significance is estimated using a permutation test. See (15) for more detail, and the overall workflow is shown in Fig. 2.

2.4. *Topological analysis of Differential Dependency Network (DDN)*

GDNs constructed for condition C_1 and C_2 are summarized into differential dependency networks (DDNs) where each edge is annotated as C1-specific, C2-specific, or common. While these condition-specific dependencies can be used to identify potential

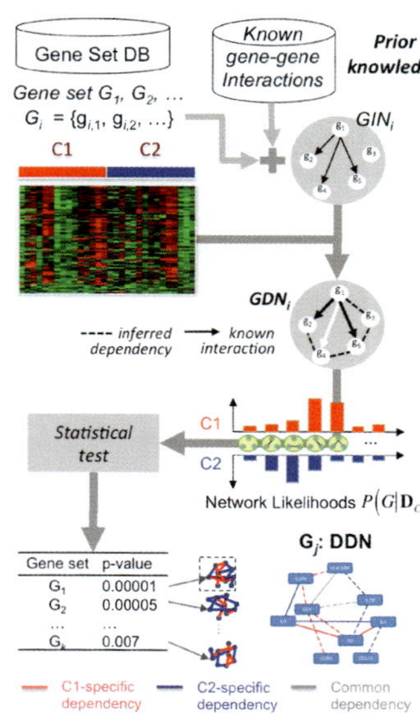

Figure 2. Workflow of knowledge-assisted EDDY

targets, the DDN often comprises hundreds of edges, rendering the prioritization of those dependencies non-trivial. We utilize the topological analysis of EDDY-derived DDNs to discern biologically important signaling nodes. These nodes could play important roles in biological signaling, hence, promising targets. For each node *i*, we will compute the normalized betweenness centrality metrics, $g(i|D_{C_1})$ and $g(i|D_{C_2})$ for GDNs, D_{C_1} and D_{C_2}, respectively (17). The regularized difference

$$\delta_{bw}(i|C_1, C_2) = \frac{g(i|D_{C_1}) - g(i|D_{C_2})}{g(i|D_{C_1}) + g(i|D_{C_2}) + \eta}, \quad (4)$$

where η is a regularization parameter, is then used to assist in prioritization of genes.

2.5. *Comparison to Knowledge-fused Differential Dependency Network (KDDN)*

The KDDN (Knowledge-fused Differential Dependency Network) model (18; 19) extends the DDN method by incorporating prior knowledge into its regularized linear regression problem with sparse constraints, where the level of prior knowledge, w_P, is a parameter taking value in [0, 1] to adjust the degree of prior-knowledge integration into the determination of differential dependency. We compare the results of knowledge-assisted EDDY against KDDN's results. KDDN does not aggregate differential dependencies of genes in a gene set and assign a score to a gene set as EDDY does, but focuses on individual differential dependencies. Hence, we focus on those pathways enriched with differential dependencies, identified by EDDY, and compare corresponding differential dependency networks between two methods.

3. Results

3.1. *Data, Gene Sets and Analysis*

We used the gene expression data of 202 glioblastoma multiforme (GBM) samples assigned with GB subtype from TCGA to identify pathways enriched with differential dependency between mesenchymal (58 samples) and non-mesenchymal samples, and between proneural (57 samples) and non-proneural samples. The gene expression data were log-transformed, standardized, and quantized prior to EDDY analysis. The gene sets queried for the analysis were 472 gene sets in REACTOME category of MSigDB. We then mined known interactions from Pathway Commons 2 (http://www.pathwaycommons.org) and matched these to all pairings in the REACTOME gene sets for prior knowledge incorporation. To investigate the effect of the degree of prior knowledge in identifying condition-specific dependencies, the prior weights w_P = 0, 0.5, and 1 were used. $w_P = 0$ specifies no influence of the known gene interactions in GDN inference and all edges in inferred GDN requires full support from the data, and $w_P = 1$ makes inferred GDN include all the known interactions unconditionally. When $w_P = 0.5$, dependencies with known interactions are added with half the support from the data.

3.2. *Pathways identified by knowledge-assisted EDDY*

Across three different prior weights (w_P = 0, 0.5, and 1.0), EDDY identified 57 pathways with statistically significant divergence between mesenchymal (MES) and non-mesenchymal for at least one of the weights, and 75 pathways between proneural (PN) and non-proneural. Table 1 presents a subset (24 pathways) of 57 mesenchymal-specific pathways, and Table 2 a subset (38 pathways) of proneural-specific 75 pathways, based on their biological interest (bold-faced) or p-value (w_P = 0.5) < 0.05. For each pathway, we include the number of genes in the pathway, p-values, P_D (the proportion of newly discovered dependencies, E_D, compared to the total number of edges in GDN, E_D+E_P) and P_C (the proportion of condition-specific dependencies, E_C, compared to total edges, E_C+E_S), for different prior weights. As w_P increases, more known interactions are added to GDN without condition-specificity, and this has three possible effects. First, condition-specific edges with weak support from data can gain support from the prior weighting, thereby increasing P_C while reducing P_D. Second, condition-specific edges with prior support can lose specificity and hence, result in reduced P_C. Finally, the loss of condition-specific edges can reduce the diversity of networks in the score distribution, having the indirect effect of increasing the influence of the surviving condition-specific edges on the divergence calculation. Indeed, we observe a consistent decrease in the number of networks in the distribution as we increase prior weight. As a result of these competing effects, p-value does not correlate with prior weight, even when examined over the finer variation of 0.1 (data not shown). However, we did note that the number of pathways with statistically significant divergence tends to decrease with prior weight – 28, 20 and 16 pathways with statistically significant divergence between mesenchymal and non-mesenchymal, and 39, 36 and 28 pathways between proneural and non-proneural, as the prior weight increases from 0 to 0.5 to 1.0.

Table 1: A subset of the REACTOME pathways with significant differential dependency between GB mesenchymal and non-mesenchymal. P_D gives the proportion of newly discovered dependencies over the total number of edges in GDN and P_C the proportion of condition-specific dependencies over total number of edges. Systematic ID from MSigDB is used instead of full pathway for shorten description. Mapping from Systematic IDs for bold-faced pathways are provided in Table 3 and Table 4, and in Appendix at the end for the rest of pathways.

Systematic ID	# genes	p-value			$P_D = E_D/(E_D+E_P)$			$P_C = E_C/(E_C+E_S)$		
		w_p=0	w_p=0.5	w_p=1	w_p=0	w_p=0.5	w_p=1	w_p=0	w_p=0.5	w_p=1
M760	27	0.0165	0.1314	0.2416	0.37			0.72		
M5113	29	0.1839	0.0173	0.4192		0.47			0.59	
M13748	34	0.1406	0.0299	0.0049		0.51	0.45		0.66	0.34
M9271	33	0.0122	0.0304	0.2399	0.77	0.66		0.75	0.68	
M506	23	0.0223	0.0478	0.1954	0.20	0.13		0.81	0.59	
M17157	19	0.0084	0.1605	0.6331	0.51			0.77		
M764	21	0.0019	0.1777	0.3609	0.73			0.83		
M571	38	0.6392	0.2754	0.0305			0.58			0.49
M9694	31	0.7833	0.0026	0.0705		0.04			0.35	
M1051	16	0.2921	0.0035			0.33			0.57	
M875	41	0.2310	0.0053	0.9018		0.58			0.76	
M612	23	0.3943	0.0104	0.8191		0.30			0.59	
M552	14	0.1828	0.0111	0.6727		0.19			0.58	
M3634	13	0.0091	0.0191		0.50	0.39		0.86	0.53	
M1062	21	0.1057	0.0222	0.1714		0.11			0.36	
M932	19	0.1187	0.0266	0.0606		0.64			0.79	
M16702	19	0.7982	0.0292	0.6791		0.39			0.61	
M1016	14	0.3862	0.0348	0.0561		0.47			0.66	
M1662	23	0.2844	0.0354	0.2397		0.33			0.64	
M6034	12	0.0568	0.0391	0.1070		0.92			0.64	
M17787	18	0.2575	0.0426	0.7349		0.69			0.33	
M7169	39	0.0082	0.0427	0.1184	0.85	0.81		0.80	0.76	

M901	35	0.0136	0.0427	0.0933	0.37	0.29		0.72	0.56	
M10122	13	0.3501	0.0433	0.6130		0.05			0.47	

Table 2: A subset of the REACTOME pathways with significant differential dependency between GB proneural and non-proneural.

Systematic ID	# genes	p-value			$P_D = E_D/(E_D+E_P)$			$P_C = E_C/(E_C+E_S)$		
		w_P=0	w_P=0.5	w_P=1	w_P=0	w_P=0.5	w_P=1	w_p=0	w_p=0.5	w_p=1
M647	16	0.0020	0.0017	0.0014	0.89	0.83	0.78	0.93	0.94	0.72
M530	37	0.0648	0.0022	0.4847		0.25			0.68	
M1092	14	0.0154	0.0071	0.0072	0.87	0.79	0.78	0.90	0.84	0.71
M549	12	0.0335	0.0114	0.8563	0.25	0.16		0.82	0.65	
M1040	19	0.0141	0.0151	0.0463	0.59	0.52	0.51	0.51	0.43	0.23
M13408	21	0.1654	0.0202	0.0242		0.43	0.40		0.65	0.33
M714	38	0.0112	0.1503	0.5874	0.56			0.75		
M570	44	0.0440	0.2321	0.5892	0.56			0.78		
M947	25	0.0045	0.0000		0.11	0.07		0.87	0.68	
M9450	12	0.3631	0.0007			0.39			0.68	
M860	28	0.1070	0.0011	0.0704		0.20			0.68	
M12967	35	0.0534	0.0013	0.0395		0.09	0.07		0.58	0.06
M936	30	0.0050	0.0020	0.0684	0.67	0.48		0.86	0.73	
M15243	10	0.0559	0.0029			0.00			0.58	
M1075	31	0.0135	0.0040	0.1367	0.39	0.29		0.88	0.74	
M846	36	0.2413	0.0052	0.5402		0.22			0.69	
M1662	23	0.0026	0.0059	0.1335	0.48	0.36		0.86	0.73	
M801	11	0.0274	0.0061	0.8040	0.50	0.38		0.75	0.58	
M899	39	0.1676	0.0073	0.1689		0.48			0.76	
M769	10	0.1899	0.0103	0.7851		0.43			0.93	
M13115	27	0.0144	0.0122	0.2782	0.03	0.02		0.77	0.64	
M12627	11	0.0001	0.0139		0.00	0.00		0.86	0.72	
M564	10	0.1861	0.0152	0.7291		0.19			0.48	
M10272	11	0.0758	0.0168	0.0001		0.54	0.50		0.72	0.40
M11184	15	0.0242	0.0180	0.0070	0.88	0.86	0.85	0.75	0.69	0.64
M719	15	0.1317	0.0190	0.1944		0.06			0.71	
M794	13	0.0326	0.0215	0.3349	0.61	0.49		0.82	0.69	
M1014	11	0.3598	0.0232			0.03			0.63	
M907	11	0.0022	0.0273	0.7901	0.63	0.52		0.68	0.65	
M837	27	0.4998	0.0273	0.4145		0.39			0.74	
M918	13	0.0023	0.0285	0.7926	0.63	0.52		0.68	0.65	
M704	44	0.1173	0.0287	0.2284		0.21			0.66	
M1016	14	0.1716	0.0359	0.2208		0.35			0.76	
M3661	22	0.0774	0.0416	0.0697		0.35			0.73	
M15195	30	0.0953	0.0432	0.0659		0.42			0.70	
M661	30	0.2166	0.0448	0.4245		0.28			0.65	
M583	18	0.0162	0.0453	0.1178	0.59	0.43		0.81	0.65	
M1825	11	0.0229	0.0488	0.0961	0.50	0.37		0.93	0.89	

3.3. *Biological Significance of Selected Signaling Pathways Identified by EDDY*

3.3.1. *Condition-specificity of Integrin αIIb β3 signaling in mesenchymal GB*

EDDY analysis of mesenchymal vs non-mesenchymal GB show significantly different (p = 0.0165 at w_P = 0.5) dependency network for INTEGRIN_ALPHAIIB_BETA3_SIGNALING (**M760**; http://bit.ly/1Dlgidx). This pathway is representative of biological mechanisms of adhesion in platelets, but there are proteins that participate in other signaling process in a diverse array of tissues and diseases. The class dependent DDNs show interesting differences in the state of this pathway's genes in mesenchymal vs. non-mesenchymal GB. DDN and GDNs in Figure 3 show that mesenchymal GB loses dependency on the cell surface integrins ITGA2B (betweenness normalized difference, δ_{bw}=-0.83[‡], rank, $R_{\delta_{bw}}$=2) and ITGB3 (δ_{bw}=-0.65, $R_{\delta_{bw}}$=7). Activation of ITGA2B/ITGB3-RAP1A-PTK2 signaling axis induces glioma cell proliferation (20). There is also a shift in the dependencies around SRC kinases between mesenchymal and non-mesenchymal GB samples with no SRC dependency evidence in mesenchymal samples but with new dependencies developed for Csk (δ_{bw}=0.12), also a member of Src-family kinase. In previous work, it is also demonstrated that Src family kinases plays very important role in migration and invasion cancer cells (21). Lastly, there is dependency shift in intracellular signaling effectors for integrins in the mesenchymal samples as evidenced by the δ_{bw} of PTPN1 (δ_{bw}=0.84, $R_{\delta_{bw}}$=1), APBB1IP (δ_{bw}=0.70, $R_{\delta_{bw}}$=6), SYK (δ_{bw}=0.43, $R_{\delta_{bw}}$=11), RAP1B (δ_{bw}=0.49, $R_{\delta_{bw}}$=9). These molecules have known roles in immunologic cell function, particularly cells of the monocytic origin (22-25). Mesenchymal GB samples have an appreciable amount of microglial (brain resident monocytic cells) cell infiltration that can be detected by RNA expression data (26), and it is interesting that EDDY appears to be detecting differential dependencies in molecules important for microglial function. In summary, this DDN demonstrates a differential wiring of ITGA2B/ITGB3 signaling network in mesenchymal vs non-mesenchymal GB. Functional validation of such differential wiring could help identifying novel nodes of vulnerability for treatment of subtype specific GB.

3.3.2. *Condition-specificity of PI3K events in ERBB2 signaling in proneural GB*

Another example of differential network dependency is illustrated in the analysis of proneural vs. non-proneural samples of GB. An example significant dependency network (p = 0.044 at w_P = 0)

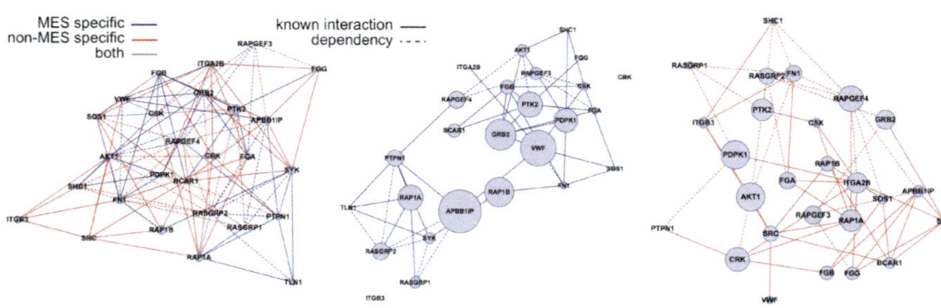

(a) DDN (b) GDN_{MES} (c) $GDN_{non\text{-}MES}$

Figure 3: (a) DDN, (b) GDN_{MES}, and (c) $GDN_{non\text{-}MES}$ of Integrin αIIb β3 signaling (M760) pathway

‡ The full data for the betweenness centrality and their difference between GDNs are not shown due to the space constraint. However, the betweenness centrality is indicated by the size of nodes in the GDNs.

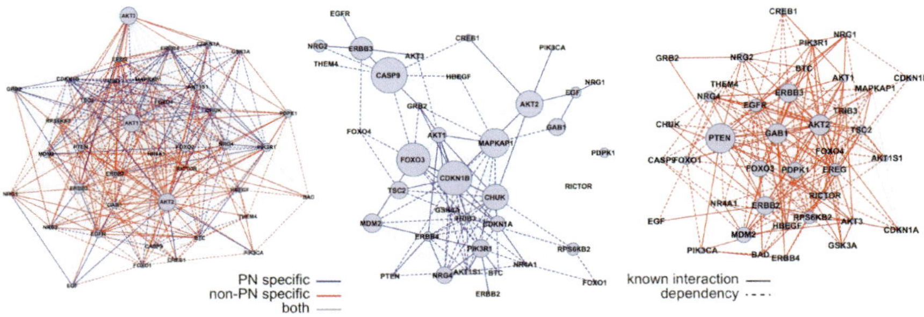

(a) DDN (b) GDN_{PN} (c) $GDN_{non\text{-}PN}$

Figure 4: (a) DDN, (b) GDN_{PN}, and (c) $GDN_{non\text{-}PN}$ of PI3K events in ERBB2 signaling (M570) pathway

is PI3K_EVENTS_IN_ERBB2 Signaling (**M570**; http://bit.ly/1I87dUt). This pathway highlights the signaling events from ERBB2, add associated family members, signal down through PIK3CA to AKT and mTOR signaling (Figure 4). There is a shift in the dependency of the ERBB signaling receptors between the proneural and non-proneural with a lessened dependency in the proneural. This is consistent with the observation that the proneural subtype of GB seems to be more reliant on PDGFRA signaling than signaling through ERBB2 (δ_{bw}=0.77, $R_{\delta_{bw}}$=4) and EGFR (δ_{bw}=0.71, $R_{\delta_{bw}}$=7) (27). However, PIK3R1 (δ_{bw}=0.60, $R_{\delta_{bw}}$=10) does show differential dependency in proneural samples, which agrees with observation of enrichment of PIK3R1 mutations in proneural samples (27). This may suggest that PIK3R1 mutations drive PIK3CA based signaling rather than PIK3CA mutations or ERBB alterations in the proneural subtype. It may also argue that PI3K signaling may needs to be targeted differently in different subtypes of GB.

3.4. *Comparison to KDDN*

Since KDDN does not aggregate score and p-value for pathway as EDDY does, we first identify pathways enriched with differential dependency, and apply KDDN to the same data set using the same prior knowledge for comparison. We used KDDN Cytoscape plug-in with parameters λ_1 set to 0.2, λ_2 to 0.05, and δ to 0.1, the default settings. The results are summarized in Tables 3 and 4.

With the default settings, kDDN identifies fewer edges than EDDY. Nevertheless, the general trend is that EDDY and kDDN find more than twice as much agreement in condition-specific edges than disagreement (selecting edges for opposite conditions). Varying λ_1 and λ_2 can increase the number of kDDN edges to approach those found by EDDY, but we sought a consistent approach to setting these parameters for fair comparison, rather than fitting agreement *ad hoc*. A key difference between the two applications is that EDDY identifies both condition-specific and shared edges for both conditions. When we include these edges, the overlap improves somewhat, but in general, the alignment between kDDN and EDDY is not substantial. We attribute this disagreement to the enhanced sensitivity of the EDDY method in assessing significance over a distribution of network scores. This might raise a concern for potential false positive discoveries by EDDY. However, our previous analysis of EDDY with simulation data indicates the false positive rate for EDDY is low, which is also supported by low P_D (< 0.5) in Table 1 and Table 2 –

majority of edges identified by EDDY are known interactions. We leave more comprehensive comparisons between EDDY and kDDN or other similar methods to our future study.

Table 3: A comparison of DDNs found by EDDY and KDDN for GB mesenchymal. EDDY queries selected specific gene sets depending on prior weight, w_P. Statistics for the two networks are common dependencies E_S and condition-specific dependencies E_C for EDDY, and condition-specific dependencies E_K for KDDN. The last column represents concordance between KDDN and EDDY DDN, specifically $|[E_{C1} \cap E_{K1}] \cup [E_{C1} \cap E_{K1}]|$ where E_{Ci}, represents C_i-specific edges identified by EDDY and E_{Ki} represents C_i-specific edges identified by KDDN.

REACTOME Pathway (PN)	ID	w_P	$\lvert E_S \rvert$	$\lvert E_C \rvert$	$\lvert E_K \rvert$	concordance
INSULIN_RECEPTOR_RECYCLING	M506	0.0	25	108	28	8
INSULIN_SYNTHESIS_AND_PROCESSING	M764	0.0	15	75	22	7
INTEGRIN_ALPHAIIB_BETA3_SIGNALING	M760	0.0	41	104	34	9
PURINE_METABOLISM	M9271	0.0	62	190	63	21
PYRUVATE_METABOLISM	M17157	0.0	16	54	53	6
GLUCONEOGENESIS	M13748	0.5	96	183	41	12
GLYCOLYSIS	M5113	0.5	105	149	35	11
INSULIN_RECEPTOR_RECYCLING	M506	0.5	80	115	28	7
PURINE_METABOLISM	M9271	0.5	94	197	63	21
GLUCONEOGENESIS	M13748	1.0	205	106	41	7
NUCLEAR_SIGNALING_BY_ERBB4	M571	1.0	185	180	65	19

Table 4: A comparison of DDNs found by EDDY and KDDN for GB proneural

REACTOME Pathway (PN)	ID	w_P	$\lvert E_S \rvert$	$\lvert E_C \rvert$	$\lvert E_K \rvert$	concordance
ACTIVATED_POINT_MUTANTS_OF_FGFR2	M647	0.0	4	57	5	3
DOWNREGULATION_OF_ERBB2_ERBB3_SIGNALING	M549	0.0	5	23	8	3
FGFR1_LIGAND_BINDING_AND_ACTIVATION	M1092	0.0	5	47	5	3
G1_S_SPECIFIC_TRANSCRIPTION	M1040	0.0	33	35	8	3
PI3K_AKT_ACTIVATION	M714	0.0	61	186	58	19
PI3K_EVENTS_IN_ERBB2_SIGNALING	M570	0.0	78	271	83	31
ACTIVATED_POINT_MUTANTS_OF_FGFR2	M647	0.5	4	61	5	3
DOWNREGULATION_OF_ERBB2_ERBB3_SIGNALING	M549	0.5	15	28	8	2
ERK_MAPK_TARGETS	M13408	0.5	53	99	27	12
FGFR1_LIGAND_BINDING_AND_ACTIVATION	M1092	0.5	9	48	5	3
G1_S_SPECIFIC_TRANSCRIPTION	M1040	0.5	44	33	8	3
NEGATIVE_REGULATION_OF_FGFR_SIGNALING	M530	0.5	130	271	48	26
ACTIVATED_POINT_MUTANTS_OF_FGFR2	M647	1.0	19	50	5	3
ERK_MAPK_TARGETS	M13408	1.0	108	54	27	5
FGFR1_LIGAND_BINDING_AND_ACTIVATION	M1092	1.0	17	41	5	3
G1_S_SPECIFIC_TRANSCRIPTION	M1040	1.0	61	18	8	2

4. Discussion

Expression profiling and whole genome sequencing from hundreds of GB specimens by TCGA has revealed a broad spectrum of genetic alterations and discrete expression signatures and subtypes (27; 28). However, the issue of how to best target these molecular subtypes using pharmacological agents remains to be addressed. An obstacle in identifying subtype-specific drug vulnerabilities is how genetic alterations and gene expression affect wiring of key signaling networks that drives tumor phenotype (29). In this work we demonstrated that using knowledge-assisted EDDY, it is possible to identify subtype specific network wiring and gene dependencies, which may be used to identify subtype specific drug vulnerabilities.

Finally, we have recently started an implementation of the EDDY algorithm on a GPU, which has shown dramatic acceleration. Besides making computations faster and allowing for the running of larger datasets, we envision a prior weight optimization over the number of condition-specific edges. Additionally, experimental validation of highlighted differences is a main priority in the future. We have access to cohort of 64 patient derived GB xenografts that include all four GBM subtypes and are available to readily deploy to test novel hypothesis indicated through EDDY analysis.

5. Acknowledgments

This work was partly funded by National Cancer Institute, National Institutes of Health (NIH) [1U01CA168397 — Center for Target Discovery and Development (CTD2)].

References

1. Subramanian A, Tamayo P, Mootha VK, Mukherjee S, Ebert BL, et al. 2005. *Proc Natl Acad Sci U S A* 102:15545-50
2. Califano A. 2011. *Molecular systems biology* 7:463
3. de la Fuente A. 2010. *Trends in genetics : TIG* 26:326-33
4. Ideker T, Krogan NJ. 2012. *Molecular systems biology* 8:565
5. Guo Z, Li Y, Gong X, Yao C, Ma W, et al. 2007. *Bioinformatics* 23:2121-8
6. Hwang T, Park T. 2009. *BMC Bioinformatics* 10:128
7. Kim Y, Kim T-K, Kim Y, Yoo J, You S, et al. 2010. *Bioinformatics*
8. Ma H, Schadt EE, Kaplan LM, Zhao H. 2011. *Bioinformatics*
9. Lai Y, Wu B, Chen L, Zhao H. 2004. *Bioinformatics* 20:3146-55
10. Hu R, Qiu X, Glazko G, Klebanov L, Yakovlev A. 2009. *BMC Bioinformatics* 10:20
11. Mentzen W, Floris M, de la Fuente A. 2009. *BMC Genomics* 10:601
12. Zhang B, Li H, Riggins RB, Zhan M, Xuan J, et al. 2009. *Bioinformatics* 25:526-32
13. Zhang B, Tian Y, Jin L, Li H, Shih Ie M, et al. 2011. *Bioinformatics* 27:1036-8
14. Choi Y, Kendziorski C. 2009. *Bioinformatics* 25:2780-6
15. Jung S, Kim S. 2014. *Nucleic acids research* 42:e60
16. Lin J. 1991. *IEEE Transactions on Information Theory* 37:145-51
17. Freeman LC. 1977. *Sociometry* 40:35-41
18. Tian Y, Zhang B, Hoffman EP, Clarke R, Zhang Z, et al. 2014. *BMC systems biology* 8:87
19. Tian Y, Zhang B, Hoffman EP, Clarke R, Zhang Z, et al. 2015. *Bioinformatics* 31:287-9
20. Sayyah J, Bartakova A, Nogal N, Quilliam LA, Stupack DG, Brown JH. 2014. *The Journal of biological chemistry* 289:17689-98
21. Guarino M. 2010. *Journal of cellular physiology* 223:14-26
22. Traves PG, Pardo V, Pimentel-Santillana M, Gonzalez-Rodriguez A, Mojena M, et al. 2014. *Cell death & disease* 5:e1125
23. Jakus Z, Fodor S, Abram CL, Lowell CA, Mocsai A. 2007. *Trends in cell biology* 17:493-501
24. Li Y, Yan J, De P, Chang HC, Yamauchi A, et al. 2007. *Journal of immunology* 179:8322-31
25. Medrano-Fernandez I, Reyes R, Olazabal I, Rodriguez E, Sanchez-Madrid F, et al. 2013. *Cellular and molecular life sciences : CMLS* 70:2395-410
26. Engler JR, Robinson AE, Smirnov I, Hodgson JG, Berger MS, et al. 2012. *PloS one* 7:e43339
27. Verhaak RG, Hoadley KA, Purdom E, Wang V, Qi Y, et al. 2010. *Cancer cell* 17:98-110
28. Cancer Genome Atlas Research N. 2008. *Nature* 455:1061-8
29. Oh YT, Cho HJ, Kim J, Lee JH, Rho K, et al. 2014. *PloS one* 9:e103327

Appendix

Systematic ID	Pathway
M10122	RETROGRADE_NEUROTROPHIN_SIGNALLING
M1014	IL_6_SIGNALING
M1016	SYNTHESIS_OF_VERY_LONG_CHAIN_FATTY_ACYL_COAS
M1016	SYNTHESIS_OF_VERY_LONG_CHAIN_FATTY_ACYL_COAS
M10272	IONOTROPIC_ACTIVITY_OF_KAINATE_RECEPTORS
M1051	INTEGRATION_OF_PROVIRUS
M1062	ANTIGEN_PRESENTATION_FOLDING_ASSEMBLY_AND_PEPTIDE_LOADING_OF_CLASS_I_MHC
M1075	INWARDLY_RECTIFYING_K_CHANNELS
M11184	ENDOGENOUS_STEROLS
M12627	DOPAMINE_NEUROTRANSMITTER_RELEASE_CYCLE
M12967	MRNA_3_END_PROCESSING
M13115	G_PROTEIN_ACTIVATION
M15195	MAPK_TARGETS_NUCLEAR_EVENTS_MEDIATED_BY_MAP_KINASES
M15243	GAP_JUNCTION_DEGRADATION
M1662	SIGNALING_BY_BMP
M1662	SIGNALING_BY_BMP
M16702	ACTIVATED_AMPK_STIMULATES_FATTY_ACID_OXIDATION_IN_MUSCLE
M17787	GLUCURONIDATION
M1825	REGULATION_OF_INSULIN_SECRETION_BY_ACETYLCHOLINE
M3634	CASPASE_MEDIATED_CLEAVAGE_OF_CYTOSKELETAL_PROTEINS
M3661	FGFR_LIGAND_BINDING_AND_ACTIVATION
M552	PROLACTIN_RECEPTOR_SIGNALING
M564	MEMBRANE_BINDING_AND_TARGETTING_OF_GAG_PROTEINS
M583	RIP_MEDIATED_NFKB_ACTIVATION_VIA_DAI
M6034	SEROTONIN_RECEPTORS
M612	CIRCADIAN_REPRESSION_OF_EXPRESSION_BY_REV_ERBA
M661	SIGNALING_BY_FGFR1_MUTANTS
M704	SIGNALING_BY_FGFR_MUTANTS
M7169	NCAM1_INTERACTIONS
M719	SHC1_EVENTS_IN_EGFR_SIGNALING
M769	ELEVATION_OF_CYTOSOLIC_CA2_LEVELS
M794	ACTIVATION_OF_CHAPERONES_BY_ATF6_ALPHA
M801	ACTIVATION_OF_CHAPERONE_GENES_BY_ATF6_ALPHA
M837	CREB_PHOSPHORYLATION_THROUGH_THE_ACTIVATION_OF_RAS
M846	FRS2_MEDIATED_CASCADE
M860	SHC_MEDIATED_CASCADE
M875	NETRIN1_SIGNALING
M899	IL1_SIGNALING
M901	GLOBAL_GENOMIC_NER_GG_NER
M907	CALNEXIN_CALRETICULIN_CYCLE
M918	N_GLYCAN_TRIMMING_IN_THE_ER_AND_CALNEXIN_CALRETICULIN_CYCLE
M932	SYNTHESIS_SECRETION_AND_INACTIVATION_OF_GLP1
M936	TRAF6_MEDIATED_IRF7_ACTIVATION
M9450	PLATELET_ADHESION_TO_EXPOSED_COLLAGEN
M947	INHIBITION_OF_VOLTAGE_GATED_CA2_CHANNELS_VIA_GBETA_GAMMA_SUBUNITS
M9694	ACTIVATION_OF_THE_PRE_REPLICATIVE_COMPLEX

INTEGRATING GENETIC AND STRUCTURAL DATA ON HUMAN PROTEIN KINOME IN NETWORK-BASED MODELING OF KINASE SENSITIVITIES AND RESISTANCE TO TARGETED AND PERSONALIZED ANTICANCER DRUGS

GENNADY M. VERKHIVKER†

Department of Computational Biosciences, Schmid College of Science & Technology, Chapman University, One University Drive, Orange CA 92866,USA

Department of Pharmacology, University of California San Diego, 9500 Gilman Drive, San Diego CA 92093, USA
Email: verkhivk@chapman.edu

The human protein kinome presents one of the largest protein families that orchestrate functional processes in complex cellular networks, and when perturbed, can cause various cancers. The abundance and diversity of genetic, structural, and biochemical data underlies the complexity of mechanisms by which targeted and personalized drugs can combat mutational profiles in protein kinases. Coupled with the evolution of system biology approaches, genomic and proteomic technologies are rapidly identifying and charactering novel resistance mechanisms with the goal to inform rationale design of personalized kinase drugs. Integration of experimental and computational approaches can help to bring these data into a unified conceptual framework and develop robust models for predicting the clinical drug resistance. In the current study, we employ a battery of synergistic computational approaches that integrate genetic, evolutionary, biochemical, and structural data to characterize the effect of cancer mutations in protein kinases. We provide a detailed structural classification and analysis of genetic signatures associated with oncogenic mutations. By integrating genetic and structural data, we employ network modeling to dissect mechanisms of kinase drug sensitivities to oncogenic EGFR mutations. Using biophysical simulations and analysis of protein structure networks, we show that conformational-specific drug binding of Lapatinib may elicit resistant mutations in the EGFR kinase that are linked with the ligand-mediated changes in the residue interaction networks and global network properties of key residues that are responsible for structural stability of specific functional states. A strong network dependency on high centrality residues in the conformation-specific Lapatinib-EGFR complex may explain vulnerability of drug binding to a broad spectrum of mutations and the emergence of drug resistance. Our study offers a systems-based perspective on drug design by unravelling complex relationships between robustness of targeted kinase genes and binding specificity of targeted kinase drugs. We discuss how these approaches can exploit advances in chemical biology and network science to develop novel strategies for rationally tailored and robust personalized drug therapies.

---

† This work is partly supported by funding from Chapman University.

1. Background

The era of significant scientific breakthroughs and technological advancements in genetics and biology has brought to clinical settings personalized health care that has the capacity to detect the onset of disease at its earliest stages and preempt the progression of disease. The comprehensive cancer genome characterization efforts have refined our understanding of specified genes responsible for development and progression of tumors[1]. Several malignancies are associated with the mutation or increased expression of protein kinases, including lung, breast, stomach, colorectal, head and neck, and pancreatic carcinomas and glioblastoma[2]. Tumor sequencing efforts have identified a rich source of naturally occurring mutations with many being simple single nucleotide polymorphisms (SNPs) in protein kinases. A subset of these SNPs occurs in the coding regions (cSNPs) of kinases and result in a change in the encoded amino acid sequence (nonsynonymous coding SNP; nscSNPs). Genome studies have revealed the importance of "driver" somatic alterations that activate crucial oncoproteins such as *EGFR, BCR-ABL,* and other kinase genes. Mutations in these protein kinases are often implicated in many cancers and exemplify the phenomenon of 'oncogene addiction,' according to which the effects of driver genomic alterations are pivotal for tumor proliferation and have a selective advantage for the formation of the tumor during somatic cell replication[3]. Oncogene dependencies induced by genetic alterations in *BCR-ABL, KIT, EGFR* and other kinase genes are well known and have provided decisive clinical proof of principle for the genomics-informed drug discovery of kinase drugs[4]. Although tumor dependencies driven by dominant oncogenes could respond to targeted therapies, clinical responses to single agents are often followed by the development of drug resistance. The tumor dependency concept is especially relevant to understand mechanisms of acquired resistance, where resistant mutations, seemingly developed due to drug treatment, may instead represent evolutionary selection of cell subpopulations which harbor preexistent somatic mutant variants which confers a primary resistance to these cells and provides them with a selective advantage. The spectrum of lung cancer EGFR mutations can induce oncogenic transformation by leading to constitutive kinase activity of EGFR and confer markedly different sensitivity to EGFR inhibitors[5]. The most common reported mutations are the deletion of five exon-19 residues and the exon-21 substitution L858R in the catalytic domain of EGFR[6]. Together, these mutations correspond to more than 90% of the activating EGFR mutations observed in non-small-cell lung cancer (NSCLC). While T790M has only a modest effect on EGFR function, a tandem of T790M and L858R mutations can result in a dramatic enhancement of EGFR activity. More than 200 activating and drug resistance EGFR mutations with different clinical responses to tyrosine kinase inhibitors have been reported[7] and molecular mechanisms of mutation-induced kinase activation have been extensively discussed[8].

Gefitinib and Erlotinib are orally effective protein-kinase targeted inhibitors that are used in the treatment of ERBB1/EGFR-mutant lung cancer. Afatinib is another EGFR-targeted kinase drug approved by the FDA for the first-line treatment of patients with metastatic NSCLC whose tumors have *EGFR* exon 19 deletions or exon 21 (L858R) substitutions. Lapatinib, a small molecule tyrosine kinase inhibitor of both EGFR and HER2/ErbB2 is now also approved for advanced HER2-amplified breast cancer[9]. Structural and biochemical studies have characterized the inhibition of intrinsic catalytic activity of EGFR and HER2/Erbb2 variants by Lapatinib using a diverse array of enzymatic and cell-based assays[10,11]. Cell-based EGFR resistance mutation screens have demonstrated that Lapatinib produced the broadest mutation spectra of any of the EGFR-targeted drugs tested in *in vitro* system, with a number of Lapatinib-specific resistant mutations clustered around the selectivity pocket and the EGFR-A-loop[12]. The association between EGFR mutations and differential drug sensitivity suggested that genetic EGFR alterations and corresponding changes in structural and interaction profiles of the EGFR kinase domain render tumors sensitive to selective inhibitors. Oncogenic kinases can adopt different mechanisms to alleviate negative regulatory processes associated with their intrinsic conformational instability. One of them is the recruitment of unstable kinase forms to the Hsp90 system that protects abnormally activated kinases in cancer cells[13]. HSP90 stabilizes viral kinases and various mutated oncogenes, including oncogenic EGFR mutants that are dependent on the chaperoning function through direct interactions to maintain their stability[14]. HSP90 inhibition reduces mutant EGFR levels and activity, suggesting a viable EGFR inhibition strategy. Crystallographic studies[15] have supported this mechanism by showing that the catalytic domains of the EGFR-L858R and EGFR-L858R/T790M oncogenic mutants can adopt flexible inactive conformations that may facilitate conformational release from the autoinhibitory state. This may be exploited by the Hsp90 chaperone to bind the unstable mutant conformations and promote an accumulation of a constitutively active form. According to the newly emerging paradigm, kinase inhibitors may exert their primary effect by "arresting" the kinase domain in the specific inactive form, thereby depriving the Hsp90 system from access to unstable conformational states and preventing uncontrollable accumulation of the active form[16].

The abundance and diversity of genetic, structural, and biochemical data underlies the complexity of mechanisms by which targeted and personalized kinase agents can combat mutational profiles in EGFR kinase. We employ a battery of synergistic computational approaches that integrate genetic, biochemical, and structural data to characterize the effect of cancer mutations in protein kinases. We show that binding specificity and drug resistance of EGFR drugs may be linked with the global network properties of key residues that are responsible for structural stability of specific targeted conformations. The results of this study offer a network-based perspective on drug design of targeted and personalized kinase drugs, showing how the efficiency and robustness of the interaction networks may be associated with kinase binding preferences and emergence of resistant mutations.

2. Methods

2.1. *Data mining*

Protein kinase sequences were obtained from Kinbase (http://kinase.com/kinbase/). Common SNPs were retrieved from PupaSNP and dbSNP using the Ensembl data mining tool, Biomart (http://www.ensembl.org/Homo_sapiens/martview). The disease causing SNPs were retrieved from OMIM, KinMutBase, and HGMD databases. We used all kinase gene entries referenced in NCBI and SwissProt database, and 7955 unique SNP entries corresponding to these kinase genes as they are referenced in NCBI. These unique SNP entries include 3722 synonymous, 3985 missense, 75 nonsense and 173 frameshift mutations. We have also gathered 780 OMIM variant entries from NCBI and 3542 SwissProt variant entries. Cancer mutations were retrieved from OMIM and COSMIC databases. Motif-based alignments of kinase sequences to the catalytic core were first generated by implementation of the Gibbs motif sampling method. This method identifies characteristic motifs for each individual subdomain of the kinase catalytic core, which are then used to generate high-confidence motif-based Markov chain Monte Carlo multiple alignments based on these motifs[17]. The nsSNPs were then mapped to the kinase catalytic domain in accordance with this alignment. Cancer driver predictions were performed by using the SVM approach as described in the earlier work[18].

2.2. *Somatic mutation distributions and driver mutation hotspots in protein kinome*

Functionally important subdomains of the kinase catalytic core were examined to determine the distribution of nsSNPs and identify structurally conserved hotspots of functionally important mutations. The number of SNPs in each of the subdomains was calculated from the structure-informed multiple sequence alignment. The expected probability E(p) of a SNP occurring in a kinase subdomain region was calculated separately for each SNP type. In brief, the average length of each region was calculated as the weighted average of the region length in each kinase considered, where weights correspond to the total number of SNPs occurring within each kinase. The probability of a SNP occurring within a particular region purely by chance was computed as its weighted average length over the sum of every region's weighted average length. The probability (p-value) of the observed total number (x) of SNPs occurring within each region, where n is the total number of SNPs considered, was calculated using the general binomial distribution. The average length of each sub-domain was calculated as the weighted average of the region length in each kinase considered, where weights correspond to the total number of SNPs occurring within each kinase. The probability of a SNP occurring within a particular region purely by chance was computed as its weighted average length divided by the sum of every region's weighted average length. The probability (p-value) of the observed total number of SNPs occurring within each region was then calculated using the general binomial distribution. Cancer mutant predictions and analysis were performed as described in previous studies[21]. A support vector machine (SVM) was trained upon common SNPs (presumed neutral) and congenital disease causing SNPs characterized by a variety of sequence, structural, and phylogenetic parameters. The threshold taken for calling a SNP a driver is 0.49 for catalytic domain mutations, and 0.53 for all other mutations.

2.3. *Network modeling of residue interaction networks in protein kinases*

Molecular dynamics (MD) simulations were carried out using NAMD 2.6 with the CHARMM27 force field[19]. The binding free energies and computational alanine scanning of kinase-drug complexes were done using MM-GBSA approach[20]. A graph-based representation of proteins was used in the protein structure network analysis, where residues were considered as nodes and edges correspond to the nonbonding residue-residue interactions. The pair of residues with the interaction strength I_{ij} greater than a user-defined cut-off $I_{\min}$ are connected by edges and produce a protein structure network graph for a given interaction strength $I_{\min}$. The strength of interaction between two amino acid side chains is

$$I_{ij} = \frac{n_{ij}}{\sqrt{(N_i \times N_j)}} \times 100 \quad (1)$$

where n_{ij} is number of distinct atom pairs between the side chains of amino acid residues i and j that lie within a distance of 4.5 Å. N_i and N_j are the normalization factors for residues i and j respectively[21]. We considered any pair of residues to be connected if $I_{\min}$ was greater than 3.0%. A weighted network representation of the protein structure is adopted that includes non-covalent connectivity of side chains and residue cross-correlation fluctuation matrix[22]. In this model, the weight w_{ij} of an edge between nodes i and j is measured as $w_{ij} = -\log(|C_{ij}|)$ where C_{ij} is the element of the covariance matrix measuring the cross-correlation residue fluctuations obtained from MD simulations. The shortest paths between two residues are determined using the Floyd–Warshall algorithm. We computed the residue-based betweenness which is defined as the sum of the fraction of shortest paths between all pairs of residues that pass through residue i:

$$C_b(n_i) = \sum_{j<k}^{N} \frac{g_{jk}(i)}{g_{jk}} \quad (2)$$

where g_{jk} denotes the number of shortest geodesics paths connecting j and k, and $g_{jk}(i)$ is the number of shortest paths between residues j and k passing through the node n_i.

3. Results

3.1. *Structural and functional signatures of cancer mutations in protein kinases*

Genetic variations in protein kinase genes are widely spread across both phylogenetic and structural space, and only a subset of all SNPs could be directly mapped to the kinase catalytic domain (Figure 1A). We constructed the distribution of various SNPs categories that could be mapped onto the 12 functional subdomains (SDs) of the kinase catalytic core (Figure 1B). Structural mapping of sSNPs resulted in a uniform coverage of kinase subdomains, showing only a weak preference towards SD II which has no obvious functional role in kinase regulation. The distribution of nsSNPs pointed to the preferential bias towards specific functional regions. Functionally important P-loop (SD I), hinge region (SD V), catalytic loop (SD VIB), and A-loop (SD VII) along with the P+1 loop region (SD VIII) are more densely populated The catalytic domain of protein kinases harbors a large number of SNPs falling into three major categories: common and neutral SNPs; inherited disease causing germline SNPs; and cancer causing SNPs. By compiling and mapping a total of 355 common SNPs, 428 inherited disease causing SNPs, and 541 cancer associated SNPs we found a statistically significant enrichment of different categories of SNPs in specific l regions of the catalytic domain (Figure 1C). Common nsSNPs are randomly distributed within the catalytic core, only sparsely populating functional segments of the catalytic core, such as the catalytic or A-loops, whereas these nsSNPs more densely occupy evolutionary unconserved regions of the C-terminal tail. The disease-causing nsSNPs primarily mapped to the regions involved in regulation and substrate binding, such as the APE-loop and the P+1 region, as well as the catalytic loop (Figure 1C). Cancer-associated nsSNPs tend to target regions directly involved in the catalytic activity that are mainly localized in the P-loop, A-loop and catalytic loop. The distribution of kinase nsSNPs across functional kinase subdomains suggested that the kinase regions that are enriched in different types of SNPs are markedly different and have only a minimal overlap. The distribution revealed a preference for cancer-causing nsSNPs to populate primarily the A-loop (SDVII) and the P-loop (SD I). The functionally important for substrate and protein binding P+1 loop are enriched largely in disease-associated mutations, but not cancer-causing mutations. These results indicated that disease-associated mutations could primarily affect the kinase regions involved in functional regulation, allosteric interactions and substrate binding[23].

Kinome-wide analysis of sequence and structure-based signatures of cancer mutations revealed that a significant number of cancer mutations could fall at structurally equivalent positions within the catalytic core. These structurally conserved mutations tend to cluster into specific mutational hotspots which may be shared by multiple kinase genes. We classified cancer mutation hotspots which had been identified as a frequent target of tumorigenic activating mutations. Cancer mutation hotspots in protein kinases are largely localized within the P-loop, hinge region, and A-loop (Figure 1).

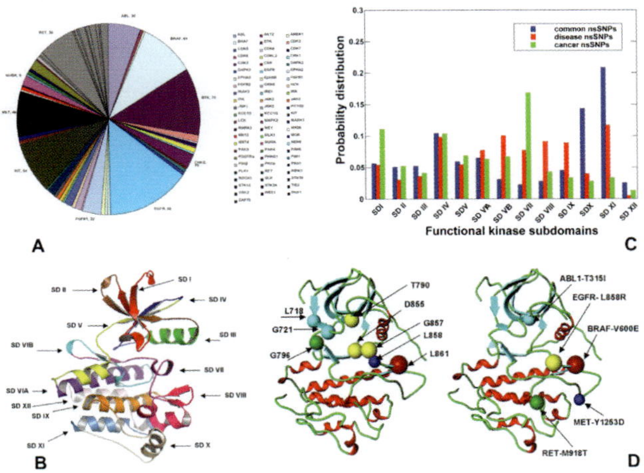

Figure 1. The distribution of nsSNPs in the catalytic core (A,C). The catalytic domain was subdivided into 12 subdomains (B) with some subdomains corresponding to functional regions : SD I (P-loop); SDIII(αC-helix); SDV (hinge region); SDVIB (catalytic loop); SDVII (A-loop) ; SDVIII (P+l loop). (B)Structural mapping is shown for common nsSNPs , disease-causing nsSNPs , and cancer-causing nsSNPs. (D) Structural localization of driver mutations is mapped onto the crystal structure of the active EGFR (pdb entry 2J6M). Structural annotation of cancer driver mutations is arranged according to their oncogenic potential. The higher the oncogenic potential of the cancer drive, the larger the ball denoting structural position of the respective mutation.

3.2 *Structural bioinformatics analysis of oncogenic kinase mutants: distinct structural signatures of Hsp90-dependent kinase clients are associated with oncogenic potential*

Oncogenic kinase mutants may rely on the Hsp90 dependence for the maintenance of stability and accumulation of the constitutively active form. In particular, Hsp90 function is essential to maintain high-level expression of mutant EGFR in lung cancer cells[14]. We performed kinome-wide structural bioinformatics analysis of chaperone-regulated kinases (Figure 2). The proteomics-based client annotation (Figure 2A) was compared against structure-based mapping of the Hsp90-Cdc37 kinase clients (Figure 2b). Structural coupling of the catalytic DFG motif and the regulatory αC-helix is recognized as central in controlling kinase activity and dynamic equilibrium between the inactive (DFG-out/αC-helix-in), the Cdk/Src-like inactive (DFG-in/αC-helix-out) and the active kinase forms (DFG-in/αC-helix-in). Although many of the Hsp90 kinase clients can occupy evolutionary different branches of the human kinome, we found they share a common Cdk/Src- type structural arrangement of their inactive functional states. The Cdk/Src-like inactive structures shared by the Hsp90 kinase clients are unified by a common structural determinant whereby the regulatory αC-helix is moved to a αC-out conformation and forms autoinhibitory clamp with the A-loop, thus preventing the formation of the catalytically competent active kinase.

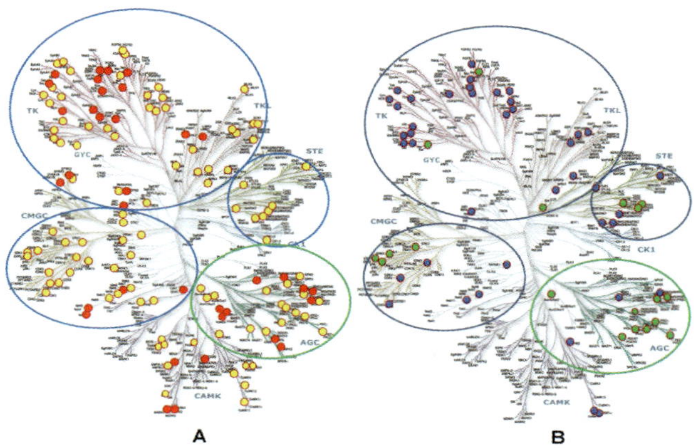

Figure 2. The distribution of the Hsp90-dependent protein kinase clients in the human kinome. (A) Kinome mapping of Hsp90-Cdc37 clients discovered in proteomic-based studies[16] is depicted. The kinases that are found to be downregulated by Hsp90 inhibition in the experimental profiling are shown in yellow (confirmed kinase clients) and red (novel kinase clients from proteomics studies[16]). (B) Structure-based mapping of the Hsp90-Cdc37 kinase clients. The Cdk/Src kinase clients are marked in blue filled spheres. A high density of the Cdk/Src clients in the TK, TKL, STE, CAMK, and CMGC groups of the human kinome tree is highlighted by blue circles. The second category of kinase clients is characterized by active structures stabilized through allosteric interactions (green spheres).

According to our analysis, oncogenic kinase mutations in the conserved hotspots (A-loop), may perturb the constraints keeping the αC-helix-out in the rigid inactive position, and allow the A-loop to assume an extended active conformation (A-loop open) that is seen in the as crystal structures of the EGFR-L858R and EGFR-L858R/T790M mutants[15]. These Cdk/Src-like active conformations that can be adopted by oncogenic mutants are far more flexible and unstable. As a result, they may be sequestered by the Hsp90 to promote uncontrollable transformation and accumulation of the constitutatively active state for kinase cancer mutants.

3.3 *Integrating genetic and structural data on oncogenic EGFR mutations: modeling of thermodynamic and networking signatures of targeted drug binding*

By using MD simulations and MM-GBSA binding free energy simulations, we evaluated the thermodynamic effect of oncogenic EGFR mutations on different conformational states of EGFR (Figure 3A). Our results showed that oncogenic mutations L747P, L747S, L858R and L861Q can destabilize the rigid autoinhibitory structure that is thermodynamically favorable in the wild-type EGFR[24]. Strikingly, oncogenic mutations L747P/S, L858R and L861Q seemed to favor a highly flexible Cdk/Src –active conformation and marginally destabilize the active conformation. As a result, EGFR mutations with a high oncogenic potential may destabilize the dormant autoinhibitory structure. These mutations may induce fast equilibrium between flexible Cdk/Src-like active conformation and active structure that could lead to uncontrollable activity, which is a "deadly" signature of cancer mutations. The major Lapatinib-resistant mutations with the high oncogenic potential occurred in the residues that do not directly contact ligand. L747 is located at a loop adjacent to αC-helix; V765 and V769 are at or near the C-terminal portion of αC-helix, and T790is at the gatekeeper position in the ATP binding site. Of the remainder, N857 is located in helix D, T854 forms the base of the ATP binding site, L858 and H870 are in the A-loop (Figure 3). To determine the thermodynamic contribution of the EGFR residues to Lapatinib binding and identify energetic hot spots susceptible to mutations, we performed free energy simulations and computational alanine scanning (Figure 3B). First, we found that only some Lapatinib-interacting residues corresponded to cancer mutation hotspots, suggesting that escaping binding interactions with the drug via mutations may not be a primary mechanism that drives emergence of Lapatinib-resistant mutations. The energetic hot spots of Lapatinib binding that corresponded to cancer mutation drug-resistant EGFR sites included L718, L777, L788, T790 (gate-keeper), and T854 residues. However, the EGFR mutations of high oncogenic potential that can render differential sensitivity to Lapatinib such as L747, L858, and L861 make fairly moderate contributions to binding energetics that could not explain high resistance. These results suggested that the mechanism of Lapatinib-induced somatic mutations may rather be associated with the intrinsic stability of the Cdk/Src inactive EGFR structure that binds Lapatinib[10-12]. Several hypotheses have suggested that the mechanism of Lapatinib-induced somatic mutations is linked with a conformation-specific mode of Lapatinib binding to an inactive EGFR structure[11,12] as drug resistant cancer mutations may stabilize the constitutively active EGFR form and thus interfere with the drug binding. To test this mechanism, we evaluated organization of the residue interaction networks and structural stability of EGFR states. The stability of the inactive EGFR conformation targeted by Lapatinib is mediated by interaction networks formed by high centrality residues F723 (P-loop), αC-helix (V765, M766, and V769), the αC-β4-loop (L774), the HRD motif (H835, D837), DFG motif (D855, F856) and L858 (A-loop) (Figure 3C, Table 1). The central result of the network analysis showed that although some somatic mutations may emerge in residues with medium centrality, Lapatinib-resistant cancer mutations can be developed in high centrality sites that determine interaction network of the specific EGFR form (Table 1). Due to their central position in the structural network, mutations of V765 and V769 (αC-helix) and L858 (A-loop) can severely compromise the integrity of the interaction network by weakening or dissolving the central autoinhibitory lock between the P-loop/A-loop interactions holding the αC-helix in the inactive position. Targeted mutations of these high centrality sites could disrupt allosteric coupling between functional regions, leading to the weakening and fragmentation of the residue interaction network. A strong network dependency on high centrality residues may explain a broad spectrum of Lapatinib-resistant mutations that are located away from the inhibitor, near the αC-helix and in the A-loop. Hence, residue centrality may be used as a metric for assessing severity of drug resistance mutations and differentiating between highly resistant and moderately resistant positions.

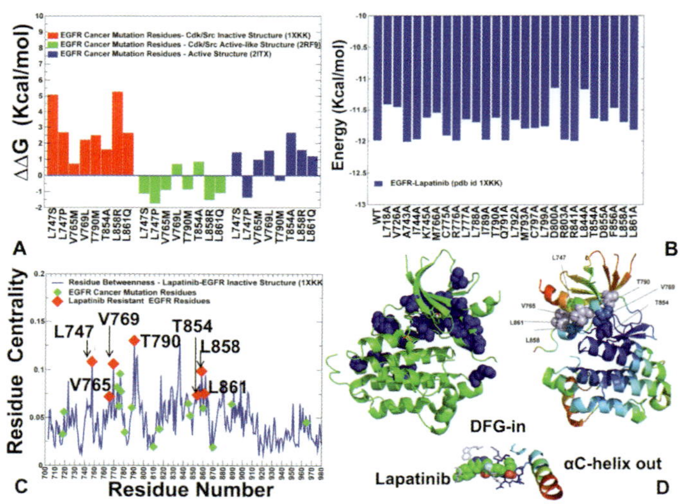

Figure 3. Structure-based network modeling of EGFR cancer mutations and drug binding. (A) Free energy changes caused by oncogenic mutations in different conformational states of EGFR. (B) Computational alanine scanning of binding site residues in the Lapatinib-EGFR complex (pdb id 1XKK). (C) The residue centrality profile of Lapatinib-EGFR complex (in blue). EGFR mutations are shown in green diamonds and Lapatinib-resistant oncogenic mutations are shown in red diamonds. (D) Structural mapping of EGFR cancer mutations (blue spheres) on the crystal structure of Lapatinib-EGFR complex (green ribbons). Mapping of Lapatinib-resistant mutations (indicated by arrows) on the crystal structure of Lapatinib-EGFR complex colored according to structural stability.

Table 1: Structure-based network analysis of the EGFR kinase domain and Lapatinib-EGFR complex. Structural region and network centrality of functional EGFR residues targeted by cancer mutations and drug resistant mutations are reported.

Residue	Residue#	Betweenness	Mutation	Exon	Kinase/segment/spine
Leu	718	0.03230	L718P	Exon 18	β1 strand
Gly	719	0.05586	G719A/C/R/S	Exon 19	Gly-rich P-loop
Leu	747	0.10818	L747S/P	Exon 19	β3-αC loop
Val	765	0.07211	V765M	Exon 20	αC-helix
Val	769	0.10593	V769L	Exon 20	αC-helix
His	773	0.08204	H773L	Exon 20	αC-β4 loop
Cys	775	0.06188	C775F/R/Y	Exon 20	αC-β4 loop
Arg	776	0.09576	R776S/C/H/P/L	Exon 20	αC-β4 loop
Leu	777	0.07761	L777Q/P/M	Exon 20	αC-β4 loop(R-spine)
Cys	781	0.03448	C781F	Exon 20	β4 strand
Leu	788	0.06048	L788V/I/F	Exon 20	β5 strand
Thr	790	0.12979	T790M/A	Exon 20	β5 strand
Gly	810	0.01939	G810S/D/A	Exon 20	αD-αE loop
Asn	816	0.03781	N816K	Exon 20	αE-helix
Val	845	0.06473	V845M/A/L	Exon 21	β7strand (C-spine)
Thr	847	0.05199	T847I/A/K	Exon 21	β7strand
Thr	854	0.07392	T854A/I/A	Exon 21	β7strand
Leu	858	0.10864	L858R/Q/K/V/M	Exon 21	Short helix/A-loop
Lys	860	0.05991	K860T/E/I	Exon21	Short helix/A-loop
Leu	861	0.07540	L861Q/R/E/F/K/P	Exon 21	Short helix/A-loop
His	870	0.01914	H870R/N/Y	Exon 21	A-loop
Arg	889	0.06413	R889S	Exon 22	A-loop
Ile	965	0.04496	I965S/N	Exon 23	αI-helix

Our study suggests that binding of selective and personalized kinase agents can be linked with the robustness of the residue networks in kinase structures. We have found that selective EGFR inhibitors with preferential binding to specific inactive conformations, such as Lapatinib, could be vulnerable to a broad spectrum of resistant mutations pointing to a "dark side" of targeted agents that reflects the inherent conflict between the efficiency and robustness of kinase drugs. The association of network properties with kinase regulation and drug binding suggests that residue interaction networks may be reorganized and specifically tailored through therapeutic agents targeting high centrality residue nodes. Integration of genetic, biochemical and structural data in the unified framework of protein structure networks and systems biology may help to understand and rationally exploit the complex relationships between robustness of targeted genes and binding specificity of personalized drugs.

References

1. E.D. Pleasance, R.K. Cheetham, P.J. Stephens, D.J. McBride et al., *Nature* **463**, 191 (2010).
2. Z.Kan, B.S. Jaiswal, J. Stinson, V. Janakiraman V et al., *Nature* **466**, 869 (2010).
3. I.B. Weinstein IB. *Science* **297**, 63 (2002).
4. W. Pao, V.A. Miller, K.A. Politi, G.J. Riely et al., *PLoS Med.* **2**, 12 (2005).
5. H. Shigematsu, T. Takahashi, M. Nomura, K. Majmudar et al., *Cancer Res.* **65**, 1642 (2005).
6. C.H. Yun, K.E. Mengwasser, A.V. Toms AV, M.S. Woo, et al., *Proc. Natl. Acad. Sci. U.S.A.* **105**, 2070 (2008).
7. E. Massarelli, F.M. Johnson, H.S. Erickson, Wistuba II, and V. Papadimitrakopoulou, *Lung Cancer* **80**, 235 (2013).
8. M.J. Eck and C.H. Yun, *Biochim. Biophys. Acta* **1804**, 559 (2010).
9. G.E. Konecny, M.D. Pegram, N. Venkatesan N, R. Finn et al., *Cancer Res.* **66**, 1630 (2006).
10. E.R. Wood, A.T. Truesdale, O.B. McDonald, D.Yuan et al., *Cancer Res.* **64**, 6652 (2004).
11. T.M. Gilmer, L. Cable, K. Alligood, D. Rusnak et al., *Cancer Res.* **68**, 571 (2008).
12. E. Avizienyte, R.A. Ward and A.P. Garner, *Biochem. J.* **415**, 197 (2008).
13. M. Taipale, I. Krykbaeva, M. Koeva, C. Kayatekin et al., *Cell* **150**, 987 (2012).
14. T. Shimamura, D. Li, H. Ji, H.J. Haringsma et al., *Cancer Res.* **68**, 5827 (2008).
15. K.S. Gajiwala, J. Feng, R. Ferre, K. Ryan et al., *Structure* **21**, 209 (2013).
16. S. Polier, R.S. Samant, P.A. Clarke, P. Workman et al., *Nat. Chem. Biol.* **9**, 307 (2013).
17. A.F. Neuwald and J.S. Liu, *BMC Bioinformatics* **5**, 157 (2004).
18. A. Torkamani and N.J. Schork *Cancer Res.* **68**, 1675 (2008).
19. J.C. Phillips, R. Braun, W. Wang, J. Gumbart et al., *J. Comput. Chem.* **26**, 1781 (2005).
20. P.A. Kollman, I. Massova, C. Reyes, B. Kuhn et al., *Acc. Chem. Res.* **33**, 889 (2000).
21. K.V. Brinda and S. Vishveshwara, *Biophys. J.* **89**, 4159 (2005).
22. A. Sethi, J. Eargle, A.A. Black and Z. Luthey-Schulten, *Proc. Natl. Acad. Sci. U.S.A.* **106**, 6620 (2009).
23. A. Dixit, L. Yi, R. Gowthaman, A. Torkamani et al., *PLoS One* 4, e7485 (2009).
24. X. Zhang, J. Gureasko, K. Shen, P.A. Cole and J. Kuriyan, *Cell* **125**, 1137 (2006).

PHENOME-WIDE INTERACTION STUDY (PheWIS) IN AIDS CLINICAL TRIALS GROUP DATA (ACTG)

SHEFALI S. VERMA[1], ALEX T. FRASE[1], ANURAG VERMA[1], SARAH A. PENDERGRASS[2], SHAUN MAHONY[1], DAVID W. HAAS[3], MARYLYN D. RITCHIE[1,2]

[1]Center for System Genomics, The Pennsylvania State University, University Park, PA 16802 USA; [2]Biomedical and Translational Informatics, Geisinger Health System, Danville, PA 17822 USA; [3]Vanderbilt Health, One Hundred Oaks 719 Thompson Lane, Suite 47183, Nashville TM 37204 USA

Association studies have shown and continue to show a substantial amount of success in identifying links between multiple single nucleotide polymorphisms (SNPs) and phenotypes. These studies are also believed to provide insights toward identification of new drug targets and therapies. Albeit of all the success, challenges still remain for applying and prioritizing these associations based on available biological knowledge. Along with single variant association analysis, genetic interactions also play an important role in uncovering the etiology and progression of complex traits. For gene-gene interaction analysis, selection of the variants to test for associations still poses a challenge in identifying epistatic interactions among the large list of variants available in high-throughput, genome-wide datasets. Therefore in this study, we propose a pipeline to identify interactions among genetic variants that are associated with multiple phenotypes by prioritizing previously published results from main effect association analysis (genome-wide and phenome-wide association analysis) based on a-priori biological knowledge in AIDS Clinical Trials Group (ACTG) data. We approached the prioritization and filtration of variants by using the results of a previously published single variant PheWAS and then utilizing biological information from the Roadmap Epigenome project. We removed variants in low functional activity regions based on chromatin states annotation and then conducted an exhaustive pairwise interaction search using linear regression analysis. We performed this analysis in two independent pre-treatment clinical trial datasets from ACTG to allow for both discovery and replication. Using a regression framework, we observed 50,798 associations that replicate at p-value 0.01 for 26 phenotypes, among which 2,176 associations for 212 unique SNPs for fasting blood glucose phenotype reach Bonferroni significance and an additional 9,970 interactions for high-density lipoprotein (HDL) phenotype and fasting blood glucose (total of 12,146 associations) reach FDR significance. We conclude that this method of prioritizing variants to look for epistatic interactions can be used extensively for generating hypotheses for genome-wide and phenome-wide interaction analyses. This original Phenome-wide Interaction study (PheWIS) can be applied further to patients enrolled in randomized clinical trials to establish the relationship between patient's response to a particular drug therapy and non-linear combination of variants that might be affecting the outcome.

Keywords: PheWAS; PheWIS; genetic interactions; Epistasis; ENCODE; Roadmap Epigenome; Pharmacogenomics; Clinical Trials; Annotations; prior biological knowledge;

1. Introduction

Investigating the precise response of antiretroviral therapies given to patients is an important area of research. Previous studies have discovered interesting single gene effects as well as genetic interaction effects associated with response to anti-retroviral medications[1,2] in the AIDS Clinical Trials Group (ACTG) data (https://actgnetwork.org/). A recently published Phenome-wide association study (PheWAS)[2] showed a number of variants associated with a list of 27 highly curated and transformed (for normal distribution) phenotypes collected in baseline model of AIDS clinical trials[3,4]. Thus, this unique clinical trials dataset and the analyses performed earlier provide a backbone for performing epistatic interactions analyses among variants and genes that might be associated with multiple drug response phenotypes.

A wealth of data are being generated from speedy advancements in genotyping and sequencing technologies, thus providing opportunities to investigate not only single gene effects but also non-linear combined genetic effects of these variants. Genome wide association studies (GWAS) have been proven to detect many SNPs associated with multiple diseases or traits. These variants discovered by GWAS can only explain small proportion of genetic risk corresponding to the problem of "missing heritability"[5]. One conceivable explanation of missing heritability is the existence of genetic interactions or epistasis[5] and the evidence for genetic interactions has been observed in both humans and model organisms[6]. Efficient identification of epistatic interactions is also an important biological problem because unlike GWAstudies, gene-gene interaction studies are not yet fully equipped to produce reproducible results most importantly due to the combinations of pairwise models that are generated from each individual study. Additionally, testing for two or multi-way interactions still remains a challenge due to overhead of computing resources and also due to correction for false positives for each test performed. Thus, filtration of variants based on prior biological knowledge is used frequently in the search for epistasis[7]. Many studies have shown that filtration of variants based on strong and marginal main effects as determined by the data can be useful in detecting interactions[8]. Combining the main effect filtration method along with filtration based on prior-biological knowledge has also been proven to increase the power to detect epistatic interactions[9–11].

The Roadmap Epigenome has provided high-resolution genome wide interaction maps based on the chromatin accessibility, histone modifications, DNA methylation and mRNA expression across 127 epigenomes[12,13]. These data can be used as a great resource of prior biological information for filtering variants based on the activity of the genomes as defined by chromatin states[14]. Annotations of variants associated with disease traits from the NHGRI GWAS Catalog[15] have shown that 81% of variants associated with a disease can be annotated into one of the functional regulatory elements using ENCODE data where functional here refers to any biochemical activity as identified from at least one of the cell lines from ENCODE [12,14]. Roadmap epigenome data is collected from an even larger list of epigenomes and thus provide an extensive and more detailed map of regulatory activity of the genome.

In this study, we intended to use this extensive knowledge about regulatory elements as criteria to filter variants based on their functional activity before performing interaction testing rather than the more traditional approach of prioritizing variants based on their activity after conducting analysis. This will reduce the multiple testing burden and increase interpretability. In the remaining sections, we explain our proposed analytic pipeline for Phenome-wide interaction study (PheWIS), its application to the pre-treatment ACTG datasets, and a series of highly significant gene-gene interactions associated with baseline clinical variables. We show that combination of biological knowledge and main effect filtering provides a high-throughput, comprehensive pipeline to address the architecture of complex traits. This method can clearly be applied to patients from on-treatment imminent clinical trial data to generate hypothesis for epistatic gene-gene interactions that could influence drug response and treatment design.

2. Materials and Methods

2.1 *Genotype and Phenotype data*

ACTG data from treatment-naïve patients has been previously reported[16–20]. We used the same dataset as described in the pilot PheWAS conducted on ACTG data that consisted of 27 pre-treatment laboratory measurements (shown in supplementary table 1 at ritchielab.psu.edu/publications/supplementary-data/psb-2016/phewis) that have been normalized by appropriate transformations. From all 27 phenotypes, 26 were used as independent variables and one phenotype (CD4 T-cell counts) was used as a covariate due to its known confounding effect in HIV patients[21]. This dataset consisted of 2547 genotyped participants which were imputed in three phases based on a separate immunogenomics project[22]. Phase I and II were combined together (Discovery dataset), which consisted of 1366 samples and Phase III consisted of 1181 samples (Replication dataset) as described in detail in pilot PheWAS[2]. **Supplementary Table 2** Lists the information on samples used in both the discovery and replication dataset along with the demographic information on these samples.

2.2 *Annotation and Filtration of variants*

The pilot PheWAS analysis reported 10,584 variants that replicated at p-value <0.01 with the same direction of effect across two datasets. We took all of these variants that passed the replication criteria in the pilot PheWAS and annotated them using Biofilter[23]. Biofilter is a unified framework that consists of data from multiple resources such as KEGG, GENCODE, RegulomeDB, etc. We added Roadmap Epigenome posterior probability data for 25 chromatin states averaged across all 127 epigenomes as a new source to Biofilter. We annotated variants with the help of Biofilter by specifying the Roadmap Epigenome as the single source to be used to annotate variants in order to remove any redundancy from similar sources such as RegulomeDB[24] or HaploReg[25] which also contain data from ENCODE project[12].

We used the 25-state chromatin models data published on the Roadmap epigenome website (http://egg2.wustl.edu/roadmap/web_portal/imputed.html#chr_imp). The roadmap epigenome posterior probability raw data is a map of the human genome where the genome is divided into 200 base pair regions (chunks) and thus there are 15,478,375 total numbers of chunks of the genome (for human genome build 37) for which probabilities for each 25 states are provided. We combined posterior probabilities from 127 epigenomes (tissues/cell types) in Roadmap Epigenome data by doing an average across all values to calculate posterior probability of each state for each 200bp region. State with highest probability was then assigned to each region. Careful investigation of these data suggested that many consecutive chunks are annotated as the same chromatin states. Thus we dynamically combined chunks together to yield a larger contiguous region of the genome, thereby reducing the total number of chunks. In order to combine the consecutive chunks, we used a rule of 80% where the two chunks were combined and annotated as the same state if the probability of the same state in consecutive chunk is 80% or greater.

To get an estimate of the total number of regions for each chromatin state in a genome-wide study, we choose to look at approximately 5M variants from Illumina Omni5 platform as that is one of the largest genotyping chips. **Table 2** provides an overall estimate of each chromatin state and the total number of regions combined dynamically for all variants genotyped on Illumina Omni5 chip (http://www.illumina.com/products/humanomni5-quad_beadchip_kit.html). We picked the Omni5 chip to show a large number of variants that can be covered with their respective chromatin states from the genotyping chips available. To get a better overview of variants on genotyping chips that are known to be associated with a disease using NHGRI GWAS catalog[15], we also mapped these variants on Omni 5 chip to GWAS catalog (accessed May 2014) using Library of Knowledge Integration (LOKI) database in Biofilter and looked at how all variants in each state are associated with one or more disease from GWAS catalog. **Table 2** also represents the number of times each chromatin state is represented in the NHGRI GWAS catalog as being associated with a disease.

10,584 variants from the pilot PheWAS were annotated using the same approach described above. **Figure 1** shows the proportion of variants in each of the 25 states. To filter these variants based on the activity of each region (corresponding to chromatin states), we removed any variants that fell in Chromatin State 25 (Quiescent/Low State) because as described in Roadmap epigenome, predominantly most of the inactive regions fall under quiescent state (approximately 40% of inactive region) and this state is represented on an average in 68% of the genome[13]. This annotation followed by filtration step resulted in 1776 variants that were further considered for association testing.

Table 2. Estimate of chromatin states from Illumina Omni5 genotyping chip and number of chromatin states in variants mapping to GWAS catalog that are associated with a disease. Here each 200 base pair region of the genome is combined together dynamically when the next region is represented as same state with at least 80% posterior probability.

State	Description	#Occurrences in Omni5 Chip	#Occurrences in NHGRI GWAS Catalog
S1	Active TSS	6803	18
S2	Promoter Upstream TSS	21901	51
S3	Promoter Downstream TSS 1	22854	65
S4	Promoter Downstream TSS 2	9007	24
S5	Transcribed 5' preferential	90330	175
S6	Strong transcription	42687	102
S7	Transcribed 3' preferential	225664	449
S8	Weak transcription	207773	404
S9	Transcribed Regulatory (Prom/Enh)	15920	47
S10	Transcribed 5' preferential and Enh	15170	40
S11	Transcribed 3' preferential and Enh	9022	25
S12	Transcribed weak Enhancer	19313	46
S13	Active Enhancer 1	7318	23
S14	Active Enhancer 2	6947	24
S15	Active Enhancer Flank	10350	23
S16	Weak Enhancer 1	8878	25
S17	Weak Enhancer 2	18104	44

S18	Primary H3K27ac possible Enhancer	895	5
S19	Primary DNAase	19959	48
S20	ZNF genes and repeats	9211	11
S21	Heterochromatin	32644	40
S22	Poised Promoter	3808	12
S23	Bivalent Promoter	12285	35
S24	Repressed Polycomb	69906	189
S25	Quiescent/Low	3289868	5565

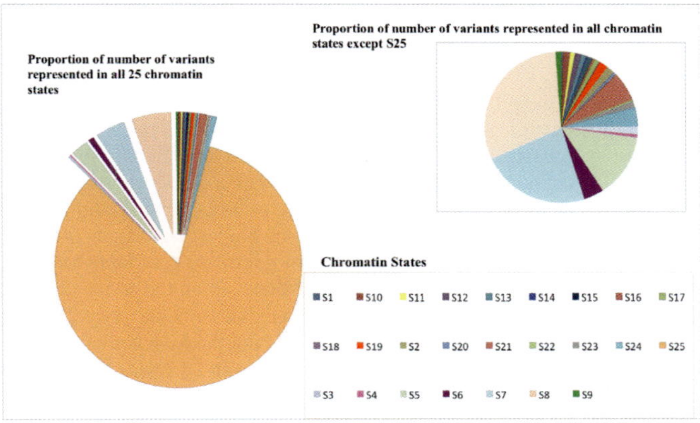

Figure 1. Distribution of all 25 chromatin states in 10,584 SNPs from the pilot PheWAS study (on left) and the proportions of variants used in PheWIS (on right)

2.3 *Statistical Analysis*

To test for pairwise interactions among 1773 annotated variants in both discovery and replication datasets, all variants were encoded as additive where risk incurred by heterozygous alternate allele is half the risk incurred by homozygous alternate alleles. We ran linear regression where a reduced model consisted of main effects of all variants adjusted by covariates and a full model consisted of main effects and an interaction term for each pairwise SNP-SNP model adjusted by covariates. A likelihood ratio test was conducted to obtain the significance of the interaction effect above and beyond the main effect of each variant. Below is the mathematical description for the reduced and full model:

$$\text{Reduced Model: } Y = \beta_0 + \beta_1\,\text{SNP1} + \beta_2\,\text{SNP2} \quad (1)$$

$$\text{Full Model: } Y = \beta_0 + \beta_1\,\text{SNP1} + \beta_2\,\text{SNP2} + \beta_3\,\text{SNP1}*\text{SNP2} \quad (2)$$

$$\text{Likelihood Ratio Test: Full Model} - \text{Reduced Model} \qquad (3)$$

We used PLATO (http://ritchielab.psu.edu/software/plato-download) to conduct PheWIS in both discovery and replication datasets where all 26 phenotypes were calculated simultaneously for each pairwise interaction model. We adjusted the analysis by age, gender, CD4 T-cell count (square root) and first 5 principal components (to account for genetic ancestry). We also calculated Bonferroni and FDR based corrected p-values[26,27] for each model tested. Here the models are adjusted for all pairwise combination of variants and all phenotypes (40,842,828 tests). We ran the regression analyses separately for discovery and replication datasets and then looked for each pairwise combination of SNPs associated with the same phenotype to determine if results were replicating across the two independent datasets.

3. Results

Annotation of all 10,584 variants from the pilot PheWAS analysis showed that the majority of variants represent state 25 (S25; Quiescent/Low) as shown in **Figure 1**.Variants detected from GWAS are highly enhanced in regulatory regions as illustrated in **Table 2** where a large number of variants are represented in all 25 states but the majority of variants associated with a disease represent the most inactive state "S25". Since a large proportion of variants known to be associated from GWA studies only represent small proportion of genetic risk[28] and one of the biggest challenges is in understanding the role of the majority of these variants[29]. Therefore, prioritizing variants based on the affect that they can impose on gene regulation is a crucial step in understanding the associations between variants and phenotypes. We aimed this study to focus on only variants that are represented in more active states (with state 1 being the most active and state 25 being the least active) with the potential for a larger proportion of variance to be explained by these variants. A total of 50,798 SNP-SNP pair and phenotype results replicate at p-value<0.01. In order to adjust for multiple testing burden and to reduce false positives, we required replication between the two datasets based on Bonferroni adjusted p-value and False Discovery Rate (FDR) adjusted p-value[26,27,30]. A total of 2,176 results replicate for just one phenotype (fasting glucose) based on Bonferroni based correction and 12,146 results replicated for two phenotypes: fasting glucose and high density lipoprotein (HDL), for FDR based correction of p-values. We used Biofilter to again annotate the position of these variants with chromatin states and then further annotate each SNP from SNP-SNP pairs with genes. SNPs are annotated as genes where the position of a SNP falls within gene boundaries. Therefore, more than one SNP can be annotated to same genes. **Table 3** presented the distribution of variants from Bonferroni and FDR based results for each of the 24-chromatin states. We also looked at the expression of top genes in various tissues using GTEx portal[31]. For HDL results, we looked for expression in adipose and liver tissue and for fasting glucose; we looked for expression in the pancreas.

We also mapped all SNP-SNP pairs to genes using Biofilter. Bonferroni significant results consisted of 212 unique genes that were mapped to 66 genes and FDR-based significant results consisted of 690 unique SNPs that represent 245 unique genes. Details of all replicated results can be found in supplementary material online (supplementary table 3 and 4 at ritchielab.psu.edu/publications/supplementary-data/psb-2016/phewis).

Figure 2 represents the top 30 results for fasting glucose that are less than Bonferroni corrected p-value 0.01. Each SNP-SNP pair and their corresponding genes are shown along with –log10 (p-value) track for both Discovery and Replication datasets. Interactions among the specific chromatin states that the SNP falls under are shown on the right side. Six unique gene-gene pairs are also expressed in the pancreas. **Figure 3** shows a circular plot for HDL providing the interaction between the SNPs in the genes and the states that the SNPs represent. The genes are colored based on the tissue that they are expressed in. **Figure 3** also represents the FDR corrected p-values for each SNP-SNP interaction pairs. For details on all results that were replicated, please refer to supplementary material online (supplementary table 3 and 4 at ritchielab.psu.edu/publications/supplementary-data/psb-2016/phewis)

Table 3. Occurrences of Bonferroni and FDR corrected results in all 24 chromatin states

State	Description	#Occurrences in Bonferoni corrected results	#Occurrences in FDR corrected results
S1	Active TSS	1	4
S2	Promoter Upstream TSS	6	15
S3	Promoter Downstream TSS 1	3	14
S4	Promoter Downstream TSS 2	3	7
S5	Transcribed 5' preferential	33	118
S6	Strong transcription	5	24
S7	Transcribed 3' preferential	38	183
S8	Weak transcription	80	292
S9	Transcribed Regulatory (Prom/Enh)	2	4
S10	Transcribed 5' preferential and Enh	2	10
S11	Transcribed 3' preferential and Enh	6	7
S12	Transcribed weak Enhancer	1	11
S13	Active Enhancer 1	6	12
S14	Active Enhancer 2	0	2
S15	Active Enhancer Flank	2	9
S16	Weak Enhancer 1	1	1
S17	Weak Enhancer 2	0	6
S18	Primary H3K27ac possible Enhancer	1	2
S19	Primary DNAase	2	8
S20	ZNF genes and repeats	2	5
S21	Heterochromatin	7	34
S22	Poised Promoter	0	1

S23	Bivalent Promoter	4	20
S24	Repressed Polycomb	10	27

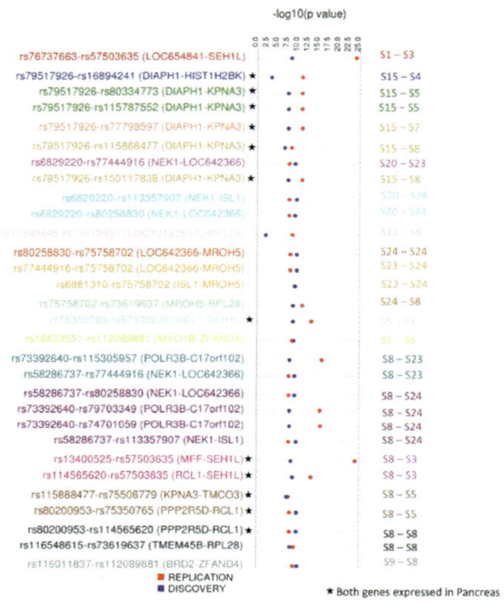

← **Figure 2.** Synthesis-view plot (http://visualization.ritchielab.psu.edu/synthesis_views/plot) illustrating interactions among top 30 SNP-SNP pair for fasting glucose phenotype. Different color for text corresponds to the combination of chromatin states that SNP-SNP pairs are mapped to as represented on the right axis.

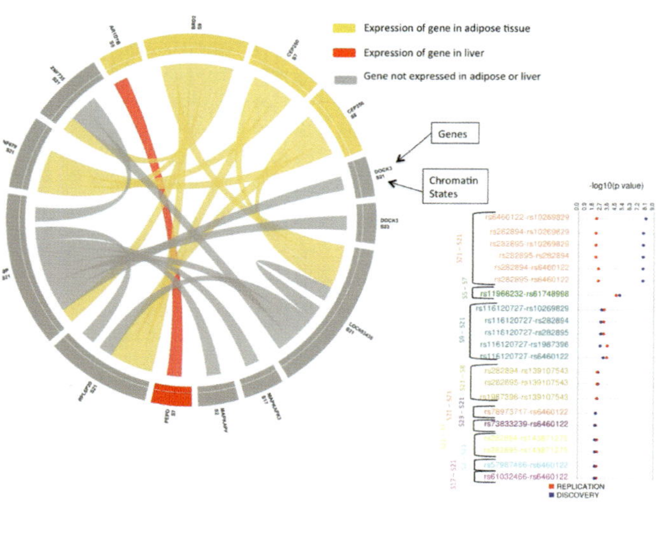

← **Figure 3.** Circular plot representing interactions of SNP-SNP pair combined based on the genes and the chromatin states represented for HDL phenotype. Yellow color corresponds to the expression of gene in adipose tissue, red color corresponds to expression of gene in liver tissue and grey color corresponds to expression on gene in neither adipose nor liver tissues. Lines show the interactions between the variants in the genes and corresponding states. On right, showing a synthesis view plot where FDR p-values of both discovery and replication dataset for each pair SNP-SNP interactions representing unique gene and chromatin state is represented. Color for SNP-SNP pair corresponds to different combinations of interactions among chromatin states

4. Discussion

This study presents a pilot Phenome-wide Interaction study (PheWIS), which is the first of its kind, in the AIDS Clinical Trials Group data. With the help of statistical methods to detect genetic interactions associated with one or multiple phenotypes, we showed significant interactions for SNPs mapped to different chromatin states. The purpose of this study is aimed at mimicking the regulatory genetic networks by showing how interactions between two different chromatin states impacted by genetic variants are associated with a trait. In this paper, we used a-priori biological information from Roadmap Epigenome data to test for variants that represent active chromatin states. Among the top associations with Bonferroni p-value<0.01 are the interactions between *SEH1L* gene and *RCL1* gene to be associated with fasting glucose. Interactions between these two genes are represented by two top-most SNP-SNP interaction pair as shown in **Figure 2**. In these interactions, the three-chromatin states represented are S3 (Promoter Downstream TSS 1), S5 (Transcribed 5' preferential) and S8 (Weak Transcription), which suggests interactions among transcribed regions that could be of potential interest. *SEH1L* gene participates in the regulation of glucose transport process (GO:0010827) and functional studies in yeast have shown that growth of yeast on glucose media requires function *RCL1*[32]. PheWIS aims at identifying interactions among variants above and beyond the main effects of individual variant. Thus, with this approach we are able to identify several known and novel interactions that could not be identified with PheWAS alone.

The majority of interactions in the FDR corrected results for HDL show interactions among chromatin state 21 (S21; Heterochromatin) and other states. In Roadmap epigenome data, heterochromatin state is mostly represented by constitutive heterochromatin and heterochromatin state is highly tissue specific[13]. Since in this analysis, we combined data from all cell lines to represent all 25 chromatin states, nothing can be said about the heterochromatin in adipose or liver cell lines. Thus, suggesting that in the future, more work would be required to look at these polymorphic regions based on the tissue that phenotype is affecting or the tissue using which the study samples are collected. For the HDL PheWIS results, one potential interesting interaction is between *ARID1B* and *PEPD* genes. Peptidase D (PEPD) and *ARID1B* genes have been known to be associated with HDL[33–35]. Both of these genes are highly expressed in adipose tissue with *PEPD* being also highly expressed in liver

There are few limitations in this study. Although after correcting for multiple testing based on Bonferroni and FDR methods, we identified many statistical interactions associated with two phenotypes; future research is required to understand these novel interaction associations. Next, all these results are based on treatment naïve patients enrolled in clinical trials, similar analysis in post-treatment quantitative phenotypes can help explore more associations that are linked to the side-effects presented by drugs as well as the benefits of the drug given to patients. Our approach is based on averaging across 127 epigenomes from Roadmap data to annotate regions of the genome. With this approach, we might have missed useful information on chromatin states that are specific to just one tissue type. Future studies can be focused on tissue specific annotation approach or a more comprehensive approach where annotations for an active region can be from any one tissue as well rather than average across all tissues. Lastly, we only excluded the variants that were mapped to state

25 from Roadmap epigenome data whereas future studies could also focus on excluding variants that are under represented in more than one states and only including the variants that map to states which are over-represented in our data.

5. Conclusions

We present the first phenome-wide SNP-SNP interaction study in a pharmacogenomics dataset. Though this study is on treatment naïve patients, it presents a great framework to look for statistical epistasis in a large number of phenotypes, which are collected post treatment. Most of the interactions associated with traits in this study are novel and would require more extensive future work to understand if any of these associations explain biological processes that are also linked to one or more phenotypes. Methods such as the one proposed for PheWIS will enable researchers to investigate more territory in the etiology of complex traits.

6. Acknowledgements

This project was supported by Award Number U01AI068636 from the National Institute of Allergy and Infectious Diseases and supported by National Institute of Mental Health (NIMH), National Institute of Dental and Craniofacial Research (NIDCR). The content is solely the responsibility of the authors and does not necessarily represent the official views of the National Institute of Allergy and Infectious Diseases or the National Institutes of Health. Grant support included AI077505, TR000445, AI054999 (DWH), HG006385, HL065962 (MDR).Study drugs were provided by Bristol-Myers Squibb Company (Princeton, NJ), Gilead Sciences (Gilead Sciences, Inc., Foster City, CA), GlaxoSmithKline, Inc (Research Triangle Park, NC), and Boehringer Ingelheim (Ridgefield, CT).Clinical Research Sites that participated in ACTG protocols ACTG 384, A5095, A5142, or A5202, and collected DNA under protocol A5128, were supported by the following grants from NIH National Institute of Allergy and Infectious Diseases (NIAID): AI069532, AI069484, AI069432, AI069450, AI069495, AI069434, AI069424, AI069439, AI069467, AI069423, AI069513, AI069477, AI069465, AI069419, AI069502, AI069474, AI069472, AI069501, AI069418, AI069494, AI069471, AI069511, AI069452, AI069428, AI069556, AI069415, AI032782, AI046376, AI046370, AI038858, AI034853, AI027661, AI025859, AI069470, AI027675, AI073961, AI050410, AI045008, AI050409, AI072626, AI069447, AI027658, AI027666, AI058740, and AI025868, and by the following grants from NIH National Center for Research Resources (NCRR): RR000046, RR000425, RR025747, RR025777, RR025780, RR024996, RR024160, RR023561, RR024156, RR024160, and RR024160.

7. References

1. Motsinger, A. A. *et al.* Multilocus genetic interactions and response to efavirenz-containing regimens: an adult AIDS clinical trials group study. *Pharmacogenet. Genomics* **16,** 837–845 (2006).
2. Moore, C. B. *et al.* Phenome-wide Association Study Relating Pretreatment Laboratory Parameters With Human Genetic Variants in AIDS Clinical Trials Group Protocols. *Open Forum Infect. Dis.* **2,** (2014).

3. Holzinger, E. R. *et al.* Genome-wide association study of plasma efavirenz pharmacokinetics in AIDS Clinical Trials Group protocols implicates several CYP2B6 variants. *Pharmacogenet. Genomics* **22,** 858–867 (2012).
4. Rotger, M. *et al.* Predictive value of known and novel alleles of CYP2B6 for efavirenz plasma concentrations in HIV-infected individuals. *Clin. Pharmacol. Ther.* **81,** 557–566 (2007).
5. Maher, B. Personal genomes: The case of the missing heritability. *Nature* **456,** 18–21 (2008).
6. Evans, D. M., Marchini, J., Morris, A. P. & Cardon, L. R. Two-stage two-locus models in genome-wide association. *PLoS Genet.* **2,** e157 (2006).
7. Ritchie, M. D. Using biological knowledge to uncover the mystery in the search for epistasis in genome-wide association studies. *Ann. Hum. Genet.* **75,** 172–182 (2011).
8. Sun, X. *et al.* Analysis pipeline for the epistasis search - statistical versus biological filtering. *Front. Genet.* **5,** 106 (2014).
9. Ma, L. *et al.* Knowledge-driven analysis identifies a gene-gene interaction affecting high-density lipoprotein cholesterol levels in multi-ethnic populations. *PLoS Genet.* **8,** e1002714 (2012).
10. Grady, B. J. *et al.* Use of biological knowledge to inform the analysis of gene-gene interactions involved in modulating virologic failure with efavirenz-containing treatment regimens in ART-naïve ACTG clinical trials participants. *Pac. Symp. Biocomput. Pac. Symp. Biocomput.* 253–264 (2011).
11. Turner, S. D. *et al.* Knowledge-Driven Multi-Locus Analysis Reveals Gene-Gene Interactions Influencing HDL Cholesterol Level in Two Independent EMR-Linked Biobanks. *PLoS ONE* **6,** (2011).
12. ENCODE Project Consortium. An integrated encyclopedia of DNA elements in the human genome. *Nature* **489,** 57–74 (2012).
13. Roadmap Epigenomics Consortium *et al.* Integrative analysis of 111 reference human epigenomes. *Nature* **518,** 317–330 (2015).
14. Schaub, M. A., Boyle, A. P., Kundaje, A., Batzoglou, S. & Snyder, M. Linking disease associations with regulatory information in the human genome. *Genome Res.* **22,** 1748–1759 (2012).
15. Welter, D. *et al.* The NHGRI GWAS Catalog, a curated resource of SNP-trait associations. *Nucleic Acids Res.* **42,** D1001–D1006 (2014).
16. Robbins, G. K. *et al.* Comparison of sequential three-drug regimens as initial therapy for HIV-1 infection. *N. Engl. J. Med.* **349,** 2293–2303 (2003).
17. Gulick, R. M. *et al.* Triple-nucleoside regimens versus efavirenz-containing regimens for the initial treatment of HIV-1 infection. *N. Engl. J. Med.* **350,** 1850–1861 (2004).
18. Gulick, R. M. *et al.* Three- vs four-drug antiretroviral regimens for the initial treatment of HIV-1 infection: a randomized controlled trial. *JAMA* **296,** 769–781 (2006).
19. Riddler, S. A. *et al.* Class-sparing regimens for initial treatment of HIV-1 infection. *N. Engl. J. Med.* **358,** 2095–2106 (2008).
20. Daar, E. S. *et al.* Atazanavir plus ritonavir or efavirenz as part of a 3-drug regimen for initial treatment of HIV-1. *Ann. Intern. Med.* **154,** 445–456 (2011).

21. Crampin, A. ., Mwaungulu, F. ., Ambrose, L. ., Longwe, H. & French, N. Normal Range of CD4 Cell Counts and Temporal Changes in Two HIVNegative Malawian Populations. *Open AIDS J.* **5,** 74–79 (2011).
22. International HIV Controllers Study *et al.* The major genetic determinants of HIV-1 control affect HLA class I peptide presentation. *Science* **330,** 1551–1557 (2010).
23. Pendergrass, S. A. *et al.* Genomic analyses with biofilter 2.0: knowledge driven filtering, annotation, and model development. *BioData Min.* **6,** 25 (2013).
24. Gronostajski, R. M., Guaneri, J., Lee, D. H. & Gallo, S. M. The NFI-Regulome Database: A tool for annotation and analysis of control regions of genes regulated by Nuclear Factor I transcription factors. *J. Clin. Bioinforma.* **1,** 4 (2011).
25. Ward, L. D. & Kellis, M. HaploReg: a resource for exploring chromatin states, conservation, and regulatory motif alterations within sets of genetically linked variants. *Nucleic Acids Res.* **40,** D930–D934 (2012).
26. Bland, J. M. & Altman, D. G. Multiple significance tests: the Bonferroni method. *BMJ* **310,** 170 (1995).
27. Benjamini, Y. Discovering the false discovery rate. *J. R. Stat. Soc. Ser. B Stat. Methodol.* **72,** 405–416 (2010).
28. Hindorff, L. A. *et al.* Potential etiologic and functional implications of genome-wide association loci for human diseases and traits. *Proc. Natl. Acad. Sci. U. S. A.* **106,** 9362–9367 (2009).
29. Visscher, P. M., Brown, M. A., McCarthy, M. I. & Yang, J. Five Years of GWAS Discovery. *Am. J. Hum. Genet.* **90,** 7–24 (2012).
30. Bhandari, M. *et al.* The risk of false-positive results in orthopaedic surgical trials. *Clin. Orthop.* 63–69 (2003). doi:10.1097/01.blo.0000079320.41006.c9
31. Lonsdale, J. *et al.* The Genotype-Tissue Expression (GTEx) project. *Nat. Genet.* **45,** 580–585 (2013).
32. Horn, D. M., Mason, S. L. & Karbstein, K. Rcl1 Protein, a Novel Nuclease for 18 S Ribosomal RNA Production. *J. Biol. Chem.* **286,** 34082–34087 (2011).
33. Global Lipids Genetics Consortium *et al.* Discovery and refinement of loci associated with lipid levels. *Nat. Genet.* **45,** 1274–1283 (2013).
34. Lin, Q.-Z. *et al.* Sex-specific association of the peptidase D gene rs731839 polymorphism and serum lipid levels in the Mulao and Han populations. *Int. J. Clin. Exp. Pathol.* **7,** 4156–4172 (2014).
35. Govindaraju, D. R. *et al.* Genetics of the Framingham Heart Study Population. *Adv. Genet.* **62,** 33–65 (2008).

A FRAMEWORK FOR ATTRIBUTE-BASED COMMUNITY DETECTION WITH APPLICATIONS TO INTEGRATED FUNCTIONAL GENOMICS

HAN YU
Biostatistics, University at Buffalo,
Buffalo, NY 14220/EST, USA
E-mail: hyu9@buffalo.edu

RACHAEL HAGEMAN BLAIR
Biostatistics, University at Buffalo,
Buffalo, NY 14220/EST, USA
E-mail: hageman@buffalo.edu

Understanding community structure in networks has received considerable attention in recent years. Detecting and leveraging community structure holds promise for understanding and potentially intervening with the spread of influence. Network features of this type have important implications in a number of research areas, including, marketing, social networks, and biology. However, an overwhelming majority of traditional approaches to community detection cannot readily incorporate information of node attributes. Integrating structural and attribute information is a major challenge. We propose a flexible iterative method; inverse regularized Markov Clustering (irMCL), to network clustering via the manipulation of the transition probability matrix (aka stochastic flow) corresponding to a graph. Similar to traditional Markov Clustering, irMCL iterates between "expand" and "inflate" operations, which aim to strengthen the intra-cluster flow, while weakening the inter-cluster flow. Attribute information is directly incorporated into the iterative method through a sigmoid (logistic function) that naturally dampens attribute influence that is contradictory to the stochastic flow through the network. We demonstrate advantages and the flexibility of our approach using simulations and real data. We highlight an application that integrates breast cancer gene expression data set and a functional network defined via KEGG pathways reveal significant modules for survival.

Keywords: KEGG pathways, logistic regression, community detection, Markov clustering, omics, survival

1. Introduction

Community structure occurs when nodes exhibit a high-degree of connectivity to each other, and a lower degree of connectivity to other groups and nodes in the network.[1,2] The community detection problem has been studied extensively in Social Network Analysis (SNA). In the areas of bioinformatics and computational biology, the problem is also referred to as module detection or graph clustering.[3,4]

In a general sense, the community detection problem can be viewed as the clustering of a network. Classical graph clustering methods inlcude Kernighan-Lin algorithm,[5] hierarchical clustering methods,[6] spectral clustering,[7,8] Newman and Girvan algorithm,[9,10] and modularity-based algorithms comprise an important class of community detection methods.[11–13] Classical approaches to community detection cannot readily incorporate information of node attributes and rely solely on network structures. The simultaneous use of attribute and connectivity information can yield more accurate results and can be leveraged in downstream analysis for prediction under attribute or network perturbations. Hanisch *et al.* derive the distance matrix by combining the structural and gene profiles distances, but require prior domain knowledge.[14] Zhou *et al.* represent attributes as additional nodes.[15] In this setting, attributes are restricted to discrete values, and consequently the size and complexity of the graph grows, and requires accounting of the different types nodes and edges.[16] Instead of graph partitioning, the algorithms of `CoPaM`[17] and `DME`[18] introduces a problem of identifying cohesive patterns or subnetworks satisfying a density threshold and cohesive constraints.

We have developed a novel community detection method that rely on stochastic flow in networks. Leveraging robust statistical classification methods, we bridge and simultaneously model the attribute and structural space. The methods that we propose are highly generalizable and flexible in their implementation. We showcase their flexibility through simulation and application that integrates breast cancer gene expression data set with KEGG ontologies and survival data.

2. Materials and Methods

Briefly, we begin by outlining *Markov CLustering (MCL)* and *regularized Markov CLustering (rMCL)* frameworks, which set the foundation of our approaches.[19,20] MCL is based on the notion that if a group of nodes belongs to the same community, then the stochastic flow from these nodes will be concentrated towards nodes in that community.[19] Performing random walks on a graph may reveal where *flows* gather, which suggests potential communities. In this setting, our focus is on undirected graphs, which have a symmetric adjacency matrix and have edge interpretations of association (not causation).

MCL algorithms depend on the iteration between two operators *expand* and *inflate*, until convergence, in order to identify communities in the network. Markov clustering utilizes a stochastic matrix that is initially derived from the adjacency matrix, $A_{adj} \in \mathbf{R}^{n \times n}$ of the graph. The stochastic matrix is defined as the matrix product, $M = A_0 \cdot D^{-1}$, where $A_0 = A_{adj} + I$, and $D \in \mathbf{R}^{n \times n}$ is the diagonal matrix containing the degree information for each node, $D(k,k) = \text{diag}\left(\sum_{i=1}^{n} A(i,k)\right)$. The operations in MCL and rMCL utilize the stochastic matrix, M, which has columns that can be interpreted as transition probabilities. In the classic MCL, the expand step at the $j + 1^{th}$ iteration requires a matrix product $\tilde{M}_{j+1} = M_j \cdot M_j$.

The inflate operator, $M_{j+1}^{inf} = Inflate(\tilde{M}_{j+1}, r)$, can be understood as the component-wise exponentiation $\tilde{m}(i,j)^r$, $\forall\, i,j = 1,\ldots,n$, where the inflation operator, r, is a constant. Following inflation, M_{j+1}^{inf} is converted to a stochastic matrix, M_{j+1}, and a new iteration is started. Importantly, the expand operator alone would give rise to a Markov Chain via a random walk on the graph. However, due to the inflation operator the process cannot be regarded as a Markov Chain. Inflation is critical to accentuate strong ties and paths, and deemphasize weak ones. The inflation constant, r, controls the degree at which this strengthening and weakening is enforced, and has a direct impact on the cluster formation. Upon convergence of MCL to steady-state, the stochastic matrix can be understood in terms of *attractors*. The matrix is sparse, and the *attractors* have at least one positive value in their row. The indices of these positive values, together with the attractor, form the community.

A regularized version of Markov Clustering, rMCL, was proposed and has been shown to overcome some fragmentation issues in the communities. The rMCL algorithm follows the same iterative approach, with an *expand* step that is replaced by a *regularization operation*, $M_{j+1} = M_j \cdot M_0$, where M_0 is the initial stochastic matrix formed from the network adjacency matrix.[20] The regularize step ensures that the original structural information is still utilized for the graph clustering process after the first iteration. Unfortunately, the regularized MCL does not naturally converge to a steady state with the same desirable interpretations in terms of community membership. In order to achieve this, at each iteration, a *prune* step is added that forces some smaller entries of the stochastic matrix to zero using a heuristic threshold. The pruning aims to eliminate entries that are small relative to other entries in the matrix.[20]

2.1. *inverse regularized Markov Clustering (irMCL)*

We propose a flexible method, inverse regularized Markov CLustering (irMCL), which utilizes the *expand* and *inflate* operators, but relies on an alternative concept of community that emphasizes the spreading of *influence* or *information* in a non-exclusive manner. Our approach relies on the following modeling assumptions:

(A1) Spreading of information/influence from Node i to Node j will not affect that from Node i to other nodes, $k \neq j$.
(A2) Nodes in the same community are influenced or share information from similar group of nodes.
(A3) Nodes with larger degrees tend to be more influential.
(A4) If an individual is highly influenced by a group of nodes, such influence tends to be self-amplified.
(A5) Spread of information between nodes with similar attributes is easier, and thus should be a function of the attributes similarity measures between nodes.

In this model, the community membership of a node is measured by *information* that flows *into* the nodes, as opposed to MCL and rMCL, where a feature is the stochastic flow that *exits* this node. Accordingly, we term this procedure "inverse regularized Markov Clustering" (irMCL). These assumptions naturally give higher weights to nodes in the network with high degrees and naturally incorporate attribute information in a flexible manner. Similar to MCL, we denote $A_{adj} \in \mathbf{R}^{n \times n}$ as the adjacency matrix of graph $\mathcal{G}$. We define a symmetric *spread matrix* as: $A = A_{adj} + I$, which defines the graph with the addition of self loops.

Algorithm 1 shows the full details of the irMCL approach. At each iteration, the initial spread matrix used to regularize. Repeated use of the spread matrix naturally puts more weight on the high degree nodes in the network (A3), and is unique to our approach. The same inflation operator as in MCL is used according to assumption (A4). Convergence is tracked empirically by examining the mean squared difference as the difference between M_j and M_{j-1}, defined as $\sum_{i=1}^{n}\sum_{k=1}^{n}\left(m_{ik}^{(j)} - m_{ik}^{(j-1)}\right)^2/n$, where $m_{ik}^{(j)}$ is the entry of M_j.

The output of this iterative method is a stochastic matrix, where the rows with high similarity are likely to belong to the same community. In our applications, we utilize complete linkage, and estimate the similarity using a euclidean distance. Silhouette plots are utilized for the determination of the number of clusters via average silhouette width.[21]

Algorithm 2.1 Feature derivation for inverse Regularized Markov Clustering (iRMCL)

Initialize:
$A_{\text{adj}} \in \mathbf{R}^{n \times n}$ Adjacency Matrix
$A_0 = A_{\text{adj}} + I$
for $k = 1$ to n **do**
  $D_0(k,k) = diag\left(\sum_{i=1}^{n} A_0\,(i,k)\right)$
end for
set: $r > 1$

Repeat until stopping criteria is met
for $j = 1$ to m **do**
  $\tilde{M}_j \leftarrow M_{j-1} \cdot A_0$
  $M_j^{\text{infl}} = \text{Inflate}(\tilde{M}_j, r)$
  for $k = 1$ to n **do**
    $D_j(k,k) = diag\left(\sum_{i=1}^{n} M_j^{\text{infl}}\,(i,k)\right)$
  end for
  $M_{\text{j}} = M_j^{\text{infl}} \cdot D_j^{-1}$
end for

Output: M_{j} for row clustering

2.2. *attribute inverse regularized Markov Clustering (airMCL)*

The irMCL algorithm is based solely on network connectivity. We propose a natural extension for clustering of networks that contain nodes with heterogenous attributes. In this setting, we use the term *attribute* to loosely to define features of the nodes. In the biological context, this could include, for example, a measurement of a phenotype, gene expression, or demographic information. The term *heterogenous* is used to describe the set of attributes defined on the network, which can be continuous or categorical. We call this method attribute inverse regularized Markov CLustering (airMCL), because it connects the inverse regularized Markov Clustering (irMCL) approach with statistical classification methods, for the purpose of community detection in attributed networks.

The link between irMCL and is achieved through use of multiple logistic regression, in which the attribute information is regressed on the vectorized structure of the network.[22] This approach gives rise to probabilistic estimate of association between network structure and attributes directly, which is embedded into the *weights* for edges in the spread matrix for Algorithm 1. Specifically, airMCL relies on vectorized versions of distance matrices, which reflect the similarity (or lack thereof) between individuals for an attribute or set of attributes. The distance matrix, $D \in R^{n \times n}$ is symmetric, and the entries $d(i,j) = d(j,i)$ convey the similarity between nodes i and j for a given set of attributes. Consequently, vectorizing the strict upper triangular portion (not including the diagonal) of these matrices maps the pairwise information between nodes and attributes into a vectorized space. This set of vectors forms the set of predictors for the logistic regression modeling.

More formally, let Z_k be the vectorized strict upper triangular regions D_k, in the same way as the vectorization of A_{adj}. The logistic model is defined as:

$$\log\left(\frac{Pr(Y=1|Z)}{1-Pr(Y=1|Z)}\right) = \beta_0 + \sum_{k=1}^{p} \beta_k Z_k, \tag{1}$$

where β_0 is an intercept term, and $\beta_1 \ldots \beta_p$ are the regression coefficients for the vectorized attributes. The left hand side of Equation 1 is the log-odds ratio. We can directly estimate the odds ratio using the estimated coefficients $\hat{\beta}$ for each pairwise-relationship: $w = \exp\left(\sum_{k=1}^{p} \hat{\beta}_k Z_k\right)$, which is embedded into the *weights* for edges in the spread matrix for Algorithm 1.

Implementations rMCL and airMCL are performed in the R programming language (https://www.r-project.org/). A library `airMCL` that implements these algorithms will be made available in the CRAN repository upon publication.

2.3. *Simulations*

We examine the performance irMCL and airMCL using a variety of network simulations following the general framework proposed by Girvan and Newman.[9] In our simulations, we consider networks containing 128 nodes that are divided into four communities of 32 nodes each. Vertices are connected independently and randomly with a probability P_{in} for those within the same community, and P_{out} for vertices in different communities ($P_{out} < P_{in}$). The probabilities are selected such that the average degree of a vertex is 16. The expected number of links to a vertex in a different community is defined as z_{out}, while the expected number of links to a vertex in the same community is defined as z_{in}. Note that the community structure is less defined (weak) when z_{out} is larger.

Within simulations of different connectivity patterns, we examined single continuous and categorical attributes, as well as their combination. Categorical attributes in the ith group were generated from a multinomial distribution:

$$p(X = x) = \begin{cases} p, & x = i \\ \dfrac{1-p}{3}, & x \in \{1, 2, 3, 4\}/i \end{cases}$$

The values of p were set to 0.9, 0.6, 0.3 to mimic strong, moderate, and weak associations to the network structure, respectively. Note that when p takes large value (0.9), the attribute X is highly homogeneous within communities. When p is small, however, it implies X has high variability within each group, and will be less informative for the purpose of community detection.

A normal distribution, $N(\mu_i, 1)$, was used for continuous attributes of group i. The difference between means of consecutive groups $\Delta\mu = \mu_{i+1} - \mu_i$ was set at 4, 2, or 0.5, to convey strong, moderate, and weak levels of association, respectively, between structural and attribute information. Within the simulation framework, we also set out to determine how sensitive our methods are to noise in network in the form of missing links. For each scenario, we performed community detection on the full network, and networks with up to 30% of their links missing at random. We compared our methods, airMCL and irMCL , with rMCL and a fast-greedy method.[11] We also examined an irMCL-adhoc method, which can be only applied to networks with single categorical attribute. In this setting, irMCL-adhoc assigns a fixed weight of 0.5 when the two nodes have different attribute values, regardless of the structural relevance.

Mixed attributes were also explored for different combinations of continuous and categorical levels of association. The mixed attribute simulations described previously were also carried out to explore performance for networks varying from well defined communities (small z_{out}) to poorly defined communities (large z_{out}). The clustering by attribute information alone is also performed. For continuous attributes, Euclidean distance and hierarchical clustering with complete linkage is used. For categorical attribute, the attribute value is directly used as cluster label. For combination of two heterogeneous attributes, the larger average performance between continuous and categorical is used, because they cannot be combined for clustering.

Performance is assessed using the Adjusted Rand Index (ARI) as a measure of agreement between two data clusterings.[23,24] Let S be a set of n elements and consider two partitions of S to compare, $X = \{X_1, \ldots, X_r\} \in S$ and $Y = \{Y_1, \ldots, Y_s\} \in S$. The ARI assumes the generalized hypergeometric distribution as the model of randomness, where the two partitions are picked at random such that the number of classes and clusters are fixed.[24] Specifically, letting n_{ij} denote the number of objects in common between X_i and Y_j and $a_i = \sum_j n_{ij}$, and $b_j = \sum_i n_{ij}$, the ARI is defined as:[24]

$$\mathrm{ARI} = \frac{\sum_{ij}\binom{n_{ij}}{2} - [\sum_i \binom{a_i}{2}\sum_j \binom{b_j}{2}]/\binom{n}{2}}{\frac{1}{2}[\sum_i \binom{a_i}{2} + \sum_j \binom{b_j}{2}] - [\sum_i \binom{a_i}{2}\sum_j \binom{b_j}{2}]/\binom{n}{2}}.$$

For each parameter setting, 100 simulated networks are tested and the standard error is calculated.

2.4. *Application to functional genomics*

We applied the airMCL method to a breast cancer microarray dataset by Van Der Vijver *et al.*[25] The data was obtained from the package `seventyGenesData` available in Bioconductor (https://www.bioconductor.org/). Our objective was to infer communities using airMCL and identify those which relate to survival. Briefly, the data consists of 295 tumor samples from a 295 women with breast cancer. Survival data was also made available for all each patient in this population. The duration for survival analysis in this study is Time To Metastasis (TTM). In this study, 101 metastasis events occurred and 194 censored data points.

The *input* to airMCL requires specification of an adjacency matrix for a corresponding network and a set of attributes. In our application, we define the network using the `KEGG` database.[26] The 24,496 transcripts in the dataset were mapped to KEGG pathways using Entrez gene identifiers with the Bioconductor annotation package `KEGG.db`. In order to obtain a 1: 1 mapping, when several transcripts mapped to a gene, the one with the most variation across the sample was retained for the modeling. After mapping, the data set consisted of 295 samples and 4,715 genes that represent nodes in the network. Transcript abundance was represented by the log10 of the ratio between each sample and the reference RNA.[25] The adjacency matrix (input) was determined through an pathway-based gene network that was formed by placing links between genes when they are present in the same KEGG pathway. The functional network consists of 4,715 nodes (genes) and 883,557 edges.

Node attributes for the airMCL are defined through a measure of dissimilarity of the gene expression data. Several dissimilarity options are feasible and we expand on this point in the discussion. The dissimilarity measure is defined as $d_{i,j} = 1 - |r_{i,j}|$, where $r_{i,j}$ is the Pearson correlation coefficient between the ith and jth genes. Logistic regression models are fit using the vectorized pairwise dissimilarity on edges (1 linked, 0 for unlinked pairs) as the predictor, and the vectorized adjacency matrix as the response variable. However, the gene network has 4,715 nodes, implying more than 11 million observations in the regression. Moreover, the sparsity of the network gives rise a severe class imbalance. To alleviate the computational complexity and address imbalance, we randomly selected the unlinked node pairs so as to have the same number as that of the edges.

Survival analysis is performed on TTM using a Cox proportional hazard model.[27] Benjamini and Hochberg method was used to control the false discovery rate.[28] A threshold of P-value< 0.05 was used to identify modules whose overall expression levels are significantly associated with the time to metastasis. Kaplan-Meier estimates were calculated for each significant module based on stratification of the 295 patients into two groups, using the median overall expression levels of the module. Specifically, $w_{kl} = \frac{1}{m_l}\sum_{i\in c_l}^{m_l} z_{ik}$, where w_{kl} is the average expression level of lth module for kth patient, c_l is the set of node index of lth module, and m_l is the number of nodes in this module.

3. Results

Each simulation was run to convergence. Some general trends persisted for the different parameter and attribute simulations (Figure 1). The overall performance of rMCL was poor, but relatively stable across missing links and different levels of association between structure and attribute. This was the case for categorical, continuous, and mixed attribute settings. When the attribute associations are moderate and weak, fast-greedy shows advantages over the other methods when the missing links is larger (Figure 1B-C,E-F).

When a categorical attribute is highly relevant to true groups ($p = 0.9$), the inclusion of attribute information significantly improved the performance (Figure 1A). In this case, the airMCL and post-hoc weighting were both useful in boosting performance. The performance for post-hoc weighting degrades as the attribute association weakens (Figures 1B-C). For continuous attributes, the airMCL is superior for strong associations across all levels of missing links (Figure 1D), and is the top-performer for moderate association with fewer missing links (Figure 1E). When the associations are weak for continuous attributes, airMCL is competitive with irMCL for scenarios with few missing links (Figure 1F). In simulations with multiple heterogeneous attributes (Figure 2G-I), the airMCL successfully extracts the structurally relevant information and improves the performance over clustering using structural information only (irMCL).

Tuning the parameter z_{out} in the simulations enables us to test the performance of our approaches in scenarios where the communities are not well defined. The performance of irMCL is comparable to fast greedy algorithm, and actually slightly outperforms fast-greedy under z_{out} ranges from 1 to 6 (Figure 2A-C). In our simulations, large z_{out} represents networks in which there is poor community structure. The airMCL's use of attributes offsets this poor structure and is the top-performing method in these extreme scenarios.

We applied the airMCL method to a breast cancer dataset using a KEGG pathway-based network and gene expression attributes.[25] A correlation-based similarity was utilized for the attributes, and the estimated coefficient for the logistic regression was -0.7624 and significant. Convergence was observed 15 iterations. The clustering of the rows of the stochastic matrix was

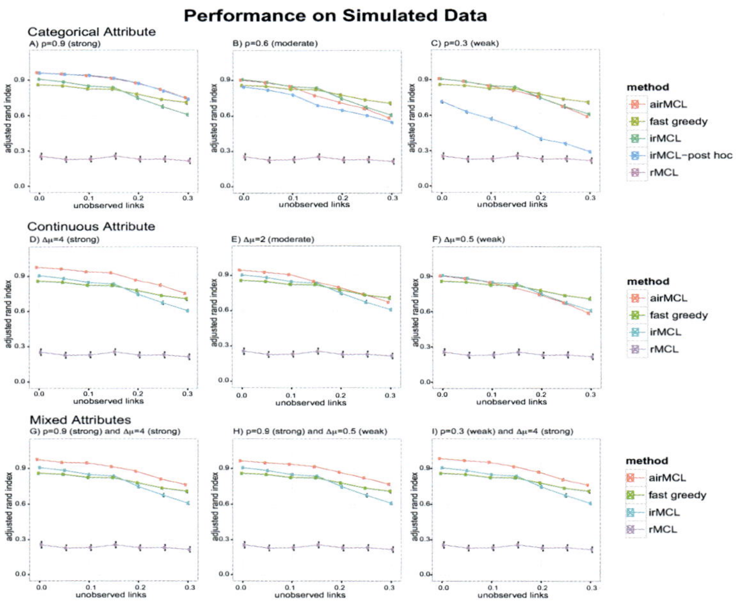

Fig. 1. Simulation results for community detection for a categorical attribute (top row), continuous attribute (second row), and a mixture of a continuous and categorical attributes (third row). Relationships between categorical attributes and community structure were simulated to be **(A)** strong, **(B)** moderate, and **(C)** weak, respectively. Likewise, for continuous attributes **(D-F)**. For the mixed attribute simulation the categorical/continuous relationships between attribute and structure considered were **(G)** strong/strong, **(H)** strong/weak, and **(I)** weak/strong.

determined using the maximum average silhouette, which was 0.85, and yielded 434 clusters. Note that the *rule of thumb* for strong structure is an average silhouette between $0.71 - 1$.[21]

Only modules with size ≥ 8 were selected for survival analysis, and the overall activation status of each module was used for the covariate (see M&M) for predicting TTM. Cox proportional hazard model was used and a multiple testing adjustment was made. A threshold criteria of P-value< 0.05, both methods yields six modules whose overall expression levels are significantly associated with the time to metastasis. Table 1 shows the summary of modules detected and a full listing of module members is available in the Supplement (posted on https://sphhp.buffalo.edu/biostatistics/news-events/workshops/). The adjusted p-values in Table 1 are from Cox regression.

In order to utilize the Kaplan-Meier product limit estimator, for each of the six modules, the 295 patients were split into two groups (low-expression and high-expression) using the median of overall expression levels as cut-off. The survival curves are shown in Figure 3. Log-rank tests were used to test the difference between survival curves of high- and low-expression

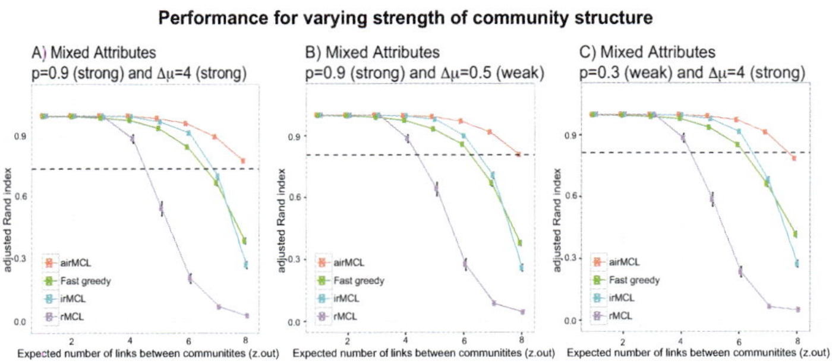

Fig. 2. Comparison of the performance of airMCL/irMCL (with/without attributes) with rMCL and fast greedy method in synthetic networks using adjusted Rand index against z_{out}. The attributes are mixed, which include attributes with **(A)** high-relevance categorical ($p = 0.9$) and high-relevance continuous ($\Delta\mu = 4$), **(B)** high-relevance categorical ($p = 0.9$) and weak-relevance continuous ($\Delta\mu = 0.5$), and **(C)** weak-relevance categorical ($p = 0.3$) and high-relevance continuous ($\Delta\mu = 4$). The horizontal black dashed line indicating the average ARI using attribute information alone.

Table 1: Breast Cancer Module Summarization

Module	Size	Pathways represented	P-value
1	8	Hedgehog signaling pathway (hsa04340)	0.02195
2	27	Pathway in cancers (hsa05200)	0.02195
		MAPK signaling pathway (hsa04010)	
		Adherens junction (hsa04520)	
		Regulation of actin cytoskeleton (hsa04810)	
		Melanoma (hsa05218)	
		Prostate cancer(hsa05215)	
		Oocyte meiosis (hsa04114)	
3	82	Ribosome pathway (hsa03010)	0.02195
4	25	Cell cycle pathway (hsa04110)	0.02195
		Non-homologous end-joining (hsa03450)	
5	19	Pathway in cancers (hsa05200)	0.03541
		Mismatch repair (hsa03430)	
		Colorectal cancer (hsa05210)	
		Small cell lung cancer (hsa05222)	
		Pancreatic cancer (hsa05212)	
		Thyroid cancer (hsa05216)	
6	35	Proteosome pathway (hsa03050)	0.03614

groups. The unadjusted p-values of log-rank tests are shown in Figure 3.

4. Discussion

The design of airMCL is such that the impact of the attributes on community formation depends on the strength of the association between attributes and network structure. Consequently, those weak associations are naturally dampened. Our approach is similar to spirit to the weighting that is done in neural network via an activation function (usually a sigmoid),

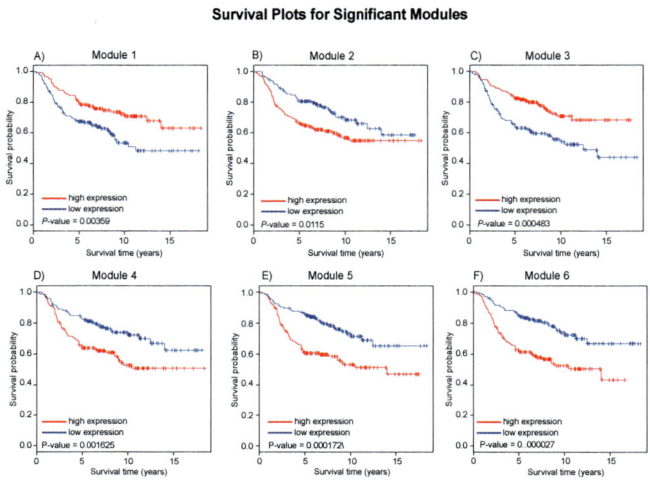

Fig. 3. **(A-F)** Kaplan-Meier survival plots for modules $1-6$. Estimate is based on the partition of the sample into two groups using median values of overall expression for each module (see methods). Red indicates higher expression, blue is for lower expression, and the unadjusted P-values for the log-rank tests are shown.

which weights the features in the input layer. In severely weak settings, the airMCL operates like the irMCL. A challenge attribute information may be irrelevant, or even contradict, the structure of the network. In our simulations, bringing in attribute with weak signals did not derail performance (Figure 1C,F,G-I). This is important as it is not up to the user to specify what attributes are important by weighting, or even eliminating them. In contrast, in the categorical case, we observed with the ad-hoc weighting can derail performance, especially in light of weak attribute associations (Figure 1C).

The fit of the logistic model itself reveals the strength of the relationship between attribute similarity network structure. Examining the regression coefficients (Equation 1) of the model can guide in model development, e.g., choice of similarity, subsets of features. For example, hypothesis testing on the coefficients (e.g., $H_0 : \beta_j = 0$) can reveal the significance of the attribute similarity as a predictor of structure. We have found this useful as a way of selecting a similarity measure for the attributes.

An important feature of the airMCL approach is that the derived inputs for the logistic regression can be handled in a flexible manner. If the set of attributes is heterogenous, one can partition the attributes into multiple subsets, and estimate distance matrices over these subsets independently. This approach enables a unique choice of similarity measure most appropriate for the given attribute or set of attributes. Differences in scales, even within variables of the same type, can also be managed by subsetting attributes. Collectively, the vectorization of the different distances would give rise to multiple predictors for the logistic regression.

In the breast cancer application, some of the identified pathways are consistent with that reported by Van't Veer *et al.*,[29] such as pathways in cell cycle regulation (Module 4) and signal transduction (Module 2). In addition, we also found that ribosome pathway is associated with breast cancer metastasis. This is consistent with the results reported by Belin *et al.*, that dysregulation of ribosome biogenesis is related to enhanced tumor aggressivity.[30] Activation of hedgehog pathway is also reported in tumors including breast cancers,[31,32] and is related to cancer metastasis.[33] Figure 3 shows that module over-expression (red) is often associated with higher hazards of metastasis. The up-regulation of Module 1 (hedgehog signaling pathway) is unexpectedly associated with better prognosis. This can be explained by the fact that up-regulated genes in this module encode inhibitors in this pathway ($GAS1$, $RAB23$, and $CK1$), which is biologically plausible.

In our simulations, we have simulated balanced communities of moderate size. However, we have also observed good performance, in terms of computational time and accuracy, in the simulation of balanced larger communities. In the case of unbalanced communities, we have achieved good performance in moderate sized simulation networks and real social networks. However, a limitation of our approach is applications to large (1000+ nodes) unbalanced networks. Addressing this form of scalability will be a direction of future research.

We have focussed on a specific application to gene expression cancer data to showcase our method. However, the `airMCL` is generalizable in the sense that it can be used in connection with data that contains a network structure and a set of attributes. The term *attribute* can be loosely defined to encompass demographic information, clinical data, omics data, and combinations of different types of data. The combination of multiple sources of data is known to be a major challenge, and our approach directly integrates them into the community detection. Framing the problem of relating the attributes to the structure via classification has several advantages. Arguably the most important of these advantages is the ability to monitor and quantify loss. Framing the connection between structure and attributes as a supervised learning problem enables the use of statistical classification methods. In this work, we outlined the framework in terms of the classic multiple logistic regression model.[22] However, several classification methods may be more or less suitable depending on the dimension of the graph and attributes, and also the correlation of predictors. Within the classification methods framework are opportunities to utilize the bias-variance tradeoff for model and feature selection. This is a direction of future research, which we anticipate will guide in elimination of extraneous attributes (and potentially nodes), and protect against overfitting.

5. Acknowledgements

HY and RHB were supported through NSF DMS 1312250 and NSF DMS 1557593.

References

1. L. Danon, A. Diaz-Guilera, J. Duch and A. Arenas, *Journal of Statistical Mechanics: Theory and Experiment* **2005**, p. P09008 (2005).

2. M. E. Newman, *The European Physical Journal B-Condensed Matter and Complex Systems* **38**, 321 (2004).
3. S. E. Schaeffer, *Computer Science Review* **1**, 27 (2007).
4. S. Horvath, *Weighted Network Analysis: Applications in Genomics and Systems Biology* (Springer Science & Business Media, 2011).
5. B. W. Kernighan and S. Lin, *Bell system technical journal* **49**, 291 (1970).
6. S. C. Johnson, *Psychometrika* **32**, 241 (1967).
7. M. Fiedler, *Czechoslovak Mathematical Journal* **23**, 298 (1973).
8. W. E. Donath and A. J. Hoffman, *IBM Journal of Research and Development* **17**, 420 (1973).
9. M. Girvan and M. E. Newman, *Proceedings of the National Academy of Sciences* **99**, 7821 (2002).
10. M. E. Newman and M. Girvan, *Physical review E* **69**, p. 026113 (2004).
11. A. Clauset, M. E. Newman and C. Moore, *Physical review E* **70**, p. 066111 (2004).
12. M. E. Newman, *Physical review E* **69**, p. 066133 (2004).
13. M. E. Newman, *Proceedings of the National Academy of Sciences* **103**, 8577 (2006).
14. D. Hanisch, A. Zien, R. Zimmer and T. Lengauer, *Bioinformatics* **18**, S145 (2002).
15. Y. Zhou, H. Cheng and J. X. Yu, *Proceedings of the VLDB Endowment* **2**, 718 (2009).
16. L. Akoglu, H. Tong, B. Meeder and C. Faloutsos, Pics: Parameter-free identification of cohesive subgroups in large attributed graphs., in *SDM*, 2012.
17. F. Moser, R. Colak, A. Rafiey and M. Ester, Mining cohesive patterns from graphs with feature vectors., in *SDM*, 2009.
18. E. Georgii, S. Dietmann, T. Uno, P. Pagel and K. Tsuda, *Bioinformatics* **25**, 933 (2009).
19. S. Van Dongen, *SIAM Journal on Matrix Analysis and Applications* **30**, 121 (2008).
20. V. Satuluri and S. Parthasarathy, Scalable graph clustering using stochastic flows: applications to community discovery, in *Proceedings of the 15th ACM SIGKDD International conference on Knowledge Discovery and Data Mining*, 2009.
21. P. J. Rousseeuw, *Journal of computational and applied mathematics* **20**, 53 (1987).
22. D. W. Hosmer Jr and S. Lemeshow, *Applied logistic regression* (John Wiley & Sons, 2004).
23. W. M. Rand, *Journal of the American Statistical Association* **66**, 846 (1971).
24. L. Hubert and P. Arabie, *Journal of classification* **2**, 193 (1985).
25. M. J. Van De Vijver, Y. D. He, L. J. van't Veer, H. Dai, A. A. Hart, D. W. Voskuil, G. J. Schreiber, J. L. Peterse, C. Roberts, M. J. Marton *et al.*, *New England Journal of Medicine* **347**, 1999 (2002).
26. M. Kanehisa, S. Goto, Y. Sato, M. Kawashima, M. Furumichi and M. Tanabe, *Nucleic acids research* **42**, D199 (2014).
27. D. R. Cox and D. Oakes, *Analysis of survival data* (CRC Press, 1984).
28. Y. Benjamini and Y. Hochberg, *Journal of the Royal Statistical Society. Series B (Methodological)* , 289 (1995).
29. L. J. Van't Veer, H. Dai, M. J. Van De Vijver, Y. D. He, A. A. Hart, M. Mao, H. L. Peterse, K. van der Kooy, M. J. Marton, A. T. Witteveen *et al.*, *Nature* **415**, 530 (2002).
30. S. Belin, A. Beghin, E. Solano-Gonzàlez, L. Bezin, S. Brunet-Manquat, J. Textoris, A.-C. Prats, H. C. Mertani, C. Dumontet and J.-J. Diaz, *PloS one* **4**, p. e7147 (2009).
31. M. Kubo, M. Nakamura, A. Tasaki, N. Yamanaka, H. Nakashima, M. Nomura, S. Kuroki and M. Katano, *Cancer research* **64**, 6071 (2004).
32. J. Taipale and P. A. Beachy, *nature* **411**, 349 (2001).
33. J. M. Bailey, P. K. Singh and M. A. Hollingsworth, *Journal of cellular biochemistry* **102**, 829 (2007).

COLLECTIVE PAIRWISE CLASSIFICATION FOR MULTI-WAY ANALYSIS OF DISEASE AND DRUG DATA

MARINKA ZITNIK

Faculty of Computer and Information Science, University of Ljubljana,
Vecna pot 113, SI-1000 Ljubljana, Slovenia
E-mail: marinka.zitnik@fri.uni-lj.si

BLAZ ZUPAN

Faculty of Computer and Information Science, University of Ljubljana,
Vecna pot 113, SI-1000 Ljubljana, Slovenia, and
Department of Molecular and Human Genetics, Baylor College of Medicine,
One Baylor Plaza, Houston, TX, 77030, USA
E-mail: blaz.zupan@fri.uni-lj.si

Interactions between drugs, drug targets or diseases can be predicted on the basis of molecular, clinical and genomic features by, for example, exploiting similarity of disease pathways, chemical structures, activities across cell lines or clinical manifestations of diseases. A successful way to better understand complex interactions in biomedical systems is to employ *collective relational learning* approaches that can jointly model diverse relationships present in multiplex data. We propose a novel collective pairwise classification approach for multi-way data analysis. Our model leverages the superiority of latent factor models and classifies relationships in a large relational data domain using a pairwise ranking loss. In contrast to current approaches, our method estimates probabilities, such that probabilities for existing relationships are higher than for assumed-to-be-negative relationships. Although our method bears correspondence with the maximization of non-differentiable area under the ROC curve, we were able to design a learning algorithm that scales well on multi-relational data encoding interactions between thousands of entities. We use the new method to infer relationships from multiplex drug data and to predict connections between clinical manifestations of diseases and their underlying molecular signatures. Our method achieves promising predictive performance when compared to state-of-the-art alternative approaches and can make "category-jumping" predictions about diseases from genomic and clinical data generated far outside the molecular context.

Keywords: Collective classification, multi-relational learning, three-way model, drug-drug interactions, drug-target interactions, symptoms-disease network, gene-disease network

1. Introduction

Collective relational learning is concerned with data domains where entities like drugs, diseases and genes are interconnected through multiple relations, such as drug-drug and drug-target interactions or disease comorbidity.[1–4] Since these approaches promote leaps across different data contexts, they are particularly well suited to model large-scale heterogeneous collections of biomedical data and have proven especially attractive for estimating binary relations, such as drug-drug interactions. These approaches take advantage of the relational effects in the data by relying on relationships within one set of entities when estimating relationships for the other entity set. For example, when predicting drug-target interactions relational approaches can consider the fact that drugs with similar pharmacological effects are likely to interact with proteins with similar genomic sequences.[1,2,5–7] Another example is mining of disease data, where relational approaches can benefit from observation that diseases caused by dysregulation of related pathways are likely to have similar clinical manifestation and show sensitivity to similar chemical compounds.[3]

State-of-the-art collective relational learning methods rely on latent factor modeling and typically measure the fit of the models to the data through a regression metric, such as the root mean-squared error, one-sided linear error or square penalty.[3,8–12] The use of this metric in the search for best model parameters is especially appealing due to the well explored theory with many statistical guarantees about the quality of least-squares solutions, efficient procedures for model estimation, and, in some cases, even the ability to find the optimal estimates. However, it is now widely recognized that approaches optimizing the error rate, such as the root mean-squared error, can perform poorly with respect to ranking of the relationships.[13,14] This situation gets exacerbated in practice where life scientists focus their attention on only a small number of predicted relationships between entities, effectively ignoring all but a short list of most promising predicted relationships. For this reason, it is better to focus on correct prediction of small but highly likely set of relations than on accurately predicting all, even the irrelevant relationships.[15]

The predictive task we need to address is ranking where the aim is to rank the relationships according to their relevance. At first it may appear that learning a good regression model is sufficient for this task, as a model that achieves perfect regression will also give perfect ranking. However, a model with near-perfect regression performance may have arbitrarily poor ranking performance. The vice versa also holds true: a perfect ranking model may give very poor regression estimates.[16] The development of prediction models that optimize for a ranking metric and can accommodate heterogeneous biomedical relations is therefore a crucial step towards accurate identification of the most promising relationships.

Taking insights from the research reviewed above, we propose a general statistical method that can estimate relationships between entities, e.g., drugs and diseases, from multi-way data, e.g., drug-drug interactions and shared human disease symptoms. Our proposed method uses pairwise classification scheme to directly optimize a ranking metric. It estimates a latent data model, which serves to make predictions about pairwise entity relationships. The contributions in this work are:

- We present a generic collective pairwise classification (COPACAR) model for multi-way data analysis.[a] We derive COPACAR model from the maximum posterior estimator for optimal collective pairwise classification on multi-relational data. We show the analogies between COPACAR and the maximization of area under the ROC curve.
- For minimizing the loss function of COPACAR, we propose a learning algorithm that is based on stochastic gradient descent with bootstrap sampling of training triplets. The *in silico* experimental results show that our algorithm has favorable convergence results w.r.t. the number of required algorithm iterations and the size of subsampled data. COPACAR can be easily parallelized, which can further increase its scalability.
- We show how to apply COPACAR to two challenges arising in personalized medicine. In studies on multi-way disease and drug data we demonstrate that our method is capable of making *category-jumping inferences*,[17] i.e. *it can make predictions within and across informational contexts.*
- Our experiments show that for the task of collective learning on multi-relational disease and drug data, learning a model with COPACAR outperforms approaches based on tensors and their decompositions.

Below we first overview related approaches for multi-relational learning and tensor decomposition. We then formulate a novel collective pairwise classification model and discuss the model fitting procedure. We present two case studies where we (1) investigate the connections between clinical manifestations of diseases and their molecular interactions, and (2) study the interactions between drugs based on drug-drug and drug-target relationships, structural similarities of the compounds, known pharmacological effects and interaction information extracted from the literature.

2. Related Work

Collective learning[11] is an umbrella term for the mechanisms that exploit information, such as that on related classes, additional attributes or relationships between related entities, to support various learning

[a]The online repository http://github.com/marinkaz/copacar includes the data and the source code used in this paper as well as additional material for experiments in a non-biological domain.

tasks on multi-relational data, like classification, link prediction in networks and association mining. The literature on relational learning is vast, hence we only give a very brief overview.

Relational learning approaches[18] assume that relations between entities arise from the interactions between *intrinsic latent attributes* of these entities.[10] Until recently, these approaches focused mostly on modeling a single relation as opposed to trying to consider a collection of similar relations. However, recently made observations that relations can be highly similar or related[3,10–12,19] suggested that superimposing models learned independently for each relation would be ineffective, especially because the relationships observed for each relation can be extremely sparse. We here approach this challenge by proposing a collective learning approach that jointly models many data relations.

Probabilistic modeling approaches for relational (network) data often translate into learning an embedding of the entities into a low-dimensional manifold. Algebraically, this corresponds to a *factorization of an appropriately defined data matrix.*[3] A natural extension to modeling of many relations is to stack data matrices and regard them as a tensor.[10,11,20] Another extension to simultaneously learning many relations is to *share a common embedding or the entities* across different relations via *collective matrix factorization.*[9,21] An extensive review of tensor decompositions and other relational learning approaches can be found in Nickel *et al.*[19]

Several clustering-based approaches have been proposed for multi-relational learning. These include classical *stochastic blockmodels*, which associate a latent class to each entity in a domain; *mixed membership stochastic blockmodels*, which allow entities to have a mixed clusters membership;[22] non-parametric Bayesian models, which automatically infer the number of latent clusters;[8,23] and neural network architectures, which embed symbolic data representations into a flexible continuous vector space.[24] Many network modeling approaches[25–27] try to detect local dependencies among the entities, i.e. nodes, and accordingly group the nodes from a multiplex network into densely interconnected groups.

Unlike clustering-based approaches, COPACAR has classification capabilities, which come from model inference based on a pairwise ranking loss. Furthermore, COPACAR uses a factorized model to estimate interactions between entities, so that we can apply our approach to large data domains. Our approach also differs from the matrix factorization approach in terms of estimation method: while matrix factorization models rely on likelihood training, we explicitly try to make the probability for existing relationships to be larger than for assumed-to-be-negative relationships.

3. Relational Data Modeling

We consider relational data consisting of triplets where each triplet encodes a relationship between two entities that we call the subject and the object. A triplet $\langle E_i, \mathcal{R}^{(k)}, E_j\rangle$ indicates that relation $\mathcal{R}^{(k)}$ holds between subject E_i and object E_j. We represent a triplet as a matrix element $\mathbf{X}_{ij}^{(k)}$, where matrix $\mathbf{X}^{(k)}$ encodes relation $\mathcal{R}^{(k)}$. We model dyadic multi-relational data as a three-way tensor where two modes are identically formed by the concatenated entities and the third dimension corresponds to the relations.

Fig. 1 illustrates our modeling method. We assume the data is given as a collection of m partially observed matrices each of size $n \times n$, where n is the number of entities and m is the number of relations[b]. A matrix element $\mathbf{X}_{ij}^{(k)} = 1$ denotes existence of a relationship $\langle E_i, \mathcal{R}^{(k)}, E_j\rangle$. Otherwise, for non-existing relationships, the associated matrix elements are set to zero. Unknown relationships can have a designated value so that they are ignored during model estimation.

We refer to a triplet also as a *relationship.* A typical example, which we discuss in greater detail in the following sections, is in pharmacogenomics, where a triplet $\langle E_i, \mathcal{R}^{(1)}, E_j\rangle$ might correspond to the interaction between drug i and drug j, and a triplet $\langle E_i, \mathcal{R}^{(2)}, E_j\rangle$ might represent the association of drug i and drug j through a shared target protein. The goal is to learn a single model of all relations, which can

[b]Note that unlike established techniques in multi-relational modeling,[11] our model does not need a homogeneous data domain. That is, entities of the first two modes can each be of different type, such as drugs, patients, diseases, etc.

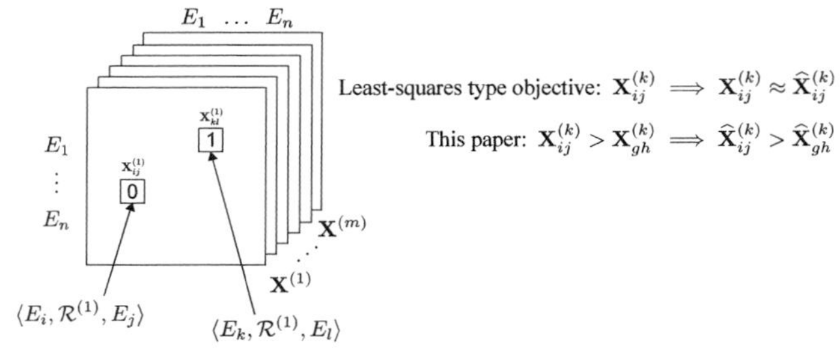

Fig. 1. A multi-relational data model for collective learning. $E_1, \dots, E_n$ denote the entities, while $\mathbf{X}^{(1)}, \dots, \mathbf{X}^{(m)}$ encode the relations in the domain.

reliably predict unseen triplets. For example, one might be interested in finding the most likely relation $\mathcal{R}^{(k)}$ for a given subject-object pair (E_i, E_j). Or, given a relation $\mathcal{R}^{(k)}$, one might like to know the most likely relationships $\langle E_i, \mathcal{R}^{(k)}, E_j \rangle$.

4. Model Description and Theoretical Aspects

Next, we formulate a generic method for collective pairwise classification on multi-relational data. It consists of the optimization criterion, which we derive by Bayesian analysis using the likelihood function for the pairwise ranking and the prior probability for model parameters. We also highlight the analogy between our model and the well known ranking statistic.

We begin with the intuition that a desirable collective learning model, which aims to identify *the most relevant relationships* in multi-relational data, should exhibit the property illustrated in Fig. 1 (right, bottom). The model should aim to *rank the relationships rather than to score the individual relationships* as ranking better represents learning tasks to which these models are applied in life and biomedical sciences. We later demonstrate that accounting for this property is important.

However, a common theme of many multi-relational models is that all the relationships a given model should predict in the future are presented to the learning algorithm as non-existing (negative) relationships during training. The algorithm then fits a model to the data and optimizes for *scoring of single relationships* with respect to a least-squares type objective[8,9,11,21,23,28] (Fig. 1, right, top). This means the model is optimized to predict the value 1 for the existing relationships and 0 for the rest. In contrast, we here consider *relationship pairs* as training data and optimize for *correctly ranking relationship pairs.*

4.1. *Collective Pairwise Classification Model for Multi-Way Data* (COPACAR)

To find the correct pairwise ranking of the relationships for all entity pairs and all relations in the domain we would like to maximize the following posterior probability:

$$p(\widehat{\mathbf{X}}^{(k)} | >_k) \propto p(>_k | \widehat{\mathbf{X}}^{(k)}) p(\widehat{\mathbf{X}}^{(k)}), \tag{1}$$

where $\widehat{\mathbf{X}}^{(k)}$, $k = 1, 2, \dots m$, denote the latent data model. Here, the notation $>_k$ indicates the relational structure for kth relation. For now, we assume that all relations act independently of each other; we will later discuss how to achieve category-jumping between the considered relations. We also assume the ordering of each relationship pair $(\langle E_i, \mathcal{R}^{(k)}, E_j \rangle, \langle E_g, \mathcal{R}^{(k)}, E_h \rangle)$ is independent of the ordering of every other relationship pair. Hence, we rewrite the above relation-specific likelihood function $p(>_k | \widehat{\mathbf{X}}^{(k)})$ as a product of single densities and then combine it for all relations $k = 1, 2, \dots, m$ as:

$$\prod_k p(>_k |\widehat{\mathbf{X}}^{(k)}) = \prod_k \prod_{i,j,g,h} p(\widehat{\mathbf{X}}_{ij}^{(k)} >_k \widehat{\mathbf{X}}_{gh}^{(k)})^{\delta(\mathbf{X}_{ij}^{(k)} >_k \mathbf{X}_{gh}^{(k)})} (1 - p(\widehat{\mathbf{X}}_{ij}^{(k)} >_k \widehat{\mathbf{X}}_{gh}^{(k)})^{\delta(\mathbf{X}_{ij}^{(k)} \not>_k \mathbf{X}_{gh}^{(k)})}, \quad (2)$$

where δ is the indicator function, $\delta(x)$ is 1 if x is true and is 0 otherwise. Assuming that the properties of a proper pairwise ranking scheme hold, we can further simplify the expression from Eq. (2) into:

$$\prod_k p(>_k |\widehat{\mathbf{X}}^{(k)}) = \prod_k \prod_{i,j,g,h} p(\widehat{\mathbf{X}}_{ij}^{(k)} >_k \widehat{\mathbf{X}}_{gh}^{(k)})^{\delta(\mathbf{X}_{ij}^{(k)} >_k \mathbf{X}_{gh}^{(k)})}. \quad (3)$$

So far it not guaranteed that the model produces a total ordering of the relationships in each relation. To achieve this we need to satisfy the requirements for a total ordering. We do so by defining the probability that relationship $\langle E_i, \mathcal{R}^{(k)}, E_j \rangle$ is more relevant than relationship $\langle E_g, \mathcal{R}^{(k)}, E_h \rangle$ as:

$$p(\widehat{\mathbf{X}}_{ij}^{(k)} >_k \widehat{\mathbf{X}}_{gh}^{(k)}) \triangleq \sigma(\widehat{\mathbf{X}}_{ij}^{(k)} - \widehat{\mathbf{X}}_{gh}^{(k)}), \quad (4)$$

where $\sigma(\cdot)$ is the logistic function, $\sigma(x) = 1/(1 + \exp(-x))$.

Until now we delegated the task of modeling the relationship $\langle E_i, \mathcal{R}^{(k)}, E_j \rangle$ to a yet unspecified latent model $\widehat{\mathbf{X}}^{(k)}$, $k = 1, 2, \dots m,$. We describe the model that can consider the intrinsic structure of multi-relational data. We build on the intuition from the RESCAL[11,12] tensor decomposition and introduce the following rank-r factorization, where each relation is factorized as:

$$\widehat{\mathbf{X}}_{ij}^{(k)} = \mathbf{A}_i^T \mathbf{R}^{(k)} \mathbf{A}_j, \text{ for } k = 1, 2, \dots, m. \quad (5)$$

Here, $\mathbf{A}$ is a $n \times r$ matrix of latent components, where n represents the number of entities in the domain and r is dimensionality of the latent space. The rows of $\mathbf{A}$, i.e., $\mathbf{A}_i^T$ for $i = 1, 2, \dots, n$, model the latent component representation of entities in the domain. Matrix $\mathbf{R}^{(k)}$ is an asymmetric $r \times r$ matrix that contains the interactions of the latent components in kth relation.

When learning a large number of relations, i.e., when k is large, the number of observed relationships for each relation can be small, leading to a risk of overfitting. To decrease the overall number of parameters, the model in Eq. (5) encodes relation-specific information with the latent matrices $\mathbf{R}^{(k)}$ and embeds the entities into the latent space spanned by $\mathbf{A}$. The effect of $r \ll n$ is *the automatic reuse of latent parameters across relations.* Collectivity of COPACAR is thus given by the structure of its model.

Thus far we discussed the likelihood function $p(>_k |\widehat{\mathbf{X}}^{(k)})$. To determine the Bayesian approach from Eq. (1), we propose a prior $p(\widehat{\mathbf{X}}^{(k)})$, which is a normal distribution with a zero mean and a covariance matrix Σ:

$$p(\mathbf{A}) \sim \mathcal{N}(\mathbf{0}, \Sigma_{\mathbf{A}}), \qquad p(\mathbf{R}^{(k)}) \sim \mathcal{N}(\mathbf{0}, \Sigma_{\mathbf{R}}), \text{ for } k = 1, 2, \dots, m. \quad (6)$$

We further reduce the number of unknown parameters by setting $\Sigma_{\mathbf{A}} = \lambda_{\mathbf{A}}\mathbf{I}$ and $\Sigma_{\mathbf{R}} = \lambda_{\mathbf{R}}\mathbf{I}$. We derive the optimization criterion for our collective pairwise classification via the maximum posterior estimator:[29]

$$\begin{aligned}
\text{OPT-COPACAR} &\triangleq \log p(\widehat{\mathbf{X}}^{(k)}| >_k) \\
&= \log p(>_k |\widehat{\mathbf{X}}^{(k)}) p(\widehat{\mathbf{X}}^{(k)}) \\
&= \log \prod_k p(>_k |\widehat{\mathbf{X}}^{(k)}) p(\widehat{\mathbf{X}}^{(k)}) \\
&= \log \prod_k \prod_{i,j,g,h} \sigma(\widehat{\mathbf{X}}_{ij}^{(k)} - \widehat{\mathbf{X}}_{gh}^{(k)})^{\delta(\mathbf{X}_{ij}^{(k)} >_k \mathbf{X}_{ij}^{(k)})} p(\widehat{\mathbf{X}}^{(k)}) \\
&= \sum_k \sum_{i,j,g,h} \ell(\widehat{\mathbf{X}}_{ij}^{(k)} - \widehat{\mathbf{X}}_{gh}^{(k)}, \mathbf{X}_{ij}^{(k)} - \mathbf{X}_{gh}^{(k)}) + \lambda_{\mathbf{A}} \|\mathbf{A}\|^2 + \lambda_{\mathbf{R}} \sum_k \|\mathbf{R}^{(k)}\|_{\text{Fro}}^2, \quad (7)
\end{aligned}$$

where $\lambda_{\mathbf{A}}$ and $\lambda_{\mathbf{R}}$ are regularization parameters and *pairwise classification loss function* ℓ is formulated as:

$$\ell(\widehat{\mathbf{X}}_{ij}^{(k)} - \widehat{\mathbf{X}}_{gh}^{(k)}, \mathbf{X}_{ij}^{(k)} - \mathbf{X}_{gh}^{(k)}) = (\mathbf{X}_{ij}^{(k)} - \mathbf{X}_{gh}^{(k)}) \log \sigma(\mathbf{A}_i^T \mathbf{R}^{(k)} \mathbf{A}_j - \mathbf{A}_g^T \mathbf{R}^{(k)} \mathbf{A}_h). \tag{8}$$

The COPACAR model rewards estimates of the model parameters that are in accordance with the input data. Intuitively, the semantics of the loss ℓ is as follows: (1) If $\mathbf{X}_{ij}^{(k)} > \mathbf{X}_{gh}^{(k)}$ then $\langle E_i, \mathcal{R}^{(k)}, E_j \rangle$ should rank higher than $\langle E_g, \mathcal{R}^{(k)}, E_h \rangle$, since it is assumed that the first relationship has greater relevance than the latter. Therefore, a model in which $\widehat{\mathbf{X}}_{ij}^{(k)} > \widehat{\mathbf{X}}_{gh}^{(k)}$ holds, scores better on OPT-COPACAR than a model with the two relationships ranked in the reversed order of their scores. (2) For relationships that are both considered relevant, i.e. $\mathbf{X}_{ij}^{(k)} = 1$ and $\mathbf{X}_{gh}^{(k)} = 1$, or both considered irrelevant, i.e. $\mathbf{X}_{ij}^{(k)} = 0$ and $\mathbf{X}_{gh}^{(k)} = 0$, we cannot infer any preference for their degree of relevance and the loss is unaffected by them.

4.2. *Connection to the AUC Optimization*

We now show the analogy between OPT-COPACAR and area under the ROC curve (AUC). The AUC under the ROC curve corresponds to the probability that a random existing (positive) relationship will be scored higher than a random non-existing (negative) relationship. The maximization of the AUC statistic is especially attractive in biomedical data domains, where the real objective is to optimize the sorting order, for example, to sort the relationships into a list so that relevant relationships are concentrated towards the top of the list.[30] However, the problems with using the AUC statistic as an objective function are that it is non-differentiable, and of complexity $O(mn^4)$ in the number of entities n, i.e., $O(n^2)$ relationships need to be compared with themselves, and relations m in the domain. The AUC for relation k is usually defined across all pairwise comparisons of the relationships:

$$\text{AUC}(k) = \frac{1}{N_1(k) N_0(k)} \sum_{\substack{i,j \\ \mathbf{X}_{ij}^{(k)}=1}} \sum_{\substack{g,h \\ \mathbf{X}_{gh}^{(k)}=0}} \delta(\widehat{\mathbf{X}}_{ij}^{(k)} - \widehat{\mathbf{X}}_{gh}^{(k)} > 0), \tag{9}$$

where δ denotes the indicator function, and $N_1(k)$ and $N_0(k)$ count the existing (positive) and non-existing (negative) relationships in kth relation, respectively.

It is easy to see the analogy between the above formula and the maximum likelihood estimator in Eq. (7). They differ in the normalization constant $1/(N_1(k)N_0(k))$ and the definition of the loss function. In contrast to the non-differentiable stepwise δ function used by the AUC, we employ the smooth loss $\log \sigma(x)$ in Eq. (8). Unlike many algorithms, which select a differentiable counterpart of a non-differentiable loss function in a heuristic manner,[30] the COPACAR adopts the AUC statistic as its objective function and specifies the loss function in Eq. (8) based on the maximum likelihood estimation.

4.3. *Related Tensor Factorizations*

The factorization scheme specified in Eq. (5) builds on the RESCAL tensor decomposition[11] and is related to other tensor decompositions. Specifically, it can be regarded as a generalization of the established DEDICOM, or an asymmetric extension of IDIOSCAL.[11] The DEDICOM tensor model is given as $\mathbf{X}^{(k)} \approx \mathbf{A}\mathbf{D}^{(k)}\mathbf{R}\mathbf{D}^{(k)}\mathbf{A}^T$ for $k = 1, 2, \ldots, m$. Here, the model assumes there is one *global model of interactions* between the latent components, i.e. an $r \times r$ latent matrix $\mathbf{R}$. Notice that its variation across relations is described by the $r \times r$ diagonal factors $\mathbf{D}_k$. The diagonal matrices $\mathbf{D}_k$ contain memberships of the latent components in the $k$tk relation. This is in contrast to Eq. (5) where we allow the *relation-specific interactions* for the latent components. While DEDICOM has been successfully applied to many domains, for example to model the changes in the corporate communication and international trade over time, our results suggest that its assumptions appear to be too stringent for multi-relational biological data, which is aligned with the observations made by Nickel *et al.*[11]

Furthermore, the model in Eq. (5) is also different from traditional multi-way factor models, such as the Tucker decomposition[31] and CANDECOMP/PARAFAC (CP).[32] The Tucker family defines a multilinear form for a tensor $\mathbf{X} \in \mathbb{R}^{n \times n \times m}$ as $\mathbf{X} = \mathbf{R} \times_1 \mathbf{A}^{(1)} \times_2 \mathbf{A}^{(2)} \times_3 \mathbf{A}^{(3)}$, where $\times_k$ denotes the mode-k tensor-matrix multiplication. Here, $\mathbf{R}$ is the global $r_1 \times r_2 \times r_3$ tensor, and $\mathbf{A}^{(k)}$ models the participation of the latent components in the kth relation. The CP family is restricted form of the Tucker-based decompositions. The definition of rank-r CP for a tensor $\mathbf{X} \in \mathbb{R}^{n \times n \times m}$ is given as a sum of component rank-one tensors, $\mathbf{a}_l \in \mathbb{R}^n$, $\mathbf{b}_l \in \mathbb{R}^n$ and $\mathbf{c}_l \in \mathbb{R}^m$, for $l = 1, \ldots, r$. Elementwise, the CP decomposition is written as $\mathbf{X}_{ijk} \approx \sum_{l=1}^{r} \mathbf{a}_{il}\mathbf{b}_{jl}\mathbf{c}_{kl}$ for $i = 1, \ldots, n$, $j = 1, \ldots, n$ and $k = 1, \ldots, m$. The model in Eq. (5) can be seen as a constrained variation of the CP model.[11]

One major difference of the COPACAR model in Eq. (7) to the existing tensor decompositions is the objective criterion used for finding the latent matrices. Other tensor decompositions are restricted to least-squares regression and cannot solve classification tasks, whereas COPACAR optimizes for a latent model with respect to ranking based on pairwise classification.

5. COPACAR Learning Algorithm

So far we derived the optimization criterion for collective pairwise classification on multi-relational data. As the criterion in Eq. (7) is differentiable, gradient descent based algorithms are a natural choice for its optimization. However, standard gradient descent is not the most effective choice for our problem due to the complexity of OPT-COPACAR (see Sec. 4.2). Instead, we propose a stochastic gradient descent algorithm based on bootstrap sampling of training triplets.

Our aim is to find the latent matrices $\mathbf{A}$ and $\mathbf{R}^{(k)}$ for $k = 1, 2, \ldots, m$ that optimize for:

$$\min_{\substack{\mathbf{A},\, \mathbf{R}^{(k)} \\ k=1,2,\ldots,m}} -\text{OPT-COPACAR}. \tag{10}$$

The gradients of the pairwise loss from Eq. (8), the integral part of OPT-COPACAR, with respect to the model parameters are:

$$\frac{\partial}{\partial \mathbf{A}} \ell(\widehat{\mathbf{X}}^{(k)}_{ij;gh}, \mathbf{X}^{(k)}_{ij;gh}) = -\frac{\partial}{\partial \mathbf{A}} \mathbf{X}^{(k)}_{ij;gh} \log \sigma(\widehat{\mathbf{X}}^{(k)}_{ij;gh}) = (\sigma(\widehat{\mathbf{X}}^{(k)}_{ij;gh}) - 1)\mathbf{X}^{(k)}_{ij;gh} \frac{\partial}{\partial \mathbf{A}} \widehat{\mathbf{X}}^{(k)}_{ij;gh} + \lambda_{\mathbf{A}} \mathbf{A} \tag{11}$$

$$\frac{\partial}{\partial \mathbf{R}^{(k)}} \ell(\widehat{\mathbf{X}}^{(k)}_{ij;gh}, \mathbf{X}^{(k)}_{ij;gh}) = -\frac{\partial}{\partial \mathbf{R}^{(k)}} \mathbf{X}^{(k)}_{ij;gh} \log \sigma(\widehat{\mathbf{X}}^{(k)}_{ij;gh}) = (\sigma(\widehat{\mathbf{X}}^{(k)}_{ij;gh}) - 1)\mathbf{X}^{(k)}_{ij;gh} \frac{\partial}{\partial \mathbf{R}^{(k)}} \widehat{\mathbf{X}}^{(k)}_{ij;gh} + \lambda_{\mathbf{R}} \mathbf{R}^{(k)},$$

where for simplicity of notation we write $\widehat{\mathbf{X}}^{(k)}_{ij;gh} = \widehat{\mathbf{X}}^{(k)}_{ij} - \widehat{\mathbf{X}}^{(k)}_{gh}$.

Let S_k denote observed relationships in kth relation and let I_k represent non-edges in kth relation. If kth relation corresponds to the human disease symptoms network, then S_k contains all disease pairs with shared symptoms and I_k holds disease pairs for which shared disease symptoms have not been recorded. To achieve descent in a correct direction, the full gradient shall be computed over all training data in each iteration and model parameters updated. However, since we have $O(\sum_k |S_k||I_k|)$ training triplets in the data, computing the full gradient in each iteration is not feasible.

Furthermore, optimizing OPT-COPACAR with a full gradient descent can lead to poor convergence due to skewness of the training data. Consider for a moment a disease i with high symptom-based similarity to many other diseases. We have many terms for triplets of the form $\langle E_i, \mathcal{R}^{(\text{symptom})}, E_j \rangle$ in the loss because for many diseases j the disease i is compared against all diseases to which a particular disease j is not related. Therefore, the gradients would be largely dominated by the terms depending on disease i. This means that very small learning rates would need to be chosen and also regularization would be difficult because the gradients would differ substantially.

To address the above issues we propose to use a stochastic gradient descent, which subsamples entity pairs (E_i, E_j) randomly (uniformly distributed) and forms an appropriately scaled gradient. In each iteration we use a bootstrap sampling without replacement to pick entity combinations, and the Armijo-Goldstein step size control to determine the maximum amount to move along a given direction of descent. The chance of picking the same entity combination in consecutive update steps is hence small.

6. Evaluation

Next, we test our algorithm for collective pairwise classification on two highly multi-relational data domains. First, we apply it to the collection of relations between drugs, where we aim to predict different types of drug relationships. We then study human disease data retrieved from the molecular and clinical contexts. We compare our method to tensor-based relational learning methods from Sec. 4.3.

6.1. *A Case Study on Pharmacogenomic Data*

6.1.1. *Data and Experimental Setup*

We obtained a list of 1,451 drugs with known pharmacological actions from the DrugBank database.[33] Examples of considered drugs include ospemifene, riluzole, chlormezanone and podofilox. Vast majority of considered drugs contained links to the corresponding chemicals in the PubChem database,[34] where we obtained information on similarity of their chemical structures. We also included information on drug-target interactions[33] and drug interaction data extracted from the literature through co-occurrence text mining.[35] Due to space constraints we refer to Kuhn *et al.*[35] for a detailed description of relationships derived from text. We also mined the drug-drug interaction network, where we connected two drugs if they are known to interact, interfere or cause adverse reactions when taken together.[33] The preprocessed dataset consisted of four drug-drug relations $\mathbf{X}^{(k)} \in \{0,1\}^{1451\times 1451}$ for $k = 1,\ldots,4$ and contained 59,990 text associations, 2,602 interactions based on chemical structures, 1,315 interactions based on shared target proteins and 48,614 drug-drug interactions based on adverse effects.

We performed 10-fold cross-validation using $\langle E_i, \mathcal{R}^{(k)}, E_j\rangle$ triplets as statistical units. Model parameters, i.e. regularization strength and factorization rank, were selected using the grid search on a random data subsample that was later excluded from performance evaluation. For kth relation, we partitioned all drugs into ten folds and deleted the kth relation-specific information of the drugs in the test fold. We then estimated the CP, DEDICOM, RESCAL and COPACAR models, and recorded the area under the ROC curve (AUC-ROC) and the area under the precision-recall curve (AUC-PR). Values of the performance metrics that are closer to one indicate better performance.

6.1.2. *Results and Discussion*

Fig. 2 shows the results of our evaluation. It can be seen that COPACAR gives better results than RESCAL, CP and DEDICOM on all data relations. The results of COPACAR and RESCAL outperform CP and DEDICOM by a large margin and show clearly the usefulness of our approach for relational drug data domain where collective learning is an important feature. A significant performance difference between the results of DEDICOM and COPACAR indicate that the constraints imposed by DEDICOM (see Sec. 4.3) are too restrictive. Another important aspect of the results in Fig. 2 is the good performance of COPACAR relative to RESCAL, which has been shown to achieve state-of-the-art performance on several relational datasets.[11,19] One possible explanation is that RESCAL is restricted to least-squares regression, which limits its ability to solve classification tasks, whereas COPACAR is designed to optimize the parameters with respect to pairwise classification.

6.2. *A Case Study on Human Disease Data*

6.2.1. *Data and Experimental Setup*

We related diseases through three dimensions. We considered the comprehensive map of disease-symptoms relationships,[36] the map of molecular pathways implicated in diseases,[37] and the map of dis-

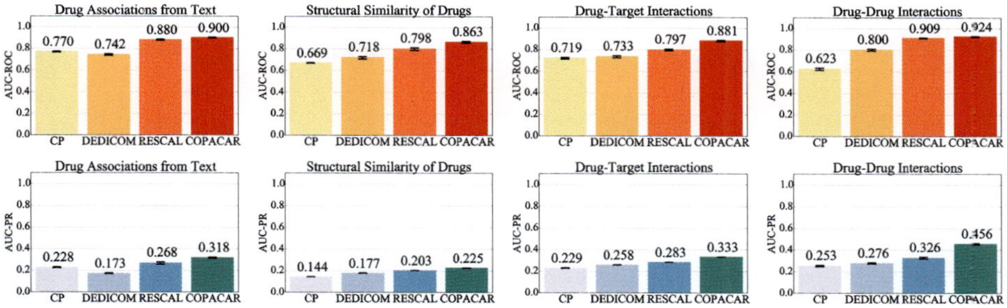

Fig. 2. The area under the ROC and the precision-recall (PR) curves via 10-fold cross-validation on drug data.

eases affected by various chemicals from the Comparative Toxicogenomics Database.[37] We used the recent high-quality disease-symptoms data resource of Zhou *et al.*[36] to generate a symptom-based relation of 1,578 human diseases, where the link between two diseases indicated significant similarity of their respective symptoms. The details of the network construction based on large-scale medical bibliographic records and the related Medical Subject Headings (MeSH) metadata are described in Zhou *et al.*[36] Examples of considered diseases are Hodgkin disease, thrombocytosis, thrombocythemia and arthritis. The preprocessed dataset consisted of three disease-disease relations $\mathbf{X}^{(k)} \in \{0,1\}^{1578 \times 1578}$ for $k = 1, 2, 3$ and contained 117,021 relationships based on significant symptom similarity, 446,488 disease relationships derived from disease pathway information and 770,035 disease connections related to drug treatment.

In the evaluation we followed the experimental protocol described in Sec. 6.1.1.

6.2.2. *Results and Discussion*

Results in Fig. 3 show the good capabilities of our COPACAR method for predicting any of the three considered disease dimensions. We see that COPACAR achieves comparable or better results than CP, DEDICOM and RESCAL models. The RESCAL and COPACAR models, which can perform collective learning, considerably boost the predictive performance of the less expressive CP and DEDICOM models by more than 20% (AUC-ROC) across all three relations. These results highlight an advantage of applying collective learning to this dataset.

The results also bear evidence that shared clinical manifestations of diseases indicate shared molecular interactions, e.g., genetic associations and protein interactions, as has already been recognized in systems medicine.[36] It should be noted that when predicting disease phenotypes (left panel in Fig. 3) the models were trained solely based on molecular-level disease components, i.e. relationships based on disease pathways and disease-chemical associations (middle and right panels in Fig. 3). Hence, the extent to which collective learning of the COPACAR has improved the quality of modeling is especially appealing. Furthermore, this result is interesting because it is known that the relations between genotype and phenotype components remain unclear and highly entangled despite impressive progress on the genetic and proteomic aspects of human disease.[36] The phenotype map[36] we use in the experiments strictly considers symptom features, excluding particular disease terms themselves, anatomical features, congenital abnormalities, and includes all disease categories rather than only monogenic diseases. Our results therefore provide robust evidence that interactions at the chemical and cellular pathway levels are also connected to similar high-level disease manifestations.

At last we want to briefly demonstrate the link-based clustering capabilities of COPACAR. We computed a rank-30 decomposition of the disease dataset and applied hierarchical clustering to the matrix

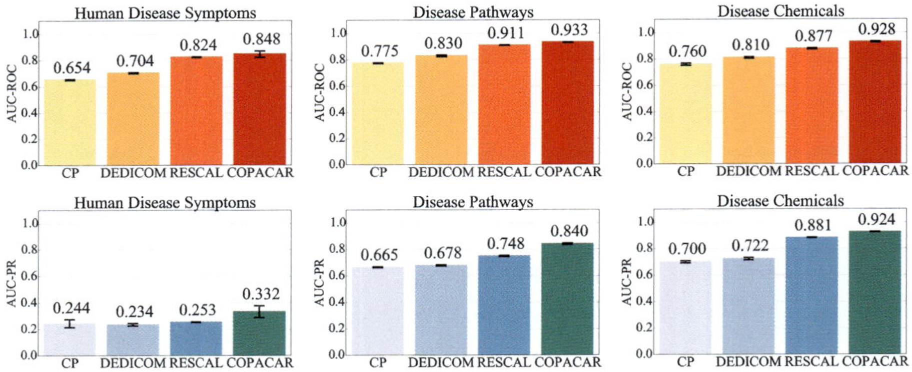

Fig. 3. The area under the ROC and the precision-recall (PR) curves via 10-fold cross-validation on disease data.

A (Fig. 4b). Diseases from the six randomly chosen clusters in Fig. 4a illustrate that we obtained a meaningful partitioning of the diseases and suggest that low-dimensional embedding of the data found by COPACAR can be a useful resource for further data modeling. Here, we were especially interested in the diseases grouped within the white bands in Fig. 4b (middle, right). Diseases therein have extremely sparse, if any at all, data profiles at the molecular or chemical levels. On the other hand, it can be seen from Fig. 4b (left) that these diseases have many common clinical phenotypes. Interestingly, COPACAR was able to make a leap across the three modeled disease dimensions and assigned poorly characterized diseases to clusters with richer molecular knowledge, such as phenylketonuria to the cluster centered around Parkinson's disease. Even when not category-jumping, COPACAR grouped diseases, such as seborrheic dermatitis and herpes, based on their symptom similarity.

6.3. *Runtime Performance and Technical Considerations*

We recorded the runtime of CP, DEDICOM, regularized RESCAL and COPACAR on various datasets and for different factorization ranks (exact times are not shown due to the space limit). The COPACAR shows training times below 3 minutes per fold on the disease data and below 5 minutes per fold on the drug data. In comparison to CP and DEDICOM, it is the case that COPACAR as well as RESCAL often give a huge improvement in terms of runtime performance on real data.

In comparison to COPACAR, we observed that RESCAL can run up to three times faster on the same data and using the same rank. We believe this is the case because RESCAL is optimized using the alternating least squares, which is possible due to its squared loss objective. In contrast, COPACAR is optimized by a stochastic gradient descent due to the nature of its optimization criterion: in each iteration, it constructs a random data subsample and makes the update. The COPACAR algorithm has two important advantages over RESCAL. First, the algorithm naturally allows for parallelization of the gradient computation on a data subsample, which further increases scalability of COPACAR. Furthermore, we do not need to have collected the entire data relations to run the algorithm. Because COPACAR operates on subsamples, it gives a natural approach for interleaving data collection and model estimation.

We also studied the technical aspects of the COPACAR learning algorithm. Specifically, we were interested in (1) the stability of algorithm performance w.r.t. the data subsample size, (2) its empirical convergence rate, and (3) its sensitivity to model parameters. Fig. 5 shows the results of this evaluation. In our experiments the algorithm typically required less than 100 iterations to converge and operated on

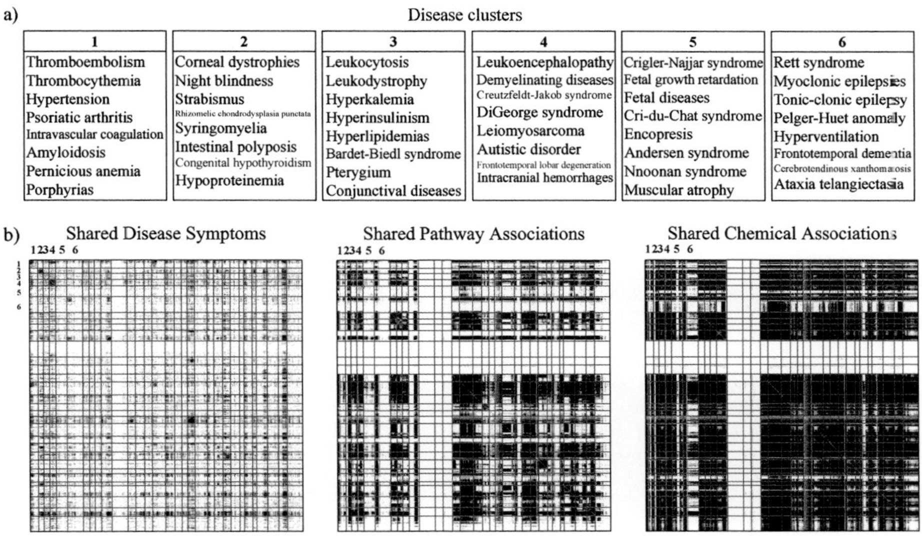

Fig. 4. (a) Disease clusters found by collective learning via COPACAR on the disease data. We labeled the clusters and shown only eight members of each. (b) Adjacency matrices for the three relations, where the rows and columns are sorted according to the disease partitioning. Black squares indicate existing disease relationships, white squares are unknown relationships.

subsamples of size at most 10% of the total number of data triplets. This means that discarding the idea of performing full cycles through the data may be useful because often only a fraction of a full cycle is sufficient for convergence. We also note that its performance is stable with regard to the wide range of values for factorization rank and regularization strength.

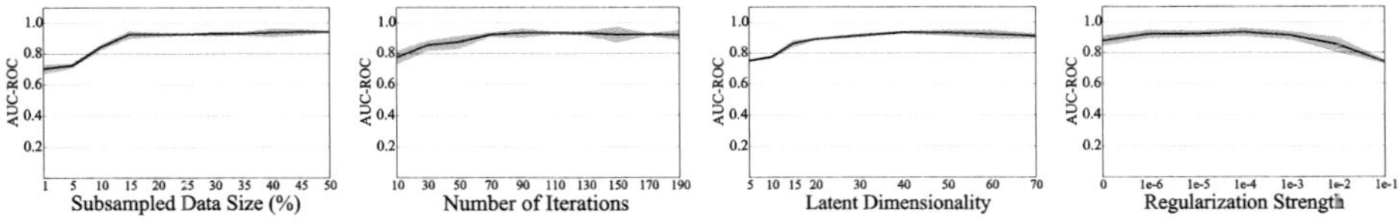

Fig. 5. The results for the area under the ROC curve (AUC-ROC) obtained by 10-fold cross-validation on the disease data. The bands indicate performance variation across folds. Shown is performance of COPACAR as a function of the subsampled data size, the number of iterations, the latent dimensionality and the regularization strength (from left to right).

7. Conclusions

Methods that can accurately estimate different types of relationships from multi-relational and multi-scale biomedical data are needed to better search through the hypothesis space and identify hypotheses that should be pursued in a laboratory environment. Towards this end, we have attempted here to address a significant limitation of current approaches for collective relational learning by developing a method for collective classification that is designed to optimize for a pairwise ranking metric. Our method achieves favorable performance in resolving which entity pairs (e.g., drugs) are most likely to be associated through a given type of relation (e.g., adverse effects or shared target proteins) by appropriately formulating a probabilistic model for pairwise classification of relationships.

Most likely, the most substantial advantage of our proposed approach is "category-jumping," which we exemplify in a case study with several relations about diseases. Category-jumping has helped us to make predictions about disease interactions at the molecular level that stem from clinical phenotype data collected far outside the molecular contexts. The implications for utility of such inference are profound. Predictions that arise from category-jumping may reveal important relationships between biomedical entities that are withheld from today-prevailing models that are trained on data of a single relation type.

Acknowledgments

This work was supported by the ARRS (P2-0209, J2-5480) and the NIH (P01-HD39691).

References

1. Y. Yamanishi, M. Kotera, M. Kanehisa and S. Goto, *Bioinformatics* **26**, i246 (2010).
2. X. Chen, M.-X. Liu and G.-Y. Yan, *Molecular BioSystems* **8**, 1970 (2012).
3. M. Zitnik, V. Janjic, C. Larminie, B. Zupan and N. Przulj, *Scientific Reports* **3** (2013).
4. F. Cheng and Z. Zhao, *Journal of the American Medical Informatics Association* **21**, e278 (2014).
5. M. Campillos, M. Kuhn, A.-C. Gavin, L. J. Jensen and P. Bork, *Science* **321**, 263 (2008).
6. J. Huang, C. Niu, C. D. Green, L. Yang, H. Mei and J. Han, *PLoS Computational Biology* **9**, p. e1002998 (2013).
7. S. V. Iyer *et al.*, *Journal of the American Medical Informatics Association* **21**, 353 (2014).
8. Z. Xu, V. Tresp, K. Yu and H.-P. Kriegel, Learning infinite hidden relational models, in *UAI*, 2006.
9. A. P. Singh and G. J. Gordon, Relational learning via collective matrix factorization, in *KDD*, 2008.
10. R. Jenatton *et al.*, A latent factor model for highly multi-relational data, in *NIPS*, 2012.
11. M. Nickel, V. Tresp and H.-P. Kriegel, A three-way model for collective learning on multi-relational data, in *ICML*, 2011.
12. M. Nickel *et al.*, Reducing the rank in relational factorization models by including observable patterns, in *NIPS*, 2014.
13. A. Gunawardana and G. Shani, *Journal of Machine Learning Research* **10**, 2935 (2009).
14. P. Cremonesi *et al.*, Performance of recommender algorithms on top-n recommendation tasks, in *RecSys*, 2010.
15. Y. Shi *et al.*, GAPfm: Optimal top-n recommendations for graded relevance domains, in *ICKM*, 2013.
16. D. Sculley, Combined regression and ranking, in *KDD*, 2010.
17. E. Horvitz and D. Mulligan, *Science* **349**, 253 (2015).
18. S. Dzeroski, *Relational data mining* (Springer, 2010).
19. M. Nickel, K. Murphy, V. Tresp and E. Gabrilovich, *arXiv:1503.00759* (2015).
20. B. W. Bader, R. Harshman, T. G. Kolda *et al.*, Temporal analysis of semantic graphs using ASALSAN, in *ICDM*, 2007.
21. M. Zitnik and B. Zupan, *IEEE Transactions on Pattern Analysis and Machine Intelligence* **37**, 41 (2015).
22. E. M. Airoldi, D. M. Blei, S. E. Fienberg and E. P. Xing, Mixed membership stochastic blockmodels, in *NIPS*, 2009.
23. C. Kemp *et al.*, Learning systems of concepts with an infinite relational model, in *AAAI*, 2006.
24. A. Bordes, J. Weston, R. Collobert and Y. Bengio, Learning structured embeddings of knowledge bases, in *AAAI*, 2011.
25. P. J. Mucha, T. Richardson, K. Macon, M. A. Porter and J.-P. Onnela, *Science* **328**, 876 (2010).
26. Y. Sun *et al.*, PathSim: Meta path-based top-k similarity search in heterogeneous information networks, in *VLDB*, 2011.
27. M. Zitnik and B. Zupan, *Bioinformatics* **31**, 230 (2015).
28. P. Hoff, Modeling homophily and stochastic equivalence in symmetric relational data, in *NIPS*, 2008.
29. S. Rendle *et al.*, BPR: Bayesian personalized ranking from implicit feedback, in *UAI*, 2009.
30. A. Herschtal and B. Raskutti, Optimising area under the ROC curve using gradient descent, in *ICML*, 2004.
31. L. R. Tucker, *Psychometrika* **31**, 279 (1966).
32. R. A. Harshman, *UCLA Working Papers in Phonetics* **16**, 1 (1970).
33. V. Law *et al.*, *Nucleic Acids Research* **42**, D1091 (2014).
34. Y. Wang, J. Xiao, T. O. Suzek, J. Zhang, J. Wang and S. H. Bryant, *Nucleic Acids Research* **37**, W623 (2009).
35. M. Kuhn *et al.*, *Nucleic Acids Research* **40**, D876 (2012).
36. X. Zhou, J. Menche, A.-L. Barabási and A. Sharma, *Nature Communications* **5** (2014).
37. A. P. Davis *et al.*, *Nucleic Acids Research* **43**, D914 (2015).

INNOVATIVE APPROACHES TO COMBINING GENOTYPE, PHENOTYPE, EPIGENETIC, AND EXPOSURE DATA FOR PRECISION DIAGNOSTICS

MELISSA A. HAENDEL
Library and Department of Medical Informatics and Epidemiology, Oregon Health and Science University
Portland, OR 97239, USA
Email: haendel@ohsu.edu

MARICEL G. KANN
Department of Biological Sciences, University of Maryland, Baltimore County
Baltimore, MD 21250, USA
Email: mkann@umbc.edu

NICOLE L. WASHINGTON
Genomics Division, Lawrence Berkeley National Laboratory
Berkeley, CA 94530, USA
Email: nlwashington@lbl.gov

1. Introduction

Despite the increasing prevalence of Whole Genome Sequencing (WGS) and Whole Exome Sequencing (WES) in clinical settings, it is still very difficult to determine the causal variant for any given disease and most such explorations fall into research contexts rather than routine clinical diagnostics. There are literally thousands of studies and talks at human genetics conferences regarding the determination of causality, and a plethora of techniques available for statistical association of variants to disease phenotypes (e.g. GWAS). However, for rare diseases, the small number of individuals prevents statistical correlation techniques used for more common diseases. For complex diseases without Mendelian inheritance patterns, the challenge is even greater. Because we know the phenotypic consequences of mutation in approximately less than 40% of human coding genes, it is necessary to utilize a diversity of other data sources and algorithms to help determine causality. What is clinically actionable is an even more difficult assessment. Further, the methods and provenance of the data by which determinations of causality and actionability are often lacking and/or produce conflicting results. Finally, to realize truly precision medicine, we must embrace the idea that all diseases are rare in that each person has their own diversity of genotypic, phenotypic, and environmental variation.

Recent work has highlighted some of the exciting new possibilities to inform rare disease diagnostics. For example, use of model organism phenotyping data, interactome data, orthology, phylogenetic inference, and epigenomics can help fill some of the gaps. Further, methods that utilize semantic inference and probabilistic modeling have also been shown to aid diagnostics Such methods combine standard WES or WGS prioritization techniques with an increasing diversity of phenotyping data and approaches. However, all of these combined approaches depend upon highly curated data and a diversity of software tools and algorithms, all of which provide the provenance for making any sort of causal or actionability judgment, the conclusions of which may change over time as the data and algorithms change. In addition, the quality of the phenotyping data varies widely and is not always accessible via clinical notes. Finally, few such combined approaches attempt to include life history data such as exposures and chronological representation of disease progression.

2. Session Summary

This session includes an invited talk, five reviewed oral presentations, and two additional accepted papers. The studies presented in this session explore problems in combining genotype and phenotype data to support rare disease and/or precision diagnostics and treatment, and spanning multiple types of data. In particular, we selected contributions from those whose methodologies leveraged multiple data modalities in their analysis of genetic variation, such as clinical measures, imaging, natural language processing, semantics, homology, mined electronic health records (EHR) and manually curated data.

2.1. *Invited Talk*

The invited talk is given by Elissa Chessler, Ph.D. an Associate Professor of Bioinformatics and Computational Biology at the Jackson Laboratory, whose work spans a diversity of biological, genomic, and behavioral data toward identifying the biological basis for the relationships among behavioral traits, particularly in mouse models of disease. The resemblance of objectively measured phenotypic characteristics across species is limited by the extent to which the phenotypic inferences supported by these assays are relevant to the disease under investigation and reflect similar characteristics across species. 'Construct validity' is a more important criterion for the matching of phenotypes across species, and to the matching of phenotypes to disease. Construct-valid assays are expected to be associated with similar molecular and other biological characteristics across species, even when the external manifestation of the disease related phenotypes is quite different in humans and model organisms. There is a wealth of relevant data consisting of gene-phenotype associations obtained through high throughput, whole genome experimentation, including genetic mapping, expression correlation, differential expression, systems genetics, mutant screens, proteomic assays and curated functional genomics experiments. A variety of statistical and combinatorial approaches may then be applied to match data from various experiments and known gene-disease or gene phenotype associations. This approach to data driven inference of the relationships among the biological characteristics of animal models, assays and disease features has been implemented in the GeneWeaver.org system, a web service consisting of a database and analytic tools for collaborative integration of functional genomic experiments.

2.2. *Papers*

In *Discovering Patient Phenotypes Using Generalized Low Rank Models,* **Schuler et al**. develop a methodology for capturing phenotypic information within EHRs. The authors show that inherited challenges on the analysis of EHRs for phenotype discovery, such as missing data, sparsity, and data heterogeneity, can be overcome by using the generalized low ranking model framework for such analysis.

In *Diagnosis-guided method for identifying multi-modality neuroimaging biomarkers associated with genetic risk factors in Alzheimer's Disease*, **Hao et al** present a novel, diagnosis-oriented, framework for selecting multi-modality quantitative traits associated with SNPs in the context of Alzheimer's Disease. This method has the potential to improve classification of patients with respect to their likelihood of developing Alzheimer's, by leveraging new data types and variables in their analysis algorithms.

In *Metabolomics Differential Correlation Network Analysis of Osteoarthritis*, **Hu et al.** describe a differential network approach to analyzing the metabolomics of an osteoarthritis (OA) cohort. The authors identified key metabolites that differ in OA and subsequently the cellular processes in which they are involved, with the goal of eventually leveraging these markers for the development of targeted therapies.

In *Integrating Clinical Laboratory Measures and ICD-9 Code Diagnoses in Phenome-wide Association Studies*, **Verma et al** describe a workflow that associates SNPs with clinical lab measures extracted from EHRs as well as ICD-9 codes. The suggested workflow would enable the use of clinical measures and their association with disease toward bringing clinical diagnoses and treatment to the level of individuals in the clinic for precision medicine.

In *Investigating the importance of anatomical homology for cross-species phenotype comparisons using semantic similarity*, **Manda et al** studies the influence of anatomical homology information on gene semantic similarity measures for phenotypic comparisons across species. Their findings are relevant to merging functional and anatomy-based gene homologue analyses.

In *Personalized Drug Targets via Network Propagation,* **Shnaps et al** present a computational strategy to simulate drug treatment in a personalized setting. The method is based on integrating patient mutation and differential expression data with a protein-protein interaction network.

In *Testing population-specific quantitative trait associations for clinical outcome relevance in a biorepository linked to electronic health records: LPA and myocardial infarction in African Americans* **Dumitrescu et al** combine genomic variant assessment (variants in LPA) and EHR phenotyping to determine risk in an unevaluated population, African Americans. This is important from the perspective of understanding how quantitative trait studies differ in different populations and highlights the challenges for complex clinical outcomes such as myocardial infarction.

2.3. *Acknowledgements*

We would like to thank all of the reviewers who provided valuable feedback for the authors of this session.

TESTING POPULATION-SPECIFIC QUANTITATIVE TRAIT ASSOCIATIONS FOR CLINICAL OUTCOME RELEVANCE IN A BIOREPOSITORY LINKED TO ELECTRONIC HEALTH RECORDS: *LPA* AND MYOCARDIAL INFARCTION IN AFRICAN AMERICANS

LOGAN DUMITRESCU

Center for Human Genetics Research, Vanderbilt University, 519 Light Hall, 2215 Garland Avenue, Nashville, TN 37232, USA
Email: logandumitrescu@gmail.com

KIRSTEN E. DIGGINS

Cancer Biology, Vanderbilt University, 742 Preston Research Building, 2220 Pierce Avenue, Nashville, TN 37232, USA
Email: kirsten.e.diggins@vanderbilt.edu

ROBERT GOODLOE

Center for Human Genetics Research, Vanderbilt University, 519 Light Hall, 2215 Garland Avenue, Nashville, TN 37232, USA
Email: robert.goodloe@gmail.com

DANA C. CRAWFORD

Institute for Computational Biology, Department of Epidemiology and Biostatistics, Case Western Reserve University, Wolstein Research Building, 2103 Cornell Road, Suite 2527, Cleveland, OH 44106, USA
Email: dana.crawford@case.edu

Previous candidate gene and genome-wide association studies have identified common genetic variants in *LPA* associated with the quantitative trait Lp(a), an emerging risk factor for cardiovascular disease. These associations are population-specific and many have not yet been tested for association with the clinical outcome of interest. To fill this gap in knowledge, we accessed the epidemiologic Third National Health and Nutrition Examination Surveys (NHANES III) and BioVU, the Vanderbilt University Medical Center biorepository linked to de-identified electronic health records (EHRs), including billing codes (ICD-9-CM) and clinical notes, to test population-specific Lp(a)-associated variants for an association with myocardial infarction (MI) among African Americans. We performed electronic phenotyping among African Americans in BioVU $\geq$40 years of age using billing codes. At total of 93 cases and 522 controls were identified in NHANES III and 265 cases and 363 controls were identified in BioVU. We tested five known Lp(a)-associated genetic variants (rs1367211, rs41271028, rs6907156, rs10945682, and rs1652507) in both NHANES III and BioVU for association with myocardial infarction. We also tested *LPA* rs3798220 (I4399M), previously associated with increased levels of Lp(a), MI, and coronary artery disease in European Americans, in BioVU. After meta-analysis, tests of association using logistic regression assuming an additive genetic model revealed no significant associations ($p<0.05$) for any of the five *LPA* variants previously associated with Lp(a) levels in African Americans. Also, I4399M rs3798220 was not associated with MI in African Americans (odds ratio

= 0.51; 95% confidence interval: 0.16 – 1.65; p=0.26) despite strong, replicated associations with MI and coronary artery disease in European American genome-wide association studies. These data highlight the challenges in translating quantitative trait associations to clinical outcomes in diverse populations using large epidemiologic and clinic-based collections as envisioned for the Precision Medicine Initiative.

1. Introduction

Labs ordered in a clinical setting provide valuable diagnostic and prognostic data at the individual patient level. In a research setting, labs can be studied to better understand the biological basis of clinical outcomes. As an example, lipid labs such as low-density lipoprotein cholesterol (LDL-C) are frequently ordered in a clinical setting to monitor the cardiovascular disease risk in patients. In turn, these labs or quantitative traits have been extensively studied in genomic research settings to identify genetic variants predictive of extreme LDL-C levels and cardiovascular disease risk [1].

A major advantage of quantitative trait genetic studies compared with case-control outcome studies is sample size resulting in statistical power [2]. As a result, there are more or larger genome-wide association studies (GWAS) and significant findings for lipid traits compared with cardiovascular disease outcomes [1], particularly for diverse populations. The emergence of electronic health records (EHRs) linked to biorepositories, however, provides contemporary opportunities to apply quantitative trait genetic variants to assess clinical relevance with an eye towards precision medicine.

We describe here the application of *LPA* genetic variants, previously associated with Lp(a) levels [3], to assess myocardial infarction associations in both an epidemiologic and clinical African American population. Lipoprotein (a) [Lp(a)] is considered an emerging biomarker or risk factor for cardiovascular disease [4-6] whose relationship with cardiovascular disease varies across races/ethnicities. Elevated plasma Lp(a) levels have been reported to be associated with cardiovascular disease in European Americans but have not been clearly documented in African Americans [7]. Paradoxically, among participants with no previous history of cardiovascular disease, the mean Lp(a) level is two- to three-fold higher in African Americans compared with European Americans [8,9]. The underlying cause(s) for this difference has not yet been determined.

Recent studies have identified common SNPs in *LPA* as strongly associated with Lp(a) levels, explaining up to 36% of the trait variance in populations of European-descent [10,11]. In a recent epidemiologic study conducted in the Third National Health and Nutrition Examination Survey (NHANES III), we demonstrated that *LPA* common genetic variants were associated with Lp(a) levels in a population-specific manner [3]. *LPA* SNP rs3798220 (I4399M) has also been associated with cardiovascular disease [11-14] and severe cardiovascular disease [15] in several European-descent populations. Thus, common genetic variants in *LPA* are strong predictors of both Lp(a) levels and cardiovascular disease risk in at least one population. We test here whether

LPA variants associated with Lp(a) levels in African Americans are associated with myocardial infarction in African Americans ascertained from epidemiologic and clinical settings.

2. Methods

2.1. *Study population*

The study populations presented here include the epidemiologic Third National Health and Nutrition Examination Survey (NHANES III) and the clinical BioVU, Vanderbilt University Medical Center's biorepository linked to de-identified electronic health records. NHANES III is a cross-sectional survey conducted between 1988 and 1994 by the National Center for Health Statistics at the Centers for Disease Control and Prevention. NHANES ascertained non-institutionalized Americans regardless of health status. Demographic, health, and lifestyle data were collected on NHANES participants through surveys, labs, and physical exams in the Mobile Examination Center (MEC). DNA is available on consenting phase 2 participants (ascertained between 1991 and 1994). The present study was approved by the CDC Ethics Review Board. Because the study investigators did not have access to personal identifiers, this study was considered non-human subjects research by the Vanderbilt University Internal Review Board.

BioVU operations [16] and ethical oversight [17] have been previously described. Briefly, DNA is extracted from discarded blood drawn for routine clinical care at Vanderbilt outpatient clinics in Nashville, Tennessee and surrounding areas. The DNA samples are linked to a de-identified version of the patient's EHR. The de-identified version of the EHR is referred to as the "Synthetic Derivative." The data in this study were de-identified in accordance with provisions of Title 45, Code of Federal Regulations, part 46 (45 CFR 46); therefore, this study was considered non-human subjects research by the Vanderbilt University Internal Review Board.

2.2. *Phenotyping*

Race/ethnicity in NHANES III is self-identified, which is concordant with global genetic ancestry for non-Hispanic whites and non-Hispanic blacks [18]. Myocardial infarction (MI) case status in NHANES III was based on data collected from a physical examination, administered by a physician, in the MEC. A continuous cardiac infarction/injury score (CIIS) was calculated based on 12 lead electrocardiogram (ECG) multiplied by 10. Those participants with a CIIS ≥ 20 were considered to have probable infarction/injury and those with a CIIS < 20 but ≥ 15 were considered to have possible infarction/injury. These thresholds correspond to an estimated specificity level of 98% and 95% [19], respectively. Our NHANES III MI case definition included participants classified as having possible or probable infarction/injury.

Race/ethnicity in BioVU is administratively assigned, which is highly concordant with genetic ancestry for European Americans and African Americans [20,21]. The de-identified EHR in BioVU contains both structured (International Classification of Diseases, Ninth Revision, Clinical

Modification billing codes [ICD-9-CM]; current procedural terminology codes; problems lists; labs) and unstructured (clinical free text) data that are accessible for electronic phenotyping. We explored five different electronic phenotyping strategies to identify cases of MI using mentions of ICD-9-CM codes (Table 1) among African American adults ≥ 40 years of age. MI case review was performed in 2013 using the browser search function in the Synthetic Derivative user interface to find the following keywords in the patient's clinical notes: myocardial infarction, MI, infar, STEMI, and NSTEMI. If none of the keywords were found in the record, the case reviewer searched for ICD-9-CM code 410 in the record and extracted the clinic visit date associated with the ICD-9-CM code. The case reviewer then searched the remainder of the patient's records on that clinic visit date for evidence of an MI. The ECG records of all possible cases were also accessed for review. Patients were considered unconfirmed for MI if EHR review failed to identify evidence of MI in the patient's medical history. Unconfirmed cases of MI were excluded from genotyping as cases. Positive predictive values (PPVs) were calculated as the total number of confirmed cases divided by the total number of potential cases. A total of 311 MI cases were identified for genotyping in BioVU.

Controls in BioVU were defined as African American adults ≥ 40 years of age with no mention of ICD-9-CM codes of MI (410) or any other codes relating to ischemic heart disease (ICD-9-CM 411-414). A total of 5,883 potential controls were identified in BioVU. Controls were frequency matched to cases by age and sex prior to selection for genotyping.

2.3. *Genotyping*

Genotyping in NHANES III was performed using the Illumina GoldenGate assay (as part of a custom 384 OPA) by the Center for Inherited Disease Research (CIDR) through the National Heart Lung and Blood Institute's Resequencing and Genotyping Service, as previously described [3]. Vanderbilt Technologies for Advanced Genomics (VANTAGE) genotyped BioVU samples for six *LPA* SNPs (rs3798220, rs41271028, rs6907156, rs10945682, rs1652507, and rs1367211) using Sequenom. Genotyping quality control for NHANES III was performed using SAS version 9.2 and for BioVU using PLINK [22].

2.4. *Statistical methods*

Tests of association in NHANES III were performed using SAS version 9.2 (SAS Institute, Cary, NC). Each *LPA* variant associated with Lp(a) levels in non-Hispanic blacks [3] was tested for association with MI status (dependent variable) using logistic regression assuming an additive genetic model adjusting for 1) age and sex and 2) age, sex, and ln(Lp[a]+1). Data was accessed remotely from the CDC's Research Data Center (RDC) in Hyattsville, Maryland using Analytic Data Research by Email (ANDRE) [23]. In BioVU, tests of association between MI (the dependent variable) and *LPA* SNPs were performed with PLINK using logistic regression assuming an additive genetic model and adjusting for age and sex (Lp[a] levels are not available in

BioVU). Meta-analyses were performed with METAL using a fixed-effects inverse-variance weighted approach [24], and results were visualized using Synthesis-View [25,26].

3. Results

As noted in the Methods section, cases of MI in NHANES III were identified using a continuous cardiac infarction/injury score applied to ECGs, an exam administered by public health professionals as part of the survey.

Table 1. Phenotyping criteria and case review results for five definitions of myocardial infarction based on mentions of billing codes. Overall, a total of 311 individual cases of confirmed MI were identified and 297 had sufficient DNA for genotyping. Abbreviations: International Classification of Diseases, Ninth Revision, Clinical Modification (ICD-9-CM); positive predictive value (PPV).

Case definition	Phenotyping criteria	Potential cases	Confirmed cases	PPV
1	ICD-9-CM code 410.* on 3 consecutive days	108	107	99.1%
2	ICD-9-CM code 410.* on 2 consecutive days	159	158	99.4%
3	≥ 3 ICD-9-CM codes 410.* ever	159	158	99.4%
4	≥ 2 ICD-9-CM codes 410.* ever	209	205	98.1%
5	≥ 1 ICD-9-CM codes 410.* ever	355	311	87.6%

In contrast, we used electronic phenotyping approaches to extract cases of MI from EHRs of African American patients. We used ICD-9-CM billing codes in various combinations in an attempt to achieve the largest samples size possible with acceptable PPV. As might be expected, the most stringent case definitions (Table 1; definitions 1 and 2) where codes for MI were required on three and two consecutive days identified the fewest number of cases at PPVs >99% after manual review. These cases of MI likely represent incident inpatient cases of MI in BioVU at the time of data extraction. Equally high in PPV but low in case count was case definition 3 where three or more ICD-9-CM codes were required. The least stringent case definitions 4 and 5 yielded the most confirmed cases (205 and 311, respectively) at acceptably high PPVs (Table 1). The high PPVs observed here are consistent with other studies examining the accuracy of using ICD-9-CM codes to identify cases of acute MI [27-30]. Of the 311 total cases identified in the Synthetic

Derivative, only 265 passed quality control after genotyping (12 had insufficient DNA for genotyping; two were compromised samples; 14 failed genotyping).

Table 2. Association between myocardial infarction and Lp(a)-associated SNPs in non-Hispanic blacks from NHANES III. A total of 19 SNPs were tested for an association with Lp(a) levels [3] and MI in non-Hispanic blacks from NHANES III. *LPA* SNPs associated with MI at $p < 0.05$ are shown here. MI case status was defined as participants with possible or probable cardiac infarction/injury (CIIS score ≥ 15). Associations with MI and transformed Lp(a) levels were performed unweighted using logistic and linear regression, respectively. [1]Adjusted for age and sex. [2]Adjusted for age, sex, and Lp(a) levels. [3]Associations between Lp(a) and *LPA* in non-Hispanic blacks in NHANES III as reported in Dumitrescu et al 2011 [3]. Abbreviations: confidence interval (CI); odds ratio (OR).

SNP	Lp(a) levels[1,3] n = 1,711		MI[1] $n_{cases} = 93$ $n_{controls} = 522$		MI[2] $n_{cases} = 91$ $n_{controls} = 498$	
	β (95% CI)	**p-value**	**OR (95% CI)**	**p-value**	**OR (95% CI)**	**p-value**
rs41271028	-0.06 (-0.18, 0.07)	0.3608	2.12 (1.01, 4.46)	***0.0470***	2.12 (1.01, 4.45)	***0.0476***
rs6907156	0.15 (0.05, 0.25)	0.0031	0.53 (0.30, 0.92)	***0.0231***	0.53 (0.30, 0.92)	***0.0241***
rs10945682	-0.14 (-0.21, -0.06)	0.0003	1.41 (1.00, 1.98)	***0.0481***	1.37 (0.97, 1.93)	0.0725
rs1652507	-0.45 (-0.59, -0.32)	1.06×10^{-10}	1.88 (1.06, 3.34)	***0.0308***	1.79 (1.00, 3.2)	0.0510
rs1367211	-0.27 (-0.34, -0.20)	3.67×10^{-14}	1.46 (1.04, 2.07)	***0.0306***	1.42 (1.00, 2.01)	0.0518

We previously tested 19 *LPA* SNPs for an association with Lp(a) levels in non-Hispanic blacks in NHANES III, 12 of which were associated at $p < 0.0001$ [3]. We tested here the same 19 *LPA* SNPs for an association with MI in non-Hispanic blacks from NHANES III. Despite the limitation of small sample size ($n_{cases} = 93$), five *LPA* variants (rs1367211, rs1652507, rs6907156, rs10945682, and rs41271028) were associated with risk of MI in non-Hispanic blacks at $p < 0.05$ (Table 2; Figure 1). Interestingly, four of the five alleles associated with *increased* risk of MI were also associated with *decreased* Lp(a) levels at $p < 0.003$.

To determine if the association of *LPA* variants and MI in non-Hispanic blacks from NHANES III was due to the putative causal role of increased Lp(a) levels in coronary artery disease, we adjusted for Lp(a) level (Table 2). This adjustment attenuated the associations for three of the five single-SNP associations with MI ($p = 0.05$, 0.05, and 0.07 for rs1367211, rs1652507, and rs10945682, respectively). In contrast, two *LPA* SNPs (rs6907156 and rs41271028) remained associated with MI at $p < 0.05$ after adjustment for Lp(a) levels. The first SNP, rs6907156, was originally associated with decreased Lp(a) levels ($p = 0.0031$) while the second, rs41271028, was not ($p = 0.36$).

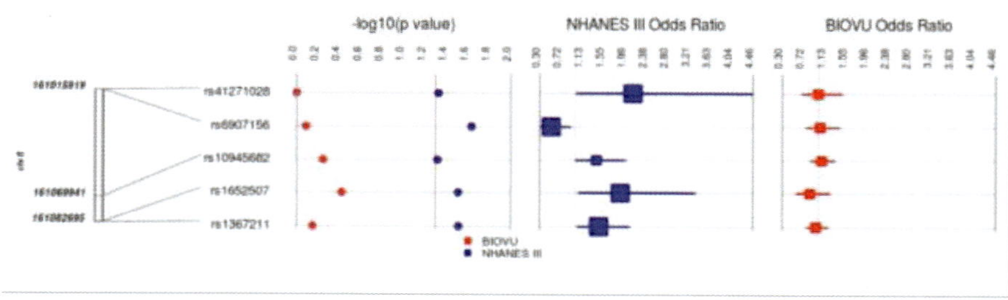

Figure 1. Synthesis-View plot of associations between Lp(a)-associated variants and myocardial infarction in non-Hispanic blacks from NHANES III and African Americans from BioVU. Tests of association were performed in non-Hispanic blacks NHANES III (n = 93 cases; n = 522 controls) and African Americans from BioVU (n = 265 cases; n = 343 controls) for myocardial infarction and each of five *LPA* SNPs previously associated with Lp(a) levels in NHANES III non-Hispanic blacks. Analyses were performed using logistic regression assuming an additive genetic model adjusted for age and sex. Odds ratios, 95% confidence intervals, and $-\log_{10}$(p-values) are plotted by study (NHANES III in blue and BioVU in red) in a forest plot generated by Synthesis-View. The red line denotes a level of significance at $p = 0.05$. Significant odds ratios are denoted by the larger squares.

To replicate the NHANES III findings, we genotyped all five *LPA* SNPs from Table 2 in BioVU African American cases (n = 265) of MI and controls (n = 343). None of the five tests of association were significant at $p < 0.05$ (Figure 1). We meta-analyzed NHANES III and BioVU tests of association and found two (rs10945682 and rs41271028) of the five associations had consistent directions of effect. However, none of the five *LPA* SNPs were associated with MI in the meta-analysis at $p < 0.05$.

Previous studies have suggested that rs3798220 (I4399M) is associated with higher Lp(a) levels and coronary artery disease risk [31,32] in European-descent populations. In the present study, we genotyped rs3798220 in BioVU African American MI cases and controls to determine if the association generalized to African-descent populations. In unadjusted tests of association, rs3798220 was not associated with MI in BioVU African Americans (odds ratio = 0.51; 95% CI: 0.16 – 1.65; p-value = 0.26).

4. Discussion

Genome-wide association studies have identified thousands of common variants significantly associated with quantitative traits, and a fraction of these, in turn, are associated with risk for specific clinical outcomes [33]. The emergence of EHRs linked to biorepositories is enabling larger clinical outcome association studies, an important step in translating quantitative trait associations into precision medicine efforts. In the present study, we tested genetic variants associated with Lp(a) levels, an emerging risk factor for cardiovascular disease, for associations with MI in African Americans ascertained from epidemiologic and clinic-based collections. Overall, Lp(a)-associated genetic variants were not associated with MI in this small sample of African Americans, highlighting the challenges of translating strong genetic associations identified for quantitative traits to clinically relevant outcomes such as cardiovascular disease.

The lack of statistical associations may be due to a combination of imprecise phenotyping and small sample size. Indeed, precise phenotyping and phenotype harmonization across studies is a major challenge for genetic association studies. We used both an epidemiologic cross-sectional survey and an EHR-based biorepository to identify cases and controls of MI. Prevalent NHANES III cases were based on ECG scores (as opposed to self-report), and BioVU cases were likely a mixture of prevalent and incident cases identified using primarily billing codes. As detailed above, we deliberately applied the least stringent MI case definition to identify the largest number of cases for manual review, a strategy that identifies "silver standard" cases that can then be combined with gold standard cases to potentially boost statistical power [34]. This silver standard strategy is likely to play a major role in the Precision Medicine Initiative as it is anticipated that the one million participants will be a combination of prevalent and incident cases of various common disease drawn from epidemiologic and clinic-based collections [35].

In the meta-analysis, we assumed that the case and control definitions were roughly equivalent. While there were a few tests of heterogeneity with p-values < 0.05 (such as *LPA* rs1652507 at p = 0.03), none were statistically significant when accounting for multiple tests. Despite the lack of evidence for gross differences in case-control definitions between the two collections, it is likely that subtle differences and possible case misclassification impacted statistical power. The NHANES III case-control definition has been shown to have high specificity, most likely resulting in good control classification (i.e., ruling out MI). Conversely, the BioVU case-control definition has high PPV or precision. Neither case-control definition addresses the underlying genetic and environmental heterogeneity typical of complex, common human diseases that, like misclassification of case-control status, decreases statistical power [36].

The present study had a small sample size, which in combination with imprecise phenotyping, led to low statistical power. The small sample size available for MI cases and controls in both the epidemiologic and EHR-based collection was disappointing given that both were drawn from relatively large collections of DNA linked to demographic and health information. Although there were 2,108 African American participants with biospecimens, NHANES III had relatively few cases of MI. At the time this study was conducted, BioVU contained DNA samples linked to EHRs for ~12,000 African Americans. However, from among these patients, only 311 cases of MI were identified through billing codes and of these, 265 were available for genotyping. It is possible that additional cases could have been identified using the clinical notes and more sophisticated natural language processing techniques, but it is doubtful that a sufficient number of additional cases would have been identified for a substantial increase in power. It is widely noted that genome-wide studies in general are not conducted in diverse populations [37], and those that are available generally have smaller sample sizes compared with their European counterparts [38]. This trend is unlikely to change with the rise in EHRs linked to biorepositories, and even concerted efforts such as the proposed Precision Medicine Initiative with one million participants [35] will be hard-pressed to muster sufficient sample sizes for clinically relevant outcomes in diverse populations.

In addition to imprecise phenotyping and low case-control counts available for study, low allele frequencies also contributed to low statistical power for specific tests of association. Overall, *LPA* rs3798220 is not common in African Americans, with a minor allele frequency of 2% in the present study. This minor allele frequency is similar to African Americans in third phase of the International HapMap Project as well as to European-descent cases and controls of coronary artery disease meta-analyzed by the CARDIoGRAM consortium [31]. The CARDIoGRAM consortium reported an odds ratio of 1.51, and assuming the same coded allele (C) and the same minor allele frequency (2%), we had ~17% power to detect an association with rs3798220 at $p < 0.05$.

Despite the numerous limitations, this study had several strengths. This study accessed both epidemiologic and clinic-based collections to identify cases and controls for MI among African Americans. Continued case-control collection for this and other clinically relevant outcomes is sorely needed to better translate genetic associations identified using quantitative traits to prevention, diagnosis, and treatment options for MI and other forms of cardiovascular disease at the bedside.

5. Acknowledgments

Genotyping services were provided by the Johns Hopkins University under federal contract number (N01-HV-48195) from NHLBI. We would like to thank Dr. Geraldine McQuillan and Jody McClean for their help in accessing the Genetic NHANES III data. The findings and conclusions in this report are those of the authors and do not necessarily represent the official position of the Centers for Disease Control and Prevention. The dataset(s) used for the analyses described were also obtained from Vanderbilt University Medical Center's BioVU, supported by institutional funding and by the Vanderbilt CTSA grant funded by the NCRR Grant UL1 RR024975-01 (now the NCATS, Grant 2 UL1 TR000445-06).

References

1. Global Lipids Genetics Consortium: **Discovery and refinement of loci associated with lipid levels.** *Nat Genet* 2013, **45:** 1274-1283.
2. Turner SD, Crawford DC, Ritchie MD: **Methods for optimizing statistical analyses in pharmacogenomics research.** *Expert Rev Clin Pharmacol* 2009, **2:** 559-570.
3. Dumitrescu L, Glenn K, Brown-Gentry K, Shephard C, Wong M, Rieder MJ *et al.*: **Variation in LPA Is Associated with Lp(a) Levels in Three Populations from the Third National Health and Nutrition Examination Survey.** *PLoS ONE* 2011, **6:** e16604.
4. Bennet A, Di Angelantonio E, Erqou S, Eiriksdottir G, Sigurdsson G, Woodward M *et al.*: **Lipoprotein(a) Levels and Risk of Future Coronary Heart Disease: Large-Scale Prospective Data.** *Arch Intern Med* 2008, **168:** 598-608.
5. Berglund L, Ramakrishnan R: **Lipoprotein(a): An Elusive Cardiovascular Risk Factor.** *Arterioscler Thromb Vasc Biol* 2004, **24:** 2219-2226.
6. Danesh J, Collins R, Peto R: **Lipoprotein(a) and coronary heart disease. Meta-analysis of prospective studies.** *Circulation* 2000, **102:** 1082-1085.
7. Heiss G, Schonfeld G, Johnson JL, Heyden S, Hames CG, Tyroler HA: **Black-white differences in plasma levels of apolipoproteins: the Evans County Heart Study.** *Am Heart J* 1984, **108:** 807-814.
8. The Emerging Risk Factors Collaboration: **Lipoprotein(a) Concentration and the Risk of Coronary Heart Disease, Stroke, and Nonvascular Mortality.** *JAMA: The Journal of the American Medical Association* 2009, **302:** 412-423.
9. Guyton JR, Dahlen GH, Patsch W, Kautz JA, Gotto AM, Jr.: **Relationship of plasma lipoprotein Lp(a) levels to race and to apolipoprotein B.** *Arterioscler Thromb Vasc Biol* 1985, **5:** 265-272.
10. Ober C, Nord AS, Thompson EE, Pan L, Tan Z, Cusanovich D *et al.*: **Genome-wide association study of plasma lipoprotein(a) levels identifies multiple genes on chromosome 6q.** *J Lipid Res* 2009, **50:** 798-806.
11. Clarke R, Peden JF, Hopewell JC, Kyriakou T, Goel A, Heath SC *et al.*: **Genetic variants associated with Lp(a) lipoprotein level and coronary disease.** *N Engl J Med* 2009, **361:** 2518-2528.
12. Luke MM, Kane JP, Liu DM, Rowland CM, Shiffman D, Cassano J *et al.*: **A polymorphism in the protease-like domain of apolipoprotein(a) is associated with severe coronary artery disease.** *Arterioscler Thromb Vasc Biol* 2007, **27:** 2030-2036.
13. Shiffman D, O'Meara ES, Bare LA, Rowland CM, Louie JZ, Arellano AR *et al.*: **Association of gene variants with incident myocardial infarction in the Cardiovascular Health Study.** *Arterioscler Thromb Vasc Biol* 2008, **28:** 173-179.
14. Shiffman D, Kane JP, Louie JZ, Arellano AR, Ross DA, Catanese JJ *et al.*: **Analysis of 17,576 Potentially Functional SNPs in Three Case-Control Studies of Myocardial Infarction.** *PLoS ONE* 2008, **3:** e2895.

15. Luke MM, Kane JP, Liu DM, Rowland CM, Shiffman D, Cassano J *et al.*: **A Polymorphism in the Protease-Like Domain of Apolipoprotein(a) Is Associated With Severe Coronary Artery Disease.** *Arterioscler Thromb Vasc Biol* 2007, ATVBAHA.
16. Roden DM, Pulley JM, Basford MA, Bernard GR, Clayton EW, Balser JR *et al.*: **Development of a Large-Scale De-Identified DNA Biobank to Enable Personalized Medicine.** *Clin Pharmacol Ther* 2008, **84:** 362-369.
17. Pulley J, Clayton E, Bernard GR, Roden DM, Masys DR: **Principles of Human Subjects Protections Applied in an Opt-Out, De-identified Biobank.** *Clinical and Translational Science* 2010, **3:** 42-48.
18. Oetjens M, Brown-Gentry K, Goodloe R, Dilks HH, Crawford DC. **Population stratification in the context of diverse epidemiologic surveys** (in preparation).
19. Rautaharju PM, Warren JW, Jain U, Wolf HK, Nielsen CL: **Cardiac infarction injury score: an electrocardiographic coding scheme for ischemic heart disease.** *Circulation* 1981, **64:** 249-256.
20. Dumitrescu L, Ritchie MD, Brown-Gentry K, Pulley JM, Basford M, Denny JC *et al.*: **Assessing the accuracy of observer-reported ancestry in a biorepository linked to electronic medical records.** *Genet Med* 2010, **12:** 648-650.
21. Hall JB, Dumitrescu L, Dilks HH, Crawford DC, Bush WS: **Accuracy of Administratively-Assigned Ancestry for Diverse Populations in an Electronic Medical Record-Linked Biobank.** *PLoS ONE* 2014, **9:** e99161.
22. Purcell S, Neale B, Todd-Brown K, Thomas L, Ferreira MA, Bender D *et al.*: **PLINK: a tool set for whole-genome association and population-based linkage analysis.** *Am J Hum Genet* 2007, **81:** 559-575.
23. Bush WS, Boston J, Pendergrass SA, Dumitrescu L, Goodloe R, Brown-Gentry K *et al.*: **Enabling high-throughput genotype-phenotype associations in the Epidemiology Architecture for Genes Linked to Environment (EAGLE) project as part of the Population Architecture using Genomics and Epidemiology (PAGE) study.** *Pac Symp Biocomput* 2013, **18:** 373-384.
24. Willer CJ, Li Y, Abecasis GR: **METAL: fast and efficient meta-analysis of genomewide association scans.** *Bioinformatics* 2010, **26:** 2190-2191.
25. Pendergrass S, Dudek SM, Roden DM, Crawford DC, Ritchie MD: **Visual integration of results from a large DNA biobank (BioVU) using synthesis-view.** *Pac Symp Biocomput* 2011, 265-275.
26. Pendergrass S, Dudek S, Crawford D, Ritchie M: **Synthesis-View: visualization and interpretation of SNP association results for multi-cohort, multi-phenotype data and meta-analysis.** *BioData Mining* 2010, **3:** 10.
27. Varas-Lorenzo C, Castellsague J, Stang MR, Tomas L, Aguado J, Perez-Gutthann S: **Positive predictive value of ICD-9 codes 410 and 411 in the identification of cases of acute coronary syndromes in the Saskatchewan Hospital automated database.** *Pharmacoepidem Drug Safe* 2008, **17:** 842-852.
28. Pladevall M, Goff DC, Nichaman MZ, Chan F, Famsey D, Ortiz C *et al.*: **An Assessment of the Validity of ICD Code 410 to Identify Hospital Admissions for Myocardial infarction:**

The Corpus Christi Heart Project. *International Journal of Epidemiology* 1996, **25:** 948-952.

29. Petersen LA, Wright S, Normand SL, Daley J: **Positive Predictive Value of the Diagnosis of Acute Myocardial Infarction in an Administrative Database.** *J Gen Intern Med* 1999, **14:** 555-558.
30. Thygesen S, Christiansen C, Christensen S, Lash T, Sorensen H: **The predictive value of ICD-10 diagnostic coding used to assess Charlson comorbidity index conditions in the population-based Danish National Registry of Patients.** *BMC Medical Research Methodology* 2011, **11:** 83.
31. Schunkert H, Konig IR, Kathiresan S, Reilly MP, Assimes TL, Holm H *et al.*: **Large-scale association analysis identifies 13 new susceptibility loci for coronary artery disease.** *Nat Genet* 2011, **43:** 333-338.
32. Tregouet DA, Konig IR, Erdmann J, Munteanu A, Braund PS, Hall AS *et al.*: **Genome-wide haplotype association study identifies the SLC22A3-LPAL2-LPA gene cluster as a risk locus for coronary artery disease.** *Nat Genet* 2009, **41:** 283-285.
33. Welter D, MacArthur J, Morales J, Burdett T, Hall P, Junkins H *et al.*: **The NHGRI GWAS Catalog, a curated resource of SNP-trait associations.** *Nucleic Acids Research* 2014, **42:** D1001-D1006.
34. McDavid A, Crane PK, Newton KM, Crosslin DR, McCormick W, Weston N *et al.*: **Enhancing the Power of Genetic Association Studies through the Use of Silver Standard Cases Derived from Electronic Medical Records.** *PLoS ONE* 2013, **8:** e63481.
35. Collins FS, Varmus H: **A New Initiative on Precision Medicine.** *N Engl J Med* 2015, **372:** 793-795.
36. Manchia M, Cullis J, Turecki G, Rouleau GA, Uher R, Alda M: **The Impact of Phenotypic and Genetic Heterogeneity on Results of Genome Wide Association Studies of Complex Diseases.** *PLoS ONE* 2013, **8:** e76295.
37. Rosenberg NA, Huang L, Jewett EM, Szpiech ZA, Jankovic I, Boehnke M: **Genome-wide association studies in diverse populations.** *Nat Rev Genet* 2010, **11:** 356-366.
38. Kaufman JS, Dolman L, Rushani D, Cooper RS: **The Contribution of Genomic Research to Explaining Racial Disparities in Cardiovascular Disease: A Systematic Review.** *American Journal of Epidemiology* 2015, **181:** 464-472.

DIAGNOSIS-GUIDED METHOD FOR IDENTIFYING MULTI-MODALITY NEUROIMAGING BIOMARKERS ASSOCIATED WITH GENETIC RISK FACTORS IN ALZHEIMER'S DISEASE

XIAOKE HAO[1,2], JINGWEN YAN[2], XIAOHUI YAO[2], SHANNON L. RISACHER[2], ANDREW J. SAYKIN[2], DAOQIANG ZHANG[1], LI SHEN[2], FOR THE ADNI

[1]*College of Computer Science and Technology, Nanjing University of Aeronautics and Astronautics, Nanjing, 210016, China*

[2]*Department of Radiology & Imaging Science, School of Medicine, Indiana University, Indianapolis,46202,USA*

Email: [1,2]robinhc@163.com, [2]jingyan@iupui.edu, [2]yao2@umail.iu.edu, [2]{srisache, asaykin}@iupui.edu, [1]dqzhang@nuaa.edu.cn, [2]shenli@iu.edu

Many recent imaging genetic studies focus on detecting the associations between genetic markers such as single nucleotide polymorphisms (SNPs) and quantitative traits (QTs). Although there exist a large number of generalized multivariate regression analysis methods, few of them have used diagnosis information in subjects to enhance the analysis performance. In addition, few of models have investigated the identification of multi-modality phenotypic patterns associated with interesting genotype groups in traditional methods. To reveal disease-relevant imaging genetic associations, we propose a novel diagnosis-guided multi-modality (DGMM) framework to discover multi-modality imaging QTs that are associated with both Alzheimer's disease (AD) and its top genetic risk factor (i.e., APOE SNP rs429358). The strength of our proposed method is that it explicitly models the priori diagnosis information among subjects in the objective function for selecting the disease-relevant and robust multi-modality QTs associated with the SNP. We evaluate our method on two modalities of imaging phenotypes, i.e., those extracted from structural magnetic resonance imaging (MRI) data and fluorodeoxyglucose positron emission tomography (FDG-PET) data in the Alzheimer's Disease Neuroimaging Initiative (ADNI) database. The experimental results demonstrate that our proposed method not only achieves better performances under the metrics of root mean squared error and correlation coefficient but also can identify common informative regions of interests (ROIs) across multiple modalities to guide the disease-induced biological interpretation, compared with other reference methods.

1. Introduction

Neuroimaging genetics emerges as one of the hottest research topics in recent studies, which identifies genetic variant associations with imaging phenotypes such as structural or functional imaging measures. Since neuroimaging plays an important role in characterizing the neurodegenerative process of many brain disease such as Alzheimer's disease (AD) [1], the quantitative imaging phenotypes can provide valuable information so that it holds great promise for revealing the complex biological mechanisms of the disease.

Genome-wide association studies (GWAS) have been widely used to identify the associations between single nucleotide polymorphisms (SNPs) and the quantitative traits (QTs) such as neuroimaging measures. To address the high dimensionality of the GWAS data and small effect size of individual SNPs, in recent imaging genetic studies, researchers have developed several generalized multivariate linear regression analysis methods by considering the priori knowledge such as inherent structural information to boost the detection power [2, 3]. Although those methods may have the potential to help discover phenotypic imaging markers related to some candidate risk SNPs [4], another problem of existing methods in imaging genetics is that the subjects' diagnosis information (e.g., class labels such as patients or healthy controls) is not fully used for revealing disease-specific imaging genetic associations. More recently, some diagnosis induced methods have been proposed to solve the imaging genetics problem [5, 6]. A two-step strategy was adopted by [5]: 1) initially, the authors identified the voxels that could provide an imaging signature of the disease with high classification accuracy using penalized linear discriminant analysis; 2) then they detected the SNPs associated with the multivariate phenotypic markers discovered in the first step. Moreover, a Bayesian framework for detecting genetic variants associated with a disease while exploiting imaging as an intermediate phenotype was proposed in [6], which was designed to jointly identify relevant imaging and genetic markers simultaneously. In addition, most of imaging genetic studies focus on discovering the associations between single imaging modality (e.g., magnetic resonance imaging (MRI)) and SNPs, while ignoring the underlying interacting relationships among multiple modalities.

With these observations, our general motivation is to identify multimodal imaging phenotypes serving as intermediate traits between a given AD genetic marker and disease status, where we hope to design a simple and powerful model to maximize disease-relevant imaging genetic associations. Accordingly, the ideas introduced in [7, 8] can be adopted and incorporated into the imaging genetics studies. Specifically in [7, 8], subjects' similarity has been successfully used for designing more powerful multi-modal models on AD classification and clinical score regression solutions, which are inspired by multi-task modeling integrated with the priori relationship between sample data and the corresponding labels in machine learning community [9].

In this study, we propose a novel diagnosis-guided multi-modality (DGMM) framework that considers robust and common regions of interests (ROIs) as well as diagnosis labels such as patients or healthy controls to handle the multi-modality phenotype associations with an AD genetic risk factor. We evaluate our DGMM method on two modalities of phenotypes, i.e., voxel-based measures extracted from structural MRI and fluorodeoxyglucose positron emission tomography (FDG-PET)) scans, as well as apolipoprotein E (APOE) SNP rs429358 (the best known AD genetic risk factor [10, 11]) data from the Alzheimer's Disease Neuroimaging Initiative (ADNI) cohort. The empirical results show that our method not only yield improved performances under the metrics of correlation coefficient and root mean squared error, but also detect a compact set of consistent and robust ROIs across two imaging modalities which are relevant to the studied genetic risk marker.

2. Method

2.1. *Genotype and Phenotype Association*

In this section, we systematically develop our computational models to explore the association between a candidate AD risk SNP and multimodal imaging phenotypes. That is, our proposed method mainly addresses the problem based on the general linear (least square) regression approach. Given imaging phenotypes $\mathrm{X} = [x_1, \dots, x_\mathrm{n}, \dots, x_\mathrm{N}]^T \in \mathrm{R}^{\mathrm{N}\times\mathrm{d}}$ as input and a candidate risk SNP $y = [\mathrm{y}_1, \dots, \mathrm{y}_\mathrm{n}, \dots, \mathrm{y}_\mathrm{N}]^\mathrm{T} \in \mathrm{R}^\mathrm{N}$ as output in the regression model, where N is the number of participants (sample size) and d is the number of imaging phenotype ROIs (feature dimensionality). The association model is designed to solve:

$$\min_{w} \frac{1}{2} ||y - Xw||^2 + \lambda R(w) \quad (1)$$

where $R(w)$ is a regularization term and λ is the corresponding parameter. The weight vector w measures the relative importance of the imaging phenotypes (i.e., ROI measures) in predicting the response of the SNP.

In the work, the goal of the learned regression model is not to discover relevant SNPs, but to select biologically meaningful imaging phenotypes that are associated jointly with a given risk SNP and the disease status. Using the linear general regression model formulated by Eq (1), we aim to identify interesting imaging phenotypes that can serve as intermediate traits on the pathway from an AD genetic risk factor to the clinical diagnosis.

2.2. *Diagnosis-Guided Single-modality Phenotype Association*

In this study, we consider the relationship between imaging phenotypes and the diagnosis information among subjects which are not fully used in conventional association analysis methods. More specifically, we will utilize the relationship information among subjects with diagnosis labels, i.e., AD, mild cognitive impairment (MCI) or healthy controls (HC). That is, if subjects are similar to each other in the original diagnosis feature space, their respective response values should be also similar. To solve this problem, we induce a new regularization term that can preserve the class level diagnosis information:

$$\min_{w} \sum_{i,j}^{N} ||f(x_i) - f(x_j)||_2^2 S_{ij} = 2w^T X^T L X w \quad (2)$$

where $S = [S_{ij}] \in R^{n \times n}$ denotes a similarity matrix that measures the similarity between every pair of samples. $L = D - S$ represents a Laplacian matrix, where D is the diagonal matrix with element defined as $D_{ii} = \sum_{j=1}^{N} S_{ij}$. Then, the similarity matrix can be defined as:

$$S_{ij} = \begin{cases} 1, & \text{if } x_i \text{ and } x_j \text{ are from the same class} \\ 0, & \text{otherwise} \end{cases} \quad (3)$$

The penalized term Eq. (2) enforces that, after being mapped into the label space, the distance between the within-class data will be small, which preserves the local neighborhood structure of the same class. We induce the diagnosis labels constraint into the single modality phenotypic solution and then formulate a diagnosis-guided single modality (DGSM) phenotype association model as follows:

$$\min_{w} \frac{1}{2} ||y - Xw||^2 + \alpha w^T X^T L X w \quad (4)$$

The strength of DGSM method is that it explicitly models the priori diagnosis information among subjects in the objective function that minimize distance within each diagnosis class for selecting the disease-relevant QT associated with the SNP. Especially, the DGSM model can generalize and handle the progressive disease with multi-diagnosis status, comparing to the binary diagnosis analysis methods that were adopted in [5, 6].

2.3. *Multi-modality Phenotype Associations*

We assume that there are N training subjects or samples, with each represented by M modalities of phenotypes. Denote $X^m = [X_1^m, \dots, X_n^m, \dots, X_N^m]^T \in R^{N\times d}$ as the data matrix of the m-th modality, and $Y = [Y_1, \dots, Y_2, \dots, Y_n]^T \in R^N$ be the corresponding response values (i.e. APOE SNP rs429358). Let $w^m \in R^d$ be the linear discriminant function corresponding to the m-th modality. Then the multi-modality phenotype association model can be formulated as follows:

$$\min_{W} \frac{1}{2} \sum_{m=1}^{M} ||Y - X^m w^m||^2 + \beta ||W||_{2,1} \tag{5}$$

where $W = [w^1, w^2, \dots, w^M] \in R^{d\times M}$ is the weight matrix whose row w_j is the vector of coefficients assigned to the j-th feature across different modalities, and $||W||_{2,1} = \sum_{j=1}^{d} ||w_j||_2$ is penalize all coefficients in the same row of matrix W for joint feature selection. First, the $l_{2,1}$-norm regularization term is a "group-sparsity" regularizer, which forces only a small number of features being selected from different modalities [12]. Second, the parameter β is a regularization parameter that is used to balance the relative contributions of the two terms in Eq (5). Finally, it is worth noting that our objective function Eq (5) is formatted as a multi-task learning framework, where each imaging modality is used to predict the same response independently (i.e., $Y_1 = Y_2 = \dots = Y_n$), but the feature selection is regularized jointly by the second term in Eq (5) to identify a set of consistent ROIs.

2.4. *Diagnosis-Guided Multi-modality Phenotype Association*

In this study, we try to develop a novel diagnosis-guided multi-modality (DGMM) framework to discover the multi-modality phenotypic associations with an AD genetic risk factor, where it explicitly models the priori diagnosis information among subjects in the objective function for selecting disease-relevant and robust multi-modality QTs associated with the SNP. We induce the diagnosis label constraint into the multi-modality phenotypic solution and design a diagnosis-guided multi-modality (DGMM) phenotype association model as follows:

$$\min_{W} \frac{1}{2} \sum_{m=1}^{M} ||Y - X^m w^m||^2 + \lambda_1 ||W||_{2,1} + \lambda_2 \sum_{m=1}^{M} (w^m)^T (X^m)^T L^m X^m w^m \tag{6}$$

where $S = [S_{ij}^m] \in R^{n\times n}$ denotes a similarity matrix that measures the similarity between every pair of samples on the m-th modality across different subjects. Here, $L^m = D^m - S^m$ represents a combinational Laplacian matrix for the m-th modality, where D_m is the diagonal matrix with element defined as $D_{ii}^m = \sum_{j=1}^{N} S_{ij}^m$. λ_1 and λ_2 denote control parameters of the regularization terms, respectively. Their values can be determined via inner cross-validation on training data. It is promising to find the better solution that is robust to noises or outliers via considering both multimodalities and the rich information inherent in the observations. The objective function can be efficiently solved using the Nesterov's accelerated proximal gradient optimization algorithm which was used in [7], which is shown in the Algorithm 1.

Firstly, we separate the objective function into the smooth part Eq (7) and non-smooth part Eq (8) as following:

$$f(W) = \frac{1}{2}\sum_{m=1}^{M} ||Y - X^m w^m||^2 + \lambda_2 \sum_{m=1}^{M} (w^m)^T (x^m)^T L^m w^m x^m \tag{7}$$

$$g(W) = \lambda_1 ||W||_{2,1} \tag{8}$$

We define the approximation function Eq (9) as following, which is composited by the above smooth part and non-smooth one:

$$\Omega(W, W_i) = f(W_i) + \left(W - W_i, \nabla f(W_i)\right) + \frac{l}{2}||W - W_i||_F^2 + g(W) \tag{9}$$

where$|| \cdot ||_F^2$ denotes the Frobenius norm, $\nabla f(W_i)$ denotes the gradient of $f(W)$ on point W_i at the i-th iteration, and l is the step size. Then, the update step of Nesterov's APG is defined as:

$$W_{i+1} = \arg\min_{W} \frac{1}{2}\left|\left|W - V\right|\right|_F^2 + \frac{1}{l} g(W) = \arg\min_{w_1, w_2, \ldots, w_d} \frac{1}{2}\sum_{j=1}^{d} ||w_j - v_j||_2^2 + \frac{\lambda_2}{l}||w_j||_2 \tag{10}$$

where w_j and v_j denote the j-th row of the matrix W and V, respectively. NAGP performs a simple step of gradient descent to go from W_i to V, and then it slide a little bit further than

$$V = W_i - \frac{1}{l}\nabla f(W_i) \tag{11}$$

Therefore, through Eq (9), this problem can be decomposed into d separate sub-problems. The key of APG algorithm is how to solve the update step efficiently. The analytical solutions of those sub-problems can be easily obtained:

$$w_j^* = \begin{cases} \left(\frac{\left\|v_j\right\|_2 - \frac{\lambda_2}{l}}{\left\|v_j\right\|_2}\right) v_j, & \text{if } \left\|v_j\right\|_2 > \frac{\lambda_2}{l} \\ 0, & \text{otherwise} \end{cases} \tag{12}$$

Instead of performing gradient descent based on W_i, we compute the search point as:

$$Z_i = (1 + \alpha_i)W_i - \alpha_i W_{i-1} \tag{13}$$

where $\alpha_i = \frac{\rho_{i-1}-1}{\rho_i}$ and $\rho_i = \frac{1+\sqrt{1+4\rho_{i-1}^2}}{2}$.

Algorithm 1: to minimize J in Equation (6)

Input: APOE genotype y= $[y_1, \dots, y_n, \dots, y_N]^T \in R^N$,
Multimodal imaging data $X^m = [X_1^m, \dots, X_n^m, \dots, X_N^m]^T \in R^{N\times d}$,
Subject diagnosis information (i.e., AD, MCI or HC)
Output: W_i, J^*
Initialization: $\lambda_1 > 0, \lambda_2 > 0, l_0 > 0, \sigma > 1, W_0 = W_1 = 0, \rho_0 = 1$
Repeat (For i=1 to max_iteration I)
1. Computed the search point Qi according to Eq (13)
2. $l = l_{i-1}$
3. while $(f(W_{i+1}) + g(W_{i+1})) > \Omega(W_{i+1}, Q_i), l = \sigma l$;
Here is computed by Eq. (10)
4. Set $l_i \leftarrow l$
Until Converges
Calculate J^*

---

3. Experiments

In this section, we evaluate the effectiveness of the proposed method on the ADNI-1 database. For up-to-date data access information, see http://adni.loni.usc.edu/data-samples/access-data/. One goal of ADNI is to test whether serial MRI, positron emission tomography, other biological markers, and clinical and neuropsychological assessment can be combined to measure the progression of mild cognitive impairment (MCI) and early AD. For more details, see www.adni-info.org. In our experiments, baseline structural MRI, FDG-PET scans, the top AD risk SNP APOE rs429358, another AD risk SNP CD33 rs386544 and non-risk SNP rs56283507 (for comparison purpose) are included. This yields a total of 357 subjects, including 87 AD, 182 MCI and 88 HC participants. Table 1 shows the numbers for each diagnosis code and each SNP.

Table 1. Diagnostic distributions on APOE SNP rs429358 and CD33 rs386544 and random non-risk SNP rs56283507

Diagnosis	APOE rs429358 Code			CD33 rs386544 Code			non-risk rs56283507 Code		
Label	0	1	2	0	1	2	0	1	2
AD	29	45	13	41	34	12	37	37	13
MCI	83	74	25	87	82	13	79	82	21
HC	66	21	1	40	36	12	39	40	9

3.1. *Imaging Phenotype Data*

The SPM Statistical Parametric Mapping software package (SPM version 12, for more details, see www.fil.ion.ucl.ac.uk/spm/software/) was used to: (1) create normalized gray matter density maps from MRI data in the standard MNI space, and (2) register the FDG-PET scans into the same space. The MarsBaR ROI toolbox [13] was used to extract mean gray matter density and FDG-PET glucose utilization values for each of 116 MarsBaR ROIs. These measures were pre-adjusted for age, gender, handedness and education.

3.2. *Genotype Data*

APOE (located on chromosome 19) has a key role in coordinating the mobilization and redistribution of cholesterol, phospholipids, and fatty acids, and it is implicated in mechanisms such as neuronal development, brain plasticity, and repair functions [14]. In imaging genetics research experiments, several whole-brain studies focused on mapping this risk genetic variable [10, 11]. In this work, we focused on studying the susceptibility SNP rs429358, which was determined using APOE ε2/ε3/ε4 status information from the ADNI clinical database for each participant. We also selected another AD risk SNP CD33 rs386544 and a random non-risk SNP rs56283507 for the comparison purpose to evaluate the performance of the proposed model.

3.3. *Experimental Settings*

In our experiment, for the input of multimodal imaging phenotypes, we normalized the FDG and VBM whose ranges are -5.29 to 6.49 and -5.34 to 4.73, respectively. For the outcome, each SNP value is coded in an additive fashion as 0, 1 or 2, indicating the number of minor alleles. We have inserted this information in our revised manuscript. 5-fold cross-validation strategy was adopted to evaluate the effectiveness of our proposed method. As for parameters of regularization, we determined the values by nested 5-fold cross-validation on the training set. In current studies, we used SM (denoting single modality based method with Lasso [15] to detect a sparse significant subset from imaging phenotypic features (i.e., ROIs)), MC (denoting modalities concatenation with Lasso to detect a sparse subset from imaging phenotypes), MM (denoting multi-modality method to detect imaging phenotypes from a sparse subset of common ROIs), DGSM, DGMC and DGMM (the standard SM, MC and MM with DG, respectively, where DG denotes the diagnosis-guided strategy).

3.4. *Results*

We compare our proposed diagnosis-guided based methods (including DGSM, DGMC and DGMM) with conventional methods (including SM, MC and MM), respectively. The performance on each dataset is assessed with root mean squared error (RMSE) and correlation coefficient (CC) between actual and predicted response values, which are widely used in measuring performances of regression and association analysis. The average results of RMSE and CC among the 5-fold test on MRI-VBM and FDG-PET modalities are calculated respectively as shown in Table 2 and Table 3. The corresponding values on the whole test data entirety (denoted Ent for short) are included in both tables, where predicted values from all cross-validation trials are pulled together for calculating a single RMSE or CC.

As shown in Table 2, the proposed DG based methods consistently outperform their non-DG based methods in the RMSE performance measure. This demonstrates that diagnosis-guided information can help improve regression performance from imaging phenotypes to genotype. DGMM and DGMC methods yield the best RMSE values of 0.9097 and 0.9096. Compared with the DG strategy, the joint regularization across multiple modalities showed negative effects on the RMSE performance in some cases (e.g., from SM to MM). Regarding the CC results in Table 3, our proposed method shows the best CC of 0.1499 with the MRI-VBM modality. The best CC of 0.1471 is obtained by DGMM in terms of the FDG measure while the second best performance is 0.1140 by our DGMM method. These results demonstrate the proposed methods can take advantage of consistent and robust multimodality information to find more important associations. Compared with the joint regularization across multiple modalities, the DG strategy had very limited contributions in most cases except the DGMM on MRI-VBM (compared with MM).

Table 2. Comparison of regression performances of the competing methods in terms of Root Mean Square Error (RMSE)

Method	MRI-VBM		FDG-PET	
	(Mean ± Std)	5-fold Ent	(Mean ± Std)	5-fold Ent
SM	1.0103±0.1123	1.0185	0.9538±0.0549	0.9569
DGSM	**0.9097±0.0342**	**0.9107**	0.9205±0.0446	0.9225
MC	0.9547±0.1088	0.9635	0.9127±0.0364	0.9138
DGMC	**0.9096±0.0342**	0.9635	**0.9096±0.0342**	**0.9106**
MM	1.3358±0.1081	1.3417	1.2267±0.0400	1.2280
DGMM	**0.9097±0.0342**	**0.9107**	**0.9097±0.0342**	**0.9106**

Table 3. Comparison of regression performances of the competing methods in terms of Correlation Coefficient (CC)

Method	MRI-VBM		FDG-PET	
	(Mean ± Std)	5-fold Ent	(Mean ± Std)	5-fold Ent
SM	-0.0154±0.1015	-0.0997	-0.1307±0.1323	-0.0557
DGSM	0.0090±0.1326	0.0039	-0.0322±0.0857	0.0363
MC	-0.0913± 0.1609	0.0345	0.0164±0.0605	-0.1037
DGMC	-0.0241±0.1318	-0.0650	-0.0354±0.1251	0.0525
MM	0.0928±0.0796	0.0886	**0.1471±0.0804**	**0.1492**
DGMM	**0.1499±0.0384**	**0.1465**	0.1140±0.0780	0.1002

We also selected another AD risk SNP CD33 rs386544 and a random SNP rs56283507 as the comparison to evaluate the performance on the proposed model. As shown in Table 4, the DGMM method with APOE rs429358 yield the best RMSE and CC performance measures, which outperform the same method involved the CD33 rs386544 or the random SNP. This matches our expectation, since the APOE SNP has a larger effect size than the CD33 SNP and the random SNP. The originality of the work is to make full use of the risk genotype and corresponding disease samples to find the intermediate phenotype between an AD genetic marker and the disease status. For evaluation purpose, it is desired to select the top AD risk SNP to demonstrate our proposed model.

Table 4. Comparison performances (RMSEs and CCs) in our proposed model with top risk SNP APOE rs429358, another risk SNP CD33 rs386544, and a random non-risk SNP rs56283507.

Candidate SNPs	MRI-VBM		FDG-PET	
	RMSE	CC	RMSE	CC
APOE-rs429358	**0.9097±0.0342**	**0.1499±0.0384**	**0.9097±0.0342**	**0.1140±0.0780**
CD33-rs386544	0.9123±0.0779	0.0582±0.1134	0.9123±0.0779	0.0960±0.0823
rs56283507	0.9628±0.0346	0.0677±0.1495	0.9628±0.0346	0.0125±0.0686

Besides the improved performances, one major goal of this study is to identify some significant and robust phenotypes that are highly correlated to risk genotype marker to capture imaging genetics associations in AD research.

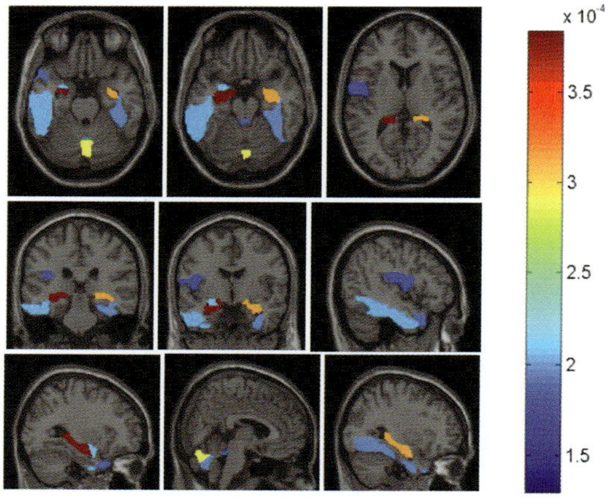

Fig. 1. Visualization of the top 10 VBM ROIs selected by the proposed method.

The top 10 selected MRI-VBM imaging features, as well as their average regression coefficients on 5-fold test, are visualized in Fig. 1 by mapping them onto the human brain. The colors of the selected brain regions indicate the regression coefficients of the corresponding MRI-VBM markers. As expected, Hippocampus_Left, Hippocampus_Right and Amygdala_Left have been detected on top 10 ROIs associated with risk genotype biomarker by the proposed DGMM method. It's worth noting that these stable markers are in accordance with the existing findings. For example, the reduction of hippocampal gray matter has been correlated with APOE SNP rs429358 [16]. The APOE polymorphism is the best established genetic risk factor for pathological changes that is also associated with anatomical brain changes.

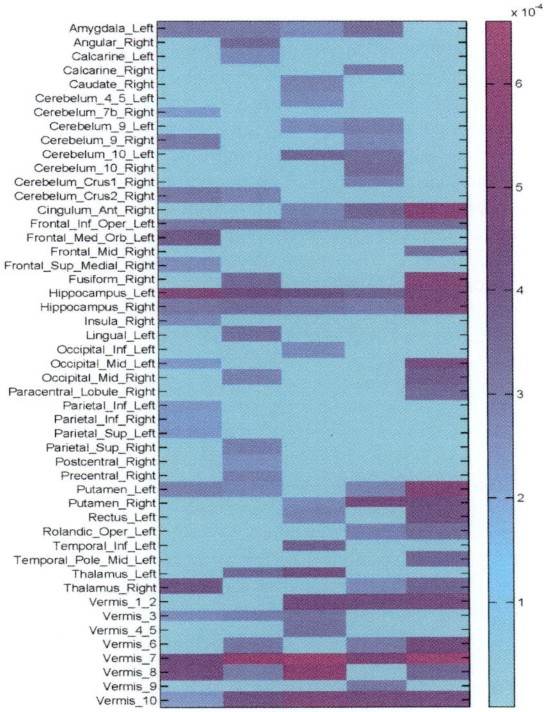

Fig. 2. Heat map of the top VBM and FDG ROI associations with APOE SNP rs429358 learned by the proposed method.

The weights of the top 20 ROIs by every fold DGMM test on the heat map are plotted in Fig. 2. Our proposed method tends to select the stable ROIs such as Vermis_7, Vermis_10, Hippocampus_Left, Hippocampus_Right and Frontal_Inf_Oper_Left that span across five cross-validation trials. The APOE SNP is the best established genetic risk factor for pathological changes that is also associated with reductions of hippocampal gray matter and glucose metabolism [10, 16, 17]. It also demonstrates the robust and consistent ROIs should be selected among the independent and different modalities, which discovers the imaging genetic associations through biological interpretation. Although reduced volume of cerebellar vermis has been associated with dementia [18], the imaging genetic finding of Vermis_7 warrants further investigation.

4. Conclusion

In this study, we have developed a diagnosis-guided multi-modality (DGMM) framework for identifying neuroimaging phenotype associations with risk genetic biomarkers. This approach explicitly models the priori diagnosis information among subjects in the objective function for selecting the most relevant and robust multi-modality QTs (i.e., MRI-VBM and FDG-PET) associated with top risk SNP (i.e., APOE rs429358). Experimental results on the ADNI database showed that our proposed DGMM method not only achieved better prediction performances under the metrics of correlation coefficient and root mean squared error compared with other single modality and non-diagnosis-guided methods, but also detected a compact set of robust and consistent ROIs across the multimodal phenotypes among the populations to guide the disease-induced biological interpretation. The similar model can be also extended to the investigation of association analyses between multi-modal brain imaging measures and any other biomarkers such as those in cerebrospinal fluid. Furthermore, the DGMM framework can be applied to other genetic associated diseases to investigate the complex biological mechanisms from genetics to intermediate traits to diagnostic outcome. An interesting future direction is to improve the efficiency of our implementation and apply it to larger scale studies such as analyzing high dimensional voxel based imaging data as well as a comprehensive set of genetic risk factors.

Acknowledgements

Data collection and sharing for this project was funded by the Alzheimer's Disease Neuroimaging Initiative (ADNI) (National Institutes of Health Grant U01 AG024904) and DOD ADNI (Department of Defense award number W81XWH-12-2-0012). ADNI is funded by the National Institute on Aging, the National Institute of Biomedical Imaging and Bioengineering, and through generous contributions from the following: AbbVie, Alzheimer's Association; Alzheimer's Drug Discovery Foundation; Araclon Biotech; BioClinica, Inc.; Bio-gen; Bristol-Myers Squibb Company; CereSpir, Inc.; Eisai Inc.; Elan Pharmaceuticals, Inc.; Eli Lilly and Company; EuroImmun; F. Hoffmann-La Roche Ltd and its affiliated company Genentech, Inc.; Fujirebio; GE Healthcare; IXICO Ltd.; Janssen Alzheimer Immunotherapy Research & Development, LLC.; Johnson & Johnson Pharmaceutical Research & Development LLC.; Lumosity; Lundbeck; Merck & Co., Inc.; Meso Scale Diagnostics, LLC.; NeuroRx Research; Neurotrack Technologies; Novartis Pharmaceuticals Corporation; Pfizer Inc.; Piramal Imaging; Servier; Takeda Pharmaceutical Company; and Transition Therapeutics. The Canadian Institutes of Health Research is providing funds to support ADNI clinical sites in Canada. Private sector contributions are facilitated by the Foundation for the National Institutes of Health (www.fnih.org). The grantee organization is the Northern California Institute for Research and Education, and the study is coordinated by the Alzheimer's Disease Cooperative Study at the University of California, San Diego. ADNI data are disseminated by the Laboratory for Neuro Imaging at the University of Southern California. This research is supported by the NNSFC (Nos. 61422204, 61473149), the JNSFDYS (No. BK20130034), the SRFDPHE (No. 20123218110009), the NFRF (No. NE2013105), the Jiangsu Qinglan Project of China and NUAA Ph.D student short-term visiting scholar project. At Indiana University, this work was supported by NIH R01 LM011360, U01 AG024904, RC2 AG036535, R01 AG19771, P30 AG10133, UL1 TR001108, R01 AG 042437, and R01 AG046171; NSF IIS-1117335; DOD W81XWH-14-2-0151, W81XWH-13-1-0259, and W81XWH-12-2-0012; NCAA 14132004; and CTSI SPARC Program.

References

1. D. C. Glahn, P. M. Thompson and J. Blangero,Human Brain Mapping, 28, 488, (2007).
2. D. P. Hibar, O. Kohannim, J. L. Stein, M. C. Chiang and P. M. Thompson,Front Genet, 2, 73, (2011).
3. T. Ge, G. Schumann and J. Feng,Quantitative Biology, 1, 227, (2013).
4. H. Wang, F. P. Nie, H. Huang, J. W. Yan, S. Kim, K. Nho, S. L. Risacher, A. J. Saykin, L. Shen and A. s. D. N. Initi,Bioinformatics, 28, I619, (2012).
5. M. Vounou, E. Janousova, R. Wolz, J. L. Stein, P. M. Thompson, D. Rueckert, G. Montana and A. D. N. Initia,Neuroimage, 60, 700, (2012).
6. N. K. Batmanghelich, A. V. Dalca, M. R. Sabuncu and G. Polina,Inf Process Med Imaging, 23, 766, (2013).
7. B. Jie, D. Q. Zhang, B. Cheng and D. G. Shen,Medical Image Computing and Computer-Assisted Intervention (Miccai 2013), Pt I, 8149, 275, (2013).
8. X. F. Zhu, H. I. Suk and D. G. Shen,Medical Image Computing and Computer-Assisted Intervention - Miccai 2014, Pt Iii, 8675, 401, (2014).
9. M. Belkin, P. Niyogi and V. Sindhwani,Journal of Machine Learning Research, 7, 2399, (2006).
10. Y. Liu, J. T. Yu, H. F. Wang, P. R. Han, C. C. Tan, C. Wang, X. F. Meng, S. L. Risacher, A. J. Saykin and L. Tan,J Neurol Neurosurg Psychiatry, 86, 127, (2015).
11. N. Filippini, A. Rao, S. Wetten, R. A. Gibson, M. Borrie, D. Guzman, A. Kertesz, I. Loy-English J. Williams, T. Nichols, B. Whitcher and P. M. Matthews,Neuroimage, 44, 724, (2009).
12. M. Yuan and Y. Lin,Journal of the Royal Statistical Society Series B-Statistical Methodology 68, 49, (2006).
13. N. Tzourio-Mazoyer, B. Landeau, D. Papathanassiou, F. Crivello, O. Etard, N. Delcroix, B Mazoyer and M. Joliot,Neuroimage, 15, 273, (2002).
14. R. W. Mahley,Science, 240, 622, (1988).
15. R. Tibshirani,Journal of the Royal Statistical Society: Series B (Statistical Methodology), 73, 273, (2011).
16. H. A. Wishart, A. J. Saykin, T. W. McAllister, L. A. Rabin, B. C. McDonald, L. A. Flashman, R. M. Roth, A. C. Mamourian, G. J. Tsongalis and C. H. Rhodes,Neurology, 67, 1221, (2006).
17. E. M. Reiman, R. J. Caselli, L. S. Yun, K. Chen, D. Bandy, S. Minoshima, S. N. Thibodeau and D. Osborne,N Engl J Med, 334, 752, (1996).
18. L. Baldacara, J. G. Borgio, W. A. Moraes, A. L. Lacerda, M. B. Montano, S. Tufik, R. A. Bressan, L. R. Ramos and A. P. Jackowski,Rev Bras Psiquiatr, 33, 122, (2011).

METABOLOMICS DIFFERENTIAL CORRELATION NETWORK ANALYSIS OF OSTEOARTHRITIS

Ting Hu*

Department of Computer Science
Memorial University, St. John's, NL, Canada
**E-mail: ting.hu@mun.ca*

Weidong Zhang

Discipline of Genetics, Faculty of Medicine
Memorial University, St. John's, NL, Canada

Zhaozhi Fan

Department of Mathematics and Statistics
Memorial University, St. John's, NL, Canada

Guang Sun

Discipline of Medicine, Faculty of Medicine
Memorial University, St. John's, NL, Canada

Sergei Likhodi

Department of Laboratory Medicine, Faculty of Medicine
Memorial University, St. John's, NL, Canada

Edward Randell

Department of Laboratory Medicine, Faculty of Medicine
Memorial University, St. John's, NL, Canada

Guangju Zhai

Discipline of Genetics, Faculty of Medicine
Memorial University, St. John's, NL, Canada

Osteoarthritis (OA) significantly compromises the life quality of affected individuals and imposes a substantial economic burden on our society. Unfortunately the pathogenesis of the disease is till poorly understood and no effective medications have been developed. OA is a complex disease that involves both genetic and environmental influences. To elucidate the complex interlinked structure of metabolic processes associated with OA, we developed a differential correlation network approach to detecting the interconnection of metabolite pairs whose relationships are significantly altered due to the diseased process. Through topological analysis of such a differential network, we identified key metabolites that played an important role in governing the connectivity and information flow of the network. Identification of these key metabolites suggests the association of their underlying cellular processes with OA and may help elucidate the pathogenesis of the disease and the development of novel targeted therapies.

Keywords: Differential correlation; Osteoarthritis; Metabolomics; Urea cycle abnormal; Obesity; Cardiovascular diseases; Differential networks; Dynamical networks; Interaction mapping.

1. Introduction

Osteoarthritis (OA) is the most common form of arthritis. It causes a substantial morbidity and disability in the elderly populations, and imposes a great economic burden on our society.[1,2] Despite high prevalence and societal impact, there is no medication that can cure it, or reverse or halt the disease progression, partly because that its pathogenesis is still unclear and there is no reliable method that can be used for early OA diagnosis.

Recent developments in the field of metabolomics provide an array of new tools for the study of OA. A large number of small-molecule metabolites from body fluids or tissues can be quantitatively detected simultaneously, which promises an immense potential for early diagnosis, therapy monitoring and understanding the pathogenesis of complex diseases.[3] Metabolites are intermediate and end products of various cellular processes and their levels of concentration serve as a good indicator of a sequence of biological systems in response to genetic and environmental influences.

In the reported studies on metabolomics analysis of OA case-control population data, the mostly adopted methodology is to test and identify metabolites that are significantly associated with the disease class using principal component analysis (PCA),[4,5] partial least square discriminant analysis (PLS),[6,7] or other individual testing techniques, and then to deduce their likely biological interrelationship with OA. Testing correlations of the concentrations of metabolites has not seen wide adoption likely due to the limited availability of methodologies. However, these correlations likely exist because metabolites are intermediate or end products of interconnected cellular processes. Analyzing their correlations provides an avenue capturing the relationships of their represented cellular processes and biological reactions associated with OA, and thus holds a great potential in OA metabolomics research.

Meanwhile, many biological systems are increasingly viewed and analyzed as highly complex networks of interlinked molecular or cellular entities or metabolites,[8] and network science has been applied to capture the interactome maps of gene-gene or protein-protein interactions[9–13] as well as transcriptional and metabolic data.[14–16]

The interaction maps of proteins, genes, metabolites or diseases can reveal the overall physical and functional landscape of a biological system, and these networks have been mostly generated under a particular static condition. More recently, differential network analysis has been promoted as a powerful framework for analyzing biological interaction maps when biological systems are considered undergoing differential changes that are dependent on the environment, tissue type, disease state, development or speciation.[17,18]

Recent interaction mapping studies have demonstrated the power of differential correlation analysis for elucidating the re-wiring of the interaction architecture of fundamental biological responses in adaptation to changing conditions.[19–25] Analyzing the rewiring of biological networks across disease conditions provides a unique insight into the dynamic response of a biological system. Instead of looking at the absolute properties of a system, differential network analysis emphasizes on the characteristics that are the most affected by genetic or environmental influences.

In this study, we proposed a differential network approach to analyzing the metabolomics population-based data of OA. We used differential analysis to quantify the variation of pairwise correlation of metabolites across case and control populations, and used networks to characterize the global interconnecting structure of such differentially correlated metabolites. Our methodology is distinct from most existing metabolomics analyses of OA in that we investigated the correlations of metabolite concentrations, and more importantly the variations of such correlations by comparing different disease status, to help elucidate the underlying biological processes specifically associated with the pathogenesis of OA. Using topological analysis of such a differential correlation network, we identified key metabolites and subsequently their represented cellular processes that may play an important role in the clinical development of OA. Our findings could be very helpful in designing novel and more targeted therapies for OA.

2. Methods

2.1. *Osteoarthritis metabolomics data*

In the current study, we used a two-stage case-control design with a discovery phase and a validation phase. For both phases, knee OA patients were selected from the Newfoundland Osteoarthritis Study (NFOAS) initiated in 2011.[26] The NFOAS aimed at identifying novel genetic, epigenetic, and biochemical markers for OA. The NFOAS recruited OA patients who underwent a total knee replacement surgery due to primary OA between November 2011 and December 2013 at the St. Clare's Mercy Hospital and Health Science Centre General Hospital in St. John's, the capital city of Newfoundland and Labrador (NL), Canada. Healthy controls for both phases were selected from the CODING study (The Complex Diseases in the Newfoundland population: Environment and Genetics), where participants were adult volunteers.[27]

Both cases and controls were from the same source population of Newfoundland and Labrador. Knee OA diagnosis was made based on the American College of Rheumatology clinical criteria for classification of idiopathic OA of the knee[28] and the judgment of the attending orthopedic surgeons. Controls were individuals without self-reported family doctor diagnosed knee OA based on their medical information collected by a self-administered questionnaire. We collected 64 OA cases and 45 healthy controls in the discovery phase and 72 cases and 76 controls in the replication phase.

Blood samples were collected after at least 8 hour fasting and plasma was separated from blood using the standard protocol. Metabolic profiling was performed on plasma using the Waters XEVO TQ MS system (Waters Limited, Mississauga, Ontario, Canada) coupled with Biocrates AbsoluteIDQ p180 kit, which measures 186 metabolites including 90 glycerophospholipids, 40 acylcarnitines (1 free carnitine), 21 amino acids, 19 biogenic amines, 15 sphingolipids and 1 hexose (above 90 percent is glucose). The details of the 186 metabolites and the metabolic profiling method were described in the previous publication.[29] Over 90% of the metabolites (167/186) were successfully determined in each sample.

Age and BMI are known factors correlated with OA. Therefore, the residual of a linear regression using attributes age and BMI was applied to remove any partial correlations as a result of those two factors, and to adjust the data for our metabolomics differential correlation analysis of OA.

2.2. *Differential analysis of metabolite correlations*

Metabolite concentrations in plasma may be correlated as a result of their represented biological processes, and the correlation may change in different phenotypic or disease conditions. Such a dynamic correlation was quantified by a differential correlation statistic in our study.

The correlation of a pair of metabolites was calculated using Pearson's correlation coefficient r in the two phenotypically distinguished samples, i.e. cases and controls. The correlation coefficients r_{case} and r_{control} were then used to compute the change of the correlation between two metabolites across two different disease classes. Specifically, for metabolites i and j, their differential correlation $r_{\text{diff}}(i,j)$ is calculated as the normalized difference of Fisher's z-transformations of $r_{\text{case}}(i,j)$ and $r_{\text{control}}(i,j)$,

$$r_{\text{diff}}(i,j) = \sqrt{\frac{n_{\text{case}}-3}{2}} \times z_{\text{case}}(i,j) - \sqrt{\frac{n_{\text{control}}-3}{2}} \times z_{\text{control}}(i,j), \tag{1}$$

where z is the Fisher's z-transformation of correlation coefficient r,

$$z_{\text{case}}(i,j) = \frac{1}{2}\ln\Big[\frac{1+r_{\text{case}}(i,j)}{1-r_{\text{case}}(i,j)}\Big],\ z_{\text{control}}(i,j) = \frac{1}{2}\ln\Big[\frac{1+r_{\text{control}}(i,j)}{1-r_{\text{control}}(i,j)}\Big]. \tag{2}$$

We used n_{case} and n_{control} to denote the total numbers of samples in cases and controls. This differential correlation statistic captures the change of the normalized correlation across two distinguishing conditions, and we used it to test if two metabolites are differentially correlated by comparing diseased and healthy populations. Note that r_{diff} describes the change of correlations by subtracting the correlation in controls from that in cases, and can take either positive or negative values.

The significance levels of differential correlations were assessed using a 1000-fold permutation test. For each permutation, we randomly shuffled the disease status of all samples combining both cases and controls to remove the association among metabolite correlations and the disease outcome. By repeating this process 1000 times, we were able to generate a null distribution under the assumption that the pairwise correlations of metabolites were not statistically distinguishing in cases and in controls. Then for each pair of metabolites, the significance (p-value) of their differential correlation was estimated as the proportion of permuted differential correlations that were greater than the observed value calculated using the original real data.

2.3. *Differential correlation network*

Network is a powerful tool to characterize the properties of entities and their complex relationships. In this study, we used networks to represent the global structure of differentially correlated metabolites by comparing OA cases and healthy controls.

Pairs of metabolites that had significant differential correlations were included to build the network. In such a differential correlation network, each node stood for a metabolite, and edges linking two metabolites represented the significant differential correlations between them. The differential correlation of a metabolite pair could be either positive or negative, meaning that their correlation in cases are significantly stronger than their correlation in controls or vice versa.

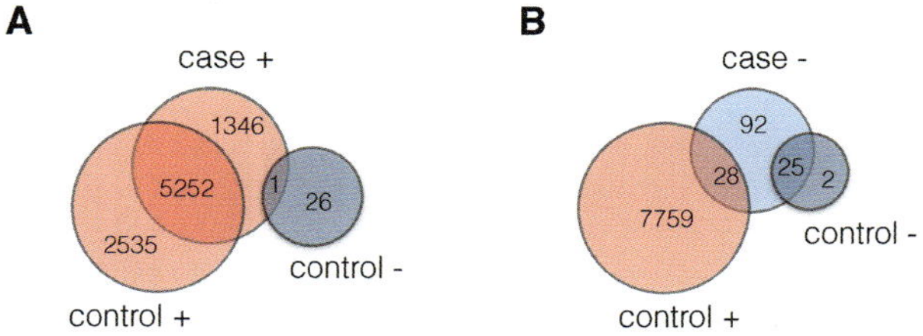

Fig. 1. Comparison of pairwise metabolite correlations (red for positive; blue for negative) in case and control populations. Only significant correlations (Pearson's correlation coefficient p-value cutoff 0.05 with Bonferroni multiple-testing correction) were included in this comparison. (**A**) For the total of 6599 (= 1346 + 5252 + 1) pairs of positively correlated metabolites in cases, the majority of them were also found positively correlated in controls. (**B**) For the total of 145 (= 92 + 28 + 25) pairs of negatively correlated metabolites in cases, a third of them were found positively correlated in controls and another third of them were found negatively correlated in controls.

3. Results

3.1. *Metabolite correlations in case and control populations*

The pairwise Pearson's correlations of 167 metabolites were calculated in both case and control samples in the discovery dataset. Of all 13,861 pairs, the majority of them were positively correlated in both cases and controls. We used a p-value threshold 0.05 and Bonferroni multiple-testing correction to define the statistical significance of pairwise correlations.

About 80% of the positively correlated pairs in cases were found also positively correlated in controls (Fig. 1**A**), and a similar link was observed for negatively correlated pairs as well (Fig. 1**B**). This large overlapping of metabolite correlations from the two phenotypic conditions suggests that the majority of the observed correlations were a result of "housekeeping" biological reactions and were not related to the disease of OA.

3.2. *Differentially correlated metabolites*

We calculated the differential correlations of all pairs of metabolites by comparing their correlations in cases and controls as described in the section of Methods. By subtracting correlations in controls from correlations in cases, metabolite pairs that were differentially correlated across these two conditions were magnified, while the persistent correlations in both conditions were removed. This differential correlation method allowed us to focus on the dynamic correlations that were specifically associated with the disease.

In the discovery dataset, 232 pairs of metabolites had significant positive differential correlations and 1060 pairs had significant negative differential correlations (permutation testing $p < 0.05$). The strongest and most significant pair of metabolites that has a positive differential correlation is Ala and Sarcosine ($r_{\text{diff}} = 9.33$, $p < 0.001$), and that has a negative differential correlation is lysoPCaC24:0 and PCaaC40:2 ($r_{\text{diff}} = -5.40$, $p < 0.001$). Fig. 2**A** shows a scatter plotting of all the pairs of metabolites with their correlations in the case population (x-axis) and the control population (y-axis). In addition, positive and negative differential correla-

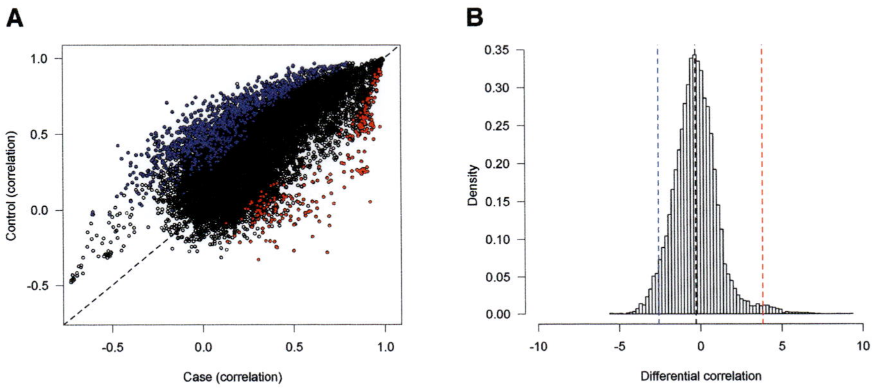

Fig. 2. (**A**) Scatter of metabolite pair correlations in cases (x axis) and controls (y axis) and identification of significant (p-value cutoff 0.05 using a 1000-fold permutation test) pairs with positive differential correlations (red) and with negative differential correlations (blue). (**B**) Distribution of all pairwise differential correlations, with a mean value of -0.283 (black dashed line). The means of significant differential correlations are also shown using dashed lines. The average significant positive correlations was 3.820 (red) and the average significant negative correlations was -2.589 (blue).

tions were shown as colored points. They represented the metabolite pairs whose correlations significantly changed across the two phenotypic conditions.

The distribution of the differential correlations of all metabolite pairs is shown in Fig. 2**B**. It follows a normal distribution approximately with a mean of -0.283. The shift of this distribution towards the negative values explained the observation that there were more significant negative differential correlations (1060 pairs) than positive ones (232 pairs). However, positive differential correlation distribution has a longer tail towards larger values, and the mean of significant positive differential correlations, i.e. 3.820, was greater than the absolute of the mean of negative ones, i.e. -2.589.

3.3. *Differential correlation network of OA*

We applied differential correlation analysis to both the discovery and replication datasets. We used the set of metabolite pairs that were significantly differentially correlated (permutation testing significance cutoff $p < 0.05$) in both datasets to build the differential correlation network of OA.

A total of 100 pairwise differential correlations were statistically significant in both datasets, including 71 metabolites. The network was comprised of four connected components and the largest component included 63 metabolites and 95 edges (Fig. 3). The remaining three components had only two or four nodes and were not included in the network visualization.

As seen in the figure, the majority of metabolite pairs were negatively differentially correlated, denoted by blue edges in the graph. Positive differential correlations, however, were

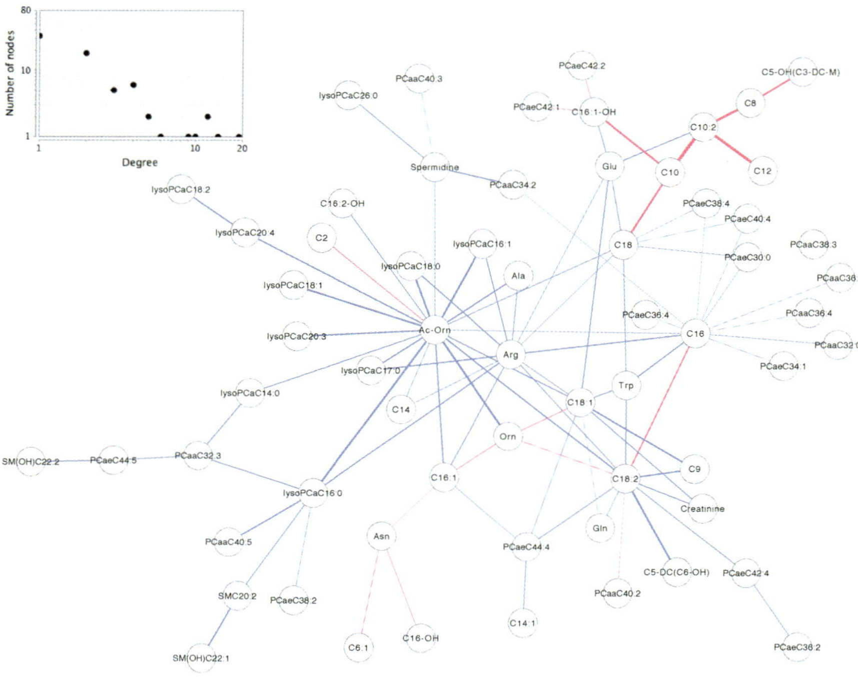

Fig. 3. The differential correlation network by comparing the discovery and replication data. Only pairs of metabolites that have significant differential correlations in both datasets are shown. There is one major connected component of the network, which has 63 nodes and 95 edges. The network is visualized using the force-directed layout with a closer node layout distance representing a stronger pairwise correlation. Edge width is proportional to the corresponding correlation strength and edge color codes for positive (red) and negative (blue) differential correlations. This network visualization was generated using Cytoscape.[30]

less observed and clustered together in sub-structures of the network. The node degree of this network had a mean of 3.02 and a heavy-tail distribution (inset of Fig. 3), showing that the majority of nodes have a very low degree but a few of them were considerably more connected than the others. This property suggests the robustness of connectivity and information flow in the network.

3.4. *Identification of key metabolites in the osteoarthritis differential correlation network*

In network science, the importance of an individual node in a network is captured by measuring its *centrality*. Besides the most commonly used centrality measure, node degree, there are more sophisticated metrics on node importance that characterize not only the number of connections

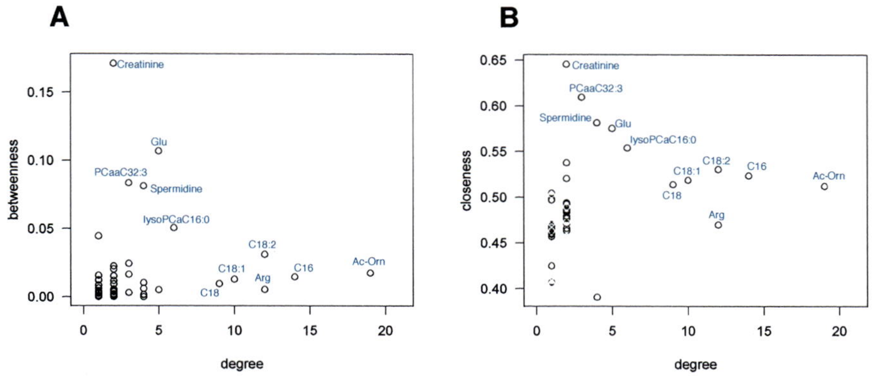

Fig. 4. Node importance characterized by (**A**) betweenness centrality and (**B**) closeness centrality in relation to node degree. Key nodes, either with high degrees, or high betweenness/closeness, or both, are identified and labeled with their represented metabolite names.

a node has, but also on how important those connections are in the global structure of an entire network. *Betweenness* centrality quantifies the number of times a node v is part of the shortest path between any pair of nodes,[31] represented as $\sum_{s \neq v \neq t \in V} \frac{\sigma_{st}(v)}{\sigma_{st}}$, where σ_{st} is the total number of shortest paths from node s to node t and $\sigma_{st}(v)$ is the number of those paths that pass through node v. Betweenness captures how important a given node acts on the connectivity of all other pairs of nodes. *Closeness* centrality is defined as $\frac{1}{\sum_{s \neq v} d_{vs}}$, where d_{vs} is the distance between nodes v and s.[32,33] This metric describes how easily a given node can reach all other nodes in a network. In the context of differential correlation networks, those centrality measures were used to identify key metabolites that play an essential role in the global interconnected structure.

Nodes with high degrees are usually referred to as "hubs" since they have more connections than the rest of the nodes in the network, and nodes with high betweenness or closeness are often referred to as "bottlenecks" since they are crucial in controlling the information flow in the network. Fig. 4 shows metabolites that are hubs, or bottlenecks, or both. The betweenness and closeness centralities are shown in relation to node degrees in the figure. The same set of 11 metabolites were identified as key nodes in both centrality measures (Fig. 4**A** and **B**).

4. Discussion

Identification of metabolic markers associated with OA holds a great potential to better understand the cellular processes in response to genetic and environmental influences that lead to the clinical outcome of the disease. The identified metabolites and their represented cellular processes will in turn help us to develop targeted therapies for OA. In this study, we developed a differential network approach to characterizing the variations of metabolite correlations in relation to different phenotypic conditions.

In our methodology, we used networks to represent the global inter-connected structure of metabolites that showed significant correlation variations in case and control populations. By exploring the topological properties of such a differential correlation network, we identified a set of key metabolites for modulating connectivity and information flow in the network, and thus hypothesized the association of their represented cellular processes with the disease.

When metabolite correlations were analyzed separately in cases and controls, we saw a large overlap of correlated metabolite pairs (Fig. 1), an observation indicating that most of the metabolite associations are not specifically related to OA. The differential analysis took a unique route by subtracting correlation coefficient of a metabolite pair in controls from that in cases, such that all the persistent pairwise correlations across the two phenotypic classes were removed and the pairs with significant variations were magnified. These differentially correlated metabolites are expected to provide useful insights into the underlying biological processes of the clinical development of OA. We observed considerably more significant negative differential correlations than positive ones (Fig. 2), which indicates that important biological processes might be compromised in OA patients.

By comparing the independent discovery and replication datasets, we built a differential correlation network of metabolites associated to OA (Fig. 3) The network included 63 metabolites and 95 pairwise differential correlations. The majority of the differential correlations were negative while the positive ones were clustered together around certain metabolites. The metabolites that have positive differential correlations are mainly coming from the same class of acylcarnitines, e.g. C18, C10, C10:2, C8, C5-OH(C3-DC-M), C12 and C16:1-OH; C18:2 and C16; C6:1, C16-OH and C16:1. From the view point of physiology function, the relationship between these metabolites is more likely a parallel relation rather than a causality.

The node degrees of this differential network had a heavy tail distribution (Fig. 3 inset), which suggests a robust property of connectivity and information flow subject to random perturbations. That is, random removal of nodes will have a very limited impact on the global connectivity of the network, a property that has been found in many biological systems including metabolic networks,[14] protein-protein interaction networks,[34] gene-regulatory networks[8] and gene-gene interaction networks.[35] In the context of OA metabolite differential correlation networks, this robustness property indicates the complexity of the molecular and cellular processes underlying the pathogenesis of OA.

Topological analysis on the node importance using centrality measures revealed a set of key metabolites that play an essential role modulating the connectivity and information flow in the network (Fig. 4). They were identified as "hubs", i.e. nodes that connect to many other nodes, and "bottlenecks", i.e. nodes that are located on major information flow paths in the network. Identification of these key metabolites may provide important insights into the pathogenesis of OA. Based on the node centrality measures, the metabolites in the network can be roughly classified into three categories. The hub-and-bottleneck metabolites Ac-Orn and Arg with their close neighbors Ala and Orn comprise the core of the network. On the network peripheral, metabolites are mostly glycerophospholipids (PC and LysoPC). Between the core and peripheral of the network is where acylcarnitines mixed with glycerophospholipids are located.

Ac-Orn, Arg, Ala and Orn have a close relationship with urea cycle in the body. Previous studies have proposed that urea cycle disorders may be related to the OA pathogenesis.[36,37] Glycerophospholipids form the essential lipid bilayer of all biological membranes and are closely involved in signal transduction, regulation of membrane trafficking and many other membrane-related phenomena.[38,39] It has been suggested that alterations in phospholipid composition and concentrations are associated with the development of OA.[40]

Acylcarnitines are related to energy metabolism. Carnitine and its acyl esters acylcarnitines are essential compounds for the metabolism of fatty acids. Carnitine can assist in the transport and metabolism of fatty acyl-CoA from the cytosol to the mitochondrial matrix, where the enzymes of oxidation are located and fatty acids are oxidized as a major source of energy. Acylcarnitine abnormal have been detected in obesity, type-2 diabetes, and cardiovascular diseases.[41,42]

The clustering of metabolites in the differential correlation network based on their centralities and the observation of urea cycle related metabolites locating on the core cluster of the network suggest that urea cycle abnormality may be a driving cause for metabolic disorders and may have a significant influence on OA development.

Acknowledgements

This work is partially supported by Newfoundland and Labrador Research & Development Corporation (RDC) Ignite grant to TH. ZF is supported by the Natural Science and Engineering Research Council (NSERC) of Canada. GZ acknowledges grants from Canadian Institute of Health Research (CIHR), Newfoundland and Labrador Research & Development Corporation (RDC) and Memorial University. We thank all the study participants who made this study possible and all the Operation Room staff in Eastern Health General Hospital and St. Clare's Hospital who helped for collecting samples.

References

1. WHO Scientific Group, *WHO Technical Report Series* **919**, p. 218 (2003).
2. J. Y. Reginstèr, *Rheumatology* **41**, 3 (2004).
3. G. N. Gowda, S. Zhang, H. Gu, V. Asiago, N. Shanaiah and D. Raftery, *Expert Review of Molecular Diagnostics* **8**, 716 (2008).
4. K. J. Deluzio and J. L. Astephen, *Gain and Posture* **25**, 86 (2007).
5. B. J. de Lange-Brokaar, A. Ioan-Facsinay, E. Yusuf, A. W. Visser, H. M. Kroon, G. J. van Osch, A. M. Zuurmond, V. Stojanovic-Susulic, J. L. Bloem, R. G. Nelissen, T. W. Huizinga and M. Kloppenburg, *Arthritis and Rheumatology* **67**, 733 (2015).
6. L. M. Gierman, S. Wopereis, B. van El, E. R. Verheij, B. J. Werff-van der Vat, Y. M. Bastiaansen-Jenniskens, G. J. van Osch, M. Kloppenburg, V. Stojanovic-Susulic, T. W. Huizinga and A. M. Zuurmond, *Arthritis and Rheumatology* **65**, 2606 (2013).
7. C. Wu, R. Lei, M. Tiainen, S. Wu, Q. Zhang, F. Pei and X. Guo, *Experimental Cell Research* **326**, 240 (2014).
8. A.-L. Barabasi and Z. N. Oltvai, *Nature Review Genetics* **5**, 101 (2004).
9. P. F. Jonsson and P. A. Bates, *Bioinformatics* **22**, 2291 (2006).
10. O. Vanunu, O. Magger, E. Ruppin, T. Shlomi and R. Sharan, *PLoS Computational Biology* **6**, p. e1000641 (2010).

11. S. Navlakha and C. Kingsford, *Bioinformatics* **26**, 1057 (2010).
12. T. Hu, Y. Chen, J. W. Kiralis and J. H. Moore, *Genetic Epidemiology* **37**, 283 (2013).
13. T. Hu, Q. Pan, A. S. Andrew, J. M. Langer, M. D. Cole, C. R. Tomlinson, M. R. Karagas and J. H. Moore, *BioData Mining* **7**, p. 5 (2014).
14. H. Jeong, B. Tombor, R. Albert, Z. N. Oltvai and A. L. Barabasi, *Nature* **407**, 651 (2000).
15. E. Ravasz, A. L. Somera, D. A. Mongru, Z. N. Oltvai and A. L. Barabasi, *Science* **297**, 1551 (2002).
16. P. Braun, E. Rietman and M. Vidal, *Proceedings of the National Academy of Sciences* **105**, 9849 (2008).
17. A. de la Fuente, *Trends in Genetics* **26**, 326 (2010).
18. T. Ideker and N. J. Krogan, *Molecular Systems Biology* **8**, p. 565 (2012).
19. Q. Zhong, N. Simonis, Q.-R. Li, B. Charloteaux, F. Heuze, N. Klitgord, S. Tam, H. Yu, K. Venkatesan, D. Mou, V. Swearingen, M. A. Yildirim, H. Yan, A. Dricot, D. Szeto, C. Lin, T. Hao, C. Fan, S. Milstein, D. Dupuy, R. Brasseur, D. E. Hill, M. E. Cusick and M. Vidal, *Molecular Systems Biology* **5**, p. 321 (2009).
20. I. W. Taylor, R. Linding, D. Warde-Farley, Y. Liu, C. Pesquita, D. Faria, S. Bull, T. Pawson, Q. Morris and J. L. Wrana, *Nature Biotechnology* **27**, 199 (09).
21. S. Bandyopadhyay, M. Mehta, D. Kuo, M.-K. Sung, R. Chuang, E. J. Jaehnig, B. Bodenmiller, K. Licon, W. Copeland, M. Shales, D. Fiedler, J. Dutkowski, A. Guenole, H. van Attikum, K. M. Shokat, R. D. Kolodner, W.-K. Huh, R. Aebersold, M.-C. Keogh, N. J. Krogan and T. Ideker, *Science* **330**, 1385 (2010).
22. J.-H. Chu, R. Lazarus, V. J. Carey and B. A. Raby, *BMC System Biology* **5**, p. 89 (2011).
23. B. Valcarcel, P. Wurtz, N.-K. S. al Basatena, T. Tukiainen, A. J. Kangas, P. Soininen, M.-R. Jarvelin, M. Ala-Korpela, T. M. Ebbels and M. de Iori, *PLoS One* **6**, p. e24702 (2011).
24. A. Fukushima, T. Nishizawa, M. Hayakumo, S. Hikosaka, K. Saito, E. Goto and M. Kusano, *Genome Analysis* **158**, 1487 (2012).
25. N. M. Penrod and J. H. Moore, *BMC Medical Genomics* **6**, p. S2 (2013).
26. G. Zhai, E. Aref-Eshghi, P. Rahman, H. Zhang, G. Martin, A. Furey, R. C. Green and G. Sun, *Journal of Orthopedics and Rheumatology* **1**, p. 5 (2014).
27. B. Fontaine-Bisson, J. Thorburn, A. Gregory, H. Zhang and G. Sun, *The American Journal of Clinical Nutrition* **99**, 384 (2014).
28. R. Altman, G. Alarcon, D. Appelrouth, D. Bloch, D. Borenstein, K. Brandt, C. Brown, T. D. Cooke and et al., *Arthritis and Rheumatology* **34**, 505 (1991).
29. W. Zhang, S. Likhodii, E. Aref-Eshghi, Y. Zhang, P. E. Harper, E. Randell, R. Green, G. Martin, A. Furey, G. Sun, P. Rahman and G. Zhai, *The Journal of Rheumatology* **42**, 859 (2015).
30. P. Shannon, A. Markiel, O. Ozier, N. S. Baliga, J. T. Wang, D. Ramage, N. Amin, B. Schwikowski and T. Ideker, *Genome Research* **13**, 2498 (2003).
31. L. C. Freeman, *Sociometry* **40**, 35 (1977).
32. A. Bavelas, *Journal of the Acoustical Society of America* **22**, 725 (1950).
33. G. Sabidussi, *Psychometrika* **31**, 581 (1966).
34. H. Jeong, S. P. Mason, A. L. Barabasi and Z. N. Oltvai, *Nature* **411**, 41 (2001).
35. T. Hu, N. A. Sinnott-Armstrong, J. W. Kiralis, A. S. Andrew, M. R. Karagas and J. H. Moore, *BMC Bioinformatics* **12**, p. 364 (2011).
36. S. B. J. Adams, L. A. Setton, E. Kensicki, M. P. Bolognesi, A. P. Toth and D. L. Nettles, *Osteoarthritis Cartilage* **20**, 64 (2012).
37. S. Kim, J. Hwang, J. Xuan, Y. H. Jung, H.-S. Cha and K. H. Kim, *PLoS One* **9**, p. e97501 (2014).
38. M. J. Berridge and R. F. Irvine, *Nature* **341**, 197 (1989).
39. A. A. Farooqui, L. A. Horrocks and T. Farooqui, *Chemistry and Physics of Lipids* **106**, 1 (2000).

40. B. A. Hills, *Internal Medicine Journal* **32**, 242 (2002).
41. V. A. Zammita, R. R. Ramsayb, M. Bonominic and A. Arduini, *Advanced Drug Delivery Reviews* **61**, 1353 (2009).
42. M. A. Ramos-Roman, L. Sweetman, M. J. Valdez and E. J. Parks, *Metabolism Clinical and Experimental* **61**, 202 (2012).

INVESTIGATING THE IMPORTANCE OF ANATOMICAL HOMOLOGY FOR CROSS-SPECIES PHENOTYPE COMPARISONS USING SEMANTIC SIMILARITY. - Accepted at Pacific Symposium on Biocomputing, 2016

PRASHANTI MANDA*

Department of Biology, University of North Carolina at Chapel Hill
Chapel Hill, NC 27599, USA
Email: manda.prashanti@gmail.com

CHRISTOPHER J. MUNGALL

Lawrence Berkeley National Laboratory
Berkeley, CA 94720, USA
Email: cjmungall@lbl.gov

JAMES P. BALHOFF

RTI International
Research Triangle Park, NC 27709, USA
Email: jbalhoff@rti.org

HILMAR LAPP

Center for Genomic and Computational Biology, Duke University
Durham, NC 27708, USA
Email: hilmar.lapp@duke.edu

TODD J. VISION

Department of Biology, University of North Carolina at Chapel Hill
Chapel Hill, NC 27599, USA
Email: tjv@bio.unc.edu

There is growing use of ontologies for the measurement of cross-species phenotype similarity. Such similarity measurements contribute to diverse applications, such as identifying genetic models for human diseases, transferring knowledge among model organisms, and studying the genetic basis of evolutionary innovations. Two organismal features, whether genes, anatomical parts, or any other inherited feature, are considered to be homologous when they are evolutionarily derived from a single feature in a common ancestor. A classic example is the homology between the paired fins of fishes and vertebrate limbs. Anatomical ontologies that model the structural relations among parts may fail to include some known anatomical homologies unless they are deliberately added as separate axioms. The consequences of neglecting known homologies for applications that rely on such ontologies has not been well studied. Here, we examine how semantic similarity is affected when external homology knowledge is included. We measure phenotypic similarity between orthologous and non-orthologous gene pairs between humans and either mouse or zebrafish, and compare the inclusion of real with faux homology axioms. Semantic similarity was preferentially increased for orthologs when using real homology axioms, but only in the more divergent of the two species comparisons (human to zebrafish, not human to mouse), and the relative increase was less than 1% to non-orthologs. By contrast, inclusion of both real and faux random homology axioms preferentially increased similarities between genes that were initially more dissimilar in the other comparisons. Biologically meaningful increases in semantic similarity were seen for a select subset of gene pairs.

Overall, the effect of including homology axioms on cross-species semantic similarity was modest at the levels of divergence examined here, but our results hint that it may be greater for more distant species comparisons.

Keywords: homology; phenotype; semantic similarity; Uberon; EQ annotation

1. Introduction

1.1. *Cross-species phenotype matching*

Organisms exhibit similarities with each other in their genetic content, anatomical structures, and other biological features due in large part to common evolutionary descent. This similarity is what allows non-human organisms to serve as models for human diseases and for biological knowledge to be transferred from model organisms to related species. In the area of biomedical informatics, an important recent application is the use of cross-species phenotype matching algorithms to generate candidate gene lists for rare and undiagnosed diseases.[1,2] Given a phenotypic profile for a human disease (e.g. a list of terms from The Human Phenotype Ontology,[3] cross-species profile matching tools generate a ranked list of candidate genes based on matches to the phenotypic profiles of orthologous genes in mutant models. This process can be automated by using phenotype ontologies and semantic similarity methods that quantify the degree of similarity.[4,5] A number of methods make use of the Uberon anatomy ontology to connect phenotype terms across species.[6,7] For example, the human phenotype "Abnormality of the upper limb" (HP_0002817) is connected to the mouse phenotype "abnormal forelimb morphology" (MP_0000550) via the Uberon class "forelimb" (UBERON_0002102).

1.2. *Homology*

Two organismal features, whether genes, anatomical parts, or another inherited feature, are considered to be homologous when they are evolutionarily derived from a single feature in a common ancestor. Orthologous genes are a particular class of homologous features, ones that are found in two different organismal lineages and that split evolutionarily into two genetic lineages during a speciation event. It is a foundational premise for much of comparative genomics that orthologous genes retain comparable functions even in distantly related organisms.[8] For example, in chick, *Tbx5* and *Tbx4* genes control early development of wing and hindlimb buds respectively, and the orthologs of these genes in zebrafish control development of *anatomically homologous* structures, the pectoral and pelvic fins[9] (Figure 1). Thus, it appears that these two gene lineages were distinct in the common ancestor of fish and birds and were deployed similarly in the development of the ancestral fore and hind appendages.

Recognizing similar phenotypes grows increasingly challenging as the evolutionary distance increases between species and anatomical features diverge in structure. Comparative anatomists have given a great deal of attention to identifying homologous anatomical structures among distantly related species.[10] Uberon does not contain explicit homology relationships,[11] such as between a hindlimb and a pelvic fin, or between the mammalian adrenal gland and the zebrafish interrenal gland. Instead, these classes are grouped according to similar structure, function or cellular composition. For example, both *forelimb* and *hindlimb* are grouped under the more general class *limb* based on their shared morphology, and limbs and fins are grouped under the general *paired appendage* (Figure 1). The situation is complicated by the fact that homology assumptions do necessarily leak into the construction of the ontology. The fact that the forelimb and hindlimb are similar morphologically is no accident if it is accepted that these are anatomical serial homologs. In fact, Uberon includes a grouping class *paired limb/fin* (UBERON_0004708) based on homology. Despite the above, homologous structures may sometimes be placed relatively distant to each other within Uberon when structural similarities are not as apparent (e.g. as is the case for certain bones in the jaw of fish that are homologous to the inner ear bones of mammals). Phenotypes affecting anatomical features that are homologous, but distantly placed within the ontology, will appear artificially dissimilar to one another.

We wish to quantify the extent to which the accuracy of cross-species phenotype matching is increased by including assertions of homology, such as those compiled by Bgee,[12] into Uberon. We do this by assessing how measures of semantic similarity are affected for orthologous relative to non-orthologous gene pairs. The underlying assumption is that orthologous genes are more often expressed in, and thus contribute to phenotypes in, homologous anatomical structures than non-orthologous genes, independently of how close the anatomical structures are within Uberon. If this assumption is correct, and homologies do indeed contribute to accuracy, we would expect to see a relatively greater increase in semantic similarity for orthologous genes relative to non-orthologous genes when non-trivial homologies are added to Uberon.

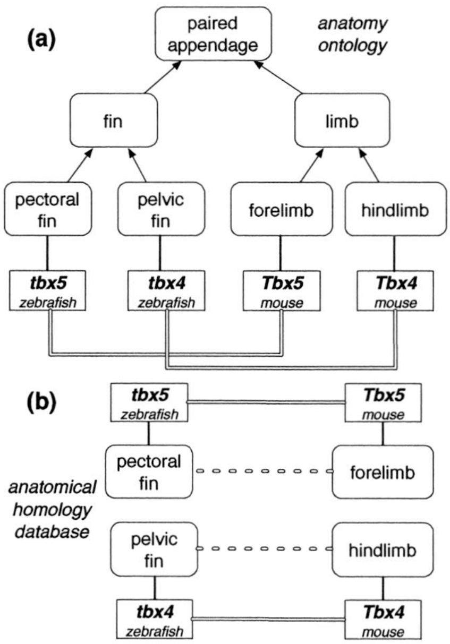

Fig. 1. The role of an orthologous gene pair in the development of different appendages. (a) genes (square boxes) are expressed in anatomical structures (rounded boxes), which are organized hierarchically in a subclass hierarchy (arrows). The two ortholog pairs are anatomically similar only at the level of "paired appendage". (b) Adding anatomical homology (dotted lines) increases anatomical similarity between orthologous pairs.

2. Methods

2.1. *Phenotype Annotation Data*

Orthologous gene pairs in zebrafish, mouse, and human were obtained from PANTHER (06/19/2014 release, v9.0).[13,14] Phenotype annotations for these genes were obtained from the Monarch Initiative v1.0 release (`https://github.com/monarch-initiative/monarch-owlsim-data`),

which aggregates data from the Human
Phenotype Ontology (HPO), Mouse Genome Informatics (MGI), and Zebrafish Information Network (ZFIN) (see `http://monarchinitiative.org` for details).

Gene pairs in which one or both genes lacked phenotype annotations were not included in the final analysis. We removed 503 mouse genes whose annotations indicate a lack of phenotypic assay (MP_0003012: no phenotypic analysis) and/or the absence of an abnormal phenotype (MP_0002169: no abnormal phenotype detected). Zebrafish and human genes for which no phenotypes are presently annotated were prefiltered by the source model organism databases (ZFIN, HPO). Anatomical homology axioms for Uberon classes were obtained from the GitHub repository of Bgee v0.2 (`https://github.com/BgeeDB/anatomical-similarity-annotations/blob/master/release/raw_similarity_annotations.tsv`).[15] Non-orthologous gene pairs across zebrafish, mouse, and human were randomly sampled with a uniform distribution from the set of gene pairs not asserted to be orthologous by PANTHER.

There are a variety of different semantic similarity measures used in the bioinformatics literature.[16] Here, we present results for a commonly used measure, Sim_{IC}, which is based on the concept of *Information Content* (IC), or the specificity of the match between two annotations relative to a chosen annotation corpus.[17] We also examined another commonly used measure, Jaccard similarity (Sim_J) which measures the ontological graph overlap between two annotations.[18] They differ in that Sim_{IC} takes into account the distribution of annotations among ontology terms while Sim_J considers ontology structure independent of annotation density. These metrics were compared because of their prior use as measures of phenotypic similarity between orthologous genes.[2]

We also have a choice in how to summarize the set of pairwise semantic similarities between two genes, both of which typically have multiple annotations. We refer to the union of the individual phenotype annotations for all alleles of one gene as a *phenotype profile*. Here, we evaluated two summary statistics for semantic similarity between two phenotype profiles, as detailed below, which we call Best Pairs and All Pairs. We only report full results for one combination of statistics, Sim_{IC} with Best Pairs, based on a test for which combination best discriminated between orthologs and non-orthologs (see Results).

The IC of ontology graph node N in an annotation corpus with Z genes is defined as the negative logarithm of the probability of a gene being annotated to N.

$$IC(N) = -\log(Z_N/Z)$$

where Z_N is the number of genes annotated to N. The IC for a pair of annotations, K and L, is defined as the IC of their *most informative common ancestor* (MICA), which is their most specific common subsumer in the ontology. Raw IC scores range from $[0, IC_{max}]$, with 0 being the score of the root node of the ontology graph, and $IC_{max} = -log(1/Z)$ the score of a node with only one gene annotation in the dataset. To obtain an IC score with a range of $[0, 1]$, the IC-based similarity measure, $Sim_{IC}(K, L)$, is normalized as follows.

$$Sim_{IC}(K, L) = IC(K, L)/ - log(1/Z)$$

The Jaccard similarity for a pair of annotations K, L is defined as the ratio of the number of nodes in the intersection of their subsumers in the ontology graph over the number of nodes in the union of their subsumers.[18]

$$Sim_J(K, L) = |A_K \cap A_L| / |A_K \cup A_L|$$

where A_K and A_L are the sets of subsumers of K and L, respectively.

2.1.1. *Similarity between phenotype profiles*

To compute the Best Pairs score between two phenotype profiles X, Y, for each annotation in X, the best scoring match in Y is determined, and the median of the $|X|$ values is taken. Similarly, for each annotation in Y, the best scoring match in X is determined, and the median of the $|Y|$ values is taken. The Best Pairs score $S_{BP}(X, Y)$ is the mean of these two medians.

$$S_{BP}(X, Y) = (1/2)[Sim(X, Y) + Sim(Y, X)]$$

where

$$Sim_{IC}(A, B) = \text{median}\Big\{Sim_{IC}(A_i, B_j) \mid i \in \{1 \ldots |A|\}, j = \underset{j=1 \ldots |B|}{\arg\max}\, Sim_{IC}(A_i, B_j)\Big\}$$

To compute the All Pairs score, one instead takes the median of of all pairwise phenotype similarities between X and Y.

$$S_{AP}(X, Y) = \text{median}\Big\{Sim(X_i, Y_j) \mid i \in \{1 \ldots |X|\}, j \in \{1 \ldots |Y|\}\Big\}$$

For both S_{BP} and S_{AP}, similarity may be measured using either Sim_{IC} or Sim_J.

2.2. *Construction of ontologies*

We constructed three ontologies for computing semantic similarity: one without homology axioms (R), one with valid homology axioms (H) and one with a random set of homology axioms (H'). Figure 2 illustrates the process by which these were built. Following the approach of Kohler et al.,[19] R, H and H' were seeded with the ontologies used by the gene phenotype annotations for all three species in the corpus: the mammalian phenotype,[20] zebrafish Phenotype,[19] and human phenotype[21] ontologies, as well as the cross-species Uberon anatomy ontology[11] and the phenotypic quality ontology PATO.[22]

There already exist a number of *homology_grouping* classes in the Uberon ontology that bundle morphologically or functionally distinct subclasses based entirely on homology. As we are seeking to determine the effect of anatomical homology on cross-species phenotype similarity of orthologous genes, we removed a number of *homology_grouping* classes from R, H, and H' (Table 1).

Next, 1,836 homology axioms from Bgee[15] relating homologous anatomical structures were added to H. For example, "pectoral fin" is asserted as being homologous to "forelimb" and "pelvic fin" to "hindlimb". These axioms restored the relations indicated by the excluded *homology_grouping* classes in Table 1. We generated a set of 1,836 'random' homology axioms by sampling anatomy terms from a permuted list of those used in the real homology axioms; these were then added to H'.

Table 1. *homology_grouping* classes in Uberon excluded from R, H, and H'.

Name	UBERON_ID	Name	UBERON_ID
adrenal/interrenal gland	0006858	paired limb/fin bud	0004357
limb/fin segment	0010538	paired limb/fin cartilage	0007389
paired limb/fin skeleton	0011582	pelvic appendage	0004709
paired limb/fin	0004708	pectoral appendage	0004710
paired limb/fin field	0005732	bone of free limb or fin	0004375

We then created *grouping classes* to subsume annotations based on different ontology class properties. These grouping classes are classified by an OWL reasoner into the pre-existing phenotype class hierarchy by virtue of subsumption reasoning and equivalence axioms. These axioms follow the standard Entity–Quality (EQ) template.[7,19] In their simplest form, EQ expressions describe a phenotype in terms of a quality (Q) and an entity (E) that is the bearer of the quality. These EQ expressions are represented in OWL as "Q and 'inheres in' some E".[7,23] The following three EQ expressions were created to serve as templates for equivalence axioms of grouping classes.

- EQ_1: Q and 'inheres in' some E
- EQ_2: Q and 'inheres in' some ('homologous to' some E)
- EQ_3: Q and 'inheres in' some (E or 'homologous to' some E)

In the above expressions, Q is the root of the PATO ontology and E can be any entity from the Uberon ontology. One class of the form EQ_1 was created for each entity class (E) in Uberon. These classes were added to R, H, and H'.

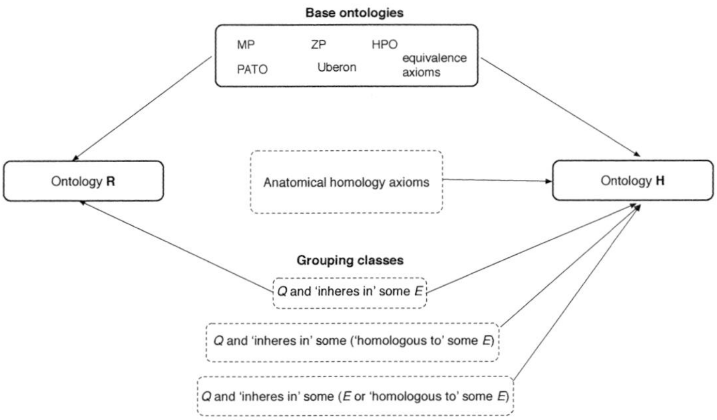

Fig. 2. Construction of ontologies for computing semantic similarity without or with supplemented knowledge of anatomical homology. First, ontology R is created by adding building block phenotype ontologies and equivalence axioms. H (and H') are created by adding anatomical homology axioms to R. Finally, one set of grouping classes is added to R and three sets of grouping classes are added to H (and H').

The templates EQ_2 and EQ_3 generate classes that group annotations whose anatomical structures are related via homology. For each entity class (E) in Uberon, we added to ontologies H and H' one class of form EQ_2 and one of form EQ_3. It is the presence of these grouping classes that control whether anatomical homology axioms are or are not used to infer common subsumers for phenotype annotations. Figure 3 illustrates the difference between ontologies for which the EQ_2 and EQ_3 classes are or are not included. When using R, which has EQ_1-template grouping classes only, fin and limb phenotypes are grouped at the high level of *paired appendage*. When using H, the phenotypes are grouped by an EQ_2-template grouping class for a more specific and thus more informative class defined by anatomical homology.

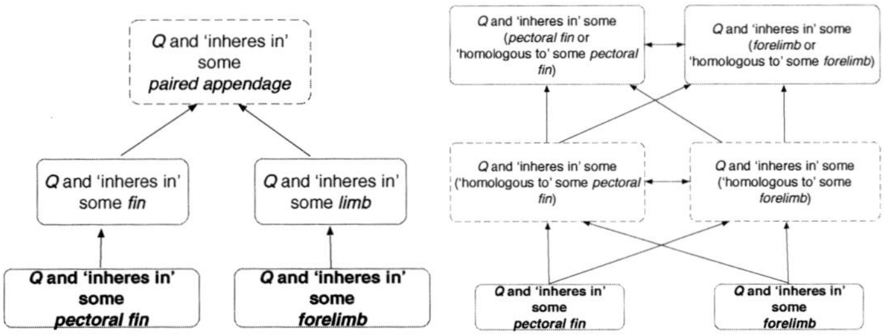

(a) Portion of ontology R (b) Portion of ontology H

Fig. 3. An example of subsumption hierarchies without and with anatomical homology. Least common subsumers for annotations of the pectoral fin and forelimb (dashed boxes) can be seen to differ for the R ontology, without homology (a) and the H ontology, with homology (b). Arrows denote subsumption relationships. It can be seen here that knowledge of homology enables the inference of more informative common subsumers for annotations with homologous anatomical structures.

2.3. *Assessing the impact of homology*

Our overall goal is to assess how the semantic similarity between the phenotypic profiles of two genes is affected by the addition of explicit homology statements in H lacking in R. Specifically, we hypothesized that there would be a greater increase in the similarity score for orthologous than non-orthologous genes when real homologies were included, and no differential increase when random homology assertions were included. We tested this hypothesis for two species pairs: mouse-human and zebrafish-human, with the expectation that the effect of homology on semantic similarity scores would be greater for the more distant evolutionary comparison.

To carry out this test, we measured the difference in similarity using R versus using either H or H'. We performed unpaired, one-sided t-tests for the null hypothesis that the distribution of differences was identically distributed for orthologs and non-orthologs. The alternate hypothesis is that the difference would be greater for orthologs. We performed four such tests, for both the zebrafish-human and mouse-human comparisons and for both H and H'.

3. Results

We obtained 1,253 orthologous gene pairs between zebrafish and mouse, 640 between zebrafish and human, and 2,034 between mouse and human, from PANTHER. Equal numbers of non-orthologous gene pairs were obtained for each species pair by sampling from permuted lists of the genes included in the orthologous pairs and requiring that the sampled pair not be included in the PANTHER orthology list. 10,055 grouping classes were added to ontology R and 30,165 to ontologies H and H'. As noted above, 1,836 anatomical homology axioms were obtained from Bgee for inclusion in H; they come from a wide variety of sources (Table 2). The same number of random homology axioms were added to H'.

Table 2. Sources of Bgee homology axioms classified by Evidence Code.[24]

Evidence Code	No. axioms	Evidence Code	No. axioms
Used in automatic assertion	709	Morphological similarity	66
Curator inference	361	Traceable author statement	55
Developmental similarity	213	Positional similarity	36
Phylogenetic distribution	197	Gene expression similarity	32
Non-traceable author statement	137	Compositional similarity	30

One of three confidence codes, "high confidence" (34.13% of axioms), "medium confidence" (61.72%), or "low confidence" (4.15%) was associated with each homology axiom by Bgee. 1,680 of the axioms assert a class to be a homolog of itself, while only 156 of the homology axioms, belonging to 12 taxonomic groups, assert homology between pairs of anatomical structures (Table 3). Thus, only a fraction of the homology axioms would be relevant for the taxonomic comparisons being made here; for example, there are only 10 non-self homology axioms that would affect comparisons between mammals.

Table 3. Distribution of Bgee homology axioms among taxa, excluding self-homologies.

Taxon name	No. axioms	Taxon name	No. axioms
Vertebrata	38	Mammalia	10
Tetrapoda	28	Metazoa	6
Bilateria	16	Sarcopterygii	8
Chordata	14	Eumetazoa	6
Amniota	10	Gnathostomata	8
Euteleostomi	10	Dipnotetrapodomorpha	2

We calculated phenotype semantic similarity for each orthologous and non-orthologous gene pair using the four combinations of semantic similarity measures described in the Methods above and for each of the three different ontologies, R, H, and H'. In order to select one semantic similarity measure for subsequent analyses, we determined which one best distinguished orthologous from non-orthologous gene pairs, reasoning that this would be an informative indicator of biological accuracy. We calculated the difference in median rank between orthologs and non-orthologs for the zebrafish-human comparison using H. We found that the combination of Sim_{IC} and S_{BP} gave the greatest discrimination between orthologous and non-orthologous gene pairs and so report results for that statistic in what follows. Full results for all four statistics, together with analysis scripts used in this study, are available from Zenodo (doi:/10.5281/zenodo.31833).

Our hypothesis that orthologs would experience a disproportionate increase in similarity when real homology axioms were used was supported by the t-tests in the case of the zebrafish-human comparison (Table 4). A one-tailed unpaired t-test found a significantly greater difference for orthologs than non-orthologs with real homology axioms but no significant difference with random axioms. However, this pattern was not seen in the mouse-human comparison, where orthologs were not significantly different than non-orthologs for H. In fact, the reverse trend was seen in all other comparisons; the mean similarity was preferentially increased for non-orthologs in the zebrafish-human H', mouse-human H, and mouse-human H' comparisons (Table 4). The underlying profile similarity values can be seen in Figure 4.

Table 4. Differences in similarity between orthologs and non-orthologs upon adding either real or random homology axioms to R. t: one-tailed, unpaired t-statistic; df: degrees of freedom; ns: not significant; δ_O, δ_{NO}: mean percent increase $\pm$ 2 standard errors relative to R for orthologs and non-orthologs, respectively.

species pair		real homology (H)	random homology (H')
zebrafish-human	t, df=1278	$t = 2.36$, $p = 0.009$	$t = 1.46$, ns
	δ_O	5.81 ± 0.70	4.85 ± 0.98
	δ_{NO}	4.88 ± 0.72	7.99 ± 3.11
mouse-human	t, df=4066	$t = -3.17$, ns	$t = -2.16$, ns
	δ_O	2.44 ± 0.27	3.22 ± 0.61
	δ_{NO}	4.39 ± 0.53	5.83 ± 0.81

4. Discussion

We wished to measure the extent to which addition of homology axioms to an anatomy ontology affects the semantic similarity of phenotypes between distantly related species. The pattern whereby orthologs between distantly related species (zebrafish and human) show a greater increase in similarity than non-orthologs when real homology axioms are added provides evidence that the inclusion of homology improves biological accuracy. However, there are three caveats. One, the relative difference in score between orthologs and non-orthologs, while significant, is less then $< 1\%$. Two, there was, unexpectedly, a larger increase in the similarity score for non-orthologs in the mouse-human comparison using real homology axioms. Third, non-orthologs had a greater increase in similarity for both species pairs when random homology axioms were added.

These results may be due to a combination of the hypothesized biological trend and a countervailing methodological artifact. First, the significant result for zebrafish-human with real homology axioms is consistent with the idea that the strength of the effect of including real homology axioms is in proportion to the evolutionary distance between the species pair. Second, the greater response of non-orthologs than orthologs in the three other comparisons, may stem from both real and faux homology axioms having a greater effect on semantic

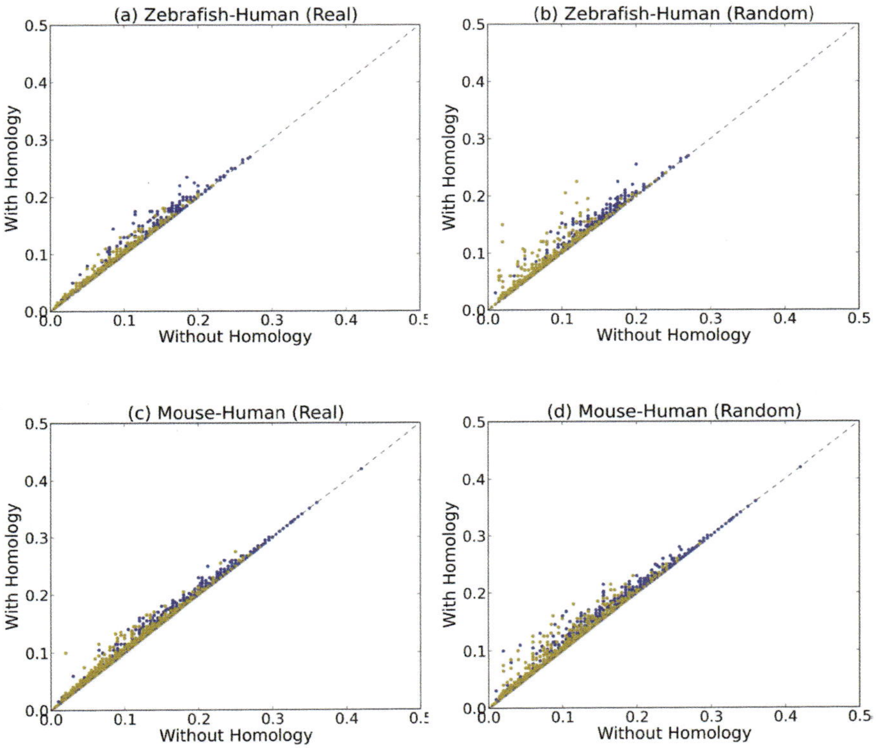

Fig. 4. $S_{IC,BP}$ for orthologs (blue) and non-orthologs (yellow) for the zebrafish-human (a,b) and mouse-human (c,d) species comparisons. The x-axis shows the scores without homology axioms (R) and the y-axis shows the scores for real (H) homology axioms (a and c) or random axioms (b and d).

similarity when phenotypes are dissimilar, as can be seen in Figure 4. When the species are closely related and orthologs are already highly similar, or when the axioms are random, then non-orthologs, which are less similar to begin with, preferentially experience the gain in similarity.

Despite the noisiness of the trends overall, we can see examples of individual gene pairs for which homology axioms have a large effect that makes biological sense. One such pair is the human gene TFAP2A (NCBI:gene:7020), which is annotated to "Fusion of middle ear ossicles", and the zebrafish gene *tfap2a* (ZFIN:ZDB-GENE-011212-6), annotated to "abnormal(ly) decreased length quadrate". The homology between the quadrate, part of the jawbone of basal vertebrates, and the incus, a middle ear ossicle in mammals, is a textbook example of vertebrate evolution.[25] When homology was excluded, "abnormal(ly) decreased length quadrate" was matched to "Micrognathia" and grouped under the relatively uninformative grouping class "Q and 'inheres in' some *bone of jaw*" with an Sim_{IC} score of 0.32. When homology assertions were included, these annotations were subsumed under the grouping class "Q and 'inheres in' some ('homologous to' some *auditory ossicle*)" with an Sim_{IC} score of 0.56.

Despite examples such as this, the modest effect of homology overall was unexpected. One explanation could be that so much anatomical homology is already implicit within the Uberon ontology that homology axioms are only needed in rare cases. In practice, it is difficult to extricate groupings in the ontology that are based on characteristics such as morphology, function, and shared development from those based on homology, potentially rendering some homology axioms redundant. Another explanation for the modest effect of homology is the relatively low number of homology assertions added to H that are not self-homologies, and the fact that only a subset of those assertions are relevant to the taxonomic groups compared here. It is not clear to what extent the results might be affected by homologies known in the literature that have yet to be curated by Bgee.

Our analysis focused on humans and two vertebrate model organisms for which abundant mutant phenotype data and a convenient set of anatomical homology statements are available. Given that the effect of homology seemed to be more pronounced in the zebrafish-human comparison than that of mouse, it would be of interest to examine species pairs with even more divergent body plans. Unfortunately, there are relatively few anatomical homology axioms linking vertebrates with model organisms outside the vertebrates, such as fruitflies and nematodes. Nonetheless, these results suggest that it would be worthwhile to explore the impact of "deeper" homology statements, either those sourced from the literature, or those derived computationally, such as by the phenolog approach.[26] In future work, we intend to explore the impact of homology reasoning on measurement of semantic similarity for phenotypes that vary naturally among vertebrate lineages, such as those in the Phenoscape Knowledgebase.[27] Independent of the use of homology axioms, some of the semantic similarity statistics that we examined showed relatively poor discrimination between orthologs and non-orthologs, suggesting the need to take a critical look at the biological accuracy of different phenotype semantic similarity measures.

References

1. P. N. Robinson and C. Webber, *PLoS Genetics* **10**, e1004268 (2014).
2. N. L. Washington, M. A. Haendel, C. J. Mungall, M. Ashburner, M. Westerfield and S. E. Lewis, *PLoS Biology* **7**, e1000247 (2009).
3. S. Kohler, S. C. Doelken, C. J. Mungall, S. Bauer, H. V. Firth, I. Bailleul-Forestier, G. C. M. Black, D. L. Brown, M. Brudno, J. Campbell, D. R. FitzPatrick, J. T. Eppig, A. P. Jackson, K. Freson, M. Girdea, I. Helbig, J. A. Hurst, J. Jahn, L. G. Jackson, A. M. Kelly, D. H. Ledbetter, S. Mansour, C. L. Martin, C. Moss, A. Mumford, W. H. Ouwehand, S.-M. Park, E. R. Riggs, R. H. Scott, S. Sisodiya, S. V. Vooren, R. J. Wapner, A. O. M. Wilkie, C. F. Wright, A. T. Vulto-van Silfhout, N. de Leeuw, B. B. A. de Vries, N. L. Washingthon, C. L. Smith, M. Westerfield, P. Schofield, B. J. Ruef, G. V. Gkoutos, M. Haendel, D. Smedley, S. E. Lewis and P. N. Robinson, *Nucleic Acids Research* **42**, D966 (2014).

4. P. N. Robinson, S. Köhler, A. Oellrich, K. Wang, C. J. Mungall, S. E. Lewis, N. Washington, S. Bauer, D. Seelow, P. Krawitz, C. Gilissen, M. Haendel and D. Smedley, *Genome Research* **24**, 340 (2014).
5. P. N. Schofield, J. P. Sundberg, R. Hoehndorf and G. V. Gkoutos, *Briefings in Functional Genomics* **10**, 258 (2011).
6. M. A. Haendel, J. P. Balhoff, F. B. Bastian, D. C. Blackburn, J. A. Blake, Y. Bradford, A. Comte, W. M. Dahdul, T. A. Dececchi, R. E. Druzinsky, T. F. Hayamizu, N. Ibrahim, S. E. Lewis, P. Mabee, A. Niknejad, M. Robinson-Rechavi, P. C. Sereno and C. J. Mungall, *Journal of Biomedical Semantics* **5**, 21 (2014).
7. C. J. Mungall, G. V. Gkoutos, C. L. Smith, M. A. Haendel, S. E. Lewis and M. Ashburner, *Genome Biology* **11**, R2 (2010).
8. R. L. Tatusov, E. V. Koonin and D. J. Lipman, *Science* **278**, 631 (1997).
9. E. K. Don, P. D. Currie and N. J. Cole, *Journal of Anatomy* **222**, 114 (2013).
10. B. K. Hall (ed.), *Homology: The hierarchical basis of comparative biology* (Academic Press, 1994).
11. C. J. Mungall, C. Torniai, G. V. Gkoutos, S. E. Lewis and M. A. Haendel, *Genome Biology* **13**, p. R5 (2012).
12. F. Bastian, G. Parmentier, J. Roux, S. Moretti, V. Laudet and M. Robinson-Rechavi, Bgee: integrating and comparing heterogeneous transcriptome data among species, in *Data Integration in the Life Sciences*, 2008.
13. P. D. Thomas, M. J. Campbell, A. Kejariwal, H. Mi, B. Karlak, R. Daverman, K. Diemer, A. Muruganujan and A. Narechania, *Genome Research* **13**, 2129 (2003).
14. H. Mi, B. Lazareva-Ulitsky, R. Loo, A. Kejariwal, J. Vandergriff, S. Rabkin, N. Guo, A. Muruganujan, O. Doremieux, M. Campbell, H. Kitano and P. D. Thomas, *Nucleic Acids Research* **33**, D284 (2005).
15. G. Parmentier, F. B. Bastian and M. Robinson-Rechavi, *Bioinformatics* **26**, 1766 (2010).
16. C. Pesquita, D. Faria, A. O. Falcao, P. Lord and F. M. Couto, *PLoS Computational Biology* **5**, e1000443 (2009).
17. P. Resnik, *Journal of Artificial Intelligence Research* **11**, 95 (1999).
18. M. Mistry and P. Pavlidis, *BMC Bioinformatics* **9**, 327 (2008).
19. S. Köhler, S. C. Doelken, B. J. Ruef, S. Bauer, N. Washington, M. Westerfield, G. Gkoutos, P. Schofield, D. Smedley, S. E. Lewis *et al.*, *F1000Research* **2**, 30 (2013).
20. C. L. Smith and J. T. Eppig, *Mammalian Genome* **23**, 653 (2012).
21. P. N. Robinson and S. Mundlos, *Clinical Genetics* **77**, 525 (2010).
22. G. V. Gkoutos, E. C. Green, A.-M. Mallon, J. M. Hancock and D. Davidson, *Genome Biology* **6**, R8 (2004).
23. G. V. Gkoutos, C. Mungall, S. Dolken, M. Ashburner, S. Lewis, J. Hancock, P. Schofield, S. Köhler and P. N. Robinson, Entity/quality-based logical definitions for the human skeletal phenome using PATO, in *Conference Proceedings of the IEEE Engineering in Medicine and Biology Society*, 2009.
24. M. C. Chibucos, C. J. Mungall, R. Balakrishnan, K. R. Christie, R. P. Huntley, O. White, J. A. Blake, S. E. Lewis and M. Giglio, *Database* **2014**, bau075 (2014).
25. E. F. Allin and J. A. Hopson, Evolution of the auditory system in synapsida (mammal-like reptiles and primitive mammals) as seen in the fossil record, in *The evolutionary biology of hearing*, (Springer, 1992) pp. 587–614.
26. J. O. Woods, U. M. Singh-Blom, J. M. Laurent, K. L. McGary and E. M. Marcotte, *BMC Bioinformatics* **14**, p. 203 (2013).
27. P. Manda, J. Balhoff, H. Lapp, P. Mabee and T. J. Vision, *Genesis* **53**, 561 (2015).

DISCOVERING PATIENT PHENOTYPES USING GENERALIZED LOW RANK MODELS

ALEJANDRO SCHULER
Center for Biomedical Informatics Research, Stanford University, 1265 Welch Road Stanford, CA, 94305. USA
Email: aschuler@stanford.edu

VINCENT LIU
Center for Biomedical Informatics Research, Stanford University, 1265 Welch Road Stanford, CA, 94305. USA
Email: vinliu@stanford.edu

JOE WAN
Computer Science, Stanford University, 353 Serra Mall Stanford, CA, 94305. USA
Email: joewan@stanford.edu

ALISON CALLAHAN
Center for Biomedical Informatics Research, Stanford University, 1265 Welch Road Stanford, CA, 94305. USA
Email: acallaha@stanford.edu

MADELEINE UDELL
Center for the Mathematics of Information, California Institute of Technology, Pasadena, CA 91125. USA
Email: udell@caltech.edu

DAVID E. STARK
Center for Biomedical Informatics Research, Stanford University, 1265 Welch Road Stanford, CA, 94305. USA
Email: dstark@stanford.edu

NIGAM H. SHAH
Center for Biomedical Informatics Research, Stanford University, 1265 Welch Road Stanford, CA, 94305. USA
Email: nigam@stanford.edu

The practice of medicine is predicated on discovering commonalities or distinguishing characteristics among patients to inform corresponding treatment. Given a patient grouping (hereafter referred to as a *phenotype*), clinicians can implement a treatment pathway accounting for the underlying cause of disease in that phenotype. Traditionally, phenotypes have been discovered by intuition, experience in practice, and advancements in basic science, but these approaches are often heuristic, labor intensive, and can take decades to produce actionable knowledge. Although our understanding of disease has progressed substantially in the past century, there are still important domains in which our phenotypes are murky, such as in behavioral health or in hospital settings. To accelerate phenotype discovery, researchers have used machine learning to find patterns in electronic health records, but have often been thwarted by missing data, sparsity, and data heterogeneity. In this study, we use a flexible framework called Generalized Low Rank Modeling (GLRM) to overcome these barriers and discover phenotypes in two sources of patient data. First, we analyze data from the 2010 Healthcare Cost and Utilization Project National Inpatient Sample (NIS), which contains upwards of 8 million hospitalization records consisting of administrative codes and demographic information. Second, we analyze a small (N=1746), local dataset documenting the clinical progression of autism spectrum disorder patients using granular features from the electronic health record, including text from physician notes. We demonstrate that low rank modeling successfully captures known and putative phenotypes in these vastly different datasets.

1. Introduction

1.1. *Learning phenotypes from the electronic health record*

With the advent and proliferation of electronic health records, *phenotyping* has become a popular mechanism with which to define patient groups based on shared characteristics- typically for conducting observational studies, defining quality metrics, or targeting clinical interventions. Current phenotyping methods vary: some rely on rules crafted from domain knowledge, others relying on statistical learning, and some employ hybrid approaches.[1,2] Regardless of the method, phenotyping has clear utility when the resulting groups are well defined, but may fail when the situation is unclear. Instead of presupposing phenotypes, recent work has leveraged advances in unsupervised learning to discover phenotypes from the data.[3,4]

A major barrier to applying machine learning approaches to phenotype discovery using health records data is that these data are often sparse, biased by non-random missingness, and heterogeneous.[3] An emerging framework, Generalized Low Rank Modeling (GLRM), offers a potential solution to address these limitations. Specific low rank models have already been successfully applied to various biomedical problems.[4,5,6] However, no prior study has considered low rank modeling as an overarching framework with which to perform phenotype discovery via models tailored to the qualities of the dataset at hand. Here, we demonstrate the use of this flexible framework to discover phenotypes in two datasets of different quality, granularity, and which represent diverse clinical situations.

1.2. *Standardizing hospital care using phenotype discovery has high impact*

Each year, Americans are admitted to hospitals over 37 million times, in aggregate spending more than 175 million days as inpatients.[7] In addition, hospitalizations cost the US economy $1 3 trillion dollars annually.[8] In light of this enormous impact, improvements in hospital care can yield dramatic results. For example, the Institute of Medicine estimated that up to 98,000 patients die each year from preventable medical errors.[9] Recent coordinated efforts to improve safety resulted in a staggering 1.3 million fewer patients harmed, 50,000 lives saved, and $12 billion in health spending avoided.[10] These efforts shared a simple premise: uncovering common phenotypes bridging diverse inpatient cohorts can drive substantial improvements in care and outcomes.[10] Given that phenotype discovery is such a critical step towards improving hospital care, existing methods for subgroup discovery are often slow and labor-intensive. For example, the codification of sepsis has taken decades[11], despite the fact that it contributes to as many as 1 out of every 2 hospital deaths[12] and is the single most expensive cause of US hospitalization.[13]

1.3. *Autism spectrum disorder phenotypes are poorly defined and badly needed*

Autism spectrum disorder (ASD) is a leading cause of mental illness in children, with an estimated 52 million cases globally.[14] In the United States, its prevalence has been estimated to be as high as 1 in 68, resulting in $11.5 billion in social costs[15,16]. ASD has eluded precise characterization of either its biological underpinnings or its clinical presentation, leading to substantial challenges in diagnosis and treatment, particularly in light of a wide range of heterogeneous phenotypes and comorbidities[17]. Although symptoms of the disorder are commonly present by age 18 months, ASD is typically not diagnosed until age 4 or later, after significant irreversible impairments in learning and neurodevelopment have already occurred[15]. Even after diagnosis, the progression of ASD is different across individuals, which has led to efforts to define subgroups that are at differential risk of comorbidities.[18] A systematic and data-driven approach for phenotype discovery can precisely characterize this heterogeneous disorder and its progression over time.

2. Methods

We analyze two datasets of different sizes, feature granularity, data-types, domains, and timelines. Instead of taking a one-size-fits-all approach, we create a tailored low rank model within the generalized low rank model framework to account for the specific qualities of each dataset and then fit the model to discover hidden phenotypes.

2.1. *Generalized low rank models*

The idea behind low rank models is to represent high-dimensional data in a transformed lower-dimensional space. Generalized low rank models[19] begin with a matrix or data table A that is populated with n samples or observations (rows) of m different features (columns; Figure 1). These features may take values from different sets (e.g. some may be real numbers, others true/false, enumerated categories, etc.) and each observation may have missing values for some features. The number of features in the dataset is referred to as its dimensionality.

We approximate A by XY, where $X \in \Re^{n\times k}$ and $Y \in \Re^{k\times m}$ (Figure 1). We interpret the rows of this "tall and skinny" X as observations from A represented in terms of the k new latent features. We interpret each row of the "short and wide" Y as a representation of one of the k latent features in terms of the m original features. In a sense, Y encodes a transformation from the original features into the latent features.

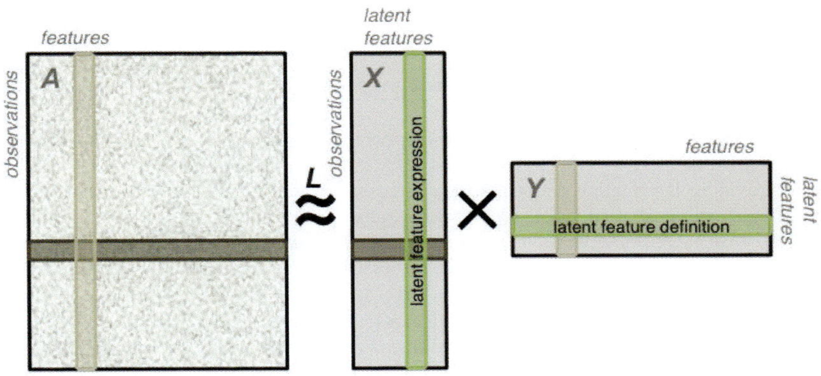

FIGURE 1: A data matrix is approximated as the product of two matrices. By construction, the resulting approximation is of lower algebraic rank. The data matrix A may contain features of different data-types and missing entries, as illustrated here. Each row of X is an encoding of an observation in A in the latent feature space. Each column of Y is an encoding of a feature of A in the latent feature space.

To find X and Y, we pose the following optimization problem:

$$\min_{X,Y} \sum_{i,j\in\Omega} l_{ij}(A_{ij}, (XY)_{ij}) + r_X(X) + r_Y(Y) \tag{1}$$

This expression consists of two parts: a loss function and regularizers. The loss

$$L = \sum_{i,j\in\Omega} l_{ij}(A_{ij}, (XY)_{ij}) \tag{2}$$

is a measure of the accuracy of our approximation of the data. Different losses may be more or less appropriate for different types of data (to reflect different noise models), so we allow the loss to be decomposed over the different elements of the dataset to account for heterogeneity in the types of features present. In addition, we only calculate the loss over the set Ω, which represents the non-missing entries in our dataset. This strategy allows us to 'borrow' statistical power from partially-filled or incomplete observations where other methods would discard the entire observation. The regularizers r_x and r_y constrain or penalize the latent feature representation. Using appropriate regularization can prevent overfitting and improve model interpretability.

To impute missing or hidden values, we solve: $\hat{A}_{ij} = arg\min_{a\in\alpha} l_{ij}(a, (XY)_{ij})$, where α represents the set of possible values that a can take (e.g. if a is a boolean feature, $\alpha = \{1, -1\}$).

Particular choices of losses and regularizations result in many well known models. For instance, using $L(A,XY) = \|A - XY\|_2^2$ and no regularization is mathematically equivalent to principal components analysis (PCA). A well-written and detailed description of GLRM and the kinds of models that can be created using this framework can be found in the seminal work by Udell et. al.[19]

2.2. *Hospitalization dataset and phenotype discovery model*

We used data from the 2010 National Inpatient Sample, the largest all-payer nationally representative dataset of US hospitalizations.[20] Each hospitalization record includes a variety of fields providing information about patient diagnoses (up to 25 different ICD-9-CM codes) and procedures (up to 15 ICD-9-CM procedure codes), as well as demographics, admission/discharge/transfer events, and comorbidities (a set of 30 AHRQ comorbidity measures, e.g. AIDS). For efficiency, we processed the dataset to consolidate the ~18,000 ICD-9-CM diagnosis and procedure codes into a total of 516 Clinical Classification Software (CCS) codes.[21] Additionally, we used 44 variables regarding patient demographics, admission circumstances, hospitalization outcome, and patient comorbidity. We expanded all categorical variables into sets of boolean dummy variables (one for each possible value) to yield a total of 557 boolean, continuous, and ordinal features. We focused specifically on adult hospitalizations (age $\geq$ 18 years) as the causes, demographics, and outcomes of pediatric hospitalizations differ substantially. To speed computation, we selected a random subsample of 100,000 hospitalizations to fit our models to.

Hospitalization records contain a diversity of data-types. We measured the accuracy of the approximation for different data elements by data-type appropriate loss functions, e.g. quadratic loss for real-valued variables such as age, hinge losses for boolean variables such as presence or absence of procedures. Real, categorical, ordinal, and boolean, and periodic data-types are familiar to most researchers, and appropriate losses for these kinds of variables are known in the machine learning and optimization communities.[19]

We defined an *epistemic boolean* variable as a boolean variable where we have a lopsided confidence about whether a true value actually indicates truth or a false value actually indicates falsehood. For example, consider diagnoses: if a clinician codes a patient for a diagnosis, it is highly likely that that patient experienced the condition that the code represents -- in other words, we are confident that "True" means true. On the other hand, if a patient did not receive a particular diagnosis, that variable would simply be missing in that patient's hospitalization record. In reality, we are less sure that the patient did not experience that condition because it may have escaped diagnosis, remained unrecognized, or simply gone uncoded. We developed a loss function to account for lopsided epistemic uncertainty of this sort. Correct predictions are not penalized regardless. Our loss function for epistemic booleans is a generalization of the boolean hinge loss and is defined as follows:

$$l(a,u) = (w_F 1_{\{-1\}}(a) + w_T 1_{\{1\}}(a)) * \max(1 + au,\ 0) \qquad (3)$$

where $1_A(x)$ is an indicator function for $x \in A$. When $w_T > w_F$, this loss function penalizes false negatives more than false positives, reflecting our greater certainty about observations labeled as "True" compared to those labeled as "False".

In light of the divergent scales and domains of the features, all loss functions and regularizers were adjusted for scaling and offsets.[19]

2.3. *Autism spectrum disorder dataset and phenotype discovery model*

We used data from the Stanford Translational Research Integrated Database Environment (STRIDE), a de-identified patient dataset that spans 18 years and more than 1.2 million patients who visited Stanford Hospital & Clinics. From all patients in STRIDE, we identified 1746 patients with at least 2 autism spectrum disorder (ASD) related visits (visits assigned a 299.* ICD9 code). For these patients, we analyzed billing data from all visits (ICD9 and CPT codes), prescribed drugs, as well as mentions of clinical concepts in their medical notes found using our previously described text annotation pipeline.[22] We restricted our analysis to data recorded when these 1746 patients were at most 15 years old because we are interested in modeling ASD phenotypes in children and adolescents. We generated a feature vector for each patient by calculating the frequency of occurrence of each visit-associated ICD9 code, prescribed drug, and medical concept mentioned in any note of that patient, binned by 6 month intervals (Figure 2). To capture the nature of this data, we used a Poisson loss over all elements in the dataset. This low rank model specification is mathematically equivalent to Poisson PCA.[23]

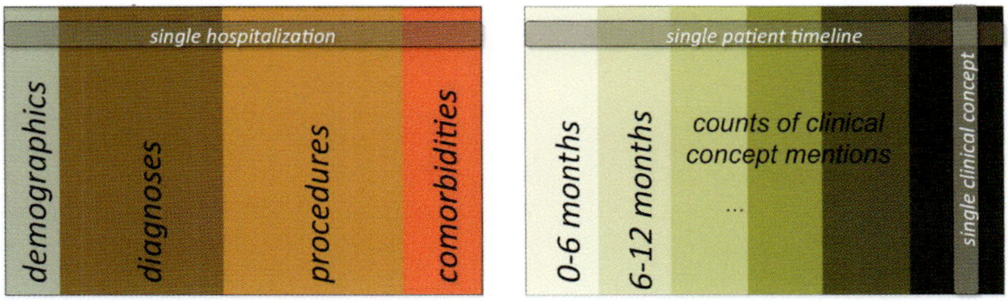

FIGURE 2: Illustration of the hospitalization (left) and ASD (right) datasets. For each ASD patient, we created a vector from the frequency of occurrence of each concept (C-1, C-2...) mentioned in their medical notes, ICD9 codes associated with a visit (ICD9-1, ICD9-2...) and medications prescribed (DRUG-1, DRUG-2...) within each 6 month period of their medical history captured in our database.

2.4. *GLRM implementation*

To fit our models, we used the Julia package *LowRankModels*[24], which implements the algorithm described by Udell et. al.[19] This software employs a general purpose, fast, and effective procedure called alternating proximal gradient descent to solve a broad class of optimization problems. Although model-specific solvers (i.e. algorithms that take advantage of the structure of a particular GLRM) could be faster, this general-purpose software allowed us to rapidly iterate through model design decisions and test choices of parameters and robustness.

The Julia *LowRankModels* package is still under active development. We dedicated substantial effort to learning and clarifying the code, contributing bug fixes, adding needed features, and optimizing performance. Our contributions will accelerate our future work and the work of other researchers using low rank models.

3. Results

3.1. *Tailored low rank models outperform PCA*

As an intrinsic evaluation, we benchmarked our tailored models against naive low rank models (PCA) of equal rank by artificially hiding a portion of the elements in the dataset and judging each model's ability to correctly impute the missing values. This procedure was repeated in a 5-fold cross validation for each model (Figure 3). In both datasets, the tailored models outperformed PCA in terms of the imputation error for held out values. Imputation errors are evaluated using a *merit function* specific to the data, not the model. While minimizing the merit function is the ultimate goal, models are fit using loss functions because the merit function is generally nonsmooth and nonconvex.

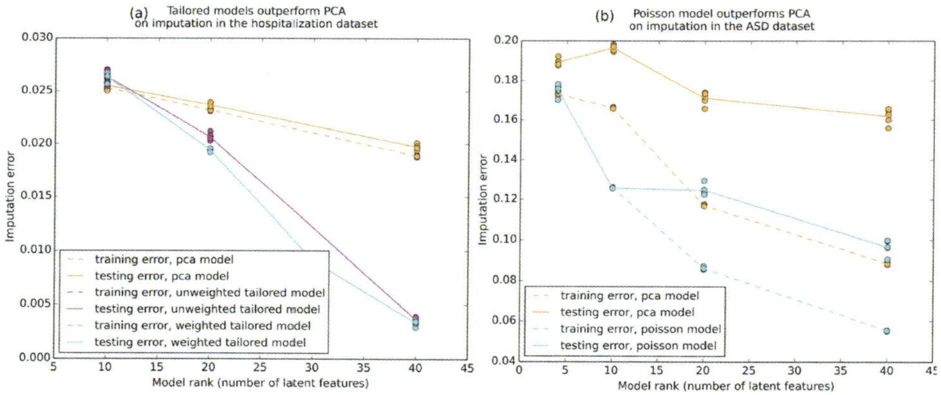

FIGURE 3: Training and testing imputation error in 5-fold cross validation of each model across a range ranks. The tailored models perform better than their naive counterpart (PCA). Imputation error is mean-normalized within each feature and by the number of data entries tested over.

3.2. *Low rank models discover hospitalization phenotypes*

We began our analysis of our hospitalization model by inspecting the latent features. Recall that each latent feature in a low rank model is a row vector in the computed matrix Y. Each entry in this vector corresponds to the influence of an original feature within this latent feature. We examined the representation of the original features in terms of the latent features by clustering the latent feature representations of the original features (the columns of the matrix Y). Hierarchical clustering clearly reproduced known associations between diagnoses, procedures, comorbidities, and demographics (not shown).

Sample Clusters

	1	2	3	4	5	6	7	8	
	n = 759 (9.2%)	*n* = 527 (6.4%)	*n* = 2,807 (33.9%)	*n* = 629 (7.6%)	*n* = 302 (3.6%)	*n* = 2,311 (27.9%)	*n* = 177 (2.1%)	*n* = 779 (9.4%)	
Phenotype	**ORTHO**	**PSYCH**	**INFECT**	**EASY OB**	**HARD OB**	**SEVERE**	**AMI**	**COMPLEX**	
									p-value
Age	63 (17)	45 (13)	62 (19)	27 (6)	30 (6)	61 (18)	63 (12)	71 (16)	<0.001
Female	58.4%	44.4%	56.7%	100%	99.3%	53.6%	31.6%	48.3%	<0.001
LOS	3.6 (3.1)	4.2 (4.2)	4.1 (4.6)	2.3 (1.4)	3.5 (3.0)	6.9 (9.0)	2.9 (2.7)	5.4 (4.4)	<0.001
Mortality	0%	0.2%	0.2%	0%	0%	8.1%	1.7%	0.5%	<0.001
Cost	48.8 (50.4)	16.0 (18.0)	23.4 (23.9)	10.5 (6.6)	18.8 (19.3)	55.2 (81.7)	65.6 (41.9)	36.0 (38.3)	<0.001
Principal Diagnosis									<0.001
#1	OA (29.4%)	Mood d/o (17.8%)	PNA (5.8%)	OB traum. (26.6%)	Prior Csxn (23.3%)	Sepsis (5.2%)	AMI (48.6%)	CHF (11.7%)	
#2	Back pain (16.2%)	EtOH d/o (14.2%)	COPD (4.6%)	Preg. Comp (12.6%)	Birth Comp (13.9%)	Biliary dz (3.7%)	CAD (42.4%)	AKI (5.5%)	
#3	LL Fxr (7.1%)	Substance (8.7%)	Sepsis (3.7%)	Birth comp (12.2%)	Breach (10.9%)	Rehab (3.4%)	Dev. Compl (4.5%)	HTN compl (5.3%)	
#4	Hip Fxr (6.2%)	Schizo. (8.0%)	Angina (3.7%)	Prol. Preg. (10.7%)	Fetal distr. (8.9%)	Complic. (2.7%)	Conduction (0.6%)	Dev compl (4.6%)	
#5	Dev Comp (5.4%)	Pancreatic (4.6%)	Dysrhythm (3.6%)	Nml. Preg. (8.0%)	Preg HTN (7.6%)	CVA (2.6%)	Dysrhytm (0.6%)	Sepsis (4.5%)	
Principal procedure									<0.001
#1	Knee arth. (22.4%)	None (64.7%)	None (64.9%)	Deliv. Assist (53.1%)	C-sxn (88.4%)	None (19.0%)	PTCA (91.0%)	None (40.8%)	
#2	Spinal fus (12.5%)	Etoh detox (16.7%)	Card. Cath (3.5%)	OB lac rep (26.2%)	Deliv Assist (4.3%)	EGD (5.4%)	Other heart (3.4%)	Dialysis (11.4%)	
#3	Hip replac (11.9%)	Ventilation (3.4%)	Other cath (2.4%)	C-sxn (6.2%)	Tubes tie (2.3%)	Ventilation (4.4%)	Card cath (1.7%)	Blood Tx (6.0%)	

Ortho: orthopedics; Psych: psychiatric; Infect: infection; Easy OB: uncomplicated obstectrics; Hard OB: complex obstetrics; Severe: severe medical illness; AMI: acute myocardial infarction; Complex: complex medical illness; LOS: length of stay; OA: osteoarthritis; LLL lower extremity; Fxr: fracture; Dev Comp: device complication; D/O: disorder; EtOH: alcohol; Schizo: schizophrenia; PNA: pneumonia; COPD: chronic obstructive pulmonary disease; OB: obstetrical; Preg: pregnancy; Comp: complication; Csxn: Ceasarean section; Fetal distr: detal distress; HTN: hypertension; Dz: disease; Complic: complication; CVA: cerebrovascular disease; CAD: coronary artery disease; CHF: congestive heart failure; AKI: acute kidney injury; Arth: arthroscopy; Spinal fus: spinal fusion; Hip replac: hip replacement; Detox: detoxification; Card cath: cardiac catheterization; OB lac rep: laceration repair; EGD: esophagoduodenoscopy; PTCA: percutaneous coronary angioplasty; Blood tx: blood transfusion

TABLE 1: Hospitalization phenotypes closely mirror common reasons for hospitalization.

To discover phenotypes, we clustered the low rank representation of our subsample of the NIS dataset (the matrix X). We chose a hierarchical cluster cutpoint for eight clusters of hospitalizations and compared cluster characteristics (Table 1) in terms of the original feature space. The eight clusters had widely divergent baseline characteristics and could be well defined within recognizable hospital phenotypes. For example, patients in clusters 4 and 5 were nearly all young females who were hospitalized for pregnancy and childbirth. They differed in that patients in cluster 5 had a slightly longer length of stay, likely associated with the marked difference in the need for C-sections (6.2% for cluster 4 vs 88.4% for cluster 5). Cluster 1 appeared to contain patients hospitalized for orthopedic procedures, while cluster 2 largely included patients with psychiatric or substance abuse hospitalization -- the most common procedure was alcohol detoxification (16.7%). Cluster 7 was nearly exclusively patients undergoing procedures for acute myocardial infarction with a 91.0% rate of cardiac percutaneous transluminal coronary angioplasty. Clusters 6 and 8 included medically complex patients with cluster 6 having a high mortality (8.1%) with a younger mean age of 61 years.

3.3. *Phenotypes discovered from a low rank model of ASD progression*

To discover ASD phenotypes we first examined the composition of the latent features in our models. Regardless of the parameterization or the rank, the primary effect that we observed in our latent feature vectors was differential enrichment for original features coming from different 6-month time-bins (Figure 4). For instance, the second latent feature shown has relatively high weights on original features corresponding to medical concepts observed in patients during the second time bin, while the first has relatively high weights on original features corresponding to clinical concepts observed in patients during the fourth time bin. This "useage timeline" effect was evident in all models we considered, regardless of rank.

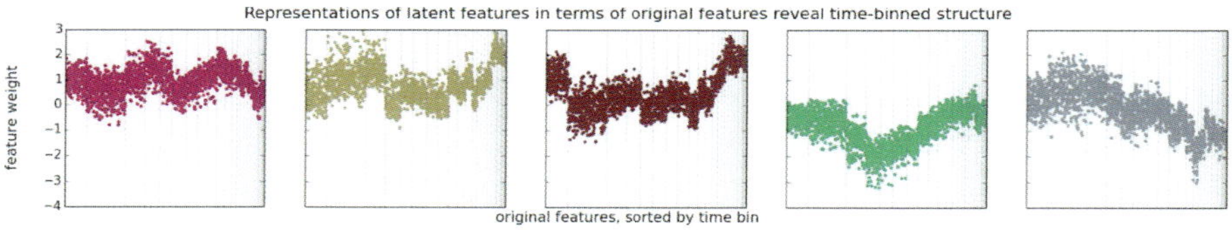

FIGURE 4: Each panel represents one latent feature vector in the matrix Y. Vertical gray lines are manually overlaid boundaries between time bins. Time bins are ordered temporally from left to right. The weights of the original features in each latent feature representation are predominantly associated according to the time bin in which each original feature was recorded, and not by clinical similarity between the original features.

To discover phenotypes, we clustered the low rank representations of our ASD dataset (the matrix X). Using k-means clustering, we derived cluster centroids (phenotypes) in terms of their latent feature representations (each phenotype is a vector in $\Re^k$). To inspect these in terms of our original features, we multiplied each phenotype vector by the matrix Y . The derived phenotypes were not differentially enriched for specific clinical concepts. Instead, these "temporal phenotypes" grouped patients by the timings of their interactions with the healthcare system.

4. Discussion

4.1. *Hospital phenotypes suggest streamlining or compartmentalizing hospital organization*

Our analysis of a nationally-representative hospitalization administrative dataset revealed that low rank modeling could identify clinically distinguishable hospital phenotypes. These phenotypes are immediately familiar to clinicians and hospital administrators with each cluster representing recognizable 'wards' or 'service lines' provided by hospitals. For example, it distinguished patients primarily admitted for orthopedic surgeries from those admitted for substance abuse or psychiatric diagnoses, essentially rediscovering hospitals' 'orthopedics' and 'psychiatric' wards. Our approach also identified sub-phenotypes within larger classes. For example, hospitalizations for childbirth are the most common reason for US inpatient stays, and our results revealed two subtypes within the obstetric population differentiated by their need for procedural intervention. Our current results establish the validity of using the low rank modeling approach for identifying known hospital phenotypes with the hope that extending this approach will yield the discovery of new phenotypes for which streamlined care pathways can be implemented.

4.2. *Time-binning masks phenotype signals in ASD dataset*

In our ASD model, we saw that the discovered phenotypes were not differentially enriched for specific clinical concepts. However, the phenotypes were not the product of random noise--they succeeded in capturing the primary source of variation in the data, which was temporality. Analysis of the latent features revealed that mentions of different clinical concepts within a time bin are more associated with each other than mentions of the same clinical concept with itself in another time bin. The model remarkably learned these associations without any *a priori* knowledge that the features represented time-binned counts. The model successfully detected a clear structure in the data, although that structure reflects an artifact of featurization and the clinical challenges associated with early diagnosis in ASD. There may be clinically relevant phenotypes present in the data, but our analysis shows that this signal is masked by time-binning. Our result is emblematic of what lies at the crux of low-rank models: the algorithm will discover the clearest and most robust signals, whether or not these signals are meaningful to the user's research interest or insight. Thus low rank models should be used to understand the profile of the dataset in order to inform future data collection or featurization. In our case, our result suggests that we should employ a different featurization method in future studies or that we should incorporate time explicitly, perhaps using tensor factorization or a convolutional approach.

4.3 *Summary*

In this study, we applied a novel and flexible machine learning method -- generalized low rank modeling -- to two very different datasets. Instead of forcing the same model onto different datasets or creating specific methods with little hope of reuse, we built two unique models united by one overarching framework and software package. Furthermore, we demonstrated different approaches to analyzing low rank models and used these techniques to discover phenotypes present in the data.

Accelerating the process of phenotype discovery has high potential to improve care and outcomes for patients, but additional work in the validation and care standardization of such phenotypes is still required. Nonetheless, using such a high-throughput approach for finding patient subgroups could dramatically shorten the time necessary to make new discoveries, especially when applied to massive datasets documenting poorly understood phenomena.

References

1. Pathak J, Kho AN, Denny JC. Electronic health records-driven phenotyping: challenges, recent advances, and perspectives. *Journal of the American Medical Informatics Association : JAMIA*. 2013;20(e2):e206-e211. doi:10.1136/amiajnl-2013-002428.
2. Shivade C, Raghavan P, Fosler-Lussier E, et al. A review of approaches to identifying patient phenotype cohorts using electronic health records. *Journal of the American Medical Informatics Association : JAMIA*. 2014;21(2):221-230. doi:10.1136/amiajnl-2013-001935.
3. Lasko TA, Denny JC, Levy MA (2013) Computational Phenotype Discovery Using Unsupervised Feature Learning over Noisy, Sparse, and Irregular Clinical Data. PLoS ONE 8(6): e66341. doi:10.1371/journal.pone.0066341
4. Ho, Joyce C., et al. "Limestone: High-throughput candidate phenotype generation via tensor factorization." *Journal of biomedical informatics* 52 (2014): 199-211.
5. Zhou, Jiayu, et al. "From micro to macro: Data driven phenotyping by densification of longitudinal electronic medical records." Proceedings of the 20th ACM SIGKDD international conference on Knowledge discovery and data mining. ACM, 2014.
6. Devarajan K (2008) Nonnegative Matrix Factorization: An Analytical and Interpretive Tool in Computational Biology. PLoS Comput Biol 4(7): e1000029. doi:10.1371/journal.pcbi.1000029
7. Weiss AJ (Truven Health Analytics), Barrett ML (M.L. Barrett, Inc.), Steiner CA (AHRQ). Trends and Projections in Inpatient Hospital Costs and Utilization, 2003–2013. HCUP Statistical Brief #175. July 2014. Agency for Healthcare Research and Quality, Rockville, MD.http://www.hcup-us.ahrq.gov/reports/statbriefs/sb175-Hospital-Cost-Utilization-Projections-2013.pdf.
8. Gonzalez JM. National health care expenses in the U.S. civilian noninstitutionalized population, 2011. MEPS Statistical Brief #425. November 2013. Agency for Healthcare Research and Quality, Rockville, MD. http://meps.ahrq.gov/data_files/publications/st425/stat425.pdf. Accessed March 28, 2014.
9. Kohn LT, Corrigan JM, Donaldson MS (editors). To Err is Human: Building a Safer Health System. National Academy Press; Washington DC. 2000.
10. Efforts To Improve Patient Safety Result in 1.3 Million Fewer Patient Harms: Interim Update on 2013 Annual Hospital-Acquired Condition Rate and Estimates of Cost Savings and Deaths Averted From 2010 to 2013. December 2014. Agency for Healthcare Research and Quality, Rockville, MD. http://www.ahrq.gov/professionals/quality-patient-safety/pfp/interimhacrate2013.html
11. Balk RA, Bone RC. The septic syndrome. Definition and clinical implications. Crit Care Clin. 1989; 5(1): 1-8.
12. Liu V, Escobar GJ, Greene JD, et al. Hospital deaths in patients with sepsis from 2 independent cohorts. JAMA. 2014; 212(1): 90-2.
13. Sutton J (Social & Scientific Systems, Inc.), Friedman B (AHRQ). Trends in Septicemia Hospitalizations and Readmissions in Selected HCUP States, 2005 and 2010. HCUP Statistical Brief #161. September 2013. Agency for Healthcare Research and Quality, Rockville, MD. http://www.hcup-us.ahrq.gov/reports/statbriefs/sb161.pdf.
14. Baxter AJ, Brugha TS, Erskine HE, Scheurer RW, Vos T, Scott JG. The epidemiology and global burden of autism spectrum disorders. Psychological medicine. 2014:1-13.
15. Prevalence of autism spectrum disorder among children aged 8 years - autism and developmental disabilities monitoring network, 11 sites, United States, 2010. Morbidity and mortality weekly report. Surveillance summaries (Washington, D.C. : 2002). 2014;63(2):1-21.

16. Lavelle TA, Weinstein MC, Newhouse JP, Munir K, Kuhlthau KA, Prosser LA. Economic burden of childhood autism spectrum disorders. Pediatrics. 2014;133(3):e520-529.
17. Jeste SS, Geschwind DH. Disentangling the heterogeneity of autism spectrum disorder through genetic findings. Nat Rev Neurol. 2014;10(2):74-81.
18. Doshi-Velez, Finale, Yaorong Ge, and Isaac Kohane. "Comorbidity Clusters in Autism Spectrum Disorders: An Electronic Health Record Time-Series Analysis." *Pediatrics* 133.1 (2014): e54–e63. *PMC*. Web. 22 July 2015.
19. Udell, Madeleine, et al. "Generalized Low Rank Models." *arXiv preprint arXiv:1410.0342* (2014).
20. Overview of the National (Nationwide) Inpatient Sample (NIS). HCUP Databases. Healthcare Cost and Utilization Project (HCUP). November 2014. Agency for Healthcare Research and Quality, Rockville, MD. www.hcup-us.ahrq.gov/nisoverview.jsp.
21. HCUP CCS. Healthcare Cost and Utilization Project (HCUP). June 2015. Agency for Healthcare Research and Quality, Rockville, MD. www.hcup-us.ahrq.gov/toolssoftware/ccs/ccs.jsp.
22. LePendu, P et al. "Pharmacovigilance Using Clinical Notes." *Clinical pharmacology and therapeutics* 93.6 (2013): 10.1038/clpt.2013.47. *PMC*. Web. 22 July 2015.
23. Collins, Michael, Sanjoy Dasgupta, and Robert E. Schapire. "A generalization of principal components analysis to the exponential family." *Advances in neural information processing systems*. 2001.
24. Udell, Madeleine, et al. "LowRankModels.jl: a julia package for modeling and fitting generalized low rank models" https://github.com/madeleineudell/LowRankModels.jl

INFERENCE OF PERSONALIZED DRUG TARGETS VIA NETWORK PROPAGATION

ORTAL SHNAPS[♦]

School of Computer Science, Tel Aviv University
Tel Aviv 69978, Israel
Email: ortalits@gmail.com

EYAL PERRY[♦]

School of Computer Science, Tel Aviv University
Tel Aviv 69978, Israel
Email: eyal.perry88@gmail.com

DANA SILVERBUSH

School of Computer Science, Tel Aviv University
Tel Aviv 69978, Israel
Email: dana.silverbush@gmail.com

RODED SHARAN*

School of Computer Science, Tel Aviv University
Tel Aviv 69978, Israel
Email: roded@post.tau.ac.il

We present a computational strategy to simulate drug treatment in a personalized setting. The method is based on integrating patient mutation and differential expression data with a protein-protein interaction network. We test the impact of in-silico deletions of different proteins on the flow of information in the network and use the results to infer potential drug targets. We apply our method to AML data from TCGA and validate the predicted drug targets using known targets. To benchmark our patient-specific approach, we compare the personalized setting predictions to those of the conventional setting. Our predicted drug targets are highly enriched with known targets from DrugBank and COSMIC ($p < 10^{-5}$), outperforming the non-personalized predictions. Finally, we focus on the largest AML patient subgroup (~30%) which is characterized by an FLT3 mutation, and utilize our prediction score to rank patient sensitivity to inhibition of each predicted target, reproducing previous findings of in-vitro experiments.

[♦] These authors contributed equally to this work.
* To whom correspondence should be addressed.

1. Introduction

Precision medicine, an approach in which medical treatment is tailored for a specific group of patients, is an arising paradigm in medical research and practice. Indeed, it is well known that some drugs affect only a specific subgroup of patients, while even harming other patients suffering from the same disease [1-2]. In recent years, computational tools have emerged to stratify diseases into informative subtypes [3] and to predict drug sensitivity per subtype in order to optimally match patients with existing medical treatments [4].

In spite of these advances, the development of new treatments in the context of precision medicine is still scarce. Consequently, there is an increasing interest in computational prediction of drug targets. Previous works [6-9] used similarity among diseases to employ drugs designed for one disease to medicate another, as well as to prioritize new compounds as potential drugs. Lamb et al. [7] created a database containing ranked drug response gene expression profiles, allowing to query the database with a disease-specific genetic signature to identify drug response profiles that correlate with it. GBA [9] predicts novel associations between drugs and diseases by assuming that if two diseases are treated by the same drug, alternative drugs treating only one of them might treat also the other. Finally, Gottlieb et al. [6] predict novel associations between drugs and diseases by utilizing multiple drug–drug and disease–disease similarity measures for the prediction task. Some of the methods, such as [6-7] could be extended for personalized prediction of drugs, yet to this date efforts for personalized design of drugs had focused on experimental work [10] or small scale networks tailored for specific condition [11-12].

As drugs often act by inhibiting their targets, attempts were also made to predict candidates for drug targets by predicting the effect of gene knockouts. These attempts focused on metabolic drugs and used metabolic network models, testing the impact of in-silico deletion of genes on the network's fluxes. For example, Fatumo et.al. [13] simulated knockouts by deleting reactions from a metabolic network to identify enzymes essential for the malaria parasite Plasmodium falciparum. Papp et al. [14] used a metabolic flux model to predict the knockout fitness effect of nonessential genes in Saccharomyces cerevisiae. In their review of current paradigms for predicting inhibitory effects, Csermely et al. [15] conclude with the need for approaches allowing the examination of multi-targets inhibition, as our new approach allows.

In this work we present a novel approach to tackle the drug target inference problem from a personalized perspective using in-silico knockouts based on propagation methods in a protein-protein interaction (PPI) network. Figure 1 provides an overview of the method: we start from a general PPI network and personal disease-related data. We rely on the framework described by Vanunu et al. [16] to prioritize casual genes by network propagation. We perform multiple network propagations in order to simulate the current patient state, the patient state after gene knockouts (by removing the corresponding nodes from the network) and an estimated "healthy" state. We use these different states in order to rank the gene knockouts and retrieve a list of candidate drug targets.

The framework we present is general and could potentially be applied to any personalized disease-related data, with cancer being a pronounced candidate for application. Cancer is a wildly heterogeneous disease, in which a group of patient phenotypically categorized into the same cancer type (or even subtype) may have only little overlap in the underlying genotype. This is especially true in acute myeloid leukemia (AML), which has striking heterogeneity in gene mutations and expression aberrations across samples [17-19]. We therefore evaluate our performance by applying it to patients suffering from AML, based upon data generated by the TCGA Research Network: http://cancergenome.nih.gov/ of mutated and differentially expressed genes [19].

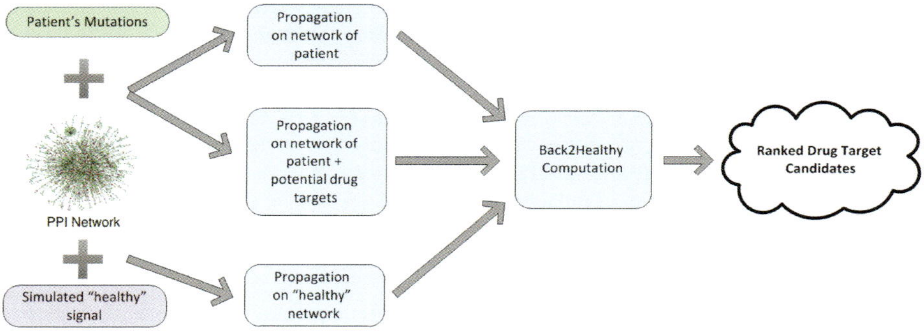

Fig. 1. An overview of the algorithmic pipeline.

2. Results

We present a novel approach to tackle the drug target inference problem from a personalized perspective using in-silico knockouts in a PPI network. As described in Figure 1, we start from a general PPI network and individual-specific disease-related data. We perform multiple network propagations in order to simulate the current patient state, the patient state after gene knockouts (by removing the gene's node from the network) and an estimated "healthy" state (see Methods). We use these different states in order to rank the gene knockouts and retrieve a list of potential drug targets.

To evaluate our performance we applied our method to TCGA gene expression and mutation data of patients suffering from acute myeloid leukemia (AML, see Methods for dataset description). First, we show that we can identify AML causal genes by synthesizing the individual propagations. Second, we show that by integrating results from a personalized knockout process we can infer potential drug targets and rank their efficacy in a patient or a subgroup of patients.

Our algorithm relies on network propagations to rank the relevance of different genes to a prior set. In order to set its parameters, we first tested the algorithm's performance in retrieving known causal genes for AML. The algorithm has two parameters (see Methods): α, determining the relative weight of the prior knowledge vs. the network in the scoring; and P, the prior set, according to which the propagation is carried out. We executed the algorithm using different settings for these parameters. To evaluate the results, we used three sets of known AML causal genes from KEGG and COSMIC, varying in confidence and size (see Methods). The application of the method to each patient resulted in a propagation score for each gene (excluding the prior set, to focus on novel discoveries). We aggregated the rank of each gene over all patients to yield a gene-based score, retaining the top 10% *affected* genes in the network. We then computed the hypergeometric enrichment of this set of genes with the different sets of known causal genes. All choices of α resulted in significant and similar p-values ($p < 10^{-5}$), which shows that the results are robust to the choice of α, as shown in Figure 2A. We use $\alpha = 0.9$ in the sequel. For the prior set, we tried four settings, defining P based on (i) mutated genes; (ii) differentially expressed genes; (iii) both, but running them separately and averaging the propagation scores obtained; and (iv) same as (iii) but taking the maximum scores rather than averages. Note that all types of mutations within coding genes were considered (missense, nonsense and silent). All prior knowledge variants resulted in significant p-values ($p < 10^{-5}$). The best variant was the first – setting P to be the set of mutated genes in each patient (Figure 2B), a choice which we use in the sequel. The mutated genes all belong to AML patients, but they are not limited only to AML-related genes.

The causal genes are thought to trigger malignant behavior by perturbing signaling pathways that regulate three core cellular processes: cell fate, cell survival, and genome maintenance [23]. In AML, cell survival and proliferation are enhanced through an aberrant signal pathway [24] represented in the KEGG database [21]. We computed the hypergeometric enrichment of the top 10% affected genes within the AML KEGG pathway (ID: hsa05221) and found that the affected genes comprise 15 out of 21 pathway components with a significant p-value ($p < 10^{-11}$), exceeding that achieved by using common mutated genes ($p < 10^{-7}$, mutations appearing in at least two patients) and capturing its downstream effect (Figure 3). It is interesting to note that although FLT3 is mutated in approximately 30% of the patients, it is not included in the top 10% affected genes after aggregation, underscoring the importance of a personalized approach.

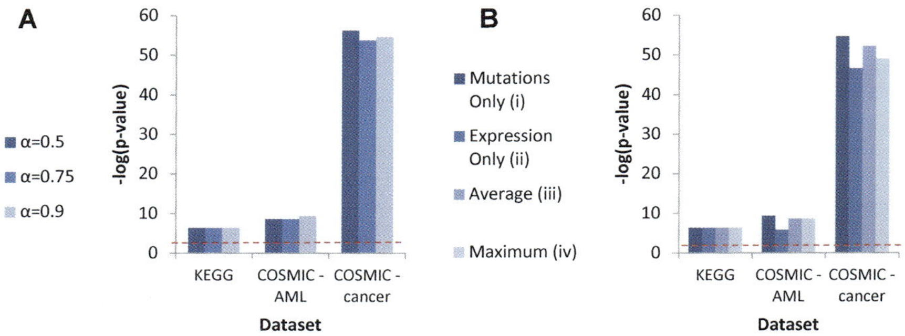

Fig. 2. Performance evaluation under different parameter (A) and prior knowledge set (B) choices. The red line denotes a p-value of 0.01.

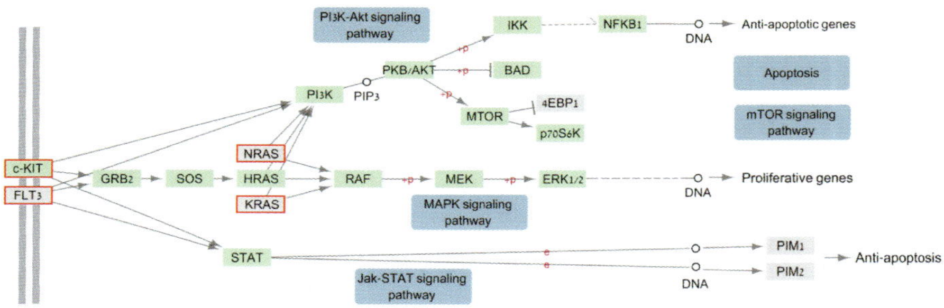

Fig. 3. The AML KEGG pathway, with top 10% affected genes (as predicted by our method) highlighted in green and commonly mutated genes framed in a red box.

The previous results imply that our propagation based scores are able to infer disease-related genes and agree with observations made by Rufallo et al. [25]. We hypothesized that good drug targets for the disease could be genes whose knockout is predicted to reverse the disease-related effects [7]. To identify such genes in-silico, we rerun the propagation based scoring while removing each gene in turn from the network, assessing the similarity between the obtained scores and those that characterize a "healthy" state. To this end, we use a Back2Healthy distance score (B2H; See Methods), taking the top scoring genes as our candidates for potential personalized drug targets. As above, we focus on non-trivial targets by excluding the patient's mutated genes from our ranking.

The process above infers drug targets for each patient individually. As information about personalized drug targets is very scarce and hard to validate, we aggregated the results over all patients, evaluating the results using known AML drug targets derived from the DrugBank database [26-29] and COSMIC [20]. The top 10% scoring genes were highly enriched with known drug targets from both sources (Figure 4A, DrugBank: $p < 10^{-5}$, COSMIC: $p < 10^{-10}$). In comparison, a naïve approach that focuses on common mutations (appearing in at least two patients), yields a set of candidate targets containing only one of the known targets ($p = 0.18$). To assess the personalized approach we took, we generated a "consensus patient", using common (appearing in at least five patients) mutated and differentially expressed genes, and applied our method to the "consensus patient". Applying the enrichment test described above, the results were insignificant (Figure 4B, DrugBank: $p = 0.22$, COSMIC: $p = 0.23$), underscoring the utility of a personalized approach.

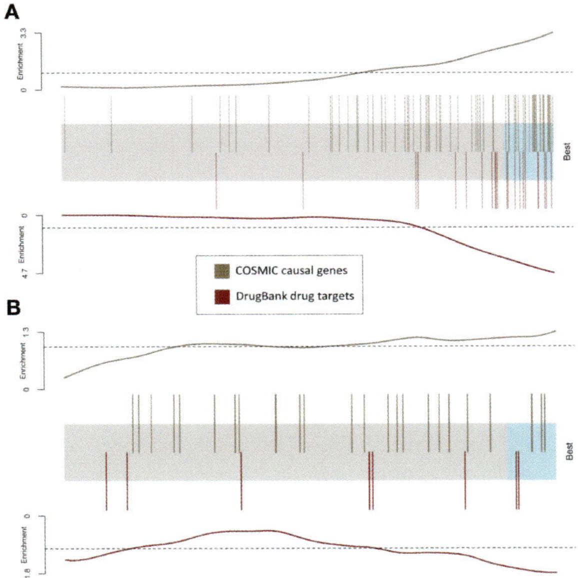

Fig. 4. Performance in drug target prediction. The candidate genes are represented by a shaded rectangle, where the top 10% are shaded cyan. Every overlaid bar stands for a single gene in a collection of known or potential drug targets. The bars are located according to their position in the candidate list generated by our method, where the rightmost bars represent the best candidates. Traces above/below the bar represent relative enrichment. (A) The barcode plot was generated by running our method on each AML patient independently and aggregating the results. (B) The barcode plot was generated by running a similar single pipeline on a "consensus" patient.

To further show the utility of our method, we used it to predict the sensitivity of the largest subgroup of AML patients – carriers of the FLT3 mutation – to known inhibitors. The following inhibitors were experimentally examined as potential drug targets and their influence on FLT3 mutated cell lines was carefully documented: Jin et al. [30] tested PI3K inhibitor and found FLT3 mutated cell lines to be poorly responsive to it; Nishioka et. al [31] showed that the MEK inhibitor caused those cell lines to respond moderately by leading to decreased abnormal proliferation, nearly resembling a healthy cell phenotype, yet showing unchanged abnormal levels of apoptosis; and Keeton et al. [32] demonstrated how PIM inhibitor caused FLT3 mutated cell lines to respond with high sensitivity, which led to development of the PIM inhibiting drug AZD1208. Our method shows in-silico sensitivity to PIM knockout, intermediate sensitivity to MEK knockout, and low sensitivity to PI3K knockout (Figure 5). These results corroborate the findings of [30-32].

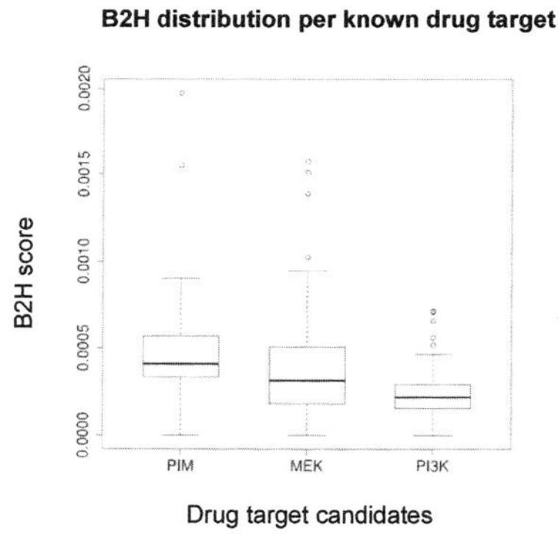

Fig 5. Sensitivity of FLT3 mutated cell lines, as predicted by B2H scores, corroborating the findings of [30-32] via in-vitro experiments.

3. Methods

3.1. *Computing propagation scores*

We use the network propagation method described in Vanunu et al. [16]. In the following we briefly describe it for the sake of completeness. The input consists of a network $G = (V, E, w)$ over a set V of proteins, where E represents the set of protein-protein interactions, and $w(u, v)$ represents the reliability of the interaction between u and v. In addition, a prior knowledge protein set P is given. The propagation process computes a scoring function $F: V \to \mathcal{R}$ that is both smooth over the network and accounts for the prior knowledge about each node.

To run the propagation process the weights are first normalized. Let W be the $|V| \times |V|$ matrix of initial weights, and let D be a diagonal matrix with $D[i,i] = \Sigma_j W[i,j]$. The normalized edge weight matrix is computed as $W' = D^{-1/2} W D^{-1/2}$. We further define a prior knowledge function $Y: V \rightarrow \{0,1\}$ such that:

$$\forall v \in V : Y(v) = \begin{cases} 1 & v \in P \\ 0 & v \notin P \end{cases}$$

We use the iterative procedure described by Law et al. [27] to compute F. Namely, starting with $F^{(0)} = Y$, we update F at iteration t as follows:

$$F^{(t)} = \alpha W' F^{(t-1)} + (1 - \alpha)Y$$

The procedure is repeated iteratively until convergence, i.e., when:

$$\left\| F^{(t)} - F^{(t-1)} \right\|_2 < 10^{-4}$$

The final propagation score for each gene is its rank among all genes, where lower ranks mean higher $F(v)$ values. In case of ties, the ranks of the corresponding genes are averaged. The genes of the prior set are assigned the highest ranks to focus the algorithm on novel discoveries.

3.2. *The Back2Healthy distance score*

Let S_{before} , S_{after} be vectors of propagation scores for a chosen gene set (here, the set of differentially expressed genes of some patient) A, where S_{before} was generated by propagating on the original PPI network, while S_{after} was generated by propagating on a "knockout" network, where one of the genes was removed. We define the Back2Healthy (B2H) distance between S_{before} and S_{after} as follows:

Let k be the size of the prior gene set of the patient (the patient's set of mutated genes). For $1 \leq i \leq n$ $(n = 1000)$, we generate a score vector S_i for A by propagating the original PPI network and setting the prior knowledge set P to be k random nodes (disjoint from A) in order to simulate a "healthy" distribution of propagation scores for A.

Next, for $a \in A$, define

$$Q_{before_a} = \frac{|\{1 \leq i \leq n | S_i[a] < S_{before}[a]\}|}{n}$$

$$Q_{after_a} = \frac{|\{1 \leq i \leq n | S_i[a] < S_{after}[a]\}|}{n}$$

Hence, Q_{before_a} represents the quantile of $S_{before[a]}$ in our simulated distribution, and similarly for Q_{after_a}. Finally, $B2H(S_{before}, S_{after})$ is defined as:

$$B2H(S_{before}, S_{after}) = \frac{\Sigma_{a \in A} |Q_{before_a} - Q_{after_a}|}{|A|}$$

3.3. Data Sets

3.3.1. *Patient, network and drug target data*

The TCGA data portal [7] contains information on 200 clinically annotated adult cases of AML (updated to 29/04/2015). The data include whole-genome sequencing of the primary tumor and matched normal skin samples from 50 patients and exome capture and sequencing for another 150 paired samples of AML tumor and skin [19]. For 174 of the patients both mutation and expression data were collected. Genes exhibiting significant expression changes were determined by the COSMIC methodology [20], by computing their z-scores based on the sequencing platform.

To construct individual-specific networks, we projected the mutations and differentially expressed genes of an individual on a human PPI network taken from HIPPIE [33], which contains 186,217 interactions among 15,029 proteins. The projected networks have on average 7.6 mutated and 340 differentially expressed genes.

We retrieved the known targets of AML drugs from the DrugBank version 4.3 database [27], obtaining 22 drug targets overall.

3.3.2. *Known causal genes*

We use three sets of known AML causal genes, varying in confidence and size. 10 causal genes were collected from the KEGG database [21,22], 94 causal genes were taken from COSMIC (72 of which are in our PPI network), and a third set of 533 cancer causal genes were collected from COSMIC (363 are in the network).

4. Conclusions

The approach we presented succeeds in predicting known drug targets for AML and could potentially be applied to other diseases with mutation and expression information, such as other cancer types recorded in TCGA. It should be noted that our method is limited to mutations that affect proteins that are part of the PPI network. More careful consideration of mutations in non-coding regions could improve its sensitivity.

Acknowledgments. RS was partially supported by a Naomi Kadar award and the National Science Foundation grant III-1217615. This study was supported in part by a fellowship from the Edmond J. Safra Center for Bioinformatics at Tel-Aviv University.

References

[1] P. I. Poulikakos, C. Zhang, G. Bollag, K. M. Shokat, and N. Rosen, "RAF inhibitors transactivate RAF dimers and ERK signalling in cells with wild-type BRAF," *Nature*, vol. 464, no. 7287, pp. 427–430, Mar. 2010.

[2] P. M. Rothwell, "Can overall results of clinical trials be applied to all patients?," *The Lancet*, vol. 345, no. 8965, pp. 1616–1619, Jun. 1995.

[3] M. Hofree, J. P. Shen, H. Carter, A. Gross, and T. Ideker, "Network-based stratification of tumor mutations," *Nat. Methods*, vol. 10, no. 11, pp. 1108–1115, Nov. 2013.

[4] M. Niepel, M. Hafner, E. A. Pace, M. Chung, D. H. Chai, L. Zhou, B. Schoeberl, and P. K. Sorger, "Profiles of Basal and Stimulated Receptor Signaling Networks Predict Drug Response in Breast Cancer Lines," *Sci. Signal.*, vol. 6, no. 294, Sep. 2013.

[5] J. A. DiMasi, R. W. Hansen, and H. G. Grabowski, "The price of innovation: new estimates of drug development costs," *J. Health Econ.*, vol. 22, no. 2, pp. 151–185, Mar. 2003.

[6] A. Gottlieb, G. Y. Stein, E. Ruppin, and R. Sharan, "PREDICT: a method for inferring novel drug indications with application to personalized medicine," *Mol. Syst. Biol.*, vol. 7, p. 496, Jun. 2011.

[7] J. Lamb, E. D. Crawford, D. Peck, J. W. Modell, I. C. Blat, M. J. Wrobel, J. Lerner, J.-P. Brunet, A. Subramanian, K. N. Ross, M. Reich, H. Hieronymus, G. Wei, S. A. Armstrong, S. J. Haggarty, P. A. Clemons, R. Wei, S. A. Carr, E. S. Lander, and T. R. Golub, "The Connectivity Map: using gene-expression signatures to connect small molecules, genes, and disease," *Science*, vol. 313, no. 5795, pp. 1929–1935, Sep. 2006.

[8] G. Hu and P. Agarwal, "Human Disease-Drug Network Based on Genomic Expression Profiles," *PLoS ONE*, vol. 4, no. 8, Aug. 2009.

[9] A. P. Chiang and A. J. Butte, "SYSTEMATIC EVALUATION OF DRUG-DISEASE RELATIONSHIPS TO IDENTIFY LEADS FOR NOVEL DRUG USES," *Clin. Pharmacol. Ther.*, vol. 86, no. 5, pp. 507–510, Nov. 2009.

[10] P. P. Zarrinkar, R. N. Gunawardane, M. D. Cramer, M. F. Gardner, D. Brigham, B. Belli, M. W. Karaman, K. W. Pratz, G. Pallares, Q. Chao, K. G. Sprankle, H. K. Patel, M. Levis, R. C. Armstrong, J. James, and S. S. Bhagwat, "AC220 is a uniquely potent and selective inhibitor of FLT3 for the treatment of acute myeloid leukemia (AML)," *Blood*, vol. 114, no. 14, pp. 2984–2992, Oct. 2009.

[11] R. Chuang, B. A. Hall, D. Benque, B. Cook, S. Ishtiaq, N. Piterman, A. Taylor, M. Vardi, S. Koschmieder, B. Gottgens, and J. Fisher, "Drug Target Optimization in Chronic Myeloid Leukemia Using Innovative Computational Platform," *Sci. Rep.*, vol. 5, Feb. 2015.

[12] M. F. Ciaccio, V. C. Chen, R. B. Jones, and N. Bagheri, "The DIONESUS algorithm provides scalable and accurate reconstruction of dynamic phosphoproteomic networks to reveal new drug targets," *Integr. Biol. Quant. Biosci. Nano Macro*, vol. 7, no. 7, pp. 776–791, Jul. 2015.

[13] S. Fatumo, K. Plaimas, J.-P. Mallm, G. Schramm, E. Adebiyi, M. Oswald, R. Eils, and R. König, "Estimating novel potential drug targets of Plasmodium falciparum by analysing the metabolic network of knock-out strains in silico," *Infect. Genet. Evol. J. Mol. Epidemiol. Evol. Genet. Infect. Dis.*, vol. 9, no. 3, pp. 351–358, May 2009.

[14] B. Papp, C. Pál, and L. D. Hurst, "Metabolic network analysis of the causes and evolution of enzyme dispensability in yeast," *Nature*, vol. 429, no. 6992, pp. 661–664, Jun. 2004.

[15] P. Csermely, V. Ágoston, and S. Pongor, "The efficiency of multi-target drugs: the network approach might help drug design," *Trends Pharmacol. Sci.*, vol. 26, no. 4, pp. 178–182, Apr. 2005.

[16] O. Vanunu, O. Magger, E. Ruppin, T. Shlomi, and R. Sharan, "Associating genes and protein complexes with disease via network propagation," *PLoS Comput. Biol.*, vol. 6, no. 1, p. e1000641, Jan. 2010.

[17] G. Marcucci, T. Haferlach, and H. Döhner, "Molecular Genetics of Adult Acute Myeloid Leukemia: Prognostic and Therapeutic Implications," *J. Clin. Oncol.*, p. JCO.2010.30.2554, Jan. 2011.

[18] H. Wang, H. Hu, Q. Zhang, Y. Yang, Y. Li, Y. Hu, X. Ruan, Y. Yang, Z. Zhang, C. Shu, J. Yan, E. K. Wakeland, Q. Li, S. Hu, and X. Fang, "Dynamic transcriptomes of human myeloid leukemia cells," *Genomics*, vol. 102, no. 4, pp. 250–256, Oct. 2013.

[19] "Genomic and Epigenomic Landscapes of Adult De Novo Acute Myeloid Leukemia," *N. Engl. J. Med.*, vol. 368, no. 22, pp. 2059–2074, May 2013.

[20] S. A. Forbes, N. Bindal, S. Bamford, C. Cole, C. Y. Kok, D. Beare, M. Jia, R. Shepherd, K. Leung, A. Menzies, J. W. Teague, P. J. Campbell, M. R. Stratton, and P. A. Futreal, "COSMIC: mining complete cancer genomes in the Catalogue of Somatic Mutations in Cancer," *Nucleic Acids Res.*, vol. 39, no. Database issue, pp. D945–D950, Jan. 2011.

[21] M. Kanehisa and S. Goto, "KEGG: Kyoto Encyclopedia of Genes and Genomes," *Nucleic Acids Res.*, vol. 28, no. 1, pp. 27–30, Jan. 2000.

[22] M. Kanehisa, S. Goto, Y. Sato, M. Kawashima, M. Furumichi, and M. Tanabe, "Data, information, knowledge and principle: back to metabolism in KEGG," *Nucleic Acids Res.*, vol. 42, no. Database issue, pp. D199–205, Jan. 2014.

[23] B. Vogelstein, N. Papadopoulos, V. E. Velculescu, S. Zhou, L. A. Diaz, and K. W. Kinzler, "Cancer Genome Landscapes," *Science*, vol. 339, no. 6127, pp. 1546–1558, Mar. 2013.

[24] C. Scholl, D. G. Gilliland, and S. Fröhling, "Deregulation of signaling pathways in acute myeloid leukemia," *Semin. Oncol.*, vol. 35, no. 4, pp. 336–345, Aug. 2008.

[25] M. Ruffalo, M. Koyuturk, and R. Sharan, "Network-Based Integration of Disparate Omic Data To Identify 'Silent Players' in Cancer," *submitted*, 2015.

[26] C. Knox, V. Law, T. Jewison, P. Liu, S. Ly, A. Frolkis, A. Pon, K. Banco, C. Mak, V. Neveu, Y. Djoumbou, R. Eisner, A. C. Guo, and D. S. Wishart, "DrugBank 3.0: a comprehensive resource for 'omics' research on drugs," *Nucleic Acids Res.*, vol. 39, no. Database issue, pp. D1035–1041, Jan. 2011.

[27] V. Law, C. Knox, Y. Djoumbou, T. Jewison, A. C. Guo, Y. Liu, A. Maciejewski, D. Arndt, M. Wilson, V. Neveu, A. Tang, G. Gabriel, C. Ly, S. Adamjee, Z. T. Dame, B. Han, Y. Zhou, and D. S. Wishart, "DrugBank 4.0: shedding new light on drug metabolism," *Nucleic Acids Res.*, p. gkt1068, Nov. 2013.

[28] D. S. Wishart, C. Knox, A. C. Guo, S. Shrivastava, M. Hassanali, P. Stothard, Z. Chang, and J. Woolsey, "DrugBank: a comprehensive resource for in silico drug discovery and exploration," *Nucleic Acids Res.*, vol. 34, no. Database issue, pp. D668–672, Jan. 2006.

[29] D. S. Wishart, C. Knox, A. C. Guo, D. Cheng, S. Shrivastava, D. Tzur, B. Gautam, and M. Hassanali, "DrugBank: a knowledgebase for drugs, drug actions and drug targets," *Nucleic Acids Res.*, vol. 36, no. Database issue, pp. D901–906, Jan. 2008.

[30] L. Jin, Y. Tabe, H. Lu, G. Borthakur, T. Miida, H. Kantarjian, M. Andreeff, and M. Konopleva, "Mechanisms of apoptosis induction by simultaneous inhibition of PI3K and FLT3-ITD in AML cells in the hypoxic bone marrow microenvironment," *Cancer Lett.*, vol. 329, no. 1, pp. 45–58, Feb. 2013.

[31] C. Nishioka, T. Ikezoe, J. Yang, and A. Yokoyama, "Inhibition of MEK/ERK signaling induces apoptosis of acute myelogenous leukemia cells via inhibition of eukaryotic initiation factor 4E-binding protein 1 and down-regulation of Mcl-1," *Apoptosis Int. J. Program. Cell Death*, vol. 15, no. 7, pp. 795–804, Jul. 2010.

[32] E. K. Keeton, K. McEachern, K. S. Dillman, S. Palakurthi, Y. Cao, M. R. Grondine, S. Kaur, S. Wang, Y. Chen, A. Wu, M. Shen, F. D. Gibbons, M. L. Lamb, X. Zheng, R. M. Stone, D. J. Deangelo, L. C. Platanias, L. A. Dakin, H. Chen, P. D. Lyne, and D. Huszar, "AZD1208, a potent and selective pan-Pim kinase inhibitor, demonstrates efficacy in preclinical models of acute myeloid leukemia," *Blood*, vol. 123, no. 6, pp. 905–913, Feb. 2014.

[33] M. H. Schaefer, J.-F. Fontaine, A. Vinayagam, P. Porras, E. E. Wanker, and M. A. Andrade-Navarro, "HIPPIE: Integrating Protein Interaction Networks with Experiment Based Quality Scores," *PLoS ONE*, vol. 7, no. 2, p. e31826, Feb. 2012.

INTEGRATING CLINICAL LABORATORY MEASURES AND ICD-9 CODE DIAGNOSES IN PHENOME-WIDE ASSOCIATION STUDIES

ANURAG VERMA[1,3], JOSEPH B. LEADER[2], SHEFALI S. VERMA[1,3], ALEX FRASE[3], JOHN WALLACE[3], SCOTT DUDEK[3], DANIEL R. LAVAGE[2], CRISTOPHER V. VAN HOUT[4], FREDERICK E. DEWEY[4], JOHN PENN[4], ALEX LOPEZ[4], JOHN D. OVERTON[4], DAVID J. CAREY[5], DAVID H. LEDBETTER[1], H. LESTER KIRCHNER[2], MARYLYN D. RITCHIE[1,3], SARAH A. PENDERGRASS[1]

Biomedical and Translational Informatics[1], Geisinger Health System, Danville, PA; Center for Health Research[2], Geisinger Health System, Danville, PA; Center for Systems Genomics[3], The Pennsylvania State University, University Park, PA; Regeneron Genetics Center, Tarrytown NY[4]; Weis Center for Research[5], Geisinger Health System, Danville, PA

Electronic health records (EHR) provide a comprehensive resource for discovery, allowing unprecedented exploration of the impact of genetic architecture on health and disease. The data of EHRs also allow for exploration of the complex interactions between health measures across health and disease. The discoveries arising from EHR based research provide important information for the identification of genetic variation for clinical decision-making. Due to the breadth of information collected within the EHR, a challenge for discovery using EHR based data is the development of high-throughput tools that expose important areas of further research, from genetic variants to phenotypes. Phenome-Wide Association studies (PheWAS) provide a way to explore the association between genetic variants and comprehensive phenotypic measurements, generating new hypotheses and also exposing the complex relationships between genetic architecture and outcomes, including pleiotropy. EHR based PheWAS have mainly evaluated associations with case/control status from International Classification of Disease, Ninth Edition (ICD-9) codes. While these studies have highlighted discovery through PheWAS, the rich resource of clinical lab measures collected within the EHR can be better utilized for high-throughput PheWAS analyses and discovery. To better use these resources and enrich PheWAS association results we have developed a sound methodology for extracting a wide range of clinical lab measures from EHR data. We have extracted a first set of 21 clinical lab measures from the de-identified EHR of participants of the Geisinger MyCode™ biorepository, and calculated the median of these lab measures for 12,039 subjects. Next we evaluated the association between these 21 clinical lab median values and 635,525 genetic variants, performing a genome-wide association study (GWAS) for each of 21 clinical lab measures. We then calculated the association between SNPs from these GWAS passing our Bonferroni defined p-value cutoff and 165 ICD-9 codes. Through the GWAS we found a series of results replicating known associations, and also some potentially novel associations with less studied clinical lab measures. We found the majority of the PheWAS ICD-9 diagnoses highly related to the clinical lab measures associated with same SNPs. Moving forward, we will be evaluating further phenotypes and expanding the methodology for successful extraction of clinical lab measurements for research and PheWAS use. These developments are important for expanding the PheWAS approach for improved EHR based discovery.

1. Introduction

Precision medicine aims to find clinical treatments based on the phenotypic and genetic makeup of each individual. Electronic health records (EHR) are a powerful resource for the investigation of common and rare disease, with the potential for discovery that will lead to meaningful and data-driven individualized patient care. Accessing de-identified EHR data linked to DNA biorepositories has already proved useful for a wide range of genetic association discovery efforts, such as through the Electronic Medical Records and Genomics (eMERGE) network [1].

In PheWAS, the association between thousands of phenotypes and any number of single nucleotide polymorphisms (SNPs) are evaluated in a high-throughput manner to identify new hypotheses, biologically relevant associations, and the identification of potential pleiotropy, highlighting important connections between networks of phenotypes and genetic architecture [2–4]. To date, de-identified EHR data coupled with genetic data have been used for multiple PheWAS, primarily through using International Classification of Disease, Ninth Edition (ICD-9) based case/control status for identifying significant associations between medical record diagnoses and genetic data [5–8].

There are other data within the EHR that can also be used for high-throughput PheWAS research, with one of the most readily available additional sources of data being clinical lab measures. Clinical lab measures are an important part of clinical decision-making, providing clues and measures of a variety of conditions as well as important reflections of health. Many of these lab measures are found in multiple diagnoses, for example, blood cell count information is important for a variety of clinical conditions and diagnoses. To date, high-throughput use of clinical lab measures from the EHR have been underutilized for multiple reasons. These include the variability and error in the units recorded that can occur across measurements, error that can occur in the collected laboratory result, change in laboratory assays, sensitivity of different assays, lack of documentation for fasting, and changes in biological function due to treatment of injury or disease (e.g. medication use). Even with these challenges there is an opportunity for further discovery by using more of the comprehensive clinical lab data available within the EHR for both high-quality phenotype algorithm development as well as expanding EHR based PheWAS beyond the use of ICD-9 based case/control status. The clinical lab measures of the EHR can more closely reflect the impact of genetic variation on phenotype, and some phenotypes observed from the clinical lab data collected in EHR are not reflected at all in case/control diagnoses or common to multiple case/control diagnoses. Using a wide range of clinical lab measures within the PheWAS framework also creates a series of results to compare and contrast with the findings of ICD-9 based PheWAS, providing a complementary set of information pertinent to health and disease and genetic association studies, enriching the interpretation and exploration of ICD-9 based PheWAS. There is also the potential for improved power for association analyses, as case numbers for ICD-9 based PheWAS can be very low depending on the ICD-9 based diagnosis compared to larger sample sizes for quantitative clinical lab measures.

We describe here our preliminary algorithmic development for high-throughput extraction of clinical lab measures from de-identified data linked to genetic data from the Geisinger Health System (GHS) MyCode^{TM} Biorepository. For these analyses we used our approach to extract 21 clinical lab measurements with some of the largest sample sizes within the EHR from, or derived from, blood: alanine amino transferase (ALT), albumin aspartate aminotransferase (AST), carbon dioxide (CO2), cholesterol, creatinine, free T3, free T4, glucose, hemoglobin A_{1C} (Hb-A_{1C}), high density lipoprotein (HDL), insulin-like growth factor (IgF-1), low density lipoprotein (LDL), platelets, urine protein, red blood cell counts (RBC), thyroglobulin antibody (TgAb), thyroid peroxidase antibody (TPO), thyroid stimulating hormone (TSH), triglycerides (TG), white blood cell counts (WBC). We also extracted body mass index measurements (BMI). We calculated the median value for each of 12,039 individuals and performed comprehensive genome-wide association analyses (GWAS) with these measurements in European-Americans within the MyCode Biorepository, and then explored associations with highly-significant SNPs from the GWAS with an ICD-9 diagnosis code based PheWAS. These preliminary analyses show the success of our approach, and the ultimate success possible in high-throughput extraction of a wide range of clinical lab measurements from the EHR.

2. Methods

2.1 Study Participants

In this study we used de-identified genetic and phenotypic data from MyCode^{TM} biorepository of Geisinger Heath System (GHS). MyCode is a biorepository that stores blood samples and Electronic Health Record (EHR) data from consented individuals for research to improve patient healthcare. GHS is located in central Pennsylvania, which is a primarily European American (EA) population with 95.7% of individuals in the study of European decent. Thus we only focused on individuals from EA ancestry for these analyses.

2.2 Genotypic data and Quality Control

GHS MyCode subjects were genotyped using the Illumina HumanOmniExpressExome Bead Chips, with coverage of a total of 964,193 SNPs. We performed Genotype Quality Control (QC) procedures to account of genotyping error prior to association testing using the R programming statistical package [9] and PLINK software [10]. We filtered out the missing data using 99% genotype and sample call rates and minor allele frequency (MAF) threshold of 1%. Also, relatedness between the individuals was calculated by Identity by Descent (IBD) and related samples were dropped using kinship coefficient of 0.125. After these QC steps and MAF filter, the genotypic data consisted of 635,525 SNPs and 12,278 samples. While individuals within GHS are primarily from EA populations, we calculated principle components to further correct for global ancestry in our associations using EIGENSOFT [11].

2.3 Clinical Lab Extraction

We extracted a total of 21 clinical lab measurements from, or derived from, blood of participants in the study. We selected an initial set of lab measurements to extract by choosing the measures with the large sample sizes that we have commonly used for other phenotype algorithm development using data from the EHR. All the summary information on the clinical lab measurements is provided in Table 1. We also extracted data to calculate body mass index (BMI). BMI is known to have confounding effects on various metabolic traits and many of the clinical lab measurements could be affected by BMI of study participants. We extracted height and weight to calculate BMI. We calculated the median value for each of these clinical lab measures over the course of all visits of each individual.

We extracted these clinical lab measurements as follows: First, we extracted each clinical laboratory measures from the de-identified EHR and log base ten transformed the results. We then standardized units within each clinical lab measure. Different Geisinger Health laboratories and Point of Care devices can have differing units of measure within the same Logical Observation Identifiers Names and Codes (LOINC) code and thus standardization and transformation of individual values needed to be performed so that all units were consistent across each clinical lab measure. We excluded measurements where the unit of measure reported on the result was different than the suggested unit of measure from LOINC when conversion was not possible. We then excluded results that were identified as implausible through a process of comparing individual level and population level medians greater than a deviation threshold determined by each LOINC code, +/- 3 standard deviations from the median. After the process of excluding these results, we transformed the results back to their original values. For TG, HDL, LDL, glucose levels and cholesterol, we omitted all observations that were not known to be fasting, i.e., observations with non-fasting or unknown fasting state. When calculating medians and standard deviation we accounted for the number of results over a patient's lifetime to adjust where exclusions are applied. Using this approach, as proof-of-principle, we extracted 21 different clinical lab tests from the entire GHS cohort, ~1.25 million people.

After these clinical lab measurements were extracted and prepared for further analyses, we calculated the lifetime median value from each individual for each of the lab measures for those individuals we had genetic data for, for association testing. We then created histograms and calculated the population median and max values for each clinical lab measure. This identified any measurements with non-normal distributions, and identified some of the most extreme outliers. We only removed outliers for white blood cell counts, values > 20K cells per/uL. We used natural-log transformation to improve the normality of the distributions for glucose and platelet measures. Our summary information of the median and mean of each measure, and whether or not each lab measure was transformed before association testing, is listed in Table 1.

Table 1: Summary of clinical lab measures, and any transformation of the variable before analysis

Phenotype	*Median*	*Mean*	*Min/Max*	*SD*	*% Male*	*% Female*
Alanine Amino Transferase (ALT) (Log Transform)	3.09	3.10	1.79/4.48	0.43	41.47	58.53
Albumin	4.30	4.26	3.45/5.30	0.26	41.19	58.81
Aspartate Aminotransferase (AST) (Log Transform)	3.13	3.16	2.35/4.04	0.26	41.23	58.77
Carbon Dioxide (CO2)	27.00	26.97	21.29/34	1.98	42.06	57.94
Cholesterol	183.00	184.55	110/320	33.56	43.13	56.87
Creatinine	57.03	57.45	21.07/105.76	13.58	63.36	36.64
Free T3	2.90	2.94	1.64/5.19	0.53	24.48	75.52

Free T4 (Log Transform)	0.19	0.22	0.009/0.63	0.14	29.95	70.05
Glucose (Log Transform)	4.60	4.64	4.14/5.12	0.18	41.83	58.17
Hemoglobin A1C (Hb-A_{1C})	6.20	6.50	4.09/10.80	1.17	43.01	56.99
High Density Lipoprotein (HDL) (Log Transform)	3.87	3.88	3.13/4.70	0.27	43.31	56.69
Insulin-like Growth Factor (IgF-1) (Log Transform)	4.96	4.90	3.08/6.08	0.62	41.67	58.33
Low Density Lipoprotein (LDL)	102.00	104.08	48/237.5	28.44	43.17	56.83
Platelets	238.50	243.80	112/526	60.16	40.67	59.33
Urine Protein (Log Transform)	2.40	2.60	0.09/6.53	1.24	44.26	55.74
Red Blood Cell Count (RBC)	4.48	4.48	3.27/6.09	0.43	40.75	59.25
Thyroglobulin antibody (TgAb)	20.00	26.83	1.70/305	31.95	21.12	78.88
Thyroid Peroxidase Antibody (TPO) (Log Transform)	2.30	3.27	0.09/7.94	1.74	18.38	81.62
Thyroid Stimulating Hormone (TSH) (Log Transform)	0.75	0.77	0.009/2.04	0.41	37.43	62.57
Triglycerides (TG) (Log Transform)	4.88	4.88	3.46/6.25	0.47	43.32	56.68
White Blood Cell Count (WBC) (Log Transform)	7.42	7.68	3.15/16.84	2.00	40.80	59.20

2.4 ICD-9 Based Case-Control Status

We used *International Classification of Diseases, Ninth Revision* (ICD-9) codes as the phenotypic data to define case-control status for PheWAS. Patients in MyCode were diagnosed with 7,039 different ICD-9 codes, these codes have been used at least once during clinic visits at GHS facilities. We defined cases as individuals with three or more visits of a specific ICD-9 code at the 5-digit code level (e.g. 250.12), no visits of an ICD-9 code were defined as a control. If an individual had one to three visits of an ICD-9 code, they were not included as a case or control (i.e. excluded from analysis). A total of 200 or more case subjects per ICD-9 code were required for inclusion in our association study. Using our criteria for inclusion/exclusion for cases and controls there were a total of 165 ICD-9 codes used for the case/control association testing.

2.5 Genetic Associations

To evaluate the association between SNPs and the 21 clinical lab measures, we used linear regression with an additive encoding for the SNPs. We used the software Platform for the Analysis, Translation and Organization of large-scale data (PLATO), freely available here: http://ritchielab.psu.edu/software/plato-download. We have implemented PLATO in DNANexus (https://www.dnanexus.com/) to use cloud-computing resources for analyses. Covariates in the models included for each association were gender, age, age^2 (age-squared), body mass index (BMI), and the first 4 principal components. We made Manhattan plots of the results of these associations, as well as a table of the results passing our Bonferroni correction based p-value available in supplementary materials (http://ritchielab.psu.edu/publications/supplementary-data/psb-2016/clinical-measure-phewas). After filtering the clinical lab measure association results by our Bonferroni threshold (described further in methods), and performing a PheWAS using ICD-9 based case/control status for these SNPs, we compared and contrasted the results of the clinical lab measurements with results of case/control ICD-9 based PheWAS. Further, we annotated the results of the clinical lab measure association testing using Biofilter [12], to add information about any genes that the SNPs from the p-value filtered results were in or near, as well as to annotate the SNPs with any known results from the NHGRI GWAS catalog. The NHGRI GWAS catalog contains results from published GWAS in the literature reaching genome-wide significance [13]. To have the magnitude and direction of effect values more comparable across the association results, we divided each beta by the standard deviation of the respective clinical lab measure for the association.

2.6 Multiple Hypothesis Testing

In terms of independent SNPs used for multiple testing adjustment, independence of SNPs varies across different populations and appropriate measures are necessary [14]. In this study, we used a linkage-disequilibrium (LD) pruning approach to identify the number of independent SNPs used in our association testing. We used PLINK to prune the 635,525 SNPs based on pairwise linkage disequilibrium (LD) at $r^2 = 0.3$ and that resulted in total of 174,401 SNPs. An r^2 of 0.3 is estimated to be a reasonable threshold for finding independent SNPs based on pairwise LD [14]. For our clinical lab measures PheWAS, our Bonferoni threshold was α value divided by the number of independent tests: $0.05/(174{,}401 \times 21) = 1.37 \times 10^{-8}$. We used the same approach for multiple hypothesis correction for the ICD-9 based PheWAS, where we calculated LD between the 286 SNPs resulting in a total of 61 independent SNPs and thus a Bonferoni threshold: $0.05 / (61 \times 165) = 4.9 \times 10^{-6}$.

3. Results

3.1 Clinical Lab Measure GWAS

In this study, we first calculated the association between 635,525 SNPs and 21 clinical lab measurements using linear regression. Figure 1 shows the Manhattan plots from each of the clinical lab measure genome-wide association studies (GWAS), points indicated in red, for p-values < 0.01. We provide higher-resolution copies of the figures of this paper, as well as Quantile-Quantile plots for the associations in supplementary materials. A total of 286 SNPs were found significantly associated with our Bonferroni defined p-value threshold of 1.37×10^{-8}, with a total of 344 SNP-clinical lab measure associations. A total of 163 out of 347 associations were found to be associated with same or similar previously reported phenotypic traits in the literature. Several associations were also for SNPs in high LD with SNPs for previously reported associations with the same or similar previously reported phenotypic trait. We observed that almost half of associations related to a previously reported association were with triglyceride and HDL-cholesterol levels, where we found 109 SNPs associated with triglycerides and 66 SNPs associated with HDL. For example, the association between SNP rs247616 downstream of *CETP* and HDL was the most significant association of all the GWAS we performed at $p = 5.25 \times 10^{-53}$, $\beta = -0.22$, this association has been previously reported in the literature[15]. *CETP* is protein-coding gene involved in cholesterol ester transfer from HDL to other lipoprotein [18]. The most significant association for TG was with SNP rs964184 downstream from *ZPR1* with $p = 1.9 \times 10^{-41}$, $\beta = -0.27$, that has also been previously reported in the literature[16,17].

A novel GWAS association found in this study was for *DPP4* SNP rs2302872 associated with thyroid globulin antibody (TgAb). TgAb is a diagnostic measure used for thyroid related autoimmune disorders like Hashimoto's disease and a measure to evaluate the treatment effectiveness of thyroid cancer. The SNP rs2302872 was associated with TgAb at $p = 1.48 \times 10^{-8}$, $\beta = -1.28$. *DPP4* has known expression in cancerous thyroid tissue in comparison to no expression in a healthy thyroid tissue[19]. TgAb as a tumor marker for thyroid cancer has been controversial where many studies suggest no association as a tumor marker[20,21] and others suggest TgAb levels can be used to identify increased risk of thyroid cancer[22]. In our case we see a potentially protective effect of this SNP in the direction of the association. In another thyroid measure, thyroid stimulating hormone (TSH), we report 4 novel loci on chromosome 5 mapped to *PDE8B*. There are other variants in *PDE8B* with known association with TSH[23], but we found 4 polymorphisms (rs1351283, rs13158164, rs6885813, rs9686502) not in LD with the previously known variants with p-values of $p = 1.16 \times 10^{-18}$, $\beta = -0.13$; $p = 2.60 \times 10^{-14}$, $\beta = -0.16$; $p = 4.74 \times 10^{-14}$, $\beta = -0.13$; $p=9.10 \times 10^{-9}$, $\beta = -0.08$. We also found 7 novel SNPs associated with aspartate amino transferase (AST) levels, where all 7 variants mapped to *MRC1,* where the most significant association was for SNP rs35038329 with $p = 2.87 \times 10^{-19}$, $\beta = 0.12$. AST levels are used for the diagnosis of various liver diseases like hepatitis and cirrhosis. In a recent study variants in *MRC1* were reported to be associated with treatment outcomes for hepatitis C [24].

3.2 Targeted ICD-9 PheWAS

We selected the SNPs from the top associations of the clinical lab measure PheWAS (p-value < 1.37×10^{-8}), resulting in 286 SNPs. We then performed an ICD-9 code based PheWAS with these SNPs, performing comprehensive associations testing between these SNPs and the ICD-9 based case/control status using logistic regression. Figure 1 shows in blue the ICD-9 based diagnosis associations with p-values less than 0.01. We found 39 associations passing our Bonferoni p-value threshold adjusted for the smaller number of SNPs for the ICD-9 base analyses, these additional association results are reported in detail within the supplementary materials. The most significant association was between SNP rs9273363 and the ICD-9 diagnosis 250.01 "Diabetes mellitus, Type I" with $p = 4.39 \times 10^{-26}$, $\beta = -0.8$. This SNP is located in the HLA region that is known to have high susceptibility for type 1 diabetes, and is one of the most high-risk polymorphisms in HLA region for Type I diabetes[25]. There were other significant associations with ICD-9 diagnoses such as SNPs associated with the diagnosis of 250.00 "Type II Diabetes" (original associated clinical lab measures of Glucose and Hb-A_{1C}), 272.4 "Hyperlipidemia" (original associated clinical lab measure: Cholesterol, LDL, TG), and 244.9 "Hypothyroidism" (clinical lab measure: TSH).

3.3 Comparing Clinical Lab Measure GWAS and ICD-9 Based PheWAS

PheWAS frequently identifies cross-phenotype associations, where one SNP is associated with more than one phenotype. These cross-phenotype associations highlight potential relationships between the phenotypes, and can also identify pleiotropy. Thus, one of the potential benefits of using clinical lab measures in addition to ICD-9 codes is the addition of a complementary set of phenotypic information for exploring the multiple cross-phenotype associations that arise in PheWAS. In this study we started with SNPs highly associated with clinical lab measures, thus have compared and contrasted what the ICD-9 diagnoses were also associated with these clinical lab measures. Table 2 shows what the clinical lab measures were, and for SNPs associated with those specific clinical lab measures what the ICD-9 codes diagnoses were also associated those SNPs.

First we identified results between the two clinical lab measures and ICD-9 diagnoses were highly related between the two sets of associations. For instance, for lipid related phenotypes and diagnoses we found the SNP rs445925 in *APOC1* associated with LDL levels ($p = 1.75 \times 10^{-37}$, $\beta = 0.30$) also associated with the ICD-9 diagnosis 272.4 "Hyperlipidemia" ($p = 1.5 \times 10^{-11}$, $\beta = 0.30$). We also found the SNP rs602633 associated with LDL levels ($p= 8.03 \times 10^{-10}$, $\beta = 0.12$) and the ICD-9 diagnosis 272.4 ($p = 4.55 \times 10^{-6}$, $\beta = 0.18$). A SNP in LD with rs599839 (r^2=1) has known associations with LDL concentrations [26] and coronary artery disease [27].

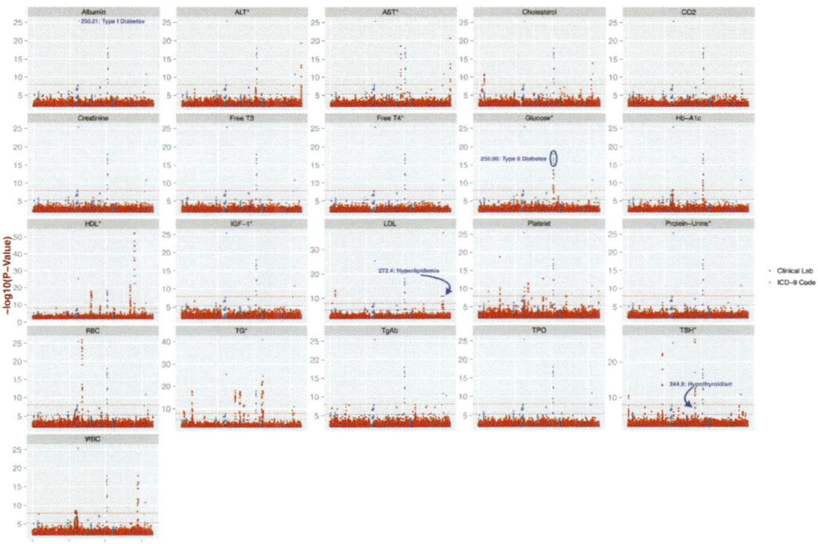

Figure 1. Manhattan plots of all 21 GWAS for clinical lab measures and the results of the following ICD-9 based PheWAS. For each of 21 clinical lab measures, the results of associations are marked as -log(10) of the p-value in red, with the abbreviation of each clinical lab measure indicated above each plot, abbreviations explained in Table 1. Plotted in blue are –log10 (p-value) from the associations of distinct ICD-9 code based case/control diagnoses. All results are from p-values < 0.01. The red dashed line in each Manhattan plot is at the Bonferroni corrected p-value of 1.37×10^{-8} for the clinical lab GWAS, and the blue dashed line is the Bonferroni corrected p-value 4.9×10^{-6} for the ICD-9 diagnoses based PheWAS.

Table 2. Phenotypes for SNPs significantly associated with clinical lab measures also significantly associated with highly related ICD-9 diagnoses

Clinical Lab Measure	***ICD-9 Diagnoses***
Cholesterol	272.4: Hyperlipidemia
Glucose	250.00: Type II Diabetes Mellitus

Hemoglobin A_{1C}	250.01: Type I Diabetes Mellitus; 250.00: Type II Diabetes Mellitus
Low Density Lipoprotein (LDL)	272.4: Hyperlipidemia
Thyroid Stimulating Hormones	244.9: Hypothyroidism
Triglycerides	272.4: Hyperlipidemia
White Blood Cell Count	250.01: Type I diabetes mellitus

For thyroid diagnoses and TSH levels we also found related phenotypic associations for specific SNPs. For example, a cluster of variants in *PTSC2* were associated with thyroid stimulating hormone levels (TSH) and also significantly associated with the ICD-9 diagnosis 244.9 "Hypothyroidism". Another *PTSC2* SNP rs10759944 was associated with TSH levels ($p = 4.12 \times 10^{-26}$, $\beta = 0.16$) and the ICD-9 diagnosis 244.9 "Hypothyroidism" ($p = 9.48 \times 10^{-8}$, $\beta = 0.21$). The SNP is in LD with rs965513 ($r^2 = 0.9$), a SNP with a known association with thyroid cancer [28,29].

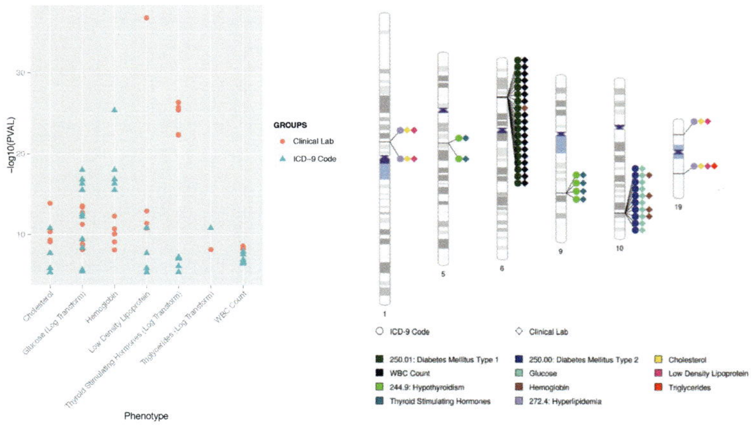

Figure 2. (a) Comparison of significant SNPs between clinical lab measures and ICD-9 code PheWAS. The x-axis has the clinical lab measures and y-axis shows its association p-value with the SNP, where red dots are the top SNPs from clinical lab PheWAS and blue triangle are the same SNPs associated with ICD-9 diagnoses. Table 2 lists what the ICD-9 diagnoses were for each of the clinical lab measures. (b) In this chromosomal ideogram, lines link SNP chromosomal locations to colored diamonds (representing clinical lab measures) or circles (representing ICD-9 diagnoses) showing the cross-phenotype associations for the SNPs identified first with associations with clinical lab measures.

For Type 2 diabetes we had another significant finding with the known Type II diabetes risk gene *TCF7L2*, with variants in *TCF7L2* associated with the ICD-9 diagnosis 250.00 "Type II diabetes" and clinical lab measures related to Type II diabetes including glucose levels and Hb-A_{1C} levels.

We did find potentially novel pleiotropic associations with a cluster of SNPs in LD on chromosome 6. These SNPs were associated with WBC as well as Hb-A_{1C}, and also associated with the ICD-9 based diagnosis of 250.01 "Type I diabetes". While diabetes has an impact on Hb-A_{1C} levels, the associations with WBC are more novel. White blood cell counts have been found to impact insulin sensitivity and diabetes development [30].

Using a less stringent cutoff for p-value for ICD-9 codes, which are often under powered associations due to the number of cases, Figure 3 shows an example of the ICD-9 PheWAS results for *LDLR* SNP rs6511720. The associated clinical lab measure was LDL levels, a previously reported association in the literature, and Figure 3 shows the spectrum of PheWAS results associated with this SNP with $p < 0.01$, a series of comorbidities related to cholesterol levels.

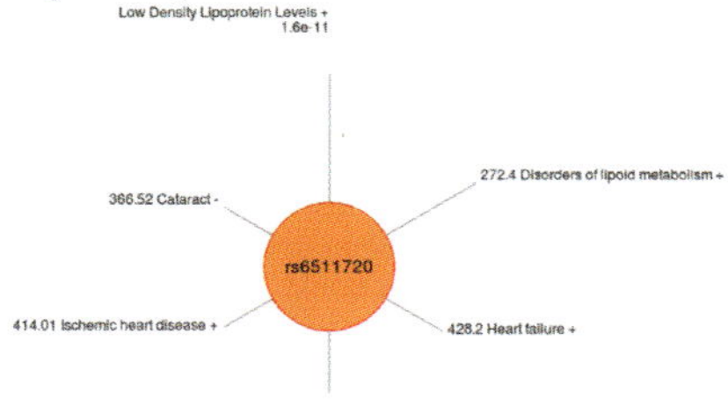

Figure 3. Spectrum of phenotypic associations for ***LDLR*** SNP rs6511720, for PheWAS p-values < 0.01. This SNP was originally associated in our study with the clinical lab measure of LDL.

4. Discussion

The goal of this study was preliminary work in the process of accessing clinical lab measurements in a high-throughput way and developing algorithms, methodologies, and ultimately an analysis pipeline to be able to use a wide range of clinical lab measure for PheWAS. We have shown here that we can successfully extract clinical lab measures for association research and use these measures for association testing. The process of extracting and preparing these clinical lab measurements has been informative. Some of the challenges with these measurements, surmounted through our research, may provide information to inform better practices for collecting these data within the clinic in terms of standardization, which could benefit patients and clinicians as well as researchers. Facing the challenges of using clinical lab measurements is also providing preliminary information for how to address the challenge of accessing medication information within the EHR for use in research, which has many of the same issues as clinical lab measurements but additional challenges for research use.

We did identify association results replicating previously published associations indicating our clinical lab measure extraction is functional, as well as a number of novel associations. We intend moving forward to do an expanded study including additional clinical lab measurements, including additional measures that have been previously studied to continue expanding our proof-of-principle results, as well as a wide array of additional measures little studied in genetic association testing. The development of algorithms for obtaining summary information about these measures, when moving to hundreds of measures, will be important for quick evaluation of these phenotypes, and sub-setting of data based on specific criteria. Of further importance will be better use of the longitudinal nature of these clinical lab measures, in health and disease, as we are currently using median values in our association testing.

A challenge within PheWAS is to understand if we find associations due to correlated phenotypes (such as glucose levels and diabetes), or if we find associations related to the impact of genetic variation on more than one phenotype (pleiotropy). A future direction is to expand our use of clinical lab measurements in addition to the use of ICD-9 codes in PheWAS to help provide more insight into the findings we have for both clinical lab measurements and ICD-9 code based case/control status to begin to understand more of the complex relationship between genetic architecture and the complex networks of signaling and phenotypic outcomes. Further, with the longitudinal nature of the EHR, we can leverage more of these data for longitudinal analyses. Clinical laboratory measures fluctuate for an individual in health and disease, and also with medication usage and age, and this complexity can be leveraged in future association testing for further discovery.

Clinical lab measurements provide an important area of exploration for PheWAS. The results of using more phenotypic measurements in a high-throughput way can enrich and expand our results of PheWAS based on ICD-9 code case/control status. Further, potential pleiotropy identified through cross-phenotype associations could show new important relationships between phenotypes through an expansion of phenotypic data available for PheWAS. Identifying a wide range of standardized and "cleaned" clinical lab measurements can also be used in the future to subset individuals based on clinical lab measure criteria before association testing. These approaches will also open the door to using more of the longitudinal nature of clinical lab measurements in future PheWAS analyses. These clinical lab measurements could also prove useful for continued development of high-quality phenotypic algorithms. The discoveries with these expanded PheWAS could prove important for discovery that leads to improvements in precision medicine as well as drug development.

References

1. Crawford, D. C. *et al.* eMERGEing progress in genomics-the first seven years. *Front. Genet.* **5,** 184 (2014).
2. Tyler, A. L., Crawford, D. C. & Pendergrass, S. A. Detecting and Characterizing Pleiotropy: New Methods for Uncovering the Connection Between the Complexity of Genomic Architecture and Multiple phenotypes. *Pac. Symp. Biocomput. Pac. Symp. Biocomput.* 183–187 (2014). doi:10.1142/9789814583220_0018
3. Pendergrass, S. A. *et al.* Phenome-Wide Association Studies: Embracing Complexity for Discovery. *Hum. Hered.*
4. Hebbring, S. J. The challenges, advantages and future of phenome-wide association studies. *Immunology* **141,** 157–165 (2014).
5. Denny, J. C. *et al.* PheWAS: demonstrating the feasibility of a phenome-wide scan to discover gene-disease associations. *Bioinforma. Oxf. Engl.* **26,** 1205–1210 (2010).
6. Hebbring, S. J. *et al.* A PheWAS approach in studying HLA-DRB1*1501. *Genes Immun.* **14,** 187–191 (2013).
7. Namjou, B. *et al.* Phenome-wide association study (PheWAS) in EMR-linked pediatric cohorts, genetically links PLCL1 to speech language development and IL5-IL13 to Eosinophilic Esophagitis. *Front. Genet.* **5,** 401 (2014).

8. Denny, J. C. *et al.* Systematic comparison of phenome-wide association study of electronic medical record data and genome-wide association study data. *Nat. Biotechnol.* **31,** 1102–1110 (2013).
9. R Development Core Team. R: A Language and Environment for Statistical Computing. (2008). at <http://www.R-project.org>
10. Purcell, S. *et al.* PLINK: a tool set for whole-genome association and population-based linkage analyses. *Am. J. Hum. Genet.* **81,** 559–575 (2007).
11. Patterson, N., Price, A. L. & Reich, D. Population Structure and Eigenanalysis. *PLoS Genet.* **2,** e190 (2006).
12. Pendergrass, S. A. *et al.* Biofilter 2.0 – Using Biological Knowledge for Advanced Filtering, Annotation, and Model Development for Genomic Analysis.
13. Hindorff, L. A. *et al.* Potential etiologic and functional implications of genome-wide association loci for human diseases and traits. *Proc. Natl. Acad. Sci.* **106,** 9362–9367 (2009).
14. Sobota, R. S. *et al.* Addressing Population-Specific Multiple Testing Burdens in Genetic Association Studies: Population-Specific Genome-Wide Thresholds. *Ann. Hum. Genet.* **79,** 136–147 (2015).
15. Smith, E. N. *et al.* Longitudinal Genome-Wide Association of Cardiovascular Disease Risk Factors in the Bogalusa Heart Study. *PLoS Genet.* **6,** e1001094 (2010).
16. Ko, A. *et al.* Amerindian-specific regions under positive selection harbour new lipid variants in Latinos. *Nat. Commun.* **5,** 3983 (2014).
17. Johansen, C. T. *et al.* Excess of rare variants in genes identified by genome-wide association study of hypertriglyceridemia. *Nat. Genet.* **42,** 684–687 (2010).
18. Hirano, K. *et al.* Disease-associated marked hyperalphalipoproteinemia. *Mol. Genet. Metab. Rep.* **1,** 264–268 (2014).
19. Ozóg, J. *et al.* [Expression of DPP4 gene in papillary thyroid carcinoma]. *Endokrynol. Pol.* **57 Suppl A,** 12–17 (2006).
20. Rago, T. *et al.* Clinical features of thyroid autoimmunity are associated with thyroiditis on histology and are not predictive of malignancy in 570 patients with indeterminate nodules on cytology who had a thyroidectomy. *Clin. Endocrinol. (Oxf.)* **67,** 363–369 (2007).
21. Holm, L.-E., Blomgren, H. & Löwhagen, T. Cancer Risks in Patients with Chronic Lymphocytic Thyroiditis. *N. Engl. J. Med.* **312,** 601–604 (1985).
22. Kim, E. S. *et al.* Thyroglobulin Antibody Is Associated with Increased Cancer Risk in Thyroid Nodules. *Thyroid* **20,** 885–891 (2010).
23. Arnaud-Lopez, L. *et al.* Phosphodiesterase 8B Gene Variants Are Associated with Serum TSH Levels and Thyroid Function. *Am. J. Hum. Genet.* **82,** 1270–1280 (2008).
24. Peng, C.-Y. *et al.* Association of MRC-1 and IL-28B with the treatment outcome of hepatitis C: a case control study. *BMC Gastroenterol.* **14,** 113 (2014).
25. Nguyen, C., Varney, M. D., Harrison, L. C. & Morahan, G. Definition of High-Risk Type 1 Diabetes HLA-DR and HLA-DQ Types Using Only Three Single Nucleotide Polymorphisms. *Diabetes* **62,** 2135–2140 (2013).
26. Sandhu, M. S. *et al.* LDL-cholesterol concentrations: a genome-wide association study. *The Lancet* **371,** 483–491 (2008).
27. Deloukas, P. *et al.* Large-scale association analysis identifies new risk loci for coronary artery disease. *Nat. Genet.* **45,** 25–33 (2012).
28. Ai, L. *et al.* Associations between rs965513/rs944289 and papillary thyroid carcinoma risk: a meta-analysis. *Endocrine* **47,** 428–434 (2014).
29. He, H. *et al.* Multiple functional variants in long-range enhancer elements contribute to the risk of SNP rs965513 in thyroid cancer. *Proc. Natl. Acad. Sci.* **112,** 6128–6133 (2015).
30. Vozarova, B. *et al.* High White Blood Cell Count Is Associated With a Worsening of Insulin Sensitivity and Predicts the Development of Type 2 Diabetes. *Diabetes* **51,** 455–461 (2002).

METHODS TO ENHANCE THE REPRODUCIBILITY OF PRECISION MEDICINE

ARJUN K MANRAI
Department of Biomedical Informatics
Harvard Medical School, Boston, MA
Email: Manrai@post.harvard.edu

CHIRAG J PATEL
Department of Biomedical Informatics
Harvard Medical School, Boston, MA
Email: Chirag_Patel@hms.harvard.edu

NILS GEHLENBORG
Department of Biomedical Informatics
Harvard Medical School, Boston, MA
Email: Nils@hms.harvard.edu

NICHOLAS P TATONETTI
Department of Biomedical Informatics
Columbia University, New York, NY
Email: Nick.Tatonetti@columbia.edu

JOHN P.A. IOANNIDIS
Department of Medicine
Stanford University School of Medicine, Stanford, CA
Email: Jioannid@stanford.edu

ISAAC S KOHANE
Department of Biomedical Informatics
Harvard Medical School, Boston, MA
Email: Isaac_Kohane@hms.harvard.edu

During January 2015, President Obama announced the Precision Medicine Initiative [1], strengthening communal efforts to integrate patient-centric molecular, environmental, and clinical "big" data. Such efforts have already improved aspects of clinical management for diseases such as non-small cell lung carcinoma [2], breast cancer [3], and hypertrophic cardiomyopathy [4]. To maintain this track record, it is necessary to cultivate practices that ensure reproducibility as large-scale heterogeneous datasets and databases proliferate. For example, the NIH has outlined initiatives to enhance reproducibility in preclinical research [5], both *Science* [6] and *Nature* [7] have featured recent editorials on reproducibility, and several authors have noted the issues of utilizing big data for public health [8], but few methods exist to ensure that big data resources motivated by precision medicine are being used reproducibly. Relevant challenges include: (1) integrative analyses of heterogeneous measurement platforms (e.g. genomic, clinical, quantified self, and exposure data), (2) the tradeoff in making personalized decisions using more targeted (e.g. individual-level) but potentially much noisier subsets of data, and (3) the unprecedented scale of asynchronous observational and population-level inquiry (i.e. many investigators separately mining shared/publicly-available data).

In this session of the Pacific Symposium on Biocomputing (PSB) 2016, we feature manuscripts that explore and propose solutions to some of the challenges of reproducibility in the era of precision medicine.

Two submissions to the session address challenges to reproducibility in observational (e.g., Electronic Health Record [EHR]) and clinical trial settings. Chen et al. [9] study the stability of predicting clinical practice patterns by varying the duration of EHR data used in training clinical order association rules, finding that larger longitudinal datasets may not improve, and might worsen, some predictions given the importance of secular practice trends. Ma et al. [10] provide a method for finding questionable exclusion criteria commonly used in clinical trials for mental disorders deposited in ClinicalTrials.gov.

Another challenge for the implementation of precision medicine involves novel methods for assessing data quality. Koire et al. [11] study threats to reproducibility when repurposing publicly available genome sequencing data, using data from The Cancer Genome Atlas [12] to study false positive variant calls and systematically evaluate variant call quality.

Software that enables analysts to transparently document analysis protocols can also help ensure reproducibility. Callahan et al. [13] create a reproducible workflow for microbiome studies using the Bioconductor [14] and knitr [15] *R* packages, providing a principled way to share protocols and explore how a multiplicity of analysis choices can sway results [16], [17]. Further, Manrai et al. [18] develop a shareable computational framework for quantifying widely-used pathogenicity assertions that relate genetic variation to disease, enabling users to identify how genetic model parameters influence risk estimates for genetic variants used in clinical practice.

These manuscripts address aspects of maintaining reproducibility as large-scale and heterogeneous datasets become increasingly common in the era of precision medicine. Concerted community-wide efforts will be critical to ensure that our ability to collect diverse types of patient-centric data is tantamount to our ability to distill reproducible findings from these data.

References

[1] F. S. Collins and H. Varmus, “A New Initiative on Precision Medicine.,” *N. Engl. J. Med.*, vol. 372, no. 9, pp. 793–5, Jan. 2015.

[2] W. Pao and N. Girard, “New driver mutations in non-small-cell lung cancer.,” *Lancet. Oncol.*, vol. 12, no. 2, pp. 175–80, Feb. 2011.

[3] S. M. Domchek, T. M. Friebel, C. F. Singer, D. G. Evans, H. T. Lynch, C. Isaacs, J. E. Garber, S. L. Neuhausen, E. Matloff, R. Eeles, G. Pichert, L. Van t’veer, N. Tung, J. N. Weitzel, F. J. Couch, W. S. Rubinstein, P. A. Ganz, M. B. Daly, O. I. Olopade, G. Tomlinson, J. Schildkraut, J. L. Blum, and T. R. Rebbeck, “Association of risk-reducing surgery in BRCA1 or BRCA2 mutation carriers with cancer risk and mortality.,” *JAMA*, vol. 304, no. 9, pp. 967–75, Sep. 2010.

[4] H. L. Rehm, “Disease-targeted sequencing: a cornerstone in the clinic.,” *Nat. Rev. Genet.*, vol. 14, no. 4, pp. 295–300, May 2013.

[5] F. S. Collins and L. A. Tabak, "Policy: NIH plans to enhance reproducibility.," *Nature*, vol. 505, no. 7485, pp. 612–3, Jan. 2014.
[6] M. McNutt, "Reproducibility.," *Science*, vol. 343, no. 6168, p. 229, Jan. 2014.
[7] "Journals unite for reproducibility.," *Nature*, vol. 515, no. 7525, p. 7, Nov. 2014.
[8] M. J. Khoury and J. P. A. Ioannidis, "Medicine. Big data meets public health.," *Science*, vol. 346, no. 6213, pp. 1054–5, Nov. 2014.
[9] J. H. Chen, M. K. Goldstein, S. M. Asch, and R. B. Altman, "Dynamically evolving clinical practices and implications for predicting medical decisions," *Pac Symp Biocomput.*, 2016.
[10] H. Ma and C. Weng, "Identification of questionable exclusion criteria in mental disorder clinical trials using a medical encyclopedia," *Pac Symp Biocomput.*, 2016.
[11] A. Koire, P. Katsonis, and O. Lichtarge, "Repurposing germline exomes of the cancer genome atlas demands a cautious approach and sample-specific variant filtering," *Pac Symp Biocomput.*, 2016.
[12] J. N. Weinstein, E. A. Collisson, G. B. Mills, K. R. M. Shaw, B. A. Ozenberger, K. Ellrott, I. Shmulevich, C. Sander, and J. M. Stuart, "The Cancer Genome Atlas Pan-Cancer analysis project.," *Nat. Genet.*, vol. 45, no. 10, pp. 1113–20, Oct. 2013.
[13] B. Callahan, D. Proctor, D. Relman, J. Fukuyama, and S. Holmes, "Reproducible research for the fine scale analyses of personalized human microbiome data," *Pac Symp Biocomput.*, 2016.
[14] R. C. Gentleman, V. J. Carey, D. M. Bates, B. Bolstad, M. Dettling, S. Dudoit, B. Ellis, L. Gautier, Y. Ge, J. Gentry, K. Hornik, T. Hothorn, W. Huber, S. Iacus, R. Irizarry, F. Leisch, C. Li, M. Maechler, A. J. Rossini, G. Sawitzki, C. Smith, G. Smyth, L. Tierney, J. Y. H. Yang, and J. Zhang, "Bioconductor: open software development for computational biology and bioinformatics.," *Genome Biol.*, vol. 5, no. 10, p. R80, Jan. 2004.
[15] Y. Xie, *Dynamic Documents with R and knitr*. 2014.
[16] S. S. Young and A. Karr, "Deming, data and observational studies."
[17] C. J. Patel, B. Burford, and J. P. A. Ioannidis, "Assessment of vibration of effects due to model specification can demonstrate the instability of observational associations," *J. Clin. Epidemiol.*, Jun. 2015.
[18] A. K. Manrai, B. L. Wang, C. J. Patel, and I. S. Kohane, "Reproducible and shareable quantifications of pathogenicity," *Pac Symp Biocomput.*, 2016.

REPRODUCIBLE RESEARCH WORKFLOW IN R FOR THE ANALYSIS OF PERSONALIZED HUMAN MICROBIOME DATA.

BENJAMIN CALLAHAN
Statistics Department, Stanford, CA 94305, USA

DIANA PROCTOR, DAVID RELMAN
Departments of Microbiology & Immunology, and Medicine
Stanford University, Stanford, CA 94305 and VA, Palo Alto, CA 94304, USA

JULIA FUKUYAMA, SUSAN HOLMES*
Statistics Department, Stanford University,
Stanford, CA 94305, USA
**E-mail:susan@stat.stanford.edu,* statweb.stanford.edu/~susan/

This article presents a *reproducible research* workflow for amplicon-based microbiome studies in personalized medicine created using Bioconductor packages and the `knitr` markdown interface. We show that sometimes a multiplicity of choices and lack of consistent documentation at each stage of the sequential processing pipeline used for the analysis of microbiome data can lead to spurious results. We propose its replacement with reproducible and documented analysis using `R` packages `dada2`, `knitr`, and `phyloseq`. This workflow implements both key stages of amplicon analysis: the initial filtering and denoising steps needed to construct taxonomic feature tables from error-containing sequencing reads (`dada2`), and the exploratory and inferential analysis of those feature tables and associated sample metadata (`phyloseq`). This workflow facilitates reproducible interrogation of the full set of choices required in microbiome studies. We present several examples in which we leverage existing packages for analysis in a way that allows easy sharing and modification by others, and give pointers to articles that depend on this reproducible workflow for the study of longitudinal and spatial series analyses of the vaginal microbiome in pregnancy and the oral microbiome in humans with healthy dentition and intra-oral tissues.

Keywords: Illumina; amplicon; DADA2; phyloseq; microbiota; microbiome; microbial ecology; longitudinal data; spatial; personalized medicine; random effects models.

1. Introduction

High-throughput (HT) DNA sequencing is allowing major advances in microbial studies; our understanding of the presence and abundance of microbial species relies heavily on the observation of their nucleic acids in a "culture independent" manner.[1] At present, the most common and cost-effective method for characterizing microbes and their communities is *amplicon sequencing*: PCR amplification of a small (100–500 bp) fragment of a conserved gene (phylogenetic marker) for which there are taxonomically-informative reference sequences available. The standard phylogenetic marker gene for bacteria is the small subunit ribosomal RNA (16S rRNA) gene,[1] for which there are also convenient tools and large reference databases.[2–4] 16S rRNA amplicon sequencing provides a census of the personalized bacterial communities present in a sampled individual.

After obtaining the amplicon sequences, a standard series of bioinformatic and statistical analyses are used to evaluate these data: filtering out low quality sequences and samples, constructing a taxonomic feature table of observations from each sample, incorporating the sample metadata, transforming and normalizing the feature table, and performing exploratory and inferential analyses. Here we explore the multiplicity of choices made during this process, show examples of their consequences, and motivate the need for better and easier reproducibility of the standard analytic workflow on amplicon data[a].

We focus here on two Bioconductor[5] packages — dada2[6] and phyloseq[7,8] — created specifically to analyze amplicon sequencing data within the R environment, and show how they enable reproducible research in several microbiome studies. We begin by illustrating the need for reproducible research workflows in microbiome studies with a typical workflow example.

2. A Case study in multiple outcomes: the Enterotypes

A few years ago, Arumugam, M. et al.[9] published an article in Nature that concluded that humans could be grouped into intestinal gut **types**. In fact, some bioinformatic forensics (presented in detail in the supplementary material) shows that during the course of the analysis, the following choices were made.

- A preliminary choice of data transformation from the original counts to proportions was made, although the authors could have chosen to take logarithms, variance stabilizing transformations[10] (here proportions replaced the original counts),
  choose between `log, rlog, subsample, prop, orig (5)`
- Nine points were dropped from the study as they were considered outliers; of course the authors could have chosen to `.. leave out 0, 1, 2 ,..,9, + criteria (100)`
- Certain taxa were filtered out as they were considered too rare or unlabeled.
  filter taxa `... remove rare taxa, ie threshold at 0.01%, 1%, 2%,...(10)`
- A distance was chosen (Jensen-Shannon, JSD) to quantify similarities between samples.
  Distances `... 40 choices in vegan/phyloseq (40)`
- An ordination method and number of coordinates has to be chosen.
  Ordination and axes `... MDS, NMDS, DCA, k=2,3,4,5.. (16)`
- A clustering method and a number of clusters
  is chosen `... PAM, KNN, hclust ... (16)`
- The authors chose an underlying continuous variable, an alternative could have been a linear or curved latent variable or group of variables. `latent variable choices... (4)`

According to this rough list, there are more than 200 million possible ways of analyzing these data [b]. Thus, there is a combinatorial explosion of the number of possible choices that an investigator makes. Some choices can impact conclusions drawn from microbiome studies; it becomes necessary for experimentalists to develop and adopt pipelines documenting choices used in these analyses with the intention of providing an assessment of the robustness and

[a]Throughout this article we use regular or `texttype` font for packages/applications with names that are capitalized or uncapitalized, respectively. We use a `courier` style font for R code, including function and class names. The supplementary material contains `Rmd` files of the complete R code that enable the reproduction of all the figures in the article.

[b]$5 \times 100 \times 10 \times 40 \times 16 \times 16 \times 4 = 204800000$

reproducibility of the analyses. In fact, an errata was published to the paper,[9] substantially weakening the original conclusions. Figure 1 shows graphical representations made after the Jensen-Shannon distance was computed on the data. The authors made an inappropriate supplementary step using the data based clusters as labels in a supervised classification between group analysis that separated the clusters more than they actually appear in the middle figure. There is unfortunately no way to use multiple hypothesis testing corrections for this number of

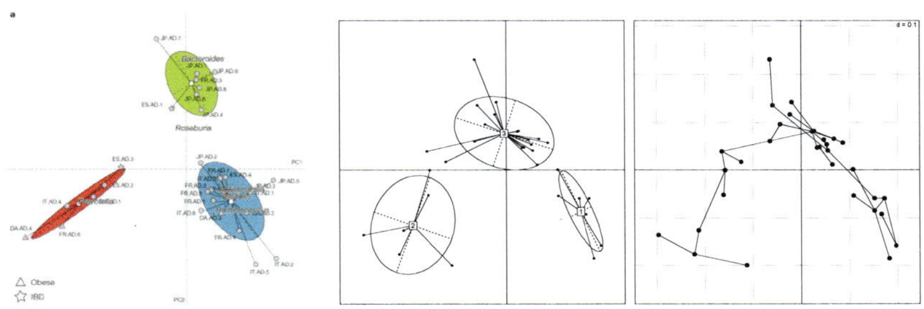

Fig. 1: On the left we show the analysis as done in[9] , in the middle we have done the same analysis with the Jensen Shannon distance but without the extra (invalid) supervised separation and on the right we have the minimum spanning tree exhibiting a clear gradient in the data.

possible analyses, thus the only way of ensuring robustness of the conclusions is to repeat the analyses with many different settings. In the supplementary material we include the output showing the ordination with 40 different distances. In particular, the clustering is not always as obvious; different choices of distance such as chisquare or Jaccard give very different results.

As this re-analysis demonstrates, access to reproducible analysis workflows is necessary for the interpretation of modern microbiome studies. In this example, in *just one stage of the analysis* (clustering of samples based on taxonomic features), the reported outcome was one out of millions of analogous alternatives, many of which differed qualitatively. Other parts of standard amplicon analysis, such as the construction of OTU tables and the evaluation of differential abundances, are accompanied by a similar myriad of choices. For this reason it is crucial that the analysis of amplicon data be made accessible — sharing the data alone is not enough.

3. A reproducible workflow in R

Here we present a workflow for the analysis of amplicon data within R (Figure 2). This workflow takes as input the amplicon sequencing reads and associated sample metadata, and provides as output exploratory and inferential statistical analyses as well as sharable analysis scripts and data files that fully reproduce those analyses. Here we focus on two particular packages developed by our group for the analysis of amplicon data within the R environment: `dada2` and `phyloseq`.

3.1. *Inferring sample sequences and abundances using DADA2*

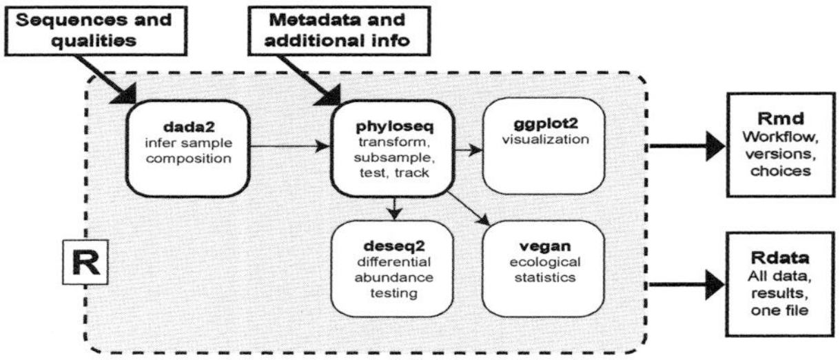

Fig. 2: Diagram of the new reproducible workflow including denoising, data integration and statistical analyses.

The DNA sequence errors introduced by PCR and sequencing complicate the interpretation of amplicon data, and present different challenges than the more well known problem of resequencing. When re-sequencing a diploid organism (like a human being) it is known that there exist either 1 or 2 variants at every position in the genome. Thus increasing depth eventually trivializes the problem of making genotype calls by overwhelming the error rate with data. However, when amplicon sequencing microbial communities the number of variants and their associated frequencies are unknown, which fundamentally changes the inference problem. When increasing sampling depth reveals new sequence variants, these might represent rare errors or rare members of the community. In addition, the PCR amplification step introduces chimeras and additional errors with a different structure than sequencing errors.

Most current studies use two methods to deal with amplicon errors, reducing their incidence by filtering out low quality reads, and lumping similar sequences together into Operational Taxonomic Units (OTUs). However, there are a significant number of choices made during this process: the type and stringency of quality filtering, the minimum abundance threshold, the size of the OTUs, the OTU construction method, and more. All of these choices can have significant downstream consequences for later analysis.[11]

This has led to serious problems for the reproducibility of amplicon-based studies. The methods used to filter sequences, construct OTUs and then perform analysis are often performed in separate environments (e.g., shell scripts vs. `python` vs. `R`). This makes the creation of a single coherent record of the analysis from input data to final product difficult and time-consuming. In practice few studies can be reproduced from the original raw data.

We have addressed this shortcoming by developing the `dada2` package[6] for R which performs the crucial filtering and sample inference steps that turn a set of raw amplicon sequences into a feature table of the types observed in each sample (e.g., an OTU table). Because `dada2` shares the R environment with downstream analysis methods already present in R, such as those in the `phyloseq` package, the publication of reproducible workflows encompassing the entirety of the analysis is far easier. One unified R script, and one unified Rdata data object, can provide a complete record of the published analysis, and allows interrogation of the full set of choices made in that process.

3.2. *Performing exploratory and inferential analysis with phyloseq*

`Phyloseq` allows the user to import a species by sample contingency table matrix (aka, an OTU Table) and data matrices from metagenomic, metabolomic, and or other −omics type experiments into the R computing environment. Phyloseq is unique in that it allows the user to integrate the OTU Table, the phylogenetic tree, the "representative sequence" fasta file, and the metadata mapping file into a single "phyloseq-class" R object. The microbial ecologist can then harness all the statistical and graphical tools available in R, including `knitr`, ggplot2 to generate reproducible research reports with beautiful graphics, as detailed in our supplementary `.Rmd` file and in the case studies below.

Combining this environment with a number of other important R packages (e.g., `vegan`, `ade4`, `DESeq2`, `multtest` ...) allows for powerful and specific analyses to be performed on amplicon-sequenced microbiome data. We share several such examples, along with the data and code necessary to reproduce them.

4. Examples: Longitudinal data analysis

Tackling the challenges involved in longitudinal patient-dependent data requires methods specifically tailored for the human body sites studied. For instance, the vaginal community is the one human body habitat that has been shown to robustly cluster into discrete community state types (CSTs).[12] This feature allows the complex information about community composition to be simplified by projecting into a small number of CSTs. Combined with longitudinal sampling information, this simplified projection is then amenable to analysis as a Markov chain.

In a 2015 study[12] we used this Markov chain representation to analyze the dynamics of the vaginal community during pregnancy. Transition rates were estimated from a set of 652 pairs of samples collected one week apart during the pregnancies of 40 women, producing an estimation of the dynamics of the vaginal community as illustrated in Figure 3. These results reproduced previous qualitative and semi-quantitative observations,[13] such as the high stability of *Lactobacillus crispatus* communities, but also provided a more detailed quantification of the stability of each CST as well as the connectivity between them.

Markov chain analysis is a powerful way to quantify and visualize dynamics, but it can only be applied to systems that are representable by a set of discrete states (a property which is often not trivial to establish as in Section 2). In the context of microbial communities this is a substantial limitation, as few communities can be so represented. A second concern, especially when applied to human-associated communities, is how the estimation of the transition rates should be performed across subjects. If the community dynamics are subject-independent, then an average over the observed transitions in each subject is appropriate, and this was the method used in our analysis of pregnant vaginal communities. However, it is possible that subject-specific factors (eg. host genetics) may influence transition rates, in which case the set of states should be expanded to include the subject effect.

Finally, the uncertainties that exist in mapping community states to discrete CSTs can also have significant consequences for the estimation of transition rates between those states, as in the case where a rare and unmodeled community state exists intermediate between the centers of two CSTs and is sometimes assigned to one or the other.

Even in the relatively very simple case of the vaginal community, this set of concerns cannot be comprehensively addressed within a single manuscript. Thus the need for access to the analytical workflow. In this study, we used the reproducible R workflow, and `Rmd` and `Rdata` files, to make our analysis easily accessible and modifiable and have deposited the data and code in a permanent repository (`permanent url`) maintained by the Stanford Digital Repository at `http://purl.stanford.edu/hg140kw6221`.

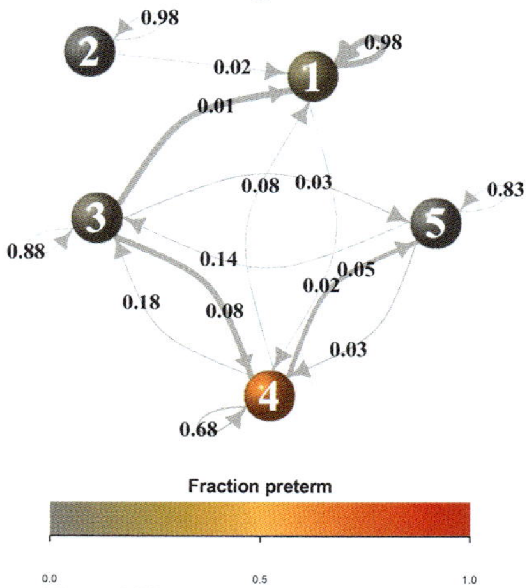

Fig. 3: Markov chain modeling of CST states across pregnancy and preterm birth.[12] Numbers indicate the one-week self-transition rate for each state. The high-diversity, low Lactobacillus class 4 is the least stable and most connected to the other CSTs. A more complete version of this figure appears in the aforementioned PNAS paper.[12]

5. Example: Spatial data analysis

Patterns of diversity and community composition across human body-sites have been well characterized.[14] When comparing human-associated microbial communities across different anatomic sites, skin and gut for instance, dramatic differences in the acquisition, development, and maintenance of microbial community composition are observed. Few studies have yet examined the extent to which microbial communities vary across fine scale spatial gradients on the human body, such as between and across individual tooth surfaces in the oral cavity. Datasets that have attempted to examine the spatial variation of oral microbial communities have shown interpersonal variation to be the strongest effect with secondary effects exerted by tooth position.[15,16] We have extended the current exploratory approaches through the use of the statistical packages available in R specifically tailored to analyze spatial or longitudinal data.

In this study, we demonstrate the usability of the `phyloseq`[7] package for applied spatial analysis of microbial communities in the oral cavity. As a test case, we generated data for this demonstration, collecting 186 independent samples from the facial (cheek-facing) and lingual (tongue-facing) surfaces of every tooth (excluding the third molars) of one adult female on each of two non-consecutive days. We extracted DNA from each sample, amplified the V3-V5 region of the 16S rRNA gene using golay-barcoded primers, and sequenced the amplicons using the 454-Titanium platform, generating 216,965 sequences with a median sequencing depth of 2,479. We use two methods to examine the spatial variation of oral microbial communities: a Between Class Analysis and a Principal Components Analysis with respect to Instrumental Variables.

5.1. *Between Class Analysis*

When dealing with a priori classes in which we know teeth communities segregate, we want to highlight differences in a supervised analysis. The segregation of supragingival communities might arise because teeth are situated along an ecological gradient. As a first examination of the spatial relationship between the oral communities, after filtering and preprocessing the data we used `ade4` to perform a Between Class Analysis (BCA) in which tooth class was used as the spatial partition.

Using teeth groupings, we found that 12% of the total variance could be explained by Tooth-Class with the first component accounting for 56.5% of the explained variance. The BCA revealed that not only can communities be distinguished from one another based on tooth class, but that these communities may exist along an ecological gradient. Molar and Premolar communities appeared to be associated with positive scores along the first BCA axis while the Central Incisors and Lateral Incisors appeared to be associated with more negative scores along the first BCA axis, regardless of whether we examine communities in the top (Maxillary) or bottom (Mandibular) jaws or whether we examined the buccal or the lingual aspect of teeth. The distribution of teeth along the first and second components suggested that supragingival plaque varies along a gradient in the mouth. Interestingly, the variance of community scores along CS1 varied according to tooth class (Figure 4) with Central Incisor communities appearing to be the most variable community class especially if communities were sampled from the lingual aspect of the teeth or from the buccal aspect of teeth in the lower jaw. This is interesting given the relative proximity of these sites to the secretions of the submandibular/sublingual glands, which may give rise to the observed structure.

5.2. *Principal Components with Instrumental Variables*

It can be challenging to incorporate covariates such as spatial variation into studies of microbial communities. Here, we explicitly model the spatial variation of microbial communities in the human mouth using a Principal Components Analysis with respect to Instrumental Variables (PCA-IV)[17] where a third-order polynomial function of the geographic coordinates was used as the constraint.

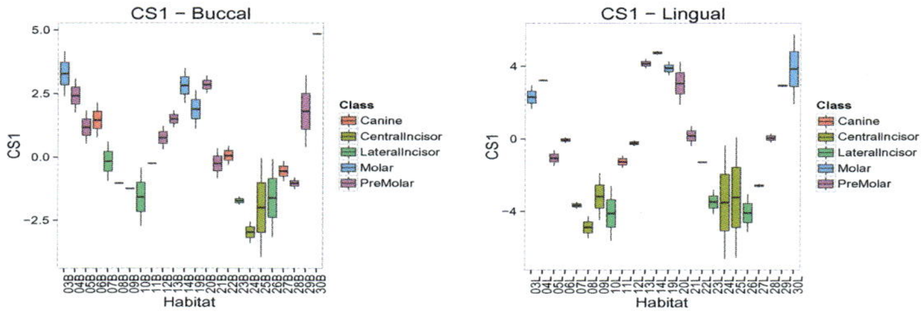

Fig. 4: Comparison of Buccal and Lingual sides of five different teeth types.

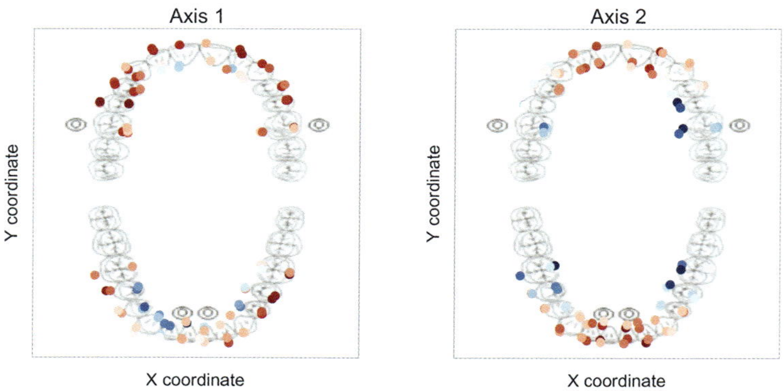

Fig. 5: A representation of the output from PCA-IV. The left and right panels show scores along the first and second PCA-IV axes, respectively. Each dot represents a sample, and its position with respect to the image of the teeth shows the area it was sampled from. Color represents whether the sample scored high (blue) or low (red) on the first and second PCA-IV axes. The first PCA-IV axis reveals a complex interaction between tooth aspect, the front vs. the back of the mouth, and jaw. Communities on the lingual aspect of most anterior teeth share negative scores along the first coordinate with a more pronounced difference between tooth aspect in the bottom jaw. The second axis describes a posterior to anterior gradient.

The PCA-IV accounts for 27% of the variance. The first principal coordinate separates buccal from lingual communities with lingual communities being associated with more positive scores along Axis 1 compared to buccal communities, especially for communities in the bottom jaw. Examining multiple individuals should confirm whether this pattern is consistent across multiple subjects and may reveal the sum of factors that structure the spatial variation of microbial communities. For now, we speculate that in this subject the relative proximity of the the submandibular glands to the upper anterior sites, the lower anterior, and posterior sites are important factors contributing to variation along Axis 1.

Axis 2, on the other hand, accounts for 23.4% of the total variance and appears to separate molar from incisor communities with molar communities being associated with positive scores along Axis 1 and incisors being associated with negative values along Axis 1.

While the data presented here pertain to just a single subject, and therefore our ability to make population level inferences limited, in this analysis we generated and provided an `.Rmd` script that can be used by other expeirmentalists to test hypotheses related to the spatial structure of host-associated communities in the oral cavity or at other body sites.

6. Relevance to Precision Medicine

The vaginal and oral community examples provided above both have relevant applications to personalized medicine. In the first example, the state of the vaginal community during pregnancy has been shown to be related to the likelihood of preterm birth outcomes in some women, with higher relative abundances of particular taxa such as *Gardnerella vaginalis* and *Ureaplasma* implicated as specific risk factors.[12] Furthermore, the duration of time spent in the high-risk states further stratified preterm risk. This suggests that longitudinal monitoring of the vaginal community during pregnancy, concomitant with the standard schedule of prenatal care, might provide a biomarker for preterm birth risks early in pregnancy allowing for pre-emptive intervention and education.

In the second example, we show an unpublished example of a spatial analysis of the oral microbiome which will have broad applicability in the dental clinic. Dentists have long known that most people experience greater difficulty brushing the lingual aspects of posterior teeth as compared to the buccal aspects of those same teeth, and that whether an individual is right or left handed impacts brushing efficacy. Differences in brushing technique may therefore give rise to differences in microbial community composition in different areas of the mouth, and these differences may be highly specific to individual patients and may relate to the incidence of dental disease. One possible application of this type of analysis in the dental clinic would be to provide patients with customized diagrams of the microbial inhabitants of their oral cavities in order to help them to better understand the impact of brushing technique on their oral health. If analyses were conducted at every semi-annual dental exam the relative impact that different dental interventions have on the spatial structure of communities could be elucidated on a per-patient basis. By wrapping up this type of metadata with phylogenetic sequence analysis and depositing the information into public repositories, we increase our ability to make this type of inference. Analysis of spatial and temporal series may also have relevance to dental disease, which is patchy. Cavities (which are not generalized) are known to form on discrete, localized surfaces of dental enamel, and in the healthy individual a single cavity typically takes years to develop. Our ability to decompose the spatial and temporal components of health-associated supragingival communities would enable clinicians to develop models that detect deviations from health, such as incipient caries, which are reversible. These types of models might then allow clinicians to detect and reverse early carious lesions before the need for costly and invasive dental restorations arises.

These examples demonstrate the need for reproducibility in microbiome research and the relevance of the reproducibility problem to the precision medicine initiative.

7. Difficulties encountered in the production of Reproducible Research

There are many biological and technical choices, which are challenging to standardize across projects, institutions, and investigators, that make experimental and analytical findings difficult to replicate. Studies that examine the ecological structure of human-associated microbial communities are tough to compare when data are generated using different sequencing technologies (e.g., 454 vs. Illumina), or even when investigators simply sequence different regions of the 16S rRNA gene. The rapid increase in the number of sequencing technologies makes it difficult to standardize the highest level technical factors that give rise to irreproducible data. In contrast, how we make use of data that have already been generated is one of the easiest ways to increase the reproducibility of experimental findings. The difficulty that any investigator faces in replicating the steps and choices of another investigator in executing their data analysis, when provided with the same raw data, is a more subtle, often-overlooked problem that is easily addressed. When considering this question in the context of precision medicine it is easy to recognize that a clinician trying to provide an individualized report on, say, oral health must be able to generate a record exactly as intended by the investigators who developed the report. To facilitate the use of these analyses in the clinic, reports could be developed to run with the Shiny Phyloseq[8] interface, which places the computational power of R in the hands of individuals less customized to working with scripts. Experimentalists will benefit from considering their data analysis pipelines from this point of view. While most bench scientists use a laboratory notebook to document choices that guide their decisions at the bench, fewer bench scientists have adopted the tools (e.g., LaTex, RMarkdown, etc.) widely used in mathematics and statistics to document the analytic choices used during data analysis. The pipelines presented here provide microbial ecologists with a single platform with which to analyze 16S rRNA gene amplicon data, providing bench scientists with the ability to generate scripts that can be executed by other scientists (or eventually by clinicians in the clinic) and enabling the genesis of figures and findings that are precisely replicable and usable in a variety of other related contexts. Here we have provided examples of three statistical analyses using `R` and in particular the `Rmd` format, easily available in `RStudio`.

7.1. *Open Data Access Barriers to Reproducibility*

The NIH Open Data Access Policy should dramatically increase our ability to reproduce findings from published datasets in addition to allowing researchers to leverage existing findings in analyses of new experiments. Nonetheless, researchers face several other barriers when trying to access these files. Non-standardized or non-existent mapping files make it difficult to analyze data that have been deposited in public repositories as multiple short read archives.

We advocate the use of platforms (e.g., Github, Bioconductor, CRAN) used by statisticians and bioinformaticians in which experimentalists can deposit not just their sequencing data but also packages containing their complete *metadata* mapping files, taxonomy files, reference sequence files, which can all be wrapped into a single `phyloseq` object within `R`.

7.2. *Advantages and weaknesses of the R workflow*

While rarely achieved in practice, the need for reproducible analysis in amplicon studies has been recognized and there are several existing approaches. The two most common pieces of pipeline software – mothur[18] and `qiime`[19] – allow analysis workflows to be shared as batch files or IPython notebooks[20] respectively. Also, restricted versions of a number of amplicon analysis tools are available through cloud based platforms such as Galaxy.[21]

However, an amplicon analysis workflow in R provides several advantages over these existing approaches. The most compelling advantage is R's access to a constellation of state-of-the art statistical methods. While common statistical tests have been implemented by pipelines like mothur and QIIME, there is no other platform that has a suite of statistical tools as broad as that implemented in R. In practice, many studies use the popular amplicon pipelines for the initial stages of their analysis, and then transition into R for deeper analysis, with the result being that even when analysis scripts are shared, they only cover part of the full workflow. Additionally, R markdown, as implemented by the `knitr` package, allows the construction of R analysis for which both the underlying code and the full output of that code can be easily shared.

There are also weaknesses in the workflow we present. One of those weaknesses is that our current R workflow does not include the optimal storage of raw sequence data. A second weakness is in the taxonomic assignment of amplicon sequences or OTUs. While one R package does exist for this purpose (`clstutils`), it requires parallel computation in an external program (`pplacer`[22]), preventing full reproducibility from the R script alone. Taxonomic assignment fully within R is an area of particular interest for future development.

8. Conclusions

We describe reproducible research in the context of studies of the human microbiome. Bioconductor packages, `dada2`, `phyloseq`, allow for denoising, handling, filtering, and analyzing high-throughput phylogenetic sequencing data. The `phyloseq` package provides extensions for leveraging analysis from other ecology-related packages, such as `spatstat` for spatial data analyses, `ade4` for multiway data analyses and `lme4` for mixed model analyses. We believe that this workflow provides a useful way to document choices in the analyses of phylogenetic sequencing data, its quality control, filtering, processing and its inferential validation.

9. Acknowledgements

We would like to thank the developers of all the open source packages we used including Joey McMurdie for his continuing dedication to `phyloseq`, Hadley Wickham for `ggplot2`, Yihui Xie for `knitr`. We are thankful to the team at RStudio, their IDE and the Shiny platform has made reproducible research much easier. Dan DiGuilio worked with us on the pregnancy study, Elies Bik, Les Dethlefsen, Daniel Sprockett and Jessica Grembi provided useful feedback. This work was partially supported by NSF DMS-1162538 (BC,SPH), NIH R01AI112401(SPH, DAR), NIH R01DE023113 (DAR), NIH R01GM099534 (DAR, SPH) and a Microbiome Seed Grant from Stanford-BioX. DAR is supported by the Thomas C. and Joan

M. Merigan Endowment at Stanford University. The supplementary Rmd and html files for this paper can be found at `http://statweb.stanford.edu/~susan/papers/PSBRR.html`.

References

1. N. R. Pace, *Science* **276**, 734 (1997).
2. T. Z. DeSantis, P. Hugenholtz, N. Larsen, M. Rojas, E. L. Brodie, K. Keller, T. Huber, D. Dalevi, P. Hu and G. L. Andersen, *Applied and Environmental Microbiology* **72**, 5069 (2006).
3. J. R. Cole, Q. Wang, E. Cardenas, J. Fish, B. Chai, R. J. Farris, A. S. Kulam-Syed-Mohideen, D. M. McGarrell, T. Marsh, G. M. Garrity and J. M. Tiedje, *Nucleic Acids Research* **37**, D141 (2009).
4. E. Pruesse, C. Quast, K. Knittel, B. M. Fuchs, W. Ludwig, J. Peplies and F. O. Glöckner, *Nucleic Acids Research* **35**, 7188 (2007).
5. W. Huber, V. J. Carey, R. Gentleman, S. Anders, M. Carlson, B. S. Carvalho, H. C. Bravo, S. Davis, L. Gatto, T. Girke, R. Gottardo, F. Hahne, K. D. Hansen, R. A. Irizarry, M. Lawrence, M. I. Love, J. MacDonald, V. Obenchain, A. K. Ole, H. Pages, A. Reyes, P. Shannon, G. K. Smyth, D. Tenenbaum, L. Waldron and M. Morgan, *Nat. Methods* **12**, 115 (Feb 2015).
6. B. J. Callahan, P. J. McMurdie, M. J. Rosen, A. W. Han, A. J. Johnson and S. P. Holmes, *bioRxiv* (2015).
7. P. J. McMurdie and S. Holmes, *PLoS ONE* **8**, p. e61217 (2013).
8. P. J. McMurdie and S. Holmes, *Bioinformatics* **31**, 282 (2015).
9. M. Arumugam, J. Raes, E. Pelletier, D. Le Paslier, T. Yamada, D. Mende, G. Fernandes, J. Tap, T. Bruls, J. Batto *et al.*, *Nature* **473**, 174 (2011).
10. P. J. McMurdie and S. Holmes, *PLoS Comput. Biol.* **10**, p. e1003531 (Apr 2014).
11. T. S. Schmidt, J. F. Matias Rodrigues and C. Mering, *Environmental microbiology* (2014).
12. D. B. DiGiulio, B. J. Callahan, P. J. McMurdie, E. K. Costello, D. J. Lyell, A. Robaczewska, C. L. Sun, D. S. Aliaga-Goltsman, R. J. Wong, G. M. Shaw, D. K. Stevenson, S. P. Holmes and D. A. Relman, *Proceedings of the National Academy of Sciences* **112**, 11060 (2015).
13. P. Gajer, R. M. Brotman, G. Bai, J. Sakamoto, U. M. Schütte, X. Zhong, S. S. Koenig, L. Fu, Z. S. Ma, X. Zhou *et al.*, *Science translational medicine* **4**, 132ra52 (2012).
14. E. K. Costello, C. L. Lauber, M. Hamady, N. Fierer, J. I. Gordon and R. Knight, *Science* **326**, 1694 (2009).
15. Y. Sato, J. Yamagishi, R. Yamashita, N. Shinozaki, B. Ye, T. Yamada, M. Yamamoto, M. Nagasaki and A. Tsuboi, *PloS one* **10**, p. e0131607 (2015).
16. A. Haffajee, R. Teles, M. Patel, X. Song, N. Veiga and S. Socransky, *Journal of periodontal research* **44**, 511 (2009).
17. S. Holmes, Multivariate data analysis: the french way, in *Probability and statistics: Essays in honor of David A. Freedman*, (Institute of Mathematical Statistics, 2008) pp. 219–233.
18. P. D. Schloss, S. L. Westcott, T. Ryabin, J. R. Hall, M. Hartmann, E. B. Hollister, R. A. Lesniewski, B. B. Oakley, D. H. Parks, C. J. Robinson *et al.*, *Applied and environmental microbiology* **75**, 7537 (2009).
19. J. G. Caporaso, J. Kuczynski, J. Stombaugh, K. Bittinger, F. D. Bushman, E. K. Costello, N. Fierer, A. G. Pena, J. K. Goodrich, J. I. Gordon *et al.*, *Nature methods* **7**, 335 (2010).
20. B. Ragan-Kelley, W. A. Walters, D. McDonald, J. Riley, B. E. Granger, A. Gonzalez, R. Knight, F. Perez and J. G. Caporaso, *The ISME journal* **7**, p. 461 (2013).
21. D. Blankenberg, J. Taylor, I. Schenck, J. He, Y. Zhang, M. Ghent, N. Veeraraghavan, I. Albert, W. Miller, K. D. Makova *et al.*, *Genome research* **17**, 960 (2007).
22. F. A. Matsen, R. B. Kodner and E. V. Armbrust, *BMC bioinformatics* **11**, p. 538 (2010).

DYNAMICALLY EVOLVING CLINICAL PRACTICES AND IMPLICATIONS FOR PREDICTING MEDICAL DECISIONS

JONATHAN H CHEN
Center for Innovation to Implementation (Ci2i), Veteran Affairs Palo Alto Health Care System
Palo Alto, CA 94304 USA
Center for Primary Care and Outcomes Research (PCOR), Stanford University
Stanford, CA 94305 USA
Email: jonc101@stanford.edu

MARY K GOLDSTEIN
Geriatrics Research Education and Clinical Center, Veteran Affairs Palo Alto Health Care System
Palo Alto, CA 94304 USA
Center for Primary Care and Outcomes Research (PCOR), Stanford University
Stanford, CA 94305 USA
Email: mary.goldstein@va.gov

STEVEN M ASCH
Center for Innovation to Implementation (Ci2i), Veteran Affairs Palo Alto Health Care System
Palo Alto, CA 94304 USA
Division of General Medical Disciplines, Department of Internal Medicine, Stanford University
Stanford, CA 94305 USA
Email: sasch@stanford.edu

RUSS B ALTMAN
Departments of Bioengineering and Genetics, Stanford University, Stanford, CA 94305 USA
Department of Medicine, Stanford University, Stanford, CA 94305 USA
Email: russ.altman@stanford.edu

Automatically data-mining clinical practice patterns from electronic health records (EHR) can enable prediction of future practices as a form of clinical decision support (CDS). Our objective is to determine the stability of learned clinical practice patterns over time and what implication this has when using varying longitudinal historical data sources towards predicting *future* decisions. We trained an association rule engine for clinical orders (e.g., labs, imaging, medications) using structured inpatient data from a tertiary academic hospital. Comparing top order associations per admission diagnosis from training data in 2009 vs. 2012, we find practice variability from unstable diagnoses with rank biased overlap (RBO)<0.35 (e.g., pneumonia) to stable admissions for planned procedures (e.g., chemotherapy, surgery) with comparatively high RBO>0.6. Predicting admission orders for future (2013) patients with associations trained on recent (2012) vs. older (2009) data improved accuracy evaluated by area under the receiver operating characteristic curve (ROC-AUC) 0.89 to 0.92, precision at ten (positive predictive value of the top ten predictions against actual orders) 30% to 37%, and weighted recall (sensitivity) at ten 2.4% to 13%, ($P<10^{-10}$). Training with more longitudinal data (2009-2012) was no better than only using recent (2012) data. Secular trends in practice patterns likely explain why smaller but more recent training data is more accurate at predicting future practices.

1. Introduction

Variability and uncertainty in medical practice compromise quality of care and cost efficiency, with overall compliance with evidence-based guidelines ranging from 20-80%.[1] Clinical decision support (CDS) tools, like order sets and alerts, reinforce best-practices by distributing information on relevant clinical orders (e.g., labs, imaging, medications),[2-5] but production is limited in scale by knowledge-based manual authoring of one intervention at a time by human experts.[6] If medical knowledge were fixed, manual approaches might eventually converge towards a comprehensive set of effective clinical decision support content from the top-down. The reality is instead a perpetually evolving body of knowledge that responds to new evidence, technology, and epidemiology that requires ongoing content maintenance to adapt to changing clinical practices.[7]

The meaningful use era of electronic health records (EHR)[8] creates an opportunity for data-driven clinical decision support (CDS) to reduce detrimental practice variability through the collective expertise of many practitioners in a learning health system.[9-13] Specifically, one of the "grand challenges" in CDS[14] is automated production of CDS from the bottom-up by data-mining clinical data sources. Such algorithmic approaches to clinical information retrieval could greatly expand the scope of medical practice addressed with effective decision support, and automatically adapt to an ongoing stream of evolving clinical practice data. This would fulfill the vision of a learning health system to continuously learn from real-world practices and translate them into usable information for implementation back at the point-of-care. The Big Data[13,15] potential of EHRs makes this vision possible, but the dynamic nature of clinical practices over time calls into question the presumption that learning from historical clinical data will inform future clinical practice. To fulfill the potential of real-time clinical prediction, we need to better understand how far back in time to mine EHRs while retaining predictive value for future decision making.

2. Background

To understand clinical practice patterns and inform potential decision support, we focus on the clinical orders (e.g., labs, imaging, medications) that concretely manifest point-of-care decision making. Prior research into data-mining for clinical decision support content includes use of association rules, Bayesian networks, and unsupervised clustering of clinical orders and diagnoses.[16-23] This prior research has largely ignored the temporal relationships between clinical data elements when training predictive models, treating individual patients or encounters as an unordered collection of items. In our own prior work, inspired by analogous information retrieval problems in recommender systems, collaborative filtering, and market basket analysis, we automatically generated clinical decision support content in the form of a clinical order recommender system[24] analogous to Netflix or Amazon.com's "Customer's who bought A also bought B" system.[25] This prior work[26] first examined the importance of matching the temporal relationship between clinical data elements to the respective timing of evaluation outcomes. For example, orders co-occuring within a short time period, such as the antibiotics vancomycin and piperacillin-tazobactam being ordered within one hour of each other, inform a more useful association than orders separated by several days of time. The impact of the temporal relationship between training and validation data has not been explored in this prior research (including our own). Instead, any evaluation of these predictive models was conducted by separating patients into random train-test subsets. This is not representative of realistic applied scenarios however, where we would have to learn from historical clinical data to inform recommendations and predictions towards future patient encounters that have never previously occurred.

In this work, we seek to determine how varying longitudinal historical training data usage can impact prediction of future clinical practices. Furthermore, we seek to quantify which inpatient admission diagnoses exhibit the most stability vs. variability of clinical practice patterns over time.

3. Materials and Methods

We extracted deidentified patient data from the (Epic) electronic medical record for all inpatient hospitalizations at Stanford University Hospital via the STRIDE clinical data warehouse.[27] The structured data covers patient encounters from their initial (emergency room) presentation until hospital discharge. With five years of data spanning 2008-2014, the dataset includes >74K patients with >11M instances of >27K distinct clinical items. The clinical item elements include >7,800 medication, >1,600 laboratory, >1,100 imaging, and >1,000 nursing orders. Non-order items include >1,000 lab results, >7,800 problem list entries, >5,300 admission diagnosis ICD9 codes, and patient demographics. Medication data was normalized with RxNorm mappings[28] down to active ingredients and routes of administration. Numerical lab results were binned into categories based on "abnormal" flags established by the clinical laboratory. To compress the sparsity of diagnosis items, we duplicated ICD9 codes up to the three digit hierarchy, such that an item for code 786.05 would have additional items replicated for code 786.0 and 786. The above pre-processing models each patient as a timeline of clinical item instances, with each instance mapping a clinical item to a patient at a discrete time point.

With the clinical item instances following the "80/20 rule" of a power law distribution,[29] most clinical items may be ignored with minimal information loss. In this case, ignoring rare clinical items with <256 instances reduces the effective item count from >27K to ~3K (11%), while still capturing 10.8M (95%) of the 11.4M item instances. After excluding common process orders (e.g., vital signs, notify MD, regular diet, transport patient, as well as most nursing and all PRN medications), 1,270 clinical orders of interest remained.

Using our previously described method,[24] we algorithmically mined association rules for clinical item pairs from past clinician behavior. Based on Amazon's product recommender,[25] we collected patient counts for all clinical item instance pairs co-occurring within 24 hours of each other to build time-stratified item association matrixes.[26] Each matrix defines a 2x2 contingency table for each pair of clinical items, from which various association statistics are derived (e.g., odds ratio (OR), positive predictive value (PPV), baseline prevalence, and Fisher's P-value).[30] To assess the varying impact of historical training data time, separate item association matrix models were built from training data from 2009, data from 2012, and data from 2009 through 2012.

We identify clinical order associations that reflect practice patterns by using query items (e.g., admission diagnosis or first several clinical orders and lab results) to score-rank all candidate clinical order items by an association statistic relative to the query items. Score-ranking by PPV (positive predictive value) prioritizes orders that are *likely* to occur after the query items, while score-ranking by Fisher's P-value for items with odds ratio > 1 prioritizes orders that are *disproportionately associated* with the query items.[26]

To find clinical orders associated with different admission diagnoses, we generated a score-ranked list of the 1,270 candidate clinical orders for each of the most common admission diagnoses (those with at least 36 instances per year), sorted by Fisher's P-value. To assess for stability in clinical order patterns, we generated two such clinical order lists for each admission diagnosis, one from the matrix built on 2009 data and the other from the 2012 data. Traditional measures of list agreement like Kendall's τ[31] are not ideal here, as they often require identically sized, finite lists, and weigh all ranks equivalently. To compare ranked clinical order lists, we instead calculate their agreement by Rank Biased Overlap (RBO).[32] When comparing two ranked lists, we define I_k as the intersection of the top k items in each list, and X_k as the size of the overlap at rank depth = $|I_k|$. The ratio of X_k to the maximum possible value (k) is the fractional overlap agreement $A_k = (X_k/k)$. RBO is a weighted summation of these agreements where the weight $w_k = (1-p)*p^{k-1}$, based on the "persistence" parameter p that reflects the probability that an observer reviewing the top k items will continue to observe the ($k+1$)-th items. The fixed $(1-p)$ factor normalizes the sum of weights to 1. For our calculations, we used a default implementation p parameter of 0.98.[33]

$$RBO = \sum_{k=1}^{\infty} w_k \cdot A_k$$

The geometric weighting scheme of RBO serves to emphasize items at the top of the list and to ensure numerical convergence regardless of list length. RBO values range from 0.0 (disjoint lists) to 1.0 (identical lists).

To assess the utility of historical clinical item associations towards predicting future practices, we performed a variation of our prior experiments to predict hospital admission orders.[24] Specifically, using association matrices trained on data from 2009, 2012, or 2009 through 2012, we used the first four hours of clinical items from every future patient admitted to the hospital in 2013 to query for a ranked list of associated clinical orders. We compared these generated order lists against the actual next 24 hours of subsequent clinical orders (that did not already occur within the query time) by area under the receiver operating characteristic (ROC-AUC), precision (positive predictive value) at 10 items, and inverse frequency weighted recall[26] (sensitivity) at 10 items. Statistical tests (*t*-tests, Pearson's correlation) were calculated with the SciPy Python package.[34]

4. Results

Table 1 reports patient demographics and the flux of new and departing ordering providers in the clinical data over the years studied. Table 2a,b,c illustrate examples of the top associated clinical orders for different admission diagnoses based on 2009 vs. 2012 data, with corresponding calculations of ranked item overlap that define the Rank Biased Overlap (RBO) score for each pair of lists. Figure 1 depicts the Rank Biased Overlap (RBO) between 2009 vs. 2012 for each of the most common admission diagnoses. Figure 2 depicts the correlation between diagnosis stability (RBO) and accuracy towards predicting future order patterns (weighted recall). Table 3 reports the overall average accuracy metrics for predicting future (2013) clinical order patterns based on association matrices trained on different subsets of historical data (2009, 2012, or 2009 through 2012).

Table 1 – Patient demographics and provider flux over the evaluation period. New providers reflect those authorizing clinical orders during a given year, but not in the prior year. Similarly, departing providers reflect those from a given year, that are not found in the subsequent year.

Metric	2009	2010	2011	2012	2013
Patients	13,493	18,459	19,070	19,327	19,523
Age (mean)	58.4	58.1	58.6	58.5	58.6
Age (std dev)	18.5	18.8	18.7	18.7	18.6
Female	52%	52%	52%	51%	51%
White	60%	63%	62%	60%	58%
Hispanic/Latino	12%	13%	13%	13%	14%
Asian	11%	11%	11%	12%	12%
Black	5%	5%	5%	5%	5%
Providers	1,709	1,892	1,917	1,798	1,821
New Providers	...	41%	33%	29%	34%
Departing Providers	34%	32%	33%	33%	...

Table 2a – Top associated clinical orders for admission diagnosis of "Encounter for Chemotherapy" (ICD9: V58.11) based on 2009 and 2012 data, score-ranked by Fisher's P-value. At each rank *k*, the intersection of the top *k* items from each list (I_k) defines the "Overlap at Rank Depth" (X_k). The ratio of overlap to rank yields a "Fractional Overlap" agreement (A_k). For the full list of 1,270 candidate clinical orders, averaging the Fractional Overlap column with a geometric weighting scheme emphasizes the importance of the top items and ensures numerical convergence. The Rank Biased Overlap (RBO) score uses a weight for each A_k term, $w_k = (1-p)*p^{k-1}$, where *p* represents a "persistence" parameter reflecting the probability that the observer of *k* items is willing to continue to inspect the *k+1* items. RBO = 0.67 for this diagnosis, indicating relatively stable rankings compared to other diagnoses. This reflects standardized practices that have not significantly changed, including chemotherapeutic agents (cyclophosphamide, rituximab) and anticipatory co-medications for side effects (filgrastim for neutropenia; ondansetron, dexamethasone, aprepitant, and diphenhydramine for nausea).

2009 Top Items	Overlap at Rank Depth	Rank	Fractional Overlap	2012 Top Items
Cyclophosphamide (IV)	0	1	0.00	Ondansetron + Dexamethasone (IV)
Ondansetron + Dexamethasone (IV)	1	2	0.50	Aprepitant (Oral)
BMT Panel 1	1	3	0.33	Filgrastim (Subcutaneous)
Ondansetron (Oral)	2	4	0.50	Cyclophosphamide (IV)
BMT Panel 2	3	5	0.60	Ondansetron (Oral)
Rituximab (IV)	3	6	0.50	Dexamethasone (Oral)
Dexamethasone (Oral)	4	7	0.57	Diphenhydramine (Intravenous)
Aprepitant (Oral)	6	8	0.75	Rituximab (IV)
Filgrastim (Subcutaneous)	7	9	0.78	D5NS KCl NaAcetate Furosemide (IV)
Diphenhydramine (Intravenous)	8	10	0.80	D5NS KCl NaAcetate (IV)
...	...	...	...	...

Table 2b – Top associated clinical orders for admission diagnosis of "Pneumonia" (ICD9: 486) based on 2009 and 2012 data, score-ranked by Fisher's P-value. Rank Biased Overlap (RBO) = 0.35 between the two lists, indicating a substantial shift in the item rankings between the two lists. A dynamic change in practice patterns is evident in response to external, epidemiologic factors as 2009 saw much more testing (Respiratory DFA Panel, Influenza A PCR) and empiric treatment (Respiratory Isolation, Oseltamavir) for the H1N1 swine flu pandemic.[35,36] The viral pandemic dissipated by 2012, with the most prominent orders shifting towards empiric treatment for community acquired pneumonia[37] (azithromycin, ceftriaxone, levofloxacin) and antibiotic resistant organisms causing health care associated pneumonia[38] (vancomycin, piperacillin-tazobactam).

2009 Top Items	Overlap at Rank Depth	Rank	Fractional Overlap	2012 Top Items
Levofloxacin (IV)	0	1	0.00	Azithromycin (IV)
Blood Culture (2x Aerobic)	1	2	0.50	Levofloxacin (IV)
Blood Culture ((An)Aerobic)	1	3	0.33	Vancomycin (IV)
Respiratory DFA Panel	1	4	0.25	Piperacillin-Tazobactam (IV)
Respiratory Isolation	1	5	0.20	Ceftriaxone (IV)
Oseltamivir (Oral)	1	6	0.17	Azithromycin (Oral)
Vancomycin (IV)	2	7	0.29	Albuterol-Ipratropium (Inhalation)
Respiratory Culture	2	8	0.25	Sodium Chloride (Inhalation)
Albuterol-Ipratropium (Inhalation)	4	9	0.44	Blood Culture (2x Aerobic)
CBC w/ Diff	5	10	0.50	Blood Culture ((An)Aerobic)
Influenza A PCR	5	11	0.45	Ipratropium (Inhalation)
...	...	...	...	...

Table 2c - Top associated clinical orders for admission diagnosis of "Joint Pain" (ICD9: 719.4) based on 2009 and 2012 data, score-ranked by Fisher's P-value. Rank Biased Overlap (RBO) = 0.29 between the two lists, indicating a substantial shift in the item rankings between the two lists. Prominent orders in 2009 reflect diagnostic workup of arthritis (including fluid cell count and culture) while 2012 reveals more prominent symptomatic treatment with intravenous opioids (hydromorphone) that concomitantly require laxatives (sennosides, polyethylene glycol, magnesium citrate) to manage the predictable constipating side effects of opioids. The 2012 prominence of "Consult Orthopedics" suggests a shift in primary treatment teams from surgical to medical services since 2009.

2009 Top Items	Overlap at Rank Depth	Rank	Fractional Overlap	2012 Top Items
Overhead Bed Frame & Trapeze	0	1	0.00	Sennosides (Oral)
XR Pelvis 1V	0	2	0.00	Polyethylene Glycol (Oral)
Prothrombin TIME (PT/INR)	1	3	0.33	XR Pelvis 1V
CBC w/ Diff	1	4	0.25	Consult Orthopedics
Metabolic Panel, Basic	2	5	0.40	Overhead Bed Frame & Trapeze
XR Femur RT	2	6	0.33	Magnesium Citrate (Oral)
XR Shoulder 1V RT	2	7	0.29	Enoxaparin (Subcutaneous)
Cell Count, Synovial Fluid	2	8	0.25	XR Hip 2V LT
Fluid Culture and Gram Stain	2	9	0.22	Hydromorphone (Intravenous)
Bupivacaine (Nerve Block)	2	10	0.20	XR Femur LT
...	...	...	...	...

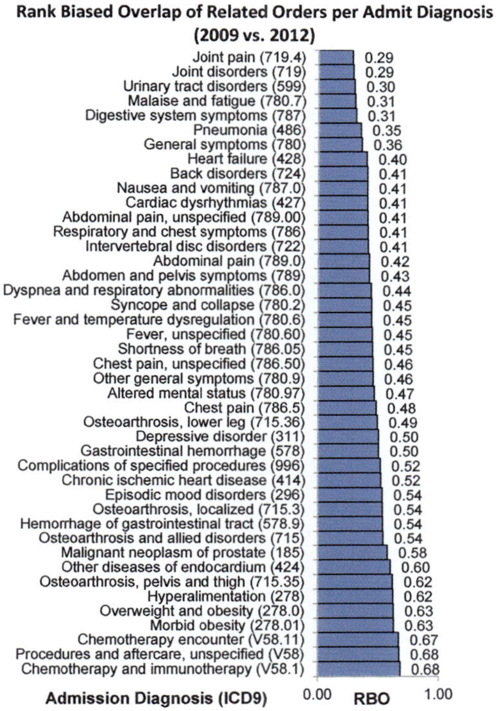

Figure 1 - Rank Biased Overlap (RBO) assessment of similarity of the lists of orders associated with the most common admission diagnoses in 2009 vs. 2012. Qualitative patterns reveal more stable ordering patterns (higher RBO) for elective hospital admissions with specific treatment plans and protocols like chemotherapy, obesity (sleep apnea and bariatric surgery), and osteoarthrosis (orthopedic surgery). Greater variability in ordering patterns (lower RBO) is seen for admission diagnoses with dynamically evolving practice patterns and less specific syndromes that may result in variable management, such as joint pain, malaise, and digestive symptoms.

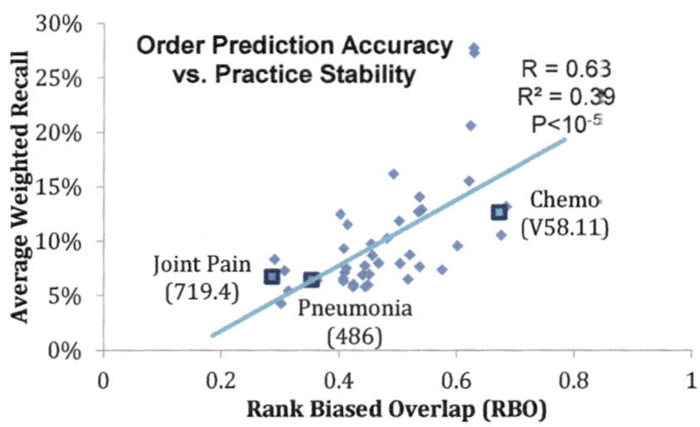

Figure 2 - Average weighted recall per admission diagnosis when predicting 2013 admission orders based on 2009-2012 training data by rank biased overlap. Using the association matrix trained on 2009-2012 data, the first clinical items from every admission in 2013 was used to query for the top ten associated clinical orders score-ranked by Fisher's P-value. Associated orders were compared against actual subsequent orders to yield a weighted recall score.[26] Each point represents the average weighted recall for one admission diagnosis vs. the respective rank biased overlap (RBO) score of order stability for 2009 vs. 2012. A linear trendline with Pearson correlation coefficient and two-tailed P-value illustrate a positive association between practice stability (higher RBO) and accuracy (weighted recall) towards predicting future order patterns.

Table 3 – Accuracy measures for predicting 2013 admission orders when using training data from different subsets of prior years. For ~15K patients with ~26K hospital admissions in 2013, data from the first four hours for each admission was used to query an association matrix trained on prior year(s) data for a list of clinical orders. The list of generated orders is score-ranked by PPV (positive predictive value ~ post-test probability) to identify orders *likely* to occur or by P-value to prioritize orders *disproportionately associated* with the query items. Generated order lists were compared against the subsequent 24 hours of clinical orders that actually occurred in each 2013 admission. Full list ranking is evaluated by the area under the receiver operating characteristic curve (ROC-AUC), while precision at ten evaluates only the top ten items. Inverse frequency weighted recall identifies methods most effective at retrieving less common, but specifically relevant orders.[26] Compared against the 2013 results, all bolded average results differed with $P<10^{-5}$ by two-tailed paired *t*-tests.

Training Data Year(s)	Training Patients	Average ROC-AUC PPV Associations	Average Precision at Ten PPV Associations	Average Weighted Recall at Ten P-value Associations
2009	10,727	0.888	**29.8%**	**2.4%**
2012	12,503	0.922	36.8%	**13.4%**
2011-2012	21,901	0.921	36.1%	**12.9%**
2009-2012	34,812	0.919	**35.4%**	**12.2%**
2013	11,278	0.924	38.0%	16.5%

5. Discussion

These results support the general supposition that clinical practices dynamically change over time (Figure 1). Elective admissions for planned procedures like chemotherapy and surgeries appear to exhibit relatively less variability over time with higher RBOs. This could of course be disrupted if future practices shifted in response to newly discovered different chemotherapy or surgical regimens, though the identified associations could still be reasonably used to suggest co-medications that are not enforced through a strict protocol. Diagnoses subject to epidemiologic shifts (i.e., pneumonia) and medical admissions for non-specific symptoms (e.g., joint pain, malaise) may trigger variable approaches to workup, represented by their lower RBOs. This method provides a quantitative assessment of clinical practice areas with the most dynamic changes, with respective implications on the reproducibility and reliability of predicting future clinical practice patterns based on historical data. It also has implications for ongoing debates on the appropriate interval for continuing medical education and maintenance of certification for individual clinicians.[39,40] For example, it could be used to identify areas where frequent education is required to adapt to rapidly shifting standards of practice vs. areas with years of stable practices that diminish the value of repetitious education maintenance.

Table 3 Table 3 reports the accuracy of models trained on different subsets towards predicting future practices by multiple measures. The area under the ROC curve (ROC-AUC) assesses discrimination accuracy for the full ranked list of candidate orders. Precision at ten items pays particular attention to the top items that a human user could realistically be expected to review. Weighted recall highlights retrieval of more "interesting" and specifically relevant suggestions over common, but potentially mundane, suggestions.[26] As might be expected, clinical order recommenders trained on more recent (2012) data are more accurate at predicting future (2013) practices than older (2009) data by all measures. The more compelling question answered is whether training on a larger longitudinal dataset (2009-2012) yields better results than just using the most recent data (2012). In this case, the extended data set is no better to slightly worse than just using the most recent data. While larger datasets are generally expected to improve the power of statistical learning methods, the correlation with RBO in Figure 2 suggests the changing clinical practice patterns over time makes older data less relevant when predicting future events.

This study focuses on the relevance of learned clinical order patterns towards predicting future events, but provides no assurance that common or strongly associated behaviors actually reflect "good" decisions. Short of randomized trials, we are evaluating our order associations against the external standards-of-care established in clinical practice guidelines.[43] With the results of this study however, it is not surprising that practice guidelines themselves must undergo regular revision, resulting in an ambiguous and moving target of clinical decision making quality that defies the existence of a fixed gold standard for clinical decision support.

A potential limitation in our evaluation of clinical practice pattern stability is the presumption that changing patterns reflect changes in clinical decision making at the management and treatment level. The nature of the EHR data source likely results in changing order patterns due to non-clinical data changes, such as shifts in diagnosis coding practices from pneumonia to sepsis.[41] Administrative infrastructure changes are expected to occur despite having little semantic difference for clinical decision making, such as the hospital orders for Respiratory Virus DFA (direct fluorescent antibody) panels being replaced with Respiratory Virus PCR (polymerase chain reaction) panels. Related work we are undertaking on probabilistic topic models of clinical data could provide opportunities to detect and resolve such "semantic" differences by noting that both such respiratory virus tests are related to "respiratory infection" scenarios, even though the two are never found together for a single patient. There may also be a substantial shift in patient characteristics insufficiently captured by admission diagnosis stratification, such as patient admissions for "joint pain" that might represent anything from elective orthopedic surgery admissions, workup for suspected septic arthritis, to pain management for a rheumatoid arthritis flare. Using more robust cohort identification methods than admission ICD9 codes, such as through natural language processing of clinical notes or SNOMED-CT codes could help normalize such factors. Individual patients could be hospitalized multiple times within each evaluation period, which could bias the association statistics without clustering statistics to mitigate internally correlated data. With all data deriving from a single medical center, significant cultural shifts in practice patterns could also be unduly influenced by the large flux of providers noted in Table 1 or even a small number of prominent clinicians.

Even if learned clinical practice patterns change for "non-clinical" reasons above, the overarching caution of depending on historical data to predict future clinical events remains relevant. The evolving clinical patterns reinforce the challenge of manually producing clinical decision support and knowledge guides for order entry, as they must be followed by ongoing manual effort to maintain them against new clinical evidence and standards that may substantially shift within just a few years. Automated algorithms to learn clinical decision support are thus even more important to not only cover the breadth of medical knowledge efficiently, but to automatically adapt to continuous streams of new information. While historical data will not predict the advent of new therapeutics or diseases, incorporating a continuous stream of data could allow automated methods to rapidly detect and adapt to shifting practice changes and alert authors to dynamic areas in need of additional decision support, just as Google Flu Trends can detect local flu activity more rapidly than conventional methods.[42] The results above inform such an approach, indicating that using the most recent data may be more important than simply accumulating a massive repository of historical data whose interpretation does not even remain internally consistent. Future opportunities could explore weighted or online learning algorithms that emphasize the relevance of recent data without completely ignoring the older data that may still capture useful information.

6. Conclusions

Clinical practice patterns for hospital admission diagnoses (automatically) learned from historical EHR data can vary substantially across years, particularly for non-specific symptom-based diagnoses and those influenced by external epidemiology (e.g., pneumonia). Elective admissions for planned procedures (e.g., chemotherapy, surgery) demonstrate more stable practice patterns over time. If the goal is predicting relevant future practices, using more recent training data is more accurate than using older data, likely due to secular trends in changing practice. Consequently, using a larger longitudinal data set from many years may be no better, and possibly worse, than using a smaller but more recent data set. Decision support and predictive analytic models should take these patterns into account.

7. Acknowledgments

Project supported by the Stanford Translational Research and Applied Medicine (TRAM) program in the Department of Medicine (DOM) and the Stanford Learning Healthcare Systems Innovation Fund and the Stanford Clinical and Translational Science Award (CTSA) to Spectrum (UL1 TR001085). The CTSA program is led by the National Center for Advancing Translational Sciences (NCATS) at the National Institutes of Health (NIH).

J.H.C supported in part by VA Office of Academic Affiliations and Health Services Research and Development Service Research funds.

R.B.A. is supported by NIH/National Institute of General Medical Sciences PharmGKB resource, R24GM61374, as well as LM05652 and GM102365. Additional support is from the Stanford NIH/National Center for Research Resources CTSA award number UL1 RR025744.

Patient data extracted and de-identified by Lee Ann Yasukawa and Susan Weber of the STRIDE (Stanford Translational Research Integrated Database Environment) project, a research and development project at Stanford University to create a standards-based informatics platform supporting clinical and translational research. The STRIDE project described was supported by the National Center for Research Resources and the National Center for Advancing Translational Sciences, National Institutes of Health, through grant UL1 RR025744.

Views expressed are those of the authors and do not necessarily represent the official views of the NIH, Department of Veteran Affairs (VA) or other affiliated institutions.

References

1. Richardson, W. C. *et al. Crossing the Quality Chasm: A New Health System for the 21st Century. Natl. Acad. Press* (Institute of Medicine, Committee on Quality of Health Care in America Committee on Quality of Health Care in America, 2001). doi:10.1136/bmj.323.7322.1192
2. Kaushal, R., Shojania, K. G. & Bates, D. W. Effects of computerized physician order entry and clinical decision support systems on medication safety: a systematic review. *Arch. Intern. Med.* **163,** 1409–1416 (2003).
3. Overhage, J. & Tierney, W. A randomized trial of 'corollary orders' to prevent errors of omission. *J. Am. Med. Informatics Assoc.* **4,** 364–75 (1997).
4. Tierney, W. M. *et al.* Computerizing Guidelines to Improve Care and Patient Outcomes: The Example of Heart Failure. *J. Am. Med. Informatics Assoc.* **2,** 316–322 (1995).
5. Chen, J. H. *et al.* Why providers transfuse blood products outside recommended guidelines in spite of integrated electronic best practice alerts. *J. Hosp. Med.* (2014). doi:10.1002/jhm.2236
6. Bates, D. W. *et al.* Ten commandments for effective clinical decision support: making the practice of evidence-based medicine a reality. *J. Am. Med. Inform. Assoc.* **10,** 523–30 (2003).
7. Goldstein, M. K. *et al.* Implementing clinical practice guidelines while taking account of changing evidence: ATHENA DSS, an easily modifiable decision-support system for managing hypertension in primary care. *Proc. AMIA Symp.* 300–4 (2000). at <http://www.pubmedcentral.nih.gov/articlerender.fcgi?artid=2243943&tool=pmcentrez&rendertype=abstract>
8. ONC. Health information technology: standards, implementation specifications, and certification criteria for electronic health record technology, 2014 edition; revisions to the permanent certification program for health information technology. Final rule. *Fed. Regist.* **77,** 54163–292 (2012).
9. Longhurst, C. a., Harrington, R. a. & Shah, N. H. A 'Green Button' For Using Aggregate Patient Data At The Point Of Care. *Health Aff.* **33,** 1229–1235 (2014).
10. Frankovich, J., Longhurst, C. A. & Sutherland, S. M. Evidence-based medicine in the EMR era. *N. Engl. J. Med.* **365,** 1758–9 (2011).
11. Smith, M., Saunders, R., Stuckhardt, L. & McGinnis, J. M. *Best care at lower cost: the path to continuously learning health care in America.* (Institute of Medicine, Committee on the Learning Health Care System in America, 2012). doi:10.5860/CHOICE.51-3277
12. Krumholz, H. M. Big Data And New Knowledge In Medicine: The Thinking, Training, And Tools Needed For A Learning Health System. *Health Aff.* **33,** 1163–1170 (2014).
13. De Lissovoy, G. Big data meets the electronic medical record: a commentary on 'identifying patients at increased risk for unplanned readmission'. *Med. Care* **51,** 759–60 (2013).
14. Sittig, D. F. *et al.* Grand challenges in clinical decision support. *J. Biomed. Inform.* **41,** 387–92 (2008).
15. Bates, D. W., Saria, S., Ohno-Machado, L., Shah, a. & Escobar, G. Big Data In Health Care: Using Analytics To Identify And Manage High-Risk And High-Cost Patients. *Health Aff.* **33,** 1123–1131 (2014).
16. Doddi, S., Marathe, a, Ravi, S. S. & Torney, D. C. Discovery of association rules in medical data. *Med. Inform. Internet Med.* **26,** 25–33 (2001).
17. Klann, J., Schadow, G. & Downs, S. M. A method to compute treatment suggestions from local order entry data. *AMIA Annu. Symp. Proc.* **2010,** 387–91 (2010).
18. Klann, J., Schadow, G. & McCoy, J. M. A recommendation algorithm for automating corollary order generation. *AMIA Annu. Symp. Proc.* **2009,** 333–7 (2009).
19. Wright, A. & Sittig, D. F. Automated development of order sets and corollary orders by data mining in an ambulatory computerized physician order entry system. *AMIA Annu. Symp. Proc.* **2006,** 819–823 (2006).
20. Zhang, Y., Padman, R. & Levin, J. E. Paving the COWpath: data-driven design of pediatric order sets. *J. Am. Med. Inform. Assoc.* **21,** e304–e311 (2014).
21. Klann, J. G., Szolovits, P., Downs, S. M. & Schadow, G. Decision support from local data: creating adaptive order menus from past clinician behavior. *J. Biomed. Inform.* **48,** 84–93 (2014).
22. Wright, A. P., Wright, A. T., McCoy, A. B. & Sittig, D. F. The use of sequential pattern mining to predict next prescribed medications. *J. Biomed. Inform.* **53,** 73–80 (2014).
23. Wright, A., Chen, E. S. & Maloney, F. L. An automated technique for identifying associations between medications, laboratory results and problems. *J. Biomed. Inform.* **43,** 891–901 (2010).

24. Chen, J. H. & Altman, R. B. Mining for clinical expertise in (undocumented) order sets to power an order suggestion system. *AMIA Summits Transl. Sci. Proc.* **2013,** 34–8 (2013).
25. Linden, G., Smith, B. & York, J. Amazon.com recommendations: item-to-item collaborative filtering. *IEEE Internet Comput.* **7,** 76–80 (2003).
26. Chen, J. H. & Altman, R. B. Automated Physician Order Recommendations and Outcome Predictions by Data-Mining Electronic Medical Records. in *AMIA Summits Transl. Sci. Proc.* 206–210 (2014).
27. Lowe, H. J., Ferris, T. a, Hernandez, P. M. & Weber, S. C. STRIDE--An integrated standards-based translational research informatics platform. *AMIA Annu. Symp. Proc.* **2009,** 391–5 (2009).
28. Hernandez, P., Podchiyska, T., Weber, S., Ferris, T. & Lowe, H. Automated mapping of pharmacy orders from two electronic health record systems to RxNorm within the STRIDE clinical data warehouse. *AMIA Annu. Symp. Proc.* **2009,** 244–8 (2009).
29. Wright, A. & Bates, D. W. Distribution of Problems, Medications and Lab Results in Electronic Health Records: The Pareto Principle at Work. *Appl. Clin. Inform.* **1,** 32–37 (2010).
30. Finlayson, S. G., Lependu, P. & Shah, N. H. Building the graph of medicine from millions of clinical narratives. *Sci. Data* **1,** 140032 (2014).
31. Kendall, M. G. A New Measure of Rank Correlation. *Biometrika* **30,** 81–93 (1938).
32. Webber, W., Moffat, A. & Zobel, J. A similarity measure for indefinite rankings. *ACM Trans. Inf. Syst.* **28,** 1–38 (2010).
33. Agrawal, R. Comparing Ranked List. (2013). at <https://ragrawal.wordpress.com/2013/01/18/comparing-ranked-list/>
34. Jones, E., Oliphant, T., Peterson, P. & Al, E. SciPy: Open source scientific tools for Python. at <http://www.scipy.org>
35. Kerr, J. R. Swine influenza. *J. Clin. Pathol.* **62,** 577–578 (2009).
36. Who. WHO Guidelines for Pharmacological Management of Pandemic Influenza A(H1N1) 2009 and other Influenza Viruses. *WHO Guidel. Pharmacol. Manag. Pandemic Influ. A(H1N1) 2009 other Influ. Viruses* 1–32 (2010). at <http://scholar.google.com/scholar?hl=en&btnG=Search&q=intitle:WHO,2010#7>
37. Mandell, L. a *et al.* Infectious Diseases Society of America/American Thoracic Society consensus guidelines on the management of community-acquired pneumonia in adults. *Clin. Infect. Dis.* **44 Suppl 2,** S27–72 (2007).
38. Focaccia, R. & Gomes Da Conceicao, O. J. Guidelines for the management of adults with hospital-acquired, ventilator-associated, and healthcare-associated pneumonia. *Am. J. Respir. Crit. Care Med.* **171,** 388–416 (2005).
39. Teirstein, P. Boarded to Death — Why Maintenance of Certification Is Bad for Doctors and Patients. *N Engl J Med* **372,** 106–8 (2015).
40. Irons, M. B. & Nora, L. M. Maintenance of Certification 2.0 — Strong Start, Continued Evolution. *N. Engl. J. Med.* **372,** 104–106 (2015).
41. Rhee, C., Gohil, S. & Klompas, M. Regulatory mandates for sepsis care--reasons for caution. *N. Engl. J. Med.* **370,** 1673–1676 (2014).
42. Ginsberg, J. *et al.* Detecting influenza epidemics using search engine query data. *Nature* **457,** 1012–1014 (2009).
43. Chen, J. H. & Altman, R. B. Data-Mining Electronic Medical Records for Clinical Order Recommendations : Wisdom of the Crowd or Tyranny of the Mob ? *AMIA Summits Transl. Sci. Proc.* (2015).
44. Brin, S. & Page, L. The anatomy of a large-scale hypertextual Web search engine BT - Computer Networks and ISDN Systems. **30,** 107–117 (1998).
45. Excell, D. Bayesian inference - The future of online fraud protection. *Comput. Fraud Secur.* **2012,** 8–11 (2012).

REPURPOSING GERMLINE EXOMES OF THE CANCER GENOME ATLAS DEMANDS A CAUTIOUS APPROACH AND SAMPLE-SPECIFIC VARIANT FILTERING

AMANDA KOIRE[†]

Program in Structural and Computational Biology and Molecular Biophysics
Baylor College of Medicine, Houston, Texas 77030, USA
Email: koire@bcm.edu

PANAGIOTIS KATSONIS[†]

Department of Molecular and Human Genetics
Baylor College of Medicine, Houston, Texas 77030, USA
Email: katsonis@bcm.edu

OLIVIER LICHTARGE

Department of Molecular and Human Genetics
Baylor College of Medicine, Houston, Texas 77030, USA
Email: lichtarg@bcm.edu

When seeking to reproduce results derived from whole-exome or genome sequencing data that could advance precision medicine, the time and expense required to produce a patient cohort make data repurposing an attractive option. The first step in repurposing is setting some quality baseline for the data so that conclusions are not spurious. This is difficult because there can be variations in quality from center to center, clinic to clinic and even patient to patient. Here, we assessed the quality of the whole-exome germline mutations of TCGA cancer patients using patterns of nucleotide substitution and negative selection against impactful mutations. We estimated the fraction of false positive variant calls for each exome with respect to two gold standard germline exomes, and found large variability in the quality of SNV calls between samples, cancer subtypes, and institutions. We then demonstrated how variant features, such as the average base quality for reads supporting an allele, can be used to identify sample-specific filtering parameters to optimize the removal of false positive calls. We concluded that while these germlines have many potential applications to precision medicine, users should assess the quality of the available exome data prior to use and perform additional filtering steps.

1. Introduction

Although the costs of whole-exome sequencing continue to decrease [1], the resources needed to identify, enroll, and sequence an entire cohort of interest will remain significant for the foreseeable future. This process is especially cumbersome when investigating rare phenotypes, including certain cancers and tumor subtypes. A more convenient alternative path is to identify and then repurpose publicly accessible datasets in order to test new hypotheses or to reproduce findings of studies performed on independent cohorts. Federal policies explicitly promote data sharing and repurposing,

† Co-first authors

by supporting public repositories like the database of Genotypes and Phenotype (dbGaP) and the Sequence Read Archive (SRA) [2,3]. The challenge, however, is that diverse datasets each developed with different goals in mind will often have unique features that require special care before they can be pooled together for repurposing. Clearly, the quality of exome variant calls varies by platform and depth of the sequencing [4,5] and also depends on the stringency of downstream pipelines for SNV identification and variant filtering [6]. Currently most whole-exome quality assessment tools focus on evaluating the quality of the raw input data [7,8] rather than on the output calls; moreover, approaches that do assess the output generally limit themselves to comparing calls to 1000 Genomes or dbSNP variants [9,10] without providing recommendations for filtering or even clear conclusions on whether the data is acceptable for use. Yet if a dataset is repurposed inappropriately, systematic biases and variability in noise levels may slant results, lower reproducibility, yield artifacts, or prevent confirmation of prior findings [11]. This presents a major problem for precision medicine in particular, since targeting a falsely called variant may result in ineffective treatment.

In order to probe the impact that dataset and variant filtering choices can have on the quality of repurposed data, we assessed in detail germline exomes from The Cancer Genome Atlas (TCGA) [12]. TCGA currently gathers diverse information from more than 11,000 patient samples across 34 cancer types. Final germline variant calls for some cancer types are available through the TCGA Data Portal, with additional lower level sequence data also available from the CGHub repository (https://cghub.ucsc.edu/). However, the primary goal of sequencing cancer patient germline samples was to provide the background information that will enable the recognition of somatic variants unique to the tumor. Secondary use of these germline exomes to further precision medicine has thus far been uncommon but shows the promise of using these germlines to predict response to treatment within a cancer cohort, detect genetic differences in individuals who develop cancer, and identify germline contributions to the process of tumorigenesis [13,14,15].

Here, we evaluated the quality of TCGA germline single nucleotide variation (SNV) calls in a given exome by testing whether two features of their collected variant calls followed the known biology of substitution and purifying selection or whether these features were lost and suggested that the variant calls were of non-biological origin.

The first feature, called Ti/Tv, has been previously described and is based on the biology of spontaneous base substitutions. In the germline, these are more often transitions (from purine to purine, or from pyrimidine to pyrimidine) than transversions (from purine to pyrimidine or pyrimidine to purine) so the Ti/Tv ratio is normally >3 across an exome, whereas for random base changes as one might produce computationally Ti/Tv is equal to 0.5 [10]; this difference can then serve as a proxy for germline variant call quality [10,16,17].

The second and novel feature, called λ, is based on the biology of purifying selection of germline mutations. Fisher's geometric model [18] predicted in 1930 that the distribution of fitness effect of germline mutations would follow a decaying exponential. For the first time, a recent study of the Evolutionary Action (EA) of human polymorphisms [19] provides a measure for the fitness effect of mutations and hence for selection constant λ decay constant of their distribution, and this selection constant is much larger for biological germline mutations than for random mutations.

Therefore, we hypothesized that Ti/Tv and λ related to the substitution and selection processes, respectively, should be complementary measures of variant call quality. Our findings support this view and also emphasize the importance of using multiple, orthogonal quality measures. Using these measures we estimate the fraction of false positive variant calls in TCGA exomes and reveal substantial variability in quality by sample, cancer subtype, and sequencing source. The methods described in this study provide an easy way to assess the quality of germline exome data and suggest, for the first time to our knowledge, the importance and feasibility of sample-specific filtering parameters.

2. Methods

2.1. *Exome data acquisition*

Gold standard germline exome variant callsets were obtained from http://www.illumina.com/platinumgenomes/ [20]. The NA12877 and NA12878, v7 (released December 2014) merged callsets were downloaded, and only 'platinum' calls limited to those with a 'PASS' designation in the filter field were considered confident calls in subsequent analysis. 1000 Genomes Project [21] Phase 3 exomes were downloaded from http://www.1000genomes.org/ on 5/11/15. All cancer germline exomes available on 10/23/14 were downloaded from TCGA [12] and separated by cancer type and institution.

2.2. *Quality Assessment*

The Ti/Tv ratios were calculated for each exome variant callset by counting all coding purine to purine and pyrimidine to pyrimidine (transition) SNV mutations, and dividing this value by the number of purine to pyrimidine and pyrimidine to purine (transversion) SNV mutations.

To measure the selection decay λ of variant callsets, the corresponding vcf files were annotated with ANNOVAR [22] and the Evolutionary Action (EA) was computed to measure the fitness effect of all missense mutations. Histograms of the distribution of EA scores, binned by deciles, were fitted to an exponential curve (Eq. 1) using a least-squares-fit to estimate λ. In Eq. 1 the 'x' values represent Evolutionary Action, 'y' represents proportion of mutations, and A and λ are constants.

$$\ln(y) = \ln(A) - \lambda x \tag{1}$$

2.3. *Simulated Variant Callsets*

Both Ti/Tv and λ were calculated for 100 datasets with different fractions of computer-generated false positive variant calls. The fraction varied from 0 to 1, in increments of 0.01. Each dataset consisted of 1000 simulated exome files, each of which contained a total of 10000 missense mutations. The mutations were drawn either from the 'true positive' pool of variants from gold standards NA12877 and NA12878, or randomly from the 'false positive' pool of all possible human SNVs in order to create the proper ratio. The average and standard deviations of λ and Ti/Tv for each fraction of false positives were calculated, and their correlations with percent noise fitted with exponential.

2.4. *Application of λ to TCGA cohorts*

For each of 21 cancer types, all germline datasets were separated by institutional source, curation, and sequencing platform. When multiple variant callsets were available for a given cancer type, files marked 'curated' were chosen over those which were not marked curated. If multiple curated sets were available, λ values were calculated for each and the set with the highest average λ was chosen. If no curation was noted on any of the files, λ values were calculated for all available sets and the set with the highest average λ was chosen. For each sample, the predicted percentage of false positive calls was then calculated from Eq. 2.

3. Results

3.1. *Calculating quality measures for 'gold standard' germline exomes*

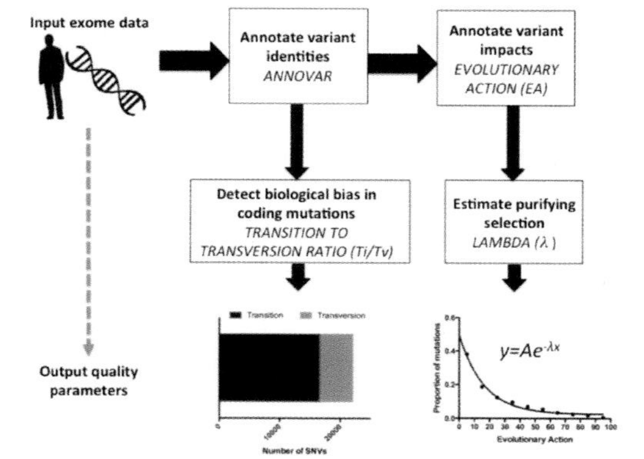

Fig. 1. Illustration depicting the steps taken to calculate λ and Ti/Tv parameters from exome data.

First, we calculated the quality scores Ti/Tv and λ (Figure 1) for gold standard germline exome calls: Illumina Platinum v7.0 variant calls for samples NA12877 and NA12878 [9]. These samples were sequenced to 200x depth on a HiSeq 2000 system, and 15 other members of their pedigree were sequenced to 50x depth on the same system. Variant calls deemed 'platinum' take into account inheritance constraints in the pedigree as well as concordance of variant calls across multiple aligners and callers [24,25,26,27,16]. As described in Methods, the transition to transversion ratio (Ti/Tv) detects a biological bias in the types of mutations that occur in the call set, while the λ shows the purifying pressure against coding mutations as a function of their Evolutionary Action (EA) [19]. For the platinum whole-exome calls for NA12877 and NA12878, we found the Ti/Tv ratios to be 3.046 and 3.036 for whole-exome variant calls and 1.97 and 1.96 for missense variant calls respectively, while λ was 0.0379 and 0.0380 respectively. These values were also reasonably consistent with exome data from The 1000 Genomes Project [21] and were independent of ethnic background; the average λ is 0.0379±0.0008 and the average Ti/Tv is 3.11±0.05 for whole-exome variants and 2.06±0.04 for missense variants. These data show a gold standard 'target' for these quality parameters that were used to assess other datasets.

3.2. *Quality measures λ and Ti/Tv decrease in a predictable fashion as false positives are added to a variant call set*

We next calculated the λ and Ti/Tv for SNV sets with varying fractions of false positive calls. Using the Illumina Platinum calls as a pool of true positives (TP) and all possible human SNVs generated by random nucleotide changes as a pool of false positives (FP), 1000 simulations of 10000 missense mutations were produced for each of 100 TP:FP ratios. As the fraction of the variant calls composed of random noise increases, both λ and Ti/Tv decrease in an exponential fashion ($R^2 > 0.99$) such that

$$\lambda = 0.038e^{-0.013*percentFP} \quad (2)$$

(Figure 2A). In addition, the two quality measurements were strongly correlated with one another (Pearson R = 0.9993, $p < 0.0001$) (Figure 2B). Use of this simulated data enables us to estimate, for a set of real exome variant calls, the degree of contamination by false positives in the sample.

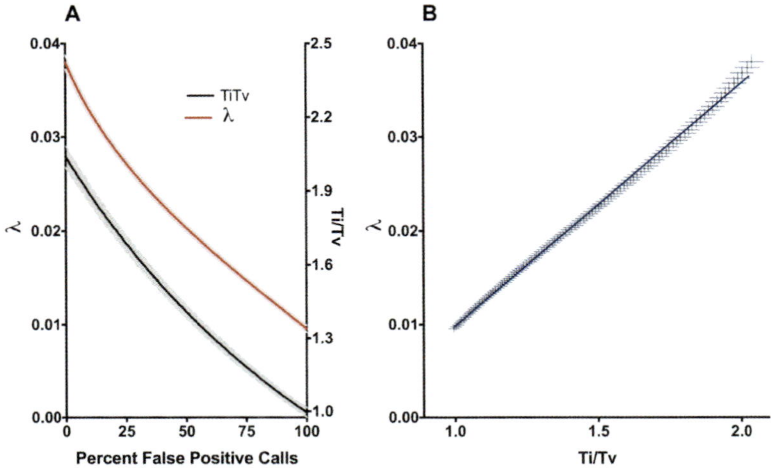

Fig. 2. Simulated noise in exome SNV calls. (a) Effect of increased noise on λ and Ti/Tv values. Shaded regions indicate the standard deviation around the mean. (b) Correlation between λ and Ti/TV.

3.3. *TCGA germline datasets vary in the amount of noise they contain*

Next, we estimated the fraction of false positive variant calls in whole-exome germline data from TCGA and assessed whether the fraction is consistent within each cancer subtype. Germline variants were available through TCGA for 21 cancer types, where they were organized by sequencing institution, sequencing platform, and curation level. Some tumors were sequenced by multiple institutes, though the extent of this overlap depended on the cancer type. We focused on data marked 'curated', rather than those marked 'automated', and when there were still multiple versions we chose the (presumably best) dataset with the largest λ (see Methods). The predicted percent of false positive calls for each exome was calculated based on Eq. 2, and the distribution for each cancer type is shown in Figure 3A. At the extremes, pheochromocytoma and paraganglioma (PCPG) exomes were predicted to uniformly contain less than 5% false calls, while most cervical cancer (CESC) exomes

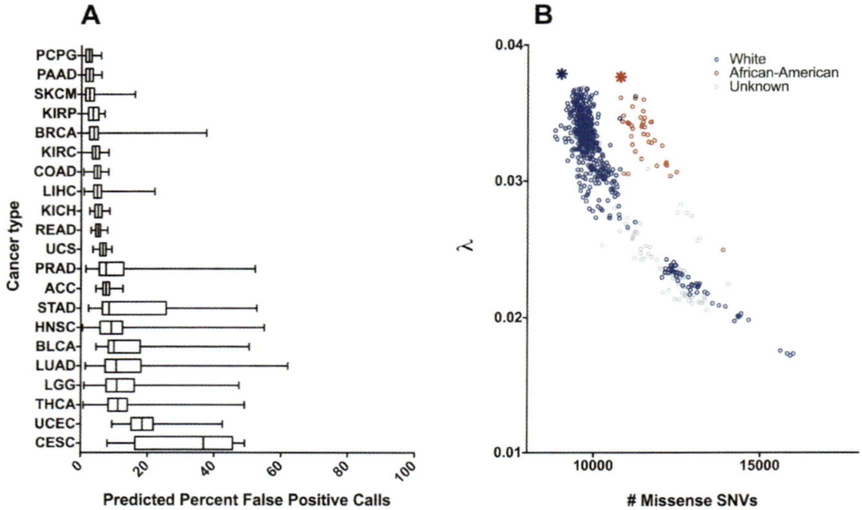

Fig. 3. Application of λ to TCGA cohorts. (a) Predicted noise across 21 TCGA cancer types. The data are represented in a box-and-whiskers plot that uses the center line to indicate median, the box to indicate quartiles, and the whiskers to indicate range. Cancer types are ordered by median. (b) Exponential relationship between λ and number of missense SNVs in Lung Adenocarcinoma. Associated open-access clinical data provided by TCGA was used to separate patients by their self-identified race. The average lambda/number of missense mutations for the 1000 Genomes Project Caucasian (CEU) and African-American (ASW) cohorts are noted with a blue and red star, respectively.

were predicted to contain more than 35% false calls. The datasets also differed greatly in variance, with rectal adenocarcinoma (READ) exomes having less than 10%, and lung adenocarcinoma (LUAD) having more than 60% variances. These results do not necessarily reflect intrinsic differences between cancer types. Large numbers of variant calls in an exome corresponded to lower λ scores and indicated an excess of false positives, as shown for LUAD in Figure 3B. However, using the number of mutations as a proxy for the false positive rate may be misleading when the cohort consists of individuals with diverse ethnic backgrounds; for example, exomes from patients of African ancestry consistently had more variants when compared to other exomes with the same λ values (Figure 3B), in agreement with data from the 1000 Genomes Project. The estimated number of 'true' missense SNV calls was consistent between samples of the same ethnic background and fit the numbers of missense mutations seen in the 1000 Genomes cohorts. These data show the marked heterogeneity of false variant call rates in TCGA germline exomes and highlight the hazards of using these datasets as-is.

3.4. *Data quality is not consistent across calling centers*

In order to test reproducibility of data across sequencing centers, we focused our analysis on the chromophobe renal cell carcinoma (KICH) dataset, which consisted of SNV germline calls from three separate institutions for the same patients. For each sequencing center, we calculated the average Ti/Tv ratio as well as the λ selection decay constant for the germline variants of each exome, as shown in Figure 4A. The center with the highest average Ti/Tv ratio also had the highest λ, which corresponded to an average of ~5% false positive calls per sample. For the second center, both the Ti/Tv ratio and λ were lower, and λ predicted an average of ~12% false positive calls per sample. For Center 3, although the average Ti/Tv ratio was nearly as good as the other two centers (2.60 compared to 2.66 and 3.1, respectively), the average λ was radically different (0.014 versus 0.032 and 0.036, respectively) and suggested that on average 76% of the calls in each sample were false positives. Indeed, the average number of missense SNV calls from this center (31000±5000) was over 3 times higher than the Illumina platinum exomes defined above as gold standard, which further supported a false positive rate of at least 70%. The different sensitivity of Ti/Tv and λ to lower quality variant calls in this case may be due may be due to technical aspects of the calling methods themselves; if the known biological bias toward transition mutations was built into the calling algorithm used by Center 3 and it was used as a factor in deciding whether to report a variant, even false positive mutations will have a high Ti/Tv. In this case, λ detected noise whereas Ti/Tv was equivocal, stressing the importance of using multiple quality measures in exome data assessment. For centers 1 and 3, additional internal filters separated the reported calls into those that 'pass' and those that do not. Restricting our analysis only to passing calls improved quality but was not sufficient to eliminate either the detected noise or the center-specific differences. These data show that the germline variant calls of TCGA patients made by different sequencing platforms and calling pipelines are sometimes very different and require careful examination by multiple, orthogonal quality measures.

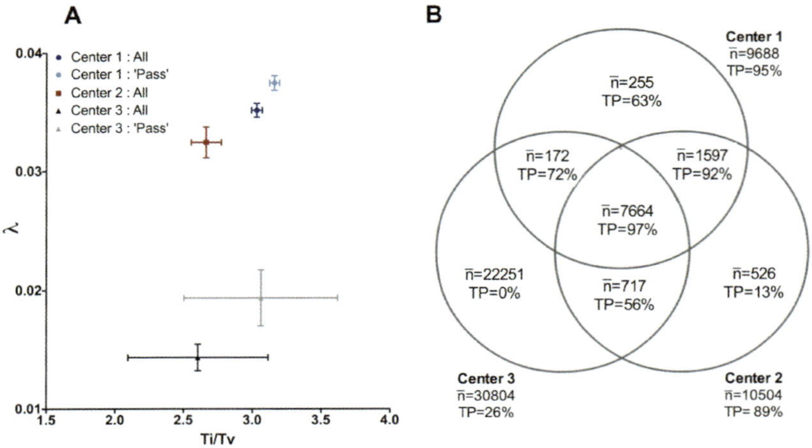

Fig. 4. KICH SNV calls for three centers (a) λ and Ti/Tv of calls. For each patient assessed by each center, λ and Ti/Tv were calculated and the average and standard deviations of these values are displayed by institution. For centers 1 and 3, internal 'pass' filters were available and are displayed as well. (b) Predicted percentage of true calls for calls agreed upon by 1, 2, or 3 institutions. For 65 KICH patients assessed by all three centers, all calls regardless of internal filtering were separated by the institution(s) that identified them. The average number of missense mutations per patient, as well as the predicted percentage of true positive calls derived from the λ value of the call set, is shown for each possible combination of sites.

A common practice to address such inconsistency in reproducibility is either to merge the available data or to use their intersection. In the first case, combining the calls from all centers would add substantial noise from Center 3. Using the intersection of all three centers, on the other hand, would result in high quality but roughly two thousand true positive calls per exome would be left out. Restricting to calls made by at least two of the centers may seem like a reasonable middle ground, but even this may not be the optimum solution. Figure 4B shows that calls made by Center 1 alone were still predicted to be of higher quality than calls agreed upon by Centers 2 and 3. These data demonstrate the caveats of 'common sense' filtering and highlight the importance of examining data quality carefully before integrating information from multiple sources.

3.5. *Appropriate filtering parameters for SNV calls are sample-dependent*

Having used λ to detect and quantify the presence of noise within these datasets, we next explored whether λ can be used for filtering false positive SNV calls. For illustration, we used exome data from two head and neck squamous cell carcinoma (HNSC) cancer patients. Each SNV in these data was associated with an average base quality for reads supporting alleles (BQ value) and a Phred-scaled quality score (QUAL value), amongst other features. For each patient, missense mutations were partitioned by BQ value and QUAL value in turn and each bin was assessed for λ. We found that λ depended on the BQ score with a sigmoidal relationship ($R^2 > 0.9$), which indicated that below a BQ cutoff the SNV calls became random (Figure 5A). Strikingly, this cutoff is specific to each exome, even those sequenced on the same date, by the same center, on the same sequencing platform, and using the same SNV calling pipeline. For example, SNVs with a BQ of 25 appeared to retain high quality in one patient but to be comprised entirely of noise in the other (see blue curve in Fig. 5A). We also found that λ depended on the QUAL value of SNVs, such that the fraction of true positive calls was lower for QUAL values near zero and gradually increased with QUAL value till it reached a plateau at about QUAL=40 (Figure 5B). For the two exomes of Figure 5b the QUAL value did not correspond to the same fraction of false positive data, since the two exomes reached different maximum λ values. These sample-dependent differences suggested that BQ and QUAL values should only be interpreted as relative measures within the context of a given exome, and that filtering parameters should be customized for each exome using λ in order to achieve optimal separation of true and false positive calls.

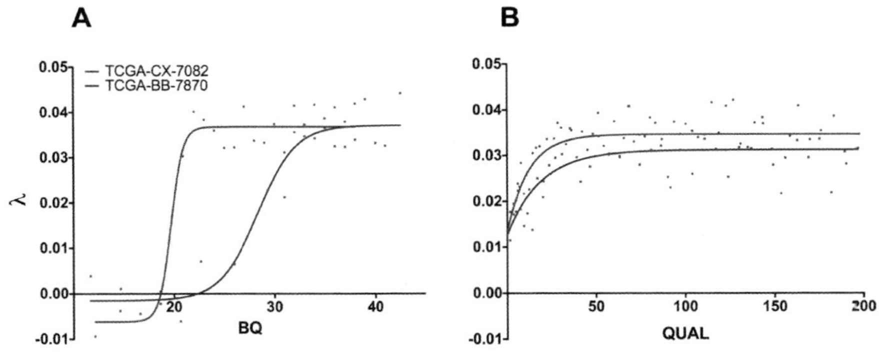

Fig. 5. Relationship between SNV features and λ for two HNSC patients. (a) Relationship between BQ and λ. For each patient, all missense SNVs were partitioned by BQ value such that every bin contained at least 50 calls; points represent the λ and average BQ of the bin. Solid lines represent sigmoidal fits. (b) Relationship between QUAL and λ. For each patient, all missense SNVs were partitioned by QUAL value such that every bin contained at least 50 calls; points represent the λ and average QUAL of the bin. Solid line represents fit to equation $y = Ae^{-kx} + b$. For display purposes values of QUAL higher than 200 were not shown.

4. Discussion

Assessing the quality of genetic variant calls has great practical importance to precision medicine, since various sequencing platforms, coverage depths, and bioinformatics pipelines to call variants result in the inclusion of an unknown number of false positive calls. This becomes a major concern when naïve users access and repurpose publically available exome data assuming all reported calls are reliable. Here, we call attention to the hazards of this assumption by applying two measures of exome data quality, the transitions-to-transversions ratio and the purifying selection pressure (λ) of variant calls, to publically accessible data. As a test system we used germline exomes of 21 cancer types available through TCGA, which were generated with the primary purpose of being a reference for calling somatic mutations. We found considerable variation in data quality between and within cancer types, such that repurposing these data as-is may mislead scientists to conclude a lack of reproducibility and unsuccessful validation of previous findings, which would hinder the progress of precision medicine as a field.

As a gold standard of true germline variants we used the Illumina Platinum samples, NA12877 and NA12878, which are the current state-of-the-art high-confidence variant calls. However, this gold standard may still include some false calls, and future developments may allow for even more accurate sequencing and variant calling. These estimates, however, have great practical importance in comparing the calling confidence of two or more exomes. Indeed, the large differences in the average fraction of false positive calls between TCGA germline exomes of different cancer types, as well as the surprising variability within a cancer cohort, underscore the need to examine all data carefully before reuse in order to improve the reproducibility of results.

When we compared the germline variants called from three different sequencing centers for the same patients, we found a considerable lack in reproducibility between centers. However, classifying variant calls by their concordance across centers was revealing. Variants called by all three centers were assessed to contain only about 3% of false positive calls; in contrast, variants called by just one of the three centers had deviant λ values that matched the simulated introduction of up to 100%, 87% and 37% false positives. For each given center, unique variant calls were predicted to contain more false positives than those also called by at least one of the other centers; still, we found that depending on the relative data quality between centers even the unique variants of a single center may contain less noise than variants agreed upon by the other two centers. This calls into question the common practices of merging data or using the overlap of calls from different centers, which may include many false positive calls or exclude true positive calls, respectively. While it was useful to exclude variants that were not annotated with "PASS" in vcf files, this filter was not able remove all or even most false positive calls. Thus, the use of λ selection pressure analysis presents a rational, quantitative approach to determining which data should be used in association studies.

Features of the SNVs, such as quality scores, can also be used to filter out false positive calls. This basic principle is already established in post-processing variant calls, but many users apply 'hard filters' to all samples and express uncertainty regarding the appropriate filters to use. Using arbitrary cutoffs for all TCGA exomes without considering quality assessment will cause some samples to retain substantial numbers of false calls and others to lose many true calls. Our results suggested that BQ value produced the most effective separation between predicted true and false positives, and that the appropriate BQ value cutoff was different for each exome, even when the data were produced by identical procedures. This analysis allows users to leverage the relationship between λ and BQ in order to choose for each exome the optimum cutoff, allowing them to repurpose these datasets with confidence and improve the reproducibility of their results.

Assessing the quality of germline variant calls is a pressing issue both to improve their intended use as well as to facilitate their repurposing for secondary goals, and since an increasing amount of exome data is being deposited in public databases. Here, we show that elementary evolutionary considerations provide a general and simple approach to detect random sequencing errors. Whereas high quality data contain variant calls that follow an invariant and known distribution of Evolutionary Action, false positive variant calls recognizably distort this Action distribution. Remarkably, this distortion can classify sequenced genomes by quality and also separate variant calls by quality within single exomes on a case-by-case basis. This work reveals wide quality disparities in sequencing data but also demonstrates how this can be overcome through the use of the Evolutionary Action concept. In the future it should therefore be possible to apply, pool and repurpose public genome sequencing data with full confidence in their quality leading to better correlations with clinical phenotypes and enhancing reproducibility in precision medicine.

Acknowledgements

Funding: National Institutes of Health R01 grants GM079656-8; U01 grant DE025181; National Science Foundation grants DBI1356569. Conflict of interest: none declared.

References

1. E.C. Hayden. *Nature* **507**(7492):294-5. (2014).
2. M.D. Mailman, M. Feolo, Y. Jin, M. Kimura, K. Tryka, R. Bagoutdinov, L. Hao, A. Kiang, J. Paschall, L. Phan, N. Popova, S. Pretel, L. Ziyabari, M. Lee, Y. Shao, Z.Y. Wang, K. Sirotkin, M. Ward, M. Kholodov, K. Zbicz, J. Beck, M. Kimelman, S. Shevelev, D. Preuss, E. Yaschenko, A. Graeff, J. Ostell, S.T. Sherry. *Nat Genet* **39**(10):1181-6 (2007).
3. R. Leinonen, H. Sugawara, M. Shumway. *Nucleic Acids Res* **39**(Database issue):D19-D21.
4. M.J. Clark, R. Chen, K.J. Karczewski, R. Chen, G. Euskirchen, A.J. Butte, M. Snyder. *Nat Biotechnol* **29**(10):908-14 (2011).
5. A.M. Meynert, L.S. Bicknell, M.E. Hurles, A.P. Jackson, M.S. Taylor. *BMC Bioinformatics* **14**:195 (2013).
6. Y. Guo, F. Ye, Q. Sheng, T. Clark, D.C. Samuels. *Brief Bioinform* **15**(6):879-89 (2014).

7. S. Pabinger, A. Dander, M. Fischer, R. Snajder, M. Sperk, M. Efremova, B. Krabichler, M.R. Speicher, J. Zschocke, Z. Trajanoski. *Brief Bioinform* **15**(2):256-78 (2014).
8. R. Bao, L. Huang, J. Andrade, W. Tan, W.A. Kibbe, H. Jiang, G. Feng. *Cancer Inform* **13**(Suppl 2):67-82 (2014).
9. V. Heinrich, T. Kamphans, J. Stange, D. Parkhomchuk, J. Hecht, T. Dickhaus, P.N. Robinson, P.M. Krawitz. *Genome Med* **5**(7):69 (2013)
10. Q. Liu, Y. Guo, J. Li, B. Zhang, Y. Shyr. *BMC Genomics* **13**(Suppl 8):S8 (2012).
11. K.J. Hoff. *BMC Genomics* **12**(10):520 (2009).
12. TCGA Research Network: http://cancergenome.nih.gov/.
13. N.J. Birkbak, B. Kochupurakkal, J.M. Izarzugaza, A.C. Eklund, Y. Li, J. Liu, Z. Szallasi, U.A. Matulonis, A.L. Richardson, J.D. Iglehart, Z.C. *PLoS One* **8**(11):e80023 (2013).
14. K.L. Kanchi, K.J. Johnson, C. Lu, M.D. McLellan, M.D.M Leiserson, M.C. Wendl, Q. Zhang, D.C. Koboldt, M. Xie, C. Kandoth, J.F. McMichael, M.A. Wyczalkowski, D.E. Larson, H.K. Schmidt, C.A. Miller, R.S. Fulton, P.T. Spellman, E.R. Mardis, T.E. Druley, T.A. Graubert, P.J. Goodfellow, B.J. Raphael, R.K. Wilson, L Ding. *Nature Communications* **5**:3156 (2014).
15. J. Ngeow, Y. Ni, R. Tohme, F. Song Chen, G. Bebek, C. Eng. *J Clin Endocrinol Metab.* **99**(7):E1316-21 (2014).
16. A. Rimmer, H. Phan, I. Mathieson, Z. Igbal, S.R. Twigg; WGS500 Consortium, A.O. Wilkie, G. McVean, G. Lunter. *Nat Genet* **46**(8):912-8 (2014).
17. H.Y. Lam, M.J. Clark, R. Chen, R. Chen, G. Natsoulis, M. O'Huallachain, F.E. Dewey, L. Habegger, E.A. Ashley, M.B. Gerstein, A.J. Butte, H.P. Ji, M. Snyder. *Nat Biotechnol* **30**(1):78-82 (2011).
18. R.A. Fisher (1930)
19. P. Katsonis, O. Lichtarge. *Genome Research* **24**(12):2050-8 (2014).
20. Illumina Platinum Genomes. http://www.illumina.com/platinumgenomes/ (2015).
21. The 1000 Genomes Project Consortium. *Nature* **491**:56-65 (2012).
22. K. Wang, M. Li, H. Hakonarson. *Nucleic Acids Research* **38**(16):e164 (2010).
23. H.A. Orr. *Nat Rev Genet* **6**(2):119-27 (2005).
24. H. Li, R. Durbin. *Bioinformatics* **25**:1754-60 (2009)
25. C. Racz, R. Petrovski, C.T. Saunders, I. Chorny, S. Kruglyak, E.H. Margulies, H.Y. Chuang, M. Kallberg, S.A. Kumar, A. Liao, K.M. Little, M.P. Stromberg, S.W. Tanner. *Bioinformatics* **29**(16):2041-3 (2013).
26. Z. Igbal, M. Caccamo, I. Turner, P. Flicek, G. McVean. *Nat Genet* **44**(2):226-32 (2012).
27. M. DePristo, E. Banks, R. Poplin, K. Garimella, J. Maguire, C. Hartl, A. Philippakis, G. del Angel, M.A. Rivas, M. Hanna, A. McKenna, T. Fennell, A. Kernytsky, A. Sivachenko, K. Cibulskis, S. Gabriel, D. Altshuler, M. Daly. *Nat Genet* **43**:491-498 (2011).

IDENTIFICATION OF QUESTIONABLE EXCLUSION CRITERIA IN MENTAL DISORDER CLINICAL TRIALS USING A MEDICAL ENCYCLOPEDIA*

HANDONG MA

Department of Biomedical Informatics, Columbia University, 622 West 168 Street, PH-20
New York, NY, 10032, USA
Email: handongma.work@gmail.com

CHUNHUA WENG

Department of Biomedical Informatics, Columbia University, 622 West 168 Street, PH-20
New York, NY, 10032, USA
Email: cw2384@cumc.columbia.edu

Precision medicine requires precise evidence-based practice and precise definition of the patients included in clinical studies for evidence generalization. Clinical research exclusion criteria define confounder patient characteristics for exclusion from a study. However, unnecessary exclusion criteria can weaken patient representativeness of study designs and generalizability of study results. This paper presents a method for identifying questionable exclusion criteria for 38 mental disorders. We extracted common eligibility features (CEFs) from all trials on these disorders from ClinicalTrials.gov. Network Analysis showed scale-free property of the CEF network, indicating uneven usage frequencies among CEFs. By comparing these CEFs' term frequencies in clinical trials' exclusion criteria and in the PubMed Medical Encyclopedia for matching conditions, we identified unjustified potential overuse of exclusion CEFs in mental disorder trials. Then we discussed the limitations in current exclusion criteria designs and made recommendations for achieving more patient-centered exclusion criteria definitions.

* This study was funded by National Library of Medicine grant R01LM009886 (Bridging the semantic gap between clinical research eligibility criteria and clinical data).

1. Introduction

Randomized controlled trials (RCT) produce high-quality evidence but often lack patient representativeness of the real-world population. Clinical research eligibility criteria define the characteristics of a research volunteer for study inclusion or exclusion. Typically, exclusion reasons relate to age, gender, ethnicity, complex comorbidities, conflicting interventions, or patient preference[1]. Although exclusion criteria do not bias the comparison between intervention and control groups, which reflects a trial's internal validity, exclusion criteria can impair the external validity of a trial[2,3]. It has been shown in various disease domains that clinical trial participants are often not representative of the real-world patient population to which an RCT is intended to apply, and that the lack of patient representativeness has impaired the generalizability of clinical trials[3,4].

Thus, it is imperative to develop methods for justifying the exclusion criteria in clinical trials. However, this task is fraught with challenges. First, many eligibility criteria are vague and complex[1] and cannot be easily represented in a computable format that allows for automated screening of unjustifiable exclusion criteria[5]. Second, clinical researchers often do not have a sufficiently precise picture of the real-world patient population to make informed decisions about exclusion criteria. Although the wide adoption of Electronic Health Record (EHR) make this idea more promising than ever[6-9], aggregating EHR data to profile the real-world patient population is a nontrivial exercise, due to common data fragmentation and data quality problems[10]. Therefore, it is worthwhile to explore alternatives to the EHR-based data-driven approach, especially through combining different data sources in order to increase patient representativeness of clinical trial eligibility criteria. This paper presents the feasibility of such a knowledge-based approach, using PubMed Health Medical Encyclopedia knowledge. PubMed Health Medical Encyclopedia (hereinafter, PubMed Encyclopedia) is a service created by the National Center for Biotechnology Information (NCBI), and made accessible by the U.S. National Library of Medicine (NLM), to provide summaries of diseases and conditions[11]. Such a meta-analysis with automatic data-mining methods across different data sources provides us new insights into clinical trial design and can inform precise evidence-based practice.

2. Methods

We chose mental disorder clinical trials for a proof of principle but the method should generalize to other fields of medicine. We hypothesized that the occurrence of a term in PubMed Encyclopedia for a symptom, a medication, or a chemical compound could be used to indicate its relevance to the mental disorder (condition) under consideration. For each term in each mental disorder, we compared the term frequencies in the exclusion criteria of all the clinical trials on that condition in ClinicalTrials.gov and the term's occurrence in PubMed Encyclopedia. On this basis we identified terms that occur frequently in both exclusion criteria and PubMed. We further hypothesized that a term with a certain level of frequency of use in PubMed Health Encyclopedia about a mental disorder should be deemed relevant to that disorder. Thus, its frequent use in excluding patients with this trait from clinical trials on that disorder could be questionable.

We built an exclusion criteria network including all mental disorders based on the method from Boland and Weng et al.'s previous work[12]. Using that network, we identified the common exclusion criteria for mental disorders and assessed their appropriateness of use. We identified clinical trials for 84 mental disorders in the category of "Behaviors and Mental Disorders" in ClinicalTrials.gov. For each condition, using our published tag-mining algorithm[13], we extracted all common eligibility features (CEFs) that each occurred in at least 3% of all clinical trials related to each condition in ClinicalTrials.gov. This method is capable of automatically deriving frequent UMLS tags from clinical text using part-of-speech (POS) tagger, N-grams model, and UMLS unique concept identifier. For example, we found the UMLS concept "ethanol", which belongs to the "organic chemical - pharmacologic substance" semantic type, occurred in 74.7% of the alcoholism clinical trials while occurred in only 26.8% of depression trials. For each mental disorder, we were able to generate a list of UMLS concepts with their frequencies of use in inclusion and exclusion criteria section.

We calculated the frequencies of use aggregated across all mental disorders for inclusion and exclusion purposes, respectively, for each of these CEFs. We also analyzed the frequency distribution of these CEFs by their UMLS semantic types. We constructed a two-mode network for all the 84 mental disorders and their top 20 CEFs, based on the disorder-CEF associations. Then we projected this network to a one-mode network based on CEFs using the Newman (2001) method (tnet)[14], a classic method used in detecting communities in networks. The process worked by selecting one set of nodes (i.e., CEFs), and linking two nodes, if they were connected to the same node in the other set of nodes (i.e., conditions). For each mental disorder, we analyzed the distribution of the degree of all the nodes in this network to assess the usage of the CEFs in the mental disorder trials. Since most CEFs occurred equally in inclusion and exclusion criteria section, we used a mutual information filter to identify distinctive CEFs, regardless inclusion or exclusion, because Mutual Information is one of the commonly used quantities that measure independence between variables. We calculated the Mutual Information (MI) for each CEF. The formula is as follows (1):

$$I(U;C) = \sum_{e_t \in \{0,1\}} \sum_{e_c \in \{0,1\}} P(U = e_t, C = e_c) \log_2 \frac{P(U=e_t, C=e_c)}{P(U=e_t)P(C=e_c)}, \quad (1)$$

For each mental disorder, U is a random variable indicating the presence (number 1) or absence (number 0) of a CEF in every eligibility criterion (et) and C is a random variable representing the inclusion (number 0) or exclusion (number 1) status of the eligibility criterion (ec). We used additive smoothing to make sure CEF unique to only one section were included in the analysis. Since we aimed to target the most informative CEFs used as exclusion criteria, we chose the CEFs with positive MI scores in exclusion criteria as candidates for future comparisons. The cutoff of MI score retained the CEFs that are more frequently used as exclusion criteria rather than inclusion criteria. We used these CEFs to represent the common confounder patient characteristics excluded by clinical trials on each condition.

To generate the PubMed dataset, due to the heterogeneous condition names, we used a semi-automatic method to match the condition names in PubMed Health with the condition names in ClinicalTrials.gov. For example, Alzheimer disease in ClinicalTrials.gov was manually matched with

Alzheimer's disease in PubMed Health database. A total of 38 mental disorders were matched and manually validated. We processed the PubMed Encyclopedia's website content[11] for each matched mental disorder and used the same tag-mining algorithm[13] to extract all the terms for risk factors, causes, symptoms, signs, exams and tests, treatment options, and complications, and obtained their aggregated frequencies across all 38 mental disorders. For each of the 38 identified mental disorders, we aligned and ranked their CEF terms by their frequencies in ClinicalTrials.gov and their occurrences in PubMed Health Medical Encyclopedia, respectively, and compared their relative importance in each content source according to the ranks. Questionable CEFs were identified with high frequencies of use in both clinical trials and PubMed Encyclopedia. The entire workflow was shown in Fig. 1(a).

3. Results

3.1. CEF overlap between exclusion and inclusion criteria for mental disorder trials

We extracted 1304 exclusion CEFs and 1155 inclusion CEFs for all of the clinical trials for mental disorders (Fig. 1 (a)). A total of 1403 unique CEFs were identified, 1056 of which were present in both inclusion and exclusion criteria. The large overlap necessitated CEF selection to identify the most informative exclusion criteria. The top three frequent semantic types were *disease or syndrome, pharmacological substance,* and *finding*, respectively (Fig. 1(b)). Table 1 shows the top 10 most frequently used CEFs for *Alzheimer's disease*.

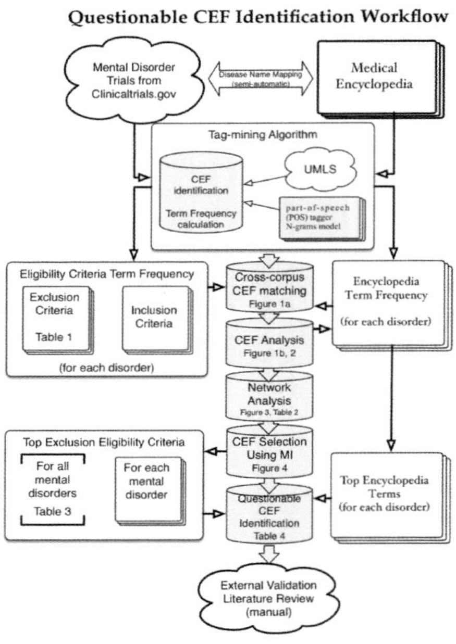

(a)

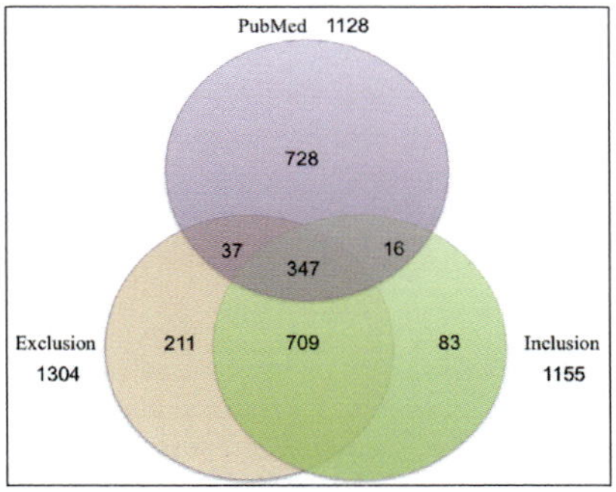

(b)

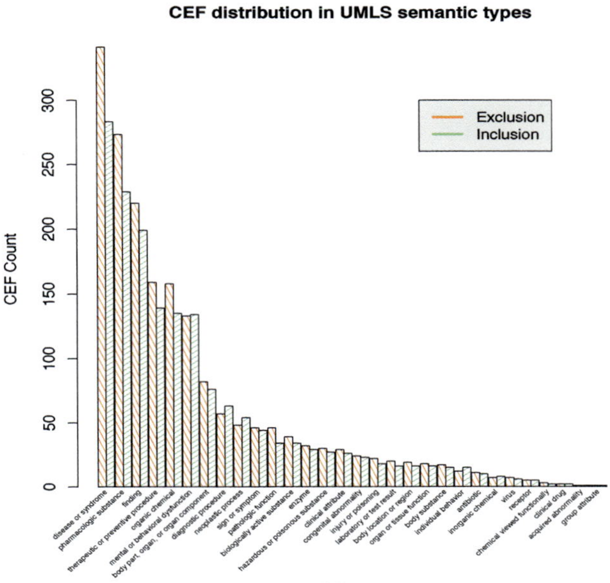

(c)

Fig 1. (a) Workflow of Identifying Questionable CEF (b) CEF Overlap between Exclusion and Inclusion CEF (c) Semantic Components between Exclusion and Inclusion CEF

Table 1. The top 10 most used exclusion CEFs for Alzheimer's diseases trials

Mostly Used Exclusion CEF	Frequency	UMLS Semantic Type
Mental disorders	29%	Mental or behavioral dysfunction
Allergy severity - severe	25%	Finding
Ethanol	23%	Organic chemical; pharmacologic substance
Depressed mood	23%	Finding; mental or behavioral dysfunction
Unstable status	21%	Finding
Cerebrovascular accident	21%	Disease or syndrome; therapeutic or preventive procedure
Magnetic resonance imaging	20%	Diagnostic procedure
Active brand of pseudoephedrine-triprolidine	17%	Organic chemical; pharmacologic substance
Pharmaceutical preparations	16%	Pharmacologic substance
Substance abuse problem	16%	Mental or behavioral dysfunction

3.2. CEF distribution among mental disorders trials

Out of the total 1403 unique CEFs, very few were used in a large number of clinical trials, while most of other CEFs were unique to one or several disorders (Fig. 2 (a)). On average, a CEF was present in 7.49 mental disorders for exclusion purposes and 6.28 for inclusion purposes. Each condition had 125.1 exclusion CEFs and 104.8 inclusion CEFs. Most of the frequently used CEFs were general factors for mental disorders (such as hypersensitivity or pharmacologic substance). Some were regularly used in clinical trials on other conditions (such as excluding gravidity, unstable states and allergy severity - severe). The top five mental disorders with the most exclusion CEFs were: *Restless Legs Syndrome* (226), *Substance Withdrawal Syndrome* (192), *Pick Disease of the Brain* (186), *Tic Disorders* (184) and *Front-temporal Dementia* (180) (Fig. 2 (b)).

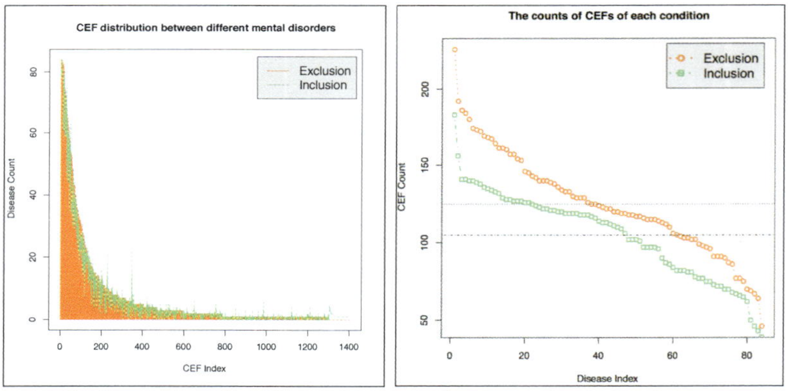

(a) (b)

Fig 2. CEF distribution between mental disorders (a) CEFs are indexed and ranked based on disease count in exclusion section. (b) Diseases are indexed and ranked based on exclusion CEF count.

3.3. Network construction and analysis for CEFs in mental disorder trials

We built a two-mode network for all mental disorders and CEFs based on the Disease-CEF linkages (Fig. 3(a)). In this network, there were two groups of nodes, the diseases and CEFs. The top 20 CEFs for each mental disorder were represented as orange ellipses and mental disorders were represented by blue round rectangles. The diseases were connected with different sets of CEFs and could be clustered based on the similarities of those connections. The edges were weighted as the frequency for a CEF to be associated with a mental disorder. Edges were red (for CEFS used only for exclusion), or orange (for CEFs used for both inclusion and exclusion), while edges of inclusion CEFs were green and dark green, respectively. In the network, we identified some hubs with higher degrees than other nodes, which indicated that a small portion of CEFs was frequently used for patient selection in most mental disorder trials. For example, diseases like *amnesia* and *bipolar disorder* shared more common CEFs with other mental disorders compared to diseases such as *associative disease* and *restless legs syndrome,* etc. In the network, most of the related disorders were clustered together using their CEF similarities such as *panic disorder* and *phobic disorder, Tourette syndrome* and *Tic disorder.* From the network, we also found some of the mental disorders, while not pathologically related, shared similar CEF sets. We also projected each of the three two-mode networks (Inclusion, Exclusion and PubMed) into one-mode network based on CEFs. Our analysis showed that three networks all display features of a scale-free network (Fig. 3(b)), which was similar to that of many of the real-world giant networks. The attributes of the network are listed in Table 2.

Table 2. Attributes of the One-Mode CEF Network

Data	CEF in network	Degree	One mode degree	Closeness	Betweenness
Inclusion	1155	7.60 ± 0.82	241.26 ± 12.00	6.3e-03 ± 3.6e-05	1656.99 ± 1685.7
Exclusion	1304	8.06 ± 0.84	309.10 ± 13.68	4.1e-03 ± 2.3e-05	1351.62 ± 935.4
PubMed	1128	2.27 ± 0.15	175.48 ± 8.04	8.6e-03 ± 3.9e-05	1895.01 ± 1407.6

* 95% confidence interval used

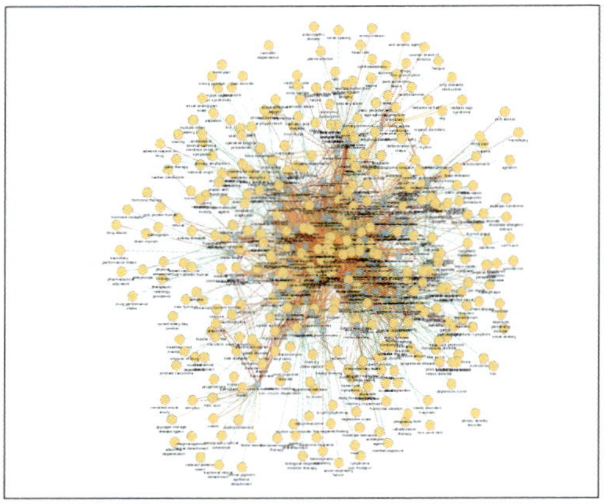

(a)

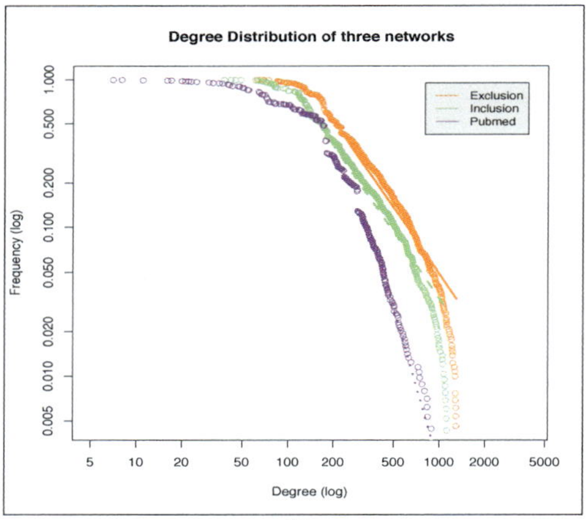

(b)

Figure 3. Network Structure (a) and Degree Distribution (b) between Mental Disorders and CEFs (b) Regression lines are plotted as solid or dashed lines.

3.4. CEF selection using Mutual Information (MI)

Some CEFs were equally used in inclusion and exclusion criteria. Using mutual information, we discarded CEFs with equal or higher occurrences in the inclusion section than in the exclusion criteria. Of a total of 1403 unique CEFs, only 632 had an MI value greater than 0, 568 had an MI value of 0 and 203 had an MI value below zero. The bigger the MI value is, the more frequently the CEF is for exclusion uses. To preserve all informative CEFs to match with the PubMed dataset, we selected all CEFs with a MI scores greater than 0 for further analysis. The benchmark analysis for CEFs and their MI distributions are in Fig. 4. Through this selection step, many common but non-discriminative CEFs were discarded, such as *pharmacologic substance*, *physical assessment findings*, and *intravenous infusion procedures*. In contrast, discriminative CEFs (e.g., *suicidal, unstable status, psychotic disorders*) were retained. However, it should be noted that some discriminative CEFs (such as *pregnancy tests, multiple endocrine neoplasia*, etc.) might be missed by this selection step.

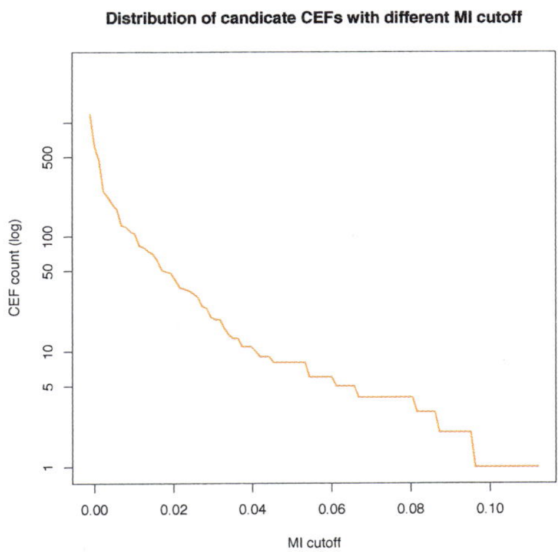

Figure 4. Mutual Information Score Distribution for CEFs

3.5. Aggregated cross-condition occurrence comparison for retained CEFs

We contrasted the aggregated occurrences of exclusion CEFs (N=1422) across the 38 matched disorders with partial results displayed in Table 3. For example, ethanol was a CEF present in all 38 mental disorders' exclusion criteria, implying 100% prevalence, and was present in the PubMed descriptions for 21 disorders (55.2%).

Table 3. A contrast of exclusion and PubMed occurrences across 38 mental disorders for top exclusion CEFs

Exclusion CEFs	Exclusion	PubMed	UMLS Semantic Types
Pharmaceutical preparations	38	38	Pharmacologic substance
Ethanol	38	21	Organic chemical; pharmacologic substance
Depressed mood	38	8	Finding; mental or behavioral dysfunction
Psychotic disorders	38	3	Mental or behavioral dysfunction
Hypersensitivity	37	3	Clinical attribute; finding; pathologic function
Hepatic	36	4	Body location or region
Antipsychotic agents	35	9	Pharmacologic substance
Unipolar depression	31	4	Mental or behavioral dysfunction
Anti-depressive agents	30	12	Pharmacologic substance
Screening for cancer	30	6	Diagnostic procedure
Benzodiazepines	30	5	Organic chemical; pharmacologic substance

The average CEF prevalence among the 38 mental disorders in the exclusion criteria and PubMed were 7.33 and 1.86, respectively, so that CEFs occurred less often in PubMed than in exclusion criteria. Among the top exclusion CEFs for the 38 mental disorders, we found some candidate CEFs that simultaneously had frequent PubMed occurrences (i.e., questionable CEFs), such as *ethanol, malignant neoplasms, anti-depressive agents,* and *depressed mood.*

We also analyzed the condition-specific CEF rankings between PubMed and exclusion criteria and identified questionable CEFs that had high PubMed rankings. Some example questionable CEFs for specific sleep disorder are listed in Table 4. *Hepatic* is associated with *sleep disorder* according to PubMed Health but was frequently used for excluding patients from 6.82% of sleep disorder clinical trials. Another questionable CEF is *sleep apnea syndromes*, whose frequency in exclusion criteria of all sleep disorder trials was as high as 24.6%, was ranked as top two relevant PubMed description for sleep disorder; therefore, we should be cautious when frequently using it as exclusion criteria. Another example is hypersensitivity (to treatment), which is common in the real-world population but is frequently excluded in randomized controlled trials (i.e., frequently as high as 13.6%).

Table 4. Questionable CEFs for excluding patients in sleep disorder clinical trials

Questionable CEF	Frequency	PubMed Rank	UMLS Semantic Type
Hepatic	6.82%	1	Body location or region
Sleep apnea syndromes	24.6%	2	Disease or syndrome
Sleep apnea obstructive	12.5%	3	Biologically active substance; disease or syndrome
Narcolepsy	8.05%	3	Disease or syndrome
Caffeine	5.10%	3	Organic chemical; pharmacologic substance
Hypersensitivity	13.6%	4	Clinical attribute; finding; pathologic function
Malignant neoplasms	9.52%	4	Finding; neoplastic process
Psychotic disorders	7.56%	4	Mental or behavioral dysfunction

4. Discussion

We investigated the exclusion criteria commonly used in mental disorder trials. The top four UMLS semantic types that contained the most questionable CEFs were *pharmacologic substance, mental or behavioral dysfunction, disease or syndrome,* and *finding*. Although some exclusion criteria of these semantic types have been used for years, their use in exclusion remains unexplained especially given their high prevalence among the real-world patients, most of who have several mental comorbidities or take multiple medications concurrently. Most of the drugs are for treating depressed mood or alcohol consumption or are anti-depressive drugs. If we exclude patients with those traits, we may generate a "pure" but not "typical"[15] test population, which may weaken the generalizability of these trials.

For a single mental disorder, the method proposed herein also detected several questionable CEFs. A recent study shows at least 50% of bipolar patient populations are excluded by at least one major exclusion criterion[15]. Using our method, we not only identified most of the exclusion criteria for bipolar disorder aggregated from previous studies (*drug abuse, alcohol abuse, significant medical conditions, pregnancy or lactation, suicidal risk* and *psychotropic medications*), but also retrieved information about which medical condition or medication was frequently used to exclude patients. *Ethanol, antipsychotic agents, and antidepressive agents* were questionable for excluding patients. This prediction corresponds to previous findings[15] that drug and alcohol abuse represent the most exclusion for bipolar trials, and provides more details for locating questionable exclusion criteria.

Although this study only focused on mental disorders for the detailed analysis, this pipeline can be easily applied to other disease domains. Most parts of the analyzing pipeline are fully automatic. The clinical trial eligibility criteria and medical encyclopedia for other diseases exist in similar format as used in this study, and can be processed in a large scale. However, considering the possible uniqueness of mental disorder domain, it is necessary to clarify the predictive power of this pipeline on a larger scale and different clinical settings, especially given poorly matched corpuses between clinical trial eligibility criteria and disease encyclopedia.

Several findings of this study shed light on future eligibility criteria designs. First, the scale-free feature of disease-CEF network suggests that a small number of exclusion criteria can be standardized and reused for most mental disorder trials. Second, trials for different conditions shared similar exclusion criteria, implying that some cohort selection criteria can be reused across conditions with little modification. Third, the power of exclusion for a single clinical trial should be quantified to avoid sampling biases in clinical trial designs.

5. Conclusion

This study demonstrates the promising value of applying a knowledge-based approach to assessing the patient-centeredness of clinical trial exclusion criteria by linking different data sources, including ClinicalTrials.gov and PubMed Medical Encyclopedia. In the future, proactive analyses like this could be conducted during clinical research designs to optimize clinical research eligibility criteria design and study participant selection to better achieve precise evidence definition[9].

6. Acknowledgments

We thank Dr. Riccardo Miotto for sharing eTACT methods for n-gram extraction.

References

1. Ross J, Tu S, Carini S, Sim I. Analysis of eligibility criteria complexity in clinical trials. *AMIA Summits on Translational Science Proceedings.* 2010;2010:46.
2. Friedman LM, Furberg C, DeMets DL. *Fundamentals of clinical trials.* Vol 4: Springer; 2010.
3. Elting LS, Cooksley C, Bekele BN, et al. Generalizability of cancer clinical trial results: prognostic differences between participants and nonparticipants. *Cancer.* 2006;106(11):2452-2458.
4. Heiat A, Gross CP, Krumholz HM. Representation of the elderly, women, and minorities in heart failure clinical trials. *Archives of Internal Medicine.* 2002;162(15).
5. Weng C, Tu SW, Sim I, Richesson R. Formal representation of eligibility criteria: a literature review. *Journal of biomedical informatics.* 2010;43(3):451-467.
6. Jensen PB, Jensen LJ, Brunak S. Mining electronic health records: towards better research applications and clinical care. *Nature Reviews Genetics.* 2012;13(6):395-405.
7. Hoffman S, Podgurski A. Improving health care outcomes through personalized comparisons of treatment effectiveness based on electronic health records. *The Journal of Law, Medicine & Ethics.* 2011;39(3):425-436.
8. Weng C, Li Y, Ryan P, et al. A Distribution-based Method for Assessing The Differences between Clinical Trial Target Populations and Patient Populations in Electronic Health Records. *Applied clinical informatics.* 2014;5(2):463.
9. Weng C. Optimizing Clinical Research Participant Selection with Informatics. *Trends in Pharmacological Sciences.* 2015:In Press.
10. Weiskopf NG, Weng C. Methods and dimensions of electronic health record data quality assessment: enabling reuse for clinical research. *Journal of the American Medical Informatics Association.* 2013;20(1):144-151.
11. Ozdas A, Shiavi RG, Silverman SE, Silverman MK, Wilkes DM. Investigation of vocal jitter and glottal flow spectrum as possible cues for depression and near-term suicidal risk. *IEEE Trans Biomed Eng.* 2004;51(9):1530-1540.
12. Boland MR, Miotto R, Weng C. A Method for Probing Disease Relatedness Using Common Clinical Eligibility Criteria. *Studies in health technology and informatics.* 2013;192:481.
13. Miotto R, Weng C. Unsupervised mining of frequent tags for clinical eligibility text indexing. *Journal of biomedical informatics.* 2013;46(6):1145-1151.
14. Newman ME. Scientific collaboration networks. II. Shortest paths, weighted networks, and centrality. *Physical review E.* 2001;64(1):016132.
15. Hoertel N, Le Strat Y, Lavaud P, Dubertret C, Limosin F. Generalizability of clinical trial results for bipolar disorder to community samples: findings from the National Epidemiologic Survey on Alcohol and Related Conditions. *The Journal of clinical psychiatry.* 2013;74(3):265-270.

REPRODUCIBLE AND SHAREABLE QUANTIFICATIONS OF PATHOGENICITY

ARJUN K MANRAI
Department of Biomedical Informatics, Harvard Medical School
10 Shattuck St., Boston, MA, 02115, USA
Email: Manrai@post.harvard.edu

BRICE L WANG
Illinois Mathematics and Science Academy
1500 Sullivan Rd., Aurora, IL 60506
Email: Bwang@imsa.edu

CHIRAG J PATEL
Department of Biomedical Informatics, Harvard Medical School
10 Shattuck St., Boston, MA, 02115, USA
Email: Chirag_Patel@hms.harvard.edu

ISAAC S KOHANE
Department of Biomedical Informatics, Harvard Medical School
10 Shattuck St., Boston, MA, 02115, USA
Email: Isaac_Kohane@hms.harvard.edu

ABSTRACT: There are now hundreds of thousands of pathogenicity assertions that relate genetic variation to disease, but most of this clinically utilized variation has no accepted quantitative disease risk estimate. Recent disease-specific studies have used control sequence data to reclassify large amounts of prior pathogenic variation, but there is a critical need to scale up both the pace and feasibility of such pathogenicity reassessments across human disease. In this manuscript we develop a shareable computational framework to quantify pathogenicity assertions. We release a reproducible "digital notebook" that integrates executable code, text annotations, and mathematical expressions in a freely accessible statistical environment. We extend previous disease-specific pathogenicity assessments to over 6,000 diseases and 160,000 assertions in the ClinVar database. Investigators can use this platform to prioritize variants for reassessment and tailor genetic model parameters (such as prevalence and heterogeneity) to expose the uncertainty underlying pathogenicity-based risk assessments. Finally, we release a website that links users to pathogenic variation for a queried disease, supporting literature, and implied disease risk calculations subject to user-defined and disease-specific genetic risk models in order to facilitate variant reassessments.

Introduction

1.1. *Clinical genomics in 2015*

Just 15 years since the completion of the Human Genome Project, researchers today can sequence a whole genome for less than $1,000. Fundamental advancements in sequencing platforms [1] coupled with concerted data-sharing efforts [2] have led to widespread and diverse uses of genomic data. Decades before the advent of next-generation sequencing, clinicians and geneticists were using targeted gene testing in diagnosis and prognosis, for example in calculating the familial risk of cystic fibrosis [3]. More recently, whole-genome and whole-exome sequencing have led to the discovery of causal lesions for numerous hitherto unsolved Mendelian disorders [4]. Other common clinical uses of genomic data include familial risk stratification for diseases such as hypertrophic cardiomyopathy [5], drug targeting based on activating mutations for cancers such as non-small-cell lung carcinoma [6], and genetic counseling for disorders such as trisomy 21 using fetal DNA circulating in maternal plasma (non-invasive prenatal testing, NIPT) [7].

While these efforts have led to real gains in diagnosis and treatment, it is now a central challenge of clinical genomics to sort through an unwieldy literature of genetic associations: in aggregate, there are hundreds of thousands of genetic associations across the entire spectrum of human disease [8]. The usual scale for summarizing findings to the clinician and patient is based on "pathogenicity," [9], or the capacity of a genomic variant to cause disease. Pathogenicity is a qualitative categorical concept, and its usual clinical scale consists of the values "Benign," "Likely Benign," "Variant of Uncertain Significance," "Likely Pathogenic," and "Pathogenic" [9].

1.2. *Recent inconsistencies between pathogenicity assertions*

Although pathogenicity assertions have been in use for decades clinically, only recently have systematic reinvestigations of pathogenicity been possible due to the widespread availability of large-scale sequencing data from the general population. The typical study design involves identifying all pathogenic variants for a given disease and then assessing the frequency of this variation in the general population. If the aggregate or individual variant frequency exceeds a disease-specific threshold, then pathogenicity for a variant or group of variants is challenged. This frequency threshold depends on the mode of inheritance (e.g. autosomal dominant), age-of-onset, prevalence in the tested population, molecular heterogeneity (fraction of disease due to a given variant), and desired penetrance cutoff (probability an individual with the variant expresses disease). For example, for an autosomal dominant disease caused by highly penetrant alleles, variant pathogenicity is called into question if the aggregate pathogenic genotype frequency exceeds the prevalence of the disease.

Several recent studies have used this approach to question the quality of pathogenicity ratings and reclassify pathogenicity assertions. Testing large-scale non-diseased populations has challenged prior pathogenicity assertions for X-linked intellectual disability [10], hypertrophic cardiomyopathy [11], non-syndromic hearing loss [12], and several other diseases. However, this is a small subset of the thousands of disorders with assertions regarding pathogenic genetic variation [8]. There is a critical need to scale up both the pace and feasibility of systematic reinvestigations of pathogenic variation using large-scale sequencing data from control populations.

1.3. *The need for reproducible, shareable, and disease-specific quantitative investigations of pathogenic variation*

It is now a central challenge in clinical genomics to reassess a scattered literature of disease-associated genetic variation as well as the large burden of novel variants discovered in whole genome or whole-exome sequencing. After achieving the "$1,000 genome," we may face the "$100,000 analysis." [13]. Several specific challenges hinder robust interpretation of potentially pathogenic genetic variation. First, pathogenicity assertions are typically not quantitative risk estimates. Second, it is usually unclear how a pathogenic variant should be interpreted in distinct clinical contexts with different prior probabilities (e.g., pathogenicity in males versus females or for patients with co-morbid conditions). Third, there is no accepted "false discovery rate" for the majority of clinically utilized pathogenic variation and, further, multiple recent re-investigations suggest that it is far greater than previously appreciated [10], [12], [14]. Fourth, and relatedly, assertions are based on a fragmented literature. It remains a challenge to assimilate findings from diverse studies with different analytic and design parameters [15]. Such re-investigations have generally concentrated on a single disease or closely related set of diseases at a time [10], [12], [14], and have required considerable bioinformatics resources to subset, clean, and work with pathogenic variation and sequence data. There is a need for a new digital platform to efficiently estimate, analyze, and share quantitative disease risk estimates for pathogenic variation.

In this manuscript we develop a shareable computational framework to quantify pathogenicity assertions that have been reported in the literature. We release a reproducible "digital notebook" which integrates executable code, text annotations, and mathematical expressions to enable investigators to study how variation in the general population and genetic model parameters dictate risk estimates underneath pathogenicity assertions. This notebook is written in the interactive computing environment IPython [16]. We extend previous disease-specific reinvestigations of pathogenicity to over 6,000 diseases and 160,000 assertions in ClinVar [17]. We document how reported pathogenicity assertions can mask large uncertainty over a wide range of risk estimates, a critical consideration for clinicians and patients using such data for treatment and diagnosis. We link pathogenicity assertions to their supporting literature and current ClinVar annotations. Investigators can use this platform to carry out rapid disease-specific quantitative analyses for pathogenic variants. Disease experts, such as genetic counselors, can tune population parameters (such as prevalence and heterogeneity) to expose the determinants of pathogenicity and prioritize pathogenicity assertions for reassessment. All code is made freely available.

2. Methods

2.1. *Genetic models*

Consider a population of n individuals. For simplicity, first consider a single bi-allelic site where the reference allele frequency is p and non-reference allele frequency is $q = 1 - p$. Under Hardy-Weinberg equilibrium, genotypes AA (homozygous reference), Aa (heterozygous), and aa (homozygous alternate) have frequencies p^2, $2pq$, and q^2, respectively. Then the *genotype frequency of* q, the fraction of individuals who carry at least one q allele, denoted $G(q)$, is given by

$$G(q) = q^2 + 2q(1 - q)$$

For a locus with k distinct alleles (by state) under Hardy-Weinberg equilibrium, this equation still holds,

$$G(q) = q^2 + 2\sum_{p_i \neq q}^{k} p_i q = q^2 + 2q(1-q)$$

We define the *penetrance* of a genotype as the conditional probability of expressing disease D for an individual possessing the genotype V, $P(D \mid V) = (P(V \mid D)P(D))/P(V)$, where $h \equiv P(V \mid D)$ is an indicator of the molecular heterogeneity of the disease, $P(D)$ is the prevalence, and $P(V)$ is the genotype frequency. Penetrance is a population-specific parameter—for a given variant, penetrance can vary substantially based on clinical context (e.g. general population vs. testing laboratory population). We consider autosomal dominant, autosomal recessive, additive, and multiplicative genetic risk models. Under these risk models, we can write genotype frequencies and relative risks given a non-reference allele frequency q and per allele risk γ for a bi-allelic locus as follows:

Genetic model	Affected genotype frequencies (relative risk)
Autosomal dominant	$q^2 + 2pq\ (\gamma)$
Autosomal recessive	$q^2\ (\gamma)$
Additive	$q^2\ (2\gamma)$, $2pq\ (\gamma)$
Multiplicative	$q^2\ (\gamma^2)$, $2pq\ (\gamma)$

Table 1: Genetic risk models. q denotes the non-reference allele frequency, γ is the per allele risk.

2.2. *Clinical variant annotations*

The ClinVar database [www.ncbi.nlm.nih.gov/clinvar] aggregates genotype-phenotype assertions across human disease [17]. ClinVar assertions are summarized on a qualitative pathogenicity scale: (Benign, Likely benign, Uncertain significance, Likely pathogenic, Pathogenic). The database further includes supporting evidence where available, such as *in vitro* and *in silico* studies of pathogenicity. The database collects submissions from investigators around the world and can be used to resolve conflicts [8]. If many investigators independently assert the same relationship, this information is used to bolster the evidence for a variant-disease relationship. In this manuscript, we use the clinvar_20150629 version of the ClinVar database retrieved from ANNOVAR [18].

2.3. *Allele frequency data from the general population*

We incorporated allele frequency data from the NHLBI Exome Sequence Project (ESP) [19] and the Broad Exome Aggregation Consortium (ExAc) [20]. These data include allele frequencies from 6,503 individuals (ESP) and 60,706 individuals (ExAc). Both databases contain frequency data separated by population groups (e.g. in ESP, allele frequency data is provided separately for the 2,203 African Americans and 4,300 European Americans that constitute ESP). ExAc has been filtered for known causes of severe pediatric diseases, as it is intended for use as a "general population" resource to filter variants [20].

2.4. *Open source software stack*

The analysis in this manuscript is performed entirely in the interactive computing environment IPython [16]. IPython combines text annotations, executable code, mathematical expressions (LaTeX), and embedded HTML in a single digital notebook. We also built a D3 visualization [21] to allow users to explore pathogenicity assertions in the browser along with supporting evidence and user-controlled genetic model parameters to compute penetrance. Genomic sequence data and ClinVar annotations were retrieved using both ANNOVAR [18] and the ClinVar website [17].

3. Results

3.1. *A reproducible and shareable workflow for quantifying pathogenicity assertions*

We developed a reproducible and shareable platform for clinical genomics annotations (Figure 1). We have released a digital notebook written in the interactive computing environment IPython [16] that integrates executable code, text annotations, mathematical expressions, and embedded HTML. Investigators can freely download this IPython notebook file, reproduce all data-gathering steps, choose any disease from ClinVar, and specify the prevalence, heterogeneity, and genetic model to estimate the penetrance of all ClinVar variants for the selected disease. All sensitivity analyses described in this manuscript can be reproduced and customized in the IPython notebook. Further, investigators can add cells of their own code and text to specify different disease-specific genetic risk models and assumptions required to compute penetrance. The analysis steps and final risk summary information, whether quantitative risks or qualitative assertions, can be stored alongside supporting data and assumptions in a single document. Customized disease-specific notebooks can be shared with collaborators to be run and customized locally.

Here we perform a sensitivity analysis for penetrance using the disease hypertrophic cardiomyopathy (HCM). Making our assumptions explicit:

- The prevalence of HCM is $1/500$ [Maron et al., *Circulation* 1995]
- There are more than a thousand causal variants for HCM [Maron et al., *J Am. Coll. Cardiol.*, 2012]. A conservative estimate for h, the heterogeneity parameter, is therefore $h = 0.1$. A more realistic estimate is $h = 0.001$
- HCM is autosomal dominant: a single copy of a highly-penetrant causal allele is sufficient to cause disease

In [25]:
```
# parameters above as Python variables
import numpy as np

prev = 1.0/500
het_range = np.linspace((0.001),(0.1),10)
pen = {} # to be computed below
genetic_model = 'AD'
```

Recall:

$$\text{Penetrance} = P(D|G=j) = \frac{k \times P(G=j|D)}{P(G=j)} = \frac{k \times P(G=j|D)}{P(G=j|D) \times k + P(G=j|\overline{D}) \times (1-k)}$$

Thus, grabbing all HCM variants and computing penetrance can be accomplished by:

In [26]:
```
HCM = merged[merged.Disease.str.contains('hypertrophic_cardiomyopathy')]
HCM = HCM.drop_duplicates(subset = ['Chromosome','Start','Stop','Ref','Alt']) # 81 distinct variants
raw_names = zip(HCM.Chromosome,HCM.Start,HCM.Ref,HCM.Alt)
clean_names = [str(chrom)+':'+str(int(pos))+str(ref)+'>'+alt for (chrom,pos,ref,alt) in raw_names]
HCM[['Chromosome','Start','Ref','Alt','ExAc_Overall_Frequency','ESP_Overall_Frequency','Accession','Disease']].head(5)
```

Out[26]:

	Chromosome	Start	Ref	Alt	ExAc_Overall_Frequency	ESP_Overall_Frequency	Accession	Disease
199	1	201328373	G	A	0.0004	0.000461	RCV000013222.22	Familial_hypertrophic_cardiomyopathy_2
203	1	201337340	G	A	0.0005	0.000308	RCV000149450.1	Primary_familial_hypertrophic_cardiomyo
250	1	236911044	C	T	0.0002	0.000077	RCV000169901.2	Familial_hypertrophic_cardiomyopathy_2
448	10	69881254	A	G	0.0009	0.000923	RCV000043545.1	Familial_hypertrophic_cardiomyopathy_2
451	10	69959174	C	T	0.0030	0.002076	RCV000043542.1	Familial_hypertrophic_cardiomyopathy_2

Figure 1: A reproducible and shareable workflow for quantifying pathogenicity assertions. Screenshot from the IPython "digital notebook" that accompanies this manuscript. The interactive computing notebook combines executable code (written in blocks), mathematical expressions, and text annotations. Code is provided to retrieve ClinVar annotations, PubMed references, and frequency data for any disease in ClinVar. The user can explicitly specify genetic model assumptions to compute penetrance and perform sensitivity analyses. Available at: https://github.com/manrai/Pathogenicity_Notebook.

3.2. *A diseaseome-wide investigation of pathogenicity assertions*

We used our computational framework to perform a diseaseome-wide analysis of pathogenicity assertions in ClinVar (Figure 2). Using the clinvar_20150629 version of ClinVar retrieved from ANNOVAR, we observed 132,584 distinct variants, as defined by unique values of (Chromosome, Start Position, Stop Position, Reference Allele, Alternate Allele) tuples in hg19 coordinates. These 132,584 variants gave rise to 160,487 distinct pathogenicity assertions about disease. As such, the majority of variants—114,107 out of 132,584 variants (86%)—were included in only a single pathogenicity assertion (Figure 2a). The 160,487 total assertions spanned 6,427 distinct disease names, although 42,761 assertions (27%) had disease names of "not specified" or "not provided." Of the 117,726 remaining assertions, just five out of 6,425 diseases (Lung Cancer, Malignant Melanoma, Hereditary Cancer-Predisposing Syndrome, Familial Cancer of Breast, Lynch Syndrome) accounted for 59,829 assertions (51%). 1,524 out of 6,425 diseases (24%) had at least five assertions (Figure 2b). Of the 160,487 total assertions, 85,455 (53.2%) were either "unknown" or "untested"; 37,871 (23.6%) were "pathogenic"; 15,483 (9.6%) were "non-pathogenic"; 11,357 (7.1%) were "probable-non-pathogenic"; 6,189 (3.9%) were "probable-pathogenic"; 3,964 (2.5%) were "other"; and 168 (0.1%) were classified as "drug-response" (Figure 2c).

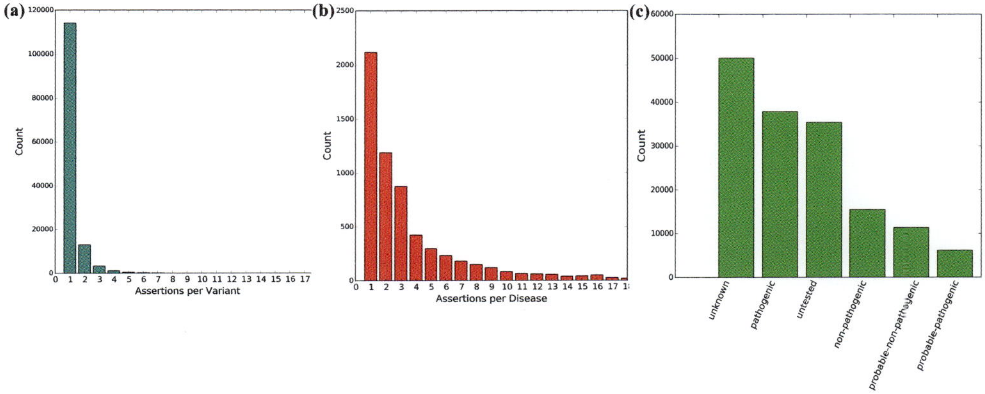

Figure 2: A diseaseome-wide investigation of pathogenicity assertions in ClinVar. (a) Distribution of 160,487 pathogenicity assertions across 132,584 distinct variants. 86% of variants had exactly one assertion. **(b)** Truncated distribution of pathogenicity assertions by disease. **(c)** Clinical significance values for assertions in ClinVar. 85,455 (53.2%) of the 160,487 total assertions were either "untested" or "unknown." "Pathogenic" assertions were the second largest overall group.

3.3. *Uncertainty in the disease risk conveyed by pathogenic variation*

The penetrance of a pathogenic variant—the probability that individuals with the variant express disease—depends on the allele frequency in both case and control individuals, mode of inheritance, age-of-onset, heterogeneity, and prevalence of the disease. To study this dependence, we analyzed the disease hypertrophic cardiomyopathy (HCM), and documented how penetrance values across all pathogenic single nucleotide variants (SNVs) for HCM vary under clinically plausible parameter values (Figure 3). We retrieved 81 distinct pathogenic SNVs with frequency data available in ExAc or ESP for HCM. We used the widely-accepted prevalence of 1:500 individuals [22] and varied the molecular heterogeneity parameter from conservative values (h = 0.1, 10% of HCM is explained by a single variant) to a more accepted model (e.g. h = 0.001) given that greater than a thousand causal variants have been identified for HCM [5]. All variants display substantial variability based on the input genetic model parameters (Figure 3), however, several pathogenic variants have consistently low penetrance due to their elevated non-reference allele frequency.

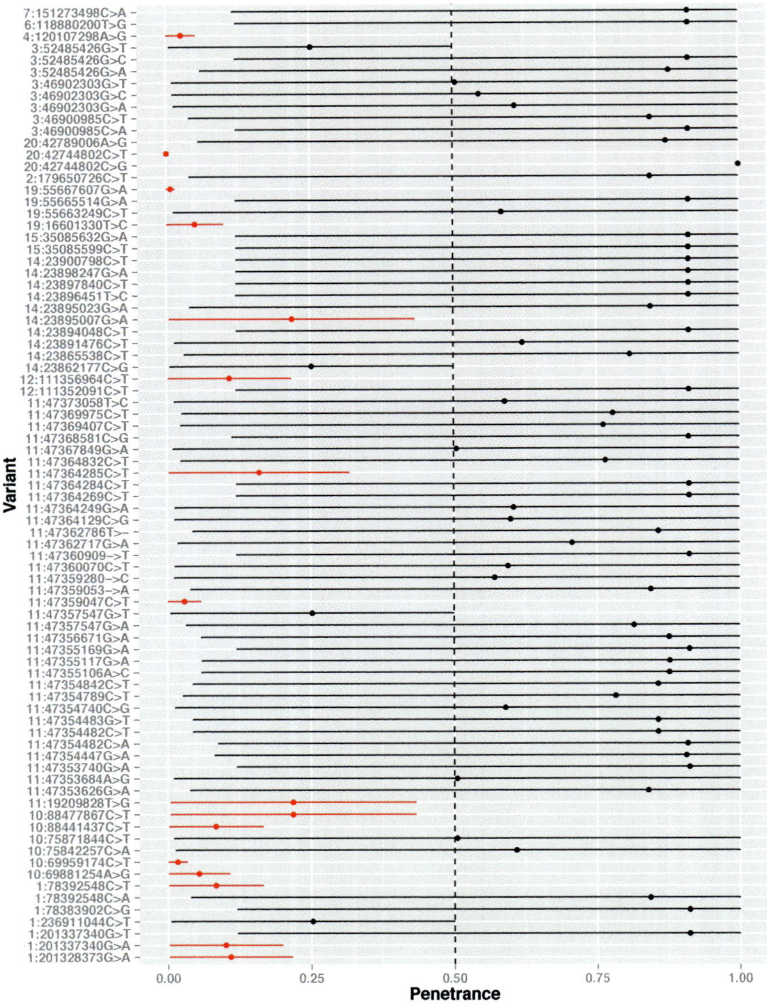

Figure 3: Uncertainty in the disease risk conveyed by pathogenic variation. Shown are the 81 pathogenic SNVs from ClinVar for hypertrophic cardiomyopathy with ExAc or ESP frequency data available. We computed a range of penetrance values for each variant by varying heterogeneity linearly in the range [0.001, 0.1]. Several variants have consistently low penetrance given their elevated non-reference allele frequency. Variants that were lower than the 50% penetrance cutoff throughout these simulations are colored in red.

3.4. *Frequency of ClinVar variants in the general population*

We studied the frequency of pathogenic variation in ClinVar by disease. Many diseases had pathogenic variants with summed minor allele frequencies that were incompatible with even moderately penetrant causal alleles (Figure 4). Considering only pathogenic SNV variation, 110 distinct disease terms in ClinVar had a summed minor allele frequency greater than 0.05 (Figure 4). The five highest frequency diseases were Neutrophil-Specific Antigens NA1/NA2, Severe Combined Immunodeficiency Autosomal Recessive T-Cell Negative B-Cell Positive NK-Cell Positive, Metachromatic Leukodystrophy, Trimethylaminuria, and Trimethylaminuria Mild.

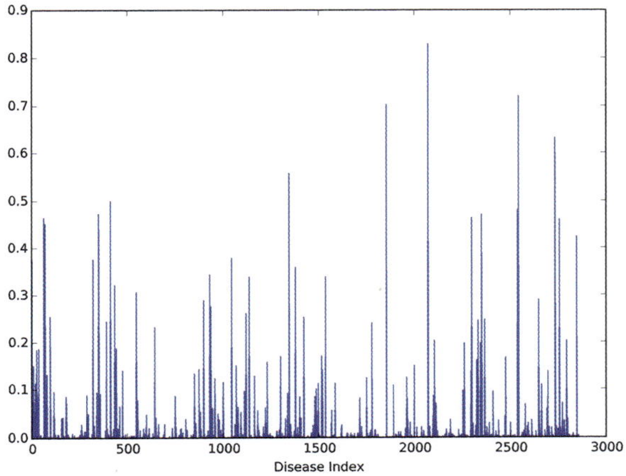

Figure 4: Summed frequency of pathogenic SNVs by disease. Many diseases have summed pathogenic SNV minor allele frequencies that far exceed the prevalence of the disease. 110 distinct disease terms have a summed minor allele frequency greater than 0.05.

3.5. *User-directed investigations of pathogenicity*

We built a website to enable investigators to conduct disease-specific analyses of pathogenic variation. After selecting a disease and specifying a genetic model, the investigator is provided with all ClinVar entries for variants with questionable pathogenicity as governed by the user-controlled parameters, as well as the supporting literature for these variants. Investigators can set genetic model parameters based on, for example, genetic testing laboratory experience from other patients with the same disease. Investigators are then provided with implied penetrance values for each variant under these assumptions as well as supporting literature references in order to efficiently prioritize pathogenic variants for reassessment.

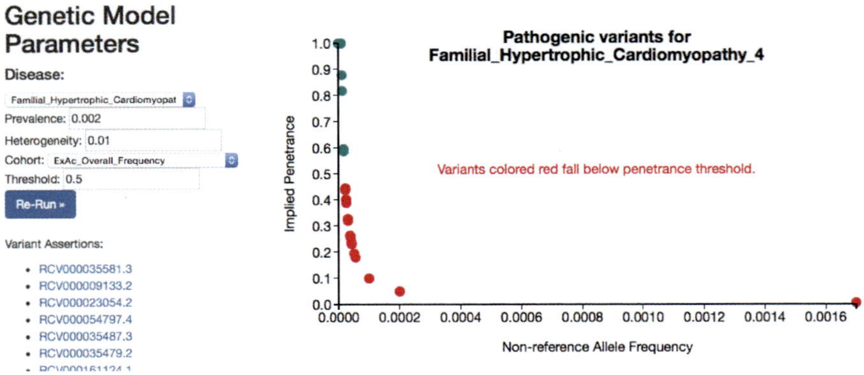

Figure 5: Exploring pathogenicity ratings. Screenshot from a website that enables users to explore disease-specific pathogenic variation. The user can select the disease, prevalence, heterogeneity, cohort used for frequency data, and penetrance threshold, and run an analysis for matching ClinVar variants. The user is linked to variant assertions in ClinVar to re-evaluate pathogenicity assertions systematically. A live version of this site can be found at http://people.fas.harvard.edu/~manrai/pathogenicity_explorer.

4. Discussion

4.1. *Summary of findings*

We developed a reproducible and shareable computational framework to quantify pathogenicity assertions across disease. We used this platform to extend previous disease-specific reinvestigations of pathogenicity to over 6,000 diseases and 160,000 assertions in ClinVar. For investigators wishing to conduct disease-specific quantitative reassessments of pathogenic variation, we released a digital notebook written in the interactive computing environment IPython that integrates executable code, text, and mathematical expressions to specify explicit genetic model assumptions and quantify pathogenicity assertions. We documented the uncertainty in disease risk estimates for pathogenic variants using, as an example, all pathogenic SNV variation for the inherited condition hypertrophic cardiomyopathy. We released a website that allows users to quickly explore pathogenic variation for individual diseases, prioritize variants for reassessment, and obtain ClinVar records and supporting literature for variants that fall below an adjustable clinical threshold for penetrance.

4.2. *Disease-specific reassessments of pathogenicity*

Bottom-up approaches to reassessing pathogenicity allow investigators to specify genetic model assumptions and filter pathogenicity assertions tailored to the individual disease in which they have expertise. The clinical utility of genomic sequence data depends heavily on prior probabilities and genetic model parameters [23], and as such it is critical to incorporate these quantities into clinical decision-making. Expertise from clinical genetic testing laboratories in measuring genetic heterogeneity and other parameters will improve reassessments going forward. It will be increasingly important to quantify our understanding of the uncertainty of pathogenicity assertions, and share these data widely to collectively improve clinical decision-making.

4.3. *The publishable unit*

Digital notebooks such as IPython/Jupyter [16] offer several advantages as a method of documenting research progress. These notebooks combine executable code divided into understandable blocks with text markup, the precision of mathematical notation, figures, and embedded HTML in an easily shareable and coherent document that lets each user tailor code and analyses for their goals. Building off of IPython, the Jupyter project (https://jupyter.org) is language agnostic, enabling users to contribute to analysis workflows such as the Pathogenicity Notebook using other popular programming languages for data analysis. Using these tools, findings can be delivered alongside the underlying data and assumptions. For pathogenicity reassessments, a digital notebook could serve as a new publishable unit of analysis.

4.4. *Future work*

It is important to stress that frequencies retrieved from ExAc and ESP are estimates of population parameters. Future work could incorporate this uncertainty into disease-specific reassessments and study the generalizability of penetrance estimates across different ethnicities using these databases. It is also important to note that using frequency data from the general population will not reclassify very rare variation that is erroneously classified as pathogenic. Additionally, a low penetrance for a particular variant does not eliminate the possibility that the variant acts in concert with other variants to impact disease. Future investigators could extend the IPython notebook published here with new data sources and genetic models for their diseases of interest. The feasibility of quantitative pathogenicity reassessments will grow both with the availability of large-scale control sequence data as well as with domain expertise to specify quantitative parameters needed to compute penetrance (e.g. heterogeneity, prevalence). The future of decision theory in clinical genomics is bright if we rigorously vet pathogenicity assertions using shared data and assumptions.

Acknowledgments

The authors thank members of the Kohane and Patel laboratories for helpful discussions. This work was supported by U54LM008748 (AKM, BLW, and ISK) and 5K99ES023504, R21ES025052, and a PhRMA Foundation award (CJP).

References

[1]S. C. Schuster, “Next-generation sequencing transforms today’s biology.,” *Nat. Methods*, vol. 5, no. 1, pp. 16–8, Jan. 2008.
[2]M. D. Mailman, M. Feolo, Y. Jin, M. Kimura, K. Tryka, R. Bagoutdinov, L. Hao, A. Kiang, J. Paschall, L. Phan, N. Popova, S. Pretel, L. Ziyabari, M. Lee, Y. Shao, Z. Y. Wang, K. Sirotkin, M. Ward, M. Kholodov, K. Zbicz, J. Beck, M. Kimelman, S. Shevelev, D. Preuss, E. Yaschenko, A. Graeff, J. Ostell, and S. T. Sherry, “The NCBI dbGaP database of genotypes and phenotypes.,” *Nat. Genet.*, vol. 39, no. 10, pp. 1181–6, Oct. 2007.
[3]B. Kerem, J. Rommens, J. Buchanan, D. Markiewicz, T. Cox, A. Chakravarti, M. Buchwald, and L. Tsui, “Identification of the cystic fibrosis gene: genetic analysis,” *Science (80-.).*, vol. 245, no. 4922, pp. 1073–1080, Sep. 1989.
[4]M. J. Bamshad, S. B. Ng, A. W. Bigham, H. K. Tabor, M. J. Emond, D. A. Nickerson, and J. Shendure, “Exome sequencing as a tool for Mendelian disease gene discovery.,” *Nat. Rev. Genet.*, vol. 12, no. 11, pp. 745–55, Nov. 2011.
[5]B. J. Maron, M. S. Maron, and C. Semsarian, “Genetics of hypertrophic cardiomyopathy after 20 years: clinical perspectives.,” *J. Am. Coll. Cardiol.*, vol. 60, no. 8, pp. 705–15, Aug. 2012.
[6]W. Pao and N. Girard, “New driver mutations in non-small-cell lung cancer.,” *Lancet. Oncol.*, vol. 12, no. 2, pp. 175–80, Feb. 2011.
[7]R. W. K. Chiu, R. Akolekar, Y. W. L. Zheng, T. Y. Leung, H. Sun, K. C. A. Chan, F. M. F. Lun, A. T. J. I. Go, E. T. Lau, W. W. K. To, W. C. Leung, R. Y. K. Tang, S. K. C. Au-Yeung, H. Lam, Y. Y. Kung, X. Zhang, J. M. G. van Vugt, R. Minekawa, M. H. Y. Tang, J. Wang, C. B. M. Oudejans, T. K. Lau, K. H. Nicolaides, and Y. M. D. Lo, “Non-invasive prenatal assessment of trisomy 21 by multiplexed maternal plasma DNA sequencing: large scale validity study.,” *BMJ*, vol. 342, no. jan11_1, p. c7401, Jan. 2011.
[8]H. L. Rehm, J. S. Berg, L. D. Brooks, C. D. Bustamante, J. P. Evans, M. J. Landrum, D. H. Ledbetter, D. R. Maglott, C. L. Martin, R. L. Nussbaum, S. E. Plon, E. M. Ramos, S. T. Sherry, and M. S. Watson, “ClinGen - The Clinical Genome Resource.,” *N. Engl. J. Med.*, vol. 372, no. 23, pp. 2235–42, May 2015.
[9]S. Richards, N. Aziz, S. Bale, D. Bick, S. Das, J. Gastier-Foster, W. W. Grody, M. Hegde, E. Lyon, E. Spector, K. Voelkerding, and H. L. Rehm, “Standards and guidelines for the interpretation of sequence variants: a joint consensus recommendation of the American College of Medical Genetics and Genomics and the Association for Molecular Pathology,” *Genet. Med.*, vol. 17, no. 5, pp. 405–423, Mar. 2015.
[10]A. Piton, C. Redin, and J.-L. Mandel, “XLID-causing mutations and associated genes challenged in light of data from large-scale human exome sequencing.,” *Am. J. Hum. Genet.*, vol. 93, no. 2, pp. 368–83, Aug. 2013.
[11]C. Andreasen, J. B. Nielsen, L. Refsgaard, A. G. Holst, A. H. Christensen, L. Andreasen, A. Sajadieh, S. Haunsø, J. H. Svendsen, and M. S. Olesen, “New population-based exome data are questioning the pathogenicity of previously cardiomyopathy-associated genetic variants.,” *Eur. J. Hum. Genet.*, vol. 21, no. 9, pp. 918–28, Sep. 2013.
[12]A. E. Shearer, R. W. Eppsteiner, K. T. Booth, S. S. Ephraim, J. Gurrola, A. Simpson, E. A. Black-Ziegelbein, S. Joshi, H. Ravi, A. C. Giuffre, S. Happe, M. S. Hildebrand, H. Azaiez, Y. A. Bayazit, M. E. Erdal, J. A. Lopez-Escamez, I. Gazquez, M. L. Tamayo, N. Y. Gelvez, G. L. Leal, C. Jalas, J. Ekstein, T. Yang, S. Usami, K. Kahrizi, N. Bazazzadegan, H. Najmabadi, T. E. Scheetz, T. A. Braun, T. L. Casavant, E. M. LeProust, and R. J. H. Smith, “Utilizing ethnic-specific differences in minor allele frequency to recategorize reported pathogenic deafness variants.,” *Am. J. Hum. Genet.*, vol. 95, no. 4, pp. 445–53, Oct. 2014.
[13]E. R. Mardis, “The $1,000 genome, the $100,000 analysis?,” *Genome Med.*, vol. 2, no. 11, p. 84, Jan. 2010.
[14]J. Jabbari, R. Jabbari, M. W. Nielsen, A. G. Holst, J. B. Nielsen, S. Haunsø, J. Tfelt-Hansen, J. H. Svendsen, and M. S. Olesen, “New exome data question the pathogenicity of genetic variants previously associated with catecholaminergic polymorphic ventricular tachycardia.,” *Circ. Cardiovasc. Genet.*, vol. 6, no. 5, pp. 481–9, Oct. 2013.
[15]C. J. Patel, B. Burford, and J. P. A. Ioannidis, “Assessment of vibration of effects due to model specification can demonstrate the instability of observational associations,” *J. Clin. Epidemiol.*, Jun. 2015.
[16]F. Perez and B. E. Granger, “IPython: A System for Interactive Scientific Computing,” *Comput. Sci. Eng.*, vol. 9, no. 3, pp. 21–29, May 2007.
[17]M. J. Landrum, J. M. Lee, G. R. Riley, W. Jang, W. S. Rubinstein, D. M. Church, and D. R. Maglott, “ClinVar: public archive of relationships among sequence variation and human phenotype.,” *Nucleic Acids Res.*, vol. 42, no. Database issue, pp. D980–5, Jan. 2014.
[18]K. Wang, M. Li, and H. Hakonarson, “ANNOVAR: functional annotation of genetic variants from high-throughput sequencing data.,” *Nucleic Acids Res.*, vol. 38, no. 16, p. e164, Sep. 2010.
[19]NHLBI GO Exome Sequencing Project (ESP), “Exome Variant Server.” [Online]. Available: http://evs.gs.washington.edu/EVS/. [Accessed: 25-Jul-2015].
[20]“Exome Aggregation Consortium (ExAC).” [Online]. Available: http://exac.broadinstitute.org. [Accessed: 05-May-2015].
[21]B. M, O. V, and H. J, “D^3: Data-Driven Documents.,” *IEEE Trans. Vis. Comput. Graph.*, vol. 17, no. 17, pp. 2301–2309.
[22]B. J. Maron, J. M. Gardin, J. M. Flack, S. S. Gidding, T. T. Kurosaki, and D. E. Bild, “Prevalence of Hypertrophic Cardiomyopathy in a General Population of Young Adults : Echocardiographic Analysis of 4111 Subjects in the CARDIA Study,” *Circulation*, vol. 92, no. 4, pp. 785–789, Aug. 1995.
[23]I. S. Kohane, D. R. Masys, and R. B. Altman, “The incidentalome: a threat to genomic medicine.,” *JAMA*, vol. 296, no. 2, pp. 212–5, Jul. 2006.

PRECISION MEDICINE: DATA AND DISCOVERY FOR IMPROVED HEALTH AND THERAPY

ALEXANDER A. MORGAN
Stanford University School of Medicine & Khosla Ventures
Stanford, CA, USA
Email: alexmo@stanford.edu

SEAN D. MOONEY
Department of Biomedical Informatics and Medical Education
Seattle, WA 98105, USA
Email: sdmooney@uw.edu

BRUCE J. ARONOW
Biomedical Informatics, Developmental Biology and Computer Science,
University of Cincinnati and Cincinnati Children's Hospital Medical Center
Cincinnati, OH 45229, USA
Email: bruce.aronow@cchmc.org

STEVEN E. BRENNER
Department of Plant and Microbial Biology, University of California, Berkeley
Berkeley, CA, USA
Email: brenner@compbio.berkeley.edu

Rapid advances in personal, cohort, and population-scale data acquisition, such as via sequencing, proteomics, mass spectroscopy, biosensors, mobile health devices and social network activity and other apps are opening up new vistas for personalized health biomedical data collection, analysis and insight. To achieve the vaunted goals of precision medicine and go from measurement to clinical translation, substantial gains still need to be made in methods of data and knowledge integration, analysis, discovery and interpretation. In this session of the 2016 Pacific Symposium on Biocomputing, we present sixteen papers to help accomplish this for precision medicine.

1. Introduction

Ultimately, precision medicine represents the significant enhancement of evidence-based medicine, where clinical guidelines gleaned from population-level studies are able to be precisely modified based on the attributes of the individual patient to both learn about new significant biological determinants of individual subtypes, and to then optimally treat that individual. The age of precision medicine is already upon us. The evolution of medicine from an art and craft to science was facilitated through the development of methods of careful data collection and statistics for clinical trials, leading to medicine guided by population level evidence. In an analogous way that industrialization in manufacturing increased production volumes with standardization and systematic improvements in quality metrics, medicine has been moving from a heuristic based craft to mechanistically-tethered measures and guidelines. However, as we learn more about heterogeneity among the strongest factors determining disease risk, progression, and response to therapies, we can now identify highly significant factors that can forge new standards and individual-level customization. In an analogous way that evidence based medicine now guides standard of care practice at the population level, newer techniques will use data to guide practice at the level of individuals. Informatics methods in this space need to take advantage of highly multiplex heterogeneous mixes of categorical and numerical data, leverage related studies taking advantage of approaches in meta-analysis and transfer learning, be robust to missing data elements and sparsity, scale with superlinear interaction complexity, and be able to deal with a feature space much greater than the number of patients/samples by using approaches such as regularization and efficient use of priors.

Major efforts to create precision medicine datasets include the new national cohort as part of the US precision medicine initiative,[1] the 100k genomes being sequenced by each of the Geisinger-Regeneron collaboration[a] and the UK 100k genomes[2] projects and linked to clinical data, the US Veteran's Administration's Million Veterans initiative,[3] the many new ongoing trials using Apple's Research Kit,[b,4] Google's ambitious Baseline Study,[c] Vanderbilt's BioVU repository,[5] Craig Venter's Human Longevity Inc.,[d] and the massive cancer molecular profiling initiatives including The Cancer Genome Atlas.[6,7] Some of these data are already publicly available, but some of these projects are clearly not intended to be made public. In its most ambitious, precision medicine will require integration of data created by clinicians, biomedical labs, and commercial devices. How the academic research community, healthcare industry, commercial device industry, diagnostic test industry, and patient advocacy groups will negotiate the challenges of collaboration and privacy in the face of sometimes conflicting interests will be a challenge. However, recent efforts by groups such as Sage Bionetworks have highlighted the value of network effects between researchers and how new collaborative frameworks can accelerate and improve the discovery and

---

[a] https://www.genomeweb.com/sequencing/regeneron-launches-100k-patient-genomics-study-geisinger-forms-new-genetics-cent

[b] http://www.apple.com/pr/library/2015/03/09Apple-Introduces-ResearchKit-Giving-Medical-Researchers-the-Tools-to-Revolutionize-Medical-Studies.html

[c] http://www.wsj.com/articles/google-to-collect-data-to-define-healthy-human-1406246214

[d] http://www.humanlongevity.com/human-longevity-inc-hli-launched-to-promote-healthy-aging-using-advances-in-genomics-and-stem-cell-therapies/

innovation process.[8,9] Importantly, there is a moral imperative to accelerate the process of healthcare innovation and improvement, as the successes are measured in lives.

The diversity of papers in this session reflect some of the exciting range of topics in precision medicine. Informatics techniques for interpreting rare variation in complex genomes are presented alongside approaches that leverage links between clinical data stores and those genomic features. Methods for quantifying and analyzing complex phenotypes in patients in their daily lives are presented along with techniques for creating patient subgroups for targeting therapies. These papers reflect a sampling of the advances in informatics that are needed as we move into the age of precision medicine. Forums such as the Pacific Symposium for Biocomputing enable researchers to share ideas and help accelerate the process of discovery.

"The future is already here — it's just not very evenly distributed."
- William Gibson, National Public Radio interview

2. Session Contributions

2.1. *Methods for managing data complexity and limited sample size*

The explosion of rich and complex data in the age of precision medicine demands fundamentally new methods of analysis. The number of discrete data points collected from a patient can easily exceed the number of patients it would be possible to enroll in a single study, and can even exceed the population of the planet[10]. **Victor Bellón and colleagues** describe using a regularized transfer learning approach using task descriptors to address this problem. In addition to the sheer volume of data, the multivariate inter-relationships and connections mean that the complexity can scale at a rate much greater than simple linearity. **Nattapon Thanitorn and colleagues** present an approach using RDF Sketch Maps to reduce representational complexity.

2.2. *Probing rare genomic variation*

Much of what drives individual differences requiring precision, personalized treatments originates in the genome. However, rare or unique variants present an exceedingly difficult challenge in genome analysis and interpretation. **Anna Okula and colleagues** present the BioBin tool which builds on previous work[11] to support variant aggregation and statistical analyses. It is being used in the Marshfield Personalized Medicine Research Project. Expanding out from SNP's and indels, **Dokyoon Kim and colleagues** develop an annotation pipeline for copy number variants to support analysis of rare CNV's, also part of the Marshfield Personalized Medicine Research project. Going beyond the genome, **Yong Fuga Li and colleagues** describe diseaseExPatho, a tool that integrates transcription data with genomic variants to develop regulatory modules to aid in the interpretation of genetic variation in rare diseases.

2.3. *Leveraging demographic and clinical data, challenges in precision medicine*

Performing studies and analyses in varying patient populations is challenging for many reasons, including biases in levels of representation and the phenotypic data collected. Three papers in this session describe how databases containing demographic and clinical data can highlight some of these challenges and provide new opportunities for research. **Sarah Laper and colleagues** discuss their inability to replicate previously well-established genetic links with cardiovascular phenotypes represented in hospital clinical records; suggesting that either the enthusiasm for the potential of clinical datasets linked to biorepositories needs to be tempered by the significant challenge of using this data for basic science research, or that this highlights the big gap between prior precision medicine results and their translation into clinical significance, or some mixture of these two. **Nophar Geifman and Atul Butte** focus their attention on a mismatch between the demographics of patients sampled in clinical outcomes studies and high molecular resolution for studies like The Cancer Genome Atlas and the general demography of cancer. **Jessica N. Cooke Bailey and colleagues** present their work looking at genetic variants associated with kidney disease in diverse ethnic subgroups, highlighting the particular challenges in investigating the genetic basis of disease pathologies that disproportionately affect particular ethnic subgroups.

2.4. *High-throughput holistic functional phenotypic profiling*

One of the most exciting aspects of precision medicine is the ability to go beyond the high-throughput molecular assays for genomics, transcriptomics, proteomics, and metabolomics, and to move into measuring phenotypes at the level of the whole individual. Rather than probing the function of a protein, we can probe the functioning of a whole human individual and how they interact with their world. Two papers in this session present efforts at phenotype profiling using mobile devices. **Elias Chaibub Neto and colleagues** describe their work profiling patients with Parkinson's disease using smartphone sensor data. **Maulik R. Kamdar and Michelle Wu** present their tool PRISM for monitoring mental wellness using a smart, sensor laden commercial wrist-based wearable. In both cases, new vistas for profiling patients are being opened up by these new data-streams and the informatics techniques to analyze them.

2.5. *Patient stratification and sample subgrouping*

Finally, our session includes five papers on subtyping patients and patient samples. An important element of clinical research has already become differentiating subgroups based on molecular/genomic level features. A recent review identified 684 registered clinical cancer trials that required genetic profiling for enrollment.[12] Disease subtyping may be done through analysis molecular/genomic level features or through a deep analysis of differences in phenotypic presentation, but either are playing an increasingly important role in clinical trials.[13,14] Importantly, for diseases like cancer, the disease may not represent just a single subtype, but may represent a population of different subtypes all coexisting simultaneously in an afflicted patient,[15-17] subtypes which need to be treated in concert, perhaps in different ways. **Vladimir Gligorijevic and colleagues** present an approach using non-negative matrix factorization for tumor stratification. **Sahand Khakabimamaghani and Martin Ester** describe a Bayesian bi-clustering approach for patient stratification using transcriptomic data. **Alex M. Fichtenholtz and colleagues** present an approach for looking at sub-groups of glial tumors to help in analysis of variants of unknown significance in a collection of 800 tumor sequences. **Subhajit Sengupta and colleagues** describe an approach for examining tumor heterogeneity using mutation pairs. Finally, **Artem Sokolov and colleagues** describe a one-class method for identifying specific cell type signatures in mixed samples.

3. Acknowledgments

We would like to thank the PSB 2016 chairs and Tiffany Murray of Stanford University for their efforts in organizing the meeting. We would like to thank UL1TR00042309 for funding (SDM).

References

1. Collins, F. S. & Varmus, H. A New Initiative on Precision Medicine. *N. Engl. J. Med.* **372,** 793–5 (2015).
2. Marx, V. The DNA of a nation. *Nature* **524,** 503–505 (2015).
3. Roberts, J. P. Million veterans sequenced. *Nat. Biotechnol.* **31,** 470–470 (2013).
4. Friend, S. H. App-enabled trial participation: Tectonic shift or tepid rumble? *Sci. Transl. Med.* **7,** 297ed10–297ed10 (2015).
5. Denny, J. C. *et al.* PheWAS: demonstrating the feasibility of a phenome-wide scan to discover gene-disease associations. *Bioinformatics* **26,** 1205–10 (2010).
6. Weinstein, J. N. *et al.* The Cancer Genome Atlas Pan-Cancer analysis project. *Nat. Genet.* **45,** 1113–20 (2013).
7. Comprehensive genomic characterization of squamous cell lung cancers. *Nature* **489,** 519–25 (2012).
8. Friend, S. H. & Norman, T. C. Metcalfe's law and the biology information commons. *Nat. Biotechnol.* **31,** 297–303 (2013).

9. Altshuler, J. S. *et al.* Opening up to precompetitive collaboration. *Sci. Transl. Med.* **2,** 52cm26 (2010).
10. Chen, R. *et al.* Personal Omics Profiling Reveals Dynamic Molecular and Medical Phenotypes. *Cell* **148,** 1293–1307 (2012).
11. Madsen, B. E. & Browning, S. R. A groupwise association test for rare mutations using a weighted sum statistic. *PLoS Genet.* **5,** e1000384 (2009).
12. Roper, N., Stensland, K. D., Hendricks, R. & Galsky, M. D. The landscape of precision cancer medicine clinical trials in the United States. *Cancer Treat. Rev.* **41,** 385–90 (2015).
13. Saria, S. & Goldenberg, A. Subtyping: What It is and Its Role in Precision Medicine. *IEEE Intell. Syst.* **30,** 70–75 (2015).
14. Röcken, C. Quality assurance in clinical trials-the role of pathology. *Virchows Arch.* (2015). doi:10.1007/s00428-015-1857-x
15. Sottoriva, A. *et al.* Intratumor heterogeneity in human glioblastoma reflects cancer evolutionary dynamics. *Proc. Natl. Acad. Sci. U. S. A.* **110,** 4009–14 (2013).
16. Navin, N. *et al.* Inferring tumor progression from genomic heterogeneity. *Genome Res.* **20,** 68–80 (2010).
17. Powell, A. A. *et al.* Single Cell Profiling of Circulating Tumor Cells: Transcriptional Heterogeneity and Diversity from Breast Cancer Cell Lines. *PLoS One* **7,** e33788 (2012).

KNOWLEDGE DRIVEN BINNING AND PHEWAS ANALYSIS IN MARSHFIELD PERSONALIZED MEDICINE RESEARCH PROJECT USING BIOBIN*

ANNA O BASILE[1], JOHN R WALLACE[1], PEGGY PEISSIG[2], CATHERINE A MCCARTY[3], MURRAY BRILLIANT[2], MARYLYN D RITCHIE[1,4]

[1]Department of Biochemistry, Microbiology and Molecular Biology, The Pennsylvania State University University Park, PA,USA; [2]Bioinformatics Research Center, Marshfield Clinic, Marshfield, WI, USA;[3]Essentia Institute of Rural Health; [4]Department of Biomedical and Translational Informatics, Geisinger Health System

Email: marylyn.ritchie@psu.edu

Next-generation sequencing technology has presented an opportunity for rare variant discovery and association of these variants with disease. To address the challenges of rare variant analysis, multiple statistical methods have been developed for combining rare variants to increase statistical power for detecting associations. BioBin is an automated tool that expands on collapsing/binning methods by performing multi-level variant aggregation with a flexible, biologically informed binning strategy using an internal biorepository, the Library of Knowledge (LOKI). The databases within LOKI provide variant details, regional annotations and pathway interactions which can be used to generate bins of biologically-related variants, thereby increasing the power of any subsequent statistical test. In this study, we expand the framework of BioBin to incorporate statistical tests, including a dispersion-based test, SKAT, thereby providing the option of performing a unified collapsing and statistical rare variant analysis in one tool. Extensive simulation studies performed on gene-coding regions showed a Bin-KAT analysis to have greater power than BioBin-regression in all simulated conditions, including variants influencing the phenotype in the same direction, a scenario where burden tests often retain greater power. The use of Madsen-Browning variant weighting increased power in the burden analysis to that equitable with Bin-KAT; but overall Bin-KAT retained equivalent or higher power under all conditions. Bin-KAT was applied to a study of 82 pharmacogenes sequenced in the Marshfield Personalized Medicine Research Project (PMRP). We looked for association of these genes with 9 different phenotypes extracted from the electronic health record. This study demonstrates that Bin-KAT is a powerful tool for the identification of genes harboring low frequency variants for complex phenotypes.

1. Introduction

Examining the genetic influence of low frequency or rare variation to complex disease susceptibility may elucidate additional trait variability and disease risk which has largely remained unexplained by traditional GWAS approaches[29]. In recent years, studies on multifactorial diseases including Alzheimer's disease and prostate cancer have provided compelling evidence that rare variants are associated with complex traits and should be further examined[9, 16]. Advances in sequencing technologies and decreases in sequencing cost have provided an opportunity for rare variant discovery. However, due to the frequency of these variants, there is often low statistical power for detecting association with a phenotype, and therefore, a necessity for prohibitively large sample sizes. Collapsing or binning methods are commonly used to aggregate variants into a single genetic variable for subsequent statistical testing, reducing the

* This work is supported by NIH grant HG006389, and is also partly funded, under a grant with the Pennsylvania Department of Health using Tobacco CURE Funds. The Department specifically disclaims responsibility for any analyses, interpretations or conclusions.

degrees of freedom in the analysis and improving power[23]. BioBin[33, 34] is an automated bioinformatics tool initially developed for the multi-level collapsing of rare variants into user-designated biological features such as genes, pathways, evolutionary conserved regions (ECRs), protein families, and regulatory regions. BioBin follows a binning approach driven by prior biological knowledge by using an internal biorepository, the Library of Knowledge Integration (LOKI)[40]. LOKI combines biological information from over a dozen public databases providing variant details, regional annotations and pathway interactions. The flexible knowledge-driven binning design of BioBin allows the user to test multiple hypotheses within one unified analysis.

Rare variant association analysis of binned variants is often performed using burden or dispersion tests. Burden methods test the cumulative effect of variants within a bin and are easily applied to case-control studies as they assess the frequency of variant counts between these phenotypic groups[24]. Burden tests assume that all variants influence the trait in the same direction and magnitude of effect, and will suffer a loss of power if a mixture of protective and risk variants is present. Standard burden tests include generalized linear model regression analyses and the weighted sum statistic(WSS)[28]. Instead of testing the cumulative effect of variants within a region, dispersion or nonburden methods will test the distribution of these variants in the cases and controls thereby maintaining statistical power in the presence of a mixture of variants. The SKAT[46] package is a dispersion test that has gained widespread use as it allows for easy covariate adjustment, analyzes both dichotomous and quantitative phenotypes, and applies multiple variant weighting options. SKAT is a score-based variance component test that uses a multiple regression kernel-based approach to assess variant distribution and test for association. Both standard burden tests and the SKAT dispersion method have been well assessed in rare variant analysis.

While various tools have been specifically developed to facilitate rare variant association analysis, many methods focus either on the creation of a relevant set of variants or on the statistical analysis of already collapsed variants. This may often lead to file conversion issues for specific tools, as well as more complicated and longer analysis time. Herein we expand the framework of BioBin by integrating select statistical tests, regression and SKAT, as well as capabilities for multiple phenotype analysis (or Phenome-wide Association Studies (PheWAS)), thereby providing a comprehensive, unified bioinformatics tool for the biological binning and association analysis of rare variants. We have evaluated the commonly used regression burden analysis and SKAT in the context of BioBin with data simulations based on individuals of European descent from 1000 Genomes Project Phase I. We have also applied a BioBin-SKAT, or Bin-KAT, test to analyze nine complex human phenotypes from the Marshfield-PMRP project[31], part of the eMERGE network[14]. Our analyses highlight the utility of BioBin as a fast, comprehensive and versatile tool for the biological binning and analysis of low frequency variants in sequence data for multiple complex phenotypes and PheWAS.

2. Methods

2.1. BioBin

2.1.1. Overview of BioBin

BioBin is a unified command line bioinformatics tool written in C++ that utilizes the LOKI database for biologically inspired binning of variants, and also provides a platform for the association analysis of rare variant bins. The framework of a BioBin analysis is to determine biological features upon which data will be binned, such as genes, pathways or intergenic regions, execute bin generation using LOKI, and apply statistical association analysis to each bin. BioBin follows an allele frequency threshold binning approach using the non-major allele frequency (NMAF), defined as 1 minus the frequency of the most common allele. As NMAF and MAF are interchangeable for biallelic markers, MAF will be used in this work. BioBin allows variants below a user-specified MAF in the case or the control group to be binned thereby facilitating the aggregation of both potential risk and protective variants. BioBin was originally developed solely for the biologically informed binning of rare variants in an automated manner. To facilitate more efficient statistical analysis, we have incorporated an extensible testing infrastructure, implementing select burden and dispersion-based tests, namely regression, wilcoxon and SKAT[46] into BioBin. These are commonly used statistical tests in rare variant association analysis, and their direct implementation into BioBin streamlines the analysis, saves time, and also avoids any potential file conversion issues. Also, if an alternate statistical test is desired, BioBin may still be utilized strictly for its biologically inspired variant collapsing function. We have also integrated multiple phenotype capabilities allowing the user to efficiently perform a binned rare variant PheWAS[35, 41, 42]. BioBin analyzes each phenotype separately and uses parallel processing to increase the speed of a PheWAS analysis through a user-specified number of processors. BioBin is open source and the code is freely available at https://ritchielab.psu.edu. It is also available on demand from the authors. All supplemental files for this manuscript are available at https://ritchielab.psu.edu/publications/supplementary-data/psb-2016/biobin-on-multiple-phenotype.

2.1.2. *Library of Knowledge Integration (LOKI)*

BioBin collapses variants into biological features by consulting the Library of Knowledge Integration (LOKI), an internal repository containing diverse knowledge from multiple sources including NCBI dbSNP and gene Entrez[38], Kyoto Encyclopedia of Genes and Genomes (KEGG)[18], Gene Ontology (GO)[11], and Pharmacogenomics Knowledge Base (PharmGKB)[32]. LOKI integrates information from these external databases into a single local repository containing knowledge from the downloaded raw data in each database. The main data types used within LOKI are position, region, group, and source. Position refers to the chromosome and base-pair position of single variants, and region represents biological features containing a start and stop position including genes and copy number variants[33]. Sources are the external databases compiled in LOKI, while groups represent various groupings of biological features such as protein interactions, protein families and pathways. While LOKI is not distributed within the BioBin code due to size constraints, tools are provided within the source distribution allowing a user to compile and perform a local installation of LOKI by downloading data directly from the external sources. The data sources within LOKI can be individually updated as necessary in order to provide the most up-to-date information.

2.2. Simulations

Simulation testing was performed in order to evaluate regression (a standard burden test) and SKAT (a dispersion test) within the framework of a BioBin variant collapsing analysis. All tests were performed using SeqSIMLA2[4] to simulate sequence data as it allowed for the simulation of common burden and dispersion test assumptions. Randomly selected protein-coding variants with a MAF<5% in individuals of European descent from the 1000 Genomes Project Phase I[8] dataset were used as the basis for our simulations. This dataset was used to obtain a distribution of allele frequencies across the whole exome for each non-monomorphic single nucleotide variant site in the represented individuals of European descent (CEU, TSI, FIN, GBR, and IBS). This allele frequency distribution was then used to create the input for SeqSIMLA2. All simulations were performed with 100 variants as we calculated this to be an approximate average number of variants expected in a median sized 24,000bp gene[12]. For this calculation, we used known gene regions in the UCSC Human Genome Browser[19] to define the total gene region length and the 1000 Genomes Project to estimate the number of SNPs identified in these gene regions.

Simulation tests and specific parameters are shown in Table 1. Our simulations focused on two main tests: altering the odds ratio (OR) and altering the proportion of risk variants, with numerous parameters tested in each of these categories. Multiple testing parameters separated by commas in Table 1 correspond to independent simulations. The proportion of causal variants represents the percentage of disease sites of the total 100 variants being simulated. Likewise, the proportion of risk variants provides the number of risk variants of these causal sites. For instance, in our altering OR test category, when simulating 40% causal variants, we had 40 disease sites, and either 40-risk variants (when testing a 100% proportion of risk variants) or 20-risk variants and 20-protective variants (when testing a 50% proportion of risk variants). The specified OR corresponds to that of the individual causal variants. Type I error was estimated with 1,000 simulated null datasets using an OR of 1. Significance was assessed using α=0.05.

Table1. Simulation tests and Parameters

Test Parameter	Altering OR	Altering Proportion of Risk Variants
Number of Simulations	1000	1000
Sample Size	1000 cases and 1000 controls	1000 cases and 1000 controls
Proportion of Causal Variants (n=100)	40%, 10%	40%
Disease Prevalence	5%	5%, 50%
Odds Ratio (OR)	1.5, 2.0, 3.0	3.0
Proportion of Risk Variants	50%, 100%	25%, 40%, 50%, 60%, 75%, 100%
Variant Weighting	No Weighting, Madsen and Browning	No Weighting, Madsen and Browning

2.3. Application of Bin-KAT to natural dataset

A Bin-KAT test was used to analyze type II diabetes (TIID) and eight diagnosis indicators in 740 de-identified European American subjects from the Marshfield Clinic Personalized Medicine

Research Project (PMRP) sequenced in the electronic Medical Records and Genomics (eMERGE) Network[15], as part of the eMERGE-PGX study[43]. Subjects were sequenced using PGRNseq[43], a next-generation sequencing platform designed for the targeted capture of selected pharmacogenes[43]. Case control status for TIID was determined using Mount Sinai's diabetes algorithm[20] from the Diabetes HTN CKD algorithm[37]. The eight diagnosis indicators analyzed are asthma, benign prostatic hyperplasia (BPH), cataracts, diverticulosis, gastroesophageal reflux disease (GERD), hypertension, hypothyroidism, and uterine fibrosis. For each diagnosis indicator, a subject was considered a case if diagnosed with one of the listed ICD-9 codes in Table 2 on two or more dates. Controls were defined as non-cases who did not meet the criteria of ICD-9 diagnosis on two or more dates.

Table 2. Analyzed Phenotypes

Phenotype	Diagnosis	Cases	Controls
TIID	Diabetes HTN CKD algorithm	99	594
Asthma	ICD-9 codes: Between '493.00' and '493.92'	90	650
(BPH)	ICD-9 codes: '600', '600.0', '600.00', '600.01', '600.09', '600.2', '600.20', '600.21', '600.9', '600.90', '600.91'	122	250
Cataracts	ICD-9 codes: '366.10', '366.12', '366.14', '366.15', '366.16', '366.17', '366.9'	202	538
Diverticulosis	ICD-9 codes: '562.00', '562.01', '562.02', '562.03', '562.10', '562.11', '562.12', '562.13'	134	606
GERD	ICD-9 codes: '530.81','530.11'	204	536
Hypertension	ICD-9 codes: Between '401.00' and '401.99'	374	366
Hypothyroidism	ICD-9 codes: '244', '244.8', '244.9', '245', '245.2', '245.8', '245.9'	98	642
Uterine Fibroids	ICD-9 codes: '218.0', '218.1', '218.2', '218.9', '654.10', '654.11', '654.12', '654.13', '654.14'	58	313

To highlight the multiple variant collapsing functions within BioBin, we binned variants having a MAF less than 0.05 by three features: gene, biological pathway and SNPEff[5] functional predictions with a minimum bin size of 5 variants. Gene binning analysis was performed on the 82 targeted pharmacogenes that passed QC. SNPEff functional predictions were used as a secondary collapsing strategy following gene binning. Variants annotated as having intergenic and intragenic effects by SNPEff were excluded from the analysis. Biological pathway variant binning was achieved using all pathway sources currently in the LOKI biorepository[40]. Overall Madsen and Browning[28] weighting was used to weigh binned variants inversely proportional to their MAF. SKAT was used to test for association between binned variants and each phenotype while adjusting for sex, year of birth, and median BMI.

3. Results

3.1. Simulations

We evaluated regression and SKAT within a BioBin coupled collapsing analysis using data simulations of 100 variants based on the allele frequencies of European subjects from the 1000 Genomes Project. All simulated conditions are shown in Table 1 and aim to test the assumptions of burden and dispersion methods. Table 3 displays that Type I error was well controlled in the analyses and was not being sacrificed in the regression or SKAT analysis.

Table 3. Type I Error Results, standard error is in parentheses.

Variant Weighting	SKAT Type I Error Rate	Regression Type I Error Rate
None	0.045 (±0.011)	0.061(±0.011)
Madsen-Browning	0.037(±0.005)	0.039(±0)

A key limitation of burden tests is loss of statistical power in the presence of a mixture of variant effects. We simulated the direction of effect by testing 100% risk variants and 50% risk, 50% protective variants. We evaluated the impact of differing directions of effect on statistical power in a Bin-KAT and BioBin-regression analysis over a varying OR range from 1.5 to 3.0. These results are shown with 10% and 40% causal variants in Figure 1 and 2, respectively. Both figures highlight the influence of variant weighting by displaying results with and without Madsen and Browning weighting.

Figure 1. Power plot of Bin-KAT and BioBin-regression analyses with a causal variant proportion of 10%. SKAT results are represented by a dashed line; regression results have a solid line. Simulations of 100% risk variants are in grey while 50% risk variants are black.

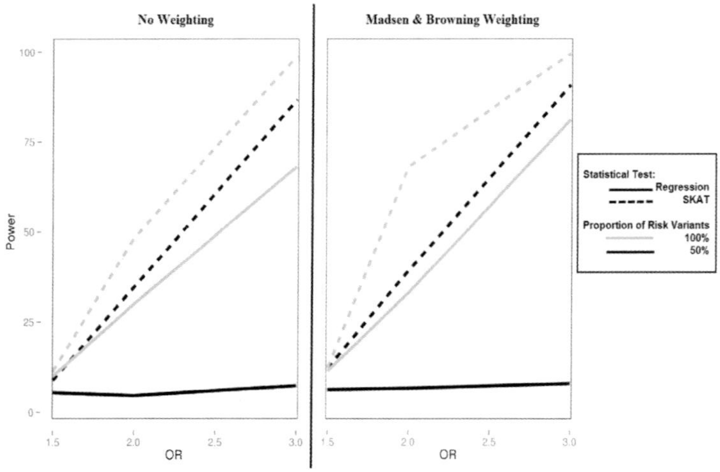

To further explore the impact of a mixture of variant effects on statistical power, we simulated data altering the proportion of risk variants over a wide range, from 25% to 100%, as seen with a disease prevalence of 5% in Figure 3. We increased this disease prevalence to 50% and present these results in Supplementary Figure 1. While a disease prevalence of 50% is high, it allowed us to create a balance in the case to control ratio and thereby symmetry in the results with comparable statistical power between 25%-75%, and 40%-60%, and a significant loss of power at 50%. This is not seen with a lower disease prevalence of 5% (Figure 3) as we are oversampling our population, so that symmetry is likely shifted.

Figure 2. Power plot of Bin-KAT and BioBin-regression analyses with a causal variant proportion of 40%. SKAT results are represented by a dashed line; regression results have a solid line. Simulations of 100% risk variants are in grey while 50% risk variants are black.

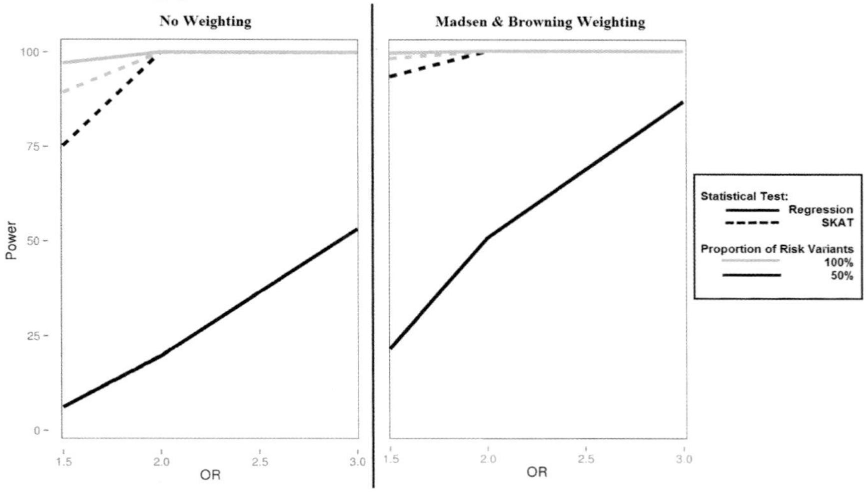

3.2 Application of Bin-KAT to natural dataset

As Bin-KAT consistently maintained greater power than a BioBin-regression, we applied this method coupled with variant weighting to simultaneously analyze 9 phenotypes in subjects of European descent from the Marshfield cohort of eMERGE-PGX project. These subjects were target sequenced for 82 pharmacogenes. We found numerous association results with p-values less than 0.05 in our gene, pathway, and SNPEff functional prediction analysis. Due to the hypothesis generating nature of this method we present all results with a p-value less than 0.05 or 0.01. As sequencing was performed on specific, targeted genes, the statistical tests are highly correlated, and therefore do not meet the independence assumptions of Bonferroni correction, which would prove too stringent in our analysis[7]. In addition, this study is exploratory in nature and all findings should be replicated in independent datasets in the future.

A full list of the results may be found in Supplementary Tables 1 and 2. Table 4 shows the number of results per phenotype and binned biological feature below a p-value cutoff of 0.05 for genes and SNPEff annotations, and an additional 0.01 cutoff for pathway analysis. We found significant associations with binned variants in 59 of the 82 targeted pharmacogenes. Figure 4 shows a Phenogram plot of all significant results collapsed by gene and SNPEff functional prediction displayed by chromosomal location of the gene. Details on the specific annotated SNPEff effect and impact can be found in Supplementary Table 1.

Figure 3. Power plot of a Bin-KAT and BioBin-regression analysis performed when altering the proportion of risk variants between 25% and 100% with a disease prevalence of 5%. SKAT results are represented by a dashed line; regression results have a solid line.

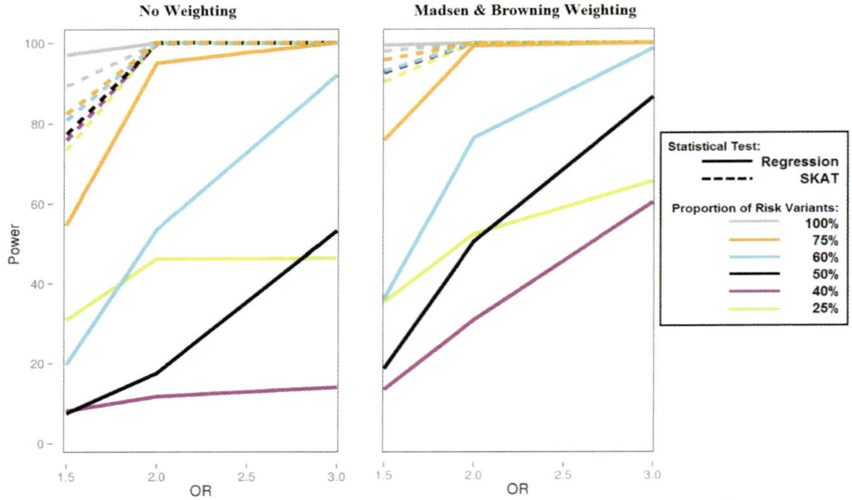

Table 4. Number of association results per phenotype and biological feature at the specified p value cutoff. Total number of bins in each biological feature is noted in parentheses.

Phenotype	Gene (p-value < 0.05)	Pathway (p-value<0.05)	Pathway (p-value<0.01)	SNPEff annotation (p-value <0.05)
Type II Diabetes	4 (82)	233 (8911)	13	17 (458)
Cataracts	5 (82)	777 (8964)	17	8 (458)
Hypothyroidism	6 (82)	324 (8991)	6	19 (458)
Hypertension	2 (82)	234 (8964)	62	1 (458)
Diverticulosis	2 (82)	248 (8964)	148	14 (458)
Asthma	6 (82)	297 (8984)	135	16 (458)
GERD	2 (82)	177 (8964)	19	3 (458)
BPH	2 (82)	102 (8964)	18	4 (458)
Uterine Fibroids	10 (82)	390 (8991)	102	18 (458)

4. Discussion

In this work, we sought to expand the framework of BioBin by integrating statistical tests to provide a tool for the automated, biologically-driven binning and association analysis of rare variants. The choice of binning algorithm is often research specific, and BioBin supports this by providing variant collapsing on multiple biological levels, as well as supporting user-customized analysis. BioBin also includes multiple variant weighting schemes outside of those within a SKAT analysis, including minimum and maximum variant weighting, as well as weighting based on

Figure 4. Phenogram plot of significant association results (p-value<0.05) in a binned gene and SNPEff functional prediction Bin-KAT analysis. The biological features are designated with different shapes, and each phenotype is represented by a different color. The target capture of the PGRNseq platform is shown by blue horizontal bands across the chromosome. The specific SNPEff effect can be found in Supplementary Table 1.

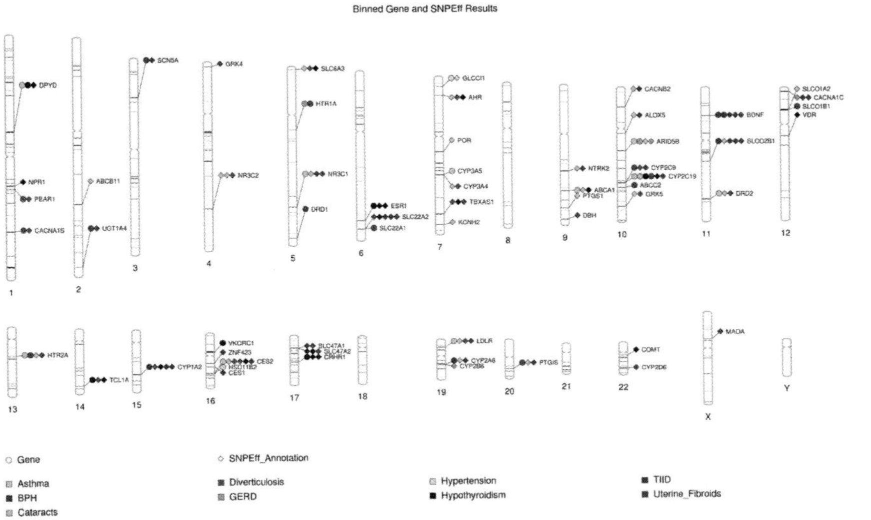

allele frequencies only within our phenotypic controls. Further, BioBin supports polyallelic variant sites and will incorporate all allelic information from these sites, a characteristic that is not supported by all tools. While multiple studies have performed exhaustive comparisons of burden and dispersion methods[2,6,10], we specifically chose to focus on regression and SKAT. Regression is a commonly used burden test, and several popular rare variant methods use a regression framework[1, 26, 27, 36]. SKAT was chosen due to its vast popularity as a dispersion method, its ease of covariate adjustment, and application to binary or quantitative phenotypes. Regression and SKAT have previously been compared in rare variant analysis[2, 10, 22] and here, are evaluated within the context of a biologically inspired binning method.

Simulation testing shows a Bin-KAT analysis maintains greater overall statistical power than BioBin-Regression. We found SKAT to outperform regression even in conditions where a burden analysis is assumed to have greater power than a dispersion test, such as variants influencing the phenotype in the same direction, as is presented in Figure 1 with 10% causal variants. In the 40% causal variant simulations (Figure 2), regression maintains higher power over SKAT in both weighted and unweighted tests. This suggests that the power of regression may be affected by the proportion of causal variants having the same direction of effect. However, when we encounter a mixture of both risk and protective variants, regression suffers a significant loss of power. In fact, SKAT maintains high power regardless of the proportion of risk variants simulated, and is held at 100% from an OR 2.0-3.0 (Figure 3). Our results also highlight that applying Madsen and Browning variant weighting to the binning analysis increases power.

We performed a Bin-KAT test with Madsen and Browning weighting to analyze 9 different phenotypes from Marshfield-PGX subjects who were target sequenced for specific pharmacogenes. We, and others, hypothesize that pharmacogenes related to drug response may also be associated with the diseases for which the drugs are used to treat. Using Bin-KAT, a series of significant associations were found. In the gene-binning analysis, an association between *BDNF* and type II diabetes (p-val 0.000437) was identified. Literature indicates that low levels of *BDNF* may be involved in type II diabetes pathogenesis, providing a potential explanation for the clustering of dementia, depression and type II diabetes[13, 21]. *BDNF* may also play a role in blood glucose metabolism and insulin resistance, a characteristic of type II diabetes[21, 30]. A number of significant results in the pathway-binning analysis performed using asthma patients included leukotriene pathways. Leukotrienes are inflammatory chemicals that can act as lipid mediators and have been well established in the pathobiology of asthma[3, 17, 44]. Leukotriene-B4 is being further investigated for its regulatory role in the development of asthma [17].

The results of this study show indications of potential pleiotropy where gene-binned variants are associated with more than one phenotype. We see this with *CYP2C19*, which is significantly associated with asthma, cataracts, hypothyroidism, and uterine fibroids. *CYP2C19* has a highly polymorphic sequence, accounting for its variability in drug metabolism as it acts on up to 10% of clinical drugs[25]. In lung tissue, cytochrome P450 enzymes may be affected by air pollutants, and the CYP2C19*2 genotype has been implicated as a risk factor for asthma[47]. Also, linkage analysis on families with endometriosis, a disorder that may be correlated with uterine fibroids[45], indicates a potential role of *CYP2C19* in endometriosis risk[39]. Association results with *CYP2C19* present exciting connections that warrant further exploration. We have looked at the co-occurrence of these four phenotypes and the correlation is fairly low. Future work will aim to evaluate *CYP2C19* and medication usage.

Bin-KAT serves as a powerful and versatile method for the biological binning and analysis of rare variants in sequence data. This approach was successful in the identifying novel and well-studied genes and pathways harboring low frequency variants in a multiple complex phenotype analysis. Studying the influence of low frequency variants has the potential to identify underlying risk factors, and uncover complex genotype-phenotype associations in multifactorial diseases.

5. Acknowledgments

We would like to thank Steven J. Schrodi, PhD, of the Bioinformatics Research Center, Marshfield Clinic in Marshfield, WI for his contributions.

6. References

[1] Asimit, J.L. et al. 2012. ARIEL and AMELIA: Testing for an Accumulation of Rare Variants Using Next-Generation Sequencing Data. *Human heredity*. 73, 2 (2012), 84–94.

[2] Bacanu, S.-A. et al. 2012. Comparison of Statistical Tests for Association between Rare Variants and Binary Traits. *PLoS ONE*. 7, 8 (Aug. 2012), e42530.

[3] Busse, W.W. et al. 1999. Leukotriene pathway inhibitors in asthma and chronic obstructive pulmonary disease. *Clinical and Experimental Allergy: Journal of the British Society for Allergy and Clinical Immunology*. 29 Suppl 2, (Jun. 1999), 110–115.

[4] Chung, R.-H. et al. 2015. SeqSIMLA2: simulating correlated quantitative traits accounting for shared environmental effects in user-specified pedigree structure. *Genetic Epidemiology*. 39, 1 (Jan. 2015), 20–24.

[5] Cingolani, P. et al. 2012. A program for annotating and predicting the effects of single nucleotide polymorphisms, SnpEff: SNPs in the genome of Drosophila melanogaster strain w1118; iso-2; iso-3. *Fly*. 6, 2 (Jun. 2012), 80–92.

[6] Clarke, G.M. et al. 2013. A Flexible Approach for the Analysis of Rare Variants Allowing for a Mixture of Effects on Binary or Quantitative Traits. *PLoS Genet*. 9, 8 (Aug. 2013), e1003694.

[7] Conneely, K.N. and Boehnke, M. 2007. So Many Correlated Tests, So Little Time! Rapid Adjustment of P Values for Multiple Correlated Tests. *American Journal of Human Genetics*. 81, 6 (Dec. 2007), 1158–1168.

[8] Consortium, T. 1000 G.P. 2012. An integrated map of genetic variation from 1,092 human genomes. *Nature*. 491, 7422 (Nov. 2012), 56–65.

[9] Cruchaga, C. et al. 2014. Rare coding variants in the phospholipase D3 gene confer risk for Alzheimer/'s disease. *Nature*. 505, 7484 (Jan. 2014), 550–554.

[10] Dering, C. et al. 2014. A comprehensive evaluation of collapsing methods using simulated and real data: excellent annotation of functionality and large sample sizes required. *Frontiers in Genetics*. 5, (Sep. 2014).

[11] Dimmer, E.C. et al. 2012. The UniProt-GO Annotation database in 2011. *Nucleic Acids Research*. 40, D1 (Jan. 2012), D565–D570.

[12] Fuchs, G. et al. 2014. 4sUDRB-seq: measuring genomewide transcriptional elongation rates and initiation frequencies within cells. *Genome Biology*. 15, 5 (2014), R69.

[13] Fujinami, A. et al. 2008. Serum brain-derived neurotrophic factor in patients with type 2 diabetes mellitus: Relationship to glucose metabolism and biomarkers of insulin resistance. *Clinical Biochemistry*. 41, 10–11 (Jul 2008), 812–817.

[14] Gottesman, O. et al. 2013. The Electronic Medical Records and Genomics (eMERGE) Network: Past, Present and Future. *Genetics in medicine : official journal of the American College of Medical Genetics*. 15, 10 (Oct. 2013), 761–771.

[15] Gottesman, O. et al. 2013. The Electronic Medical Records and Genomics (eMERGE) Network: past, present, and future. *Genetics in Medicine*. 15, 10 (Oct. 2013), 761–771.

[16] Gudmundsson, J. et al. 2012. A study based on whole-genome sequencing yields a rare variant at 8q24 associated with prostate cancer. *Nature Genetics*. 44, 12 (Dec. 2012), 1326–1329.

[17] Hallstrand, T.S. and Henderson, W.R. 2010. An update on the role of leukotrienes in asthma. *Current opinion in allergy and clinical immunology*. 10, 1 (Feb. 2010), 60–66.

[18] Kanehisa, M. et al. 2012. KEGG for integration and interpretation of large-scale molecular data sets. *Nucleic Acids Research*. 40, Database issue (Jan. 2012), D109–114.

[19] Kent, W.J. et al. 2002. The Human Genome Browser at UCSC. *Genome Research*. 12, 6 (Jun. 2002), 996–1006.

[20] Kho, A.N. et al. 2012. Use of diverse electronic medical record systems to identify genetic risk for type 2 diabetes within a genome-wide association study. *Journal of the American Medical Informatics Association : JAMIA*. 19, 2 (2012), 212–218.

[21] Krabbe, K.S. et al. 2007. Brain-derived neurotrophic factor (BDNF) and type 2 diabetes. *Diabetologia*. 50, 2 (Feb. 2007), 431–438.

[22] Ladouceur, M. et al. 2012. The Empirical Power of Rare Variant Association Methods: Results from Sanger Sequencing in 1,998 Individuals. *PLoS Genet*. 8, 2 (Feb. 2012), e1002496.

[23] Lee, S. et al. 2012. Optimal tests for rare variant effects in sequencing association studies. *Biostatistics (Oxford, England)*. 13, 4 (Sep. 2012), 762–775.

[24] Lee, S. et al. 2014. Rare-Variant Association Analysis: Study Designs and Statistical Tests. *American Journal of Human Genetics*. 95, 1 (Jul. 2014), 5–23.

[25] Lee, S.-J. 2013. Clinical Application of CYP2C19 Pharmacogenetics Toward More Personalized Medicine. *Frontiers in Genetics*. 3, (Feb. 2013).

[26] Li, B. and Leal, S.M. 2008. Methods for detecting associations with rare variants for common diseases: application to analysis of sequence data. *American Journal of Human Genetics*. 83, 3 (Sep. 2008), 311–321.

[27] Lin, D.-Y. and Tang, Z.-Z. 2011. A General Framework for Detecting Disease Associations with Rare Variants in Sequencing Studies. *American Journal of Human Genetics*. 89, 3 (Sep. 2011), 354–367.

[28] Madsen, B.E. and Browning, S.R. 2009. A Groupwise Association Test for Rare Mutations Using a Weighted Sum Statistic. *PLoS Genet*. 5, 2 (Feb. 2009), e1000384.

[29] Maher, B. 2008. Personal genomes: The case of the missing heritability. *Nature News*. 456, 7218 (Nov. 2008), 18–21.

[30] Marchelek-Myśliwiec, M. et al. 2015. Insulin resistance and brain-derived neurotrophic factor levels in chronic kidney disease. *Annals of Clinical Biochemistry*. 52, Pt 2 (Mar. 2015), 213–219.

[31] McCarty, C.A. et al. 2007. Informed consent and subject motivation to participate in a large, population-based genomics study: the Marshfield Clinic Personalized Medicine Research Project. *Community Genetics*. 10, 1 (2007), 2–9.

[32] McDonagh, E.M. et al. 2011. From pharmacogenomic knowledge acquisition to clinical applications: the PharmGKB as a clinical pharmacogenomic biomarker resource. *Biomarkers in Medicine*. 5, 6 (Dec. 2011), 795–806.

[33] Moore, C.B. et al. 2013. BioBin: a bioinformatics tool for automating the binning of rare variants using publicly available biological knowledge. *BMC Medical Genomics*. 6, Suppl 2 (May 2013), S6.

[34] Moore, C.B. et al. 2013. Low Frequency Variants, Collapsed Based on Biological Knowledge, Uncover Complexity of Population Stratification in 1000 Genomes Project Data. *PLoS Genet*. 9, 12 (Dec. 2013), e1003959.

[35] Moore, C.B. et al. 2014. Phenome-wide Association Study Relating Pretreatment Laboratory Parameters With Human Genetic Variants in AIDS Clinical Trials Group Protocols. *Open Forum Infectious Diseases*. 2, 1 (Dec. 2014).

[36] Morgenthaler, S. and Thilly, W.G. 2007. A strategy to discover genes that carry multi-allelic or mono-allelic risk for common diseases: A cohort allelic sums test (CAST). *Mutation Research/Fundamental and Molecular Mechanisms of Mutagenesis*. 615, 1–2 (Feb. 2007), 28–56.

[37] Nadkarni, G.N. et al. 2014. Development and validation of an electronic phenotyping algorithm for chronic kidney disease. *AMIA Annual Symposium Proceedings*. 2014, (Nov. 2014), 907–916.

[38] NCBI Resource Coordinators 2013. Database resources of the National Center for Biotechnology Information. *Nucleic Acids Research*. 41, Database issue (Jan. 2013), D8–D20.

[39] Painter, J.N. et al. 2011. High-density fine-mapping of a chromosome 10q26 linkage peak suggests association between endometriosis and variants close to CYP2C19. *Fertility and Sterility*. 95, 7 (Jun. 2011), 2236–2240.

[40] Pendergrass, S.A. et al. 2013. Genomic analyses with biofilter 2.0: knowledge driven filtering, annotation, and model development. *BioData Mining*. 6, 1 (Dec. 2013), 25.

[41] Pendergrass, S.A. et al. 2013. Phenome-Wide Association Study (PheWAS) for Detection of Pleiotropy within the Population Architecture using Genomics and Epidemiology (PAGE) Network. *PLoS Genetics*. 9, 1 (Jan. 2013).

[42] Pendergrass, S.A. et al. 2011. The Use of Phenome-Wide Association Studies (PheWAS) for Exploration of Novel Genotype-Phenotype Relationships and Pleiotropy Discovery. *Genetic epidemiology*. 35, 5 (Jul. 2011), 410–422.

[43] Rasmussen-Torvik, L.J. et al. 2014. Design and anticipated outcomes of the eMERGE-PGx project: a multicenter pilot for preemptive pharmacogenomics in electronic health record systems. *Clinical Pharmacology and Therapeutics*. 96, 4 (Oct. 2014), 482–489.

[44] Sampson, A. and Holgate, S. 1998. Leukotriene modifiers in the treatment of asthma. *BMJ : British Medical Journal*. 316, 7140 (Apr. 1998), 1257–1258.

[45] Uimari, O. et al. 2011. Do symptomatic endometriosis and uterine fibroids appear together? *Journal of Human Reproductive Sciences*. 4, 1 (2011), 34–38.

[46] Wu, M.C. et al. 2011. Rare-Variant Association Testing for Sequencing Data with the Sequence Kernel Association Test. *The American Journal of Human Genetics*. 89, 1 (Jul. 2011), 82–93.

[47] Yildirim Yaroğlu, H. et al. 2011. CYP2C19 gene polymorphism may be a risk factor for bronchial asthma. *Medical Principles and Practice: International Journal of the Kuwait University, Health Science Centre*. 20, 1 (2011), 39–42.

MULTITASK FEATURE SELECTION WITH TASK DESCRIPTORS

VÍCTOR BELLÓN*, VÉRONIQUE STOVEN AND CHLOÉ-AGATHE AZENCOTT

MINES ParisTech, PSL-Research University, CBIO-Centre for Computational Biology,
35 rue St Honoré 77300 Fontainebleau, France
Institut Curie, 75248 Paris Cedex 05,France
INSERM U900, 75248 Paris Cedex 05, France
*victor.bellon@mines-paristech.fr**

Machine learning applications in precision medicine are severly limited by the scarcity of data to learn from. Indeed, training data often contains many more features than samples. To alleviate the resulting statistical issues, the multitask learning framework proposes to learn different but related tasks joinlty, rather than independently, by sharing information between these tasks. Within this framework, the joint regularization of model parameters results in models with few non-zero coefficients and that share similar sparsity patterns. We propose a new regularized multitask approach that incorporates task descriptors, hence modulating the amount of information shared between tasks according to their similarity. We show on simulated data that this method outperforms other multitask feature selection approaches, particularly in the case of scarce data. In addition, we demonstrate on peptide MHC-I binding data the ability of the proposed approach to make predictions for new tasks for which no training data is available.

1. Introduction

A substantial limiting factor for many machine learning applications in bioinformatics is the scarcity of training data. This issue is particularly critical in precision medicine applications, which revolve around the analysis of considerable amounts of high-throughput data, aiming at identifying the similarities between the genomes of patients who exhibit similar disease susceptibilities, prognoses, responses to treatment, or immune responses to vaccines. In such applications, collecting large numbers of samples is often costly. It is therefore frequent for the number of samples (n) to be orders of magnitudes smaller than the number of features (p) describing the data. Model estimation in such $n \ll p$ settings is a major challenge of modern statistics, and the risk of overfitting the training data is high.

Fortunately, it is often the case that data is available for several related but different problems (or tasks). While such data cannot be pooled together to form a single, large data set, the *multitask* framework makes it possible to leverage all the available information to learn related but separate models for each of these problems. For example, genetic data may be available for patients who were included and followed under different but related conditions. If each condition is considered separately, we may not have enough data to detect the relevant genetic variations associated to the trait under study. Multitask learning approaches where each condition corresponds to a task can be used to circumvent this issue by increasing the number of learning examples while keeping the specificity of each dataset[1,2]. Another prevalent strategy to avoid overfitting the training data is to apply *regularization*, that is to say, to impose a penalization on the complexity of the model. One of the most common penalizations takes the form of an l_1-norm over the weights assigned to the features. In the context of least-squares regression, this is known as the Lasso[3]. This approach drives many of the regression weights to

0, resulting in sparse models, that is to say, models that involve a small number of predictors. This makes them particularly suitable for biological applications, where it is often desirable for models to not only exhibit good predictive abilities, but also to be interpretable. For example, if samples are patients encoded by genetic features, if only a small number of features are selected by the model (i.e. are assigned non-zero weights), it may be possible to relate these features to the biological pathways involved in the predicted trait. Further down the line, these features can be used to aid diagnosis or design companion tests. However, l_1-regularized methods are sensitive to small perturbations of the data, and it is therefore necessary to pay attention to their stability.

The Multitask Lasso[4] was the first approach to apply regularization in the multitask setting. It employs a block regularization over the weights that imposes to select the same features for all tasks. The coefficients assigned to these features are allowed to vary smoothly, resulting in separate models for the separate tasks. However, many problems require more flexibility. Indeed, since the tasks considered in multitasks approaches are related, but not identical, we can expect some sharper variation in the degree to which the selected features are relevant for the different tasks.

In line with this idea, the Multi-level Multitask Lasso[5] expresses each regression coefficient as the product of two factors. One factor controls the overall sparsity and captures features common to all tasks; the second factor modulates the weights of the selected features, reflecting the task specificities.

These approaches have two limitations. First, they cannot be directly applied to make predictions for new tasks for which no training data is available. This could be relevant to predict the cytotoxicity of a new drug on cells or patients, or to evaluate the prognosis of a previously unseen cancer subtype. Second, the degree of similarity between tasks is not explicitly taken into account. However, intuitively, we would like to explicitly enforce that more information should be shared between more similar tasks.

These two limitations can be addressed by defining an explicit representation of the tasks. This provides a convenient way to relate tasks and to share information between them, as is done in kernel methods[6,7]. Based on the intuition that the second factor of the MML[5] should be similar for similar tasks, we propose to characterize each task by a set of descriptor variables and re-write this factor as a linear combination of these descriptor variables.

In this paper, we start by formulating the multitask least-squares regression problem and by presenting the state of the art. We then introduce our model, give a result on the asymptotic convergence of the estimator, and present an algorithm for solving the optimization problem. Experimental results on simulated data show our approach to be competitive both in terms of prediction error and in terms of the quality of the selected features. Finally, we illustrate the validity of the proposed method for the prediction of new tasks by applying it to MHC-I binding prediction, a problem relevant to the design of personalized vaccines.

2. State of the art

In this section we present existing approaches to the problem of multitask feature selection.

2.1. *Problem formulation*

Let us assume that we want to learn K different tasks, corresponding to K datasets $(X^k, Y^k)_{k=1,\dots,K}$. Let $X^k \in \mathbb{R}^{n_k \times p}$ be the data matrix containing n_k instances of dimension p, and $Y^k \in \mathbb{R}^{n_k}$ the corresponding real-valued output data. Our objective is to find, for every $k = 1, \dots, K$ and for every $i = 1, \dots, n_k$, $\beta \in \mathbb{R}^{K \times p}$ such that

$$y_i^k = f\left(x_i^k\right) + \epsilon_i^k = \sum_{j=1}^{p} \beta_j^k x_{ij}^k + \epsilon_i^k,$$

where ϵ_i^k is the noise for the i-th instance of task k. For each feature j, β_j is a K-dimensional vector of weights assigned to this feature for each task. Direct minimization of the loss between Y and f would be equivalent to fitting K different linear regressions in a single step. Therefore, this formulation does not allow to share information across tasks.

2.2. *Multitask Lasso and Sparse Multitask Lasso*

One of the first formulations for the joint selection of features across related tasks, commonly referred to as Multitask Lasso[4] (ML), uses a method related to the Group Lasso[8]. Information is shared between tasks through a regularization term: An l_2-norm forces the weights β_j of each feature to shrink across tasks, and an l_1-norm over these l_2-norms produces a sparsity pattern common to all tasks. These penalties produce patterns where every task is explained by the same features. This results in the following optimization problem:

$$\min_{\beta \in \mathbb{R}^{K \times p}} \frac{1}{2} \sum_{k=1}^{K} \frac{1}{n_k} \sum_{i=1}^{n_k} \left(y_i^k - \sum_{j=1}^{p} \beta_j^k x_{ij}^k \right)^2 + \lambda \sum_{j=1}^{p} \|\beta_j\|_2, \tag{1}$$

A common extension of this problem is the Sparse Multitask Lasso (SL), based on the Sparse Group Lasso[9]. It consists in adding the regularization term $\lambda_s \|\beta\|_1$ to Equation 1, which generates a sparse structure both on the features as well as between tasks. These sparse optimization problems have been well studied and can be solved using proximal optimization[10].

2.3. *Multi-level Multitask Lasso*

To allow for more flexibility in the sparsity patterns of the different tasks, the authors of the Multi-level Lasso[5] (MML) propose to decompose the regression parameter β into a product of two components $\theta \in \mathbb{R}^p$ and $\gamma \in \mathbb{R}^{K \times p}$. The intuition here is to capture the global effect of the features across all the tasks with θ, while γ provides some modulation according to the specific sensitivity of each task to each feature. This results in the following optimization problem:

$$\min_{\theta \in \mathbb{R}^p, \gamma \in \mathbb{R}^{K \times p}} \frac{1}{2} \sum_{k=1}^{K} \frac{1}{n_k} \sum_{i=1}^{n_k} \left(y_i^k - \sum_{j=1}^{p} \theta_j \gamma_j^k x_{ij}^k \right)^2 + \lambda_1 \sum_{j=1}^{p} |\theta_j| + \lambda_2 \sum_{k=1}^{K} \sum_{j=1}^{p} |\gamma_j^k| \tag{2}$$

with the constraint that $\theta > 0$.

The authors prove that this approach generates sparser patterns than the so-called Dirty model[11], where the β parameter is decomposed into the sum (rather than product) of two parameters. In practice, this model also gives sparser representations than the ML, and has the advantage not to impose to select the exact same features across all tasks.

The optimization of the parameters is a non-convex problem that can be decomposed in two alternate convex optimizations. Furthermore, the optimal θ can be calculated exactly given γ.[12] This optimization, however, is much slower than that of the ML. Finally, note that in this approach, the multitask character is explicitly provided by the parameter θ, which is shared across all tasks, rather than implicitly enforced by a penalization term.

3. Multiplicative Multitask Lasso with Task Descriptors

The approaches presented above do not explicitly model relations between tasks. However, an explicit representation of the task space might be available. Inspired by kernel approaches, where task similarities are encoded in the model[6,7], we introduce a new model called Multiplicative Multitask Lasso with Task Descriptors (MMLD), where we use a vector of task descriptor variables to encode each task, and to explain the specific effect modulating each feature for each task.

Following the MML formulation[5], we decompose the parameter β into a product of two components. We keep the notation θ for the first component, which corresponds to the global feature importance common to all tasks. The second component is now a linear combination of the L-dimensional task descriptors $D \in \mathbb{R}^{L\times K}$. The L task descriptors have to be defined beforehand and depend on the application. For example, if the different tasks are sensitivity to different drugs to which cell lines are exposed, one could use molecular fingerprints[13] to describe the drugs, i.e. the tasks. The regression parameter $\alpha \in \mathbb{R}^{p\times L}$ indicates the importance of each descriptor for each feature, and controls the specificity of each task. Hence we formulate the following optimization problem:

$$\min_{\theta\geq 0,\alpha\in\mathbb{R}^{p\times L}} \frac{1}{2}\sum_{k=1}^{K}\frac{1}{n_k}\sum_{i=1}^{n_k}\left(y_i^k - \sum_{j=1}^{p}\theta_j\left(\sum_{l=1}^{L}\alpha_{jl}d_l^k\right)x_{ij}^k\right)^2 + \lambda_1\sum_{j=1}^{p}|\theta_j| + \lambda_2\sum_{j=1}^{p}\sum_{l=1}^{L}|\alpha_{jl}|, \tag{3}$$

where λ_1 and λ_2 are the regularization parameters for each component of β.

Importantly, because predictions for a new data point x are made as $\sum_{j=1}^{p}\theta_j(\sum_{l=1}^{L}\alpha_{jl}d_l^k)x_{ij}$, this formulation allows to make predictions for tasks for which no training data is available: the only task-dependent parameters are the descriptors d_l^k. This ability to extrapolate to new tasks is not shared by the existing multitask Lasso methods.

3.1. *Theoretical guaranties*

Let us define, for all $k = 1,\dots,K$, $i = 1,\dots,n_k$, $j = 1,\dots,p$, $l = 1,\dots,L$, $\xi_{ijl}^k = d_l^k x_{ij}^k$ and $\mu_{jl} = \theta_j\alpha_{jl}$. Problem 3 can be reformulated as

$$\min_{\theta\geq 0,\mu\in\mathbb{R}^{p\times L}} \frac{1}{2}\sum_{k=1}^{K}\frac{1}{n_k}\sum_{i=1}^{n_k}\left(y_i^k - \sum_{j=1}^{p}\sum_{l=1}^{L}\mu_{jl}\xi_{ijl}^k\right)^2 + \lambda_1\|\theta\|_1 + \lambda_2\sum_{j=1}^{p}\theta_j^{-1}\|\mu_j\|_1 \tag{4}$$

Following Lemma 1 in Ref. 12, it is immediate to prove that, when $\omega = 2\sqrt{\lambda_1\lambda_2}$, Problem 4 is equivalent to

$$\min_{\mu \in \mathbb{R}^{p\times L}} \frac{1}{2}\sum_{k=1}^{K}\frac{1}{n_k}\sum_{i=1}^{n_k}\left(y_i^k - \sum_{j=1}^{p}\sum_{l=1}^{L}\mu_{jl}\xi_{ijl}^k\right)^2 + \omega\sum_{j=1}^{p}\sqrt{\|\mu_j\|_1}, \tag{5}$$

with $\hat{\theta}_j = \sqrt{\frac{\lambda_1}{\lambda_2}\|\mu_j\|_1}$. Problem 5 has a convex loss function and a non-convex regularization term. The characterization of the asymptotic distribution of the estimator for this problem, as well as its $\sqrt{n}$-consistency, have been previously given by Lozano and Swirszcz[5], based on a more general result[14].

3.2. *Algorithm*

Problem 3 is non-convex. We therefore propose to adapt the algorithm of Ref. 5 and separate it in alternate convex optimization steps: the optimization of θ for a fixed α, corresponding to a nonnegative Garrote problem[15], and the optimization of α for a gixed θ, corresponding to a Lasso optimization[3]. Details can be found in Algorithm 3.1. Python code is available at: `https://github.com/vmolina/MultitaskDescriptor`

Algorithm 3.1.

Input $\{X^k, Y^k, D^k\}_{k=1,\dots,K}$, $\lambda_1, \lambda_2, \epsilon, m_{max}$.
Define $n = \sum_{k=1}^{K} n^k$, $\tilde{X} = \{x_1^1, \dots, x_{n^1}^1, x_1^2, \dots, x_{n^k}^k\}$ *and* $\tilde{Y} = \{y_1^1, \dots, y_{n^1}^1, y_1^2, \dots, y_{n^k}^k\}$
Initialize $\theta_j(0) = 1$ *and* $\alpha_j(0)$ *according to an initial estimate, for* $j = 1,\dots,p$.
For $m = 1, \dots m_{max}$:
  Solve *for* α:
    $w_{ijl}(m) = \theta_j(m-1)d_{il}\tilde{x}_{ij}$.
    $\alpha(m) = \arg\min_\alpha \frac{1}{2}\sum_{i=1}^{n}\left(\tilde{y}_i - \sum_{j=1}^{p}\sum_{l=1}^{L}\alpha_{jl}w_{ijl}(m)\right)^2 + \lambda_2\sum_{j=1}^{p}\sum_{l=1}^{L}|\alpha_{jl}|$
  Solve *for* θ:
    $z_{*j}(m) = \left[\sum_{l=1}^{L}\alpha_{jl}(m)d_l^1 x_{1j}^1, \dots, \sum_{l=1}^{L}\alpha_{jl}(m)d_l^k x_{n^k j}^k\right]$, *for* $j = 1, \dots p$
    $\theta(m) = \arg\min_{\theta\geq 0}\frac{1}{2}\sum_{i=1}^{n}\left(\tilde{y}_i - \sum_{j=1}^{p}\theta_j(m-1)z_{ij}(m)\right)^2 + \lambda_1\sum_{j=1}^{p}|\theta_j(m-1)|$
  $\beta_j^k(m) = \theta_j(m)\sum_{l=1}^{L}\alpha_{jl}(m)d_l^k$
  If $R(\beta(m-1)) - R(\beta(m)) \leq \epsilon$ *(where* $R(\beta)$ *denote the squared loss over all tasks)*
    Break
Return $\beta(m)$

4. Experiments on simulated data

In this section, we compare our method to the ML, the SL and the MML based on two different criteria. First, we compare them in terms of the *quality* of the selected features. By quality, we mean the ability to recover the true support of β (that is to say, its non-zero entries), as well as the stability of the selection upon data perturbation. Second, we evaluate the methods in terms of prediction performance.

4.1. *Simulated data*

We simulate K design matrices $X_k \in \mathbb{R}^{n_k \times p}$ according to a Gaussian distribution with mean 0 and a precision matrix $\Sigma \sim Wishart(p+20, I_p)$, where I_p is the identity matrix of dimension p. In our simulations $n_1 = n_2 = \ldots = n_K$. For each task k, we sample L descriptors d_l^k from a normal distribution with mean $\mu_{d_l} \sim \mathcal{N}(0,5)$ and variance $\sigma^2_{d_l} \sim \mathcal{G}amma(0.2,1)$. We build θ by randomly selecting $p_s < p$ indices for non-zero coefficients, which we sample from a Gamma distribution $\mathcal{G}amma(1,2)$. All other entries of θ are set to 0. We build α in the following manner: For each of the non-zero θ_j, we randomly select $L_s < L$ entries of α_j to be non-zero, and sample them from a Gaussian distribution $\mathcal{N}(0,2)$. All other α_{jl} are set to 0.

We then compute $\beta_j^k = \theta_j \left(\sum_{l=1}^{L} \alpha_{jl} D_l^k\right)$ and normalize it by dividing by $\beta^* = \max_{j,k}|\beta_j^k|$. Finally, we randomly chose with replacement S_s entries of β_k. If the chosen entry is different from 0, we set it to 0; conversely, if it was equal to 0, we set it to a new value sampled from a Gaussian distribution $\mathcal{N}(0,0.5)$. This last randomization step is performed to relax the structure of β. Finally, we simulate $Y = \beta X + \epsilon$ where ϵ is Gaussian noise with $\sigma^2 = 0.1$.

Each of our experiments consists in evaluating the different models in a 10-fold cross-validation. We create a first set of experiments containing 5 datasets generated with the parameters $K = 4$, $n_k = 100$, $p = 100$, $L = 10$, $p_s = 20$, $L_s = 4$, and $S_s = 100$. We generate a second set of experiments using $n_k = 20$ to simulate a scarce setting. We report the results of additional experiments in a scarcer setting ($p = 8000$, $n_k = 20$) in the Supplementary Materials.

In each experiment we train 4 different models: the ML[4], the SL[9], the MML[5], and the MMLD we propose here. In order to better understand the role of the task descriptor space, we use 3 variants of the MMLD: one that uses the same task descriptors as those from which the data was generated; one that uses these descriptors, perturbed with Gaussian noise ($\sigma = 0.1$); and one with a random set of task descriptors, sampled from a uniform distribution over $[0,1]$. Perturbing the task descriptors with more noise should give results in between those obtained in those last two scenarios.

Each of these 6 methods estimates a real-valued matrix $\hat{\beta} \in \mathbb{R}^{K \times p}$. We then consider as selected, for a given task k, the features j for which $\hat{\beta}_j^k$ is different from 0. For all methods, λ is set by cross-validation: Let $\lambda_{\min}$ be the value of λ that yields the lowest cross-validated RMSE $E_{\min}$. Then, we pick, amond all $\lambda > \lambda_{\min}$ resulting in a cross-validated RMSE less than one standard deviation away from E_min, the λ that yields the median cross-validated RMSE. This heuristic compromises between optimizing for RMSE and imposing more regularization.

4.2. *Feature selection and stability*

In this section, we evaluate the ability of the feature selection procedure to select the correct features, as well as the stability of the procedure, on two sets of experiments.

Stability of the feature selection In precision medicine applications, it is often critical that feature selection methods be stable: If a method selects different features under small perturbations, we cannot rely on it to identify biologically meaningful features. To evaluate the stability of the feature selection procedures, we calculate the consistency index[16] between the sets of features selected over each fold.

Figure 1(a) shows the consistencies of the different methods for the first set of experiments. We observe that the consistency of the feature selection for the proposed method is much higher than the consistency of SL and MML. By contrast, ML presents a very high consistency index, that decays when the data is scarcer. (Fig. 1(b)). The addition of small noise to the task descriptors does not have a strong effect on the stability of the selection, using random task descriptors negatively affects it, especially when data is scarce. In an even scarcer scenario the consistency presents high variation for all methods (Supp. Mat.).

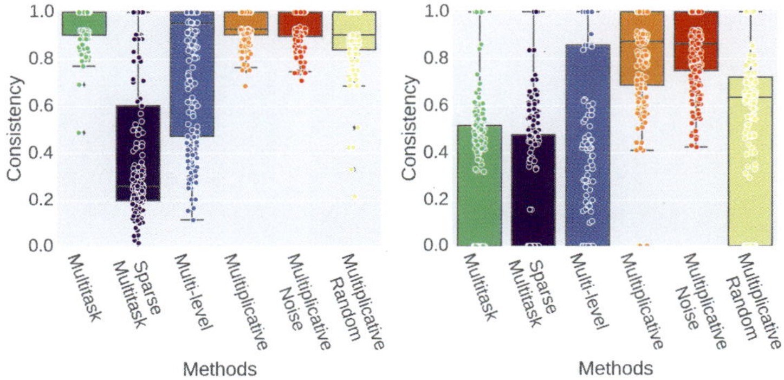

(a) $n_k = 100$ instances per task (b) $n_k = 20$ instances per task

Fig. 1. Boxplot depiction of the consistency index of the different methods for simulated data.

Number of selected features We report in Table 1 the mean number of non-zero coefficients assigned by each method in each scenario. We evaluate sparsity at the level of the β coefficients, hence the total number of coefficients is $n_k \times K$. The ML and the SL both recover more features than all other methods. The MML chooses more features than the MMLD when $n_k = 100$, but selects fewer parameters when the number of instances is reduced. Finally, the MMLD presents a much lower variation in the number of selected features than all other methods.

Table 1. Mean number of non-zero coefficients assigned by each method.

	True	ML	SL	MML	MMLD	Noisy MMLD	Random MMLD
$n_k = 100$	126.8 ± 6.8	169.28 ± 163.4	231.62 ± 121.3	83.9 ± 80.7	54.88 ± 9.8	56.88 ± 11.2	49.12 ± 56.5
$n_k = 20$	126.8 ± 3.2	80.88 ± 79.8	43.96 ± 48.8	17.82 ± 21.7	46.24 ± 15.9	48.56 ± 18	46.72 ± 34.6

Ability to select the correct features We report the Positive Predictive Value (PPV, Fig. 2) and the sensitivity (Fig. 3) of the feature selection for the different methods. The PPV is the proportion of selected features that are correct. The sensitivity is the proportion of

correct

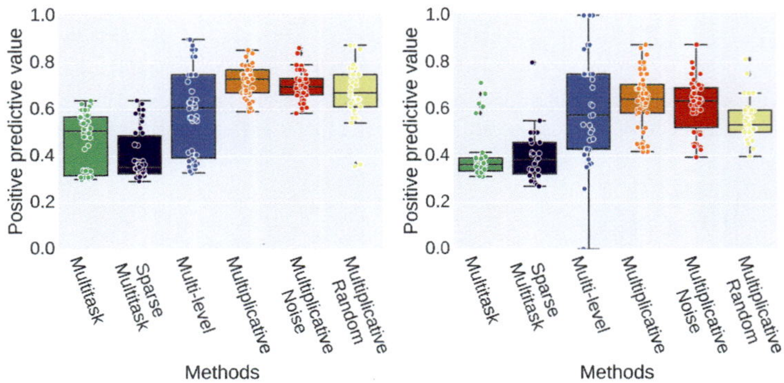

(a) $n_k = 100$ instances per task (b) $n_k = 20$ instances per task

Fig. 2. Boxplot depiction of the positive predictive value of the different methods for simulated data.

While the MML outperforms the ML and the SL in terms of PPV (Fig. 2), its sensitivity is worse (Fig. 3). Indeed, the ML and the SL select many more features: this higher sensitivity comes to the price of a large number of false positives. By contrast, the proposed MMLD performs well according to both criteria. It clearly outperforms all other methods in terms of PPV (Fig. 2), even when using noisy descriptors. In the case of random descriptors, the performance is close to that of the MML, and more degraded when the data is scarce. In terms of sensitivity (Fig. 3), the MMLD also outperforms its competitors. We observe a higher variability in performance for these other methods, due to the higher variability in the number of features they select. The ML, SL and MML suffer greater losses in sensitivity than the proposed method when data is scarce. Using task descriptors hence seems to increase the robustness of the feature selection procedure. As would be expected, using random task descriptors negatively affects the ability of the MMLD to recover the correct features. Small perturbations of the task descriptors appear to have little effect on the quality of the selected features. We report similar results for the setting where $p = 8000$ (Supp. Mat.).

4.3. *Prediction error*

The other important criterion on which to evaluate the model we propose is the quality of the predictions it makes. Figure 4 presents the 10-fold cross-validated Root Mean Squared Error (RMSE) of the different methods, for both $n_k = 100$ and $n_k = 20$. We observe that the proposed method performs better than its competitors, even with perturbed task descriptors. According to a paired Wilcoxon signed rank test (Supp. Mat.), these differences in RMSE on scarce data are significant. Interestingly, this is true even in comparison with the ML and the SL, which select more features and could hence be expected to yield lower RMSEs.

This improvement in predictive performance is particularly visible in the scarce setting(Fig. 4). In addition, the variance of the RMSE of the MMLD remains stable when the

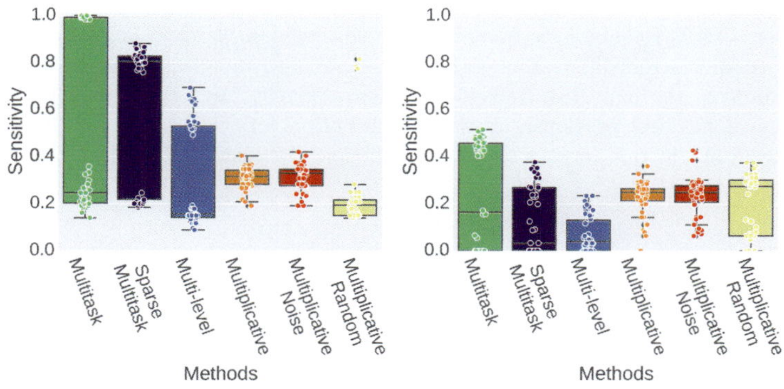

(a) $n_k = 100$ instances per task (b) $n_k = 20$ instances per task

Fig. 3. Boxplot depiction of the sensitivity of the different methods for simulated data.

number of samples decreases, while it clearly increases for the other approaches. Once again, we report similar results for the setting where $n_k = 20$ and $p = 8000$ (Supp. Mat.).

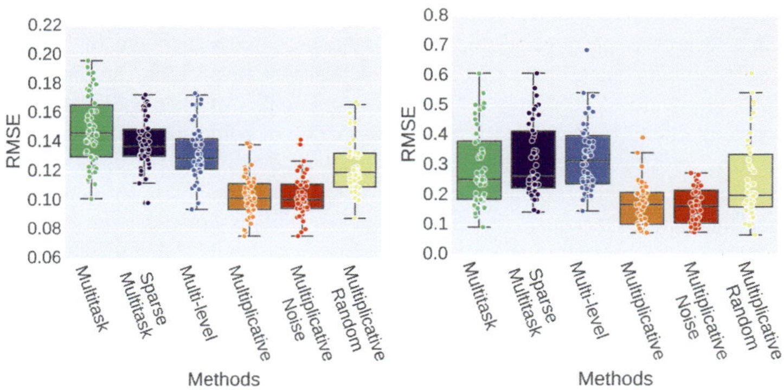

(a) $n_k = 100$ instances per task (b) $n_k = 20$ instances per task

Fig. 4. Boxplot of the 10-fold cross-validated Root Mean Squared Error (RMSE) of the different methods for simulated data. For readability, (a) and (b) are plotted on different scales.

5. Peptide-MHC-I binding prediction

The prediction of whether a peptide can bind to a given MHC-I (major histocompatibility complex class I) protein is an important tool for the development of peptide vaccines. MHC-I genes are highly polymorphic, and hence express proteins with diverse physico-chemical properties across individuals. The binding affinity of a peptide is thus going to depend on the MHC-I allele expressed by the patient. It is therefore important that predictions are allele-specific. This in turns opens the door to administering patient-specific vaccines.

While some MHC-I alleles have been well studied, others have few if any known binders. Sharing information across different alleles has the potential to improve the predictive accuracy of models. Indeed, the multitask framework, where different tasks correspond to different MHC-I proteins, has been previously shown to be beneficial for this problem[17,18]. In addition, it can be necessary in this context to make predictions for tasks (i.e. alleles) for which no training data is available.

5.1. *Data*

Following previous work[17], we test our model on three freely available benchmark datasets[19,20]. The data consists of pairs of peptide sequences and MHC-I alleles, labeled as binding or non-binding. Ref. 19 provides two datasets for the same 54 alleles, containing 1363 (resp. 282) positive and 1361 and (resp. 141784) negative examples. The dataset from Ref. 20 has 35 different alleles, 1548 positive examples and 4331 negative examples. As an example of an allele with few training data, allele B*57:01 in Ref. 20 only has 11 known binders.

The peptides are of length 9 and are classically represented by a 20-dimensional binary vector indicating which amino acid is present. While in this case $p < n$, this example allows us to evaluate the proposed method on real data, relevant for personalized medicine applications. Because the MHC-I alleles are much longer than that, we do not adopt the same representation and define task descriptors as follows: Using sequences extracted from the IMGT/HLA database[21], we keep only the amino acids located at positions involved in the binding sites of all three HLA superfamilies[17,22]. Inspired by the Linseq kernel[17], we then compute a similarity matrix between all alleles (tasks), based on the proportion of coincident amino acids at each position. We then perform a Principal Component Analysis on this matrix and keep the first 4 principal components, having observed that the structure of this matrix is not much perturbed by this dimensionality reduction. In the end, each task is represented by the 4-dimensional vector of its projections on each of these 4 components.

5.2. *Experiments*

We predict whether a peptide binds to a certain allele using the ML, the SL, the MML and the MMLD. Additionally, we compare these approaches to single task Lasso regressions.

We run cross-validation using the same folds as in the original publications[19,20]. The first Heckerman dataset[19] is divided in 5 folds and the second one in 10. Because this second dataset is highly unbalanced, we randomly keep only one negative example for each of the positive examples. The Peters dataset[20] is divided in 5 folds. We run an inner cross-validation to set the regularization parameters.

We show in Fig. 5.2 the Receiver Operator Curves (ROC) for the three datastets. Each curve corresponds to one fold. We additionally report the mean and standard deviation of the area under the ROC curve (ROC-AUC) for each approach. We observe that the ML, the SL and the MMLD perform comparatively, and consistently outperform the two other methods.

Furthermore, we evaluate the ability of the different methods to predict binding for alleles

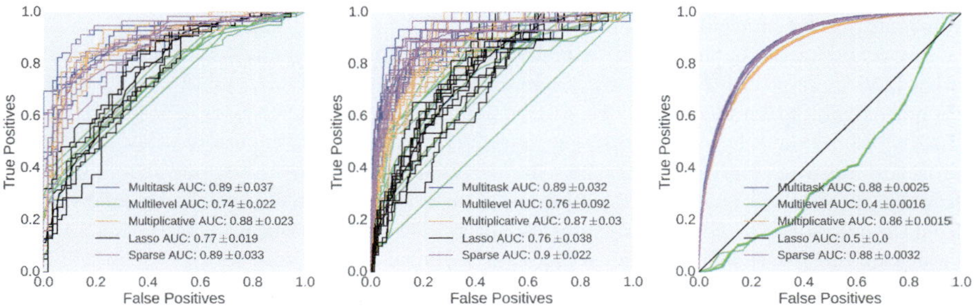

(a) First Heckerman dataset (b) Second Heckerman dataset (c) Peters dataset

Fig. 5. Cross-validated ROC curves for the prediction of MHC-I binding.

for which no training data is available. For this purpose, we use the models previously trained on the folds of the two first datasets to predict on the folds of the third dataset. When predicting for a new task with the ML, the SL and the MML, we use the mean of the predictions made by all trained models. As can be seen in Fig. 6, the proposed method is the only one that outperforms the trivial baseline (ROC-AUC=0.5), hence illustrating its ability to make predictions for previously unseen tasks, by contrast with all other methods.

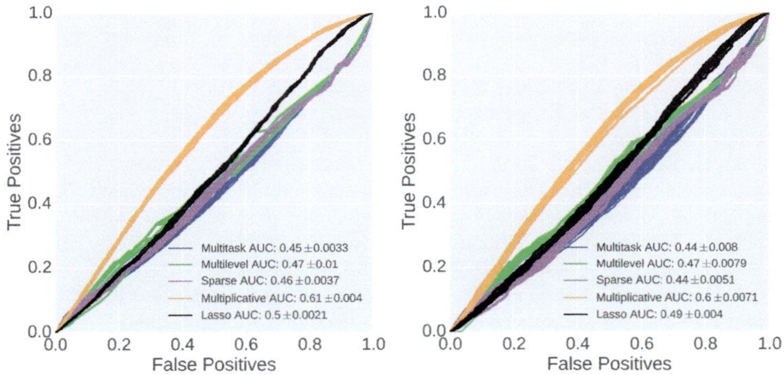

(a) Models trained on the first Heckerman dataset, evaluated on the Peters dataset
(b) Models trained on the second Heckerman dataset, evaluated on the Peters dataset

Fig. 6. ROC curves for the prediction of MHC-I binding, cross-dataset.

6. Conclusion

We have presented a novel approach for multitask least-squares regression. Our method extends the MML framework[5] to leverage task descriptors. This allows to tune how much information is shared between tasks according to their similarity, as well as to make predictions for new tasks. Multitask kernel methods[6,17,18] also allow to model relations between tasks, but do not offer the advantages of the Lasso framework in terms of sparsity and interpretability, which are key for biomedical applications.

Our experiments on simulated data show that the proposed method is more stable than other Lasso approaches. The features it selects are hence more reliable, and the resulting models more easily interpreted. In addition, true support recovery suffers less in scarce settings. Finally, the predictivity of the resulting models is competitive with that of other Lasso approaches. Unsurprisingly, performance deteriorates when task descriptors are inappropriate. However, neither the quality of the selected features nor the model predictivity suffer much from the addition of small noise to these descriptors. These results suggest that the MMLD approach we propose is well adapted to precision medicine applications, which require building stable, intepretable models from $n \ll p$ data.

Finally, our experiments on MHC-I peptide binding prediction illustrate that the method we propose is well-suited to making predictions for tasks for which no training data is available.

7. Supplementary Materials

`http://cazencott.info/dotclear/public/publications/Bellon_PSB2016_SuppMat.pdf`

Acknowledgments

This project has received funding from the European Unions Seventh Framework Programme for research, technological development and demonstration under grant agreement no. 316861.

References

1. K. Puniyani, S. Kim and E. P. Xing, *Bioinformatics* **26**, i208 (2010).
2. L. Chen, C. Li, S. Miller and F. Schenkel, *BMC Genetics* **15**, p. 53 (2014).
3. R. Tibshirani, *J Roy Stat Soc B Stat Meth* , 267 (1996).
4. G. Obozinski, B. Taskar and M. Jordan, *Statistics Department, UC Berkeley, Tech. Rep* (2006).
5. A. C. Lozano and G. Swirszcz, Multi-level lasso for sparse multi-task regression (2012).
6. T. Evgeniou, C. A. Micchelli and M. Pontil, *J Mach Learn Res* , 615 (2005).
7. E. V. Bonilla, K. M. Chai and C. Williams, *NIPS* **20**, 153 (2007).
8. M. Yuan and Y. Lin, *J Roy Stat Soc B Stat Meth* **68**, 49 (2006).
9. N. Simon, J. Friedman, T. Hastie and R. Tibshirani, *J Comput Graph Stat* **22**, 231 (2013).
10. Y. Nesterov, *Introductory Lectures on Convex Optimization* (Springer, 2004).
11. A. Jalali, S. Sanghavi, C. Ruan and P. K. Ravikumar, *NIPS* **23**, 964 (2010).
12. X. Wang, J. Bi, S. Yu and J. Sun, On multiplicative multitask feature learning (2014).
13. D. R. Flower, *J Chem Inform Comput Sci* **38**, 379 (1998).
14. G. V. Rocha, X. Wang and B. Yu, *arXiv preprint arXiv:0908.1940* (2009).
15. L. Breiman, *Technometrics* **37**, 373 (1995).
16. L. I. Kuncheva, *AIA* **25**, 421 (2007).
17. L. Jacob and J.-P. Vert, *Bioinformatics* **24**, 358 (2008).
18. C. Widmer, N. C. Toussaint, Y. Altun and G. Rätsch, *BMC Bioinformatics* **11**, p. S5 (2010).
19. D. Heckerman, C. Kadie and J. Listgarten, *J Comput Biol* **14**, 736 (2007).
20. B. Peters, H.-H. Bui, S. Frankild, M. Nielson *et al.*, *PLoS Comput Biol* **2**, p. e65 (2006).
21. J. Robinson, A. Malik, P. Parham, J. Bodmer and S. Marsh, *Tissue Antigens* **55**, 280 (2000).
22. I. A. Doytchinova, P. Guan and D. R. Flower, *J Immunol* **172**, 4314 (2004).

PERSONALIZED HYPOTHESIS TESTS FOR DETECTING MEDICATION RESPONSE IN PARKINSON DISEASE PATIENTS USING iPHONE SENSOR DATA

ELIAS CHAIBUB NETO*, BRIAN M. BOT, THANNEER PERUMAL, LARSSON OMBERG, JUSTIN GUINNEY, MIKE KELLEN, ARNO KLEIN, STEPHEN H. FRIEND, ANDREW D. TRISTER

Sage Bionetworks, 1100 Fairview Avenue North, Seattle, Washington 98109, USA

* *corresponding author e-mail: elias.chaibub.neto@sagebase.org*

We propose hypothesis tests for detecting dopaminergic medication response in Parkinson disease patients, using longitudinal sensor data collected by smartphones. The processed data is composed of multiple features extracted from active tapping tasks performed by the participant on a daily basis, before and after medication, over several months. Each extracted feature corresponds to a time series of measurements annotated according to whether the measurement was taken before or after the patient has taken his/her medication. Even though the data is longitudinal in nature, we show that simple hypothesis tests for detecting medication response, which ignore the serial correlation structure of the data, are still statistically valid, showing type I error rates at the nominal level. We propose two distinct personalized testing approaches. In the first, we combine multiple feature-specific tests into a single union-intersection test. In the second, we construct personalized classifiers of the before/after medication labels using all the extracted features of a given participant, and test the null hypothesis that the area under the receiver operating characteristic curve of the classifier is equal to 1/2. We compare the statistical power of the personalized classifier tests and personalized union-intersection tests in a simulation study, and illustrate the performance of the proposed tests using data from mPower Parkinsons disease study, recently launched as part of Apples ResearchKit mobile platform. Our results suggest that the personalized tests, which ignore the longitudinal aspect of the data, can perform well in real data analyses, suggesting they might be used as a sound baseline approach, to which more sophisticated methods can be compared to.

Keywords: personalized medicine, hypothesis tests, sensor data, remote monitoring, Parkinson

1. Introduction

Parkinson disease is a severe neurodegenerative disorder of the central nervous system caused by the death of dopamine-generating cells in the midbrain. The disease has considerable worldwide morbidity and is associated with substantial decrease in the quality of life of the patients (and their caregivers), decreased life expectancy, and high costs related to care. Early symptoms in the motor domain include shaking, rigidity, slowness of movement and difficulty for walking. Later symptoms include issues with sleeping, thinking and behavioral problems, depression, and finally dementia in the more advanced stages of the disease. Treatments are usually based on levodopa and dopamine agonist medications. Nonetheless, as the disease progresses, these drugs often become less effective, while still causing side effects, including involuntary twisting movements (dyskinesias). Statistical approaches aiming to determine if a given patient responds to medication have key practical importance as they can help the physician in making more informed treatment recommendations for a particular patient.

In this paper we propose personalized hypothesis tests for detecting medication response in Parkinson patients, using longitudinal sensor data collected by iPhones. Remote monitoring of Parkinson patients, based on active tasks delivered by smartphone applications, is an active research field.[1] Here we illustrate the application of our personalized tests using sensor data collected by the mPower study, recently launched as part of Apple's ResearchKit[2,3] mobile platform. The active tests implemented in the mPower app include tapping, voice, memory, posture and gait tests, although in this paper we focus on the tapping data only. During a tapping test the patient is asked to tap two buttons on the iPhone screen alternating between two fingers on the same hand for 20 seconds. Raw sensor data collected during a single test is given by a time series of the screen x-y coordinates on each tap. Processed data corresponds to multiple features extracted from the tapping task, such as the number of taps and the mean inter-tapping interval. Since the active tests are performed by the patient on a daily basis, before and after medication, over several months, the processed data corresponds to time series of feature measurements annotated according to whether the measurement was taken before or after the patient has taken his/her medication. Though others have investigated the feasibility of monitoring medication response in Parkinson patients using smartphone sensor data, this previous work did not focus on the individual effects that medications have, but rather focused on the classification on a population level.[4]

The first step in analyzing these data is to show that simple feature-specific tests, which ignore the serial correlation in the extracted features, are statistically valid (the distribution of the p-values for tests applied to data generated under the null hypothesis is uniform). This condition guarantees that the tests are exact, that is, the type I error rates match the nominal levels, so that our inferences are neither conservative nor liberal. In other words, if we adopt a significance level cutoff of α, the probability that our tests will incorrectly reject the null when it is actually true is given by α.

Even though the simple feature-specific tests are valid procedures for testing for medication response, in practice, we have multiple features and need to combine them into a single decision procedure. The second main contribution of this paper is to propose two distinct approaches to combine all the extracted features into a single hypothesis test. In the first, and most standard approach, we combine simple tests, applied to each one of our extracted features, into a single union-intersection test. Although simple to implement, scalable, and computationally efficient, this approach requires multiple testing correction, which might become burdensome when the number of extracted features is large. In order to circumvent this potential issue, our second approach is to construct personalized classifiers of the before/after medication labels using all the extracted features of a given patient, and test the null hypothesis that the area under the receiver operating characteristic curve (AUROC) of the classifier is equal to 1/2 (in which case the patient's extracted features are unable to predict the before/after medication labels, implying that the patient does not respond to the medication). A slight disadvantage of the classifier approach, compared to the union-intersection tests, is the larger computational cost (especially for classifiers that require tuning parameter optimization by cross-validation) involved in the classifier training. In any case, the increased computational demand is by no means a limiting factor for the application of the approach.

The rest of this paper is organized as follows. In Section 2 we present our personalized tests, discuss their statistical validity, and perform a power study comparison. In Section 3 we illustrate the application of our tests to the tapping data of the mPower study. Finally, in Section 4 we discuss our results.

2. Methods

2.1. *Notation and a few preliminary comments on the data*

Throughout this paper, we let x_{kt}, $k = 1, \ldots, p$, $t = 1, \ldots, n$, represent the measurement of feature k at time point t, and let $y_t = \{b, a\}$, represent the binary outcome variable, corresponding to the before/after medication label, where b and a stand for "before" and "after" medication, respectively.

Even though the participants where asked to perform the active tasks 3 times per day, one before the medication, one after, and one at any other time of their choice, participants did not always follow the instructions correctly. As a result, the data is non-standard, with variable number of daily tasks (sometimes fewer, sometimes greater than 3 tasks per day), and variable timing relative to medication patterns (e.g., bbabba..., aaabbb..., instead of bababa...). Furthermore, the data also contains missing medication labels, as sometimes, a participant performed the active task but did not report whether the task was taken before or after medication. In our analysis we restrict our attention to data collected before and after medication only. Hence, for each participant, the number of data points used in our tests is given by $n = n_b + n_a$, where n_b and n_a correspond, respectively, to the number of before/after medication labels.

2.2. *On the statistical validity of personalized tests which ignore the autocorrelation structure of the data*

It is common knowledge that the t-test, the Wilcoxon rank-sum test, and other two-sample problem tests, suffer from inflated type I error rates in the presence of dependency. We point out, however, that this can happen when the data within each group is dependent, but the two groups are themselves statistically independent. When the data from both groups is sampled jointly from the same multivariate distribution, the dependency of the data might no longer be an issue. Figure 1 provides an illustrative example with t-tests applied to simulated data.

The t-test's assumption of independence (within and between the groups' data) is required in order to make the analytical derivation of the null distribution feasible. It doesn't mean the test will always generate inflated type I error rates in the presence of dependency (as illustrated in Figure 1f). As a matter of fact, a permutation test based on the t-test statistic is valid if the group labels are exchangeable under the null,[5] even when the data is statistically dependent. Exchangeability[6] captures a notion of symmetry/similarity in the data, without requiring independence. On the examples presented in Figure 1, the group labels are exchangeable on panels a and c as illustrated by the symmetry/similarity of the data between groups 1 and 2 at each row of the heatmaps. For panel b, on the other hand, the lack of symmetry between the groups on each row illustrates that the group labels are not exchangeable.

In the context of our personalized tests, the before/after medication labels are exchangeable under the null of no medication response, even though the measurements of any extracted feature are usually serially correlated. Note that the exchangeability is required for the medication labels, and not for the feature measurements, which are not exchangeable due to their serial correlation. Figure 2 illustrates this point, showing the symmetry/similarity of the separate time series for the before and after medication data.

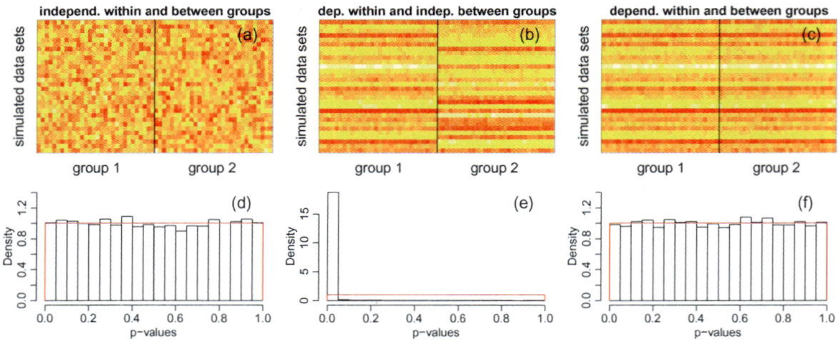

Fig. 1. The effect of data dependency on the t-test. Panels a, b, and c show heatmaps of the simulated data. Columns are split between group 1 and 2, and each row corresponds to one simulated null data set (we show the top 30 simulations only). Bottom panels show the p-value distributions for 10,000 tests applied to null data simulated according to: (i) $\boldsymbol{x}_1 \sim \mathrm{N}_{30}(\boldsymbol{\mu}_1, \boldsymbol{I})$ and $\boldsymbol{x}_2 \sim \mathrm{N}_{30}(\boldsymbol{\mu}_2, \boldsymbol{I})$ with $\boldsymbol{\mu}_1 = \boldsymbol{\mu}_2 = \boldsymbol{0}$ (panel d); (ii) $\boldsymbol{x}_1 \sim \mathrm{N}_{30}(\boldsymbol{\mu}_1, \boldsymbol{\Sigma})$ and $\boldsymbol{x}_2 \sim \mathrm{N}_{30}(\boldsymbol{\mu}_2, \boldsymbol{\Sigma})$, where $\boldsymbol{\mu}_1 = \boldsymbol{\mu}_2 = \boldsymbol{0}$ and $\boldsymbol{\Sigma}$ is a correlation matrix with off-diagonal elements equal to $\rho = 0.95$ (panel e); and (iii) $(\boldsymbol{x}_1, \boldsymbol{x}_2)^t \sim \mathrm{N}_{60}(\boldsymbol{\mu}, \boldsymbol{\Sigma})$, with $\boldsymbol{\mu} = (\boldsymbol{\mu}_1, \boldsymbol{\mu}_2)^t = \boldsymbol{0}$ and $\boldsymbol{\Sigma}$ as before (panel f). The density of the Uniform[0, 1] distribution is shown in red. Panel d shows that under the standard assumptions of the t-test, the p-value distribution under the null is (as expected) uniform. Panel e shows the p-value distribution for strongly dependent data, showing highly inflated type I error rates, even though the data was simulated according to t-test's null hypothesis that $\boldsymbol{\mu}_1 = \boldsymbol{\mu}_2$. Panel b clarifies why that is the case. For each row (i.e., simulated data set), the data tends to be quite homogeneous inside each group, but quite distinct between the groups. Because on each simulation we sample the data vectors $\boldsymbol{x}_1$ and $\boldsymbol{x}_2$ from a multivariate normal distribution with a very strong correlation structure, all elements in the $\boldsymbol{x}_1$ vector tend to be close to each other, and all elements in $\boldsymbol{x}_2$ tend to be similar to each other. However, because $\boldsymbol{x}_1$ and $\boldsymbol{x}_2$ are sampled independently from each other, their values tend to be distinct. In combination, the small variability in each group vector together with the difference in their means leads to high test statistic values and small p-values. Panel f shows the p-value distribution for strongly dependent data, when sampled jointly. In this case, the distribution is uniform. Panel c clarifies why. Now, each row tends to be entirely homogeneous (within and between groups), since the joint sampling of $\boldsymbol{x}_1$ and $\boldsymbol{x}_2$ makes all elements in both vectors quite similar to each other, so that the difference in their means tends to be small.

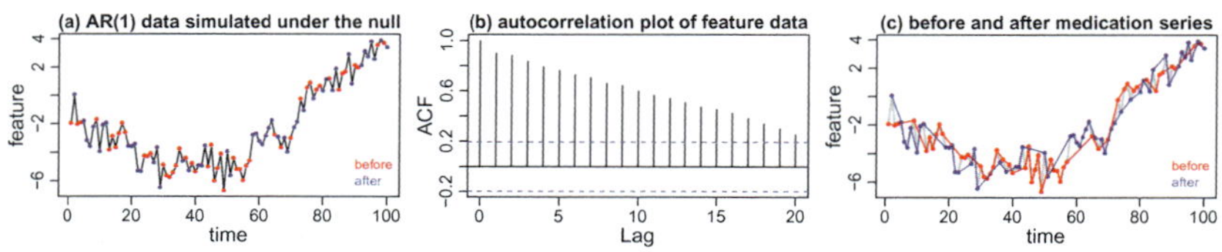

Fig. 2. Exchangeability of the before/after medication labels in time series data. Panel a shows a feature, simulated from an AR(1) process, under the null hypothesis that the patient does not respond to medication. In this case, the before medication (red dots) and after medication (blue dots) labels are randomly assigned to the feature measurements. Panel b shows an autocorrelation plot of the feature data. Panel c shows the separate "before medication" (red) and "after medication" (blue) series. Note the symmetry/similarity of the two series. Clearly, under the null hypothesis that the patient does not respond to medication, the medication labels are exchangeable, since shuffling of the before/after medication labels would not destroy the serial correlation structure or the trend of the series.

Hence, even though our longitudinal data violates the independence assumption of the t-test, the permutation test based on the t-test statistic is still valid. Of course the same argument is valid for permutation tests based on other test statistics. Figure 3 illustrates this point, with permutation tests based on the t-test and Wilcoxon rank-sum test. Panel a shows the original data. Red and blue dots represent measurements before and after medication, respectively. The grey dots represent data collected at another time or where the medication label is missing. Panel b shows one realization of a random permutation of the before/after medication labels. In order to generate a permutation null distribution, we perform a large number of random label shuffles, and for each one, we evaluate the adopted test statistic in the permuted data. Panels c and d show the permutation null distributions generated from 10,000 random permutations of the medication labels based, respectively, on the t-test and on the Wilcoxon rank-sum test statistics. The red curve on panel c shows the analytical density of a

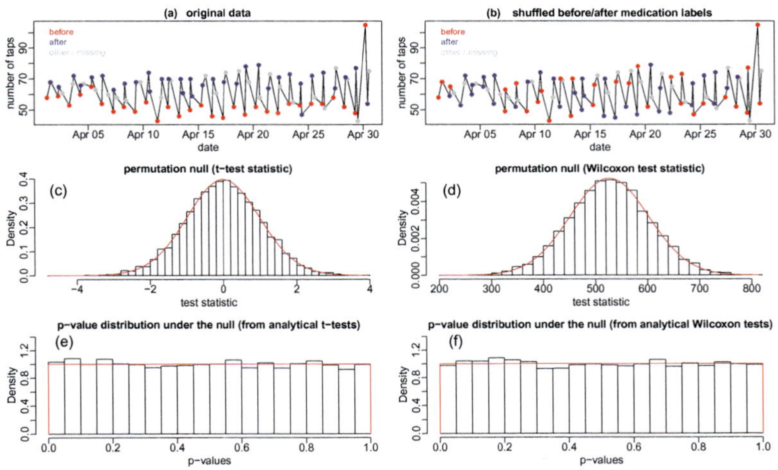

Fig. 3. Personalized tests for the null hypothesis that the patient does not respond to medication, according to a single feature (number of taps), and based on t-test and Wilcoxon rank-sum test statistics. Panel a shows the data on the number of taps for one of mPower's study participants during April 2015. Panel b shows one realization of a random permutation of the before/after medication labels. Panel c shows the permutation null distribution based on the t-test statistic. The red curve represents the analytical density of a t-test, namely, a t-distribution with $n_b + n_a - 2$ degrees of freedom. Panel d shows the permutation null distribution based on Wilcoxon rank-sum test statistic. The red curve shows the density of the normal asymptotic approximation for Wilcoxon's test, namely, a Gaussian density with mean, $n_b n_a/2$, and variance, $n_b n_a(n_b + n_a + 1)/12$. In this example $n_b = 31$ and $n_a = 34$. Panels e and f show the analytical p-value distributions under the null. Results are based on 10,000 null data sets. Each null data set was generated as follows: (i) randomly sample the number of "before" medication labels, n_b, from the set $\{10, 11, \ldots, n-10\}$, where n is the total number of measurements; (ii) compute the number of "after" medication labels as $n_a = n - n_b$; and (iii) randomly assign the "before medication" and "after medication" labels to the number of taps measurements. The plots show the histograms of the p-values derived from the application of t-tests (panel e) and Wilcoxon's tests (panel f) to each of the null data sets. The density of the Uniform[0, 1] distribution is shown in red.

t-test for the data on panel a, while the red curve on panel d shows the density of the normal asymptotic approximation for the Wilcoxon test represented in panel b (we show the normal approximation density, since the exact null distribution is discrete). The close similarity of the permutation and analytical distributions suggests that, in practice, we can use the analytical p-values of t-tests or Wilcoxon rank-sum tests instead of the permutation p-values. Panels e and f further corroborate this point, showing uniform distributions for the analytical p-values of t-tests and Wilcoxon tests respectively, derived from 10,000 distinct null data sets.

2.3. *Personalized union-intersection tests*

In order to combine the feature-specific tests, H_{0k} : the patient does not respond to the medication, according to feature k, versus H_{1k} : the patient responds to the medication, across all extracted features, we construct the union-intersection test,

$$H_0 : \cap_{k=1}^{p} H_{0k} \quad \text{versus} \quad H_1 : \cup_{k=1}^{p} H_{1k} \,, \tag{1}$$

where, in words, we test the null hypothesis that the patient does not respond to medication, for all p features, versus the alternative hypothesis that he/she responds to medication according to at least one of the features. Under this test, we reject H_0 if the p-value of at least one of the feature-specific tests is small. Hence, the p-value for the union-intersection test corresponds to the smallest p-value (across all p tests) after multiple testing correction.

We implement union-intersection tests based on the t-test and Wilcoxon rank-sum test statistics. We adopt the Benjamini-Hochberg approach[7] for multiple testing correction. As detailed in Figure 4, the union-intersection test tends to be slightly conservative when applied to correlated features.

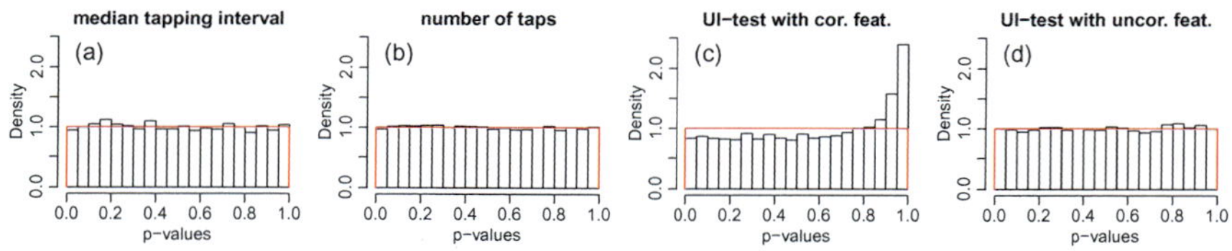

Fig. 4. P-value distributions for the union-intersection test under the null. Results are based on 10,000 null data sets generated by randomly permuting the before/after medication labels. Panels a and b show the p-value distributions from, H_{0k}, for 2 out of the 24 features combined in the union-intersection test. We observed uniform distributions for all features, but report just 2 here due to space constraints. Panel c shows the p-value distribution of the union-intersection test using Benjamini-Hochberg correction. The skewness of the distribution towards larger p-values indicates that the test is conservative, meaning that at a nominal significance level α the probability of rejecting the null when it is actually true is smaller than α. Since the p-value distributions of the features are uniform (panels a and b), the skewness of the union-intersection p-value distribution is clearly due to the multiple testing correction. We experimented with other procedures, but the Benjamini-Hochberg correction was the least conservative one. Panel d reports the p-value distribution for the union-intersection test using a shuffled version of the feature data. Note that by shuffling the data, we destroy the correlation among the features, so that the now uniform distribution suggests that the violation of the independence assumption (required by the Benjamini-Hochberg approach) seems to be the reason for the slightly conservative behavior of the union-intersection test. Results were generated using Wilcoxon's test.

2.4. *Personalized classifier tests*

An alternative approach to combine the information across multiple features into a single decision procedure is to test whether a classifier trained using all the features is able to predict the before/after medication labels better than a random guess. Adopting the AUROC as a classification performance metric, we have that a prediction equivalent to a random guess would have an AUROC equal to 0.5, whereas a perfect prediction would lead to an AUROC equal to 1. Furthermore, if a classifier generates an AUROC smaller than 0.5, we only need to switch the labels in order to make the AUROC larger than 0.5. Therefore, we can test if a classifier's prediction is better than random using the one-sided test,

$$H_0 : \text{AUROC} = 1/2 \quad \text{versus} \quad H_1 : \text{AUROC} > 1/2 \; . \tag{2}$$

It has been shown[8] that, when there are no ties in the predicted class probabilities used for the computation of the AUROC, the test statistic of the Wilcoxon rank-sum test (also known as the Mann-Whitney U test), U, is related to the AUROC statistic by $U = n_b \, n_a (1 - \text{AUROC})$ (see section 2 of reference[9] for details). Hence, under the assumption of independence (required by Wilcoxon's test) the analytical p-value for the hypothesis test in (2) is given by the left tail probability, $P(U \leq n_b \, n_a (1 - \text{AUROC}))$, of Wilcoxon's null distribution. In the presence of ties, the p-value is given by the left tail of the asymptotic approximate null,

$$U \approx \text{N}\left(\frac{n_b \, n_a}{2} \, , \; \frac{n_b \, n_a (n+1)}{12} - \frac{n_b \, n_a}{12 \, n \, (n-1)} \sum_{j=1}^{\tau} t_j (t_j - 1)(t_j + 1) \right) \, , \tag{3}$$

where τ is the number of groups of ties, and t_j is the number of ties in group j.[9] Alternatively, we can get the p-value as the right tail probability of the corresponding AUROC null,

$$\text{AUROC} \approx \text{N}\left(\frac{1}{2} \, , \; \frac{n+1}{12 \, n_b \, n_a} - \frac{1}{12 \, n_b \, n_a \, n \, (n-1)} \sum_{j=1}^{\tau} t_j (t_j - 1)(t_j + 1) \right) \, . \tag{4}$$

As before, even though the test described above assumes independence, the exchangeability of the before/after medication labels guarantees the validity of the permutation test based on the AUROC statistic. Figure 5 illustrates this point.

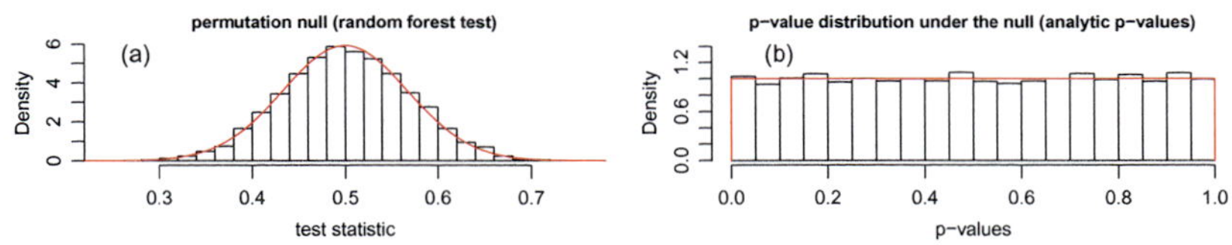

Fig. 5. Panel a shows the permutation null distribution for the personalized classifier test (based on the random forest algorithm). The red curve represents the density of the AUROC (approximate) null distribution in eq. (4). Panel b shows the p-value distribution for the (analytical) classifier test, based on 10,000 null data sets, generated by randomly permuting the before/after medication labels as described in Figure 3. The density of the Uniform[0, 1] distribution is shown in red.

2.5. *Statistical power comparison*

In this section we compare the statistical power of the personalized classifier test (based on the random forest[10] and extra trees[11] classifiers) against the personalized union-intersection tests (based on t-tests and Wilcoxon rank-sum tests). We simulated feature data, x_{kt}, according to the model,

$$x_{kt} \; = \; A \, \cos(\pi \, t) \; + \; \epsilon_{kt} \; , \tag{5}$$

where $\epsilon_{kt} \sim \mathrm{N}(0, \sigma_k^2)$ represents i.i.d. error terms, and the function $A \cos(\pi t)$ describes the periodic signal, with A representing the peak amplitude of the signal. Figure 6 describes the additional steps involved in the generation of the before/after medication labels.

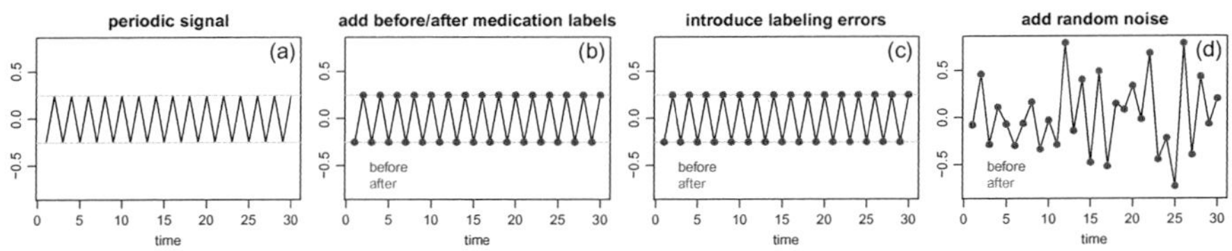

Fig. 6. Data simulation steps. First, we generate the periodic signal, $A\cos(\pi t)$, shown in panel a for $t = 1, \ldots, 30$, and $A = 0.25$. Second, we assign the "before medication" label (red dots) to the negative values, and the "after medication" label (blue dots) for the positive values (panel b). Third, we introduce some labeling errors (panel c). Finally, at the fourth step (panel d), we add $\epsilon_{kt} \sim \mathrm{N}(0, \sigma_k^2)$ error terms to the measurements.

We ran six simulation experiments, covering all combinations of sample size, $n = \{50, 150\}$, and number of features, $p = \{10, 50, 200\}$. In all simulations we adopted $A = 0.25$, mislabeling error rate of 10%, and increasing $\sigma_k = \{1.00, 1.22, 1.44, 1.67, 1.89, 2.11, 2.33, 2.56, 2.78\}$ for the first 9 features, and $\sigma_k = 3$, for $10 \leq k \leq p$. In each experiment, we generated 1,000 data sets and computed the personalized tests p-values. Figure 7 presents a comparison of the empirical power of the personalized tests as a function of the significance threshold α. For each test, the empirical power curve was estimated as the fraction of the p-values smaller than α, for a dense grid of α values varying between 0 and 1.

The results showed some clear patterns. First, the comparison of the top and bottom panels showed that for all 4 tests an increase in sample size leads to an increase in statistical power (as one would have expected). Second, the empirical power of the union-intersection tests based on Wilcoxon and t-tests was very similar in all simulation experiments, with the t-test being slightly better powered than the Wilcoxon test. This result was expected since the simulated features were generated using Gaussian errors, and the t-test is known to be slightly better powered than the Wilcoxon rank-sum test under normality. Similarly, the empirical power of the personalized classifier tests was also similar, with the extra trees algorithm tending to be slightly better powered. Third, this study shows that neither the personalized classifier nor the union-intersection approaches dominate each other. Rather, we see that for this particular data generation mechanism and simulation parameters choice, the union-intersection tests tended to be better powered than the classifier tests for smaller values of p, whereas the converse was

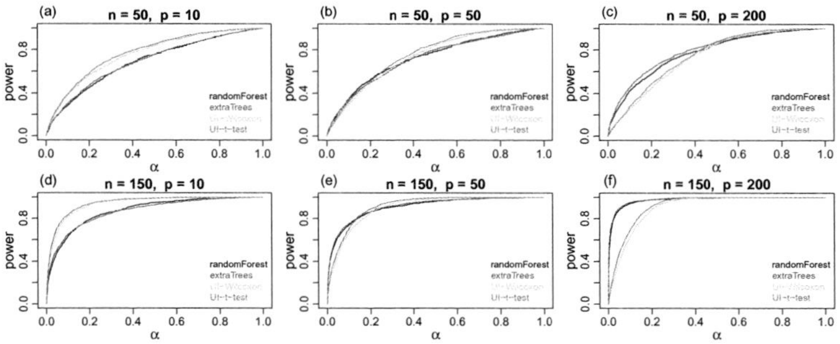

Fig. 7. Comparison of the personalized test's empirical power as a function of the significance cutoff, α.

true for larger p. Furthermore, observe that the power of the union-intersection tests tended to decrease as the number of features increased (note the slight decrease in the slope of the cyan and pink curves as we go from the panels on the left to the panels on the right). The tests based on the classifiers, on the other hand, tended to get better powered as the number of features increased (note the respective increase in the slope of the blue and black curves).

The observed decrease in power of the union-intersection tests might be explained by the increased burden caused by the multiple testing correction required by these tests. On the other hand, the personalized classifier tests are not plagued by the multiple testing issue since all features are simultaneously accounted for by the classifiers. Furthermore, in situations where none of the features is particularly informative, the classifiers might still be able to better aggregate information across the multiple noisy features.

3. Real data illustrations

In this section we illustrate the performance of our personalized hypothesis tests, based on tapping data collected by the mPower study between 03/09/2015 (date the study opened) and 06/25/2015. We restrict our analyzes to 57 patients, who performed at least 30 tapping tasks before medication, as well as 30 or more tasks after medication.

Figure 8 presents the results. Panel a shows the number of tapping tasks (before and after medication) performed by each participant. Panel b reports the AUROC scores for 4 classifiers (random forest, logistic regression with elastic-net penalty,[12] logistic regression, and extra trees). Panel c presents the p-values for the respective personalized classifier tests[a], as well as for 2 personalized union-intersection tests (based on t- and Wilcoxon rank-sum tests).

Panel b shows that the tree-based classifier tests (random forest and extra trees) showed comparable performance across all participants, whereas the regression-based approaches (elastic net and logistic regression) were sometimes comparable but sometimes strongly out-

[a]The results for the personalized classifier tests were based on 100 random splits of the data into training and test sets, using roughly half of the data for training and the other half for testing. The AUROC and p-values reported on Figure 8 b and c correspond to the median of the AUROCs and p-values across the 100 data splits.

performed by the tree-based tests. Panel c shows that the union-intersection tests, nonetheless, produced at times much smaller p-values than the classifier tests. At a significance level of 0.001 (grey horizontal line), about one quarter of the patients (leftmost quarter) respond to dopaminergic medication, according to most tests.

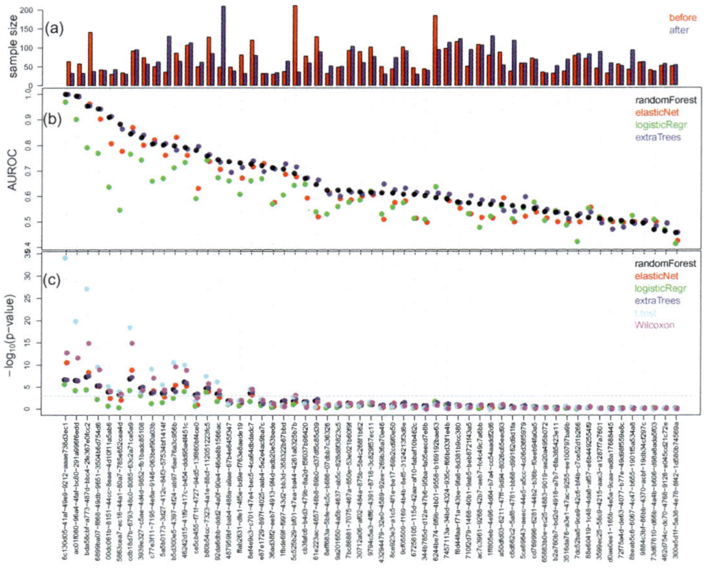

Fig. 8. Results of personalized hypothesis tests applied to the tapping data from the mPower study. Panel a shows the number of tapping tasks performed before (red) and after (blue) medication, per participant. Panel b shows the AUROC scores (across 100 random splits of the data into training and test sets) for four classifiers. Panel c shows the adjusted p-values (in negative log base 10 scale) of 4 classification tests and 2 union-intersection tests. The p-values were adjusted using Benjamini-Hochberg multiple testing correction. The grey horizontal line represents an adjusted p-value cutoff of 0.001. The participants were sorted according to the AUROC of the random forest algorithm (black dots in panel b).

In order to illustrate how our personalized tests are able to pick up meaningful signal from the extracted features, we present on Figure 9, time series plots of 2 features (number of taps and mean tapping interval) which were consistently ranked among the top 3 features for the top 3 patients on the left of Figure 8c, and compared it to the data from the bottom 3 patients on the right of Figure 8c. (For the random forest classifier test, we ranked the features according to the importance measure provided by the random forest algorithm. For the union-intersection tests, we ranked the features according to the p-values of the feature-specific tests.) Comparison of panels a to f, which show the data from patients that respond to medication (according to our tests), against panels g to l, which report the data from patients that do not respond to medication, shows that our tests can clearly pick up meaningful signal from these two extracted features.

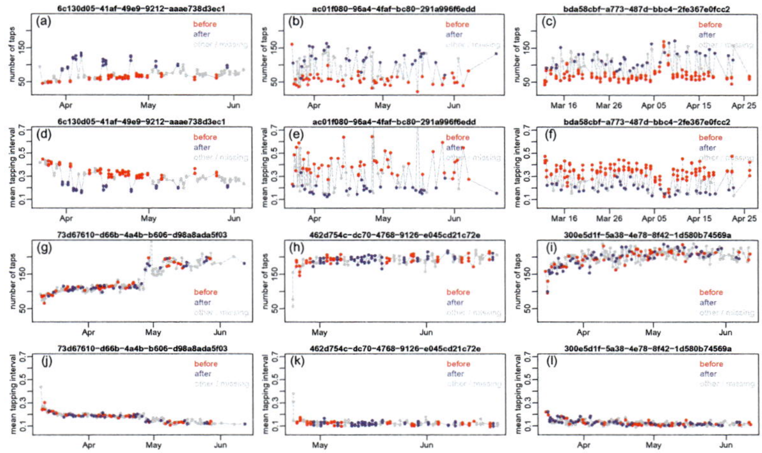

Fig. 9. Comparison of the leftmost 3 patients against the rightmost 3 patients in Figure 8c, according to 2 tapping features (number of taps and mean tapping interval). Panels a to f show the results for the top 3 patients (which respond to medication according to our tests), while panels g to l show the results for the bottom 3 patients (which don't respond to medication according to our tests).

4. Discussion

In this paper we describe personalized hypothesis tests for detecting dopaminergic medication response in Parkinson patients. We propose and compare two distinct strategies for combining the information from multiple extracted features into a single decision procedure, namely: (i) union-intersection tests; and (ii) hypothesis tests based on classifiers of the medication labels. We carefully evaluated the statistical properties of our tests, and illustrated their performance with tapping data from the mPower study. We have also successfully applied our tests to features extracted from the voice and accelerometer data collected using the mPower app, but cannot present these results here due to space limitations.

About one quarter of the patients analyzed in this study showed strong statistical evidence for medication response. For these patients, we observed that the union-intersection tests tended to generate smaller p-values than the classifier based tests. This result corroborates our observations in the empirical power simulation studies, where union-intersection tests tended to out-perform classifier tests when the number of features is small (recall that we employed only 24 features in our analyses).

Although our tests can detect medication responses, they do not explicitly determine the direction of the response, that is, they cannot differentiate a response in the expected direction from a paradoxical one. We point out, however, that their main utility is to help out the physician pinpoint patients in need of a change on their drug treatment. For instance, our tests are able to detect patients for which the drug has an effect in the expected direction but is not well calibrated (so that its effect is wore off by the time the patient takes the medication), or patients showing paradoxical responses. Note that both cases flag a situation which requires an action by the physician (calibrating the medication dosage in the first case, and stopping/changing the medication in the second one). Therefore, even though our tests cannot distinguish between these cases they are still able to detect patients which can potentially benefit from a dose calibration or a change in medication.

Even though we restricted our attention to 4 classifiers, other classification algorithms could also be easily used for generating additional personalized classifier tests. In our experience, however, tree based approaches such as the random forest and extra trees classifiers tend to perform remarkably well in practice, providing a robust default choice. Similarly, we focused on t- and Wilcoxon rank-sum tests in the derivation of our personalized union-intersection tests, since these tests show robust practical performance for detecting group differences.

Although others have investigated the feasibility of monitoring medication response in Parkinson patients using smartphone sensor data,[4] their study focused on the classification of the before/after medication response at the population level, with the classifier applied to data across all patients, and not at a personalized level, as done here.

In this paper we show that simple hypothesis tests, which ignore the serial correlation in the feature data, are statistically valid and well powered to detect medication response in the mPower study data. We point out, however, that, in theory, more sophisticated approaches able to leverage the longitudinal nature of the data could in principle improve the statistical power to detect medication response. In any case, the good practical performance of our simple personalized tests suggests that they can be used as sound baseline approaches, against which more sophisticated methods can be compared.

Code and data availability. All the R[13] code implementing the personalized tests, and used to generate the results and figures in this paper is available at `https://github.com/Sage-Bionetworks/personalized_hypothesis_tests`. The processed tapping data is available at `doi:10.7303/syn4649804`.

Acknowledgements. This work was funded by the Robert Wood Johnson Foundation.

References

1. S. Arora, et al, *Parkinsonism and related disorders* **21**, 650-653 (2015).
2. Editorial, *Nature Biotechnology* **33**, 567 (2015).
3. S. H. Friend, *Science Translational Medicine* **7**, 297ed10 (2015).
4. A. Zhan, et al, *High frequency remote monitoring of Parkinson's disease via smartphone: platform overview and medication response detection* (manuscript under review) (2015).
5. B. Efron, R. J. Tibshirani, *An introduction to the bootstrap*, Chapter 15 (Chapman and Hall/CRC, 1993).
6. J. Galambos, *Encyclopedia of Statistical Sciences* **7**, 573-577 (1996).
7. Y. Benjamini, Y. Hochberg, *Journal of the Royal Statistical Society, Series B* **57**, 289-300 (1995).
8. D. Bamber, *Journal of Mathematical Psychology* **12**, 387-415 (1975).
9. S. L. Mason, N. E. Graham, *Quarterly Journal of the Royal Meteorological Society* **128**, 2145-2166 (2002).
10. L. Breiman, *Machine Learning* **45**, 5-32 (2001).
11. P. Geurts, D Ernst, L. Wehenkel, *Machine Learning* **63**, 3-42 (2006).
12. J. H. Friedman, T. Hastie, R. Tibshirani, *Journal of Statistical Software* **33 (1)**, (2010).
13. R Core Team, *R Foundation for Statistical Computing* (Vienna, Austria. ISBN 3-900051-07-0, URL http://www.R-project.org/ 2015).

KIDNEY DISEASE GENETICS AND THE IMPORTANCE OF DIVERSITY IN PRECISION MEDICINE

JESSICA N. COOKE BAILEY

Institute for Computational Biology, Department of Epidemiology and Biostatistics, Case Western Reserve University, Wolstein Research Building, 2103 Cornell Road, Suite 2527, Cleveland, OH 44106, USA
Email: jnc43@case.edu

SARAH WILSON

Center for Human Genetics, Vanderbilt University, 519 Light Hall, 2215 Garland Avenue, Nashville, TN 37232, USA
Email: sarah.wilson928@gmail.com

KRISTIN BROWN-GENTRY

Center for Human Genetics Research, Vanderbilt University, 519 Light Hall, 2215 Garland Avenue, Nashville, TN 37232, USA
Email: kristin.gentry@healthspring.com

ROBERT GOODLOE

Center for Human Genetics Research, Vanderbilt University, 519 Light Hall, 2215 Garland Avenue, Nashville, TN 37232, USA
Email: robert.goodloe@gmail.com

DANA C. CRAWFORD

Institute for Computational Biology, Department of Epidemiology and Biostatistics, Case Western Reserve University, Wolstein Research Building, 2103 Cornell Road, Suite 2527, Cleveland, OH 44106, USA
Email: dana.crawford@case.edu

Kidney disease is a well-known health disparity in the United States where African Americans are affected at higher rates compared with other groups such as European Americans and Mexican Americans. Common genetic variants in the myosin, heavy chain 9, non-muscle (*MYH9*) gene were initially identified as associated with non-diabetic end-stage renal disease in African Americans, and it is now understood that these variants are in strong linkage disequilibrium with likely causal variants in neighboring *APOL1*. Subsequent genome-wide and candidate gene studies have suggested that *MYH9* common variants among others are also associated with chronic kidney disease and quantitative measures of kidney function in various populations. In a precision medicine setting, it is important to consider genetic effects or genetic associations that differ across racial/ethnic groups in delivering data relevant to disease risk or individual-level patient assessment. Kidney disease and quantitative trait-associated genetic variants have yet to be systematically characterized in multiple racial/ethnic groups. Therefore, to further characterize the prevalence of these genetic variants and their association with kidney related traits, we have genotyped 10 kidney disease or quantitative trait-associated single nucleotide polymorphisms (SNPs) (rs2900976, rs10505955, rs10502868, rs1243400, rs9305354, rs12917707, rs17319721, rs2467853, rs2032487, and rs4821480) in 14,998 participants from the population-based cross-sectional National Health and Nutrition Examination Surveys (NHANES) III and 1999-2002 as part of the Epidemiologic Architecture for Genes Linked to Environment (EAGLE) study. In this general adult population ascertained regardless of health status (6,293 non-Hispanic whites, 3,013 non-Hispanic blacks, and 3,542 Mexican Americans), we observed higher rates of chronic kidney disease among non-Hispanic blacks compared with the other groups as expected. We performed single SNP tests of association using linear regressions assuming an additive genetic model adjusted for age, sex, diastolic blood pressure, systolic blood pressure, and type 2 diabetes status for several outcomes including creatinine (urinary), creatinine (serum), albumin (urinary), eGFR, and albumin-to-urinary creatinine ratio (ACR). We also tested for associations between each SNP and chronic kidney disease and albuminuria using logistic regression. Surprisingly, none of the *MYH9* variants tested was associated with kidney diseases or traits in non-Hispanic blacks ($p>0.05$), perhaps attributable to the clinical heterogeneity of kidney disease in this population. Several associations were observed in each racial/ethnic group at $p<0.05$, but none were consistently associated in the same direction in all three groups. The lack of significant and consistent associations is most likely due to power highlighting the importance of the availability of large, diverse populations for genetic association studies of complex diseases and traits to inform precision medicine efforts in diverse patient populations.

1. Introduction

The kidney is an essential organ that excretes metabolic wastes from blood to maintain fluid homeostasis, osmoregulation, blood pressure, and electrolyte balance – key processes for survival [1]. The health risks and financial burden of poor kidney health are well-documented (e.g. [2]). Also well-documented are the higher prevalence and incidence of kidney disease among African Americans compared with other racial/ethnic groups in the United States [3,4]. This is a tremendous health disparity that exists even after accounting for socioeconomic status, as evidenced by reports that have evaluated varying degrees of kidney disease and have detected significant risk in African Americans compared to whites even when distinct methods are implemented and when income is taken into account [5,6]. Recent admixture studies in African-descent populations with focal segmental glomerulosclerosis [7], nondiabetic end-stage disease (ESRD)[8], and other kidney diseases have established a genetic basis that partially explains the observed racial/ethnic differences in the development and progression of these diseases [9].

Kidney disease is often symptom-free until it has significantly diminished the ability of the organ to function, and it is therefore crucial to identify genetic variants associated with biological indicators of kidney health. Kidney disease can be detected with biomarkers obtained through standardized blood tests that estimate renal function and by monitoring excretion of protein in the urine. Chronic kidney disease (CKD), estimated glomerular filtration rate (eGFR), albumin, and creatinine are clinical measures used to identify potential kidney failure. Numerous genetic variants have been implicated in studies of kidney disease and function [8,10-13]; however, not all of these variants have been evaluated in large, diverse population-based studies. To determine the utility of these variants for precision medicine settings, we asked the following: Do kidney trait-associated single nucleotide polymorphism (SNP) allele frequencies differ across racial/ethnic groups? Can kidney trait and disease associations be generalized across populations?

To answer these questions, we as the Epidemiologic Architecture for Genes Linked to Environment (EAGLE), a study site of the Population Architecture using Genomics and Epidemiology I (PAGE) study [14], accessed the National Health and Nutrition Examination Surveys to evaluate the associations between kidney-related traits and ESRD-associated genetic variants across multiple racial/ethnic groups.

2. Methods

2.1. *Study population*

The study population presented here is from the National Health and Nutrition Examination Survey (NHANES) conducted by the National Center for Health Statistics at the Centers for Disease Control and Prevention. NHANES are cross-sectional surveys of non-institutionalized Americans regardless of health status. Demographics and health data are collected via survey (self-identified), labs, and physical exams in the Mobile Examination Center by public health professionals. CDC collected biospecimens for DNA extraction from consenting participants between 1991 and 1994 (NHANES III), 1999-2000, and 2001-2002 (Continuous NHANES). All procedures were approved by the CDC Ethics Review Board and written informed consent was obtained from all participants. Because no identifying information was accessed by the investigators, Vanderbilt University's Institutional Review Board determined that this study met the criteria for a "non-human subjects" determination

Estimated glomerular filtration rate was calculated using the following equation: $175 \times (\text{standardized } S_{cr}^{-1.154}) \times (\text{age}^{-0.203}) \times (0.742 \text{ if female}) \times (1.212 \text{ if black})$, where S_{cr} is standardized serum creatinine. Albuminuria as a binary trait was defined as either 1) urinary albumin-to-urinary creatinine ratio (ACR) ≥ 30 mg/g or 2) sex-specific thresholds (urinary ACR ≥ 17 mg/g in men and ≥ 25 mg/g in women). Chronic kidney disease was defined as eGFR <60 ml/min or the presence of albuminuria. Participants were considered to have type 2 diabetes if they answered "yes" to "Ever been told you have sugar/diabetes?" and "Are you now taking insulin?" or if they had fasting blood glucose levels >126 mg/dL.

2.2. *SNP selection and genotyping*

As part of the PAGE I study [14], we as the EAGLE study site selected candidate gene and genome-wide association study (GWAS)-associated variants in late 2009 (Table 1). A total of 11

SNPs (rs2900976, rs10505955, rs10502868, rs1243400, rs9305354, rs12917707, rs17319721, rs2467853, rs2032487, rs4821480, and rs4821481) were targeted for genotyping as part of a custom 96-OPA on the Illumina BeadXpress. In addition to genotyping experimental NHANES samples, we genotyped blind duplicates provided by CDC and HapMap controls (n=360). *MYH9* rs4821481 was out of Hardy Weinberg Equilibrium in more than one NHANES III racial/ethnic group (at $p<0.001$) and was therefore dropped from subsequent analyses; all other SNPs passed quality control.

2.3. *Statistical methods*

All statistical tests were performed stratified by race/ethnicity. Race/ethnicity is self-reported in NHANES, which has been shown to be correlated with global genetic ancestry [15]. Single SNP tests of association were performed for each of the ten SNPs and the following quantitative trait outcomes among adults (17 years of age or older) using linear regression: creatinine (urinary), creatinine (serum), albumin (urinary), eGFR, and albumin-to-urinary creatinine ratio (ACR). Single SNP tests of association were also performed using logistic regression for albuminuria and chronic kidney disease. Non-normal quantitative trait distributions were natural log-transformed prior to analysis. All tests of association assumed an additive genetic model and were adjusted by age, sex, diastolic blood pressure, systolic blood pressure, and type 2 diabetes status. All analyses were performed unweighted using SAS v9.2 (SAS Institute, Cary, NC) and the Analytic Data Research by Email (ANDRE) portal of the CDC Research Data Center in Hyattsville, MD [16]. Results from quantitative trait tests of association were plotted using Synthesis-View [17,18].

3. Results

Study population characteristics are given in Table 2. Overall half of the adult participants were non-Hispanic white and female. Both non-Hispanic black and Mexican American participants were younger on average compared with non-Hispanic white participants. As expected based on the known epidemiology [2], the labs associated with kidney function were worse in non-Hispanic blacks compared with the other two groups. More cases of chronic kidney disease were identified among non-Hispanic black participants compared with the other two groups. Conversely, more cases of albuminuria were identified among Mexican American participants compared with the other two groups (Table 2).

The allele frequencies for the coded allele of each SNP are displayed in Figure 1 by race/ethnicity. Coded alleles for rs10502868, rs12917707, rs2032487, rs2467853, rs2900976, and rs4821480 were all more common in non-Hispanic blacks than non-Hispanic whites and Mexican Americans. Coded alleles for rs10505955, rs17319721, and rs9305354 were all more common in non-Hispanic whites than non-Hispanic blacks and Mexican Americans. Coded alleles for rs1243400 were more common in Mexican Americans than non-Hispanic blacks and non-Hispanic whites. Four of the SNPs characterized here (rs12917707, rs2032487, rs2467853, and rs4821480) are not included in the International HapMap Project Phase 3 [19] and therefore do not have

Table 1. SNPs selected for targeted genotyping in NHANES 1999-2002 and their previously reported associations. Abbreviations: beta (β); family-based association tests (FBAT); generalized estimating equations (GEE); not reported (NR); odds ratio (OR).

rs number (Coded allele)	**Nearest gene (Location)**	**Associated phenotype (Population)**	**Reported genetic effect (p-value)**	**PubMed ID**
rs2900976 (NR)	*DYSF-RPS20P10* (intergenic)	Albumin (Tuscans living in the Chianti region of Italy)	NR (1.4x10-6)	18464913
rs10505955 (G)	*BCAT1* (intronic)	Albumin (Korculans from Korcula, Croatia)	β = 0.10 ($9.5x10^{-6}$)	19260141
rs10502868 (G)	*SLC14A2* (intronic)	Albumin (Korculans from Korcula, Croatia)	β = -0.40 ($6.5x10^{-6}$)	19260141
rs1243400 (NR)	- (chromosome 10)	Albumin, urinary (European Americans from Framingham, MA)	NR ($4.8x10^{-6}$ based on FBAT)	17903292
rs9305354 (NR)	*LOC284825* (intergenic)	Albumin, urinary (European Americans from Framingham, MA)	NR ($8.4x10^{-6}$ based on GEE)	17903292
rs12917707 (G)	*UMOD* (5′ flanking)	Chronic kidney disease Glomerular filtration rate, estimated by serum creatinine (European-descent participants from multiple cohorts)	OR =1.25 ($2.3x10^{-12}$) β = 0.02 ($5.2x10^{-16}$)	19430482
rs17319721 (A)	*SHROOM3* (intronic)	Glomerular filtration rate, estimated by serum creatinine (European-descent participants from multiple cohorts)	β = -0.01 ($1.2x10^{-12}$)	19430482
rs2467853 (G)	*SPATA5L1* (intronic)	Glomerular filtration rate, estimated by serum creatinine (European-descent participants from multiple cohorts)	β = -0.01 ($6.2x10^{-14}$)	19430482
rs2032487 (C)	*MYH9* (intronic)	End-stage renal disease, non-diabetic (African Americans)	OR = 2.19 ($1.46x10^{-11}$, recessive genetic model)	18794854
rs4821480 (T)	*MYH9* (intronic)	End-stage renal disease, non-diabetic (African Americans)	OR = 2.29 ($7.31x10^{-11}$, recessive genetic model)	18794854
rs4821481 (T)	*MYH9* (intronic)	End-stage renal disease, non-diabetic (African Americans)	OR = 2.25 ($1.46x10^{-12}$, recessive genetic model)	18794854

reference allele frequency data available for comparison across populations. Of the remaining six SNPs, the majority of allele frequencies observed in NHANES were similar to those observed in HapMap populations with similar genetic ancestry [African Americans from Southwestern United States (ASW), European Americans from Utah (CEU), and Mexican Americans from Los Angeles, California (MEX)]. Of note is *SLC14A2* rs10502868 where the MEX allele frequency (2%) was significantly higher compared with the frequency estimated in Mexican Americans from NHANES (0.001%). Also, MEX allele frequencies are not available for *SHROOM3* rs17319721, common variant in Mexican Americans from NHANES (37%; Figure 1).

Table 2. Study population characteristics. Means (+/- standard deviation) given unless otherwise noted.

	Non-Hispanic whites	**Non-Hispanic blacks**	**Mexican Americans**
n	6,293	3,013	3,542
Female (%)	3,385 (53.8%)	1,652 (54.8%)	1,761 (49.7%)
Age, in years	53.24 (19.70)	44.08 (17.27)	44.00 (17.69)
ln(serum creatinine, mg/dL)	-0.10 (0.31)	0.003 (0.32)	-0.20 (0.34)
ln(urinary creatinine, mg/dL)	4.53 (0.75)	4.97 (0.65)	4.65 (0.73)
Albuminuria (%)	14/5944 (0.2%)	15/2779 (0.5%)	34/3377 (1.0%)
Urinary albumin-to-urinary creatinine ratio (ACR)	0.002 (0.05)	0.006 (0.07)	0.010 (0.10)
Urinary albumin, mg/mL	33.20 (221.08)	76.27 (454.86)	77.06 (583.25)
eGFR	50.78 (26.38)	73.91 (52.74)	48.24 (30.83)
CKD (%)	1734/5940 (29.2%)	1555/2796 (55.6%)	922/3378 (27.3%)

Test of association p-values and directions of genetic effect are displayed in Figure 2 for each SNP and trait across each population sample. Eight of the genotyped SNPs were associated (at $p<0.05$) with at least one trait in at least one population. Of these, three SNPs were limited to association with one trait in one population: 1) *DYSF-RPS20P10* rs2900976 was associated with natural log transformed creatinine in non-Hispanic blacks (β = -0.022), 2) *SHROOM3* rs17319721 was associated with natural log transformed creatinine in non-Hispanic whites (β = 0.011), and 3) *LOC284825* rs9305354 was associated with natural log transformed urinary creatinine in non-Hispanic blacks (β = -0.047). Three SNPs were associated with multiple traits in non-Hispanic whites: 1) *UMOD* rs12917707 was associated with natural log transformed creatinine and eGFR (β = 0.016 and 1.209, respectively), 2) *MYH9* rs4821480 was associated with natural log transformed creatinine, albumin-creatinine ratio, and CKD (β = 0.023, 0.180, and odds ratio = 1.305 and 95% confidence interval 1.053 – 1.617, respectively), and 3) *MYH9* rs2032487 was associated with natural log transformed creatinine, eGFR, albumin-creatinine ratio, and CKD (β = 0.030, 1.963, 0.196, and odds ratio = 1.340 and 95% confidence interval 1.076 – 1.669, respectively).

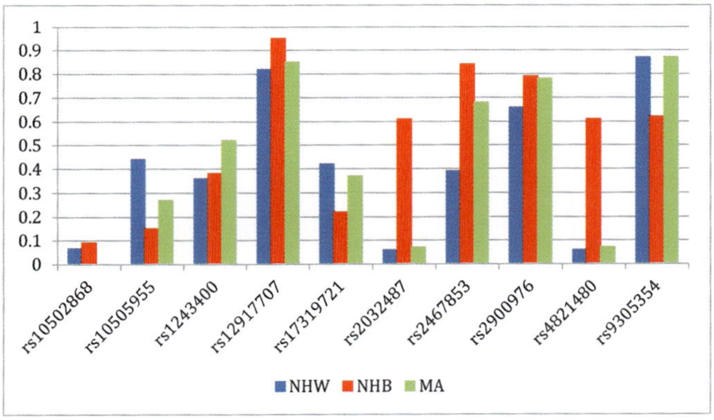

Figure 1. Coded allele frequency of kidney disease or trait-associated SNPs by race/ethnicity. Allele frequencies (y-axis) are given for each of the ten SNPs genotyped in NHANES (x-axis) for each race/ethnicity. Race/ethnicity is color-coded (blue for non-Hispanic whites, red for non-Hispanic blacks, and green for Mexican Americans). Allele frequencies displayed here were calculated based on NHANES III and NHANES 1999-2002 frequencies combined.

SNP rs1243400 on chromosome 10 was associated with CKD in non-Hispanic whites and non-Hispanic blacks, though in opposite directions of effect for the same coded allele (odds ratio = 1.182; 95% confidence interval = 1.064 – 1.313 and odds ratio = 0.851; 95% confidence interval 0.74 - 0.98, respectively). *SPATA5L1* rs2467853 was associated with several traits in all three populations, with natural log transformed creatinine in non-Hispanic whites and Mexican Americans (β = 0.016 and - 0.017, respectively), with eGFR in non-Hispanic whites and non-Hispanic blacks (β = 1.137 and 4.115, respectively), with natural log transformed urinary creatinine in non-Hispanic blacks (β = 0.050), with eGFR in Mexican Americans (β = 1.407), and with CKD in non-Hispanic whites (odds ratio = 1.135; 95% confidence interval 1.026 – 1.257).

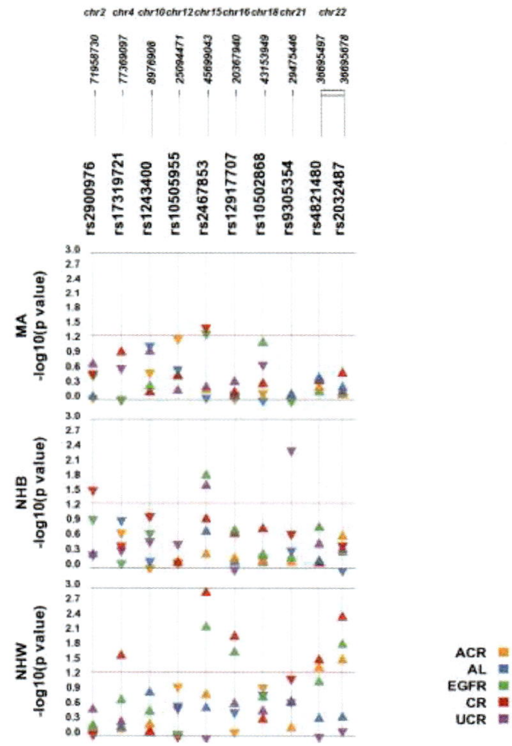

Figure 2. Results of tests of association are displayed using Synthesis-View by SNP, quantitative trait, and race/ethnicity. Plotted are the p-values (y-axis is –log of the p-value). The triangles denote the direction of the genetic effect. The red line is a p-value threshold of 0.05. Abbreviations: Albumin-creatinine ratio (ACR); albumin (AL), estimated Glomerular Filtration Rate (eGFR), serum creatinine (CR), urinary creatinine (uCR); non-Hispanic white (NHW), non-Hispanic black (NHB), Mexican American (MA).

4. Discussion

We tested ten kidney disease and trait-associated SNPs for an association with CKD and kidney traits in non-Hispanic whites, non-Hispanic blacks, and Mexican Americans ascertained regardless of health status for a national survey. As might be expected based on the SNP selection criteria, we observed eight associations with at least one trait in at least one population at $p<0.05$ in this diverse epidemiologic survey. No one SNPs was associated for the same trait or outcome across all three populations tested. However, we did observe that SNPs such as those in *MYH9* were associated with several traits and outcomes in a single population.

That the *MYH9* SNPs were associated with several traits/outcomes in non-Hispanic whites is not surprising, given the previous reports in the literature [20]. Surprising, however, is the lack of association of these SNPs in non-Hispanic blacks and Mexican Americans given the strong linkage disequilibrium between the *MYH9* SNPs and *APOL1* variants that are strongly associated with kidney disease in African Americans. In the present study, the three *MYH9* SNPs targeted for genotyping are in strong linkage disequilibrium with one another in all three racial/ethnic groups (r^2 ranging from 0.86 – 1.0). Reports have implicated *APOL1* as the driving cause of racial/ethnic disparity in kidney disease [21], though other function studies suggest *MYH9* remains relevant to kidney disease risk [22]. The lack of association may be attributable to the combination of heterogeneous kidney diseases among individuals in the Mexican American and non-Hispanic black samples.

Affordability and representation are among the ten things that must be addressed in order to achieve precision medicine [23]. Ideally, genetic variants selected for clinical genotyping are relevant to all populations tested, and therefore efficient in providing potentially healthcare-related data even at the individual patient level. To achieve this goal, it is crucial that variants selected for genotyping have relevancy to traits in multiple populations, not just European-descent individuals.

Arguably, precision medicine will also be more efficiently achieved with the addition and expansion of discovery studies that assess the impact of genetic variation in all populations and racial/ethnic backgrounds. The *MYH9-APOL1* variants, which have a much greater impact in individuals of African ancestry, are an example of precision medicine targets that will streamline the process of identifying patients at greater risk for developing kidney disease, and also identifying donor kidneys that are more likely to survive [24].

The present study had numerous weaknesses and strengths. Despite the overall large sample size of Genetic NHANES (n = 14,998), the present study was limited to adult participants with kidney traits available for analysis. As a result, the sample size of participants with CKD was modest resulting in lower statistical power to replicate known genetic associations. Additionally, not all participants with CKD will progress to end-stage renal disease requiring dialysis. An assessment of end-stage renal disease as opposed to the more general CKD may have allowed the detection and replication of genetic associations observed for *MYH9* in African Americans. Additionally, evaluating a specific subset of kidney disease (diabetic nephropathy, focal segmental glomerulosclerosis, HIV-associated nephropathy, etc.) would likely also yield more harmonized results. Likewise, protein in the urine (proteinuria or albuminuria, depending upon method of measurement) is the diagnostic hallmark and indicator of kidney dysfunction [1]; however, this was an uncommon condition in the present study resulting in low statistical power.

Despite these limitations, the present study had several strengths including the availability of multiple kidney disease and quantitative traits as well as three racial/ethnic groups from the United States. Large prospective studies and clinical-based repositories will be required to realize the vision of precision medicine particularly for health disparities across diverse populations. The current governmental support for focus on precision medicine heralds the necessity of studies such as the one presented herein. In a precision medicine setting, it is crucial to realize the different genetic effects and associations that can be observed in racial/ethnic populations. Kidney disease is a well-known example of health disparities with a strong, known genetic component influencing disease risk (*MYH9-APOL1*) and, while a genetic basis for the disparate rates of kidney diseases across racial/ethnic groups is widely recognized, research such as this is necessary to systematically characterize genome-wide and candidate gene identified genetic variants across diverse populations.

5. Acknowledgments

Genotyping for NHANES was supported in part by The Population Architecture Using Genomics and Epidemiology (PAGE) I study, which was funded by the National Human Genome Research Institute (NHGRI/NIH U01HG004798). EAGLE would like to thank Dr. Geraldine McQuillan and Jody McLean for their help in accessing the Genetic NHANES data. The Vanderbilt University Center for Human Genetics Research, Computational Genomics Core provided computational and/or analytical support for this work. The findings and conclusions in this report are those of the authors and do not necessarily represent the views of the National Institutes for Health or the Centers for Disease Control and Prevention. JNCB is supported by a PhRMA Informatics Fellowship.

References

1. Scott RP, Quaggin SE: **The cell biology of renal filtration.** *The Journal of Cell Biology* 2015, **209:** 199-210.

2. USRDS Coordinating Center. United States Renal Data System. 2015. 7-24-2015.

3. Cowie CC, Port FK, Wolfe RA, SAVAGE PJ, Moll PP, Hawthorne VM: **Disparities in Incidence of Diabetic End-Stage Renal Disease According to Race and Type of Diabetes.** *N Engl J Med* 1989, **321:** 1074-1079.

4. Williams WW, Pollak MR: **Health Disparities in Kidney Disease--Emerging Data from the Human Genome.** *N Engl J Med* 2013, **369:** 2260-2261.

5. Lipworth L, Mumma MT, Cavanaugh KL, Edwards TL, Ikizler TA, Tarone R *et al.*: **Incidence and Predictors of End Stage Renal Disease among Low-Income Blacks and Whites.** *PLoS ONE* 2012, **7:** e48407.

6. McClellan WM, Newsome BB, McClure LA, Howard G, Volkova N, Audhya P *et al.*: **Poverty and Racial Disparities in Kidney Disease: The REGARDS Study.** *American Journal of Nephrology* 2010, **32:** 38-46.

7. Kopp JB, Smith MW, Nelson GW, Johnson RC, Freedman BI, Bowden DW *et al.*: **MYH9 is a major-effect risk gene for focal segmental glomerulosclerosis.** *Nat Genet* 2008, **40:** 1175-1184.

8. Kao WHL, Klag MJ, Meoni LA, Reich D, Berthier-Schaad Y, Li M *et al.*: **MYH9 is associated with nondiabetic end-stage renal disease in African Americans.** *Nat Genet* 2008, **40:** 1185-1192.

9. Parsa A, Kao WHL, Xie D, Astor BC, Li M, Hsu Cy *et al.*: **APOL1 Risk Variants, Race, and Progression of Chronic Kidney Disease.** *N Engl J Med* 2013, **369:** 2183-2196.

10. Melzer D, Perry JRB, Hernandez D, Corsi AM, Stevens K, Rafferty I *et al.*: **A Genome-Wide Association Study Identifies Protein Quantitative Trait Loci (pQTLs).** *PLoS Genet* 2008, **4:** e1000072.

11. Zemunik T, Boban M, Lauc G, Jankovic S, Rotim K, Vatavuk Z *et al.*: **Genome-wide association study of biochemical traits in Korcula Island, Croatia.** *Croat Med J* 2009, **50:** 23-33.

12. Hwang SJ, Yang Q, Meigs JB, Pearce EN, Fox CS: **A genome-wide association for kidney function and endocrine-related traits in the NHLBI's Framingham Heart Study.** *BMC Med Genet* 2007, **8:** S10.

13. Kottgen A, Glazer NL, Dehghan A, Hwang SJ, Katz R, Li M *et al.*: **Multiple loci associated with indices of renal function and chronic kidney disease.** *Nat Genet* 2009, **41:** 712-717.

14. Matise TC, Ambite JL, Buyske S, Carlson CS, Cole SA, Crawford DC *et al.*: **The Next PAGE in Understanding Complex Traits: Design for the Analysis of Population Architecture Using Genetics and Epidemiology (PAGE) Study.** *American Journal of Epidemiology* 2011, **174:** 849-859.

15. Oetjens M, Brown-Gentry K, Goodloe R, Dilks HH, Crawford DC. **Population stratification in the context of diverse epidemiologic surveys** (in preparation).

16. Bush WS, Boston J, Pendergrass SA, Dumitrescu L, Goodloe R, Brown-Gentry K *et al.*: **Enabling high-throughput genotype-phenotype associations in the Epidemiology Architecture for Genes Linked to Environment (EAGLE) project as part of the Population Architecture using Genomics and Epidemiology (PAGE) study.** *Pac Symp Biocomput* 2013, **18:** 373-384.

17. Pendergrass S, Dudek SM, Roden DM, Crawford DC, Ritchie MD: **Visual integration of results from a large DNA biobank (BioVU) using synthesis-view.** *Pac Symp Biocomput* 2011, 265-275.

18. Pendergrass S, Dudek S, Crawford D, Ritchie M: **Synthesis-View: visualization and interpretation of SNP association results for multi-cohort, multi-phenotype data and meta-analysis.** *BioData Mining* 2010, **3:** 10.

19. **Integrating common and rare genetic variation in diverse human populations.** *Nature* 2010, **467:** 52-58.

20. Cooke JN, Bostrom MA, Hicks PJ, Ng MCY, Hellwege JN, Comeau ME *et al.*: **Polymorphisms in MYH9 are associated with diabetic nephropathy in European Americans.** *Nephrology Dialysis Transplantation* 2012, **27:** 1505-1511.

21. Quaggin SE, George AL: **Apolipoprotein L1 and the Genetic Basis for Racial Disparity in Chronic Kidney Disease.** *Journal of the American Society of Nephrology* 2011, **22:** 1955-1958.

22. Anderson BR, Howell DN, Soldano K, Garrett ME, Katsanis N, Telen MJ *et al.*: **In vivo Modeling Implicates APOL1 in Nephropathy: Evidence for Dominant Negative Effects and Epistasis under Anemic Stress.** *PLoS Genet* 2015, **11:** e1005349.

23. Kohane IS: **Ten things we have to do to achieve precision medicine.** *Science* 2015, **349:** 37-38.

24. Freedman BI, Julian BA: **Should kidney donors be genotyped for APOL1 risk alleles?.** *Kidney Int* 2015, **87:** 671-673.

PREDICTING SIGNIFICANCE OF UNKNOWN VARIANTS IN GLIAL TUMORS THROUGH SUB-CLASS ENRICHMENT

ALEX M. FICHTENHOLTZ†
afichtenholtz@foundationmedicine.com

NICHOLAS D. CAMARDA†
ndc9@duke.edu

ERIC K. NEUMANN†
eneumann@foundationmedicine.com

†Technology Innovation, Foundation Medicine Inc.
Cambridge, MA 02141, USA

Glial tumors have been heavily studied and sequenced, leading to scores of findings about altered genes. This explosion in knowledge has not been matched with clinical success, but efforts to understand the synergies between drivers of glial tumors may alleviate the situation. We present a novel molecular classification system that captures the combinatorial nature of relationships between alterations in these diseases. We use this classification to mine for enrichment of variants of unknown significance, and demonstrate a method for segregating unknown variants with functional importance from passengers and SNPs.

1. Introduction

Molecular diagnostics are increasing in importance to clinical oncology as the number of therapies targeting specific molecular alterations and pathways in cancer grows. These new drugs are accompanied by a shift in the tumor classification paradigm away from one based on histopathology to one centered on the molecular drivers of cancer. This has resulted in a proliferation of studies investigating the roles of various tumor suppressors and oncogenes in many types of tumors. The methods by which samples are interrogated have also shifted away from single gene hotspot tests to massively parallel, multiple marker integrated platforms[1]. In addition to genetic alterations in driver genes, gene expression changes, promoter mutations and methylation status have been implicated in cancer progression. This explosion in our ability to measure does not always lead to an increase in understanding, as we struggle to understand the relationships between the many markers we can now observe.

The genomic landscape of brain cancer, in particular tumors of glial origin, is a particularly difficult area for interpretation, for while large-scale sequencing studies of glioblastoma and lower grade astrocytomas have identified multiple targetable oncogenic driver alterations[2], these results have yet to meaningfully impact treatment decisions. Targeted therapies have had limited success in these tumor types, and multiple clinical trials have failed to show benefit with targeted tyrosine kinase inhibitors[3,4]. Existing molecular classification schemes are either based on gene expression[5], performed exclusively in lower grade gliomas like oligodendrogliomas[6], or focused on the 'main three' markers (*IDH* mutation, *TERT* promoter mutation and chromosome 1p/19q loss)[7]. We present a genomic classification for glial tumors based on comprehensive massively parallel sequencing of over 800 glial cancers of different grades, annotation of the resultant variant calls, and subsequent latent class analysis of the detected genetic alteration landscape.

In our classification, we take care to annotate alterations as either: 'known or likely' drivers of cancer, or 'variants of unknown significance' (VUS), as described in Methods and Materials. The motivation for this is to segregate genomic events that play a role in the tumor mechanism from innocuous alterations (i.e. SNPs and passenger mutations). While there are notable exceptions, in general it is somatic alterations that drive tumors[8]. We filter out suspected germ line variant calls using dbSNP[9], but this database is based on 1000 human genomes, meaning that the rarer SNPs will not be accounted for. 'Passenger' mutations can also confound the alteration landscape. Briefly speaking, passenger mutations are alterations accumulated during clonal expansion that do not currently confer any selective advantage onto the tumor[10]. The clinical significance of labeling these genomic events is paramount: if the oncologist elects to target a passenger mutation, the therapeutic regimen will presumably have no effect.

Any alteration that cannot be explicitly labeled as driver, passenger, or SNP is considered to be a VUS. If we look at the number of variant calls that are considered 'known and likely' versus the number we consider to be VUSs, we find that the VUSs account for the majority of what we detect. Thus we have a scenario in which there are modes of cancer unaccounted for, but this signal is mixed heavily with the noise of germ line variants and passenger mutations, making this excellent territory for variant prioritization approaches. Existing methods to evaluate the significance of unknown alterations in the context of disease range widely in their strategies. Sequence conservation based algorithms make the argument that mutations at heavily conserved residues in oncogenes and tumor suppressors are more likely to be deleterious[11]. Structural biology based methods stratify alterations based on their impact on protein folding energy and solubility[12].

We propose that a parallel method to assign significance to uncharacterized mutations is to align them to existing knowledge. Our classification of glioma samples, which is based on a small number of heavily mutated known cancer drivers, is statistically robust and well supported by existing literature. We use this classification as a reference point representing the current state of knowledge regarding the molecular landscape of glioma and examine how VUSs, which were not included in the definition of the molecular classes, distribute themselves along class partitions. We argue that genes that show skewed distributions towards a specific class participate in the mechanism driving tumors of that class in an as yet previously undescribed manner.

2. Materials and Methods

2.1 Comprehensive genomic profiling

All samples were submitted to a CLIA-certified, New York State and CAP-accredited laboratory (Foundation Medicine, Cambridge MA) for NGS-based genomic profiling, as previously described[13]. Extracted DNA was adaptor-ligated and capture was performed for all coding exons of 287 cancer-related and 47 introns of 19 genes frequently rearranged in cancer (Sup Table S1).

Captured libraries were sequenced to a median exon coverage depth of >500x, and resultant sequences were analyzed for base substitutions, insertions, deletions, copy number alterations (focal amplifications and homozygous deletions) and select gene fusions. Natural germline variants from the 1000 Genomes Project (dbSNP135)[8] were removed, and known confirmed somatic alterations deposited in the Catalog of Somatic Mutations in Cancer (COSMIC v62)[14] were highlighted as biologically significant (i.e. 'known and likely variants'). All inactivating events (i.e. truncations and deletions) in known tumor suppressor genes were also called as significant.

2.2 Data selection and filtering

Analysis was performed on a combined dataset of 847 glial tumor samples from four separate diseases: 76 oligodendrogliomas (BOD), 99 low and mid-grade astrocytomas (LGA), 101 anaplastic astrocytomas (AA), and 571 brain glioblastomas (GBM). For each gene that is altered in the data set, we count how many samples within that set carry an alteration in this gene, not taking into account alteration type (e.g. gene amplification, point mutation, etc.). If a sample has multiple alterations in a given gene (i.e. an indel and an amplification), it is still only counted once. The list of gene counts is then sorted, and all genes not altered in at least six samples are discarded in order to keep statistical power high. Because certain sets of genes tend to occur together in co-amplified vectors (e.g. *CDKN2A* and *CDKN2B*), we added a 'co-amplification' feature: if two genes occur in a data set, and their genetic co-ordinates are within 10 Mb of one another on the same chromosome, a separate variable that indicates their co-occurrence is added to the feature list. For instance, if co-mutations in *CDKN2A* and *CDKN2B* occur with sufficient frequency in a data set, there should be three features: the presence of a *CDKN2A* alteration only, the presence of a *CDKN2B* alteration only, and the presence of an alteration in both genes. This concept can be extended to three and four co-amplified genes. We estimate structure models while changing the 'class number' parameter r for each r=1,...,R, and then progressively increase r until the number of parameters to be estimated in the model exceed the number of samples (i.e. system is underdetermined). The full list of features used for clustering is given in Sup Table S2.

2.3 Latent class analysis

Latent class analysis was performed with the R package **poLCA**, version 1.4[15]. For each tumor sample i in our set, there are J manifest (observable) variables (genes), each of which can have one of two K_j outcomes (altered|not altered). The latent class model approximates the observed distribution of the manifest variables with a weighted sum of R cross-classification tables. The probability that a sample in class $r = 1,\ldots,R$ produces the kth outcome on the jth variable is represented by π_{jrk}, and the weight for a given class r is denoted by p_r. Thus, for each manifest variable within a class, $\sum_{k=1}^{K_j} \pi_{jrk} = 1$, and across all classes $\sum_r p_r = 1$. Denoting the observed value of the jth manifest variable for sample i having the kth outcome as Y_{ijk} (such that if gene j in sample i is mutated $Y_{ijk} = 1$, otherwise $Y_{ijk} = 0$), the probability that sample i in class r has any given set of mutations J is given by:

$$P(Y_i;\pi_r) = \prod_{j=1}^{J}\prod_{k=1}^{K_j}(\pi_{jrk})^{Y_{ijk}} \tag{1}$$

Across all classes 'r', this probability is given by:

$$P(Y_i|\pi,p) = \sum_{r=1}^{R} p_r \prod_{j=1}^{J}\prod_{k=1}^{K_j}(\pi_{jrk})^{Y_{ijk}} \tag{2}$$

The parameters estimated by **poLCA** are p_r and π_{jrk}. Denoting the total number of samples in the set as N, the latent class model is found by maximizing the log-likelihood function (3) with respect to p_r and π_{jrk} using expectation-maximization:

$$\ln L = \sum_{i=1}^{N} \ln \sum_{r=1}^{R} p_r \prod_{j=1}^{J}\prod_{k=1}^{K_j}(\pi_{jrk})^{Y_{ijk}} \tag{3}$$

The above equation reaches its maximum values with class partitions that best satisfy the conditional independence criterion (i.e. that manifest variables show conditional independence within a class). The number of classes in a model is a parameter adjusted by the user, and a larger value results in a higher log-likelihood score for the model. **poLCA** finds local maxima starting from initial values of p_r and π_{jrk}, thus to ensure the global maximum is found for each model we estimate the model 10,000 times with random initial parameter values each time, ultimately keeping the estimated model with the best log-likelihood score. To select the most probable of the many candidate latent structure models, we use a metric called the Akaike Information Criterion[16], defined as:

$$AIC = 2k - 2\ln L \tag{4}$$

where k is the number of estimated parameters specified by the model (a product of the number of features and the number of classes in the model), and L is the maximized value of the log-likelihood function we described earlier. Under certain conditions, an alternate information metric called the Bayesian Information Criterion[17] can be used. This is defined as:

$$BIC = -2\ln L + k\ln n \tag{5}$$

where n is the total number of sample instances being analyzed. Discussed in greater detail elsewhere[18], the BIC can be the appropriate metric for evaluating multiple models when the model space is dominated by a few major effects and contains likely nested models, which we believe to be the case for our combined dataset. Given a set of candidate models, the most appropriate model is the one that minimizes the AIC or BIC.

2.4 Feature selection approach

We performed an initial LCA on the dataset using a large number of features. For each feature, we calculated the initial entropy of the feature using Shannon's entropy formula[19], where p is the probability of seeing that feature in any given sample across the data set:

$$H(X) = -p \log_2 p \tag{6}$$

This allows us to compare the loss of entropy in each feature across classes as we increase the complexity of the models we fit the data to. We use entropy loss as a measure of how well the LCA partitions the alteration occurrences. If the probability of seeing a feature within a class is p_x and the probability of a given sample being a member of that class is given as p_y, then this entropy can be calculated across multiple classes using the conditional version of Shannon's formula[20]:

$$H(X|Y) = p_{x|y} \log_2 \frac{p_y}{p_{x|y}} \tag{7}$$

We calculate the entropy loss for each feature across all models to determine if that feature accounts for a significant portion of the entropy drop for the entire system. We then use that metric as a basis to set a lower threshold for the number of features to be included in the modeling and then performed a 'higher-resolution' LCA using this reduced feature set.

2.5 *Association of VUSs with known classes*

After computing the most likely classification for each sample using the high-resolution LCA based on a small number of known and likely features, we performed statistical testing of ***all*** of the alterations detected in the samples assigned to each class. We grouped alterations by gene, and separated those annotated as 'known and likely' from ones annotated as 'VUS.' For each feature in each class, we calculated the enrichment within the given class against all other classes of 'known and likely' alterations, 'VUS' alterations, and total alterations using a one-sided Fisher's Exact test. P-values obtained using Fisher's Exact are adjusted for multiple hypothesis testing using the Bonferroni formula[21]:

$$\alpha_{\text{adj}} = 1 - (1 - \alpha)^n \tag{8}$$

where α is the significance level of the result, and n is the total number of independent tests conducted, which in this case is 83 features × 5 tests each = 415 total tests. In the case where a gene's association with two classes simultaneously is tested, the exponent used for Bonferroni correction is 1245.

3. Results

Investigational LCA of the combined glial tumors set was performed using 83 features with up to eight classes being modeled. Analysis of entropic loss per feature during modeling reveals that the majority of information loss is accounted for by the 17 most frequently occurring features (84.1%, 84.1%, 70.8%, 70.2%, 70.2%, 69.2%, 69.1% for 2-class, 3-class, 4-class, 5-class, 6-class, 7-class, 8-class models, respectively; Figure 1). LCA was performed with these 17 features with the 6-class solution deemed most likely given BIC metrics (Table 1). Class 1 is dominated by alterations in *CDKN2A/B* (190/190 samples; p=1) and *EGFR* (161/190; p=0.84) (Figure 2). Class 2 also features alterations in *CDKN2A/B* (116/162; p=0.72), but instead of *EGFR* shows alterations in *NF1* (92/162; p=0.56). Class 3 showcases alterations in *EGFR* (64/114; p=0.56) and *CDK4/MDM2* (49/114; p=0.43). Class 4 is selective for alterations in *IDH1* (137/151; p=0.91), *TP53* (145/151; p=0.96) and *ATRX* (140/151; p=0.93). Class 5 is characterized by *IDH1* mutation (89/89; p=1) and alterations in *CIC* (31/89; p=0.35) and *TP53* (38/89; p=0.43). Class 6 is represented by alterations in *TP53* (111/141; p=0.79), *RB1* (55/141; p=0.39) and *PTEN* (68/141; p=0.48). The full model is given in Table 2.

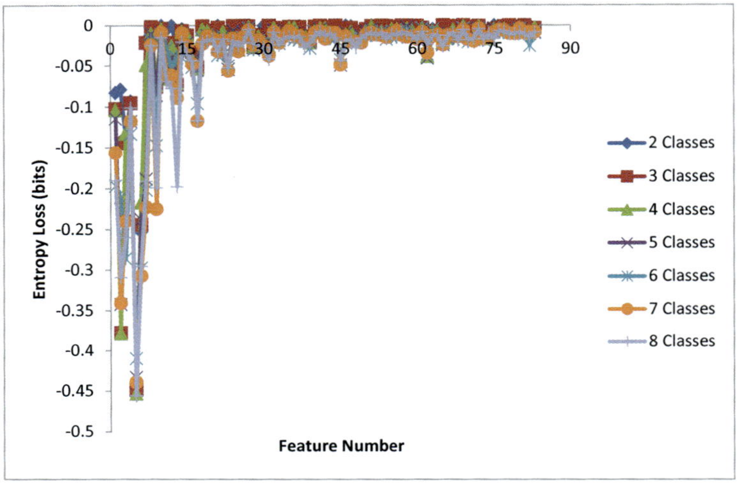

Figure 1. Information theoretic analysis of the clustering behavior of the 83 features most frequently altered in the glial tumor cohort. The bulk (~70%) of entropy loss achieved by class segregation is concentrated in the top 17 features.

Enrichment analysis of all alterations including VUSs for samples within a given class reveals notable contributions from *PTEN* (p=1.2e-8) in Class 1, *PTPN11* (p=0.018) in Class 2, and *NOTCH1* (p=0.018), *NOTCH4* (p=0.043), *ARID1A* (p=5.6e-4), and *SMARCA4* (p=0.09) in Class 4 (Table 3). The full list of enrichment results is listed in Supplemental Table S3.

Classes	LL	AIC	BIC	Parameters	Relative P (AIC)	Relative P (BIC)
1	-5370.53	10775.07	10855.68	17	0	1.20e-261
2	-4829.37	9728.731	9894.69	35	3.30e-146	5.74e-53
3	-4682.99	9471.985	9723.295	53	1.86e-90	9.48e-16
4	-4602.15	9346.296	9682.957	71	3.65e-63	5.44e-07
5	-4528.9	9235.796	9657.807	89	3.61e-39	0.16
6	-4466.37	9146.745	9654.107	107	7.84e-20	1
7	-4422.78	9095.558	9688.27	125	1.02e-08	3.82e-08
8	-4386.38	9058.758	9736.821	143	1	1.10e-18

Table 1. High resolution LCA solutions for the combined glial tumor data set, with class number parameter varying between 1 and 8. LL = log likelihood, AIC = Akaike Information Criterion, BIC = Bayesian Information Criterion. Relative probabilities for the models depend on AIC and BIC, and are calculated using the Akaike weight formula: $p_i = e^{-(\mathrm{AIC}_i - \mathrm{AIC}_{min})/2}$.

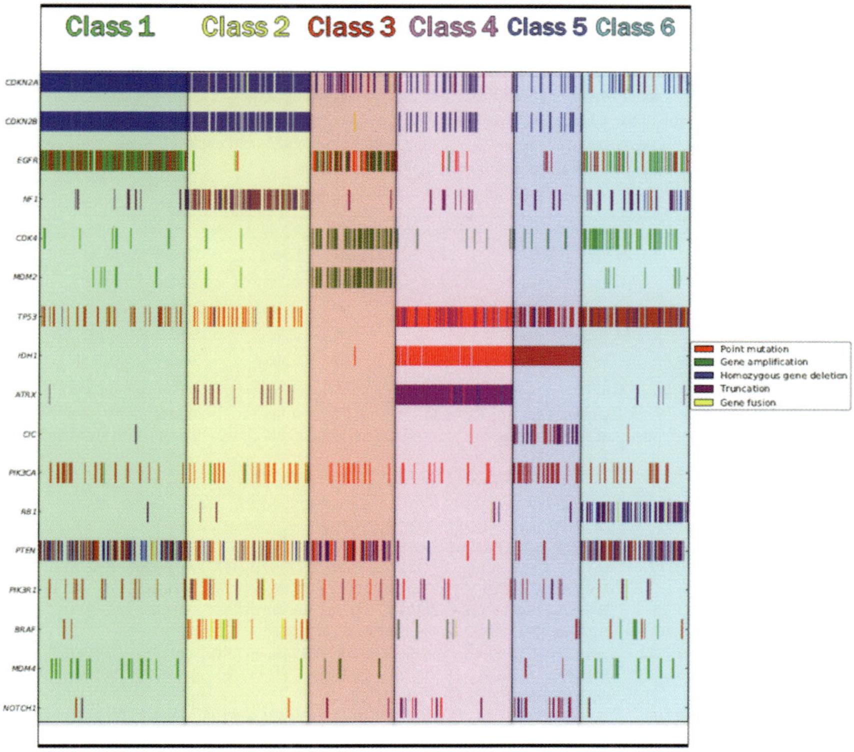

Figure 2. Latent model of the combined glial tumor cohort featuring six classes. The glial tumor cohort includes clinically confirmed cases of oligodendroglioma, low- and mixed-grade astrocytoma, anaplastic astrocytoma, and glioblastoma.

Feature	Class 1	Class 2	Class 3	Class 4	Class 5	Class 6
TP53	0.178	0.197	0	0.963	0.433	0.79
CDKN2A/B	1	0.716	0	0.142	0.106	0
EGFR	0.848	0.033	0.56	0.019	0.038	0.263
PTEN	0.44	0.29	0.37	0.028	0.036	0.483
IDH1	0	0	0.014	0.908	1	0
ATRX	0.005	0.079	0	0.929	0	0.033
NF1	0.047	0.559	0.036	0.049	0.062	0.167
PIK3CA	0.096	0.137	0.109	0.063	0.306	0.114
RB1	0.007	0.012	0	0.019	0.013	0.388
PIK3R1	0.069	0.101	0.041	0.044	0.087	0.045
CDKN2A	0	0.033	0.208	0.028	0	0.163
CDK4	0.025	0	0	0.051	0.049	0.228
CDK4/MDM2	0.007	0.012	0.425	0	0	0

BRAF	0.015	0.139	0	0.019	0.024	0.033
MDM4	0.077	0.012	0.03	0	0.026	0.088
NOTCH1	0.009	0.008	0.017	0.07	0.162	0.015
CIC	0.005	0	0	0.005	0.345	0.008

Table 2. Best LCA model of the brain glioblastoma data set according to BIC metrics presented in the form of a class-conditional probability table. Each class is a column, and the probability of observing a sample of a given class is shown at the top of the column. Each row is a feature used in the model, and each cell value is the probability of observing an alteration in that feature in a sample assigned to that class.

Gene	Class	Known/Likely	VUS	p-Value	Adj. p-Value
PTEN	1	88	15	2.9e-11	1.2e-8
PTPN11	2	13	4	4.3e-5	1.8e-2
ARID2	4/5	6	14	3.1e-5	3.8e-2
SMARCA4	4/5	11	18	1.3e-5	1.5e-2
ARID1A	5	9	12	1.4e-6	5.6e-4
NOTCH1	5	13	12	4.5e-5	1.8e-2
NOTCH4	5	1	12	1.0e-5	4.3e-2

Table 3. Enrichment of selected genes in pre-defined molecular sub-classes taking VUSs into account.

Gene	Class-Specific Variants of Unknown Significance
PTEN	R15del, D24ins, I28T, P30L, Y46C, R47S, N48K, N48S, H61Q, Y65N, V85I, V175M, Y188D, D326Y, D326H
PTPN11	D61A, F285S, T411M, A461T
ARID1A	A139T, A165_166insA, G278D, P289A, G352R, G712A, P870L, P989L, V1082G, K1124N, P1209L, G1222E, G1274R, P1275S, V1834M
ARID2	G16E, S129C, V138A, R231K, S365L, V450I, A858T, V1040I, A1107V, E1249del, P1395S, V1503L, D1703E, N1796K
SMARCA4	G10E, P16S, P24S, G53E, G206S, P324S, L754F, P913S, Q1104R, Q1104R, V1016L, R1157G, G1159R, D1177N, A1186T, R1192H, R1203L, E1287K, T1358I, E1512K
NOTCH1	C344G, F357S, P422S, P447S, C461Y, C467Y, G519S, A958V, R1114H, P1287C, C1467Y, R1664K, G1704E, S2030F, A2035_L2048del, G2169E, V2249M
NOTCH4	E33K, G47R, R113K, G216S, S312F, G337D, P631L, G642N, A647G, R807H, V1457M, G1510S, G1701E

Table 4. Class-specific VUSs in genes found to associate significantly with pre-defined classes.

4. Discussion

Oligodendroglioma (OD), low- and mixed-grade astrocytoma (LGA), anaplastic astrocytoma (AA), and brain glioblastoma (GBM) tumors are all thought to originate from glial precursor cells, and are difficult to segregate using histopathology[22]. Multiple studies have grouped various subsets of these four diseases together for the purposes of molecular profiling[23]. We sequenced 847 of these tumors on a comprehensive massively parallel sequencing platform capable of detecting alterations in several hundred cancer related genes, examined the genetic alteration landscape from an information theoretic perspective using unsupervised classification, and found the genes that contribute most to classification. After stratifying the dataset into the likely molecular classes based on known and likely somatic alterations, we examined the distribution of variants of unknown significance (VUS) within each class. The highest enrichments of VUSs within a given class as measured by p-value are considered in more detail.

4.1 Analysis of alteration landscape

Recognizing the similarities in histology between each disease type, we felt it was prudent to combine the tumor sets into one large superset consisting of 847 total samples. The initial analysis was done using 83 features. Analysis of the information content in every model showed that the entropic loss is concentrated in the 17 most frequently occurring features, which we consider sufficient to classify a large subset of tumors belonging to any of these four disease types. These features are alterations in *TP53*, *CDKN2A/B*, *EGFR*, *PTEN*, *IDH1*, *ATRX*, *NF1*, *PIK3CA*, *RB1*, *PIK3R1*, *CDKN2A*, *CDK4*, *CDK4/MDM2*, *BRAF*, *MDM4*, *NOTCH1*, and *CIC*. LCA considering just these 17 features yields a well-delineated six-way mixture model. Class 1 is driven by disruption to the cell-cycle mechanism with every sample in this class showing a co-deletion of *CDKN2A* and *CDKN2B*, as well as a hyper-activated signal transduction network with *EGFR* alterations being found in most samples in this class. Though not unique to this class, *PTEN* alterations are found in nearly half of all tumors in class 1. Class 2 seems to be related to Class 1, with both featuring *CDKN2A/B* co-deletion, though with *NF1* alterations in place of *EGFR*. Interestingly, alterations in these genes between these two classes show a nearly completely mutually exclusive relationship, suggesting that the alterations have a similar functional effect. Previous subtyping studies detected both of these classes, but were unable to segregate them, and did not allude to any mutual exclusivity between *EGFR* and *NF1*[4-6]. These two classes are significantly associated with poor prognosis, high tumor grade, and positive response to aggressive therapy.

Classes 4 and 5 are also related, and have also been discovered and confirmed to be clinically significant by prior studies. Class 4 features alterations in *IDH1*, *ATRX* and *TP53*, and is associated with positive prognosis and lower tumor grade. This class is typically found in astrocytoma, though we have seen this profile in samples assigned to both lower (e.g. oligodendroglioma) and higher grade (e.g. glioblastoma) histology categories. Class 5 shows *IDH1* mutation without *ATRX* alteration. Previous studies suggest that this class should also be associated with heterozygous deletion of chromosome arms 1p and 19q; manual inspection of sequencing data confirmed this to be the case. Class 5 should be associated with the best prognosis according to previous work, and is found primarily in oligodendrogliomas.

Class 3, which features alteration of *EGFR* in conjunction with co-amplification of *CDK4* and *MDM2* in a variation of the cell-cycle/signal transduction perturbation theme found in classes 1 and 2, and Class 6, which is driven by the combination of *TP53* and *RB1* alteration represent novel classes that have not been detected in previous cohorts, though a prior investigation of anaplastic oligodendrogliomas found simultaneous disruption of the *RB1* and *TP53* pathways in 9/20 tumors[24].

4.2 *Association of VUSs with known classes*

Learning the classification of glial tumors by known and likely somatic variants leads us to comparing the distribution of VUSs in genes across those classifications. Class 1, which is driven by the combination of *CDKN2A/B* deletion and *EGFR* alteration, shows preference for mutations in *PTEN*, and in particular, VUSs. *PTEN* alterations are known to de-regulate the PI3K pathway, and have been found to synergize with activation of *EGFR* to promote increased tumor growth[25]. The unknown variants we detected in *PTEN* within class 1 tend to cluster towards the front end of the phosphatase tensin-type domain found between residues 14 and 185 in this protein. Inspection of the COSMIC databases shows multiple confirmed somatic events around this same area, with several direct overlaps between a somatic event from COSMIC and a VUS at the residue of interest.

Class 2 features a combination between *CDKN2A/B* co-deletion and *NF1* alteration, and is enriched for *PTPN11* alterations, including four previously undetected VUSs. *PTPN11* encodes a cytoplasmic protein tyrosine phosphatase that promotes the activation of the Ras/MAPK pathway, and mutations in this protein lead to it being constitutively active[26]. The association of alterations in these genes with alterations in *NF1*, which encodes a negative regulator of the GTPase HRAS, suggests that tumors of this type rely on perturbation of the Ras and PI3K pathways for their tumorigenicity, and that targeting proteins in these pathways may be a viable treatment paradigm. Comparing the unknown variants we found to the collection of known somatic *PTPN11* variants from COSMIC reveals that there are confirmed somatic variants at D61 and at A461.

While classes 1 and 2 are variants of the cell-cycle/signal transduction co-perturbation mode, class 5 is driven primarily by *IDH1* mutation and some mutations in *CIC*. Previous studies claim that this class should be associated with mutation in *NOTCH1* and *FUBP1*, along with heterozygous deletion of chromosome arms 1p and 19q. These have all been confirmed in this dataset. This class features enrichment of both *NOTCH1* and *NOTCH4* at a statistically significant level. These proteins are transmembrane receptors known for their role in patterning during embryogenesis. Alterations in *NOTCH* genes have been found in a large number of cancers, and are thought to contribute to oncogenesis by promoting angiogenesis and modulating the EMT[27]. Less frequently discussed is the formation of complexes featuring *NOTCH* proteins and histone de-methylases. Histone modification is known to be essential for transcription of *NOTCH*, and may be related to the fact that alterations in the SWI/SNF proteins *ARID1A* and *SMARCA4* are also heavily enriched for in this class. Furthermore, *ARID2*, another known chromatin remodeler associates with classes 4 and 5. The role of the chromatin signaling network in this glioma subtype has not been previously described.

4.3 *Extending glial tumor genomics knowledge*

An obvious question regarding this approach is whether the ‘known’ LCA classes we use to explore VUS distributions are clinically relevant, or merely statistically significant. Several previous studies have discovered these classes independently, and verified their clinical significance. A study of low-grade gliomas in the TCGA data set uncovered both a *TP53/IDH1/ATRX* class (Class 4) as well as a class driven primarily by *IDH1* alterations with no concomitant *ATRX* events, with the occasional alteration in *CIC* (Class 5)[5]. These classes were shown to be clinically significant in terms of event-free and overall survival, age at diagnosis, primary tumor site, and molecular phenotype (i.e. methylation, gene expression, protein expression). Another study of GBM patients in the TCGA data found classes driven by *CDKN2A/B* co-deletion and *NF1/EGFR* alterations (Classes 1 and 2), though it did not recognize the mutual exclusivity between *NF1* and *EGFR*[4]. The researchers found this class to be significantly associated with poorer prognosis and an older age at diagnosis. The combination of *RB1* and *TP53* alterations that typify Class 2 has been found in pre-clinical studies to generate sarcomas in mesenchymal stem cells, and is generally known to be a transforming combination in cell lines[28]. Class 3, which features *EGFR* alterations in combination with *CDK4/MDM2* co-amplifications, is a novel result that has not been described before in glial tumor literature. Our results have added novel associations of tumor suppressor and oncogene alterations with these classes.

Additionally, we can learn from the distribution of variants across the length of the gene. For genes significantly associated with a given class, the pileup of alterations has a conspicuous pattern. A superb example is *SMARCA4*. This SWI/SNF protein is enriched in classes 4 and 5, and the specific VUSs associated with these classes are distributed non-randomly. Specifically, 10 of 20 unknown variants are found in a 200 aa region between AA1100 and AA1300. PROSITE[29] suggests that this area is home to a helicase domain. Further investigation should be conducted as to the role of this domain in this protein in *IDH1* driven brain tumors, as disrupting this mechanism represents a potential avenue to treating these cancers.

5. Acknowledgements

We wish to acknowledge Alex Poliakov, Bryan Lewis, Marylin Matz, and the rest of the Paradigm4 team for their technical support and enhancement of the SciDB platform used during this project.

References

1. Heuckmann JM, et al. *Annals of Oncology* **26**: 1830-7 (2015).
2. Brennan CW, et al. *Cell* **155**: 462-77 (2013).
3. Razis E, et al. *Clinical Cancer Research* **15**: 6258-66 (2009).
4. Galanis E, et al. *Clinical Cancer Research* **19**: 4816-23 (2013).
5. Verhaak RG, et al. *Cancer Cell* **17**: 98-110 (2010).
6. Cancer Genome Atlas Research Network, et al. *New England Journal of Medicine* **372**: 2481-98 (2015).
7. Eckel-Passow JE, et al. *New England Journal of Medicine* **372**: 2499-508 (2015).
8. Armitage P, et al. *British Journal of Cancer* **8**: 1-12 (1954).

[9]. Sherry ST, et al. *Nucleic Acids Research* **29**: 308-11 (2001).
[10]. Greaves M, et al. *Nature* **481**: 306-13 (2012).
[11]. Ng PC, et al. *Nucleic Acids Research* **31**: 3812-4 (2003).
[12]. Adzhubei I, et al. *Nature Methods* **7**: 248-9 (2010).
[13]. Frampton GM, et al. *Nature Biotechnology* **31**: 1023-31 (2013).
[14]. Forbes SA, et al. *Nucleic Acids Research* **43**: D805-11 (2014).
[15]. Linzer DA, et al. *Journal of Statistical Software* **42**: 1-29 (2011).
[16]. Akaike, H. *IEEE Transactions on Automatic Control* **19**: 716-23 (1974).
[17]. Schwarz GE. *Annals of Statistics* **6**: 461-4 (1978).
[18]. Burnham KP, et al. *Sociological Methods & Research* **33**: 261-304 (2004).
[19]. Shannon, CE. *Bell System Technical Journal* **27**: 379-423 (1948).
[20]. Cover, TM, et al. *Elements of Information Theory (1st Ed.)* New York: Wiley (1991).
[21]. Bland JM, et al. *British Medical Journal* **310**: 170 (1995).
[22]. Maher EA, et al. *Genes and Development* **15**: 1311-33 (2001).
[23]. Vigneswaran, et al. *Annals of Translational Medicine* **3**: 95-108 (2015).
[24].Watanabe T, et al. *Journal of Neuropathology and Experimental Neurology* **60**: 1181-9. (2001).
[25]. Pires MM, et al. *Cancer Biology and Therapy* **14**: 246-53 (2013).
[26]. Bentires-Alj M, et al. *Cancer Research* **64**: 8816-20 (2004).
[27]. Allenspach EJ, et al. *Cancer Biology and Therapy* **1**: 466-76 (2002).
[28]. Rubio R, et al. *Oncogene* **32**: 4970-80 (2013).
[29]. Sigrist CJA, et al. *Nucleic Acids Research* **41**: D344-7 (2012).

Supplementary material is hosted at http://files.fm/u/ozghtas/

DO CANCER CLINICAL TRIAL POPULATIONS TRULY REPRESENT CANCER PATIENTS? A COMPARISON OF OPEN CLINICAL TRIALS TO THE CANCER GENOME ATLAS

NOPHAR GEIFMAN

Institute for Computational Health Sciences, University of California, San Francisco
Mission Hall, 550 16th Street
San Francisco, CA 94158-2549, USA
Email: Nophar.Geifman@ucsf.edu

ATUL J. BUTTE

Institute for Computational Health Sciences, University of California, San Francisco
Mission Hall, 550 16th Street
San Francisco, CA 94158-2549, USA
Email: Atul.Butte@ucsf.edu

Open clinical trial data offer many opportunities for the scientific community to independently verify published results, evaluate new hypotheses and conduct meta-analyses. These data provide a springboard for scientific advances in precision medicine but the question arises as to how representative clinical trials data are of cancer patients overall. Here we present the integrative analysis of data from several cancer clinical trials and compare these to patient-level data from The Cancer Genome Atlas (TCGA). Comparison of cancer type-specific survival rates reveals that these are overall lower in trial subjects. This effect, at least to some extent, can be explained by the more advanced stages of cancer of trial subjects. This analysis also reveals that for stage IV cancer, colorectal cancer patients have a better chance of survival than breast cancer patients. On the other hand, for all other stages, breast cancer patients have better survival than colorectal cancer patients. Comparison of survival in different stages of disease between the two datasets reveals that subjects with stage IV cancer from the trials dataset have a lower chance of survival than matching stage IV subjects from TCGA. One likely explanation for this observation is that stage IV trial subjects have lower survival rates since their cancer is less likely to respond to treatment. To conclude, we present here a newly available clinical trials dataset which allowed for the integration of patient-level data from many cancer clinical trials. Our comprehensive analysis reveals that cancer-related clinical trials are not representative of general cancer patient populations, mostly due to their focus on the more advanced stages of the disease. These and other limitations of clinical trials data should, perhaps, be taken into consideration in medical research and in the field of precision medicine.

1. Introduction

Approximately 30,000 clinical trials are conducted each year across the globe and various market and regulatory forces are driving initiatives to publicly share patient-level data from these trials [1-3]. With the advancement of science and betterment of the human condition in mind, there are several purported benefits for the sharing of clinical trial data [4-6]. Sharing these data offers the opportunities for the scientific community to independently verify published results. Lack of availability of original research data is a known, significant, barrier against reproducibility. Availability of the data may provide opportunities to evaluate new hypotheses that were not originally formulated in the studies, either by extending the analysis of data from a clinical trial or by combining data from different trials.

The availability of clinical data from different trials makes such data an attractive source for systemic research and meta-analysis [7]. Examining disease-related patterns by meta-analysis can help gain better understanding of disease-related characteristics and lead to new discoveries and insights. Combining data from multiple clinical studies and evaluating the same disease with various outcome measures could help leverage an improvement in efficacy by suggesting possible combination of treatments. Additionally, these data may be used to identify and define subgroups of subjects who respond better to a specific treatment. The plethora of raw, individual-level clinical data should provide a real springboard for scientific advances in precision medicine and development of new techniques in clinical informatics [8-10].

Due to the complexity of the different types of cancer, and the difficulties in selecting the right therapeutic approaches, increasing efforts are being dedicated towards improving cancer care via precision medicine [10, 11]. While most of the focus in this field is on the genetics and molecular characteristics of the cancer and increasing a drug efficacy based on the properties of a given tumor, other forms of clinical data can be useful in advancing precision medicine in cancer.

Recently, several pharmaceutical clinical trial data sharing platforms such as the Project Data Sphere [1, 2] and the *clinicalstudydatarequest.com* site (developed by GlaxoSmithKline) have emerged, making raw data from clinical trials available for research. Another, well-established source for cancer-related clinical data is The Cancer Genome Atlas (TCGA), which aims to comprehensively characterize and analyze many cancer types and makes its data freely available for research [12]. While the TCGA's focus lies in the genetics of cancer, establishing a large database of cancer genome sequences and aberrations, it also holds clinical data such as patient survival, treatment and demographics.

While pharma-released data is increasingly becoming available for research, to our knowledge there are very few works that utilized these data sources. Here we present the integration of patient-level data from many cancer clinical trials, present the potential of these data and systematically evaluate whether, in the field of cancer, trial data can usefully represent patient populations.

2. Methods

2.1. *Clinical trial data from the Project Data Sphere portal*

Clinical trial data was obtained from the Project Data Sphere portal (projectdatasphere.org) which stores, shares and allows analysis of patient level phase III cancer trial data [2]. The database currently holds data from the comparator arms of 48 cancer clinical trials. Subjects assigned to the comparator arms of cancer clinical trials usually receive standard of care treatment. Following registration, submission and approval of a research proposal, the data is made available for either download or analysis on an online SAS platform (Figure 1).

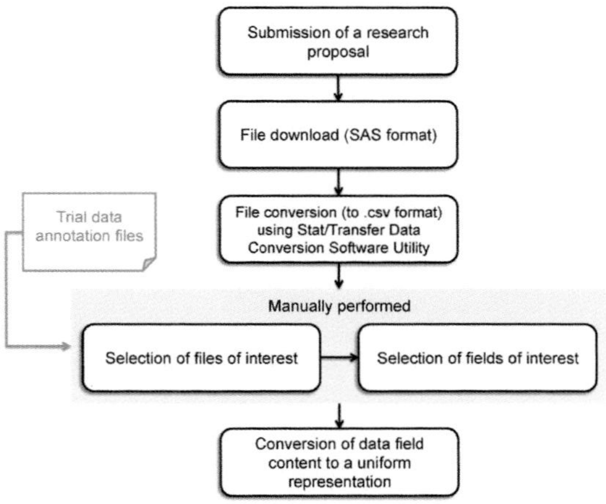

Fig. 1. Clinical trial data processing workflow.

Survival, demographic and treatment data for 10 cancer clinical trials (the first made available on the site which matched cancer types in the TCGA and for which annotation files were available) were downloaded from the Project Data Sphere database and used for the analyses presented here (Table 1).

Table 1. Clinical trial data downloaded from the Project Data Sphere portal.

	Cancer type	Study sponsor	No. of subjects*	ClinicalTrial.gov ID
1	Breast	Pfizer	166	NCT00373113
2	Colorectal	Astra Zeneca	690	NCT00384176
3	Prostate	Astra Zeneca	635	NCT00626548
4	Prostate	Astra Zeneca	248	NCT00672282
5	Prostate	Astra Zeneca	550	NCT00673205
6	Prostate	Centocor	45	NCT00385827

7	Prostate	Celgene	226	NCT00988208
8	Prostate	Pfizer	278	NCT00676650
9	Prostate	Sanofi	371	NCT00417079
10	Pancreas	Sanofi	156	NCT00417209

The majority of subjects have stage IV cancer. Approximately half (n=1439) of the prostate cancer subjects from the Astra Zeneca trials do not have a specific stage assigned. All breast cancer subjects are female; all prostate cancer subjects are male. In colorectal cancer, there are 290 females and 400 males. In the pancreatic cancer dataset, there are 64 females and 92 males. * The number of subjects for which survival data was available.

2.2. *The Cancer Genome Atlas dataset*

Data from the TCGA was selected to represent the general cancer subject population, representing a wide range of cancer stages, patient ages, ethnic groups and treatments. For each of the four cancers evaluated in this study (breast, colorectal, prostate and pancreatic), survival, treatment and demographic data were obtained from the TCGA database, April 27, 2015. These cancer types were selected due to their matching to cancer types available in our clinical trial dataset (Table 2). Only subjects for which survival data was available were included. Subcategories of cancer stages were converted to a more inclusive stage, for example: stages IIa, IIb, IIc were converted to stage II.

Table 2. The Cancer Genome Atlas dataset.

Cancer type	No. of subjects*	Gender (%)	Stage (%)
Breast	1018	Male - 12 (1.2) Female - 1006 (98.8)	I - 179 (17.6) II - 574 (56.4) III - 239 (23.5) IV - 21 (2.1) NA - 5 (0.5)
Colorectal	355	Male - 190 (53.5) Female - 165 (46.5)	I - 55 (16.74) II - 136 (38.8) III - 107 (28.4) IV - 48 (14.1) NA - 9 (2)
Prostate	492	Male - 492 (100) Female - NA	II - 337 (68.5) III - 48 (9.8) IV - 4 (0.8) NA - 103 (20.9)
Pancreas	171	Male - 94 (55) Female - 77 (45)	I - 18 (10.5) II - 142 (83) III - 5 (2.9) IV - 4 (2.3) NA - 2 (1.2)

* The number of subjects for which survival data was available.

2.2.1. *Assigning cancer stage for the prostate cancer subset*

For the prostate cancer data subset, the cancer stage was not available for any of the subjects. We therefore assigned each subject with a cancer stage based on the tumor and metastasis status for which data was available. If the assigned metastasis status is M1, stage IV is assigned to that subject (regardless of tumor status). For metastasis status M0 and tumor status T1b, T1c, T1 and T2, stage II was assigned. For metastasis status M0 and tumor status T3, stage III was assigned and for metastasis status M0 and tumor status T4, stage IV was assigned. There were no other combinations of tumor and metastasis status in the data.

2.3. *Survival analyses*

Kaplan-Meier analyses were conducted using the 'Survival' package (version 2.38-3) in R [13]. For calculation of the significance for the difference between sample sets, the log-rank test was used. For each analysis (by cancer type, cancer stage, gender etc.) subjects for which data was missing (from either dataset) were excluded.

3. Results

3.1. *Clinical trial subjects show lower survival in comparison to TCGA subjects*

We first compared the survival of subjects between the four different cancer types (breast, colorectal, prostate and pancreatic) in each of the datasets. The survival of subjects from the clinical trial dataset are overall significantly ($p<0.0001$) lower in comparison to matching cancer types from the TCGA (Figure 2). This observation can logically be explained by the generally more progressive cancer stage of subjects included in the trial dataset, in comparison to the TCGA. One interesting observation however, is the differences in the order (from best to worst survival) of cancers in the survival plots, illustrated in Figure 2A and 2B. In trial subjects, breast cancer shows lower survival than colorectal while in TCGA subjects, breast cancer has a higher survival than colorectal cancer. In both datasets, the differences in survival between breast and colorectal cancer subjects are significant ($p<0.005$). Comparison of the demographics of trial and TCGA subjects reveals that age distribution is similar in both datasets other than for the "over 75 years" age group (Figure 3A). For breast, colorectal and pancreatic cancers, the TCGA has more subjects over the age of 75 than the clinical trials; indicating that, clinical trials tend to select fewer older subjects. Gender distribution is similar in both datasets (Figure 3B) and the majority of subjects in both datasets are Caucasian (Figure 3C).

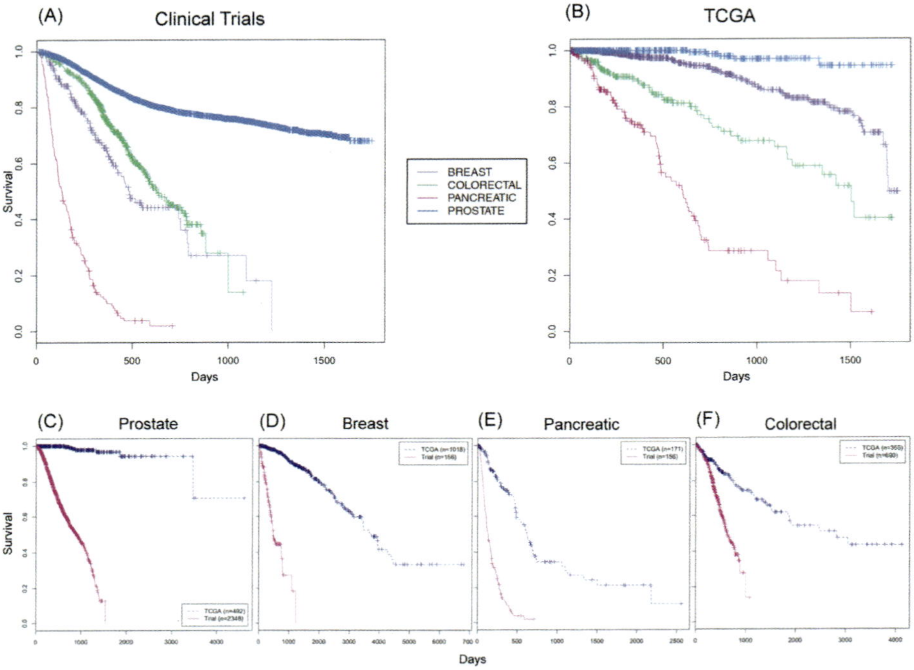

Fig. 2. Cancer patient survival in clinical trials and TCGA.

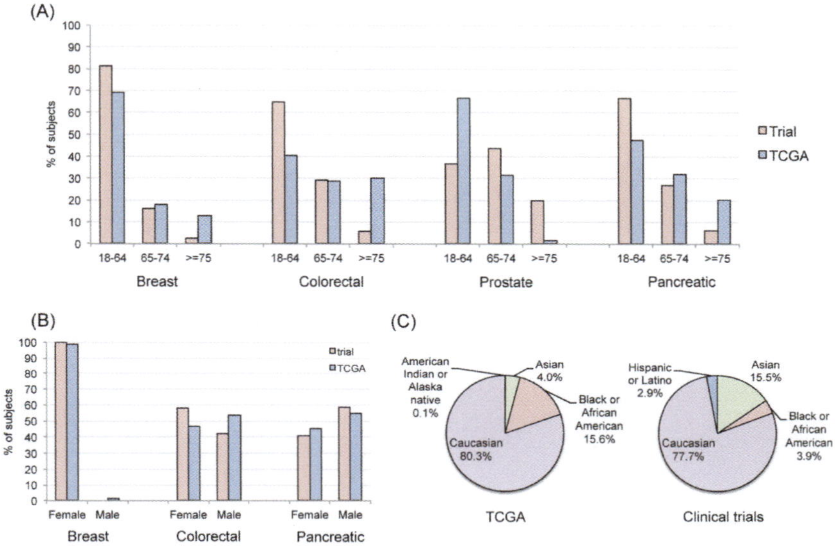

Fig. 3. Demographic variables in clinical trials and TCGA. (A) Age distribution in each cancer subset. (B) Gender distribution in each cancer subset (prostate cancer was excluded since subjects are always male). (C) Ethnic distribution in each of the datasets.

3.2. *Gender differences in clinical trial data but not TCGA*

We next compared the survival of subjects from trials to those from the TCGA on the basis of gender. When comparing subjects from all cancer types (breast, colorectal, prostate and pancreatic), significant differences in survival were found between males and females in the trial dataset (p=3.37e-12) but not in the TCGA (Figure 4). To evaluate whether it is the gender-specific cancer types (such as breast or prostate) which are driving this difference, we next compared the survival between genders in each dataset but this time limiting the analysis to only the pancreatic and colorectal cancer types which have a similar distribution of males and females in both datasets. In this instance, no significant gender differences in survival were found in either of the datasets.

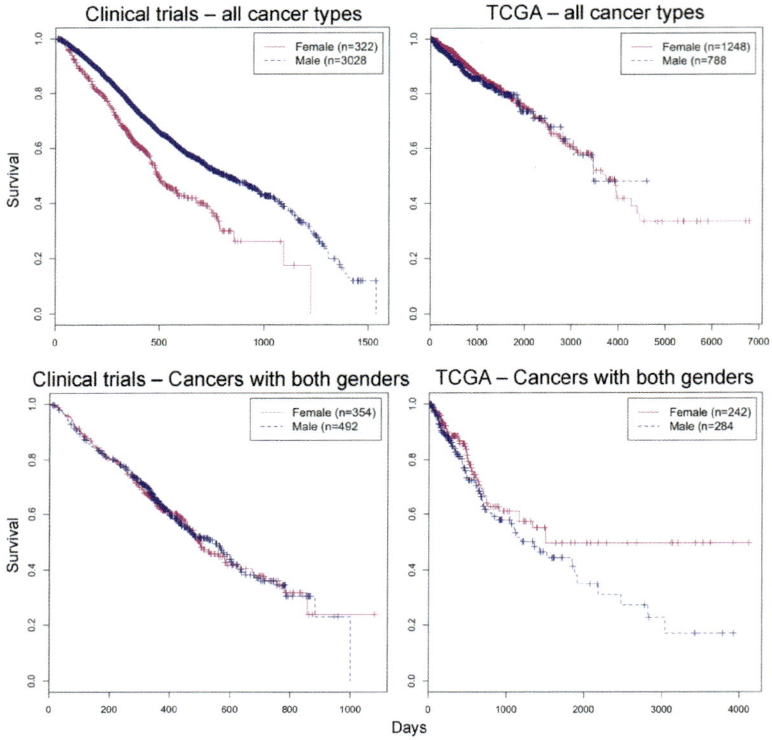

Fig. 4. Gender differences in survival in clinical trials and TCGA.

3.3. *The effect of the cancer stage*

In order to evaluate the differences in survival based on patients' cancer stages in the TCGA and trial datasets, for each cancer type all available stages from the TCGA were compared to stage IV subjects from the trial dataset (Figure 5). The trial data was limited to only stage IV (or advanced) subjects since that was the only available stage for most of the cancer types (breast, colorectal and pancreatic) in this dataset. In all four cancer types, stage IV subjects from clinical trials showed lower survival rates than subjects with other stages of cancer from the TCGA. In addition, in all four cancer types, subjects with stage IV cancer from the trials dataset showed lower survival than stage IV subjects from TCGA; significantly in breast cancer ($p<0.005$), though for colorectal cancer this was not significant ($p=0.764$) and for pancreatic and prostate cancer ($p<0.05$) the TCGA sample size of stage IV subjects was very small, see Table 2).

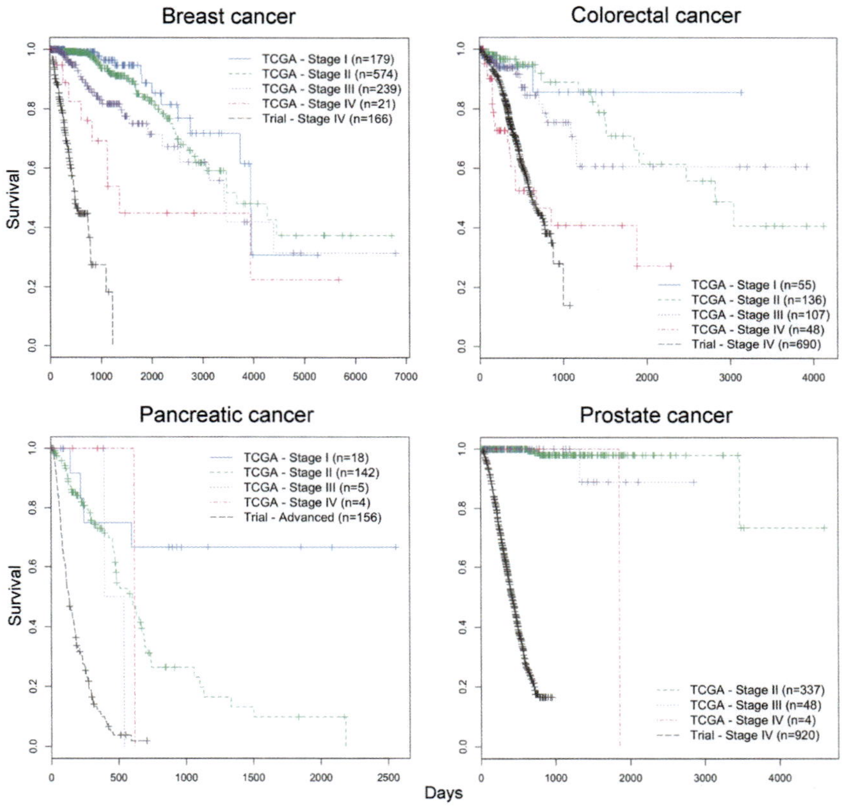

Fig. 5. Survival of different cancer stages in TCGA and clinical trials.

4. Discussion

Recently, clinical trial data sharing platforms such as the Project Data Sphere, the *clinicalstudydatarequest.com* site (developed by GlaxoSmithKline) and ImmPort [14]; have been making raw data from clinical trials available for investigation. These emerging, rich datasets offer many opportunities for research and the advancement of precision medicine. To evaluate whether cancer-related pharma-released trial data is representative and comparable to cancer patient populations, we present here the comparison of cancer-related patient-level data from a newly open clinical trial dataset to The Cancer Genome Atlas.

Our first analysis, comparing survival of different cancer types between the two datasets, revealed that clinical trial breast cancer subjects show lower survival than clinical trial colorectal cancer subjects; while in the TCGA dataset, breast cancer has a higher survival than colorectal cancer. Overall, lower survival of trial subjects in comparison to TCGA subjects can be explained by the generally more advanced stage of cancer in trial subjects. However, differences in cancer stages cannot explain the differences between the datasets in survival of breast and colorectal cancer. In both these cancers in the trial dataset, all subjects are stage IV and in both those cancers in the TCGA datasets, subjects have a mix of stages with no significant differences between the distribution of cancer stages in breast and colorectal subjects.

Based on this observation, it can be deduced that for stage IV cancer, colorectal cancer patients have a better chance of survival than breast cancer patients. On the other hand, for all other stages, breast cancer patients have better survival than colorectal cancer patients. Generally speaking, trial subjects have a worse prognosis; this is unlikely because they are in a trial, but rather because of how they are selected for trials. It should be pointed out that this observation is made based on a single breast cancer trial and a single colorectal cancer trial. Therefore, it is possible that other factors, such as practices of the company conducting the trial, differences in tumor subtypes or the population's characteristics may contribute to the differences in survival between these cancer types. Further analysis, which would include data from additional breast and colorectal cancer trials, could further elucidate this observation.

Comparison of survival by gender revealed that survival differences between genders exist in trial data, but not in the TCGA dataset. This is most likely driven by the stage of the cancer rather than the gender of the subjects. More specifically, by the lower survival of breast cancer subjects in the trial dataset in comparison to that of prostate cancer subjects. Since in the trial data, all breast cancer subjects are stage IV while the prostate cancer subjects are of a mix of stages (I, II, II and IV) it could be the cancer stage which is the cause for the difference. This gender difference is not seen in the TCGA dataset, probably because both the breast and prostate cancer subjects have a similar distribution of a mix of stages (with very few stage IV subjects).

Subjects with stage IV breast, pancreatic and prostate cancer from the trials dataset showed lower survival than matching stage IV subjects from TCGA. One possible explanation for this is that subjects included in clinical trials tend to have undergone several lines of therapy and only because those have failed, have these subjects enrolled in a trial. On the other hand, many of the subjects in the TCGA dataset are facing their first line of treatment. Therefore, the colorectal cancer subset is more comparable, since inclusion criteria for the colorectal cancer trial was that subjects have not undergone previous treatment for colorectal cancer and that the study's treatment is the first line of treatment. For this cancer type, no significant differences were found between the survival of trial and TCGA subjects; further indicating that stage IV trial subjects have lower survival rates since their cancer is less likely to respond to treatment.

One of the major shortcomings our analysis is that while we were able to compare the two datasets based on cancer type, cancer stage or gender, one factor that was not taken into consideration is the treatment given to the subjects. Matching data from TCGA to the that of a respective clinical trial dataset while taking into consideration all possible variables including treatment (type and course) is virtually impossible, likely to leave a very small number of subjects if any at all. For example, in the TCGA there are only 4 stage IV colorectal cancer patients who were treated with FOLFOX and would be comparable to the colorectal trial subjects. On the other hand, there is indication that there are few treatment-associated differences in the survival or long-term outcomes of early or localized prostate cancer subjects [15, 16]. Therefore, it could be argued that the survival differences shown here between TCGA and trial subjects in prostate cancer may be caused, at least in part, by something other than differences in treatment. As more data becomes available, integrative analyses taking treatment arms into account can be carried out and meta-analysis can be conducted across many clinical trials.

The clinical trials dataset we describe here holds a lot of potential for research. However, while getting hold of the data was a fast and relatively straightforward process, other than when using the somewhat restrictive online tools, the data were far from readily usable. One of the major issues we encountered was that for the available studies, data schemas are not uniform and table and variable names were lacking annotation which had to be manually extracted from long annotation files. Since considerable manual effort needed to be invested in the processing of these data, only ten trials (of the 48 available) were downloaded and used as proof of concept for the work presented here. Future work will include adding data from all trials available on the Project Data Sphere site. In addition, as the trend of opening clinical trials data to the research community grows, further clinical trial datasets can be included. While the results of the work described here indicate that clinical trial data does not accurately represent subject populations (for the cancer types evaluated here), this dataset can still be very useful. The availability of clinical data from different trials makes such data an attractive source for systemic research and meta-analysis. For example, one possible use, given a large enough number of studies pertaining to the same disease, would be to conduct a meta-analysis and compare the outcomes of different treatments. Examining disease-related patterns by meta-analysis can help gain better insight into disease-related characteristics and assist in finding new discoveries and insights. Moreover, by combining data from multiple clinical studies we can leverage the improvement in efficacy by possible combination of treatments.

5. Conclusions

We present here the analyses of data from a newly available pharma-released clinical trial dataset and its comparison to data from The Cancer Genome Atlas. Our comprehensive analysis reveals that clinical trials, in the field of cancer, are not representative of cancer patient populations, probably due largely to their focus on advanced stages. However, while recognizing that these data have their limitations, they hold great potential for advancing medical research in the field of precision medicine. Future research design should include consideration of the representativeness of the cancer patient population.

6. Acknowledgments

The authors would like to thank Dr. Marina Sirota and Dr. Dvir Aran for helpful discussion and useful suggestions.

7. Disclaimer

This publication is based on research using information obtained from www.projectdatasphere.org, which is maintained by Project Data Sphere, LLC. Neither Project Data Sphere, LLC nor the owner(s) of any information from the web site have contributed to, approved or are in any way responsible for the contents of this publication.

References

1. A. K. Green, et al., The Project Data Sphere Initiative: Accelerating Cancer Research by Sharing Data. *Oncologist* (2015).
2. K. Hede, Project data sphere to make cancer clinical trial data publicly available. *J Natl Cancer Inst* **105** 16 (2013).
3. P. Nisen and F. Rockhold, Access to patient-level data from GlaxoSmithKline clinical trials. *N Engl J Med* **369** 5 (2013).
4. H. M. Krumholz and E. D. Peterson, Open access to clinical trials data. *JAMA* **312** 10 (2014).
5. B. Lo, Sharing clinical trial data: maximizing benefits, minimizing risk. *JAMA* **313** 8 (2015).
6. M. Rosenblatt, S. H. Jain and M. Cahill, Sharing of clinical trial data: benefits, risks, and uniform principles. *Ann Intern Med* **162** 4 (2015).
7. J. S. Ross, R. Lehman and C. P. Gross, The importance of clinical trial data sharing: toward more open science. *Circ Cardiovasc Qual Outcomes* **5** 2 (2012).
8. in *Toward Precision Medicine: Building a Knowledge Network for Biomedical Research and a New Taxonomy of Disease*. 2011: Washington (DC).
9. N. V. Chawla and D. A. Davis, Bringing big data to personalized healthcare: a patient-centered framework. *J Gen Intern Med* **28 Suppl 3** (2013).
10. A. R. Shaikh, et al., Collaborative biomedicine in the age of big data: the case of cancer. *J Med Internet Res* **16** 4 (2014).
11. L. A. Garraway, J. Verweij and K. V. Ballman, Precision oncology: an overview. *J Clin Oncol* **31** 15 (2013).
12. N. Cancer Genome Atlas Research, et al., The Cancer Genome Atlas Pan-Cancer analysis project. *Nat Genet* **45** 10 (2013).
13. T. T. A Package for Survival Analysis in S. version 2.38. (2015); Available from: http://CRAN.R-project.org/package=survival.
14. S. Bhattacharya, et al., ImmPort: disseminating data to the public for the future of immunology. *Immunol Res* **58** 2-3 (2014).
15. M. R. Cooperberg, J. M. Broering and P. R. Carroll, Time trends and local variation in primary treatment of localized prostate cancer. *J Clin Oncol* **28** 7 (2010).
16. L. Holmberg, et al., A randomized trial comparing radical prostatectomy with watchful waiting in early prostate cancer. *N Engl J Med* **347** 11 (2002).

PATIENT-SPECIFIC DATA FUSION FOR CANCER STRATIFICATION AND PERSONALISED TREATMENT

VLADIMIR GLIGORIJEVIĆ, NOËL MALOD-DOGNIN AND NATAŠA PRŽULJ*

Department of Computing, Imperial College London,
London, SW7 2AZ, United Kingdom
**E-mail: natasha@imperial.ac.uk*
http://www.doc.ic.ac.uk/~natasha

According to Cancer Research UK, cancer is a leading cause of death accounting for more than one in four of all deaths in 2011. The recent advances in experimental technologies in cancer research have resulted in the accumulation of large amounts of patient-specific datasets, which provide complementary information on the same cancer type. We introduce a versatile data fusion (integration) framework that can effectively integrate somatic mutation data, molecular interactions and drug chemical data to address three key challenges in cancer research: stratification of patients into groups having different clinical outcomes, prediction of driver genes whose mutations trigger the onset and development of cancers, and repurposing of drugs treating particular cancer patient groups. Our new framework is based on graph-regularised non-negative matrix tri-factorization, a machine learning technique for co-clustering heterogeneous datasets. We apply our framework on ovarian cancer data to simultaneously cluster patients, genes and drugs by utilising *all* datasets. We demonstrate superior performance of our method over the state-of-the-art method, Network-based Stratification, in identifying three patient subgroups that have significant differences in survival outcomes and that are in good agreement with other clinical data. Also, we identify potential new driver genes that we obtain by analysing the gene clusters enriched in known drivers of ovarian cancer progression. We validated the top scoring genes identified as new drivers through database search and biomedical literature curation. Finally, we identify potential candidate drugs for repurposing that could be used in treatment of the identified patient subgroups by targeting their mutated gene products. We validated a large percentage of our drug-target predictions by using other databases and through literature curation.

Keywords: Data fusion; Tumor stratification; Drug repurposing; Cancer driver genes; Non-negative Matrix Factorization.

1. Introduction

Cancer is a leading cause of morbidity worldwide. It is a complex genetic disease in which the genomes of normal cells accumulate somatic mutations and other alterations that are eventually perturbing vital cellular functions. Recent advances in DNA sequencing technologies have enabled identification of somatic mutations across tumor genomes and exomes of individual patients[1,2]. These somatic mutations provide a new and rich source of data for addressing many challenges in cancer research, such as indentifying driver genes (i.e., genes whose mutations lead progression of oncogenesis), stratifying patients into biologically meaningful classes with different clinical outcomes and creating new opportunities for development of successful personalized treatment strategies[3]. Cancer is also a highly heterogeneous disease with large genetic diversity even between tumors of the same cancer type. Namely, two clinically identical tumors rarely have a large set of common mutated genes. Moreover, very few genes are frequently mutated across tumor samples. This makes the use of somatic mutations for identification of driver genes, as well as for patient stratification into subtypes, much harder[1,4,5]. However, despite this genetic diversity between tumor samples, the perturbed pathways are often similar[1]. Therefore, integration of somatic mutations with other genomic data, such as with molecular networks that contain pathways, is a promising direction for addressing these problems.

Development of computational methodologies that can efficiently integrate genome-scale molecular information and address the above mentioned challenges in cancer research is of foremost importance. A majority of previous studies do not utilise data on somatic mutations, but instead, they are mainly based on mRNA expression and methylation data. Because of noisiness of these data, the patient stratification studies for cancer types often do not produce patient subgroups that agree well with any clinical or survival data[6]. Therefore, a recent study proposed the use of somatic mutation data in combination with biological networks as a new source of information for tumor stratification[5]. However, the proposed methodology cannot account for additional data types (e.g., drug data) and cannot be used for identifying novel driver genes, nor for predicting a personalised therapy. Moreover, previous data integration methods can only be used for either cancer patient stratification[5], driver gene prediction[7] or drug repurposing[8].

Here, we present a versatile patient-specific data integration (fusion) methodology capable of: 1) uncovering patient subgroups (stratification) with prognostic survival outcome, 2) predicting novel driver genes and 3) repurposing drugs, i.e., predicting new candidate drugs for targeting mutated gene products in individual patients and that can be used in treatment of identified patient subgroups. To our knowledge, this is the first method that can address all three challenges simultaneously. Our methodology is based on Non-negative Matrix Tri-Factorization (NMTF) technique, initially proposed for dimensionality reduction and co-clustering problems in machine learning[9]. It approximates (factorises) a high-dimensional data matrix, representing relations between two data types, as a product of three non-negative, low-dimensional matrices[9]. The clustering interpretation of low-dimensional matrices and their previously established relatedness to the k-means clustering has enabled the use of NMTF in co-clustering problems[10,11]. Recently, there has been a significant development in the use of NMTF in data fusion because of its ability to extend to any number of interrelated data types by *simultaneously* decomposing their relation matrices. This has provided us with a valuable framework for fusion (integration) of any number and type of interrelated heterogeneous datasets[12,13]. NMTF has demonstrated a great potential in addressing various biological problems, such as disease association prediction[12], disease gene discovery[14], protein-protein interaction prediction[15] and gene function prediction[16].

We use NMTF to integrate somatic mutation profile (SMP) data of serous ovarian cancer patients from TCGA[4] with molecular networks (MNs) from BioGRID[17] and KEGG[18], drug-target interaction (DTI) and drug chemical similarity (DCS) data from DrugBank[19] (detailed in Sec. 2.3). We perform consensus clustering by using NMTF to simultaneously cluster patients, genes and drugs based on the evidence from *all* datasets. First, we stratify patients into three groups that we assess by using clinical data. We show significant difference in survival outcomes between these groups, as well as a good agreement with other clinical data.

Second, from clusters of genes, we identify those enriched in known driver mutations; we postulate genes strongly related to known driver genes in these clusters as potential drivers genes, i.e., genes responsible for ovarian cancer progression. Finally, we use the matrix completion property of NMTF to predict new drug-target relations and to identify new drug candidates that could be used for repurposing and treatment of identified ovarian cancer patient groups. Furthermore, we evaluate the influence of all combinations of datasets onto the accuracy of drug-target predictions by performing a 5-fold cross validation. We shown that the highest accuracy is achieved when all datasets are taken into account, proving the utility of integrating all considered datasets (detailed in Sec. 3).

2. Methods

2.1. *Patient-specific data fusion framework*

We consider there different datasets: patients, genes and drugs. Patients and genes are related to each other by somatic mutation profiles (SMPs), constructed for n_1 patients over n_2 genes and encoded in high-dimensional relation matrix, $\mathbf{R}_{12}^{n_1 \times n_2}$. Its entries are binary values, with $\mathbf{R}_{12}[p][g] = 1$ if gene g is found to be mutated in patient p, and zero otherwise. Genes and drugs are related to each other according to drug-target interactions (DTIs). DTIs between n_2 genes and n_3 drugs are encoded in relation matrix, $\mathbf{R}_{23}^{n_2 \times n_3}$. Its entries are also binary values, with $\mathbf{R}_{23}[g][d] = 1$, if the product of gene g is a target of drug d and zero otherwise. See Sec. 2.3 and Fig. 1 for details of construction of the relation matrices and for an illustration of these datasets.

We use NMTF to simultaneously decompose both relation matrices into a product of three non-negative low-dimensional matrices as follows: $\mathbf{R}_{12} \approx \mathbf{G}_1\mathbf{H}_{12}\mathbf{G}_2^T$. and $\mathbf{R}_{23} \approx \mathbf{G}_2\mathbf{H}_{23}\mathbf{G}_3^T$. The low dimensional matrices can be obtained by solving the following optimisation problem: $\min_{\mathbf{G}_i \geq 0, 1 \leq i \leq 3} J = \min_{\mathbf{G}_i \geq 0, 1 \leq i \leq 3} \left(\| \mathbf{R}_{12} - \mathbf{G}_1\mathbf{H}_{12}\mathbf{G}_2^T \|_F^2 + \| \mathbf{R}_{23} - \mathbf{G}_2\mathbf{H}_{23}\mathbf{G}_3^T \|_F^2 \right)$, where F denotes Frobenius norm and J is the objective function. The non-negativity constraints imposed on $\mathbf{G}_i$ matrices for $1 \leq i \leq 3$ provide easier interpretation of their values in the clustering assignment. Many of the data types are characterised by additional, internal connectivity structure represented by graphs (networks). In this study, genes are connected by molecular networks (MNs), while drugs are connected based on the similarity of their chemical structures, i.e., drug chemical similarity (DCS) network (illustrated in Fig. 1). We incorporate these networks (MN and DSC) into the above objective function by adding two regularisation terms to constrain the construction of $\mathbf{G}_2$ and $\mathbf{G}_3$ matrices. This approach is also known also as *Graph-regularized NMTF* (or GNMTF)[20]. Namely, the aim is to enforce two interacting genes to belong to the same cluster (similarly with drugs) and a violation of these constrains results in penalties to our objective function. Hence, the final objective function has the following form:

$$\min_{\mathbf{G}_i \geq 0, 1 \leq i \leq 3} J = \min_{\mathbf{G}_i \geq 0, 1 \leq i \leq 3} [\, \| \mathbf{R}_{12} - \mathbf{G}_1\mathbf{H}_{12}\mathbf{G}_2^T \|_F^2 + \| \mathbf{R}_{23} - \mathbf{G}_2\mathbf{H}_{23}\mathbf{G}_3^T \|_F^2 + tr(\mathbf{G}_2^T\mathbf{L}_2\mathbf{G}_2) + tr(\mathbf{G}_3^T\mathbf{L}_3\mathbf{G}_3)] \quad (1)$$

where, tr denotes the trace of a matrix, and $\mathbf{L}_2$ and $\mathbf{L}_3$ are graph Laplacians of MN and DCS networks, respectively.

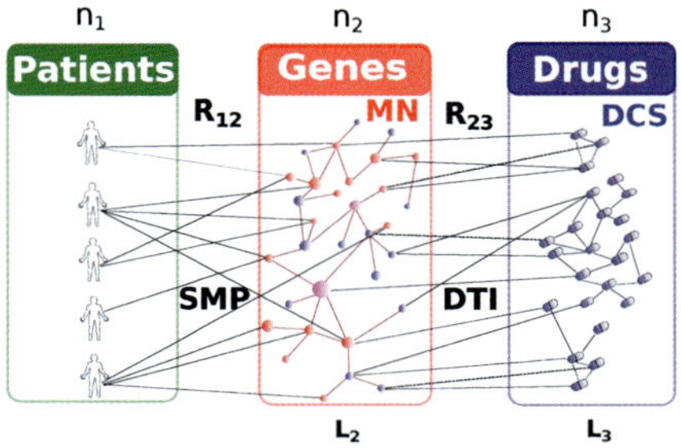

Fig. 1. Schematic illustration of datasets used in this study. Three types of objects (nodes) are represented: n_1 *ovarian cancer patients* (in green), n_2 *genes* (in red) and n_3 *drugs* (in blue). Somatic mutation profiles (SMP) for ovarian cancer patients are represented by patient-gene links denoting assignment of mutated genes (in red) to each individual patient. These connections are encoded into relation matrix, $\mathbf{R}_{12}$. Genes are connected by a molecular network (MN) obtained by merging three different interaction data types (see Sec. 2.3). Also, MN is the union of three types of genes: *mutated genes* coming from patients' SMPs (in red), *driver genes* (in pink) coming from TCGA database and *normal genes* (in blue) coming from other databases used for construction of networks (details in Sec. 2.3). Connections in this network, MN, are represented by Laplacian matrix, $\mathbf{L}_2$. Links between genes (i.e., their protein products, that are drug targets) and drugs are drug-target interactions (DTIs) and are represented by relation matrix, $\mathbf{R}_{23}$. Links between drugs are represented by drug chemical similarity (DCS) network (details in Sec. 2.3). Connections in this network, DCS, are represented by Laplacian matrix, $\mathbf{L}_3$.

The key idea of our GNMTF-based data fusion approach is in sharing low-dimensional matrix $\mathbf{G}_2$ whilst simultaneously learning from (i.e., decomposing) relation matrices, $\mathbf{R}_{12}$ and $\mathbf{R}_{23}$. Such decomposition accounts for collective influence of all data sets (along with molecular and chemical constraints effectively integrated within the same framework) onto the resulting clustering of patients, genes and drugs. This approach corresponds to the *intermediate* data fusion in which the structure of the data is preserved during the model inference. Such an approach has been shown to result in the best accuracy among all data fusion approaches[12].

Minimisation of the objective function, J, is done by *multiplicative update rules* used to compute all low-dimensional matrices; under these multiplicative rules, the objective function is non-increasing[11]. The minimisation starts by randomly initialising $\mathbf{G}_i$ matrices for $1 \le i \le 3$ and then iteratively updating their values until the convergence criterion is reached. In all our runs, we use *Random Acol* initialisation strategy[21] and the convergence criterion is reached when $\frac{|J_{n+1}-J_n|}{|J_n|} < 10^{-5}$. The multiplicative update rules, their derivation and proof of convergence can be found in Wang *et al.*[11].

Co-clustering of patients, genes and drugs. Matrices $\mathbf{G}_1^{n_1 \times k_1}$, $\mathbf{G}_2^{n_2 \times k_2}$ and $\mathbf{G}_3^{n_3 \times k_3}$ from Equation 1 above are *cluster membership indicator* matrices for patients, genes and drugs, respectively; based on their entries, n_1 patients are assigned to k_1 patient clusters, n_2 genes are assigned to k_2 gene clusters and n_3 drugs are assigned to k_3 drug clusters, respectively. In particular, following the *hard clustering* procedure of Brunet *et al.*[22], matrix $\mathbf{G}_1^{n_1 \times k_1}$, with rows representing patients and columns representing clusters, is used to place patient p into cluster k if $\mathbf{G}_1[p][k]$ is the largest entry in row p. This assignment procedure results in the binary connectivity matrix for patients, $\mathbf{C}_1^{n_1 \times n_1}$, with entry $\mathbf{C}_1[p_1][p_2] = 1$ if patients p_1 and p_2 belong to the same cluster and $\mathbf{C}_1[p_1][p_2] = 0$ otherwise. We apply this procedure for all cluster membership indicator matrices. The number of clusters (also called *rank parameters*) for each dataset are chosen to be $k_1 \ll n_1$, $k_2 \ll n_2$ and $k_3 \ll n_3$, which provides dimensionality reduction of the relation matrices. Matrices $\mathbf{H}_{12}^{k_1 \times k_2}$ and $\mathbf{H}_{23}^{k_2 \times k_3}$ in Equation 1 above represent compressed, low-dimensional versions of $\mathbf{R}_{12}$ and $\mathbf{R}_{23}$, respectively.

An important step in our methodology is estimating rank parameters, which are the numbers of clusters of patients, genes and drugs, k_1, k_2 and k_3, respectively. These parameters need to be known before factorisation is performed. The usual procedure for obtaining these parameters is by varying these parameters for each run and estimating cluster stability[22,23]. We take the values of parameters for which the most stable clustering is achieved. In particular, multiplicative update rules converge to a different solution in each run, depending on the random matrix initializations (i.e., initial clustering assignment given by the initial values in matrices $\mathbf{G}_i$, $1 \leq i \leq 3$). For example, if a clustering of patients into k_1 classes is stable, we expect small variations in the assignment to clusters from run to run. To measure this, we perform multiple factorisation runs with the same values of rank parameters. Each time, a connectivity matrix is computed (e.g., $\mathbf{C}_1$ for patients); based on these, an averaged connectivity matrix (also called the *consensus* matrix) over all runs is computed, $\hat{\mathbf{C}}_1$. If the clustering is stable, then the entries in $\mathbf{C}_1$ (also referred to as the *cluster association scores*) will be either close to zero, or close to one. Otherwise, the entries will be scattered in the interval $[0, 1]$. We use the *dispersion* coefficient, $\rho_{k_1}(\hat{\mathbf{C}}_1)$, introduced by Kim *et al.*[23], as a measure of cluster stability. The values of the dispersion coefficient range in $0 \leq \rho_{k_1}(\hat{\mathbf{C}}_1) \leq 1$, where 1 denotes a stable clustering. In our study, for each rank parameter, we perform a grid search in intervals of 1 for $1 \leq k_1 \leq 5$, $5 \leq k_2 \leq 30$ and $5 \leq k_3 \leq 30$, and compute dispersion coefficients, $\rho_{k_1}(\hat{\mathbf{C}}_1)$, $\rho_{k_2}(\hat{\mathbf{C}}_2)$ and $\rho_{k_3}(\hat{\mathbf{C}}_3)$ for patients, genes and drugs, respectively. We choose the values for k_1, k_2 and k_3 for which dispersion coefficients are of the highest values.

Matrix completion property. In addition to co-clustering of patients, genes and drugs, we model the existing and predict new drug-target interactions by using the *matrix completion* property of GNMTF. Namely, after obtaining low-dimensional matrices, the reconstructed drug-target matrix, $\hat{\mathbf{R}}_{23} \approx \mathbf{G}_2\mathbf{H}_{23}\mathbf{G}_3^T$, is more complete than the initial matrix, $\mathbf{R}_{23}$, and it can be used for extracting new, unobserved drug-target relations and therefore, finding new drug candidates for repurposing.

2.2. *Drug repurposing, patient stratification and driver gene prediction*

Drug repurposing. We use the reconstructed drug-target relation matrix, $\hat{\mathbf{R}}_{23}$, to extract new, previously, unobserved drug-gene interactions and to postulate new candidates for drug repurposing in the treatment of ovarian cancer patients. We apply a combination of row-centric and column-centric rules to extract new, strongly associated drug-gene pairs[13]. Namely, a drug-gene pair, (d, g), is considered to be predicted, if the estimated association score, $\hat{\mathbf{R}}_{23}[g][d]$, is greater than the mean association score of all relations of gene g, as well as greater than the mean association score of all relations of drug d.

Patient stratification. We stratify ovarian cancer patients into groups, according to the consensus matrix, $\hat{\mathbf{C}}_1$. We use the approach of Brunet *et al.*[22]: we use the off-diagonal entries of $\hat{\mathbf{C}}_1$ as a measure of patient similarity and apply average linkage hierarchical clustering to group patients into k_1 classes. Results and validations are shown in Sec. 3.1 below.

Cancer driver gene prediction. Similar to the patient consensus matrix, we use the gene consensus matrix, $\hat{\mathbf{C}}_2$, to extract gene clusters and identify those that are enriched in mutations and known driver genes by using *the standard model sampling without replacement test (i.e., hypergeometric test).* In clusters that are enriched in known drivers, we identify genes that are highly associated with known driver genes based on the clustering association scores from the gene consensus matrix. We postulate that these genes are new driver genes for ovarian cancer. Results and validations are presented in Sec. 3.2 below.

2.3. *Datasets, pre-processing and matrix construction*

We downloaded high-grade serous ovarian cancer somatic mutation data from TCGA data portal[4] on the 2nd of July 2015. We only consider data generated by using Illumina GAIIx platform, having the largest number of patients. Following the same procedure for data filtering as in Hofree *et al.*[5], we retain only the patients with more that 10 mutated genes. This results in $n_1 = 353$ serous ovarian cancer patients with mutations in the total of 11,148 genes. Mutated genes are mapped onto the Molecular Network (MN) that we obtain by merging three different biological networks: protein-protein interaction (PPI) and genetic interaction (GI) network from BioGRID database (version 3.4.126)[17], and metabolic interaction (MI) network from KEGG database[18]. This results in MN of 236,751 interactions among $n_2 = 19,118$ genes (mutated and normal). We represent these interactions by Laplacian matrix, $\mathbf{L}_2^{n_2 \times n_2}$, computed as: $\mathbf{L}_2 = \mathbf{D}_2 - \mathbf{A}_2$, where $\mathbf{A}_2$ is the adjacency matrix of MN and $\mathbf{D}_2$ is the diagonal degree matrix of MN (i.e., whose entries on the diagonal are row sums of $\mathbf{A}_2$ and all other entries in $\mathbf{D}_2$ are zeros). For each patient, we create an n_2- long binary $(0, 1)$ somatic mutation profile (SMP) vector, where "1" indicates the existence of a mutated gene in the patient and all other entries are "0". These mutation profiles for all n_1 patients are captured in a binary relation matrix $\mathbf{R}_{12}^{n_1 \times n_2}$ consisting of these SMP vectors. Due to the sparsity of matrix $\mathbf{R}_{12}$, we apply a network propagation technique as the pre-processing step to smooth the patient profiles, by spreading the influence of each mutation over its neighbours in MN network. We use the network propagation method proposed by Vanunu *et al.*[24], based on which the new patient-gene relation matrix is computed iteratively as follows: $\mathbf{R}_{12}^{t+1} = \alpha \mathbf{R}_{12}^{t} \bar{\mathbf{A}}_2 + (1-\alpha)\mathbf{R}_{12}^{0}$, where $\bar{\mathbf{A}}_2$ is the normalised adjacency matrix of MN network computed as $\bar{\mathbf{A}}_2 = \mathbf{A}_2 \mathbf{D}_2^{-1}$, $\mathbf{R}_{12}^{0} = \mathbf{R}_{12}$ is the initial patient-gene matrix and α is a tuning parameter that controls the distance of diffusion through MN network. In all our runs, we set $\alpha = 0.6$ (as it produced the best results), and we took the final network-smoothed, patient-gene matrix (after convergence, $|\mathbf{R}_{12}^{t+1} - \mathbf{R}_{12}^{t}| < 10^{-6}$, is achieved) as input to GNMTF. This pre-processing step has been shown to lead to much better and more robust patient stratification results in previous studies[5], hence we use it as well.

Drug-target interactions (DTIs) are downloaded from DrugBank database (version 4.3)[19]. We retrieved $n_3 = 6,620$ drugs (FDA-approved and experimental) targeting 1,385 genes in MN. These interactions are captured by DTI binary relation matrix, $\mathbf{R}_{23}^{n_2 \times n_3}$. SMILES chemical representation of the n_3 drugs are also retrieved from DrugBank database. The two-dimensional chemical similarity between drugs are computed by using Tanimoto similarity coefficient[25]. We retain only the top 5% most similar drug pairs, which results in 1,069,393 links in the drug chemical similarity (DCS) network. We represent these links by Laplacian matrix, $\mathbf{L}_3^{n_3 \times n_3}$ (computed in the same way as for MN network, described above).

2.4. *Clinical and biological validation of results*

For all patients, we also downloaded clinical follow up data from TCGA database, including the overall patients' survival information (days to the last follow-up and vital status), age, tumor grade, size and tumor position. We used these data to assess the clinical relevance of the patient clusters that we obtain after data fusion. We used Kaplan-Meier survival curves, as well as the log-rank p-value, to measure the significance of the difference in survival profiles between different patient clusters. The log-rank p-value measures the probability of the null hypothesis that patients in each cluster are drawn from the same underlying survival distribution[26]. From TCGA database, we also retrieved a list of 83 known ovarian cancer driver genes, out of which 76 are present in our set of mutated patient genes. We use this set of genes to assess gene clusters obtained after fusion and to identify clusters enriched in drivers.

3. Results

3.1. *Patient stratification*

We perform the consensus clustering, as described in Sec. 2.1, with 20 different random initialisations (initial cluster assignment) of GNMTF and compute the consensus matrices of patients, genes and drugs. We observe that rank parameters of $k_1 = 3$ (the number of patient clusters), $k_2 = 25$ (the number of gene clusters) and $k_3 = 20$ (the number of drug clusters), lead to the most stable clustering (with the largest dispersion coefficients: $\rho_{k_1=3}(\hat{\mathbf{C}}_1) = 0.56$, $\rho_{k_2=25}(\hat{\mathbf{C}}_2) = 0.91$ and $\rho_{k_3=20}(\hat{\mathbf{C}}_3) = 0.88$).

To assess the prognostic capabilities of our patient-specific data fusion approach on ovarian cancer patients, we perform clinical validation of the three obtained patient clusters. The Kaplan-Meier survival curves, shown in Fig.2 (A), reveal the low-survival group (*Cluster 2*) with 56% of death cases and the good outcome group (*Cluster 1*) with 38% of death cases. We observe that the identified clusters are highly discriminative with the log-rank p-value of 5.3×10^{-3}. The same number of clusters has been also reported in previous studies done on somatic mutation and molecular interaction data[5], and also in study done only on miRNA expression data[4]. Furthermore, the identified clusters display a good agreement with the median age of patients in clusters, with *Cluster 2* having the oldest patients. In addition, *Cluster 2* has the largest number of patients with abnormal growth of tissue (tumor), 78%, as compared to *Cluster 1* with 60% of such patients.

We compare the performance of our method with the state-of-the-art somatic mutation-based stratification method called Network-based Stratification (NBS)[5]. NBS takes as input a

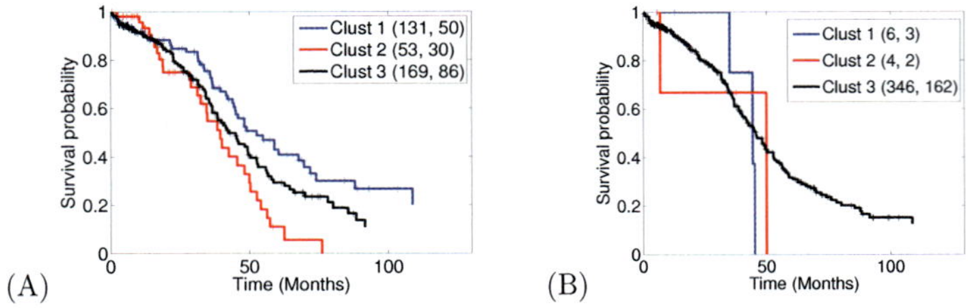

Fig. 2. Kaplan-Meier survival curves for 3 different patient groups produces by GNMTF (A) and NBS (B). The total number of patients and the number of deceased patients for each cluster are shown in the legend, the first and the second number in brackets, respectively.

patient-gene post-smoothing relation matrix and a molecular network matrix. We apply it on the same set of data described in Section 2.3, excluding drug data, which only our framework can take into account. We test NBS for different numbers of patient clusters (i.e., $k \in \{2, 3, 4, 5\}$) and compute the Kaplan-Meier survival curves[26] for the obtained patient clusters. We compare the survivability results of NBS with our method with the same number of patient clusters (Fig. 2 (A,B)). Unlike our method, which can produce clusters with significantly different survival outcomes (i.e., p-val $= 5.3 \times 10^{-3}$), NBS cannot (p-val ≥ 0.74 for all $k \in \{2, 3, 4, 5\}$). Thus, our framework is the only one able to extract personalised knowledge from somatic mutation profiles.

3.2. *Identification of driver genes*

We performed biological assessment of the $k_2 = 25$ gene clusters that we obtain from the gene consensus matrix, $\hat{\mathbf{C}}_2$. We identify 9 gene clusters that are significantly enriched in mutations and 5 gene clusters that are significantly enriched in known drivers ($p - val \leq 0.05$, see Fig. 3). Out of these clusters, cluster number 8 has the largest number of driver genes (26) and the highest enrichment in driver genes (with $p - val = 2.06 \times 10^{-4}$). To identify new driver genes, we further analyse this cluster as follows: first, based on the cluster association scores in the gene consensus matrix, we extract the mutated genes that are strongly associated with the known driver genes. In particular, we focus only on genes associated with the known driver genes with the cluster association score ≥ 0.9 (as explained below). That is, out of 20 restarts of GNMTF, we extract genes that appear 18 times in cluster 8 with other driver genes. Then, for each of these genes, we compute the average cluster score based on its associations with all driver genes. We provide the list of the top 20 genes (out of 809 predicted drivers in total) that we postulate as new driver genes of ovarian cancer progression and we sort it according to the average cluster association score, as shown in Table 1. This procedure is motivated by the observation that out of the 76 known driver genes, 67 of them are strongly related (with cluster association score ≥ 0.9) among themselves through all gene clusters.

We assess our predicted driver genes against two cancer driver gene databases, COSMIC database Cancer Gene Census[29] and IntOGen[30], as well as against a database of putative cancer driver genes, the Candidate Cancer Gene Database (CCGD)[31]. Our results show that

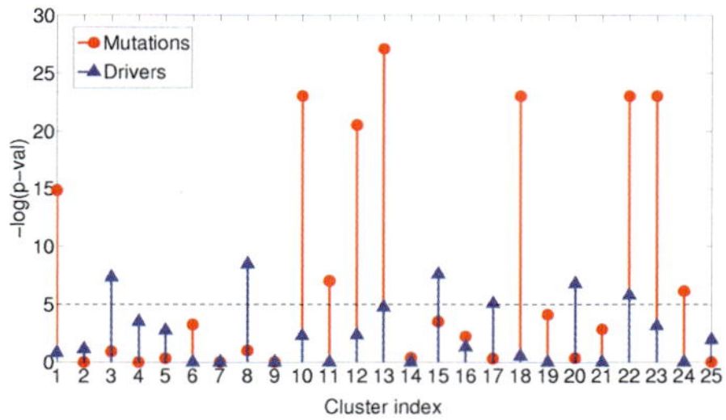

Fig. 3. Clusters enriched in mutated (red) and driver (blue) genes. For each cluster, $-log(pval)$ is plotted as a measure for enrichment (y-axis).

$\sim 40\%$ of our 809 predicted driver genes (with scores ≥ 0.9) have either been already proposed as drivers (in CCGD), or validated by experts (Census, or IntOGen). The list of our 20 top-scoring predicted cancer driver genes is presented in Table 1. Also, we investigated the literature to assess the relevance of our two top-scoring predictions that are not found in other databases and found evidence that they are biologically relevant. Our top-scoring cancer driver gene prediction is ADAM32, which is strongly clustered with driver gene BMPR2 (Table 1). The association between the two genes is biologically relevant, because both are involved with transforming growth factors (TGFs). Our prediction of ADAM32 as a cancer driver gene is also relevant, because ADAM genes are known to be responsible for cancer cell proliferation and progression[32]. The second best prediction is REG1P (from the REG family of proteins), which is strongly clustered with driver gene CLASP2. Our prediction of REG1P as a cancer driver gene is also relevant, because the REG family plays different roles in proliferation, migration, and anti-apoptosis through activating different signalling pathways; their dis-regulation is closely associated with cancer and REG proteins have been proposed as markers for prognosis of cancers[33].

3.3. *Drug-target interaction prediction*

To demonstrate the predictive power of our data fusion approach and to assess the contribution of each dataset on the drug-target interaction prediction, we perform a 5-fold cross validation for each combination of the datasets shown in Fig. 1. In all our experiments, true positives are correctly predicted DTIs, while false positives are predicted DTIs that are not present in the initial dataset.

We compute average Area Under the Receiver Operator Characteristic (ROC) and Precision-Recall (PR) curves (over 20 repetitions) to evaluate the performance of our methodology for each combination of datasets included in the integration process. The results are shown in Fig. 4. The lowest values of average AUC ROC and AUC PR are observed when only DTI dataset is used. The values increase with the inclusion of other datasets, resulting in the highest average AUC ROC when all datasets are taken into account. With all datasets taken into account by GNMTF, we use the reconstructed DTI relation matrix, $\hat{\mathbf{R}}_{23}$, to extract new drug-target interactions, as described in Sec. 2.2. We assess our prediction accuracy against two different large drug-target interaction databases, MATADOR[28] and CTD[27]. Out

Table 1. The list of the top scoring proposed driver genes (1st column) and their associated known driver genes (2nd column), with the association score (3rd column), and the confirmation of their presence in CCGD database (4th column).

New driver	Known drivers	Score	DB
ADAM32	BMPR2	1.000	–
REG1P	CLASP2	1.000	–
PCDHA2	CHD4	1.000	–
NCR1	BMPR2	1.000	–
USPL1	CLASP2	1.000	–
GDPD3	DDX5	1.000	–
LECT1	CLASP2	1.000	CCGD
IL25	CDK12, CCAR1	0.975	–
BAK1	ATRX, TFDP1, NDRG1	0.967	–
MOGAT2	ATRX, TFDP1, NDRG1	0.967	–
CHAF1A	ATRX, TFDP1, NDRG1	0.967	CCGD
PITX2	ATRX, TFDP1, NDRG1	0.967	–
SIN3B	ATRX, TFDP1, NDRG1	0.967	–
RPL30	ATRX, TFDP1, NDRG1	0.967	–
GRWD1	ATRX, TFDP1, NDRG1	0.967	–
SNAI1	ATRX, TFDP1, NDRG1	0.967	CCGD
RBMXP4	ATRX, TFDP1, NDRG1	0.967	–
CPNE7	ATRX, TFDP1, NDRG1	0.967	–
HIPK3	ATRX, TFDP1, NDRG1	0.967	CCGD
EPOR	ATRX, TFDP1, NDRG1	0.967	CCGD

Table 2. The list of predicted top scoring drug–target associations (first two columns), the association scores (third column), and the confirmation of their presence in CTD (C) or MATADOR (M) database (fifth column). All drugs are FDA-approved.

Gene	Drug	Score	Clusters	DB
KIT	ATP	0.873	1, 2, 3	–
GABRQ	Adinazolam	0.808	1	M
GABRQ	Fludiazepam	0.808	1	M
GABRQ	Cinolazepam	0.809	1	M
GABRQ	Clotiazepam	0.809	1	M
HTR2A	Dopamine	0.809	1, 3	C, M
GRIN3A	Pethidine	0.801	1, 2	–
CACNA2D1	Verapamil	0.761	1, 3	M
PDGFRB	ATP	0.724	1, 2	–
KDR	ATP	0.724	1, 3	C
HTR1A	Mirtazapine	0.720	1, 2	C ,M
GABRA6	Adinazolam	0.688	1	M
GABRA6	Fludiazepam	0.688	1	M
GABRA6	Cinolazepam	0.688	1	M
GABRA6	Clotiazepam	0.688	1	M
GABRA4	Adinazolam	0.687	1, 3	M
GABRA4	Fludiazepam	0.687	1, 3	M
GABRA4	Cinolazepam	0.687	1, 3	M
GABRA4	Clotiazepam	0.687	1, 3	M
CACNA1D	Magnesium Sulfate	0.676	1, 2, 3	M

of our 225,947 predicted DTIs, 37% have already been confirmed in MATADOR, or CTD. The list of our 20 top scoring predicted DTIs is shown in Table 2, out of which 17 are confirmed in CTD, or MATADOR database. Second, we investigated the literature to assess the relevance of our two top-scoring predicted DTIs that are not found in other databases and found evidences that they are biologically relevant. The top scoring target gene KIT (C-Kit) is particularly relevant. It is a receptor tyrosine kinase (e.g., it catalyses ATP/ADP reactions). It has been shown that unregulated activity of this gene leads to occurrence of tumors and thus, it has been proposed as a potential drug target in cancer[34]. Interestingly, we predict the drug candidate for targeting this gene to be Adenosine triphosphate (or ATP), for which a precise role in cancer is still under investigation. Increasing ATP intake is known to improve cancer patient conditions[35]. The reason could be that ATP is linked to cancer cell metabolism and either activates cell death mediated by restoration of normal mitochondrial function, or alterates the cytosolic ATP/ADP ratio, which is postulated to deactivate glycolysis (Warburg effect) in a cancer cell[36]. Another drug-target in Table 2 whose predicted drug is not present in CTD and MATADOR databases is GRIN3A. GRIN3A (NMDAR-1) is a sub-unit of NMDA receptor (a glutamate-regulated ion channel). NMDA receptor has been proposed as a target for cancer chemotherapy[37]. It has been proposed that glutamate antagonist molecules should be used as potential drug targets[37]. Interestingly, our predicted drug, Pethidine (also known as Meperidine), is a glutamate antagonist that is known to bind NMDA receptors[38], which provides evidence that our prediction of Pethidine as a drug for targeting GRIN3A is biologically relevant. However, evidence that Pethidine can bind to GRIN3A in particular has not yet been established. Furthermore, based on the mutated genes of particular patients, we propose these newly discovered drugs (see column four in Table 2) for treatment of the three patient groups described in Sec. 3.1.

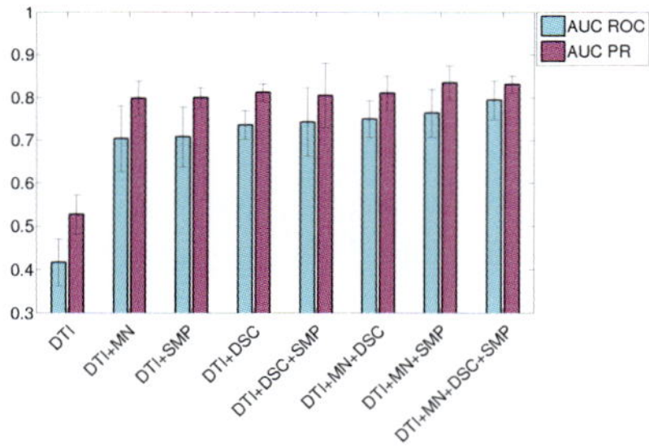

Fig. 4. Area under the ROC and PR curves for GNMTF runs done on the combination of particular datasets listed on x-axis. The results are sorted increasingly according to the AUC ROC values. See Fig. 1 for the abbreviated names of the datasets.

4. Conclusions

In this paper, we propose a data fusion framework that can effectively integrate somatic mutation data along with molecular networks and drug chemical data. It is based on GNMTF method for co-clustering heterogeneous data and it can be even further extended to incorporate any number and type of data. One important advantage of our framework is that when applied to a specific cancer, it can simultaneously perform three different tasks: patient stratification into clinically different groups, novel driver gene identification and drug-repurposing predictions for treating cancer.

We apply the GNMTF-based data fusion framework to ovarian cancer patients and identify three substantially different groups of patients with different survival outcomes. In addition, from the obtained gene clusters, we identify a list of genes that we postulate as potential drivers of ovarian cancer progression due to their strong cluster associations to known ovarian cancer driver genes. We perform biomedical literature curation for the top scoring predictions, ADAM32 and REG1P, and show that they are related to cancer cell proliferation and tumor progression, while 40% of other predictions we validate in other databases. Moreover, our framework is capable of predicting new drugs that could be used for targeting mutated genes and thus, for treatment of identified groups of ovarian cancer patients. We provide a list of predicted drug-target interactions, a good number of which is matching those reported in other databases. Other, non-validated predictions for driver genes and drug-target interactions could be true, awaiting experimental validation.

Our analysis also suggests that somatic mutation data is a valuable complement to other molecular data, whose integration with those data could lead to an improvement in the performance of data fusion methods. Our approach has a potential to enable the derivation of new hypotheses, improve drug selection and lead to improvement in patient genomics-tailored therapeutics for cancer.

Acknowledgement

This work was supported by the European Research Council (ERC) Starting Independent Researcher Grant 278212, the National Science Foundation (NSF) Cyber-Enabled Discovery and Innovation (CDI) OIA-1028394, the ARRS project J1-5454, and the Serbian Ministry of

Education and Science Project III44006.

References

1. B. Vogelstein *et al.*, *Science* **339**, 1546 (2013).
2. C. Kandoth, M. D. *et al.*, *Nature* **502**, 333 (2013).
3. C. Rubio-Perez, *et al. Cancer Cell* **27**, 382 (2015).
4. Cancer Genome Atlas Research Network, *Nature* **474**, 609 (2011).
5. M. Hofree *et al.*, *Nature Methods* **10**, 1108 (2013).
6. Cancer Genome Atlas Research Network, *Nature* **487**, 330 (2012).
7. Y. Chen *et al.*, *Scientific Reports* **3** (2013).
8. Y. Yamanishi *et al.*, *Bioinformatics* **26**, i246 (2010).
9. C. Ding *et al.*, in *KDD '06: Proceedings of the 12th ACM SIGKDD International Conference on Knowledge Discovery and Data mining*, (2006).
10. C. Ding *et al.* in *Proceedings of SIAM Data Mining Conference*, (2005).
11. F. Wang, T. Li and C. Zhang, in *SIAM Conference on Data Mining (SDM)*, (2008).
12. M. Žitnik *et al. Scientific Reports* **3** (2013).
13. M. Žitnik and B. Župan, *Pattern Analysis and Machine Intelligence, IEEE Transactions on* **37**, 41 (2015).
14. T. Hwang *et al.*, *Nucleic Acids Research* **40**, e146 (2012).
15. H. Wang *et al.*, *Journal of Computational Biology* **20**, 344 (2013).
16. V. Gligorijević, V. Janjić and N. Pržulj, *Bioinformatics* **30**, i594 (2014).
17. A. Chatr-aryamontri *et al.*, *Nucleic Acids Research* **43**, D470 (2015).
18. M. Kanehisa and S. Goto, *Nucleic Acids Research* **28**, 27 (2000).
19. D. S. Wishart, C. Knox, A. C. Guo *et al.*, *Nucleic Acids Research* **36**, D901 (2008).
20. F. Shang, L. Jiao and F. Wang, *Pattern Recognition* **45**, 2237 (2012).
21. R. Albright, J. Cox, D. Duling, A. Langville and C. Meyer, *North Carolina State University, Tech. Rep* **81706** (2006).
22. J.-P. Brunet *et al.*, *Proceedings of the National Academy of Sciences* **101**, 4164 (2004).
23. H. Kim and H. Park, *Bioinformatics* **23**, 1495 (2007).
24. O. Vanunu *et al.*, *PLoS Computational Biology* **6**, e1000641 (01 2010).
25. N. Nikolova and J. Jaworska, *QSAR & Combinatorial Science* **22**, 1006 (2003).
26. V. Bewick, L. Cheek and J. Ball, *Critical Care* **8**, 389 (2004).
27. A. P. Davis *et al.*, *Nucleic Acids Research* **43**, D914 (2015).
28. S. Günther *et al.*, *Nucleic Acids Research* **36**, D919 (2008).
29. P. A. Futreal *et al.*, *Nature Reviews Cancer* **4**, 177 (2004).
30. G. Gundem *et al.*, *Nature Methods* **7**, 92 (2010).
31. K. L. Abbott *et al. Nucleic Acids Research* **43**, D844 (2015).
32. S. Mochizuki and Y. Okada, *Cancer Science* **98**, 621 (2007).
33. J. Zhao, J. Wang, H. Wang and M. Lai, *Adv. Clin. Chem* **61**, 153 (2013).
34. J. Lennartsson and L. Ronnstrand, *Current Cancer Drug Targets* **6**, 65 (2006).
35. A. Jatoi and C. L. Loprinzi, *Journal of Clinical Oncology* **20**, 362 (2002).
36. E. N. Maldonado and J. J. Lemasters, *Mitochondrion* **19**, 78 (2014).
37. S. I. Deutsch, A. H. Tang, J. A. Burket and A. D. Benson, *Biomedicine & Pharmacotherapy* **68**, 493 (2014).
38. T. Yamakura, K. Sakimura and K. Shimoji, *Anesthesia & Analgesia* **90**, 928 (2000).

PRISM: A DATA-DRIVEN PLATFORM FOR MONITORING MENTAL HEALTH

MAULIK R. KAMDAR*

Program in Biomedical Informatics, Stanford University,
Stanford, CA 94305, USA
E-mail: maulikrk@stanford.edu

MICHELLE J. WU*

Program in Biomedical Informatics, Stanford University,
Stanford, CA 94305, USA
E-mail: mjwu@stanford.edu

Neuropsychiatric disorders are the leading cause of disability worldwide and there is no gold standard currently available for the measurement of mental health. This issue is exacerbated by the fact that the information physicians use to diagnose these disorders is episodic and often subjective. Current methods to monitor mental health involve the use of subjective DSM-5 guidelines, and advances in EEG and video monitoring technologies have not been widely adopted due to invasiveness and inconvenience. Wearable technologies have surfaced as a ubiquitous and unobtrusive method for providing continuous, quantitative data about a patient. Here, we introduce PRISM — Passive, Real-time Information for Sensing Mental Health. This platform integrates motion, light and heart rate data from a smart watch application with user interactions and text entries from a web application. We have demonstrated a proof of concept by collecting preliminary data through a pilot study of 13 subjects. We have engineered appropriate features and applied both unsupervised and supervised learning to develop models that are predictive of user-reported ratings of their emotional state, demonstrating that the data has the potential to be useful for evaluating mental health. This platform could allow patients and clinicians to leverage continuous streams of passive data for early and accurate diagnosis as well as constant monitoring of patients suffering from mental disorders.

Keywords: mental health; wearables; user interactions; visualization

1. Introduction

1.1. *The current state of mental health care*

Neuropsychiatric disorders are the leading cause of disability among non-communicable diseases worldwide and are estimated to be the cause of around 10.4% of the global burden of disease.[1,2] These disorders include mood disorders (depression, unipolar and bipolar disorders), anxiety-related disorders (anorexia and other eating disorders, obsessive-compulsive disorders, panic disorders and post-traumatic stress syndrome), schizophrenia, substance- and alcohol-use disorders, dementia, as well as neurological disorders like epilepsy, Parkinson's and multiple sclerosis. Within this group, depressive disorders accounted for the most DALYs, followed by anxiety disorders.[1] Historically, major health policy decisions have been primarily informed through mortality statistics. As a result, the impact of these pervasive and disabling neuropsychiatric disorders has been undervalued in comparison to cardiovascular diseases,

*These authors contributed equally to this work.

cancer and communicable diseases. Thus, research on these disorders has not been a priority and improvement in the management of mental health is lagging behind. The current standard practices for patients with such mental health disabilities, including those described in the APA's Diagnostic and Statistical Manual of Mental Disorders (DSM), rely on descriptive criteria and are often not evidence-based.[3,4] Unlike in other areas of medicine, very few objective clinical tests or medical devices are routinely used in mental health care.[5] The information physicians use to make diagnoses is not only subjective, but it is also episodic, capturing only occasional snapshots from patient visits and ignoring the finer dynamics of a patient's condition. EEG-based methods have been shown to be effective for quantitatively capturing a patients condition,[6] but they are invasive and thus do not permit continuous data collection. Other methods, such as video monitoring, have been shown to be useful for monitoring the mental health status of patients, but are extremely difficult to interpret.[7] These drawbacks have been a barrier to widespread adoption of these modern technologies. While the first EEG-based mental health evaluation tool has recently been approved for ADHD,[8] more comprehensive patient-friendly technologies and robust methods are needed to improve the state of mental health care.

1.2. *The applicability of web and wearable technologies*

Wearable products have emerged as a ubiquitous technology for passive collection of quantitative data. Over 10 million fitness trackers, smart watches and other such devices are sold each year and the market is growing rapidly.[9] Many such products have the ability to measure physiological data, environmental data, and activity data. In addition, other data, such as keystroke dynamics and eye tracking, can be collected based on the user's interaction with their wearables and other devices. Unlike in other emerging solutions discussed previously, these data can be collected in a passive manner, without affecting the day-to-day life of the user. Preliminary evidence has shown that these data can be leveraged for use in monitoring of mental health patients.[10,11] Variability in heart rate, activity and light exposure have been studied previously for screening in disorders like sleep apnea, dementia and epilepsy.[12,13] Patient-reported outcomes, as collected through social networks and citizen science approaches, serve as real-time evidence for determining emotional distress symptoms.[14,15] Performing sentiment analysis over social streams provides a better understanding of human emotions than occasional surveys.[16] Web and wearable technologies have been experimented with for unobtrusive data collection to detect seizures and mental fatigue.[10,11,17] However, such methods have only been applied individually in a very narrow set of test cases. We believe that the integration of these diverse passive data types will provide a robust and quantitative method for characterizing a wide range of mental health states.

Here, we describe a novel platform, PRISM (**P**assive, **R**eal-time **I**nformation for **S**ensing **M**ental Health), which allows for passive collection of many types of quantitative measures that have been shown to be related to mental health outcomes.[11,13,17] In addition, the web interface enables patients to easily access and understand their own data and thus their health condition. This system has the potential both to generate more knowledge about which treatments work best for patients and to allow informed decision making in the clinic. We believe that this integrated system for collecting and analyzing quantitative information about a patient will have huge ramifications for the improvement of the care of mental health patients.

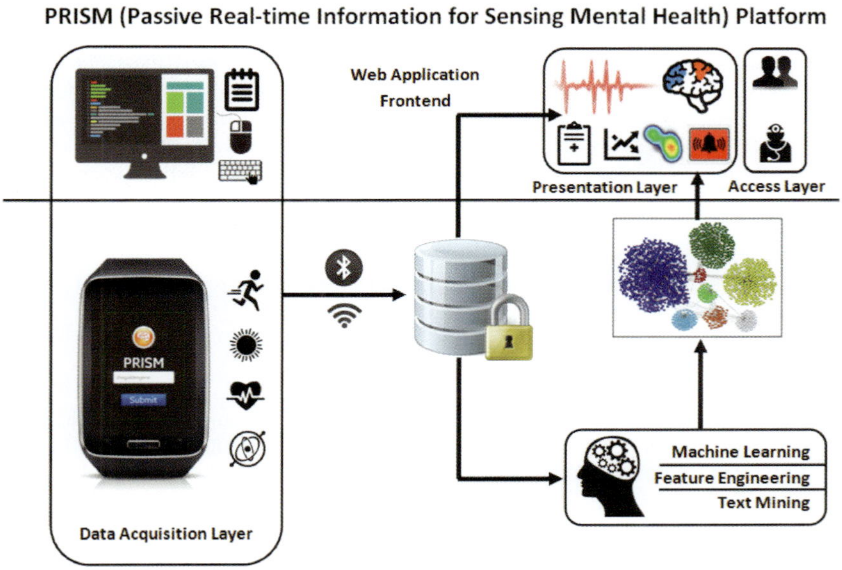

Fig. 1. The PRISM platform for data collection, integration and analysis is outlined here. Our system allows for a smooth workflow that incorporates data acquisition, data analysis and feedback to the user.

2. Materials & Methods

2.1. *Platform Development*

The first phase of our project was to develop a platform that aggregates passively recorded user data from smart watches with user interaction patterns from a web application. Our platform architecture is composed of six different layers — *i)* Data acquisition, *ii)* Text mining, *iii)* Feature Engineering, *iv)* Machine Learning, *v)* Access and *vi)* Presentation (Fig. 1). The data acquisition layer relies on the Samsung Gear S smart watches and the Tizen APIs[18] to collect physiological, environmental and activity data of the users. A separate web application incorporates a private blog-based feature for users to self-report their moods. User keystroke and mouse interaction patterns are captured as the user types and navigates across the web application. The web application also serves as a presentation layer that provides different visualizations through interactive widgets, allowing users to explore their own data. A role-based access is enabled to delineate two different categories of users of our platform: *i)* General — normal users or existing patients who wish to track their mental health and/or become aware of any dormant neuropsychiatric conditions, and *ii)* Health care providers — physicians and other providers who can gain summarized access to their patients' data.

While developing the platform, we were concerned with four main concepts — *privacy*, *connectivity*, *longevity* and *storage*. Since we are dealing with personal data, it was imperative to store this information in a secure database in an anonymized format. To ensure HIPAA compliance,[19] we do not store any patient identifiers, such as personal information (name, date of birth, mobile or email) and GPS location information. To ensure that the data collection is not affected by the connectivity of the wearable, all the data is first written to a local file and a separate backend service stores it in our secure database using a Bluetooth-paired mobile device or a WiFi connection. The battery life of the wearable can be a rate-limiting factor to our continuous data collection process. By excluding GPS information and giving preference to Bluetooth over WiFi connections, we extended the battery life of our wearable to multiple days. To ensure further longevity, no data analysis is itself carried out on the wearable and the sole purpose is just to record the data. Because sensor and interactions data is voluminous, we implemented gap encoding and stored data as text blocks for disjoint periods of times to achieve efficient space utilization.

2.2. *Data Collection*

As the second phase of our project, we conducted a pilot study with anonymized participants between the ages of 19–37 years. The study protocol was approved by the Institutional Review Board at Stanford University. As the participant engaged in his/her daily activities, the watch passively collected the following data with one second granularity:

(1) **Environmental:** Light intensity levels
(2) **Physiological:** Heart rate (beats per minute and R-R intervals)
(3) **Activity:** Device acceleration and rotation in all three axes, cumulative distance walked, total number of walk and run steps, speed, cumulative calories burnt, and walk frequency.

Using our web application, the participants entered free form text entries about how they were feeling and reported three outcomes - happiness, energy and relaxation - on a scale of 1 to 10 (Fig. 2C). These were collected at approximately three hour time intervals. As the participant typed and navigated our interface, we collected the following user interaction patterns:

(1) **Keystroke patterns:** Typing speed, number of spelling errors, key press down times, interkey latencies (time taken between typing two keys), number of times 'Enter', 'Delete' (backspace), 'Ctrl+Y' (redo) and 'Ctrl+Z' (undo) are pressed.
(2) **Mouse interaction patterns:** Total number and locations of mouse clicks, mouse hover and drag positions and times, screen width and height.

All the participants were made aware of the types of data that we were collecting before registering in this pilot study, and the days they participated in our study were selected randomly. We also asked participants to complete a post-study survey to evaluate the usability of PRISM. We used a standard 10-question system usability scale (SUS) questionnaire.[20]

2.3. *Text mining*

We generated string tokens by parsing the subjective text entries. We assigned each word with mean valence, arousal and dominance scores (VAD) retrieved from a database of VAD norms (1-10) for nearly 14,000 English lemmas.[21] We generated stemmed representation of words that did not have a VAD score using Snowball Stemmer[22] and assigned VAD scores for associated similar words, e.g. VAD score of *annoy* instead of *annoyance*. We calculated average VAD scores for each text entry.

Table 1. This table summarizes all of the features extracted from both the smart watch and user interaction data.

Data Type(s)	**Feature(s) used**
all smart watch data	mean, standard deviation, dominant frequency
pedometer data	cumulative sums
key press times, interkey latencies	mean, standard deviation
undo/redo, spelling errors, enters, backspaces	raw counts
mouse moves and drags	average distance & velocity, proportion of total time
number of clicks	total counts scaled by total time

2.4. *Feature engineering*

All features used are described in Table 1. All data was grouped into 1 hour-long time periods and each period was treated as a separate training example for our analysis, resulting in 347 training examples for all 13 users. For each period, each smart watch data type was composed into a time series with appropriate timestamps. Accelerometer data was converted into spherical coordinates for this analysis in order to capture more natural descriptors of the movements. For each time series, three summary statistics — mean, standard deviation, and dominant frequency — were computed. The dominant frequency was derived from a discrete Fourier Transform of the time series. For user interaction data types, each hour-long time period was associated with the keystroke and mouse dynamics data from the most recent blog entry. The majority of data types, including counts of typing errors and mouse move and drag information, were used directly as features. Key press times and interkey latencies were stratified by unigrams and bigrams, respectively, and the distribution of times for each unigram or bigram was represented by its mean and standard deviation in the feature set. This resulted in a total of 1658 features, 1591 of which are key press and latency statistics.

2.5. *Machine learning*

For unsupervised learning, data was scaled and centered for each feature. Hierarchical clustering of both features and training examples was performed using complete linkage with a Euclidean distance metric.

For supervised learning analysis, all missing data, composed mostly of keystroke dynamics data types, filled in by median imputation. Model selection, including tuning of model parameters, was performed using cross validation, stratified by the text entry associated with

Table 2. This table summarizes all of the models tested, along with the parameters tuned for each.

Model	Parameters
random forest	number of trees, number of features considered
gradient boosted regressor tree	learning rate, number of boosting stages
lasso, ridge, elastic net	regularization parameter
support vector machine	kernel type
k nearest neighbors	number of neighbors

the training example. For example, if the 8-9am, 9-10am, and 10-11am training examples for one participant all contain user interaction data from the 8am text entry, they are grouped together in cross validation. This ensures that training and testing are not performed on redundant datasets. Model types and parameters altered are described in Table 2. Performance was evaluated using an independent 20% evaluation dataset to avoid bias in model selection.

2.6. *Resources*

The PRISM platform can be accessed at `http://54.200.211.229/BrainHealth/`. The wearable application can be downloaded from the platform. Source code for the platform can be found at `https://github.com/wuami/PRISM`.

3. Results

3.1. *PRISM Platform*

Users can register with the PRISM platform, by providing a username, year of birth and a desired password. After installation of our application on the Samsung Gear S smart watch, the username can be entered in the GUI of the application screen (Fig. 2A). The application then starts to collect light, heart rate, accelerometer and pedometer data passively (Fig. 2B). Users have the option to stop data collection anytime by clicking on the 'Stop Collecting' button. If a Bluetooth-paired mobile device or a WiFi network is available, the wearable data is automatically uploaded to our database. They can authenticate with our web application and report their mood in free text format and their happiness, energy and relaxation on a quantitative scale (Fig. 2C, 3A). A log of users' past entries is accessible through the web application.

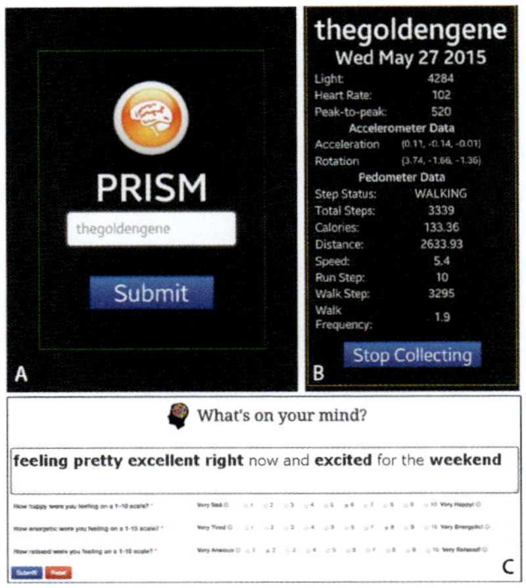

Fig. 2. PRISM data collection: (A) The Samsung Gear S wearable application welcome screen for username input. (B) The light, heart rate, accelerometer and pedometer data collected passively at every second. (C) The subjective text entry and marked tokens.

User data can be visualized by clicking on the 'Explore' tab in the navigation toolbar. For a selected day, the heart rate, light intensity, net device acceleration and rotation can be viewed as line charts and walking and running steps per minute as stacked area charts. These plots are aligned to an interactive scrollable timeline (Fig. 3D). Keystroke patterns are shown as a heatmap superimposed on the actual keys pressed. The radius indicates the number of times a key is pressed and the color scale indicates the average time on the key. Interkey latencies can be visualized as a network chart where the nodes represent the keys and thickness of the edges represent the average time (Fig. 3B). A similar heatmap, superimposed over the user screen, is generated for the mouse interaction patterns, where a single dot, a blurred region or a line indicates mouse click, move or drag patterns respectively (Fig. 3A). An 'Emotion Cuboid' visualizes the average VAD scores for each text entry as a scatter plot in a 3D space (Fig. 3C). Hovering over any point shows the associated text entry and average VAD scores. VAD scores are comparable to the happiness, energy and anxiety, hence such a spatial visualization can help easily separate the text entries and time period based on a VAD estimate of the sentiment in that text entry.

3.2. *Pilot study*

We conducted a pilot study with 13 healthy participants between 19 and 37 years of age. Each participant wore the smart watch for at least 6 hours and entered at least three text entries over the course of a single day. In addition, one participant wore the watch for 10 days in order to collect longitudinal data. Participants walked an average of 6800 steps per day and were exposed to an average of 1050 lux. Reported happiness levels ranged from 2 to 10, forming a roughly normal distribution with a mean of 6.6 out of 10. Reported energy and relaxation levels spanned the full range from 1 to 10, in a roughly uniform distribution with means of 6.8 and 4.5, respectively. Energy levels were skewed to the left, while relaxation levels were slightly skewed to the right. VAD analysis revealed that the content of the text was indicative of reported outcomes, suggesting that mining of passive sources of text data, like Twitter feeds, could replace the self-reported outcome measures. Happiness levels and estimated valence were correlated with a Pearson's correlation of 0.48. Energy level and estimated arousal had a correlation of 0.23.

The SUS questionnaire revealed a generally positive reaction to PRISM, with an average score of 74 out of 100. Participants reported that they found the system generally easy to use, without require much knowledge or training, but reported a lack of desire to use the system frequently. This may be a result of our study population consisting of healthy individuals who do not have a strong incentive to understand their mental health.

3.3. *Model development*

We began with unsupervised learning analysis in order to gain an understanding of any structure present within our data. For this analysis, we used only the 22 features where there was no missing data. As shown in Figure 4A, the data points seem to cluster somewhat according to the associated relaxation level. The top section of the heatmap shows the most variation in relaxation levels, but this is consistent with the longer branch lengths in the dendrogram, suggesting that the training examples are not as similar as in the other clusters. Watch acceleration and rotation data as well as keyboard data seemed to be the most variable across

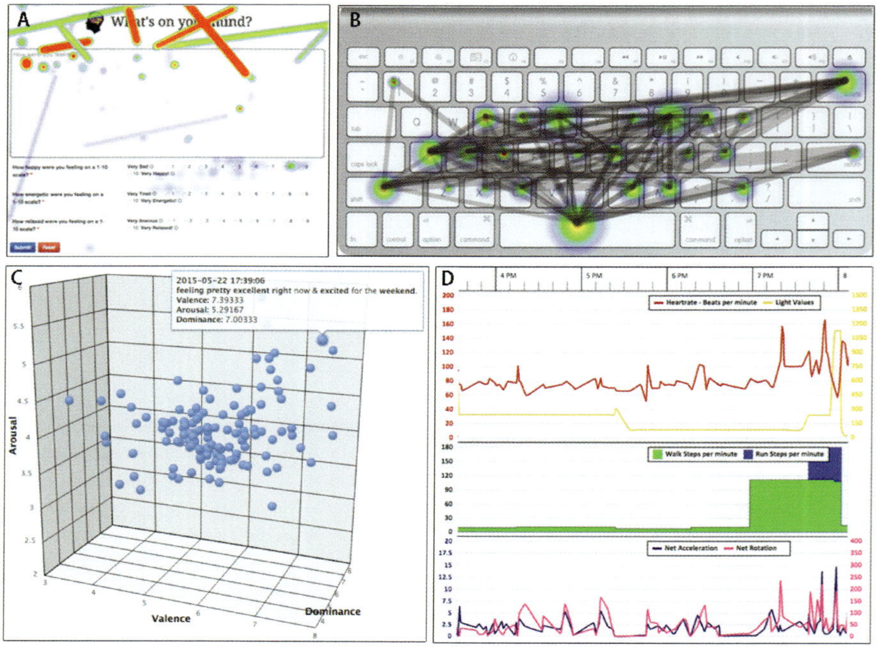

Fig. 3. PRISM Web Application: (A) The private blog-based workflow for users to input subjective free-text descriptions of their mood. A mouse interactions heatmap is superimposed over this screen. (B) Keystroke interactions heatmap and network visualizations. (C) A 3D-scatter plot (Emotion Cuboid) for visualization of average VAD scores of text entries. (D) Timeline-scrollable, exploratory interface for visualization of wearable data — [1] heart rate (beats per minute), [2] light intensity values, [3] net device acceleration, and [4] rotation, as line charts, and [5] walk steps per minute and [6] run steps per minute, as stacked area charts.

users and are the most informative for this clustering. This may have to do with the relatively high reliability of these data compared to most of the watch data types.

In the supervised learning phase, we used machine learning techniques to develop models for predicting the user-reported response variables from the smart watch and user interaction data. We evaluated a variety of methods, including decision tree-based models, generalized linear models and non-parametric models. We found that tree-based models, including random forests and gradient boosted regressor trees, resulted in the highest performance. A random forest model explained about 51% of the variance in our data, resulting in a p-value of 8.8×10^{-8}, as shown in Figure 4B. This suggests that while we are unable to accurately predict the reported relaxation levels, our data does capture information about the user's emotional state. Top features were derived primarily from user interactions and included interkey latency mean values, number of backspaces and number of mouse clicks. The selection of the interkey latency features may be an artifact of the sparsity of those features, as some bigrams may be present only in a single text entry. However, the number of backspaces and mouse clicks are more likely to reflect true changes in the user's style of interaction with their computer when they are more anxious or more relaxed.

In order to understand person-to-person variation, we also trained a user-specific model for the participant for which we collected longitudinal data. However, we found that this did not result in a significant improvement over the generalized model (data not shown). This suggests that the variation across time points in a single individual are comparable than that between individuals.

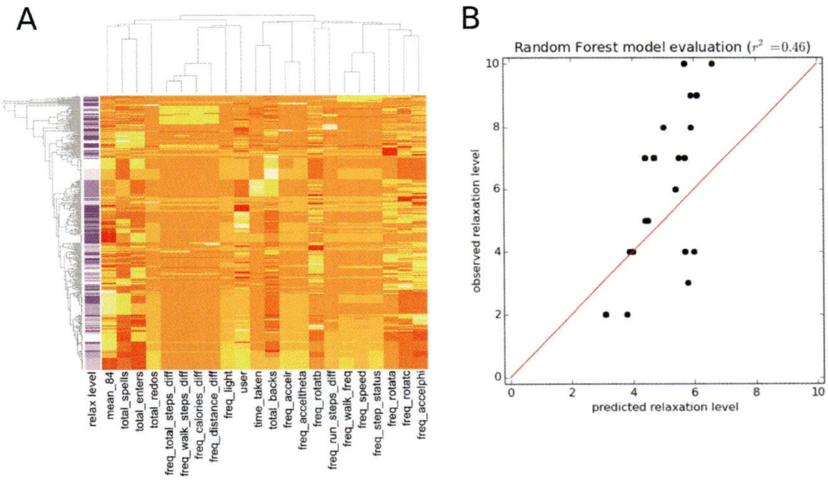

Fig. 4. Our results from machine learning analysis are shown here. (A) This heatmap shows hierarchical clustering of a subset of features, where labels beginning with "freq" indicate major frequencies and labels ending in "diff" indicate differences between time points. The associated relaxation levels, shown in purple, do cluster with the data. (B) A random forest model explains 51% of the variance in reported relaxation levels on a withheld evaluation dataset, with a p-value of 8.8×10^{-8}.

4. Discussion

We have designed, created, and tested the PRISM platform for the collection and analysis of smart watch and user interaction data. Based on our pilot study, we have shown that it is effective from both a usability and data analysis perspective.

4.1. *Usability and User Experience*

A mental health monitoring system can achieve maximum usability only if it can seamlessly integrate into the user's and physician's daily workflow. Developing an application for a smart watch wearable allows us to achieve passive, non-invasive data collection without the need for any manual intervention. While a new blog-based feature might not be appealing to the users at first, the methods developed here can easily be combined with a Wordpress-based blog or a Twitter plugin. Both of these networks have been established for user engagement and have also been used as a data source of free text for early diagnosis of neuropsychiatric conditions.[23] As it is cumbersome to enter the username in the wearable screen without a stylus, we wish to remove that requirement in the subsequent versions of the wearable application. As each Samsung Gear S smart watch can be identified with a unique International Mobile Station Equipment Identity (IMEI) key, we can ask the user to provide the IMEI key during registration with our platform to link the watch to a specific user. Providing interactive visualizations to the user for exploring the data that we collect helps increase user engagement and build trust. In the future, we will also provide a panel to present our inferred insights and map mental activity to a 3D Brain Browser visualization. We plan to extend our platform to provide feedback and recommendations to our users through the visualization panels or as notifications in the smart watch. In light of the rapid advances in this domain with the introduction of Apple Research Kit and Samsung Simband wearables, it will be interesting to develop solutions that are interoperable across different wearable platforms.

4.2. *Data Quality*

In the process of testing our watch and web applications, we observed several factors contributing to poor data quality. Many of the watch sensors are extremely sensitive to the tightness of the strap and the placement on the wrist. In particular, heart rate data is not collected properly when the sensor is not directly in contact with the skin. Further, even when the watch was fastened tightly, the measurement varied by 10 bpm or more as compared to a direct pulse measurement and was not consistent when placed on different wrists. The heart rate is derived using photoplethysmography, or optical measurements of blood volume changes at the surface of the skin. While this has been shown to be effective for heart rate monitoring in wearables,[24] it is clear that either the Gear S hardware or the algorithms used by the Tizen API are failing to capture the true signal. Similarly, pedometer data is very sensitive to hand placement. Activities like cycling or treadmill walking, where hands are often gripping a stable handle, are omitted. Additionally, pedometer data was observed to update sporadically rather than continuously, affecting our time series analysis. These factors are introducing significant noise into our data and significantly impacting our ability to create predictive models.

While there is potential for upgrading to more advanced wearables like the Samsung Simband, we will no doubt need to develop smarter ways to handle noisy data and to distinguish between true and artifactual features of the data. This will require the collection of larger datasets that will give us the power to identify predictive variation over chance variation in the data. Bigger data would also allow us to take advantage of tools like `softImpute`,[25] which take advantage of low rank structure in data to impute missing values. Similar methods, like generalized low rank models,[26] could also use this structure to smooth over noise in the data. In the future, we also plan to carry out sentiment analysis by generating a VAD-weighted term frequency matrix and using a Naive Bayes or an SVM classifier.

4.3. *Future Work*

We believe that PRISM has the potential to dramatically change the way that mental health is diagnosed, monitored and treated. While our analyses are currently severely limited by the size of the pilot study, the power of the data will increase dramatically as more data is collected. In order to fully realize this potential, further studies with neuropsychiatric patients will be necessary to validate the utility of the data. This will allow us to extract meaningful relationships between our quantitative metrics and patient condition as well as treatment outcomes. More longitudinal data will also allow us to infer user-specific patterns that can enable personalized analyses of users' data. Our system can be seamlessly integrated into a patient's daily life; however, introduction into a physician's workflow may require integration with existing EMR systems. Visualizations will enable clinicians to process both raw data and analytics rapidly in order to assist decision making, while our models will allow for the triggering of alerts to the physician when their patients are in an unstable mental state. In addition, the patient studies will enable us to link our data with actual outcomes, allowing both physicians and analytic models to learn how physiology affects treatment response.

5. Conclusions

Physiological and exogenous data (behavioral, social, environmental) is overwhelming to capture and analyze, but makes up a large portion of health determinants. In this work, we have developed the PRISM platform that can leverage heterogeneous, continuous streams of data, collected in a passive, non-invasive fashion, with the hopes of making mental health diagnosis more quantitative. We have shown the potential of such a platform by developing models that can recapitulate users' reported emotional states. While developing the platform, we have ensured that the privacy and anonymity of our users is maintained, and that data collection is not hindered by the rate-limiting factors of device connectivity and longevity. We have also developed interactive visualization panels that allow users to explore and understand their own data, and can also serve as mechanisms through which feedback can be provided.

Acknowledgments

We would like to thank Zeshan Hussain for his work in the intial stages of the project and Samsung for providing the Gear S watches and additional technical support. We would also like to acknowledge Dave Stark, Russ Altman, Steve Bagley and Tim Sweeney for providing valuable advice throughout the course of this research. M.W. was supported by the National Science Foundation Graduate Research Fellowship under Grant No. DGE-114747.

References

1. H. A. Whiteford *et al.*, Global burden of disease attributable to mental and substance use disorders: findings from the Global Burden of Disease Study 2010. *Lancet* **382**November 2013.
2. W. H. Organization, WHO global report: preventing chronic diseases: A vital investment2005.
3. A. P. Association, Diagnostic and statistical manual of mental disorders: DSM-52013.
4. R. E. Drake *et al.*, Implementing Evidence-Based Practices in Routine Mental Health Service Settings *Psychiatric Services* **52** (American Psychiatric Publishing, February 2001).

5. S. Kapur *et al.*, Why has it taken so long for biological psychiatry to develop clinical tests and what to do about it? *Molecular psychiatry* **17** (Macmillan Publishers Limited, December 2012).
6. S. M. Snyder, T. A. Rugino, M. Hornig and M. A. Stein, Integration of an EEG biomarker with a clinician's ADHD evaluation *Brain and Behavior* March 2015.
7. J. G. Ellis, W. S. Lin, C.-Y. Lin and S.-F. Chang, Predicting Evoked Emotions in Video, in *2014 IEEE International Symposium on Multimedia*, (IEEE, December 2014).
8. I. D. Delgado-Mejia *et al.*, Theta/beta ratio (NEBA) in the diagnosis of attention deficit hyperactivity disorder. *Revista de neurologia* **58 Suppl 1**February 2014.
9. J. Wei, How Wearables Intersect with the Cloud and the Internet of Things : Considerations for the developers of wearables. *IEEE Consumer Electronics Magazine* **3**July 2014.
10. M.-Z. Poh, T. Loddenkemper, C. Reinsberger *et al.*, Convulsive seizure detection using a wrist-worn electrodermal activity and accelerometry biosensor. *Epilepsia* **53**May 2012.
11. C. Epp, M. Lippold and R. L. Mandryk, Identifying emotional states using keystroke dynamics, in *Proceedings of the 2011 annual conference on Human factors in computing systems - CHI '11*, (ACM Press, New York, New York, USA, May 2011).
12. J. Jeppesen, A. Fuglsang-Frederiksen, R. Brugada, B. Pedersen *et al.*, Heart rate variability analysis indicates preictal parasympathetic overdrive preceding seizure-induced cardiac dysrhythmias leading to sudden unexpected death in a patient with epilepsy *Epilepsia* 2014.
13. S. Ancoli-Israel, M. R. Klauber *et al.*, Variations in circadian rhythms of activity, sleep, and light exposure related to dementia in nursing-home patients. *Sleep* **20**January 1997.
14. P. A. Pilkonis, S. W. Choi, S. P. Reise, A. M. Stover, W. T. Riley and D. Cella, Item banks for measuring emotional distress from the Patient-Reported Outcomes Measurement Information System (PROMIS): depression, anxiety, and anger. *Assessment* **18**September 2011.
15. R. R. Morris, S. M. Schueller and R. W. Picard, Efficacy of a Web-Based, Crowdsourced Peer-To-Peer Cognitive Reappraisal Platform for Depression: Randomized Controlled Trial *Journal of Medical Internet Research* **17** (Journal of Medical Internet Research, March 2015).
16. M. Larsen, T. Boonstra, P. Batterham, B. O'Dea, C. Paris and H. Christensen, We Feel: Mapping emotion on Twitter. *IEEE journal of biomedical and health informatics* **PP**February 2015.
17. A. Pimenta *et al.*, Monitoring Mental Fatigue through the Analysis of Keyboard and Mouse Interaction Patterns, in *Hybrid Artificial Intelligent Systems*, (Springer Berlin, 2013).
18. H. Jaygarl, C. Luo *et al.*, *Professional Tizen Application Development* (Wiley, 2014).
19. C. for Disease Control, Prevention *et al.*, HIPAA privacy rule and public health. guidance from CDC and the US Department of Health and Human Services *MMWR: Morbidity and Mortality Weekly Report* **52** (US Centers for Disease Control and Prevention, 2003).
20. J. Lewis and J. Sauro, The Factor Structure of the System Usability Scale, in *Human Centered Design*, ed. M. Kurosu, LNCS, Vol. 5619 (Springer Berlin Heidelberg, 2009) pp. 94–103.
21. A. B. Warriner, V. Kuperman and M. Brysbaert, Norms of valence, arousal, and dominance for 13,915 english lemmas *Behavior research methods* **45** (Springer, 2013).
22. M. F. Porter, Snowball: A language for stemming algorithms (2001).
23. M. L. Wilson, S. Ali and M. F. Valstar, Finding information about mental health in microblogging platforms, in *Proceedings of the 5th Information Interaction in Context Symposium on - IIiX '14*, (ACM Press, New York, New York, USA, August 2014).
24. H. Asada, P. Shaltis, A. Reisner and R. Hutchinson, Mobile monitoring with wearable photoplethysmographic biosensors *IEEE Engineering in Medicine and Biology Magazine* **22**May 2003.
25. R. Mazumder, T. Hastie and R. Tibshirani, Spectral Regularization Algorithms for Learning Large Incomplete Matrices. *Journal of machine learning research : JMLR* **11**March 2010.
26. M. Udell, C. Horn, R. Zadeh and S. Boyd, Generalized Low Rank ModelsOctober 2014.

BAYESIAN BICLUSTERING FOR PATIENT STRATIFICATION

SAHAND KHAKABIMAMAGHANI and MARTIN ESTER

School of Computing Science, Simon Fraser University, 8888 University Drive
Burnaby, BC, V5A 1S6, Canada
E-mail: {sahandk, ester}@sfu.ca

The move from Empirical Medicine towards Personalized Medicine has attracted attention to Stratified Medicine (SM). Some methods are provided in the literature for patient stratification, which is the central task of SM, however, there are still significant open issues. First, it is still unclear if integrating different datatypes will help in detecting disease subtypes more accurately, and, if not, which datatype(s) are most useful for this task. Second, it is not clear how we can compare different methods of patient stratification. Third, as most of the proposed stratification methods are deterministic, there is a need for investigating the potential benefits of applying probabilistic methods. To address these issues, we introduce a novel integrative Bayesian biclustering method, called B2PS, for patient stratification and propose methods for evaluating the results. Our experimental results demonstrate the superiority of B2PS over a popular state-of-the-art method and the benefits of Bayesian approaches. Our results agree with the intuition that transcriptomic data forms a better basis for patient stratification than genomic data.

1. Introduction

In Empirical Medicine every patient of a particular disease receives the same treatment. However, although working for simpler diseases to a degree, this approach has not been successful for more complex diseases like cancer. Therefore, the paradigm in medicine is shifting from Empirical to so called Personalized Medicine, which is a patient derived approach with the goal of providing individual treatments for each patient according to his/her particular conditions and features. As an intermediate step currently being investigated, "Stratified Medicine is an approach by which groups of patients with the same disease are subdivided into different categories depending on the underlying mechanism of disease and their probable response to a therapeutic intervention [1]."

According to the definition of stratified medicine, a cohort of patients is divided into subgroups, called subtypes, and the specific features of each subtype that constitute the disease mechanism for that subtype are identified. These features will then be used to design subtype-specific treatments. One possible approach to patient stratification is Biclustering, which is proven useful for this task [2] and is commonly in use for it. A comprehensive discussion of bi-clustering methods can be found in [3]. Most of the biclustering algorithms proposed in the literature utilize an optimization method to find the solution. They can be categorized into two main classes:

1. *Deterministic:* Examples are Singular Value Decomposition (SVD) [4] and Non-negative Matrix Factorization (NMF) [5], which try to optimize the value of latent variables indicating the clustering structure. Although these methods initialize the latent factors randomly, given the same initial random parameters, they will always produce the same final result.
2. *Probabilistic:* this family of methods models the data as a Bayesian network of variables with cluster ids being a latent variable. Examples are Plaid [6] and SAMBA [7]. These methods

also use random initialization; however, since they use stochastic optimization, they might produce different solutions in different executions given the same initial values.

Methods in the second group usually return a probabilistic assignment of objects to clusters. This is more desirable for patient stratification, because first, it provides a model-based (rather than ad-hoc) approach to predict subtypes for new patients with unknown subtypes, and second, patients in one subtype often share features with patients in other subtypes and probabilistic assignments to subtypes capture these similarities and are more informative than strict assignments [8]. Furthermore, stochastic optimization methods are less prone to get stuck in local optimums. In addition, probabilistic models allow for introduction of prior knowledge into model.

In terms of the diversity of data types used as stratification input, methods can be categorized into single-input and integrative. Hofree et al. [5] and Hochreiter et al. [9] are examples of single-input approaches. They, respectively, use somatic point mutation and gene expression data. While some (but not all) of these publications provide comparisons between their methods and existing methods, these comparisons were conducted using either synthetic data or real databases with clinically known subtypes and, as also discussed in [2] and to the best of our knowledge, no suitable metric is provided for benchmarking when the data are real and unlabeled.

Some single-input stratification methods use a different approach by finding the subtypes based on only a single data type, fixing the detected subtypes, and then integrating other data types to investigate subtype-specific features in those datasets. Examples are two prominent references Verhaak et al. [10] and Cho and Przytycka [8], both of which used gene expression data as the main datatype for patient stratification, but they did not discuss the logical reasons for this choice.

As an example of integrative methods, Shen et al. [11] proposed a Bayesian method, namely iCluster, for integrative clustering of genomic data and applied it to breast and lung cancer data. In another study, Sun et al. [4] proposed a multi-view SVD method and applied it for integrating genomic and clinical data to find disease subtypes and their associated genetic variations. We note that these publications do not compare with competitors and do not demonstrate the merits of the integrative approach compared to single-input patient stratification through benchmarking experiments. Although Sun et al. [4] used AUC scores for discussing this point, we believe that their results are not an indicator of superiority of the integrative method, but are the natural result of their experimental setup. Furthermore, they only examine combining clinical and point mutation data and do not consider other genomic, transcriptomic, or proteomic data types.

Table 1 summarizes the mentioned approaches to patient stratification and compares them according to the discussed aspects. According to our discussion, the merit of integrating different datasets for patient stratification is still an open issue. Furthermore, no systematic methods and metrics have been presented in the literature for evaluating patient clustering results and efforts have been focused rather on gene clustering (as in Prelic et. al [2]). Moreover, as also seen in Table 1, the utility of probabilistic methods in patient stratification is overlooked, although they are frequently applied for gene clustering. As discussed earlier, these methods provide potential solutions for the problems in patient stratification.

In this paper, we address these open issues by proposing a novel Probabilistic Graphical Model (PGM), which we call B2PS (Bayesian Biclustering for Patient Stratification), and appropriate evaluation metrics. To the best of our knowledge, the model provided here is the first Integrative Bayesian Biclustering model. While there are solutions for Integrative Biclustering [12] as well as Bayesian Biclustering [13] in the literature, no work so far combines integrative, Bayesian, and Biclustering concepts in one model.

Table 1. Existing and proposed methods

Method	Probabilistic or Deterministic	Clustering/ Biclustering	Stratification Input Datatypes
Verhaak et al. (2010) [10]	Deterministic (HC)	Clustering	Expression
Hochreiter et al. (2010) [9]	Deterministic (FA)	Biclustering	Expression
Hofree et al. (2013) [5]	Deterministic (NMF)	Biclustering	Mutation
Shen et al. (2009, 2012) [11, 14]	Deterministic (FA)	Clustering	Multiple
Sun et al. (2014) [4]	Deterministic (SVD)	Clustering	Multiple
Cho & Przytycka (2013) [8]	Probabilistic (PGM)	Clustering	Multiple
B2PS	Probabilistic (PGM)	Biclustering	Multiple

Abbreviations used in this table: HC (Hierarchical Clustering) – FA (Factor Analysis)

The main contributions of this paper are as follows:

- The proposed model allows for **incorporation of prior knowledge**, which is useful for dealing with noisy data. Our experimental results show that this ability is useful for processing noisy biological data and improves the stratification performance.
- The proposed method is able to **detect the natural number of clusters** for each dimension (i.e., row and column), identification of which requires an iterative trial process in deterministic methods. Measured evaluation metrics indicates that the natural sample clusters detected by our method form a better partitioning than the one detected by conventional NMF.
- Unlike conventional bi-clustering methods, **the number of row and column clusters is not assumed to be the same** in our model. This is a useful assumption that is more consistent with typical biological datasets and, according to our experimental results, provides a more informative clustering across both dimensions.
- The integrative method proposed here allows for examination of patient stratification results when using **different combinations of diverse datatypes** with no theoretical limitation on the number of data types. This makes it possible to identify the datatypes that are more useful for patient stratification. Experimental results with two TCGA datasets suggest that gene expression data is more informative than genomic data for patient stratification.

We compare the performance of B2PS against NMF, a state-of-the-art deterministic method. Experimental results demonstrate the superiority of B2PS over NMF regarding both patient stratification and feature clustering in different experimental settings. We believe that the outputs of the proposed method can be a useful basis for detecting the subtype-specific driver aberrations, which is one of the goals of stratified and personalized medicine.

2. Methods

2.1. Model

To perform patient stratification using different datatypes, an integrative probabilistic graphical model for biclustering is proposed. The model is shown in Figure 1. Observed variables are shaded and hyperparameters are in dotted circles. Table 2 includes a detailed description of the variables of the model.

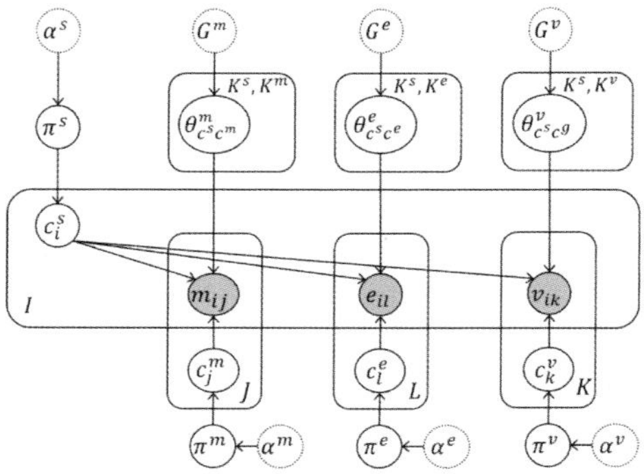

Fig. 1. The probabilistic graphical model of B2PS

Because the goal is to integrate different datatypes about the same set of patients/samples, in our model, datasets of different datatypes are assumed to have the same rows/samples but can have different columns/features. Accordingly, the row clustering is shared across different datatypes, but each dataset has its particular column clustering. However, column clusterings of different datatypes are indirectly related to each other through the shared row clustering. While, no direct dependency is assumed between sample clusters c_i^s and gene clusters c_l^e,c_j^m, and c_k^v in this model, they are indirectly dependent given the observed data variables. In terms of clustering structures discussed in [13], B2PS produces a single non-overlapping clustering, meaning that each row/column belongs to a single cluster that has no overlap with other clusters.

2.2. Parameter Learning and Inference

The Gibbs sampling method [15] is used for parameter learning and latent variable inference. After random initialization, the latent variables (see Table 2) are iteratively sampled one by one based on computed marginal conditional probabilities. Eq. 1 shows the conditional probability for sample/row clusters. Parameters π^m, π^e, and π^v and hyperparameters α^m, α^e, and α^v are not included in this equation for they are conditionally independent from c_i^s given c^m, c^e, and c^v (refer to the model in Figure 1). Other absent parameters are integrated out.

Table 2: Parameters and variables included in B2PS

Type	Name	Description	Distribution
Observed Variables	e_{il}	Expression status of gene l of sample i	$e_{il} \sim \text{Multinomial}_3\left(\theta^e_{c_i^s c_j^g}\right)$
	m_{ij}	Mutation status of gene j of sample i	$m_{ij} \sim \text{Bernoulli}\left(\theta^m_{c_i^s c_j^g}\right)$
	v_{ik}	Copy number variation of gene k of sample i	$v_{ik} \sim \text{Multinomial}_5\left(\theta^v_{c_i^s c_j^g}\right)$
Hyperparameters	α^s	The parameter of prior Dirichlet distribution for samples. K^s is the number of sample clusters. K^s and p^s are provided as input.	$\alpha^s = [\frac{p^s}{K^s} \cdots \frac{p^s}{K^s}]_{1\times K^s}$
	α^x	The parameter of prior Dirichlet distribution for fetures of data type x. K^x is the number of feature clusters. K^x and p^x are provided as input.	$\alpha^x = [\frac{p^x}{K^x} \cdots \frac{p^x}{K^x}]_{1\times K^x}$
	G^x	The parameters for prior distributions of $\theta^x_{c^s c^x}$ for data type x. β values are provided as input.	$G^m = \{\beta_0^m, \beta_1^m\}$ $G^v = \{\beta_{-2}^v, \beta_{-1}^v, \beta_0^v, \beta_1^v, \beta_2^v\}$ $G^e = \{\beta_{-1}^e, \beta_0^e, \beta_1^e\}$
Model Parameters	π^s	Distribution of the probability of belonging to different sample clusters	$\pi^s \sim \text{Dirichlet}_{K^s}(\alpha^s)$
	π^x	Distribution of the probability of belonging to different feature clusters for data type x	$\pi^x \sim \text{Dirichlet}_{K^x}(\alpha^x)$
	$\theta^x_{c^s c^x}$	Parameters for distribution of the values of the entities belonging to bicluster (c^s, c^x) datatype x	$\theta^m_{c^s c^\mu} \sim \text{Beta}(G^m)$ $\theta^v_{c^s c^v} \sim \text{Dirichlet}_5(G^v)$ $\theta^e_{c^s c^e} \sim \text{Dirichlet}_3(G^e)$
Latent Variables	c_i^s	Cluster id for ith sample (sampled variable)	$c_i^s \sim \text{Multinomial}_{K^s}(\pi^s)$
	c_l^e, c_j^m, c_k^v	Cluster id for lth, jth, and kth gene in corresponding datasets (sampled variable)	$c_r^x \sim \text{Multinomial}_{K^x}(\pi^x)$

In the above table, x can be m (point mutation), e (gene expression), or v (copy number variation).

All variables used in Eq. 1 and Eq. 2 (below) are described in Table 3. The right side of Eq. 1 has generally two terms; the first term accounts for the size of clusters (i.e., larger clusters are assigned greater probability) and the second term incorporates the similarity of row i to the members of each cluster (i.e., giving higher probability for assigning row i to clusters with more similar members). Values of the hyperparameters control the balance between these two terms. Feature clusters for different data types are sampled similarly. As an example, the Eq. 2 is the conditional probability of feature clusters according to gene expression data.

$$P(c_i^s = q | c_{-i}^s, c^m, c^e, c^v, m, e, v; \alpha^s, G^m, G^v, G^e)$$
$$\propto P(c_i^s = q, c_{-i}^s, c^m, c^e, c^v, m, e, v; \alpha^s, G^m, G^v, G^e)$$
$$\propto \frac{ns_q^{-i} + \alpha_q^s}{ns^{-i} + p^s} \times \prod_{x \in \{m,e,v\}} \left[\prod_{t=1}^{K^x} \prod_{\{r | c_r^x = t\}} \left(\frac{nx_{qt}^{x_{ir},-i} + \beta_{x_{ir}}^x}{nx_{qt}^{-i} + \beta_x} \right) \right]^{D_x} \quad (1)$$

$$P(c_j^e = q | c_{-j}^e, c^s, e; \alpha^e, G^e) \propto \frac{ne_q^{-j} + \alpha_q^e}{ne^{-j} + p^e} \times \prod_{t=1}^{K^s} \prod_{\{i | c_i^s = p\}} \left(\frac{ne_{tq}^{e_{ij},-j} + \beta_{e_{ij}}^e}{ne_{tq}^{-j} + \beta_e} \right) \quad (2)$$

Table 3: The variables included in sampling conditional probabilities

Variable	Description
c_{-i}^s	Cluster id variables for all samples except ith sample
c_{-j}^e	Cluster id variables for all features of expression datatype except jth feature
ns^{-i}	The total number of samples minus one (the ith sample)
ns_a^{-i}	The number of samples in sample cluster a excluding the ith sample
nx^{-r}	The total number of features in database x minus one (the rth feature)
nx_b^{-r}	The number of features in feature cluster b of dataset x excluding the rth feature
nx_{ab}^{-i}, nx_{ab}^{-r}	The number of elements in bicluster (a,b) in dataset x except those elements related to the ith sample or rth feature, respectively
$nx_{ab}^{x_{ir},-i}$, $nx_{ab}^{x_{ir},-r}$	The number of elements in bicluster (a,b) in dataset x whose value equals x_{ir} except those elements related to the ith sample or rth feature, respectively
β_x	$\beta_x = \sum_d \beta_d^x$, where d is the values that a data point of type x can take (e.g., for point mutation $d \in \{0,1\}$)
D_x	A binary variable indicating inclusion ($D_x = 1$) or exclusion ($D_x = 0$) of data type x in or from the conditional probability, when examining different combinations of datatypes.

In the above table, x can be m (point mutation), e (gene expression), or v (copy number variation).

The number of clusters for samples and genes are denoted respectively by K^s and K^x, where x can be m, e, or v (see Table 2). The random initialization of cluster id variables produces a uniform distribution of entities to these clusters. However, according to the terms included in above conditional probabilities, sampling tends to minimize the number of clusters such that the members of a cluster are highly similar. So, as the biclustering converges throughout the iterations, some clusters become empty with no entities assigned to them, if the values for K^s and K^x are set large enough. Accordingly, after each execution of learning algorithm (until convergence) the natural number of clusters can be determined as the number of occupied clusters.

2.3. Computing Final Clusters and Model Parameters

Due to the stochastic nature of Gibbs sampling, the results of two distinct executions can be different. Therefore, as in [5] and [16], a consensus method based on repeated execution of the learning algorithm is used to yield a more robust clustering. This method is based on a similarity matrix, where the similarity is measured as the number of times (out of several executions) that two entities (samples or genes) belong to the same cluster at the end of an execution. Then, the consensus matrices (one for each dimension) are used to perform UPGMA hierarchical clustering to identify the final sample and gene clusters. The number of clusters used for hierarchical clustering is the average of the number of clusters occupied at the end of different executions. After finding the final clustering structures, the model parameters can be estimated as maximum a posteriori probabilities.

2.4. Comparison Partner

To compare the performance of the proposed probabilistic model with deterministic methods, we use a popular method for patient stratification based on Non-negative Matrix Factorization (NMF). We used the multiplicative NMF algorithm of Lee and Seung [17]. We downloaded the MATLAB implementation by Zhang et al. [12], who modified and used the algorithm for biclustering genomic and transcriptomic data. We amended the code to produce consensus matrices for further post-processing described in section 2.6.

2.5. Evaluation

Between two main categories of internal and external measures used to evaluate clustering results, we used external measures, which are more suitable for assessing the performance of patient or gene clustering algorithms [2]. According to the goal of patient stratification, different patient groups are expected to exhibit distinctive responses to treatments. Therefore, for evaluating the patient clustering results, we use clinical data and perform survival analysis. We use the log-rank test [18] implemented in R 'survival' package. The smaller the log-rank *p*-value, the more distinctive the survival behavior of different patient clusters. This measure is a popular measure for validating stratification results, but, to the best of our knowledge, it has not been used for comparing different clustering algorithms.

Since the main goal of this study is sample stratification, we also measure the stability and robustness of sample clustering outputs regarding the Cophenetic Correlation Coefficient using the method described by Brunet et al. [16]. This is a measure between 0 and 1 and approaches 1 as results of an experiment are more repeatable and robust. Since almost all of the features of the datasets used in our experiments are genes, the Gene Ontology Term Overlap (GOTO) [19] criterion is used for evaluating the feature clustering. Larger values of this metric imply more meaningful clustering in terms of biological relationship between cluster members.

2.6. Parameter Tuning

To determine the best number of clusters for NMF, the method proposed by Brunet et al. [16] is used, which is based on the Cophenetic Correlation Coefficient briefly described in section 2.5. Similar to method described in section 2.3 for B2PS, a consensus matrix is computed throughout execution of NMF for the same number of times as for B2PS. This experiment is repeated with different numbers of clusters and the Cophenetic Correlation Coefficient is recorded for each experiment. Finally, a chart showing the trend of the Cophenetic Correlation Coefficient versus the increasing number of clusters is drawn and the number after which the coefficient value decreases considerably is chosen as the optimal number of clusters.

The parameters of B2PS are the hyperparameters of prior distributions of values for data points and cluster assignment probabilities. Sample clustering hyperparameter α^s is common among all datatypes, however, feature clustering and data value priors are distinct for different datatypes. Clustering hyperparameters are set uniformly as shown in Table 2 and depend on the values of p^s (for samples) and p^x (for features). For weak or non-informative priors, these values are set to 1 and for strong or informative priors they are set according to the number of samples and features of the dataset being analyzed. Data value prior hyperparameters are set according to their real distribution in the dataset under investigation. When weak, they are scaled such that β values (see Table 2) of the data types being analyzed sum to one. Strong priors are adjusted according to the size of the dataset under analysis.

The optimal values of hyperparameters for each datatype are selected through a trial process that optimizes for log-rank p-value. For integrated analysis of several datatypes, the prior settings of individual data types are used. For common hyperparameter α^s, the value used for the datatype producing the best sample clustering in its independent analysis is used.

3. Experiments

3.1. Data

Data for this research are obtained from The Cancer Genome Atlas (TCGA) online dataset [20]. Data include genomic data, namely somatic point mutation and genome-wide copy number variation, and transcriptomic gene expression data. Data are about Glioblastoma Multiform (GBM) and Breast Invasive Carcinoma (BRCA) patients. For each disease, data of a subset of patients/samples having records for all three datatypes mentioned above is downloaded.

To be analyzable with our method, data are preprocessed into three matrices where rows refer to samples and columns refer to features (i.e., genes or miRNAs). According to different properties of the three datatypes, different preprocessing methods are used. Final values are 0 (for genes not containing any non-silent mutation) and 1 (otherwise) for point mutation data, {-2, -1, 0, 1, 2} (the change in the normal number of copies of a gene or miRNA computed by GISTIC2.0 [21]) for CNV, and -1 (under-expression), 0, and +1 (over-expression) for gene expression data (capturing changes more than two fold). Number of features of preprocessed final datasets for somatic point mutation, CNV, and expression data were respectively 4117, 23082, 11874 for 102 GBM samples and 13776, 23082, and 17814 for 501 BRCA samples. Because NMF only accepts non-negative values, for experiments with NMF these data are further preprocessed using the method described in [12]. Clinical data were also available for the patients and contained information required for survival analysis. We retrieved gene ontology data for GOTO analysis using the 'biomaRt' R package [22].

3.2. *Results*

The experiments are designed with three goals in mind: 1) to show the benefit of the ability to incorporate prior knowledge enabled by the Bayesian approach, 2) to identify the best combination of datatypes for patient stratification, and 3) to compare the proposed method with a state-of-the-art method. In all experiments, the learning algorithm is executed 50 times for both B2PS and NMF. To set the number of iterations for each execution, the learning algorithm is first applied with a large number of iterations, the point of (relative) convergence of the objective function is detected manually, and then the algorithm is run with that number of iterations.

3.2.1. Effects of Priors

To investigate the effects of priors on performance of B2PS, different combinations of strong and weak values for hyperparameters are examined. As an example, the results of a subset of different possible settings for GBM expression dataset are shown in Table 5. Since the main goal of this research was sample stratification, final selected priors (bolded in table) favor better sample clustering over better gene clustering.

According to these and similar results for the BRCA dataset (not reported due to page limit), strong data priors increase the performance regarding the sample clustering with a slight decrease in gene clustering score. This can be explained by the fact that strong priors cancel the noise of gene expression data to a degree, which generally, is expected to increases the sizes of sample and gene clusters. For sample clusters, this effect is somewhat attenuated according to strong patterns in expression profiles of each cluster and the number of clusters remain almost the same. However for gene clusters, this effect merges more similar gene clusters resulting in fewer clusters.

Strong priors for clustering have a reverse effect on clustering structure. As the clustering priors increase, tendency to create clusters with higher similarity among their members increases. So, we should expect smaller and more precise clusters and, consequently, larger number of clusters. Once more, for the same reasons mentioned for data prior, this is more observable for gene clustering rather than sample clustering. Generally the results endorse the usefulness of ability to include prior knowledge in patient stratification.

3.2.2. Informative Datatypes for Patient Stratification

To identify the most informative datatypes for patient stratification we examined different combinations of three datatypes: somatic point mutation, copy number variation and gene expression. Results are summarized in Table 6 for GBM and BRCA datasets. Here, no results are reported for point mutation data, because, due to high heterogeneity of these data, independent experiments with point mutation dataset did not converge to any stable results and, moreover, point mutation data did not have any effects on the output of integrative experiments.

According to the results, gene expression data, when used alone, produces the best result according to both sample clustering (log-rank p-value) and gene clustering (GOTO score). For sample clustering, this can be related to the fact that gene expression profiles are closer to final phenotypes and reflect the cumulative effects of molecular aberrations occurred in earlier steps of central dogma of biology better than other mentioned data types. For gene clustering, higher

GOTO score for expression data compared to others is interpretable according to the fact that genes with similar expression patterns across different samples are more likely to share the same functions in cell than genes with similar CNV.

Table 5: Different prior settings for experiments with GBM gene expression dataset

Priors			Num. of Sample Clusters	Num. of Feature Clusters	Log-rank p-value	GOTO
Data	Sample Clustering	Gene Clustering				
weak	weak	weak	8	66	0.018	3.444
strong	**weak**	**weak**	**8**	**25**	**0.004**	**3.408**
strong	strong	weak	9	21	0.017	3.404
strong	weak	strong	8	73	0.019	3.415
strong	strong	strong	8	70	0.008	3.418

Table 6: Results of integrative and single input experiments for GBM and BRCA

Dataset	Data Types	Sample Clusters	Feature Clusters		Log-rank p-value	Cophenetic Corr. Coef.	GOTO	
			Exp.	CNV			Exp.	CNV
GBM	Exp.	8	25	NA	**0.004**	0.958	**3.408**	NA
GBM	CNV	19	NA	86	0.411	0.976	NA	1.820
GBM	Exp. and CNV	7	22	68	0.292	0.799	3.403	1.802
BRCA	Exp.	8	69	NA	**0.140**	0.935	**2.598**	NA
BRCA	CNV	20	NA	63	0.353	0.913	NA	1.854
BRCA	Exp. and CNV	11	69	68	0.535	0.897	2.580	1.857

Moreover, according to the results, combination of expression and CNV data types introduces noise and decreases the robustness (the Cophenetic Correlation Coefficient) of the results and, deteriorates performance of sample and gene clustering compared to when gene expression is used alone. This is related to the inconsistency between different data types and the fact that different genotypes can be transcribed and translated into similar phenotypes.

3.2.3. B2PS vs. NMF

Comparison between the proposed method and NMF is conducted using gene expression data, which is here detected as the most informative datatype for patient stratification. To identify the number of clusters of NMF, the method described in section 2.6 is used. The results of NMF with the selected number of clusters and B2PS with the detected number of clusters are included in Table 7 for GBM and BRCA datasets. According to the results, although NMF produces slightly more robust results (which can be related to the higher number of clusters for B2PS), B2PS produces remarkably more meaningful stratification and feature clusters.

Table 7. Comparison between B2PS and NMF

Dataset	Method	Sample Clusters	Feature Clusters	Log-rank p-value	Cophenetic Corr. Coef.	GOTO
GBM	B2PS	8	25	**0.004**	0.958	**3.408**
	NMF	3	3	0.458	0.965	2.535
	B2PS	3	29	0.047	0.967	**3.405**
	B2PS	3	6	0.217	0.999	3.392
BRCA	B2PS	8	69	**0.140**	0.935	**2.598**
	NMF	3	3	0.226	0.991	2.541
	B2PS	3	101	0.120	0.998	**2.603**
	B2PS	3	6	0.489	0.983	2.548

To see whether B2PS can also perform as well when the numbers of sample clusters are the same for both methods, in another experiment, B2PS is forced to find the clustering structure with the number of subtypes detected by NMF. Results shown in Table 7 approves that B2PS performs better stratification and, interestingly, when the number of sample clusters of B2PS is restricted, the number of detected feature clusters increases and the quality of feature clusters remain almost the same as (slightly better than) the unrestricted case. To examine if this flexibility in the number of clusters across two different dimensions is an advantage that is effective in superior performance of B2PS, the results are compared with the case when this flexibility is discarded by simulating the inflexibility of NMF. For this, the numbers of sample and feature clusters are set "logically" equal for B2PS. Since, unlike NMF, B2PS inputs consists of both negative and positive values, then "logically" equivalent setting for B2PS is when the number of feature clusters is twice the number of sample clusters. The results of these double-restricted experiments are also included in Table 7. As it can be seen, this additional restriction distorts the performance in both aspects of sample and feature clustering considerably. Accordingly, results support the hypothesis that flexibility in the number of clusters improves the performance.

4. Conclusions

We proposed a novel probabilistic graphical model, called B2PS, for Bayesian integrative biclustering of biological data for patient stratification. Our experimental results demonstrate the effectiveness of the Bayesian approach for inclusion of prior knowledge and detection of a natural number of clusters. Our experiments also show that B2PS is more effective in patient stratification than NMF, due to the probabilistic nature of B2PS and its flexibility in the number of clusters across two dimensions. In cases where gene expression data is collectible (e.g., cancer), this type of data turns out to be more informative than other genomic data for patient stratification at least for the datasets used in this study. For diseases where gene expression data cannot be gathered from the relevant tissue, methods like the one proposed in [5], which preprocess the genomic data to reduce their heterogeneity, can be useful. B2PS helps achieving the ultimate goal of stratified medicine by providing more robust subtypes and gene clusters, which can serve as a starting point to find subtype-specific gene expression profiles and consequently subtype specific pathways or subnetworks. This information together with the mutation profiles can then be employed to find the driver genetic variations for each subtype (the hallmark of stratified medicine). Future research may explore the integration of other data types (e.g., methylation, miRNA expression, and other structural variations like gene fusion) as well as increasing the resolution of the current datatypes (e.g., modeling gene expression as continuous distribution).

References

[1] J. C. D. Willis and G. M. Lord, *Nature Reviews (Immunology)* 15, 323 (2015).

[2] A. Prelic, S. Bleuler, P. Zimmermann, A. Wille, et al., *Bioinformatics* 22, 1122 (2006).

[3] A. Oghabian, S. Kilpinen, S. Hautaniemi and E. Czeizler, *PLoS ONE* 9 (2014).

[4] J. Sun, J. Bi and H. R. Kranzler, *BMC Genetics* 15 (2014).

[5] M. Hofree, J. P. Shen, H. Carter, A. Gross and T. Ideker, *Nature Methods,* 1108 (2013).

[6] L. Lazzeroni and A. Owen, *Statistica Sinica* 12, 61 (2002).

[7] A. Tanay, R. Sharan and R. Shamir, *Bioinformatics* 18, 136 (2002).

[8] D. Y. Cho and T. M. Przytycka, *Nucleic Acids Res.* 41, 8011 (2013).

[9] S. Hochreiter, U. Bodenhofer, M. Heusel, A. Mayr et al., *Bioinformatics* 26, 1520 (2010).

[10] R. G. Verhaak, K. A. Hoadley, E. Purdom, V. Wang, et al., *Cancer Cell* 17, 98 (2010).

[11] R. Shen, A. B. Olshen and M. Ladanyi, *Bioinformatics* 25, 2906 (2009).

[12] S. Zhang, C. Liu, W. Li, H. Shen et al., *Nucleic Acids Res* 40, 9379 (2012).

[13] E. Meeds and S. Roweis, UTML TR 2007–001, University of Toronto, Toronto (2007).

[14] R. Shen, Q. Mo, N. Schultz, V. E. Seshan, A. B. Olshen, J. Huse, et al., *PLoS One* 7, (2012).

[15] G. Casella and E. I. George, *The American Statistician* 46, 167 (1992).

[16] D. D. Lee and H. S. Seung, *Adv. Neural Inform. Process. Syst* 13, 556 (2001.

[17] N. Mantel, *Cancer Chemotherapy Reports* 50, 163 (1966).

[18] J. P. Brunet, P. Tamayo, T. R. Golub and P. M. Jill, *PNAS* 12, 4164 (2004.

[19] M. Mistry and P. Pavlidis, *Bioinformatics* 9 (2008).

[20] "The Cancer Genome Atlas," [Online]. Available: http://cancergenome.nih.gov/.

[21] C. Mermel, S. Schumacher, B. Hill, M. L. Meyerson, R. Beroukhim and G. Getz, *Genome Biology* 12 (2011).

[22] S. Durinck, Y. Moreau, A. Kasprzyk, S. Davis et al., *Bioinformatics* 21, 3439 (2005).

BIOFILTER AS A FUNCTIONAL ANNOTATION PIPELINE FOR COMMON AND RARE COPY NUMBER BURDEN

DOKYOON KIM[1], ANASTASIA LUCAS[1], JOSEPH GLESSNER[2], SHEFALI S. VERMA[1], YUKI BRADFORD[1], RUOWANG LI[1], ALEX T. FRASE[1], HAKON HAKONARSON[2], PEGGY PEISSIG[3], MURRAY BRILLIANT[3], MARYLYN D. RITCHIE[1,4]*

[1]*Center for Systems Genomics, Department of Biochemistry and Molecular Biology, Pennsylvania State University, University Park, Pennsylvania, USA*

[2]*Center for Applied Genomics, Children's Hospital of Philadelphia, Philadelphia, Pennsylvania, USA*

[3]*Biomedical Informatics Research Center, Marshfield Clinic Research Foundation, Marshfield, Wisconsin, USA*

[4]*Biomedical & Translational Informatics, Geisinger Health System, Danville, Pennsylvania, USA*

Email: marylyn.ritchie@psu.edu

Recent studies on copy number variation (CNV) have suggested that an increasing burden of CNVs is associated with susceptibility or resistance to disease. A large number of genes or genomic loci contribute to complex diseases such as autism. Thus, total genomic copy number burden, as an accumulation of copy number change, is a meaningful measure of genomic instability to identify the association between global genetic effects and phenotypes of interest. However, no systematic annotation pipeline has been developed to interpret biological meaning based on the accumulation of copy number change across the genome associated with a phenotype of interest. In this study, we develop a comprehensive and systematic pipeline for annotating copy number variants into genes/genomic regions and subsequently pathways and other gene groups using Biofilter – a bioinformatics tool that aggregates over a dozen publicly available databases of prior biological knowledge. Next we conduct enrichment tests of biologically defined groupings of CNVs including genes, pathways, Gene Ontology, or protein families. We applied the proposed pipeline to a CNV dataset from the Marshfield Clinic Personalized Medicine Research Project (PMRP) in a quantitative trait phenotype derived from the electronic health record – total cholesterol. We identified several significant pathways such as toll-like receptor signaling pathway and hepatitis C pathway, gene ontologies (GOs) of nucleoside triphosphatase activity (NTPase) and response to virus, and protein families such as cell morphogenesis that are associated with the total cholesterol phenotype based on CNV profiles (permutation *p-value* < 0.01). Based on the copy number burden analysis, it follows that the more and larger the copy number changes, the more likely that one or more target genes that influence disease risk and phenotypic severity will be affected. Thus, our study suggests the proposed enrichment pipeline could improve the interpretability of copy number burden analysis where hundreds of loci or genes contribute toward disease susceptibility via biological knowledge groups such as pathways. This CNV annotation pipeline with Biofilter can be used for CNV data from any genotyping or sequencing platform and to explore CNV enrichment for any traits or phenotypes. Biofilter continues to be a powerful bioinformatics tool for annotating, filtering, and constructing biologically informed models for association analysis – now including copy number variants.

Keywords: Copy number burden, functional annotation, electronic medical record, precision medicine

1. Introduction

Precision medicine, an emerging approach for prevention and treatment strategies that takes into account individual variability in genes, lifestyle, and environment for each person, has become one of the main research interests of biomedical science [1]. Recently, a precision medicine initiative was announced as a new research initiative that plans to boost progress toward a new era of personalized medicine [1]. Thus, collecting and utilizing patients' rich information through electronic health records (EHRs) is one of the most important keys in precision medicine in order to tailor disease prevention and effective treatment strategies. First, precision medicine will need to be tested in many pilot studies to guide clinical practice.

The electronic MEdical Record and GEnomics (eMERGE) is a national network organized and funded by the National Human Genome Research Institute (NHGRI) that combines DNA repositories linked with electronic medical record (EMR) systems for performing large scale, high-throughput genetic association studies [2]. Many genome-wide association studies (GWAS) have been performed for multiple phenotypes generated from the eMERGE network [3,4]. In addition, a phenome-wide association study (PheWAS) approach has been used to query genotype-phenotype associations between targeted single-nucleotide polymorphisms (SNPs) and multiple phenotypes and to detect pleiotropy [5]. Despite many efforts to investigate genotype-phenotype associations, genetic studies to date have still only identified a small fraction of the heritability of complex traits [6]. Many alternative approaches to improve the 'missing heritability' problem have been proposed such as investigating gene-gene interactions associated with phenotypes or a systems genomics approach [7,8]. In addition, an alternative explanation for the 'missing heritability' could be copy number variations (CNVs) [9].

Disease-associated rare/common CNVs have been identified through multiple studies [10,11]. However, one conclusion from the extensive CNV association studies is that there are hundreds or thousands of genes or genomic regions that contribute to disease susceptibility for certain disorders such as autism. Thus, total genomic copy number burden, as an accumulation of copy number change, is a meaningful measure of genomic change that may contribute to phenotypes that are associated with many genes/regions. Previously, we found that autism is associated with increased levels of copy number burden [12]. However, one of the current limitations of this approach is that it is difficult to interpret biological meaning based on the accumulation of copy number change genome-wide. Is it the amount of copy number change that is important or is it which genes/pathways the copy number change occurs that is important? In this study, we develop a comprehensive and systematic pipeline for annotating copy number variants into genes/genomic regions and subsequently pathways and other gene groups using Biofilter – a bioinformatics tool that aggregates over a dozen publicly available databases of prior biological knowledge [13]. Next we conduct enrichment tests of biologically defined groupings of CNVs including pathways, Gene Ontology (GO), or protein families. We applied the proposed pipeline to a CNV data set in a cholesterol phenotype from the Marshfield Clinic, a study site of the eMERGE network. We identified several significant pathways, GOs, and protein families that are associated with the cholesterol phenotype based on CNV profiles. The results discussed herein demonstrate the utility of the proposed pipeline as a novel method for annotating the results of CNV burden analysis underlying complex traits such as cholesterol.

2. Methods

2.1. *Data*

Median total cholesterol as a phenotype for this study was extracted from the EHR from the Marshfield Personalized Medicine Research Project (PMRP) [14]. Table 1 shows the descriptive statistics of the data set. High-density SNP genotyping was performed on DNA samples at the Center for Inherited Disease Research (CIDR) using the Illumina 660W-Quad. After quality controls (QC), 3,399 samples with available *median total cholesterol* phenotype from the Marshfield PMRP were selected for the present study. DNA samples from this site were genotyped using the Illumina 660W-Quad array as previously described [15]. QC is described in further detail in the *CNV Burden Analysis* section.

Table 1. Descriptive statistics on Marshfield *Median Total Cholesterol* data set. Total number of samples after QC is presented.

Phenotype	Sex		Birthdate Year* (Mean±SD)	Total
	Male	Female		
Median Total Cholesterol	1,428	1,971	3.1779±1.1772	3,399

**Birthdate Year* denotes decade of birth where 1=1910, 2=1920, 3=1930, 4=1940, 5=1950, and 6=1960.

2.2. *CNV Burden Analysis*

Figure 1 shows the illustration of the entire pipeline. In order to detect CNV, log R ratio and B Allele Frequency values were extracted from the Illumina 660W-Quad array. The PennCNV software, based on a hidden Markov model, was used for calling CNVs [16]. First, individual CNV calls were generated as raw CNV calls and then several QC steps were performed. CNVs that had a high success rate of attempted SNPs, a low standard deviation of normalized intensity, and low genomic wave artifacts passed QC thresholds. All samples had genetically inferred European ancestry and any genotypic duplicates were removed. In addition, samples with spurious large homozygous deletions were removed. After QC, 3,399 samples were analyzed for the CNV burden analysis. Linear regression models using PLATO software [17] were fit to the data to evaluate the associations between CNV burden, i.e. accumulation of duplication or deletion in each individual, or collectively, as total base pairs of altered copy number (i.e. total CNV burden), and the median total cholesterol phenotype. Analyses were adjusted for potential confounders, including age (decade of birth), sex, and the first three principal components of ancestry that were generated from the PCA analysis based on SNP data set.

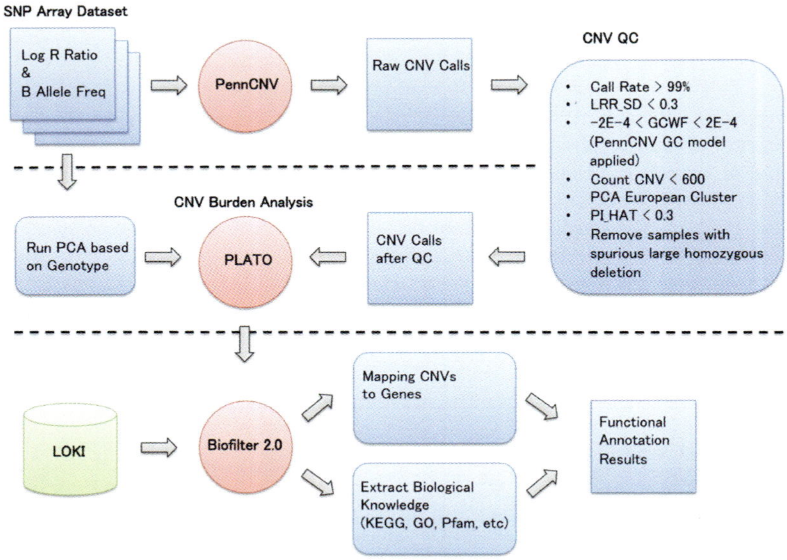

Fig. 1. Illustration of the pipeline for functional annotation based on the results of the CNV burden analyses. PennCNV is used for calling CNVs, then copy number burden analysis is performed using CNV calls after QC. A new function of Biofilter 2.0 provides functional annotation results based on copy number burden.

2.3. *Biofilter 2.0*

Biofilter 2.0 is a software tool that provides a convenient single interface for high-throughput annotation, filtering of genetic data via accessing multiple publicly available human genetic data sources, and constructing biologically informed models for association analysis [13]. This software uses a build-in database called the Library of Knowledge Integration (LOKI), which contains a number of public data resources. LOKI includes not only information about the genomic locations of SNPs and genes, but also information about biological networks, connections, and/or pathways to be used for determining relationships between genes. For more information, see: http://ritchielab.psu.edu/software/.

A new function was added in Biofilter 2.0 for CNV analyses. CNV data, which are specified by a chromosome and base pair range from any genotyping or sequencing platform, can be mapped to genes (Fig 1). These CNV regions can be mapped to genes based on percent of overlap of the genes with the CNV region or based on the number of base pairs overlapped. In addition, biological knowledge such as pathway, GO, or Pfam along with list of its gene members can be extracted using Biofilter 2.0 for the functional annotation calculation based on the results of CNV burden analyses (Fig 1). For the current study, 281 Kyoto Encyclopedia of Gene and Genomes (KEGG) pathways, 1,454 GOs, and 2,908 Pfams were used.

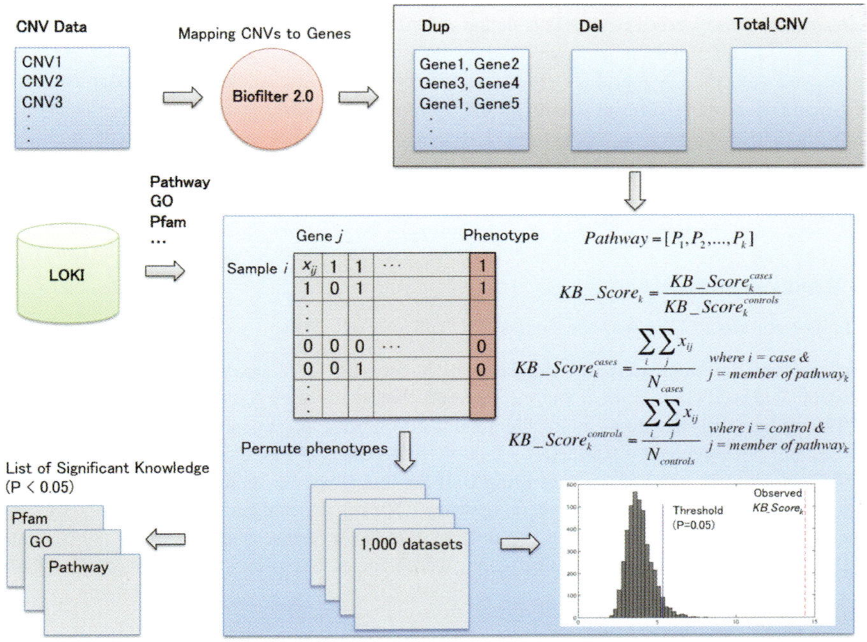

Fig. 2. Overview of the functional annotation calculation based on CNV profiles. After the CNV data set was mapped to genes using Biofilter 2.0, functional enrichment tests can be used to identify significantly enriched biological knowledge such pathway, GO or Pfam. KB, knowledgebase.

2.4. *Functional Annotation based on CNV Profiles*

Figure 2 describes the overview of the functional annotation calculation based on CNV profiles. After the CNV data set was mapped to genes using Biofilter 2.0, functional enrichment tests can be used for the functional annotation. However, an over-representation analysis (ORA) approach, which is one of the most common methods for the pathway analysis, is not appropriate for annotating the results of CNV burden analyses since it does not consider the frequency information of rare and/or common CNV across samples. Thus, we propose a new functional annotation method based on the results of CNV burden analyses in order to capture the frequency information of rare and common CNV. Knowledgebase score (*KB_Score*) was calculated via aggregating not only frequency of genes within a specific pathway but also frequency of those genes across samples (Fig 2). *KB_Scores* can be obtained from cases and controls, respectively, and then each *KB_Score* should be normalized by dividing by the number of cases/controls. Next, a final *KB_Score* can be achieved as a ratio of each *KB_Score*, which is generated from the cases/controls, respectively. After calculating *KB_Score* per pathway, we randomly permuted the phenotype 1,000 times to generate random data sets, that is, the phenotypes are randomly associated with the CNV profiles. To assess whether annotated biological knowledge (observed *KB_Score*) is more significant than expected by chance, any observed *KB_Score* higher than the 950th highest *KB_Score* in the permuted data set was recorded in a final list of significant biological knowledge (Fig 2). Even though the proposed method is suitable for case/control phenotypes, continuous phenotypes can be dichotomized based on quartiles as described more below.

3. Results and Discussion

3.1. *The Results of CNV Burden Analysis*

We assessed the significance of the association between CNV burden variables (duplication, deletion, and total CNV burden) and median total cholesterol using linear regression. Through the CNV burden analysis, duplication and total CNV burden were significantly associated with cholesterol phenotype, $P = 0.0023$, $P = 0.0099$, respectively. Thus, duplication regions and total CNV regions were mapped to genes using Biofilter 2.0. Since functional annotation results were similar between different overlap criteria between CNV regions and genes (data not shown), CNV data was mapped to genes based on 1bp overlap criteria for further analysis. From 3,399 samples 7,150 distinct genes and 9,587 distinct genes were mapped based on duplication and total CNV, respectively.

3.2. *Significant Pathways, GOs, and Pfams Associated with Cholesterol*

Since the proposed method for annotating the results of CNV burden analyses is appropriate for the case/control phenotype, the median total cholesterol, a continuous phenotype, was dichotomized in three different ways based on quartiles: (1) 4th quartile (cases) vs. 3rd, 2nd, 1st quartiles (controls); (2) 4th, 3rd quartiles (cases) vs. 2nd, 1st quartiles (controls); (3) 4th, 3rd, 2nd quartiles (cases) vs. 1st quartile (controls). We compared the annotation results between different dichotomized phenotypes (Table 2). Since total number of significant biological knowledge between dichotomized phenotypes was not too different and there were many distinct biological knowledge that were shared between at least two dichotomized phenotypes, we chose the first way

Table 2. Comparison of total number of significant knowledge features based on different dichotomized cholesterol phenotypes. Each element is the number of significant knowledge features from the functional annotation calculation (P<0.05).

Knowledge Type	CNV Type	Dichotomizing Cholesterol Phenotype			Shared Significant Knowledge*
		4th Quartile **vs.** 3rd, 2nd, 1st Quartiles	4th, 3rd Quartiles **vs.** 2nd, 1st Quartiles	4th, 3rd, 2nd Quartiles **vs.** 1st Quartile	
Pathway	Dup	32	38	36	28
	Total CNV	20	10	21	13
GO	Dup	39	62	62	43
	Total CNV	50	38	58	26
Pfam	Dup	43	46	35	22
	Total CNV	63	57	68	20

**Shared significant knowledge* denotes the number of distinct knowledge that appears in at least more than two dichotomized phenotypes.

of dichotomized phenotype, top quartile as cases vs. three bottom quartiles as controls, for further analysis. This approach is also commonly used in many epidemiology studies in order to calculate odds ratio for the continuous phenotype [18].

Through the proposed functional annotation method, significant pathways, GOs, and Pfams were obtained based on the selected dichotomized phenotype. Table 3 shows the results of pathway knowledge for duplication and total CNV burden. We restricted the significance threshold (permutation *P-value* <0.01) to remove marginally significant results. Based on a stricter threshold, 6 pathways were found from duplication burden and 2 pathways were selected from total CNV burden as pathways associated with the cholesterol phenotype (Table 3). Similarly, significant GOs and Pfams were also found (Table 4 and Table 5).

3.3. *Biological interpretation*

Previously, many studies have reported that hypercholesterolemia or lower cholesterol levels are associated with CNV [19]. In addition, hyperlipidemia is associated with many other diseases such as myocardial infarction. For example, one study found several CNVs to have a link to myocardial infarction and hyperlipidemia [20]. Most of these studies were focused on specific CNVs or genes within CNV regions. However, we found that CNV burden is associated with cholesterol level. This is the first study to identify the association between the cholesterol quantitative trait and CNV burden in the literature. This suggests that cholesterol levels may also be associated with global genetic effects of many genes/regions.

Table 3. The list of significant pathways. Significant pathways associated with cholesterol were selected based on CNV burden data set (P<0.01). Continuous cholesterol phenotype was dichotomized at the 75th percentile in order to perform the proposed functional annotation pipeline, comparing CNVs in the top quartile ('high') with those in the bottom 3 quartiles ('low').

CNV Type	Significant Pathways	Permutation *P-value*
Dup	Hepatitis C	0.001998
	Toll-like receptor signaling pathway	0.001998
	Cytokine-cytokine receptor interaction	0.006993
	Shigellosis	0.006993
	RIG-I-like receptor signaling pathway	0.007992
	Influenza A	0.00999
Total CNV	Toll-like receptor signaling pathway	0.001
	Renal cell carcinoma	0.002997

Table 4. The list of significant protein families. Significant protein families associated with cholesterol were selected based on CNV burden data set (P<0.01). Continuous cholesterol phenotype was dichotomized at the 75th percentile in order to perform the proposed functional annotation pipeline comparing CNVs in the top quartile ('high') with those in the bottom 3 quartiles ('low').

CNV Type	Significant Pfams	Permutation *P-value*
Dup	Cell morphogenesis central region	0.001998
	Cell morphogenesis C-terminal	0.001998
	Cell morphogenesis N-terminal	0.001998
	RAVE protein 1 C terminal	0.00999
	Zinc-binding domain	0.00999
Total CNV	Poly (ADP-ribose) glycohydrolase (PARG)	0.001
	Adenylate and Guanylate cyclase catalytic domain	0.001
	Thrombospondin type 1 domain	0.000999
	ADAM-TS Spacer 1	0.000999
	Cell morphogenesis central region	0.001998
	Cell morphogenesis C-terminal	0.001998
	Cell morphogenesis N-terminal	0.001998
	Reprolysin (M12B) family zinc metalloprotease	0.008991
	Zinc binding domain	0.00999

Table 5. The list of significant GOs. Significant GOs associated with cholesterol were selected based on CNV burden data set (P<0.01). Continuous cholesterol phenotype was dichotomized at the 75th percentile in order to perform the proposed functional annotation pipeline comparing CNVs in the top quartile ('high') with those in the bottom 3 quartiles ('low').

CNV Type	Significant GOs	Permutation *P-value*
Dup	Nucleoside triphosphatase activity	0.001
	GTPase activity	0.001
	Pyrophosphatase activity	0.001
	Cellular defense response	0.000999
	Hydrolase activity, acting on acid anhydrides	0.000999
	Caspase regulator activity	0.005994
	Histone acetyltransferase activity	0.006993
	Response to virus	0.007992
	Microtubule organizing center organization and biogenesis	0.008991
	Centrosome organization and biogenesis	0.008991
	Nuclear envelope	0.00999
Total CNV	Sensory organ development	0.001
	Nicotinic acetylcholine-gated receptor-channel complex	0.000999
	Nicotinic acetylcholine-activated cation-selective channel activity	0.000999
	Double stranded DNA binding	0.001998
	Cyclase activity	0.002997
	Phosphorus-oxygen lyase activity	0.002997
	Secondary metabolic process	0.005994
	Learning and/or memory	0.006993
	Serotonin receptor activity	0.007992
	Microvillus	0.008991
	Amino acid transmembrane transporter activity	0.008991

In order to better understand possible mechanisms of the association between the cholesterol phenotype and CNV burden, the proposed functional annotation test was performed based on CNV profiles. Six pathways, hepatitis C, toll-like receptor signaling pathway, cytokine-cytokine receptor interaction, shigellosis, RIG-I-like receptor signaling pathway, and influenza A were found in the annotation results based on duplication burden. In particular, toll-like receptor (TLR) signaling pathway is a well-known pathway that acts an important role in atherosclerosis [21]. A prior study found that *TLR4* can directly interfere with cholesterol metabolism in macrophages, which suggests that *TLR4* could affect disease pathology [21]. The results from a second study revealed that Hepatitis C virus entry, in cooperation with *CD81* and scavenger receptor B type I, is also partially dependent on membrane cholesterol [22]. In addition, TLR signaling pathways and renal cell carcinoma were obtained based on total CNV burden. In a recent study of patients who underwent surgery for renal cell carcinoma, preoperative serum cholesterol was implicated as an independent factor for prognosis. Lower cholesterol levels were found to be associated with advanced disease and worse survival, which may be due to cholesterol's increased storage in tumour cells and role in new membrane biosynthesis [23]. For Pfam, 5 and 9 protein families were found based on duplication and total copy number burden, respectively. Interestingly, many cell morphogenesis-related protein families were found, in line with findings that cholesterol is important for proper cell morphogenesis due to its role in maintaining membrane order [24]. Among many significant GOs, nucleoside triphosphatase activity (NTPase) was found to be associated with cholesterol. Nuclear membrane cholesterol both modulates NTPase activity and can alter activity when oxidized [25]. Furthermore, similarly to the aforementioned Hepatitis C virus, cholesterol is important for membrane fusion during virus infection into host cells, as the enrichment of cholesterol helps to maintain membrane fluidity in the cell [26]. Taken together these results demonstrate the utility of the proposed pipeline for annotating the results of CNV burden analysis underlying complex traits such as total cholesterol phenotype.

4. Conclusions

In this study, we developed a systematic pipeline for annotating copy number variants into genes/genomic regions and subsequently pathways and other biological knowledge using Biofilter 2.0. In addition, a new method that takes into account the frequency information of genes in rare/common CNVs was proposed and led to the finding of many biologically relevant pathways, GOs, and protein families associated with cholesterol. Based on the copy number burden analysis, it follows that with larger copy number changes and a greater accumulation of copy number changes, it is more likely that genes known to influence disease risk and phenotypic severity will be affected. Thus, our study suggests the proposed pipeline could improve the interpretability of copy number burden analysis where hundreds of loci or genes contribute toward disease susceptibility via biological knowledge groups such as pathways. This CNV annotation pipeline with Biofilter can be used for CNV data from any genotyping or sequencing platform and to explore CNV enrichment for any traits or phenotypes. Biofilter is open source and freely available at http://ritchielab.psu.edu/software. Biofilter continues to be a powerful bioinformatics tool for annotation, filtering, and constructing biologically informed models for association analysis – now including copy number variants.

As demonstrated by this and other studies, CNV burden analysis is a new powerful method to investigate the association between accumulated genetic effects and many traits or phenotypes. In particular, the development of an appropriate annotation pipeline for CNV burden analysis will be valuable to better understand possible mechanisms associated with phenotypes in the context of accumulated effect of rare/common CNVs. As more well-designed genetic and phenotypic data are generated based on EHR for better precision medicine, CNV burden analysis continues to demonstrate the strengths along with the proposed annotation pipeline.

Acknowledgments

This work was funded by NHGRI grant U01 HG006389, NHLBI grant U01 HL065962, and CTSI: UL1 RR033184-01. This work is also supported by a grant with the Pennsylvania Department of Health using Tobacco CURE Funds.

References

1. Collins FS, Varmus H (2015) A new initiative on precision medicine. N Engl J Med 372: 793-795.

2. Gottesman O, Kuivaniemi H, Tromp G, Faucett WA, Li R, et al. (2013) The Electronic Medical Records and Genomics (eMERGE) Network: past, present, and future. Genet Med 15: 761-771.

3. Namjou B, Keddache M, Marsolo K, Wagner M, Lingren T, et al. (2013) EMR-linked GWAS study: investigation of variation landscape of loci for body mass index in children. Front Genet 4: 268.

4. Ritchie MD, Denny JC, Crawford DC, Ramirez AH, Weiner JB, et al. (2010) Robust replication of genotype-phenotype associations across multiple diseases in an electronic medical record. Am J Hum Genet 86: 560-572.

5. Denny JC, Bastarache L, Ritchie MD, Carroll RJ, Zink R, et al. (2013) Systematic comparison of phenome-wide association study of electronic medical record data and genome-wide association study data. Nat Biotechnol 31: 1102-1110.

6. Manolio TA, Collins FS, Cox NJ, Goldstein DB, Hindorff LA, et al. (2009) Finding the missing heritability of complex diseases. Nature 461: 747-753.

7. Ritchie MD, Holzinger ER, Li R, Pendergrass SA, Kim D (2015) Methods of integrating data to uncover genotype-phenotype interactions. Nat Rev Genet 16: 85-97.

8. Hall MA, Verma SS, Wallace J, Lucas A, Berg RL, et al. (2015) Biology-Driven Gene-Gene Interaction Analysis of Age-Related Cataract in the eMERGE Network. Genet Epidemiol 39: 376-384.

9. Eichler EE, Flint J, Gibson G, Kong A, Leal SM, et al. (2010) Missing heritability and strategies for finding the underlying causes of complex disease. Nat Rev Genet 11: 446-450.

10. Prakash SK, LeMaire SA, Guo DC, Russell L, Regalado ES, et al. (2010) Rare copy number variants disrupt genes regulating vascular smooth muscle cell adhesion and contractility in sporadic thoracic aortic aneurysms and dissections. Am J Hum Genet 87: 743-756.

11. Connolly JJ, Glessner JT, Almoguera B, Crosslin DR, Jarvik GP, et al. (2014) Copy number variation analysis in the context of electronic medical records and large-scale genomics consortium efforts. Front Genet 5: 51.

12. Girirajan S, Johnson RL, Tassone F, Balciuniene J, Katiyar N, et al. (2013) Global increases in both common and rare copy number load associated with autism. Hum Mol Genet 22: 2870-2880.

13. Pendergrass SA, Frase A, Wallace J, Wolfe D, Katiyar N, et al. (2013) Genomic analyses with biofilter 2.0: knowledge driven filtering, annotation, and model development. BioData Min 6: 25.

14. Peissig PL, Rasmussen LV, Berg RL, Linneman JG, McCarty CA, et al. (2012) Importance of multi-modal

approaches to effectively identify cataract cases from electronic health records. J Am Med Inform Assoc 19: 225-234.

15. Turner S, Armstrong LL, Bradford Y, Carlson CS, Crawford DC, et al. (2011) Quality control procedures for genome-wide association studies. Curr Protoc Hum Genet Chapter 1: Unit1 19.

16. Wang K, Li M, Hadley D, Liu R, Glessner J, et al. (2007) PennCNV: an integrated hidden Markov model designed for high-resolution copy number variation detection in whole-genome SNP genotyping data. Genome Res 17: 1665-1674.

17. Grady BJ, Torstenson E, Dudek SM, Giles J, Sexton D, et al. (2010) Finding unique filter sets in PLATO: a precursor to efficient interaction analysis in GWAS data. Pac Symp Biocomput: 315-326.

18. Volk HE, Lurmann F, Penfold B, Hertz-Picciotto I, McConnell R (2013) Traffic-related air pollution, particulate matter, and autism. JAMA Psychiatry 70: 71-77.

19. Pollex RL, Hegele RA (2007) Genomic copy number variation and its potential role in lipoprotein and metabolic phenotypes. Curr Opin Lipidol 18: 174-180.

20. Shia WC, Ku TH, Tsao YM, Hsia CH, Chang YM, et al. (2011) Genetic copy number variants in myocardial infarction patients with hyperlipidemia. BMC Genomics 12 Suppl 3: S23.

21. Curtiss LK, Tobias PS (2009) Emerging role of Toll-like receptors in atherosclerosis. J Lipid Res 50 Suppl: S340-345.

22. Kapadia SB, Barth H, Baumert T, McKeating JA, Chisari FV (2007) Initiation of hepatitis C virus infection is dependent on cholesterol and cooperativity between CD81 and scavenger receptor B type I. J Virol 81: 374-383.

23. de Martino M, Leitner CV, Seemann C, Hofbauer SL, Lucca I, et al. (2015) Preoperative serum cholesterol is an independent prognostic factor for patients with renal cell carcinoma (RCC). BJU Int 115: 397-404.

24. Arita Y, Nishimura S, Ishitsuka R, Kishimoto T, Ikenouchi J, et al. (2015) Targeting cholesterol in a liquid-disordered environment by theonellamides modulates cell membrane order and cell shape. Chem Biol 22: 604-610.

25. Ramjiawan B, Czubryt MP, Massaeli H, Gilchrist JS, Pierce GN (1997) Oxidation of nuclear membrane cholesterol inhibits nucleoside triphosphatase activity. Free Radic Biol Med 23: 556-562.

26. Tanner LB, Lee B (2013) The greasy response to virus infections. Cell Host Microbe 13: 375-377.

THE CHALLENGES IN USING ELECTRONIC HEALTH RECORDS FOR PHARMACOGENOMICS AND PRECISION MEDICINE RESEARCH

SARAH M. LAPER

Eastern Virginia Medical School, Norfolk, VA, 23507, USA
Email: lapersm@evms.edu

NICOLE A. RESTREPO

Center for Human Genetics Research, Vanderbilt University, 519 Light Hall, 2215 Garland Avenue, Nashville, TN 37232, USA
Email: n.restrepo@vanderbilt.edu

DANA C. CRAWFORD

Institute for Computational Biology, Department of Epidemiology and Biostatistics, Case Western Reserve University, Wolstein Research Building, 2103 Cornell Road, Suite 2527, Cleveland, OH 44106, USA
Email: dana.crawford@case.edu

Access and utilization of electronic health records with extensive medication lists and genetic profiles is rapidly advancing discoveries in pharmacogenomics. In this study, we analyzed ~116,000 variants on the Illumina Metabochip for response to antihypertensive and lipid lowering medications in African American adults from BioVU, the Vanderbilt University Medical Center's biorepository linked to de-identified electronic health records. Our study population included individuals who were prescribed an antihypertensive or lipid lowering medication, and who had both pre- and post-medication blood pressure or low-density lipoprotein cholesterol (LDL-C) measurements, respectively. Among those with pre- and post-medication systolic and diastolic blood pressure measurements (n=2,268), the average change in systolic and diastolic blood pressure was -0.6 mg Hg and -0.8 mm Hg, respectively. Among those with pre- and post-medication LDL-C measurements (n=1,244), the average change in LDL-C was -26.3 mg/dL. SNPs were tested for an association with change and percent change in blood pressure or blood levels of LDL-C. After adjustment for multiple testing, we did not observe any significant associations, and we were not able to replicate previously reported associations, such as in *APOE* and *LPA*, from the literature. The present study illustrates the benefits and challenges with using electronic health records linked to biorepositories for pharmacogenomic studies.

1. Introduction

As the costs for genomic sequencing continue to decrease, the goals of precision medicine to use a patient's genetic makeup to calculate his or her risk of disease, prevent illness, and determine the best drug or treatment for his or her medical problem, become more feasible [1,2]. Large-scale genome-wide association studies (GWAS) and smaller sequencing studies are rapidly identifying common and rare genetic variants that contribute to human disease [3] and response to drug treatment [4]. The results from pharmacogenomic studies, the study of variants that effect drug response and efficacy, can be translated to clinical practice more easily than variants that effect disease risk. For example, knowledge that one treatment is most effective for one genotype group, while another treatment is optimal for another group, can aid in the selection of treatments [5]. There have already been successful pharmacogenomic studies that have been translated to clinical practice. A variant in *HLA-B* was identified that is associated with increased risk of a hypersensitivity reaction when using Abcavir for the treatment of HIV [6], dosing recommendations for thiopurines have been developed based on *TPMT* genotype [7], and variants in *CYP2D6* have been identified that cause patients to either be poor metabolizers or rapid metabolizers of codeine [8]. Many of the early pharmacogenomic studies focused on variants in candidate genes that code for drug-metabolizing enzymes or drug targets. However, with advances in molecular assaying technology and the increased practicality of sequencing the entire genome, variants in other regions that have a clinically important effect may be discovered [9].

The majority of genetic association studies, including pharmacogenomic studies [10,11], have been in European populations [12]. It is important to conduct GWAS in diverse populations in order to discover variants that may not be present in European populations [12]. Previous studies have already found population specific frequencies for variants that effect drug response. For example, it has been found that there are significant differences in allele frequencies between populations for genes encoding drug metabolizing enzymes [13], that variants in *CYP2C9* and *VKORC1* differ among racial/ethnic groups and effect the dosing of warfarin [14], and that African Americans have the lowest frequency of the variant near the *IL28B* gene that is associated with response to hepatitis C treatment [15].

Longitudinal epidemiological cohorts are the gold standard for genetic association studies particularly in the context of gene-environment studies [16]. Properly designed cohorts, however, require enormous resources for the study of common health outcomes and may not be feasible for the study of rare outcomes, such as adverse events in pharmacogenomics. The recent emergence of electronic health records (EHR) linked to biorepositories offers an alternative strategy for rapid and cost-effective data collection for genetic association studies. EHRs contain a large amount of patient data, and it has been shown that, when linked to biorepositories, this data source can be utilized in genetic studies [17]. The use of EHRs linked to biorepositories has advantages over the traditional cohort design, such as cost, timeliness, and the ability to select for a wide range of phenotypes [18]. Also, EHRs contain data not typically collected in a traditional epidemiological study, such as information related to drug response [5]. Extracting medication from EHRs has been found to be one of the most time-consuming processes when using EHR driven genomic studies. However, advances in natural language processing have been successful in identifying medication relevant information from clinical notes in EHRs [19]. Finally, an advantage of using EHRs is that they provide a more accurate representation of the clinical population, including minority populations, than traditional cohort studies [18].

In this study, we used EHRs linked to a biorepository to analyze drug response in an African American population of almost 12,000 patients genotyped on the Illumina Metabochip [20]. We extracted data related to two common clinical treatments: 1) the use of antihypertensive medication to lower blood pressure, and 2) the use of lipid lowering medication to lower blood levels of low-density lipoprotein cholesterol (LDL-C). Individual response to both of these treatments varies greatly, although the exact cause of this variation is unknown and likely due to many interacting factors. The availability of EHR data allowed us to study drug response in an African American population. However, this study provides an illustration of challenges that arise when using EHRs linked to biorepositories for genetic association analyses.

2. Methods

2.1. *Study population*

The data described here were obtained from BioVU, the Vanderbilt University Medical Center's biorepository linked to de-identified electronic health records. BioVU operations [21] and ethical oversight [22] have been described elsewhere. Briefly, DNA is collected from discarded blood samples remaining after routine clinical testing at Vanderbilt outpatient clinics in Nashville, Tennessee and surrounding areas, and is linked to a de-identified version of the patient's EHR termed the "Synthetic Derivative." The data were de-identified in accordance with provisions of Title 45, Code of Federal Regulations, part 46 (45 CFT 46), and this study was considered non-human subjects research by the Vanderbilt University Internal Review Board.

2.2. *Genotyping*

DNA samples from mostly non-European Americans in BioVU were genotyped on the Illumina Metabochip by the Vanderbilt University Center for Human Genetics Research DNA Resources Core as part of the Population Architecture using Genomics and Epidemiology (PAGE) I Study [23]. This dataset is herein after referred to as "EAGLE BioVU" [24]. The Metabochip is a custom array of approximately 200,000 SNPs designed for replication and fine-mapping of genome-wide association study-identified variants for metabolic and cardiovascular traits [25].

Prior to analyses, quality control procedures were performed using PLINKv1.09 [26]. SNPs were filtered to exclude those with low minor allele frequency (<1%), low genotyping frequency (<95%), and deviations from Hardy-Weinberg expectations ($p<10^{-6}$). After quality control, approximately 116,000 SNPs were available for analysis. Patients were excluded from analysis if his or her biological sex did not match the recorded sex in the EHR, if they had a low genotyping rate (<95%), or if they were determined to be related to another sample (identical-by-descent).

2.3. *Phenotyping*

BioVU de-identified EHRs contain structured (such as International Classification Disease codes or billing codes, current procedural terminology codes, vital signs, and labs) and unstructured (clinical notes) data accessible for electronic phenotyping. We extracted systolic blood pressure, diastolic blood pressure, and LDL-C from the de-identified EHRs for the present study. Prescription medication is available in the de-identified version of the EHR through MedEx [19]. Clinic dates and corresponding prescriptions for both antihypertensives and LDL-C lowering medications were extracted as previously described [24,27,28]. Blood pressure and LDL-C measurements were considered "post-medication" if a prescription for an antihypertensive or lipid lowering drug was extracted prior to the lab. "Pre-medication" values include all values available per patient prior to a prescription, including values within the normal range. Median values of measurements were calculated for both "pre-medication" and "post-medication" categories for each patient. Body mass index was extracted from the EHR and cleaned as previously described [29]. This study only considered patients who had both pre-and post-medication median measurements for systolic and diastolic blood pressure, or LDL-C.

EAGLE BioVU has a total of 15,863 patients [24]. For the present study, only patients who were identified as African American and who were over the age of 18 by the year 2010, based on date of birth, were included. There were 2,653 and 1,244 patients who had both pre- and post-medication systolic and diastolic blood pressure measurements and pre- and post-medication LDL-C lab values, respectively. After quality control, there were 2,268 and 1,028 patients analyzed for blood pressure medication response and lipid medication response, respectively.

2.4. *Statistical Methods*

Single SNP tests of association assuming an additive genetic model using linear regression were performed using PLINKv1.09 [26]. Two dependent variables were considered: 1) the difference between the median post- and pre-medication measurements and 2) the percent change between the median post- and pre-medication measurements. For each dependent variable, three different models were considered: 1) unadjusted, 2) adjusted for age and sex and 3) adjusted for age, sex, and the first three principal components of ancestry. Age was the patient's age as of 2010. The principal components of ancestry were obtained from the PAGE I Study Coordinating Center [30]. Results from tests of association for *APOE* and *LPA* gene regions were visualized using LocusZoom [31].

3. Results

3.1. *Response to Anti-Hypertensive Medication*

Study population characteristics for patients with pre- and post-antihypertensive medication blood pressure measurements are given in Table 1. A total of 2,620 adult African American patients had both a pre- and post-antihypertensive medication blood pressure measurement. The majority of the patients were female (65.5%) and born between 1940 and 1970 (70.9%). On average, the patients were overweight (BMI = 29.7 kg/m^2). For systolic blood pressure, the average median pre-medication measurement was 132.6 mm Hg, reflecting the fact that all pre-medication values over the course of a patient's care including those within the normal range were included. The average median post-medication measurement was 132.0 mm Hg, and the average change in systolic blood pressure with the use of medication was a decrease by 0.6 mm Hg. For diastolic blood pressure, the average median pre-medication measurement was 79.7 mm Hg, the average median post-medication measurement was 79.0 mm Hg, and the average change in diastolic blood pressure with the use of medication was a decrease by 0.8 mm Hg.

Table 1. Blood Pressure Medication Response Study Population Characteristics. Study population characteristics (sex, decade of birth, average median body mass index, average median pre-medication blood pressure measurement, average median post-medication blood pressure measurement, and average change in blood pressure are given for the 2,620 patients who were above the age of 18, African American, and who had both pre-medication and post-medication blood pressure measurements. Abbreviations: standard deviation (SD).

Female, %	65.5
Decade of birth, %	
1910	0.7
1920	3.1
1930	8.4
1940	16.1
1950	24.6
1960	20.5
1970	13.3
1980	12.1
1990	1.2
Mean (± SD) Body Mass Index (kg/ m^2)	29.2 (12.7)
Mean (± SD) pre-medication blood pressure measurement (systolic/ diastolic) (mm Hg)	132.6 (17.1) / 79.7 (10.6)
Mean (± SD) post-medication blood pressure measurement (systolic/diastolic) (mm Hg)	132.0 (15.0) / 79.0 (9.4)
Mean (± SD) change in blood pressure (systolic/ diastolic) (mm Hg)	-0.6 (15.2) / -0.8 (10.0)

The 116,000 SNPs on the Metabochip that passed quality control, as described in the methods, were tested for an association with change and percent change in systolic and diastolic blood pressure with the use of antihypertensive medication. No SNPs passed a strict Bonferroni corrected significance level ($p = 4 \times 10^{-7}$) for an association with either the change or percent change in systolic or diastolic blood pressure for any of the models (Figure 1). The most significant SNP for both the change and percent change in systolic blood pressure was rs8058830 on chromosome 16 (beta = -2.6 mm Hg, $p = 1.4 \times 10^{-5}$; beta = -1.9%, $p = 1.3 \times 10^{-5}$). The most significant SNP for the change in diastolic blood pressure mapped to rs183551129 on chromosome 15 (beta = 1.5 mm Hg, $p = 1.2 \times 10^{-5}$). The most significant SNP for the percent change in diastolic blood pressure was rs17672219 on chromosome 16 (beta = -2.2%, p = 6.5 x 10

[6]). These results are for the linear regression model adjusted for age, sex, and the first three principal components, and the results were not appreciably different for the other two models.

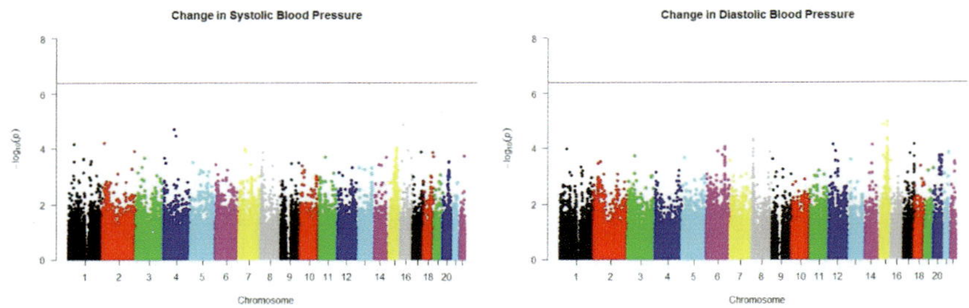

Figure 1. Single SNP association results for change in blood pressure with the use of antihypertensive medication. Single SNP tests of association were performed using linear regression assuming an additive genetic model adjusting for age, sex, and the first three principal components. The $\log_{10}(p)$ values (y-axis) were plotted using R for the association of each tested SNP (x-axis) with the change in systolic and diastolic blood pressure respectively. The red line represents the Bonferroni corrected significance level of $p = 4 \times 10^{-7}$.

3.2. *Response to Lipid Lowering Medication*

Study population characteristics for patients with pre- and post-medication LDL-C measurements are given in Table 2. A total of 1,244 patients African American adult patients had both a pre- and post-medication LDL-C lab measurement. The majority of the patients were female (61.4%) and born between 1930 and 1970 (85.8%). On average, the patients were obese (BMI = 30.9 kg/m^2). The average pre-medication LDL-C measurement was 129.9 mg/dL, the average post-medication LDL-C measurement was 103.6 mg/dL, and the average change in LDL-C with the use of lipid lowering medication was a decrease by 26.3 mg/dL.

The 116,000 SNPs on the Metabochip that passed quality control as described in the methods were tested for an association with change and percent change in LDL-C with the use of lipid lowering medication. For both the change and percent change in LDL-C, none of the SNPs passed a Bonferroni corrected significance level ($p = 4 \times 10^{-7}$) for any of the linear regression models (Figure 2). The most significant SNP for the change in LDL-C was rs12564350 located on

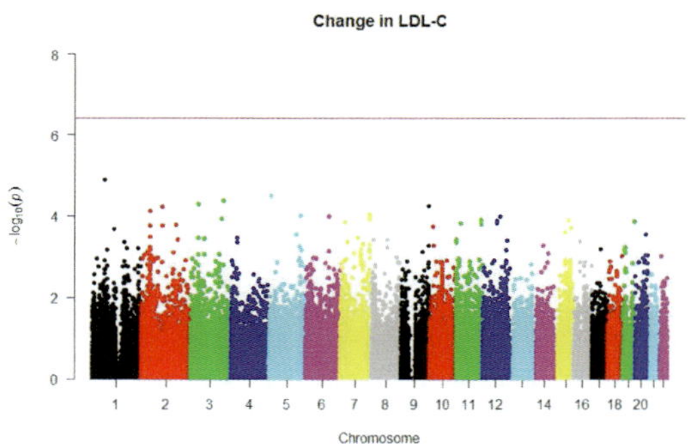

Figure 2. Single SNP association results for change in LDL-C with the use of lipid lowering medication. Single SNP tests of association were performed using linear regression assuming an additive genetic model adjusting for age, sex, and the first three principal components. The $\log_{10}(p)$ values (y-axis) were plotted using R for the association of each tested SNP (x-axis) with the change in LDL-C measurements. The red line represents the Bonferroni corrected significance level of $p = 4 \times 10^{-7}$.

chromosome 1 (beta = -14.8 mg/dL, $p = 1.3 \times 10^{-5}$). The most significant SNP for the percent change in LDL-C was rs4309741 on chromosome 3 (beta = 9.3%, $p = 5.6 \times 10^{-6}$). These results are for the linear regression model adjusted for age, sex, and the first three principal components, and the results for the other two models tested were not appreciably different.

Table 2. Statin Response Study Population Characteristics. Study population characteristics (sex, decade of birth, average median body mass index, average median pre-medication LDL-C measurement, average median post- medication LDL-C measurement, and average change in LDL-C are given for the 1,242 patients who were above the age of 18, African American, and who had both pre-medication and post-medication LDL-C measurements. Abbreviations: standard deviation (SD); low density lipoprotein cholesterol

Female, %	61.4
Decade of birth, %	
1910	1.0
1920	5.6
1930	15.7
1940	22.9
1950	27.9
1960	19.3
1970	6.0
1980	1.4
1990	0.2
Mean (± SD) Body Mass Index (kg/ m^2)	30.9 (10.9)
Mean (± SD) pre-medication blood LDL-C measurement (mg/dL)	129.9 (45.5)
Mean (± SD) post-medication blood LDL-C measurement (mg/dL)	103.6 (35.7)
Mean (± SD) change in LDL-C (mg/dL)	-26.3 (41.4)

Two genes for which variants have been located that are significantly associated with lipid response to medication and have been consistently replicated are *APOE* and *LPA* [32]. There were approximately ~250 and ~500 SNPs assayed in the *LPA* and *APOE* regions shown, respectively.

The SNPs that were tested in these regions on the Metabochip were not found to be significantly associated with the change in LDL-C with the use of lipid lowering medication (Figure 3).

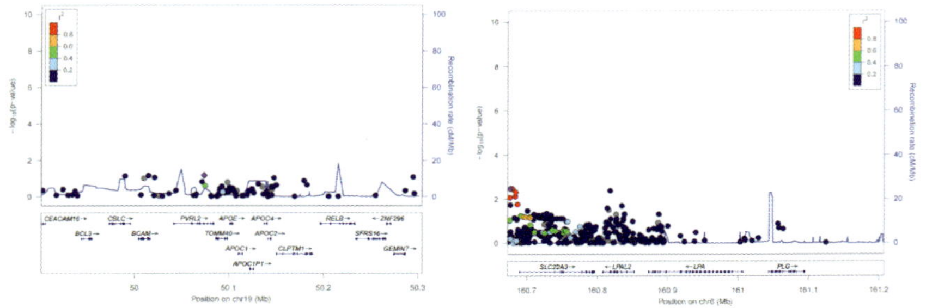

Figure 3. Locus Zoom plot of association results for *APOE* and *LPA*. Figure was generated using LocusZoom (http://csg.sph.umich.edu/locuszoom/) with linkage disequilibrium calculations from the hg18 1000 Genomes June 2010 YRI dataset. Results are shown for the association results in the *APOE* and *LPA* regions for the change in LDL-C with the use of medication, for the linear regression model adjusted for age, sex, and the first three principal components.

4. Discussion

In this study, we extracted data from BioVU related to antihypertensive and lipid lowering medication use in adult, African American patients. We tested for the association of SNPs on the Metabochip with the change in blood pressure and LDL-C measurements with the use of antihypertensive and lipid lowering medicines, respectively. Previous studies have found variants significantly associated with drug response for both of these medications [32–34]. After correction for multiple testing, we did not identify significant novel associations nor did we replicate previously reported associations for these response to treatment outcomes. There were several limitations to this study that illustrate many of the challenges for pharmacogenomics studies that use data from EHRs linked to biorepositories.

A major challenge for all pharmacogenomics studies is sample size and statistical power. The electronic phenotyping strategy outlined here balanced sample size ("lumping") versus precise phenotyping ("splitting"). In this present study, we did not distinguish by class of medication prescribed. There have been previous genome-wide association studies in African American populations that have found SNPs significantly associated with response to antihypertensive medications [33,34]. However, these studies differed from the current study in that they had much smaller sample sizes and studied specific classes of antihypertensive medication. A study on the response to thiazide diuretics had a sample size of 204 African American patients [34] and a study on angiotensin II receptor blockers used 193 African American patients [33]. Our study had more power to detect a significant association since we had 2,620 African American individuals, but this was balanced by the fact that the Metabochip did not assay any of the specific previously identified variants and our reliance on linkage disequilibrium. Also, the strategy of lumping rather splitting medication classes may have precluded our ability to identify associations in this dataset.

There have also been previous genome-wide association studies on response to lipid lowering medications that have found significant associations [35–39]. Although these studies found significant associations for variants in several genes, only variants in the *APOE* and *LPA* genes have been consistently replicated [32]. Our study found no significant SNPs in these regions (Figure 3). However, all of the previous studies were conducted using European populations. Since our study was performed in an African American population it is not entirely surprising that we were not able to replicate previous significant associations since allele frequencies and levels of linkage disequilibrium vary between populations, risk variants vary in effect size between different populations, and there may be risk variants in African American populations that do not exist in European populations [12].

One of the limitations of this study is that genotyping was performed on the Metabochip, a custom array designed for fine mapping of variants identified by genome-wide association studies for cardiovascular and metabolic traits [25]. Although we had variants in the gene regions of significant SNPs identified in previous studies on antihypertensive and lipid lowering medication response, most of the specific previously identified SNPs were not tested. Thus, we may have missed significant associations because the SNPs were not directly genotyped or in strong linkage disequilibrium with the SNPs targeted by the Metabochip. This, indeed, is another problem in pharmacogenomic studies. Variants in regions known to effect drug metabolism often do not pass genotype quality control, and their coverage on genome-wide association study genotyping platforms is limited [40,41].

As already noted, a common problem with pharmacogenomic studies is obtaining a large enough sample size since it is difficult to obtain a large population taking a certain class of drug or having a particular response [5,40]. Even though we had access to a large number of African American patients, we still only had a small sample size for patients who met our phenotype requirements. This illustrates one of the disadvantages to using EHR data in the United States. Since it is an open system, some of the patients did not come back for a post-medication measurement or entered the system after already starting a medication.

Phenotyping is a major challenge when using EHRs for pharmacogenomics studies. We used MedEx to extract data on when a prescription for a qualifying medication was entered into the system. However, we did not distinguish between the types or doses of medication because our sample size would have been too small to have the power to detect an association. Also, EHR does not include data on many environmental factors. For example, we could not access compliance, since BioVU is an opt-out model and we could not follow-up with patients, so we assumed that if a patient had a prescription for a medication than they were actually taking the medication. One of the advantages to using EHR data is that there is access to patient information over a long time span. The biorepository we used contains patient data spanning 20 years. However, this presents a challenge in our pharmacogenomics study because prescribing practices have changed for both blood pressure and lipid lowering medication.

As compared to a traditional epidemiological study with standard protocols aimed at uniform measurement, the quality of measurements recorded in EHRs can be variable. Blood pressure is hard to measure for numerous reasons [42]. It is possible that different instruments are used across the different outpatient clinics to measure blood pressure, and these differences can result in slightly higher or slightly lower measurements. In a clinical setting, patient's blood pressure measurements are taken at different times of the day depending on appointment time, and each person's blood pressure varies throughout the day. The white coat effect may also cause an artificially high measurement. Finally, there is inherent human error in taking blood pressure measurements. The lab values for LDL-C are more reliable, but we have to assume that the patient in BioVU fasted, and this is not typically documented in the EHR.

Although we experienced challenges in using EHR data linked to a biorepository to study drug response for antihypertensive and lipid lowering medications, this dataset did allow us to study a diverse population. The previous studies on response to antihypertensive medication had small sample sizes for African Americans and there have been no genome-wide association studies on response to lipid lowering medication in African Americans. Some of the challenges that have been described here would be difficult to overcome in our current health care system. However, simply combining EHR data from multiple locations such as EHR-linked biorepositories participating in the electronic MEdical Records & GEnomics (eMERGE) network [17] may allow us to distinguish between the type and dose of medication while maintaining adequate sample sizes. The use of EHR data is a promising and valuable resource for the future of pharmacogenomic studies in diverse populations as we enter the era of precision medicine.

5. Acknowledgements

This work was supported in part by NIH grant U01 HG004798 and its ARRA supplements. The dataset(s) used for the analyses described were obtained from Vanderbilt University Medical Center's BioVU which is supported by institutional funding and by the Vanderbilt CTSA grant funded by the National Center for Research Resources, Grant UL1 RR024975-01, which is now at the National Center for Advancing Translational Sciences, Grant 2 UL1 TR000445-06.

References

1. W. G. Feero, JAMA **299**, 1351 (2008).
2. F. S. Collins and H. Varmus, N. Engl. J. Med. **372**, 793 (2015).
3. T. A. Manolio, Nat. Rev. Genet. **14**, 549 (2013).
4. M. Pirmohamed, Annu. Rev. Genomics Hum. Genet. **15**, 349 (2014).
5. M. D. Ritchie, Hum. Genet. **131**, 1615 (2012).
6. M. A. Martin, T. E. Klein, B. J. Dong, M. Pirmohamed, D. W. Haas, and D. L. Kroetz, Clin. Pharmacol. Ther. **91**, 734 (2012).

7. M. V. Relling, E. E. Gardner, W. J. Sandborn, K. Schmiegelow, C.-H. Pui, S. W. Yee, *et al,* Clin. Pharmacol. Ther. **89**, 387 (2011).
8. K. R. Crews, A. Gaedigk, H. M. Dunnenberger, T. E. Klein, D. D. Shen, J. T. Callaghan, *et al*, Clin. Pharmacol. Ther. **91**, 321 (2012).
9. W. E. Evans and M. V. Relling, Science **286**, 487 (1999).
10. V. E. Ortega and D. A. Meyers, J. Allergy Clin. Immunol. **133**, 16 (2014).
11. L. H. Cavallari and M. A. Perera, Future Cardiol. **8**, 563 (2012).
12. N. A. Rosenberg, L. Huang, E. M. Jewett, Z. A. Szpiech, I. Jankovic, and M. Boehnke, Nat. Rev. Genet. **11**, 356 (2010).
13. J. Wilson, M. Weale, A. C. Smith, F. Gratix, B. Fletcher, M. G. Thomas, *et al*, Nature Genetics **29**, 265 (2001).
14. J. A. Johnson, L. Gong, M. Whirl-Carrillo, B. F. Gage, S. A. Scott, C. M. Stein, *et al,* Clin. Pharmacol. Ther. **90**, 625 (2011).
15. D. Ge, J. Fellay, A. J. Thompson, J. S. Simon, K. V. Shianna, T. J. Urban, *et al*, Nature **461**, 399 (2009).
16. T. A. Manolio, J. E. Bailey-Wilson, and F. S. Collins, Nat. Rev. Genet. **7**, 812 (2006).
17. D. C. Crawford, D. R. Crosslin, G. Tromp, I. J. Kullo, H. Kuivaniemi, M. G. Hayes, *et al*, Front. Genet. **5**, (2014).
18. I. S. Kohane, Nat. Rev. Genet. **12**, 417 (2011).
19. H. Xu, M. Jiang, M. Oetjens, E. A. Bowton, A. H. Ramirez, J. M. Jeff, et al, Am. Med. Inform. Assoc. JAMIA **18**, 387 (2011).
20. D. C. Crawford, R. Goodloe, E. Farber-Eger, J. Boston, S. A. Pendergrass, J. L. Haines, *et al*, Hum. Hered. **79**, 137 (2015).
21. D. Roden, J. Pulley, M. Basford, G. Bernard, E. Clayton, J. Balser, *et al*, Clin. Pharmacol. 38 Ther. **84**, 362 (2008).
22. J. Pulley, E. Clayton, G. R. Bernard, D. M. Roden, and D. R. Masys, Clin. Transl. Sci. **3**, 42 (2010).
23. T. C. Matise, J. L. Ambite, S. Buyske, C. S. Carlson, S. A. Cole, D. C. Crawford, *et al*, Am. J. Epidemiol. **174**, 849 (2011).
24. D. Crawford, R. Goodloe, E. Farber-Eger, J. Boston, S. Pendergrass, J. Haines, et al, Human Hereditary **79**, 137 (2015).
25. B. F. Voight, H. M. Kang, J. Ding, C. D. Palmer, C. Sidore, P. S. Chines, *et al*, PLoS Genet. **8**, e1002793 (2012).
26. S. Purcell, B. Neale, K. Todd-Brown, L. Thomas, M. A. R. Ferreira, D. Bender, J. Maller, P. Sklar, P. I. W. de Bakker, M. J. Daly, and P. C. Sham, Am. J. Hum. Genet. **81**, 559 (2007).
27. L. Dumitrescu, R. Goodloe, Y. Bradford, E. Farber-Eger, J. Boston, and D. C. Crawford, BioData Min. **8**, (2015).
28. L. Dumitrescu, R. Goodloe, E. Farber-Eger, S. Pendergrass, W. Bush, and D. Crawford (submitted).
29. R. Goodloe, E. Farber-Eger, J. Boston, D. Crawford, and W. Bush, (in preparation).
30. S. Buyske, Y. Wu, C. L. Carty, I. Cheng, T. L. Assimes, L. Dumitrescu, *et al*, PLoS ONE **7**, e35651 (2012).

31. R. J. Pruim, R. P. Welch, S. Sanna, T. M. Teslovich, P. S. Chines, T. P. Gliedt, *et al*, Bioinformatics **26**, 2336 (2010).
32. J. C. Hopewell, C. Reith, and J. Armitage, Curr. Opin. Lipidol. **25**, 438 (2014).
33. S. T. Turner, E. Boerwinkle, J. R. O'Connell, K. R. Bailey, Y. Gong, A. B. Chapman, *et al*, Hypertension **62**, 391 (2013).
34. S. T. Turner, K. R. Bailey, B. L. Fridley, A. B. Chapman, G. L. Schwartz, H. S. Chai, *et al*, Hypertension **52**, 359 (2008).
35. J. F. Thompson, C. L. Hyde, L. S. Wood, S. A. Paciga, D. A. Hinds, D. R. Cox, *et al*, Circ. Cardiovasc. Genet. **2**, 173 (2009).
36. M. J. Barber, L. M. Mangravite, C. L. Hyde, D. I. Chasman, J. D. Smith, C.A. McCarty, *et al*, PLoS ONE **5**, e9763 (2010).
37. D. I. Chasman, F. Giulianini, J. MacFadyen, B. J. Barratt, F. Nyberg, and P. M. Ridker, Circ. Cardiovasc. Genet. **5**, 257 (2012).
38. H. A. Deshmukh, H. M. Colhoun, T. Johnson, P. M. McKeigue, D. J. Betteridge, P. N. Durrington, *et al*, J. Lipid Res. **53**, 1000 (2012).
39. J. C. Hopewell, S. Parish, A. Offer, E. Link, R. Clarke, M. Lathrop, *et al*, Eur. Heart J. **34**, 982 (2013).
40. A. A. Motsinger-Reif, E. Jorgenson, M. V. Relling, D. L. Kroetz, R. Weinshilboum, N. J. Cox, *et al*, Pharmacogenet. Genomics **23**, 383 (2013).
41. M. T. Oetjens, J. C. Denny, M. D. Ritchie, N. B. Gillani, D. M. Richardson, N. A. Restrepo, *et al*, Pharmacogenomics **14**, 735 (2013).
42. D. W. Jones, L. J. Appel, S. G. Sheps, E. J. Roccella, and C. Lenfant, JAMA **289**, 1027 (2003).

SEPARATING THE CAUSES AND CONSEQUENCES IN DISEASE TRANSCRIPTOME

Yong Fuga Li[1, 2], Fuxiao Xin[3], and Russ B. Altman[1, 4, #]
[1.] Department of Bioengineering, Stanford University. [2.] Stanford Genome Technology Center, Stanford University. [3.] Machine Learning Lab, GE Global Research. [4.] Department of Genetics, Stanford University
[#] Email: russ.altman@stanford.edu

The causes of complex diseases are multifactorial and the phenotypes of complex diseases are typically heterogeneous, posting significant challenges for both the experiment design and statistical inference in the study of such diseases. Transcriptome profiling can potentially provide key insights on the pathogenesis of diseases, but the signals from the disease causes and consequences are intertwined, leaving it to speculations what are likely causal. Genome-wide association study on the other hand provides direct evidences on the potential genetic causes of diseases, but it does not provide a comprehensive view of disease pathogenesis, and it has difficulties in detecting the weak signals from individual genes. Here we propose an approach diseaseExPatho that combines transcriptome data, regulome knowledge, and GWAS results if available, for separating the causes and consequences in the disease transcriptome. DiseaseExPatho computationally de-convolutes the expression data into gene expression modules, hierarchically ranks the modules based on regulome using a novel algorithm, and given GWAS data, it directly labels the potential causal gene modules based on their correlations with genome-wide gene-disease associations. Strikingly, we observed that the putative causal modules are not necessarily differentially expressed in disease, while the other modules can show strong differential expression without enrichment of top GWAS variations. On the other hand, we showed that the regulatory network based module ranking prioritized the putative causal modules consistently in 6 diseases, We suggest that the approach is applicable to other common and rare complex diseases to prioritize causal pathways with or without genome-wide association studies.

1. Introduction

Complex diseases result from the interplay of multiple genetic variations and environment factors (*1*, *2*). The putative causal genetic variants can be identified through their associations with disease phenotypes using approaches such as genome wide association study (GWAS) (*3*). However, the genetic variants do not directly cause disease, but do so by altering cells' molecular status, as described by epigenomes, transcriptomes, etc., which then escalate to the individual level and manifest as diseases. Hundreds of GWAS studies have been carried out for diverse traits and diseases (*3*, *4*), yet our understanding of most common diseases remains fragmented and uncertain (*5*). In most cases, knowing the causal genes of diseases is far from knowing the mechanism, limiting our ability to translate the knowledge of disease genetics into prevention and treatment strategies (*6*, *7*).

High-throughput technologies based on sequencing or microarray have enabled genome-wide studies at multiple levels, from GWAS, transcriptome profiling, to meta-genomics (*8–11*). Integration and joint modeling of the complementary sources of data will enable the most complete view of disease pathogenesis (*12–14*). Transcriptomic, proteomic, and metagenomic profiling can potentially provide key insights on the pathogenesis of diseases, but the signal from the disease causes and consequences are intertwined (*4*, *15*, *16*), making it challenging to extract the causal signals. GWAS and genome sequencing provides direct evidences of genetic cause of diseases, yet variants with small effect size pose great challenges (*3*, *4*).

The gene-regulation network is a graphical summary of the regulation mechanisms of human gene transcriptions. It is composed of the binary relationships among transcription factor – target genes. Despite its simplicity, studies based on the network have revealed important properties of gene regulations (*17–20*). However there has been limited application of human gene regulatory network in the computational inference of disease causes or mechanisms due to the lack of data (*21*). With the development of ChIP-seq technology (*22*, *23*) and the coordinated effort such as

ENCODE (*20, 24*) to measure genome wide transcription factor binding profiles, increasingly higher coverage of the human gene regulation network is being achieved.

Here we propose a computational pipeline, diseaseExPatho, to infer the molecular mechanism underlying complex human diseases (**Figure 1**). It takes three types of inputs, transcriptome of a disease of interest, GWAS implicated putative disease causal genes if known, and gene regulation network, which is independent of the specific disease. DiseaseExPatho first computationally decomposes the gene expression data using independent component analysis (ICA) to obtain functional coherent gene modules. It then labels the modules as differentially expressed (DE) and/or putative causal, using a novel statistical inference method for detecting gene enrichment. Finally, it hierarchically ranks the gene modules based on the gene transcriptional regulation network in order to prioritize the putative causal modules even when the disease causal genes are unknown. We applied the method to psychiatric disorders, type II diabetes, and inflammatory bowel diseases, and demonstrated its ability to decompose and prioritize the causal signal in disease transcriptome data with or without the knowledge of putative causal genes.

2. Methods

2.1. *Transcriptome data*

Transcriptome data for psychiatric disorders and diabetes are obtained from GEO(*25*). Microarray data are preprocessed using the fRMA algorithm(*26–29*) on batches defined by experiment date. The expression values are summarized to the gene level. For RNA-seqs, the FPKM values are quantile normalized and summarized to the gene level and log2 transformed. Only protein-coding genes are retained. Multiple datasets are merged based on shared gene identifiers and further quantile normalized. Metadata for patients are manually cleaned and standardized.

For the psychiatric disorders, five studies (GEO accessions GSE21935, GSE21138, GSE35974, GSE35977, and GSE25673) are combined. The first four are transcriptomes of brain regions and the last one is a study of iPS cell derived neurons from patients and normal controls. There are 429 samples in total, covering bipolar disorder (BD), schizophrenia (SZ), and major depression (MD). For type II diabetes (T2D), 4 studies (GSE38642, GSE50397, GSE20966, and GSE41762) of pancreatic islets tissues or beta cells are selected. For inflammatory bowel diseases (IBDs), a single RNA-seq dataset (GSE57945) of pediatric IBD patients is used, total 322 samples.

2.2. *Human gene regulation network and disease genetic associations*

The gene transcriptional regulation network is computationally extracted from ChIP-seq experiments as well as low throughput studies reported in the literature (see (*30*) for more details). The dataset is comprised of 146096 direct transcriptional regulation relationships between 384 transcription factors (TFs) and 16967 target genes, and is viewed as a directed graph with edges pointing from TFs to the target genes.

Gene-disease associations from genome wide association studies (GWAS) for psychiatric disorders (bipolar disorder, schizophrenia, and major depression), type II diabetes, and inflammatory bowel diseases were retrieved from dbGAP(*31*), NHGRI(*32*) and NHLBI(*33*) catalogs and filtered with loose p-value cutoff 1×10^{-5} to retain the weak but true disease causing genes. Phenotype terms related to the same diseases are manually examined and putative causal genes are combined. For each SNP, the closest gene or two genes for inter-genic SNP, are retained.

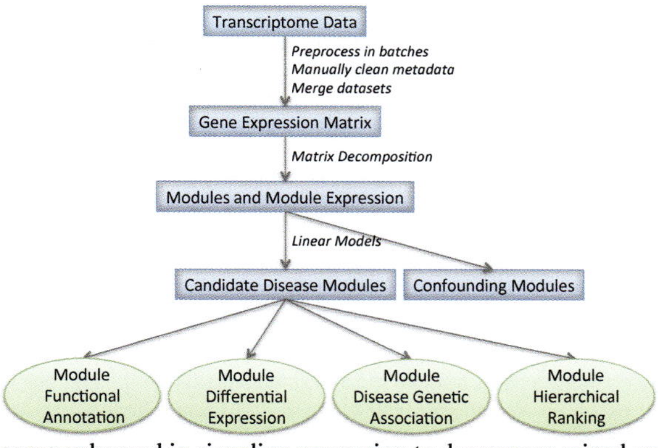

Figure 1: Overview of the diseaseExPatho pipeline for causal gene module prioritization

2.3. *Independent component analysis for learning gene modules*

Independent component analysis (ICA) is an unsupervised machine-learning algorithm for decomposing matrix into underlying simpler and potentially more meaningful component. It is commonly used in signaling processing to decompose mixed and noisy audio signals and estimate the original independent sound sources (*34*). When applied to transcriptome data, ICA decomposes gene expression into functionally coherent gene modules that correspond to cellular processes or pathways that co-express and co-vary in a biological sample (*35*). In this study, we decompose transcriptome data from patients in order to achieve mechanistic view of diseases.

Specifically, let matrix Y denotes the expression of G genes in N samples with dimension $G \times N$. ICA approximates a matrix Y as the product of two matrixes $Y \sim S \cdot A$, where S is a $G \times M$ matrix containing weights of G variables (genes) in the M independent components, while A is a $M \times N$ matrix containing the mixing coefficients of the M components in the N samples. It can be viewed as biclustering methods with the two matrixes providing the row and column clustering for original matrix Y. ICA achieves the matrix decomposition through the assumption that the M components in matrix S are statistically independent. We perform ICA using the fastICA algorithm(*34, 36, 37*) as in previous study(*38*). The M independent components learned from gene expression data are called gene modules. Each module represents a soft clustering of genes, with the dominant genes having the highest positive or negative weights. Previous study suggest that ICA provide functionally more coherent gene clusters compared to PCA(*35*). We hence interpret the modules as computationally estimated gene pathways composed of functionally related genes. In addition, we interpret each row of matrix A as the expression of a module in the N samples following previous study(*38*). For each gene expression matrix, we learn $M = 50$ gene modules.

2.4. *Differential expression of gene modules in disease*

For each gene module, we apply linear model of the form $a_i = \beta_0 + \beta_{disease} \cdot x_i^{disease} + \beta_{confound} \cdot x_i^{confound} + \varepsilon_i$ to infer the differential expression of the module in diseases versus normal. Note that a_i from matrix A is the expression of a module in sample i, $x_i^{disease}$ is the disease status variables, while $x_i^{confound}$ is the confounding variables. The significance is accessed by the Wald-T test of $\beta_{disease}$ and $\beta_{confound}$ being zero. P-values are then corrected by the BH procedure(*39*) for multiple hypothesis testing, and FDR < 0.05 is viewed significant.

Given the capability of ICA to separate signals resulting from different latent variables, we assume each gene module is associated with *one* latent variable. This will be true if the number of samples is much larger than the number of latent variables. We therefore label each module by the type of variable that is most strongly associated with it. Specifically, each gene module is marked *disease related*, if the module expression is significantly associated with the disease status, while less or not significantly associated with confounding variables; or *confounding*, if the opposite is true. The remaining modules not significant for any of the variables are of *uncertain* status. Both disease-related or the uncertain modules are retained, while the confounding modules are ignored. For psychiatric disorders and IBD, gender and ages are treated as confounding variables, while for T2D, gender, age and BMI are treated as confounding variables.

2.5. *Directed graph based ranking of gene modules*

We propose a hierarchical ranking method for gene modules based on gene regulation network. At the high level, we will assign a hierarchical ranking of each gene based on its position in the network, and then for each module, we compute its rank as the weighted average rank of genes in the modules. This approach can be extended to general gene clusters and known gene pathways without loss of generality since an ICA gene module is a weighted gene list, while gene clusters or known gene pathways are special cases of weighted gene lists taking only binary weights.

Given a directed graph and its adjacency matrix $M = (m_{ij}|m_{ij} \in \{0,1\}\, for\ i,j = 1 \dots n)$, we define a *non-negative* rank measure $r = (r_i | i = 1 \dots n)$ that is associated with nodes in the graph. We require that the rank of node j to equal (or be the best least square approximation of) the average of the ranks of all parent nodes plus 1, with 1 representing one layer downstream, i.e.,

$$r_j \sim \frac{\sum_i m_{ij}(r_i + 1)}{\sum_i m_{ij}}.$$

When the network is rooted, the rank measure can be interpreted the average distance from the root of the network to node j. Written in matrix format, the above problems are formally solved by $r^T = (\mathbf{1} \cdot M') \cdot (I - M')^{\dagger}$, where M' is the in-degree normalized adjacency matrix, $\mathbf{1}$ is a row vector of n 1s, and $\dagger$ is the pseudo-inverse. However, computing $(I - M')^{\dagger}$ directly can be intractable for large network. Alternatively, when $I - M'$ is invertible, we can numerically compute r^T iteratively through $r^T \leftarrow (1 + r^T) \cdot M'$ until convergence. For human gene regulation network, we found that $I - M'$ is invertible when we removed the self-loops. In this study, we removed the self-loops and used the iterative algorithm for computational efficiency.

Based on the gene ranks, we then calculate the ranks of gene modules. For an ICA module $\mathbf{s_m} = (s_{gm} | g = 1 \dots G)$, normalized s.t. $\sum_g {s_{gm}}^2$=1, the module's rank is calculated as $R_m = \sum_g s_{gm}^2 \cdot r_g$. Only genes in the regulatory network are included in this calculation.

2.6. *Inference of gene modules' association with genetic causes*

We propose a novel algorithm to associate a set of GWAS implicated putative causal genes with a gene module. Specifically, for each module m we built a linear model,

$$|s_{gm}| = \alpha_m + \beta_m x_g + \varepsilon_{gm},$$

where s_{gm} is the weight of gene g in module m, $x_g \in \{0,1\}$ indicates if gene g is a putative causal gene of a disease according to GWAS association p-value < 10^{-5}. When β_m is significantly greater than 0, the causal genes are significantly contributing to the gene module, thus module m is considered a *putative causal* module. Notice that since extreme positive or negative values are equally important for a gene module, we use absolute values $|s_{gm}|$. Hence, we name the approach *bidirectional* linear model (biLM).

We also note that due to the nature of this problem, x_g is binary, and the method is equivalent to performing a special T-test on the data, thus it can be called bidirectional T-test (biT-test). We use the same approach to detect the association of putative causal genes with the differential expression of genes in disease versus control, by replacing $|s_{gm}|$ with centered differential gene expression $|y_g - \overline{y_g}|$, where y_g is the differential expression of gene g.

3. Results

We apply diseaseExPatho to three major types of adult psychiatric disorders, schizophrenia (SZ), bipolar disorder (BD) and major depression (MD), as well as type II diabetes (T2D) and inflammatory bowel diseases (IBDs), Crohn's disease (CD) and ulcerative colitis (UC). The three psychiatric disorders together affect over 10% of the US population. They are widely studied with both transcriptome and GWAS approaches, allowing us to evaluate our method. We compiled transcriptome data from 5 studies of brain tissues and neuron cells. In total, there are 429 samples, including 82 BD, 27 MD, and 160 SZ patients, and 160 normal controls. For T2D we combine transcriptomic data of pancreatic islet from four studies, totally 199 samples, including 50 T2D patients and 149 normal controls. IBD data come from 1 study, total 322 samples, including 218 CD disease, 62 UC, and 42 normal controls. The putative causal genes for the disorders are manually compiled from databases of GWAS associations(*31–33*), with totally 151, 306, 87, 485, 71, and 229 putative causal genes for BD, MD, SZ, T2D, CD, and UC respectively.

3.1. *Genetic causes of diseases leave detectable signals in the transcriptome*

Despite the popularity of both GWAS and transcriptomic approach for disease study, there has been limited research on the consistency between GWAS and transcriptome approaches. A recent study reported the gene expression outliers are enriched with rare genetic variations in SZ patients(*40*). Here we examine the enrichment of GWAS-implicated putative causal genes of 6 diseases in the *two tails* of gene differential expression profiles. Statistically significant enrichment is detected for both the BD (p-value 0.00024, biLM) and MD (p-value 9.0×10^{-8}), but not SZ (p-value 0.23, see **Figure 2**). Enrichment is also observed for putative causal genes of T2D (p-value 4.3×10^{-8}), and CD (p-value 0.032), but not significant for UC (p-value 0.13).

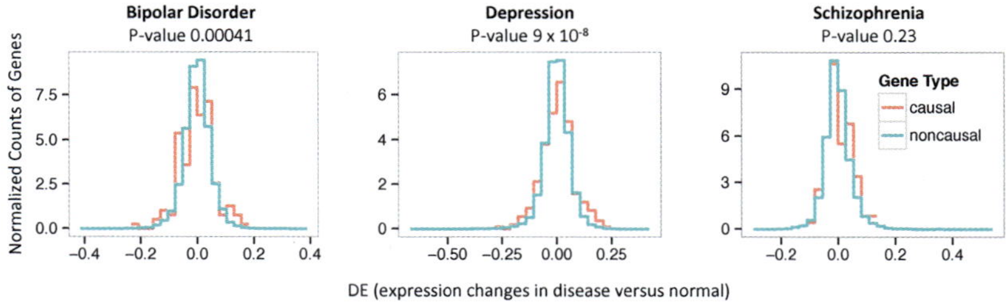

Figure 2: Overlay of the distributions (normalized counts) for differential expression (DE) of the putative causal genes (red) against DE of the other genes (cyan) in three types of psychiatric disorders. The p-values are obtained by bidirectional linear model comparing the spreads of the DE values.

3.2. *Matrix decomposition separates the causal and differentially expressed gene modules*

Diseases are generally complex processes. For complex diseases, multiple genetic and environmental factors together contribute to the disease risks. We believe for complex diseases, the causal factors, regardless of the type, cause disease through common molecular pathways of multiple genes. When we apply ICA to patient transcriptome, we expect some of the learnt gene modules to capture the underlying *causal* molecular pathways, driven by the same underlying causal factor (or a set of closely related causal factors) of the disease. The remaining modules can be downstream in disease pathogenesis or related to (possibly unknown) confounding factors.

We applied ICA and bidirectional linear model to psychiatric disorders and identified 17, 16, and 8 putative causal gene modules for BD, MD, and SZ. We refer to these significant modules the *putative causal* modules. We observed that many of the gene modules show a stronger enrichment of putative disease causal genes (**Figure 3A**) compared to the overall differential expression profiles (**Figure 2**) in terms of the association p-values. For example, 11, 6, and 8 of the putative causal modules have stronger enrichment p-values than the original differential expression profiles.

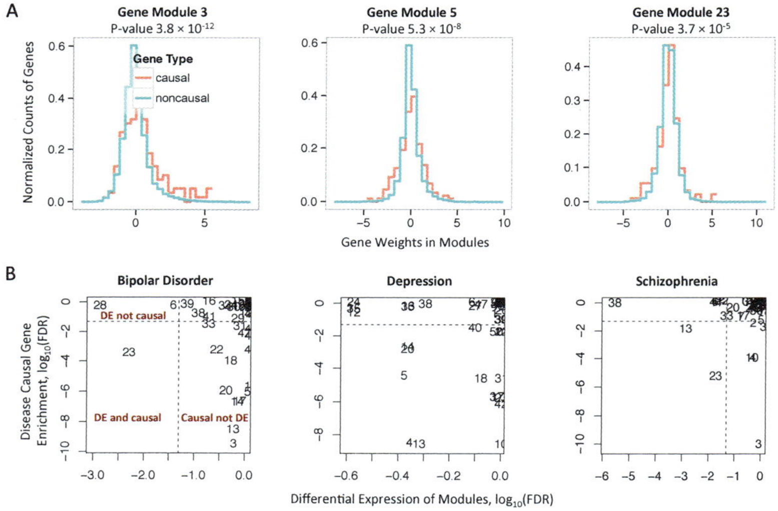

Figure 3: Disease causing and differential expression (DE) are orthogonal at the pathway level. **A**. Gene modules learnt by ICA and GWAS implicated putative disease causing genes are associated by enrichment analysis. The plots overlay the distributions of the weights of the putative causal genes of bipolar disorder (red) against the distribution of the weights of the other genes (cyan) in 3 modules. The p-values are obtained by biLM comparing the spreads of the genes' weights in two distributions. **B**. Scatter plot of the causal gene modules versus the DE gene modules for 3 psychiatric disorders. Many causal gene modules are not significantly differentially expressed, while many DE gene modules are not enriched with putative causal genes. The numbers in the plots are the IDs of gene modules. The x and y axes are FDR (the multi-testing corrected p-values) at log scale. Dashed lines correspond to FDR level 0.05.

Since the modules are derived based on gene expression data, it is important to examine if the putative causal modules are always differentially expressed (DE) in disease versus normal individuals. We calculate each module's DE as described in method section using a linear regression by removing the effects of confounding factors and correcting the p-values for multi-hypothesis testing. We then use the log of corrected p-value, FDR value here, to indicate the extent of DE. The extent of module's disease causing effect is calculated using bidirectional linear model. Surprisingly, we observed that majority of the putative causal gene modules are not differentially expressed for the psychiatric disorders (**Figure 3B**), and similar results are observed on separate analysis of type II diabetes and inflammatory bowel diseases (CD and UC). For example, module 3 is associated with putative causal genes for all 3 psychiatric disorders, but it is not differentially expressed for any of them. This however is consistent with the improved causal gene enrichment at the module level compare to the differential gene expressions, since many disease causal genes are apparently not associated with strong differential expression. In addition, we observed modules (e.g., module 38) that are only differentially expressed but not enriched with putative disease causing genes. Despite this, some putative causal modules are indeed differentially expressed (e.g. module 23, see **table 1** for details on selected modules). Overall, we identified 3 and 9 DE gene modules for bipolar disorder and schizophrenia, while 1 and 2 of them are overlapping with the putative causal modules. No DE modules are identified for major depression.

3.3. *Putative causal modules are ranked lower in the gene regulatory network*

GWAS studies are not available for all complex diseases. Given the large sample size requirement, some complex diseases may not have enough population to enable GWAS. We hence examine the possibility to infer putative causal gene modules from expression data directly.

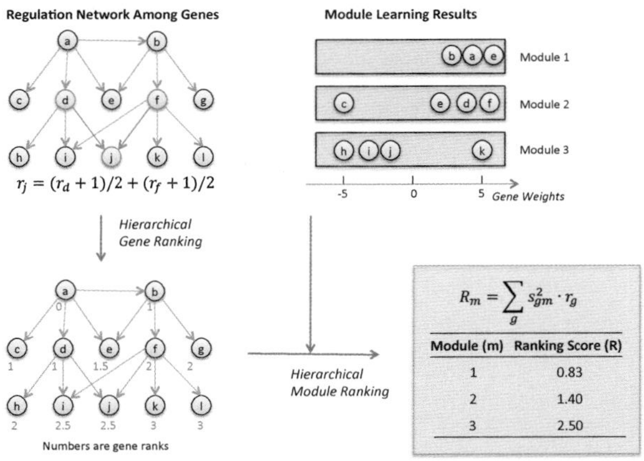

Figure 4: Directed gene regulation network based ranking of genes (letters a-l) and weighted gene lists (gene modules 1-3). The gene ranking is interpreted as the average distance of a gene to the root of the network when the network is rooted, as the example shown in this figure.

We rank the gene modules based on the directed gene regulation network to prioritize the modules. Multiple studies suggest that essential genes are less likely to be disease causing(*23*). Our basic assumption is that gene modules ranked top of the network are more essential in the cell, and are less likely to be associated with phenotypically weak variants for complex diseases.

We propose a novel and intuitive rank score for genes based on the regulatory network structure (**Figure 4**). The key property of the ranking is that a node's rank is the average of all its parent nodes' ranks plus 1 (see methods section 2.5 for details). For simple rooted graphs, we show that the resulting rank is the average distance of a node to the root of the graph (**Figure 4**). It is different from previously proposed gene ranking approaches(*17–19*) in two major ways. First, it provides an intuitive ranking of nodes that is consistent with topological sort when the graph is acyclic. Second, previous approaches focus on the transcription factors (TFs) and rank from the bottom to the top. Our approach ranks from top to bottom and both TF and non-TFs receive meaningful ranks depending on their locations in the network.

We first examine the ranking of single genes. The putative causal genes are enriched significantly in the bottom half of the network (p-value 0.0007, odd ratio 1.13 for putative causal gene obtained at p-value cutoff 1×10^{-5})*. Relatedly, GWAS implicated transcription factors also show a weak trend of favoring the bottom half of the network (p-value 0.20), despite that fact that TFs overall have higher ranks (p-value 0.0002, t-test).

We then examine the module rankings by aggregating the gene rankings based on genes' weights in the modules (see methods section 2.5). For the psychiatric disorders, we discovered that the putative causal gene modules are ranked significantly lower than the other modules (**Figure 5A** left, p-value 0.018, two-tailed t-test of ranks compared putative causal and non-causal modules, or p-value 0.028, Spearman rank correlation -0.33 between enrichment p-value and module ranking). Similarly for type II diabetes (p-value 0.16, two-tailed t-test; or p-value 0.0023, Spearman rank correlation -0.43), and inflammatory bowel diseases (p-value 0.019, two-tailed t-test; or p-value 0.00028, Spearman rank correlation -0.50). We further compared the putative causal modules with the differential expressed non-causal modules, and observed significantly lower ranking of the causal modules, this is true for 3 psychiatric disorders together (p-value 0.0019, two-tailed t-test, **Figure 5A** right), as well as for each psychiatric disorder separately (**Figure 5B**). This is however not significant for T2D and IBDs.

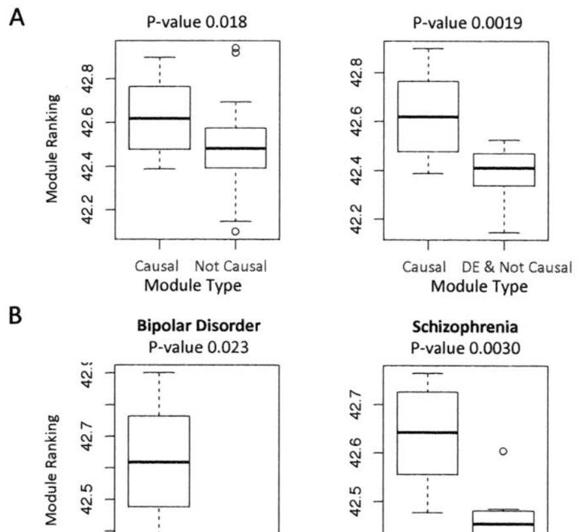

Figure 5: Gene regulation network based ranking differentiates the causal versus non-causal gene modules. The causal modules tend to be ranked lower (i.e. with higher rank values). **A**. Boxplots comparing the ranking of putative causal modules versus other modules (left) or putative causal modules versus the DE & Not Causal modules (right). A *putative causal* module is defined as a module that shows significant enrichment (FDR < 0.05) of GWAS implicated putative causal genes. A *DE & Not Causal* module is defined as a differentially expressed module (FDR < 0.05) that is not causal. For each module, p-values for 3 psychiatric disorders are combined into one p-value by Fisher's methods, for either differential expression or the enrichment of putative causal genes. The p-values shown in the figures are obtained by two-tailed two-sample t-tests. **B**. The comparison of putative causal modules versus the DE & Not Causal modules for individual psychiatric disorders. Note there are no significant DE modules for major depression.

As a control, we evaluate the ranking on the inverse network (by reversing the TF to target gene regulation directions), to obtain a bottom-

* The true causal genes will likely show stronger enrichment, given that a significant portion of the SNP-disease associations are spurious, and two candidate genes are included for intergenic SNPs.

up ranking mimicking the approach in previous studies (*17–19*). None of the results are significant based on the new ranking.

3.4. *Biological functions of the gene modules for psychiatric disorders*

We annotated the functions of gene modules for psychiatric disorders based on the enrichment of known gene functions curated in the Gene Ontology and canonical pathway databases (*41–44*). We carefully examined 8 gene modules, covering the top 5 putative causal and the top 5 differentially expressed modules (**Table 1**). The five putative causal gene modules are annotated with neural related gene functions, such as synaptic transmission and glutamate receptor activity. The three DE and non-causal gene modules are annotated mostly with functions that are not unique to neuronal systems, and are ranked top in gene regulatory network.

ID	Module Function %	Top 5 Genes	Type	Differential Expression P-value			Enrichment of Putative Causal Genes, P-value			Module Ranking
				BP $	MD $	SZ $	BP $	MD $	SZ $	
3	neuronal system; synaptic transmission; gated channel activity	SPHKAP, *GABRA6*, NEUROD1, *CADPS2*, *CNTN6*	Causal	0.23	0.42	0.49	4.E-12	2.E-07	1.E-12	34
13	axon guidance; nervous system development; glutamate receptor activity	DLK1, *ZIC1*, *PTN*, *GNAL*, DNER	Causal & DE	0.14	0.13	3.E-05	7.E-11	8.E-11	8.E-04	33
10	neuronal system; nervous system development; glutamate receptor activity	PMP2, SLC22A3, SLC17A8, *KAL1*, *CHL1*	Causal	0.90	0.72	0.09	2.E-07	7.E-11	5.E-06	42
4	neuronal system; nervous system development; voltage gated cation channel activity	*RGS4*, TESPA1, GDA, *HTR2A*, *CDH9*	Causal	0.61	0.10	0.18	7.E-04	2.E-10	4.E-06	41
23	GPCR downstream signaling; transmission of nerve impulse; receptor activity	*RELN*, *MET*, PENK, CALB1, GCNT4	Causal & DE	2.E-04	0.83	8.E-04	4.E-05	1.E-07	1.E-07	37
38	HIF1 TF pathway; signal transduction; drug binding	*FKBP5*, SLC14A1, PDK4, IL1RL1, ZBTB16	DE	9.E-03	0.14	3.E-08	0.05	0.35	0.26	13
6	oxidative phosphorylation; carbohydrate metabolic process; oxidoreductase activity	*GSTT1*, LAPTM4B, ATP6AP1, ATP6V0B, PITHD1	DE	2.E-03	0.25	1.E-03	0.21	0.62	0.43	7
28	cell cycle; cell cycle process; taste receptor activity	PI15, DLEU1, MSTN, FKBP14, SYCP2L	DE	2.E-05	0.79	0.24	0.20	0.87	0.66	4

Table 1: Function annotations of putative causal and differentially expressed modules for the psychiatric disorders. The top 5 putative causal modules (first 5 rows) and top 5 differentially expressed modules (marked DE) are included. Five highest-weighted genes are listed for each module, and those genetically associated with psychiatric disorder are underlined. [%]For each module, three functional terms are provided. They are the most significant terms in canonical pathways, gene ontology (GO) biological process, and GO molecular functions. [$]BP: bipolar disorder; MD: major depression; SZ: schizophrenia.

It is worth noting that stronger overlap among the three psychiatric disorders is observed at the module level than the gene level (**Figure 6**). We believe this is because the gene modules provide additional statistical power than single genes, and the impact of false causal genes from GWAS is minimized at the module level, as the modules are comprised of mainly functional related genes.

We also examine the functions of the disease-specific putative causal modules. Module 2 is unique to schizophrenia (and weakly for BP). Its top function annotations include GPCR ligand binding, G protein coupled receptor protein signaling pathway, and hormone activity. Module 29 is unique to bipolar disorder. Its top function annotations include 3-UTR mediated translational regulation, translation, and structural constituent of ribosome. Module 33 is unique to bipolar disorder (and weakly to schizophrenia). Its top function annotations include taste transduction, synaptogenesis, and taste receptor activity. Module 47 is unique to bipolar disorder. Its top function annotations include integrin-1 pathway, multicellular organismal development, and actin binding. Module 11 is unique to depression. Its top function annotations include cell adhesion molecules (CAMs), membrane organization and biogenesis, and phosphoric diester hydrolase activity. Module 50 is unique to depression, but it has no significant function annotation.

4. Discussion

Human complex diseases are the consequences of long-term interplay among a suite of abnormal genetic variants and environmental conditions. Previous studies have identified strong organizational patterns of human disease genes from the study of biological networks(*23*), and it is suggested that human disease genes tend to cluster into modules(*45*). Various approaches have been developed for predicting gene functions or disease genes using the guilty-by-association rule.

In this study we propose diseaseExPatho that integrates disease transcriptome and human gene regulation network to unravel the pathogenesis pathways in specific diseases. The diseaseExPatho is composed of 4 major components (**Figure 1**). A) ICA decomposition of gene expression matrix from patients; B) Module differential expression analysis; C) A novel algorithm (biLM) for associating a gene module with GWAS implicated putative causal genes of a disease; D) A novel algorithm for ranking genes and modules based on the gene regulation network.

Especially, we focus on prioritizing the disease causing pathways common to multiple patients by leveraging the gene co-expression pattern as well the hierarchical structure in the gene regulation network. We applied diseaseExPatho to 3 datasets for psychiatric disorders, type II diabetes (T2D) and inflammatory bowel diseases (IBDs), and obtained consistent and promising results.

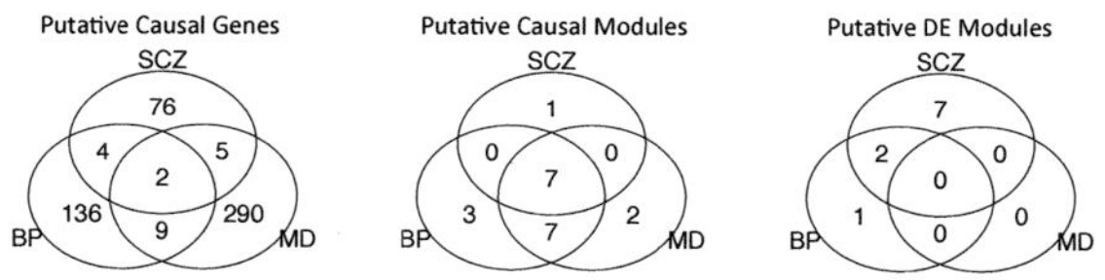

Figure 6: Overlap of putative causal genes, putative causal modules, and DE modules among psychiatric disorders. BP: bipolar disorder; MD: major depression; SCZ: schizophrenia. Significant modules for each disease are identified as those with FDR<0.05 for that disease.

4.1. *Gene co-expression, differential expression and disease causing*

The disease transcriptome data provide two key ingredients of information. First, it provides the genes that are likely active in the disease. This acts as a filter of the gene regulation network to obtain the disease-relevant sub-network. Second, it provides the gene co-expression patterns, which divides the disease-related genes into compact modules with close-related functions.

We use the independent component analysis to simultaneously extract these two types of information, through estimating a set of independent gene expression modules. After labeling the putative causal gene modules based on the enrichment of putative disease causing genes, we observed that majority of the gene modules activated/deactivated in the disease are not associated with causal genetic variations (**Figure 3B**). In fact, many putative causal modules do not show DE in the patients, while many non-causal gene modules are significantly differentially expressed (**Figure 3B** and **table 1**). Such non-causal DE modules may correspond to downstream molecular pathways, or they may be driven by disease-unrelated confounding factors.

With the help of computationally derived gene modules (pathways), we can elevate from the individual causal genes to causal pathways. This provides us with four advantages. First, we have stronger statistical power in detecting the causal mechanisms of diseases, when we aggregate the GWAS signals from the individual gene level to the pathway level. Second, we have a more systematic view of the disease mechanism as revealed by the common functions of multiple genes in a module. Third, we have higher statistical power in detecting gene module's expression changes compared to gene's expression changes, as we suffer much less from the multi-hypothesis testing issues, since there are much fewer number of modules than genes. Fourth, the identified gene expression suffers much less from confounding factors, such as patient heterogeneity due to gender and age.

It is a significant and underappreciated fact that the disease causing genes leave significant expression signals in patients' transcriptome(*40*). We observed increased expression changes of putative causal genes for psychiatric disorders (**Figure 2**), as well as T2D and IBDs. This serves as the foundation of transcriptome-based disease etiology inference.

We observed a stronger enrichment of disease causing genes in individual gene modules than using the overall DE profiles (**Figure 3A**). In the extreme cases of schizophrenia, we observed 3 modules that are significantly enriched with putative schizophrenia causal genes, while no significant enrichment is observed for the differential expression profile in patient versus normal (**Figure 2** right). This implies that, first, human disease variations gather in functionally related genes, and second, these functional related genes cluster as co-expression modules in the disease transcriptome, even when the genes do not show strong differential expression in disease.

A weak consistency has been observed between GWAS and gene expression data for prostate cancer (*46*) and schizophrenia (*40*). Our findings not only support the consistency, but also provide an explanation of the failure to observe a much stronger consistency. We suggest that many DE signals in expression data are non-causal but rather consequences or driven by confounding factors, as supported by the *DE & not causal* modules. On the other hand, DE is not a legitimate requirement for all causal genes, as supported by the *causal & non-DE* modules.

We believe the differential expression approach, although commonly used in transcriptome study, is not the best approach to extract the causal signals in expression data. A recent study observed improved GO term enrichment when selecting SNPs that are associated with gene expression changes(*47*). We support the integration of expression data and GWAS as a way to remove the noises in GWAS findings. However, given our observations, we believe requiring DE on the causal genes will remove true causal genes. We instead advocate using gene co-expression modules rather than disease differential expression for improved interpretation of GWAS results.

4.2. *Causal module prioritization without using known genetic causes*

Prior studies suggest that human essential genes (with knock-out lethality in mouse) are less likely to be disease causing (*23*). We propose a network-based module ranking, and hypothesize that the top-ranked modules are more essential, while the near bottom-ranked modules are more likely to be disease causing. This hypothesis is supported by module ranking results in psychiatric disorders (**Figure 5** and **table 1**), as well as T2D and IBDs. Consistently, GWAS implicated putative disease/phenotype causal genes also prefers the bottom-half of the regulatory network.

To our knowledge, the network we compiled for this study is the largest published network, yet it only covers 384 transcription factors, 25% of the putative 1500 transcription factors in human (*48*, *49*). Despite this, the network already provides meaningful signal for prioritizing the putative causal modules, as is observed in 3 disease datasets. We hence expect improved performance of diseaseExPatho with the accumulation of more and higher quality gene-regulation data.

Although we demonstrate the applications of diseaseExPatho to complex diseases with extensive GWAS results, we suggest the module ranking approach can be applied to prioritize putative causal modules for complex disease that are not well studied by GWAS, such as the idiopathic inflammatory myositis (*50*, *51*).

5. Acknowledgement

YFL would like to acknowledge the support of TRAM pilot grant for part of this work and Ron Davis for insightful discussions. RBA would like to acknowledge the supported by MH094267, HL117798, LM005652, GM102365 and a grant from Pfizer Inc. We thank Annie Altman-Merino for her assistance with metadata curation of patient samples.

References

1. K. T. Zondervan, L. R. Cardon, *Nat. Rev. Genet.* **5**, 89–100 (2004).
2. J. Marchini, P. Donnelly, L. R. Cardon, *Nat. Genet.* **37**, 413–417 (2005).
3. P. M. Visscher, M. a. Brown, M. I. McCarthy, J. Yang, *Am. J. Hum. Genet.* **90**, 7–24 (2012).
4. M. I. McCarthy *et al.*, *Nat. Rev. Genet.* **9**, 356–369 (2008).
5. V. K. Rakyan, T. a Down, D. J. Balding, S. Beck, *Nat. Rev. Genet.* **12**, 529–541 (2011).
6. M. J. Bamshad *et al.*, *Nat. Rev. Genet.* **12**, 745–755 (2011).
7. M. L. Freedman *et al.*, *Nat. Genet.* **43**, 513–518 (2011).
8. O. Morozova, M. a. Marra, *Genomics.* **92**, 255–264 (2008).
9. A. Kahvejian, J. Quackenbush, J. F. Thompson, *Nat. Biotechnol.* **26**, 1125–1133 (2008).
10. Z. Wang, M. Gerstein, M. Snyder, *Nat. Rev. Genet.* **10**, 57–63 (2009).
11. M. J. Heller, *Annu. Rev. Biomed. Eng.* **4**, 129–153 (2002).
12. C. Giallourakis, C. Henson, M. Reich, X. Xie, V. K. Mootha, *Annu. Rev. Genomics Hum. Genet.* **6**, 381–406 (2005).
13. D. G. MacArthur *et al.*, *Nature.* **508**, 469–76 (2014).
14. C. Auffray, Z. Chen, L. Hood, *Genome Med.* **1**, 2 (2009).
15. P. Thagard, *Minds Mach.* **8**, 61–78 (1998).
16. L. Darden, *Mechanism and Causality in Biology and Medicine* (2013; http://link.springer.com/10.1007/978-94-007-2454-9), vol. 3.
17. N. Bhardwaj, K.-K. Yan, M. B. Gerstein, *Proc. Natl. Acad. Sci. U. S. A.* **107**, 6841–6 (2010).
18. H. Yu, M. Gerstein, *Proc. Natl. Acad. Sci. U. S. A.* **103**, 14724–31 (2006).
19. K.-K. Yan, G. Fang, N. Bhardwaj, R. P. Alexander, M. Gerstein, *Proc. Natl. Acad. Sci. U. S. A.* **107**, 9186–91 (2010).
20. M. B. Gerstein *et al.*, *Nature.* **489**, 91–100 (2012).
21. R. M. Piro, F. Di Cunto, *FEBS J.* **279**, 678–696 (2012).
22. A. Valouev *et al.*, **5**, 829–834 (2008).
23. A.-L. Barabási, N. Gulbahce, J. Loscalzo, *Nat. Rev. Genet.* **12**, 56–68 (2011).
24. B. E. Bernstein *et al.*, *Nature.* **489**, 57–74 (2012).
25. T. Barrett *et al.*, *Nucleic Acids Res.* **41**, D991–5 (2013).
26. M. N. McCall, B. M. Bolstad, R. a Irizarry, *Biostatistics.* **11**, 242–53 (2010).
27. R. A. Irizarry *et al.*, *Biostatistics.* **4**, 249–64 (2003).
28. M. N. McCall, H. A. Jaffee, R. A. Irizarry, *Bioinformatics.* **28**, 3153–4 (2012).
29. B. M. Bolstad, R. . Irizarry, M. Astrand, T. P. Speed, *Bioinformatics.* **19**, 185–193 (2003).
30. Y. F. Li, R. B. Altman, *Prep.* (2014).
31. M. D. Mailman *et al.*, *Nat. Genet.* **39**, 1181–6 (2007).
32. D. Welter *et al.*, *Nucleic Acids Res.* **42**, 1001–1006 (2014).
33. J. D. Eicher *et al.*, *Nucleic Acids Res.* **43**, D799–D804 (2014).
34. A. Hyvärinen, E. Oja, *Neural Networks.* **13**, 411–430 (2000).
35. S.-I. Lee, S. Batzoglou, *Genome Biol.* **4**, R76 (2003).
36. A. Hyvärinen, E. Oja, *Neural Comput.* **9**, 1483–1492 (1997).
37. A. Hyvarinen, *IEEE Trans. Neur. Net.* **10**, 626–634 (1999).
38. J. M. Engreitz, B. J. Daigle, J. J. Marshall, R. B. Altman, *J. Biomed. Inform.* **43**, 932–44 (2010).
39. Y. Benjamini, Y. Hochberg, *J. R. Stat. Soc. Ser. B.* **57**, 289 – 300 (1995).
40. J. Duan *et al.*, *Hum. Mol. Genet.*, 1–12 (2015).
41. M. Ashburner *et al.*, *Nat. Genet.* **25**, 25–9 (2000).
42. G. Joshi-Tope *et al.*, *Nucleic Acids Res.* **33**, D428–32 (2005).
43. M. Kanehisa, *Nucleic Acids Res.* **28**, 27–30 (2000).
44. C. F. Schaefer *et al.*, *Nucleic Acids Res.* **37**, D674–9 (2009).
45. X. Wu, R. Jiang, M. Q. Zhang, S. Li, *Mol. Syst. Biol.* **4**, 189 (2008).
46. I. P. Gorlov, G. E. Gallick, O. Y. Gorlova, C. Amos, C. J. Logothetis, *PLoS One.* **4** (2009), doi:10.1371/journal.pone.0006511.
47. H. Zhong, X. Yang, L. M. Kaplan, C. Molony, E. E. Schadt, *Am. J. Hum. Genet.* **86**, 581–591 (2010).
48. J. M. Vaquerizas, S. K. Kummerfeld, S. A. Teichmann, N. M. Luscombe, *Nat. Rev. Genet.* **10**, 252–63 (2009).
49. S. K. Kummerfeld, S. a Teichmann, *Nucleic Acids Res.* **34**, D74–D81 (2006).
50. M. Jani *et al.*, *Lancet.* **381**, S56 (2013).
51. Q. Gang, C. Bettencourt, P. Machado, M. G. Hanna, H. Houlden, *Orphanet J. Rare Dis.* **9**, 88 (2014).

A BAYESIAN NONPARAMETRIC MODEL FOR RECONSTRUCTING TUMOR SUBCLONES BASED ON MUTATION PAIRS

SUBHAJIT SENGUPTA[1*], TIANJIAN ZHOU[2*], PETER MÜLLER[3], YUAN JI[1,4†]

[1]*Program for Computational Genomics and Medicine, NorthShore University HealthSystem;* [2]*Department of Statistics and Data Sciences, The University of Texas at Austin;* [3]*Department of Mathematics, The University of Texas at Austin;* [4]*Department of Public Health Sciences, The University of Chicago*

We present a feature allocation model to reconstruct tumor subclones based on mutation pairs. The key innovation lies in the use of a pair of proximal single nucleotide variants (SNVs) for the subclone reconstruction as opposed to a single SNV. Using the categorical extension of the Indian buffet process (cIBP) we define the subclones as a vector of categorical matrices corresponding to a set of mutation pairs. Through Bayesian inference we report posterior probabilities of the number, genotypes and population frequencies of subclones in one or more tumor sample. We demonstrate the proposed methods using simulated and real-world data. A free software package is available at http://www.compgenome.org/pairclone.

Keywords: Categorical Indian buffet process; Latent feature model; Local Haplotype; NGS data; Random categorical matrices; Tumor heterogeneity.

1. Introduction

1.1. *Background*

With the recent development of next-generation sequencing (NGS) technology, whole-genome or whole-exome sequencing has been used to interrogate genetic landscape of tumors within and across different patients. Using single nucleotide variants (SNVs), NGS data can reveal whether a tumor sample is composed of cell subpopulations, i.e., subclones that contain somatic mutations.[1–6] In essence, the main problem of subclone reconstruction is to identify more than two haploid genomes in a tumor sample. Since humans are diploid, a homogeneous cell population can only harbor two distinct haploid genomes, or else the cell population must be heterogeneous and contain at least two different subclones with different genomes. In NGS data, short reads are mapped to each SNV locus. Compared to the reference nucleotide base on the locus, some short reads may harbor the same reference base while others may bear a variant base. The latter are called variant reads and the proportion of variant reads among all the reads mapped to the SNV is called the observed variant allele fraction (VAF). If all the cells in a tumor sample share the same genome, i.e., they are genetically homogeneous, the VAFs must be close to 0, 0.5, or 1, reflecting the three possible genotypes at a single locus – AA, AB, or BB. For example, when all the cells in the tumor bear the heterozygous AB genotype, roughly half of the reads will harbor A and the other half B. Therefore, the observed VAF should be close to 0.5. Homozygous alleles should give rise to observed VAFs close to 0 or 1. When the VAF at the SNV is neither of 0, 0.5, or 1, the cellular genomes might

---

*have equal contributions.

†Address for Correspondence: Research Institute, NorthShore University HealthSystem, 1001 University Place, Evanston, IL 60201, USA. Email: koaeraser@gmail.com

be heterogeneous containing distinct gentoypes at the SNV. For example, a sample of 50% of cells bearing genotype AB and 50% of cells bearing AA results in 75% of A alleles and 25% B alleles. If the A allele is the reference genome, the VAF is expected to be around 25%, or 0.25, which is not close to 0, 0.5, or 1. Based on this basic logic, many methods[7–12] have been developed to infer subclones using NGS data.

1.2. *Main idea*

Inference of subclones that hinges on "unusual" VAFs is vulnerable to the noise and artifacts in the NGS data. In particular, due to the complexity and limitation of the NGS experiment, the observed VAF at an SNV can deviate from ideal values 0, 0.5, or 1 even when the cell population is homogeneous. When the population is indeed heterogeneous, noise in the NGS data can still affect the accuracy of subclone reconstruction. Currently the noise and artifacts in NGS data cannot be properly modeled and accounted for due to its complexity,[13] and therefore SNV-based subclone callers often require lengthy and ad-hoc noise filters. The effects of these noise filters on the subclone reconstruction is usually unknown.

To mitigate this problem, we consider a different approach. We assume that paired-end short reads are used in the NGS experiment. Instead of modeling reads mapped to individual SNVs, we consider a pair of them, i.e., mutation pairs. We consider proximal mutation pairs that are close enough to be phased by some of the same short reads. Such mutation pairs can be retrieved by existing tool[14] with high confidence. Since there are two loci in each mutation pair, the observed data are haplotypes (of two phased SNVs). With four possible nucleotides at each SNV, there could be up to 16 different haplotypes at each mutation pair (details in Section 2.2). Observing more than two haplotypes is evidence of tumor heterogeneity, again, due to diploidy. See Fig. 1 for an example.

We assume a total of T ($T \geq 1$) samples are obtained from a single patient, and consider intra-tumor heterogeneity as the main inference goal. Consider a finite number of K mutation pairs that are shared across the T samples, and assume that an unknown number of C subclones are present. We denote a subclone by a set of matrices $\boldsymbol{z}_{kc}$ for mutation pairs $k = 1, 2, \ldots, K$. Each $\boldsymbol{z}_{kc}$ is a 2×2 matrix that codes the two diploid genotypes of mutation pair k for subclone c. Detail of $\boldsymbol{z}_{kc}$ is given in the upcoming discussion. We also assume that the C subclones are shared by the T samples, with different population frequencies for each sample, denoted by $\boldsymbol{w}_t = (w_{t0}, w_{t1}, \ldots, w_{tC})$ for sample t, where $0 < w_{tc} < 1$ for all c and $\sum_{c=0}^{C} w_{tc} = 1$. Using the NGS data we infer $\boldsymbol{Z}$ and $\boldsymbol{w}$ based on a simple idea that that variant reads can only arise from subclones with variant genotypes.

Among exiting methods, SciClone, TrAp, Clomial, PhyloSub and PhyloWGS [(8–12)] are of relevance to this work. The main difference of our method from all the other existing methods is that we use mutation pairs as experimental units instead of unpaired SNVs. Also our model is based on latent feature allocation methods that allow overlapping mutations across subclones. This is different from cluster-based methods in the literature.

The paper is structured as follows: Sec. 2 and Sec. 3 describes the Bayesian feature allocation model and posterior inference, respectively. Sec. 4 presents two simulation studies. Sec. 5 reports analysis results for a real-world dataset. Sec. 6 concludes with a final discussion.

2. Probability Model

2.1. *Sampling Model*

We start the construction of a sampling model by considering one mutation pair k (see Fig. 1). Two loci, denoted by $r = 1, 2$ mark the mutation pair. A set of short reads are mapped to the genomic region that contain the two loci. Index short reads by d. When short reads are mapped to the region, we require that at least one of the two loci is covered, or else the short reads are excluded from our analysis since they do not provide any information on the mutation pair. Consider short read d mapped to mutation pair k in sample t. Define $\boldsymbol{s}_{tk}^{(d)} = \left\{s_{tkr}^{(d)}\right\}_{r=1,2} = \left(s_{tk1}^{(d)}, s_{tk2}^{(d)}\right)$, where $s_{tkr}^{(d)}$ takes three values of $\{0, 1, -\}$ representing that the base on read d mapped to locus r is reference, variant, or missing, respectively. For example, in Fig. 1 locus $r = 1$, $s_{tk1}^{(d)} = 0$ for read $d = 1$, $s_{tk1}^{(d)} = 1$ for read $d = 2$, and $s_{tk1}^{(d)} = -$ for read $d = 3$. Aggregating across two loci, each $\boldsymbol{s}_{tk}^{(d)}$ can take $G = 8$ possible genotypes, including the reference, variant, and missing genotypes, denoted by $\mathcal{H} = \{\boldsymbol{h}_1, \ldots, \boldsymbol{h}_G\} = \{(0,0), (0,1), (1,0), (1,1), (-,0), (-,1), (0,-), (1,-)\}$, where each $\boldsymbol{h}_g = \{h_{gr}\}_{r=1,2} = (h_{g1}, h_{g2})$ denotes the potential genotype at each locus r of a short read. Let $n_{tkg} = \sum_d I\left(\boldsymbol{s}_{tk}^{(d)} = \boldsymbol{h}_g\right)$ be the read count representing the number of short reads having genotype $\boldsymbol{h}_g$. Here $I()$ is the indicator function. The total number of reads that are mapped to the loci of the mutation pair k in sample t is then $N_{tk} = \sum_{g=1}^{G} n_{tkg}$. We assume a multinomial sampling model for n_{tkg} conditional on N_{tk}, given by

$$n_{tk1}, \ldots, n_{tkG} \mid N_{tk}, p_{tk1}, \ldots, p_{tkG} \overset{indep.}{\sim} \text{Multinomial}\left(N_{tk}; p_{tk1}, \ldots, p_{tkG}\right), \tag{1}$$

where $p_{tkg} = Pr(\boldsymbol{s}_{tk}^{(d)} = \boldsymbol{h}_g)$ is the probability that a read bears genotype $\boldsymbol{h}_g$ on mutation pair k in sample t.

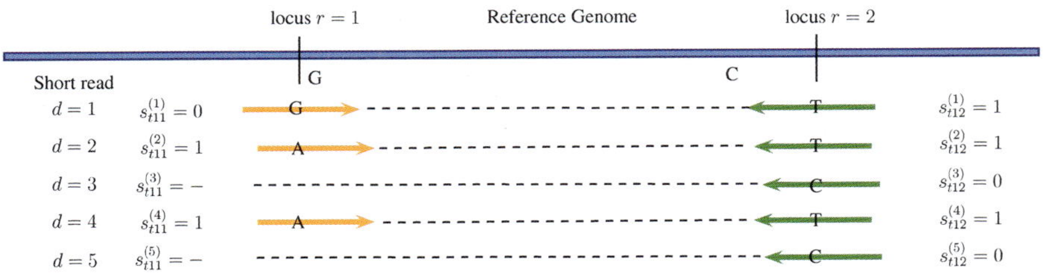

Fig. 1. Illustration of read count data for a mutation pair. There is a total of five reads mapped to the two loci that mark the mutation pair. The five reads exhibit genotypes $(0,1)$, $(1,1)$, $(-,0)$, $(1,1)$, $(-,0)$, which implies that there could be three haplotypes for the mutation pair in the sample.

2.2. *Subclone Representation using Z*

We collect all the $\boldsymbol{z}_{kc}$'s in a matrix format, denoted as a $K \times C$ matrix $\boldsymbol{Z} = [\boldsymbol{z}_{kc}]$. Technically, $\boldsymbol{Z}$ is a matrix of matrices, since each $\boldsymbol{z}_{kc}$ is itself a matrix. See Fig. 2. The total number of subclones, denoted by C, is random. The c-th column of $\boldsymbol{Z}$, $\boldsymbol{z}_c = (\boldsymbol{z}_{1c}, \ldots, \boldsymbol{z}_{Kc})$ denotes one particular subclone. Each element $\boldsymbol{z}_{kc}$ records the two alleles of a particular mutation pair k for subclone c. Let $j = 1, 2$ index the two alleles in a subclone and $r = 1, 2$ represent the two loci in a mutation pair. We write $\boldsymbol{z}_{kc} = \{z_{kcjr}\} = ((z_{kc11}, z_{kc12}), (z_{kc21}, z_{kc22}))$. See Fig. 2 for an example. Note that $z_{kcjr} = 1$ indicates that r-th locus of j-th allele of $\boldsymbol{z}_{kc}$ bears a mutation compared to the reference genome. Clearly $\boldsymbol{z}_{kc}$ can take $Q = 16$ possible values i.e. $\boldsymbol{z}_{kc} \in \{\boldsymbol{z}^{(q)}\}_{q=1}^{16} = \{\boldsymbol{z}^{(1)}, \ldots, \boldsymbol{z}^{(16)}\} = \{((0,0),(0,0)), ((0,0),(0,1)), \ldots, ((1,1),(1,1))\}$. For example, in Fig. 2 reference genome at the loci of mutation pair 1 is AT, and the corresponding genotype of subclone 3 is $((G,T),(G,C))$, which translates to $\boldsymbol{z}_{kc} = ((1,0),(1,1))$. However, we can collapse some $\boldsymbol{z}^{(q)}$ values since we do not distinguish the order of the two alleles for a mutation pair in a subclone. That is $\boldsymbol{z}_{kc} = ((z_{kc\underline{1}1}, z_{kc\underline{1}2}), (z_{kc\underline{2}1}, z_{kc\underline{2}2}))$ and $\boldsymbol{z}_{kc} = ((z_{kc\underline{2}1}, z_{kc\underline{2}2}), (z_{kc\underline{1}1}, z_{kc\underline{1}2}))$ lead to the same probability model. Therefore, the two alleles are coded invariant of their orders and we reduce the number of possible outcomes of $\boldsymbol{z}_{kc}$ to from 16 to $Q = 10$ and they are listed as: $\boldsymbol{z}^{(1)} = ((0,0),(0,0)), \boldsymbol{z}^{(2)} = ((0,0),(0,1)), \boldsymbol{z}^{(3)} = ((0,0),(1,0)), \boldsymbol{z}^{(4)} = ((0,0),(1,1)), \boldsymbol{z}^{(5)} = ((0,1),(0,1)), \boldsymbol{z}^{(6)} = ((0,1),(1,0)), \boldsymbol{z}^{(7)} = ((0,1),(1,1)), \boldsymbol{z}^{(8)} = ((1,0),(1,0)), \boldsymbol{z}^{(9)} = ((1,0),(1,1))$ and $\boldsymbol{z}^{(10)} = ((1,1),(1,1))$.

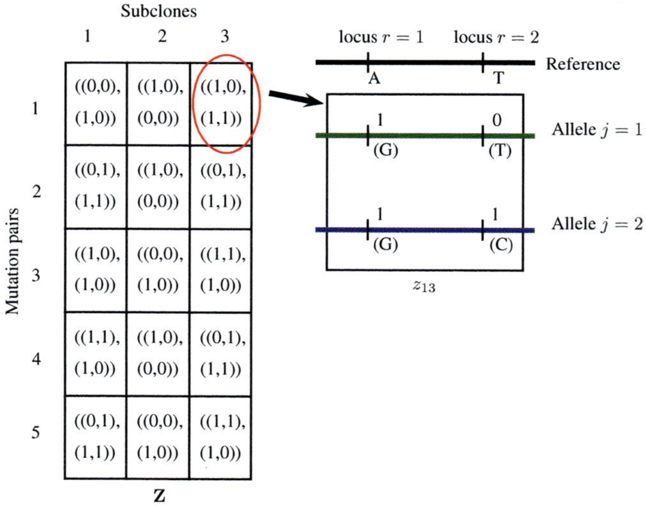

Fig. 2. Illustration of $\boldsymbol{Z}$ (left panel) for subclones in a sample and a particular subclonal genotypes for a mutation pair (right panel). Each column of Z represents a subclone, with each element representing the subclonal genotypes for a mutation pair. The genotypes for mutation 1 in subclone 3 is $((1,0),(1,1))$, which can be shown as a stylized example in the right panel.

Each sample is potentially an admixture of the subclones (columns of $\boldsymbol{Z}$), mixed in different proportions. Given $\boldsymbol{Z}$, we can denote the proportions of the C subclones by $\boldsymbol{w}_t = (w_{t0}, w_{t1}, \ldots, w_{tC})$ for sample t, where $0 < w_{tc} < 1$ for all c and $\sum_{c=0}^{C} w_{tc} = 1$. Notice that the subclones are common for all tissue samples, but the weights w_{tc} vary across samples. A background subclone, which has no biological meaning and is indexed by $c = 0$, is included to account for experimental noise (sequencing errors, mapping errors, etc.).

2.3. *Prior model*

Prior for p_{tkg}: The prior for the multinomial probabilities p_{tkg} in (1) is based on a simple idea: a short read harboring a particular haplotype $\boldsymbol{h}_g$ can only come from subclones that also harbor the same haplotype in their genomes. The probability of observing such a short read depends on the population frequencies $\boldsymbol{w}_t$ of such subclones harboring the haplotype. Therefore, we define

$$p_{tkg} \propto \sum_{c=1}^{C} w_{tc} A(\boldsymbol{h}_g, \boldsymbol{z}_{kc}) + w_{t0}\, \rho_g, \text{ for } g = 1, \ldots, 8, \tag{2}$$

where $A(\boldsymbol{h}_g, \boldsymbol{z}_{kc})$ is the expected proportion of alleles with genotype $\boldsymbol{h}_g$ at mutation pair k of subclone c. Accounting for the potential missing genotype at each of the two loci corresponding to the mutation pair, there are three ways a short read can cover the mutation pair: (i) the read maps to both loci; (ii) the read maps to the second locus but does not map to the first (left missing), and (iii) the read maps to the first locus but not the second locus (right missing). Therefore, we define

$$A(\boldsymbol{h}_g, \boldsymbol{z}_{kc}) = \begin{cases} \sum_{j=1}^{2} 0.5 \times I\left(h_{g1} = z_{kcj1}, h_{g2} = z_{kcj2}\right), & \text{for } g = 1, \ldots, 4; \\ \sum_{j=1}^{2} 0.5 \times I(h_{g2} = z_{kcj2}), & \text{for } g = 5, 6; \\ \sum_{j=1}^{2} 0.5 \times I(h_{g1} = z_{kcj1}), & \text{for } g = 7, 8. \end{cases} \tag{3}$$

In (3), the three equations correspond to the three coverage cases (i) – (iii) mentioned above. The factor 0.5 is used to reflect that any short read comes from one of the two alleles in the genome with equal probability. Quantifying the expected proportion of alleles in the genome, $A(\boldsymbol{h}_g, \boldsymbol{z}_{kc})$ can only take three values 0, 0.5 or 1. According to (3) and assuming no sequencing error, a read that covers both loci ($g = 1, 2, 3, 4$) and bears genotype $\boldsymbol{h}_g$ must be generated from a subclone having the same $\boldsymbol{h}_g$ genotype in at least one allele. When the read only covers one of the two loci, the requirement is to match the sequence on the covered locus only, and hence the equations in (3) for cases $g = 5, 6, 7, 8$.

In (2) we also include a background subclone denoted by $c = 0$ with proportion of w_{t0} to account for experimental noise. The background subclone does not exist and is only used as a mathematical device to account for noise and artifacts in the NGS data. See Ref. [15] for details.

Prior for $\boldsymbol{Z}$: We develop a latent-feature-allocation prior for the latent matrix $\boldsymbol{Z}$, the elements of which take categorical values. The prior $p(\boldsymbol{Z} \mid C)$ is constructed under fixed C. Let $\boldsymbol{\pi}_c = (\pi_{c1}, \pi_{c2}, \ldots, \pi_{cQ})$ where $p(\boldsymbol{z}_{kc} = \boldsymbol{z}^{(q)}) = \pi_{cq}$ and $\sum_{q=1}^{Q} \pi_{cq} = 1$. We use the beta-Dirichlet distribution[16] as the prior for $\boldsymbol{\pi}_c$. Conditional on C, $p(\boldsymbol{z}_{kc} = \boldsymbol{z}^{(1)}) = \pi_{c1}$ follows a beta distribution with parameters 1 and α/C, and $(\tilde{\pi}_{c2}, \ldots, \tilde{\pi}_{cQ})$, where $\tilde{\pi}_{cq} = \pi_{cq}/(1 - \pi_{c1})$ with $q = 2, \ldots, Q$, follows a Dirichlet distribution with parameters $(\gamma_2, \ldots, \gamma_Q)$. Here $\boldsymbol{z}^{(1)}$ is special because it refers to the reference genome. We write

$$\pi_c \sim \text{Beta-Dirichlet}(\alpha/C, 1, \gamma_2, \ldots, \gamma_Q).$$

As shown in Ref. [17], the marginal limiting distribution of $\boldsymbol{Z}$ follows a categorical Indian buffet process (cIBP) as $C \to \infty$.

Prior for $\boldsymbol{w}$: Next, we introduce a prior distribution for $\boldsymbol{w}_t$ as

$$\boldsymbol{w}_t \mid C \overset{iid}{\sim} \text{Dirichlet}(d_0, d, \ldots, d),$$

for $t = 1, \ldots, T$. For all practical purpose, we set $d_0 < d$ to imply that the hypothetical background subclone has a small population frequency.

Prior for $\boldsymbol{\rho}$ and C: Then we construct the prior for $\boldsymbol{\rho}$, where ρ_g is the conditional probability of observing a read with a genotype $\boldsymbol{h}_g$ due to experimental noise. We assume Dirichlet priors on ρ_g's,

$$\rho_{g_1} \sim \text{Dirichlet}(d_1, \ldots, d_1); \; \rho_{g_2} \sim \text{Dirichlet}(2d_1, 2d_1); \; \rho_{g_3} \sim \text{Dirichlet}(2d_1, 2d_1) \tag{4}$$

where $g_1 = \{1, 2, 3, 4\}$, $g_2 = \{5, 6\}$ and $g_3 = \{7, 8\}$.

Finally, we put a geometric distribution prior on number of subclones i.e. $C \sim \text{Geom}(r)$, and hence $E(C) = 1/r$ a priori.

3. Posterior Inference

3.1. *Posterior computation:*

Markov chain Monte Carlo (MCMC) simulation[18] is used to draw samples from the posterior of the unknown parameters. Let $\boldsymbol{x} = (\boldsymbol{Z}, \boldsymbol{\pi}, \boldsymbol{w}, \boldsymbol{\rho})$ denote all the parameters except C. With fixed C, sampling $\boldsymbol{x}$ from the respective posterior distribution is straightforward. Gibbs sampling transition probabilities are used to update $\boldsymbol{Z}$ and $\boldsymbol{\pi}$, and Metropolis-Hastings transition probabilities are used to update $\boldsymbol{w}$ and $\boldsymbol{\rho}$.

Updating the value of C is more challenging, since it involves change of dimension of parameter space. We use an approach similar to Ref. [19], which is a reversible jump[20] style algorithm, with a model comparison approach using modified fractional Bayes factor.[21,22] The basic idea is to consider a finite number of possible C, denoted by $\{C_{\min}, \ldots, C_{\max}\}$, split the data into a training set $\boldsymbol{n}' = b\boldsymbol{n}$ and a test set $\boldsymbol{n}'' = (1-b)\boldsymbol{n}$ (where $0 < b < 1$), and do a model comparison among those possible C. Details are given in [19].

3.2. *Estimate of Z:*

The point estimates for the parameters are determined as follows. We use the posterior mode C^* as a point estimate of C. Conditional on C^*, we follow Ref. [19] to find a point estimate of $\boldsymbol{Z}$. For any two $K \times C^*$ matrices $\boldsymbol{Z}$ and $\boldsymbol{Z}'$, $1 \le c, c' \le C^*$, let $\mathcal{D}_{cc'}(\boldsymbol{Z}, \boldsymbol{Z}') = \sum_{k=1}^{K} \|\boldsymbol{z}_{kc} - \boldsymbol{z}'_{kc'}\|_1$. Here we take the vectorized form of $\boldsymbol{z}_{kc}$ and $\boldsymbol{z}'_{kc'}$ to compute L^1 distance between them. The distance between $\boldsymbol{Z}$ and $\boldsymbol{Z}'$ is then defined as $d(\boldsymbol{Z}, \boldsymbol{Z}') = \min_{\boldsymbol{\sigma}} \sum_{c=1}^{C^*} \mathcal{D}_{c,\sigma_c}(\boldsymbol{Z}, \boldsymbol{Z}')$, where $\boldsymbol{\sigma} = (\sigma_1, \ldots, \sigma_{C^*})$

is a permutation of $\{1, \ldots, C^*\}$ and the minimum is over all possible permutations. A posterior point estimate for $\boldsymbol{Z}$ is defined as

$$\boldsymbol{Z}^* = \underset{\boldsymbol{Z}' \in \{\boldsymbol{Z}^{(l)}, l=1,\ldots,L\}}{\arg\min} \frac{1}{L} \sum_{l=1}^{L} d(\boldsymbol{Z}^{(l)}, \boldsymbol{Z}'),$$

where $\{\boldsymbol{Z}^{(l)}, l = 1, \ldots, L\}$ are posterior Monte Carlo samples of $\boldsymbol{Z}$. Finally, we report posterior point estimates $\boldsymbol{w}^*$ and $\boldsymbol{\rho}^*$ conditional on C^* and $\boldsymbol{Z}^*$ and calculate posterior point estimates $\boldsymbol{p}^*$ in order to check goodness of fit of the model.

4. Simulation

4.1. *Simulation 1*

We carry out two simulation studies to validate our proposed model. In the first simulation, we consider $K = 100$ mutation pairs for $T = 1$ sample. We assume the number of latent subclones is $C^{\text{TRUE}} = 3$, and set the subclone proportions as $\boldsymbol{w}^{\text{TRUE}} = (1 \times 10^{-7}, 0.65, 0.28, 0.07)$ (note that 1×10^{-7} refers to the proportion of the background subclone). The latent $\boldsymbol{Z}^{\text{TRUE}}$ matrix is shown in Fig. 3(a) in the form of a heatmap. For example, subclone 3 has genotype $\boldsymbol{z}^{(q)}$ with different q values. Specifically, $q = 10$ for mutation pairs 1-20, $q = 9$ for mutation pairs 21-40, $q = 6$ for mutation pairs 41-60, $q = 1$ for mutation pairs 61-80, and $q = 5$ for mutation pairs 81-100. Fig. 3(b) shows a possible lineage structure among subclones. We generate $\boldsymbol{\rho}^{\text{TRUE}}$ from its prior given in Eq. (4) with hyperparameter $d_1 = 1$. Next, we calculate multinomial probabilities p_{tkg}^{TRUE} shown in Eq. (2) and (3) from the simulated $\boldsymbol{Z}$, $\boldsymbol{w}$ and $\boldsymbol{\rho}$. We generate random numbers ranging from 400 to 600 as total read counts N_{tk}, and finally we generate read counts n_{tkg} from the multinomial distribution given N_{tk} as shown in Eq. (1).

We fit the model with hyperparameters as $\alpha = 4$, $\gamma_2 = \cdots = \gamma_Q = 0.5$, $d = 0.5$, $d_0 = 0.1$, $d_1 = 1$, and $r = 0.4$. We set $C_{\min} = 1$ and $C_{\max} = 10$ as the range of C. The choice of b needs to be calibrated. We choose b such that the test sample size $(1 - b) \sum_{t=1}^{T} \sum_{k=1}^{K} N_{tk}$ is approximately equal to $250/\sqrt{T}$. This choice leads to better posterior inference in our calibration process. We run MCMC simulation for 50,000 iterations, discarding the first 20,000 iterations as initial burn-in, and keep one sample every 10 iterations. The initial values are randomly generated from the prior.

The posterior mode $C^* = 3$ recovers the truth. Fig. 3(c) shows the point estimate of $\boldsymbol{Z}^{\text{TRUE}}$, given by $\boldsymbol{Z}^*$, which is very close to the truth. Fig. 3(d) shows the difference between $(p_{tkg}^* - p_{tkg}^{\text{TRUE}})$, which can be considered as the residual of model fitting. The histogram is centered at zero with a small variance that indicates a good model fit. The estimated subclone weights are $\boldsymbol{w}^* = (1.20 \times 10^{-168}, 0.650, 0.277, 0.073)$, which is also close to the truth. Typically in a real scenario, the number of available samples are quite low. In fact, in most of the cases data for only one sample can be obtained. We perform this simulation example in order to show that our model performs quite well even with a single sample.

We compare the performance of our model against BayClone[15] which is an SNV-based subclone caller. It chooses the model based on log pseudo marginal likelihood (LPML). According to LPML, the estimated number of subclones under BayClone is $C^* = 5$, which does not recover the truth. Fig. 3(e) shows the true subclone matrix in the form of Bay-

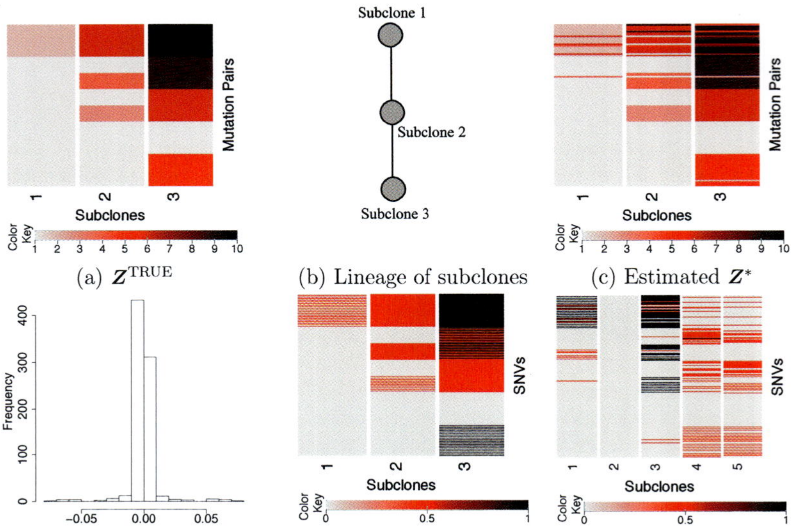

(a) $\boldsymbol{Z}^{\text{TRUE}}$ (b) Lineage of subclones (c) Estimated $\boldsymbol{Z}^*$

(d) Histogram of $(p^*_{tkg} - p^{\text{TRUE}}_{tkg})$ (e) $\boldsymbol{Z}^{\text{TRUE}}_{\text{BC}}$ (f) Estimated $\boldsymbol{Z}^*_{\text{BC}}$

Fig. 3. (a-d) Heatmap of the true subclone matrix $\boldsymbol{Z}^{\text{TRUE}}$, lineage structure and results from posterior inference. (e-f) Heatmap of the true and estimated subclone matrix using BayClone.

Clone's notation, denoted by $\boldsymbol{Z}^{\text{TRUE}}_{\text{BC}}$, and Fig. 3(f) shows the estimated matrix $\boldsymbol{Z}^*_{\text{BC}}$, where $z_{kc} = 0$, $z_{kc} = 0.5$ and $z_{kc} = 1$ refer to homozygous wild-type, heterozygous variant and homozygous variant at SNV locus k, respectively. The estimated subclone proportions are $\boldsymbol{w}^*_{\text{BC}} = (0.004, 0.364, 0.349, 0.171, 0.057, 0.054)$. From the BayClone's output, we can notice three problems. Firstly, BayClone could not recover the true number of subclones. Secondly, since BayClone infers the subclone structure by VAF of an SNV, the connection between adjacent SNVs is not modeled, and thus BayClone could not recover the $\boldsymbol{Z}$ matrix and cellular fractions accurately. For example, BayClone could not distinguish the difference between $\boldsymbol{z}_{kc} = \boldsymbol{z}^{(4)}$ and $\boldsymbol{z}_{kc} = \boldsymbol{z}^{(6)}$ in our model. Lastly, because of the noise in the data, BayClone includes a relatively larger proportion for the background subclone ($w_0 = 0.004$ in this example) which is significantly reduced for mutation pair data.

4.2. *Simulation 2*

In the second simulation study, we generate hypothetical reads data for $K = 100$ mutation pairs and $T = 5$ samples. We assume $C^{\text{TRUE}} = 4$. The subclone matrix $\boldsymbol{Z}^{\text{TRUE}}$ is shown in Fig. 4(a) and a possible lineage structure is given in Fig. 4(c). For each sample t, we generate the subclone proportions $\boldsymbol{w}^{\text{TRUE}}_t$ from $Dirichlet(0.01, \sigma(20, 14, 10, 4))$, where $\sigma(20, 14, 10, 4)$ is a random permutation of $(20, 14, 10, 4)$. The proportions $\boldsymbol{w}^{\text{TRUE}}$ which is now a matrix shown by a heatmap in Fig. 4(b). In the heatmap for $\boldsymbol{w}$, darker color indicates high abundance of a subclone in a sample, and light grey color represents low abundance. The parameters $\boldsymbol{\rho}^{\text{TRUE}}$ and N_{tk} are generated using the same approach as before. Finally, we calculate p_{tkg}^{TRUE} and generate read counts n_{tkg} from Eq. (1) similar to previous simulation.

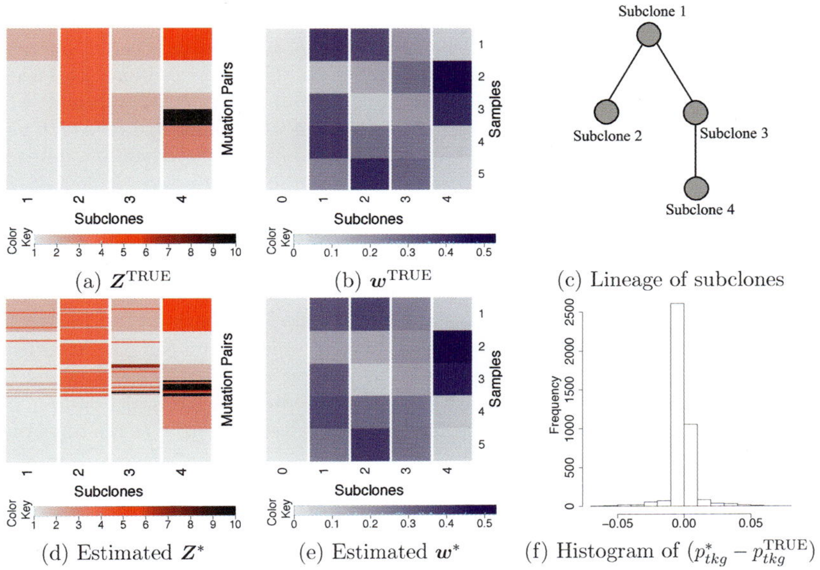

(a) $\boldsymbol{Z}^{\text{TRUE}}$ (b) $\boldsymbol{w}^{\text{TRUE}}$ (c) Lineage of subclones

(d) Estimated $\boldsymbol{Z}^*$ (e) Estimated $\boldsymbol{w}^*$ (f) Histogram of $(p_{tkg}^* - p_{tkg}^{\text{TRUE}})$

Fig. 4. Heatmap of the true subclone matrix, lineage structure and the results from posterior inference.

We fit the model with the same hyperparameters and same MCMC setting, except here we use $C_{\max} = 8$ in order to accelerate MCMC sampling. Also here due to the presence of multiple samples, we use a smaller (compared to simulation 1) test sample size. The posterior mode $C^* = 4$ recovers the truth. Fig. 4(d) shows the heatmap of $\boldsymbol{Z}^*$, and Fig. 4(e) shows the heatmap of $\boldsymbol{w}^*$. Comparing those two figures with Fig. 4(a) and 4(b), we can see that the truth is nicely recovered. Some mismatches are due to the relatively complex subclone structure. Fig. 4(f) shows the histogram of $(p_{tkg}^* - p_{tkg}^{\text{TRUE}})$ which indicates a good model fit.
We also compare our results with BayClone for this simulation. BayClone chooses the model with 5 subclones, which does not recover the truth.

5. Head and Neck Cancer Dataset

Whole exome data of 30 pairs of matched tumor (head and neck cancer) and normal samples are downloaded from the Sequence Read Archive (`http://www.ncbi.nlm.nih.gov/sra`).[23] We map the pair-end reads from the FASTQ format files to the human genome (version HG19) using BWA to generate BAM files for each individual sample. GATK's UnifiedGenotyper is used to call variants and to generate a single VCF file for all of them. Next task is to find mutation pair positions, their genotypes and number of reads supporting them. It is done by a bioinformatics tool *LocHap*[14] which searches for multiple single nucleotide variants (SNVs) that are scaffolded by the same reads. The scaffolded SNVs are referred to as local haplotypes. When a local haplotype exhibits more than two genotypes, *LocHap* calls it a local haplotype variant (LHV). Using the individual BAM file and the combined VCF file, *LocHap* generates HCF format output file.[14] HCF files contain LHV with two or three SNV locations. This whole process runs very fast as *LocHap* is an ultra-fast tool that can process an WES sample with about 30X coverage under a minute.[14] On an average we find a few hundreds LHVs with high quality in a WES sample. We select LHVs with two SNV locations as we are interested in mutation pairs only. Among those LHVs, we first remove the LHVs where the loci of two SNVs are very close to each other (within, say 50 bps) or close to other types of structural variants such as indels. We remove the LHVs where most reads were aligned to any of the SNVs at a base near the end of the reads. Also we filter out those LHVs where any of the SNVs are mapped by most reads with strand bias. At first, we find the intersection of mutation pair loci between normal and tumor samples and then we select randomly around 100 loci for each sample and record the read data from HCF files. In order to compare the underlying subclonal structure of normal and tumor samples we run our model on both separately. We run MCMC for $50,000$ iterations and discard the first $20,000$ iterations as initial burn-in. We use thinning count equals to 10. Hyperparameter settings are exactly same as the simulation 1 (Section 4.1).

Fig. 5 shows the number of subclones of a tumor sample and its matched normal for all 30 samples. Note that in almost all the samples the number of subclones in tumor is higher than the matched normal. In Fig. 6, we put subclone matrix ($\boldsymbol{Z}$) from six tumor and matched

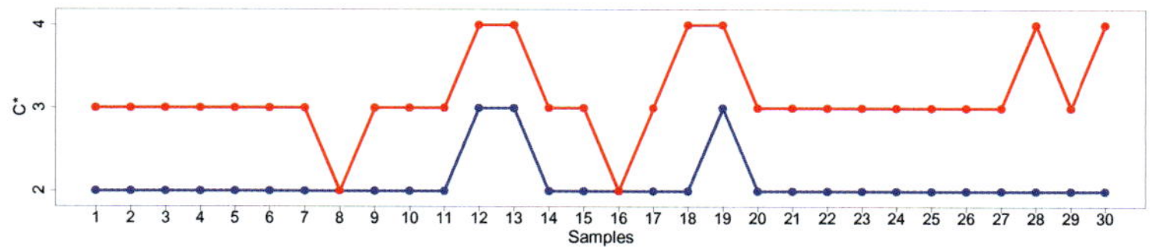

Fig. 5. Inferred number of subclones (C^*) for tumor (in red) and matched normal (in blue)

normal samples side by side. As one can notice, in tumor sample the corresponding subclonal structure is somewhat preserved with an addition of a new subclone. This indicates that tumor sample is more heterogeneous than the corresponding matching normal sample. We show the proportion of each subclone below each column of $\boldsymbol{Z}$ and columns of $\boldsymbol{Z}$ is reordered according to decreasing order of weights of the subclones.

We also run BayClone on those samples. The results look different. Due to space limitation, we omit the details since BayClone results are less reliable according to simulation studies.

Analysis of real data provides valuable clinical information. For example, one could seek potential biomarker mutation pairs for targeted therapy. These results could also be used as future diagnosis reference.

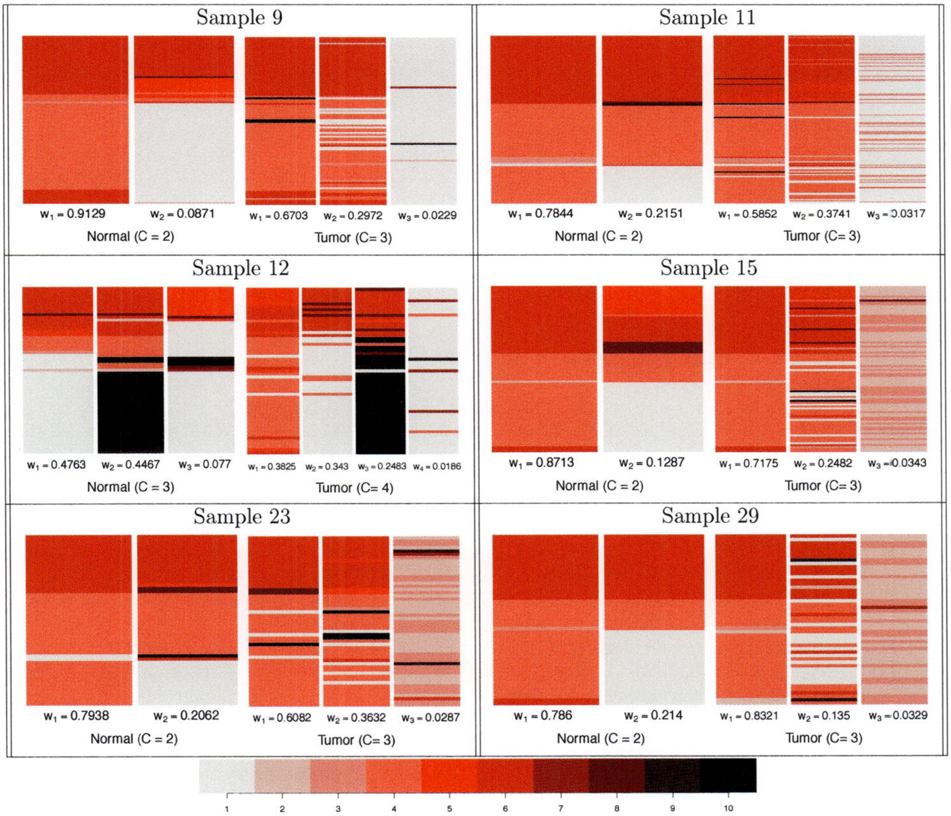

Fig. 6. Heatmap of subclone matrix $\boldsymbol{Z}$ from selected 6 samples (ordered according to age).

6. Discussion and future work

With the proposed model we infer subclonal structure and their proportions using mutation pairs data. The methods describe tumor heterogeneity in a principled manner based on a feature allocation model. It explicitly models overlapping mutation pairs across subclones. Through simulations, we show that mutation pair-based inference is more powerful than SNV-based subclone calling. This is not surprising since mutation pairs naturally provide heterogeneity of tumor samples through poly-genotypic short reads. In other words, direct evidence of having more than two haplotypes from short reads can be used to infer subclones in a tumor sample rather than indirect modeling on unusual VAFs for SNVs.

Our approach can be extended to model more than two SNVs. In order to accommodate more number of SNVs we only need to increase the number of categorical values that the $\boldsymbol{Z}$ matrix can take. Also, we are working on extensions that explicitly take into account potential phylogenetic relationship of subclones, which requires modeling the dependence among columns of the $\boldsymbol{Z}$ matrix.

References

1. N. D. Marjanovic, R. A. Weinberg and C. L. Chaffer, *Clinical chemistry* **59**, 168 (2013).
2. V. Almendro, A. Marusyk and K. Polyak, *Annual Review of Pathology: Mechanisms of Disease* **8**, 277 (2013).
3. K. Polyak, *The Journal of clinical investigation* **121**, p. 3786 (2011).
4. J. Stingl and C. Caldas, *Nature Reviews Cancer* **7**, 791 (2007).
5. M. Shackleton, E. Quintana, E. R. Fearon and S. J. Morrison, *Cell* **138**, 822 (2009).
6. D. L. Dexter, H. M. Kowalski, B. A. Blazar, Z. Fligiel, R. Vogel and G. H. Heppner, *Cancer Research* **38**, 3174 (1978).
7. L. Oesper, A. Mahmoody and B. J. Raphael, *Genome Biol* **14**, p. R80 (2013).
8. C. A. Miller, B. S. White, N. D. Dees, M. Griffith, J. S. Welch, O. L. Griffith, R. Vij, M. H. Tomasson, T. A. Graubert, M. J. Walter *et al.*, *PLoS computational biology* **10**, p. e1003665 (2014).
9. F. Strino, F. Parisi, M. Micsinai and Y. Kluger, *Nucleic acids research* **41**, e165 (2013).
10. W. Jiao, S. Vembu, A. G. Deshwar, L. Stein and Q. Morris, *BMC bioinformatics* **15**, p. 35 (2014).
11. H. Zare, J. Wang, A. Hu, K. Weber, J. Smith, D. Nickerson, C. Song, D. Witten, C. A. Blau and W. S. Noble, *PLoS computational biology* **10**, p. e1003703 (2014).
12. A. G. Deshwar, S. Vembu, C. K. Yung, G. H. Jang, L. Stein and Q. Morris, *Genome biology* **16**, p. 35 (2015).
13. H. Li, *Bioinformatics* **30**, 2843 (2014).
14. S. Sengupta, K. Gulukota, Y. Zhu, C. Ober, K. Naughton, W. Wentworth-Sheilds and Y. Ji, *Nucleic Acids Research (to appear)* (2015).
15. S. Sengupta, J. Wang, J. Lee, P. Müller, K. Gulukota, A. Banerjee and Y. Ji, *Pacific Symposium of Biocomputing* , 467 (2015).
16. Y. Kim, L. James and R. Weissbach, *Biometrika* **99**, 127 (2012).
17. S. Sengupta, J. Ho and A. Banerjee, *Two Models Involving Bayesian Nonparametric Techniques*, tech. rep., University of Florida (2013).
18. S. Brooks, A. Gelman, G. Jones and X.-L. Meng, *Handbook of Markov Chain Monte Carlo* (CRC Press, 2011).
19. J. Lee, P. Müller, Y. Ji and K. Gulukota, *A Bayesian Feature Allocation Model for Tumor Heterogeneity*, tech. rep., UC Santa Cruz (2013).
20. P. J. Green, *Biometrika* **82**, 711 (1995).
21. A. O'Hagan, *Journal of the Royal Statistical Society. Series B (Methodological)* **57**, 99 (1995).
22. G. Casella and E. Moreno, *Journal of the American Statistical Association* **101**, 157 (2006).
23. N. Stransky, A. M. Egloff, A. D. Tward, A. D. Kostic, K. Cibulskis, A. Sivachenko, G. V. Kryukov, M. S. Lawrence, C. Sougnez, A. McKenna *et al.*, *Science* **333**, 1157 (2011).

ONE-CLASS DETECTION OF CELL STATES IN TUMOR SUBTYPES

ARTEM SOKOLOV*, EVAN O. PAULL, JOSHUA M. STUART*
Department of Biomolecular Engineering,
University of California Santa Cruz
**E-mail: {sokolov,jstuart}@soe.ucsc.edu*

The cellular composition of a tumor greatly influences the growth, spread, immune activity, drug response, and other aspects of the disease. Tumor cells are usually comprised of a heterogeneous mixture of subclones, each of which could contain their own distinct character. The presence of minor subclones poses a serious health risk for patients as any one of them could harbor a fitness advantage with respect to the current treatment regimen, fueling resistance. It is therefore vital to accurately assess the make-up of cell states within a tumor biopsy. Transcriptome-wide assays from RNA sequencing provide key data from which cell state signatures can be detected. However, the challenge is to find them within samples containing mixtures of cell types of unknown proportions. We propose a novel one-class method based on logistic regression and show that its performance is competitive to two established SVM-based methods for this detection task. We demonstrate that one-class models are able to identify specific cell types in heterogeneous cell populations better than their binary predictor counterparts. We derive one-class predictors for the major breast and bladder subtypes and reaffirm the connection between these two tissues. In addition, we use a one-class predictor to quantitatively associate an embryonic stem cell signature with an aggressive breast cancer subtype that reveals shared stemness pathways potentially important for treatment.

Keywords: One-class models; Embryonic Stem Cells; Breast Cancer; Pan-Cancer

1. Introduction

Precision medicine in cancer has seen significant advances for treating patients based on molecular subtypes revealed by DNA and RNA-based analyses. Some examples include the now classic use of Gleevec to virtually cure the BCR-ABL form of Chronic Myeloid Leukemia, the more recent use of crizotinib for cancers beyond lung cancers with ALK fusions, including pediatric neuroblastoma, and the development of targeted inhibitors in breast cancer for both estrogen expressing and HER2-amplified forms. Despite these successes, many patients recur with disease as new tumor sub-populations emerge with evolved resistance or harbor even minor fractions of tumor subtypes refractory to treatment.

One promising future direction for cancer therapy is to catalog all subtypes for which such options are available. A patient's treatment can then be tailored according to their particular tumor's makeup. The problem with this approach is that many tumors consist of a heterogeneous collection of cell types, either those that have evolved through mutation and selection from the initial primary, or are "normal" cells such as those from the immune system or stroma that coexist with tumor cells in either antagonistic or synergistic ways. Tumor biopsies contain a mixture of various cell types. High-throughput data collected from the biopsy, such as RNA-sequencing data, reflects a superposition of the contributing cell sub-populations in the sample.

Several methods have been developed to deconvolute gene expression data, collected on a possibly mixed sample, into a set of distinct profiles representing separate cell types.[1] The most popular approach is to use unsupervised methods such as those based on non-negative matrix factorization[2] or other matrix decomposition techniques (e.g. independent component analysis). However, unsupervised methods attempt to identify all tumor subtypes in a single optimization, which is a difficult problem.

On the other hand, traditional supervised approaches require the presence of two or more classes to train models. In these kinds of situation, there is no definitive negative class, just a set of classes we wish to detect and some that are unknown. Often, we would like to contrast a particular subtype against all/any other subtypes, not any one in particular. One solution, albeit cumbersome, involves training $k-1$ dichotomous classifiers in which one-class is chosen as the positive set and each of the other $k-1$ classes are used separately or together as the contrasting negative set. It is unclear how the classes in the negative set should be weighted, either during training (if they are combined) or in the predictor (if $k-1$ separate classifiers are used). One drawback is that the negative classes have as much influence as the positive class on the ability to detect whether a sample represents an example from the positive set, which may be undesirable.

Our approach in this paper is to instead frame the problem as a detection task: given a particular known cell type, can we identify whether it is present at some appreciable level in a sample that contains possibly numerous cell types? This formulation fits naturally into the precision medicine framework as it can make suggestions based on disease subtypes of interest; e.g. those that are particularly aggressive, or those that have specific treatment options. Some possible approaches for one-class detection might use gene set enrichment approaches to detect if a set of genes is significantly upgregulated. However, we focus the work here on methods that provide an abstraction layer of the data to reach a higher-level understanding of the cell states under study.

We compare the ability of one-class methods against comparable two-class methods to learn a signature for a "pure" class and then detect it in possibly mixed samples. Our experiments compare two established one-class methods based on support vector machines (SVMs) against a binary SVM. We also introduce one-class logistic regression (OCLR) and measure its performance against standard binary logistic regression. We show that the one-class methods are able to outperform the standard two-class methods in simulated mixed data sets. In particular, when positive examples are among the negative examples in the training set, the one-class methods remain accurate However, the two class methods drop significantly in their performance.

We compare OCLR against SVM-based one-class predictors by training models for breast cancer subtypes. The empirical results show that OCLR achieves comparable performance while offering a more flexible formulation that can be extended to incorporate regularization schemes to, e.g., produce sparse models or integrate pathway information.

Lastly, we apply one-class models to recognize a specific molecular signal to new data where the presence of that signal is suspected. Specifically, models trained to recognize breast cancer subtypes are applied to bladder cancer samples, confirming transcriptome-level similarity between subtypes of the two diseases. We also investigate the level of de-differentiation in breast cancer subtypes by applying a one-class model trained to recognize embryonic stem cells. Our experiments reveal enrichment of a specific stemness program in breast basal tumors that illuminate the proliferative, metabolic, and developmental pathways that could suggest alternative targets.

2. Methods

We consider three one-class methods. Two of them are ν-SVM[3] and Support Vector Data Description (SVDD),[4] both based on the maximum-margin principle of SVMs. The former method aims to maximize the margin between the data and the origin. SVDD, on the other hand, finds a sphere with the smallest radius that fully encapsulates the data. Other approaches are possible, but the SVM-based approaches have been shown to perform well on a wide variety of tasks.

In addition to the two SVM-based methods, we propose the one-class logistic regression (OCLR) model. The proposed method functions similarly to ν-SVM, where it aims to identify the direction from the origin towards the data. Unlike the ν-SVM, however, logistic regression has a differentiable loss function, allowing for natural application of regularization schemes, such as group LASSO[5] and Elastic Nets,[6] to build sparse models and integrate pathway information. While some of these regularization schemes have been derived for Support Vector Machines, the general non-differentiability of the hinge loss requires the use of optimization methods that are not always straightforward.

Formally, given a set of n samples $\mathcal{X} = \{\mathbf{x}_i\}$, we define a *one-class logistic regression* model by a weight vector $\mathbf{w}$ that maximizes the log-likelihood $l(\mathbf{w}|\mathcal{X}) = \sum_{i=1}^{n} \log p(\mathbf{x}_i|\mathbf{w})$, where the likelihood is modeled with the logistic function:

$$p(\mathbf{x_i}|\mathbf{w}) = \frac{e^{\mathbf{w}^T\mathbf{x_i}}}{1 + e^{\mathbf{w}^T\mathbf{x_i}}} \tag{1}$$

By itself, the logistic function is not enough to model the data, as setting $\mathbf{w}$ to infinity gives a degenerate solution of $p(\mathbf{x}_i|\mathbf{w}) = 1.0$ for all data samples. To make the problem well-defined, we impose a regularizer, $\mathcal{R}(\mathbf{w})$, on the weights $\mathbf{w}$ to obtain the modified objective function:

$$\max_{\mathbf{w}} \frac{l(\mathbf{w}|\mathcal{X})}{n} - \lambda\mathcal{R}(\mathbf{w}), \tag{2}$$

or equivalently

$$\max_{\mathbf{w}} \frac{1}{n} \sum_{i=1}^{n} \left[\mathbf{w}^T\mathbf{x}_i - \log(1 + e^{\mathbf{w}^T\mathbf{x}_i})\right] - \lambda\mathcal{R}(\mathbf{w}), \tag{3}$$

where λ is a regularization meta-parameter that controls the tradeoff between model accuracy and complexity, and the factor of $\frac{1}{n}$ is introduced to keep the values of λ comparable across datasets of varying size.

Note the absence of a constant bias term commonly found in linear models. Similarly to the discussion above, the bias term requires regularization to avoid producing a degenerate solution. The ν-SVM formulation does utilize such a regularizer.[3] However, folding the bias term into a regularizer is equivalent to solving Equation (3) in a homogeneous coordinate space, where an auxiliary dimension is introduced to the data, and all samples are given a coordinate of 1.0 along that dimension. Because of this equivalence, we don't explicitly model the bias term.

To solve the optimization problem in Equation (3), we follow the Newton-Raphson method proposed by Friedman, *et. al.*[7] The approach constructs iteratively reweighted least squares estimates of the loss function using a Taylor series expansion. Let $\hat{\mathbf{w}}$ be the current model estimate. The second-order Taylor series approximation of the log-likelihood is given by

$$l_Q(\mathbf{w}|\mathcal{X}, \hat{\mathbf{w}}) = -\frac{1}{2}\sum_{i=1}^{n} a_i(y_i - \mathbf{w}^T\mathbf{x}_i)^2, \tag{4}$$

where the sample weights a_i and the working response y_i are computed using the current model estimate via

$$\hat{p}_i = \frac{\exp(\hat{\mathbf{w}}^T\mathbf{x}_i)}{1+\exp(\hat{\mathbf{w}}^T\mathbf{x}_i)}, \qquad a_i = \hat{p}_i(1-\hat{p}_i), \qquad y_i = \hat{\mathbf{w}}^T\mathbf{x}_i + \frac{1}{\hat{p}_i}. \tag{5}$$

To iterate on the model estimate itself, we now simply solve

$$\max_{\mathbf{w}} \frac{l_Q(\mathbf{w}|\mathcal{X}, \hat{\mathbf{w}})}{n} - \lambda\mathcal{R}(\mathbf{w}), \tag{6}$$

which is a standard regularized weighted linear regression problem. The specifics of solving this problem depend on the regularization scheme used. We stress that because the vast majority of novel regularization methods are initially derived for linear regression, their application to the proposed one-class logistic regression model is much more straightforward than to the hinge loss of ν-SVM.

One of the main draws to using Support Vector Machine methods is their generalization to reproducing kernel Hilbert spaces.[8] One-class logistic regression models maintain this advantage. Specifically, when the regularizer is a ridge penalty ($\mathcal{R}(\mathbf{w}) = \|\mathbf{w}\|_2^2$), constructing and solving the Lagrangian of the optimization problem in Equation 6 yields the following saddle point constraint: $\mathbf{w} = \sum_{i=1}^{n} \alpha_i\mathbf{x}_i$, where α_i are the Lagrange multipliers. The constraint allows us to compute the probability of any sample $\mathbf{z}$ given the model $\mathbf{w}$ as

$$p(\mathbf{z}|\mathbf{w}) = \frac{\exp\left(\sum_{i=1}^{n} \alpha_i\mathbf{x}_i^T\mathbf{z}\right)}{1+\exp\left(\sum_{i=1}^{n} \alpha_i\mathbf{x}_i^T\mathbf{z}\right)}. \tag{7}$$

Since this probability computation is at the heart of the optimization problem in Equation 6, replacing the dot products $\mathbf{x}_i^T\mathbf{z}$ with kernel computations $K(\mathbf{x}_i, \mathbf{z})$ allows us to learn a one-class logistic regression model in the Hilbert space corresponding to the kernel function K without explicitly mapping the data to that space.

The implementation of the one-class logistic regression method, including the kernel variant are available as part of our `gelnet` package in R. The package is available for download as open source from `https://cran.r-project.org/web/packages/gelnet/index.html`.

3. Results

3.1. *Detection of stemness signal in mixed populations of cells*

We tested the ability of the methods to detect the presence of a subtype of interest embedded in a mixture. The cancer stem cell hypothesis posits that a small fraction of a tumor's cells

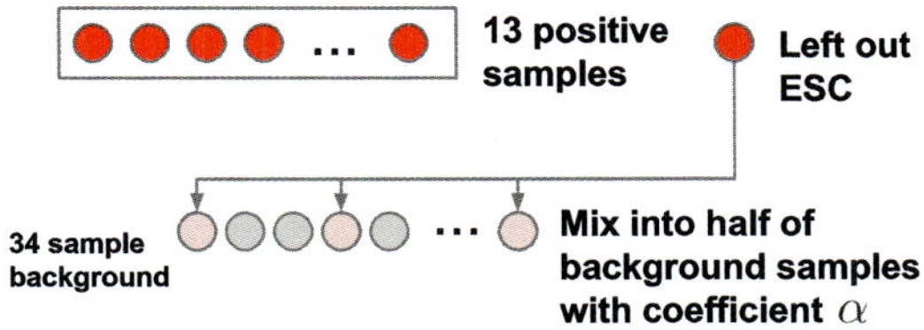

Fig. 1. A depiction of the leave-one-out experimental setup. Each of the Embryonic Stem Cells (ESCs) in turn was mixed into one half of randomly-chosen background samples with a predefined mixing coefficient α. A predictor is then given the remaining 13 ESC samples and asked to identify which of the 34 background samples contain the mixture.

harbor stem cell-like properties and that these cells may exhibit more aggressive phenotypes such as the ability to resist treatment, maintain proliferative potential even through oxidative stress conditions, and exhibit the ability to metastasize via cells of different character than the originating primary. Our simulation models the situation in which a tumor sample may contain a collection of cell types, some more or less differentiated than others. While it is possible the simulation might miss nuances present in actual patient data, for example if sub-clones mix in a non-linear fashion. However, the synthetic data offers the advantage of complete control so that the detection of latent cell states embedded into a simulated sample can be evaluated clearly.

For this experiment, we used the data from the Progenitor Cell Biology Consortium (PCBC) project on Synapse (syn1773109). The dataset contains RNAseq for 14 embryonic stem cells (ESCs) and 34 cells committed to a lineage. We performed a leave-one-out experiment by withholding each ESC sample in turn. The remaining 13 samples comprise the positive set, while the left-out sample was mixed into randomly selected half of the 34 background samples (Figure 1). The resulting machine learning task is to build a model that can correctly rank the background samples containing the stemness signal above those that do not. The accuracy is evaluated via Area under the ROC curve (AUC), which can be interpreted as the probability that the predictor correctly ranks a mixture sample above a non-mixture sample.

We evaluated the performance of ν-SVM, SVDD and our newly-proposed one-class logistic regression method. LIBSVM[9,10] was used to train ν-SVM and SVDD models using a linear kernel and the recommended parameter settings of $\nu = 0.5$ and $C = 2/n$ (where n is the number of training samples.) Note that parameters ν in ν-SVM and C in SVDD have a reciprocal relationship[3,4] and the values stated above provide exactly the same level of regularization. For consistency, we used the logistic regression model defined in Equation (3) with $\lambda = 1/4$ and $\mathcal{R}(\mathbf{w}) = \|\mathbf{w}\|_2^2$, which yields an identical regularizer to the one used by ν-SVM and SVDD.

In addition to the three one-class models, we also considered two binary predictors: logistic regression and binary SVM. Binary predictors require a negative set of samples for training, and several methods exist for identifying "true" negative examples in an unlabeled set.[11,12] Many of these methods begin by using the entire unlabeled set as the negative set to train the initial binary predictor. The initial predictor is then used to rank the unlabeled set, and the ranking is analyzed to select samples to be used as negative examples for subsequent retraining of the binary predictor. In this paper, we consider only the initial step of using the entire unlabeled background set as negative examples to highlight the issue binary predictors face in the absence of "true" negative data. LIBSVM was used to train binary SVM models, while logistic regression models were trained using the R package `gelnet`. The regularization parameter was kept at 1.0 for both types of binary predictors.

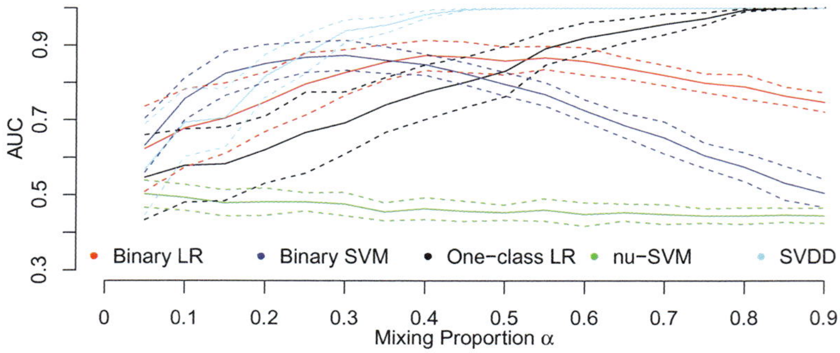

Fig. 2. The accuracy of predictors plotted against the mixture coefficient. The solid lines represent the mean performance across 30 trials. The dashed lines are one standard deviation away from the mean.

Figure 2 presents the performance of all methods as a function of the mixing coefficient α. We note the general upward trend in the performance of one-class models as α increases. This is expected, since a larger mixing proportion of the left-out ESC sample makes the mixtures look more like the positive class, yielding an easier detection task. The trend is not shared by ν-SVM models, which are unable to identify samples with mixed stemness signal from others in the background. A potential explanation for the poor performance comes from sample locality in high-dimensional feature spaces. An SVM can be viewed as a mechanical system, where the decision plane is a "stiff sheet" in mechanical equilibrium, upon which the training samples exert forces and torques.[13] Because the high-dimensional space of RNAseq is vastly undersampled by the PCBC dataset, the training data is effectively localized to a tiny fraction of that space. Thus, tiny perturbations in the training samples will create giant "swings" of the "stiff sheet" in other portions of the feature space. This effectively makes the model highly sensitive to noise and reduces its generalization to the unsampled portions of the feature space.

As mixture samples gain similarity to the positive class, it also throws off the binary predictors, as observed by their decreased performance for higher values of α. This highlights the challenge binary predictors face when presented with positive and unlabeled data: unlabeled data may contain a strong representation of the positive signal, leading to a skewed decision boundary. The challenge acts as a motivating factor for finding high-quality negative sets in the unlabeled data and iterative re-training of the binary predictors using those sets. The issue is completely side-stepped by the one-class methods, because they require positive samples only.

3.2. *One-class models distinguish breast cancer subtypes*

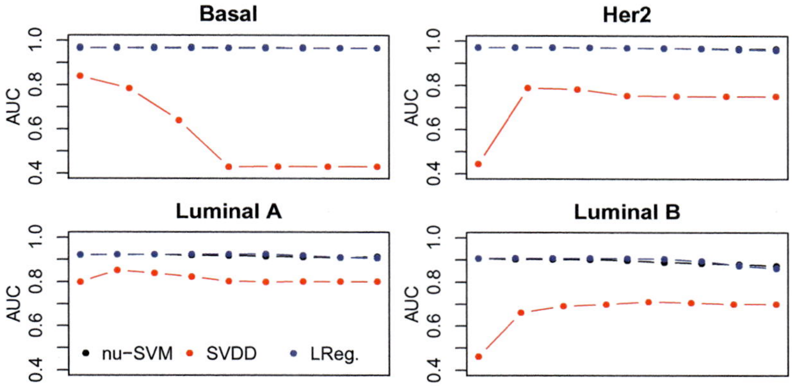

Fig. 3. The accuracy of the one-class methods plotted against the regularization parameter. Each of the four panels corresponds to a specific TCGA BRCA subtype.

We next applied the one-class predictors to an established classification task in cancer genomics – that of determining the major breast cancer subtypes from gene expression profiles. The transcription-derived subtypes in breast cancer have demonstrated prognostic value that have led to the establishment of FDA-approved tests including Oncotype-DX and the mammoprint. Defining treatment decisions based on gene expression subtypes has been shown to improve greatly over the use of pathology information alone.[14] The luminal subtype is associated with better prognosis and the expression of estrogen and progesterone receptors that can be targets of therapy (e.g. tamoxifen or aromatase inhibitors). The luminals can be further divided into three subclasses including the luminal-As, luminal-Bs, and the HER2-amplified sets. Luminal B tumors display somewhat more basal-like characteristics and tend to have higher levels of TP53 mutations. HER2-amplified tumors, which have a greater number of genomic copies of the amplicon on which the HER2 growth receptor gene resides, respond effectively to agents that block the receptor. The basal subtype, on the other hand, exhibits a much more aggressive character. Basal tumors are often further subdivided into tumors that either do or do not express the gene Claudin as the Claudin-low group display an even more severe outcome than the other basal tumors.

Most importantly, new evidence has revealed that primary tumors can be comprised of many different sub-populations of cells, exhibiting different subtype characters.[15] Indeed, it has been postulated that cancer cells have the ability to transdifferentiate from one subtype to another such as adenocarcinomas of the lung or prostate into neuroendocrine-like cells.[16] Taken together, the accurate assessment of a primary tumor's subtype or its mixture of possibly many subtypes, is currently perhaps the most important step in planning treatment for breast cancer patients.

We therefore applied the one-class predictors to the task of defining gene expression-based signatures of the four major breast cancer subtypes: Basal, Her2-amplified (Her2), Luminal A, and Luminal B. For every subtype, the one-class methods were evaluated via leave-one-out cross-validation, and the AUC score was computed to capture the probability that a sample withheld from the positive class was scored higher than a sample from another subtype. We investigated the effect of regularization on performance by sweeping across the meaningful values of the regularization parameters: $\nu \in (0,1)$ for ν-SVMs,[3] $C \in [1/n, 1]$ for SVDD[10] and $\lambda = 10^k$ with $k \in [-4, 4]$ for one-class LR. As seen in Figure 3, the level of regularization had a marginal impact on performance of ν-SVM and one-class LR, while SVDD was more sensitive to the parameter value choice.

As expected, all of the methods achieved high levels of accuracy for an interval of parameter choices, but the ν-SVM and logistic regression approaches outperformed SVDD in this prediction setting (Figure 3). The logistic regression-based approach performed as well as the top SVM-based strategy in both simulation and in this real-tumor application (SVDD in the former and ν-SVM in the latter). Because logistic regression has comparable performance to the SVM-based method but can be used to identify sparse and interpretable sets of features due to its differentiable loss, we elected to use it for the remainder of this study.

3.3. *Breast cancer one-class models detect molecular similarity in bladder cancer*

While the location in the body of a primary tumor contributes a dominant influence on gene expression signatures, the disease subtype can be revealed through Pan-Cancer comparisons that reflect cell-of-origin commonalities across tissues.[17] Recently, transcriptome- and genome-wide analyses from three independent groups have revealed the similarity between bladder and breast subtypes.[18–20] In particular, muscle-invasive bladder cancers can also be distinctly grouped into Claudin-low, basals, P53-enriched luminals, and non-P53-enriched luminals.

We asked if one-class predictors could connect cancer subtypes across tissues. Specifically, we investigated the hypothesis that bladder cancer subtypes share common cell-of-origin signatures with breast cancer subtypes. The subtype assignment of TCGA bladder carcinoma (BLCA) samples was taken from the molecular characterization literature,[21] and the corresponding RNAseq data was obtained from the Broad Institute's Firehose pipeline (2014-10-17 run). Indeed, the one-class predictors confirm the connection of major subtypes between bladder and breast cancers (Figure 4). Strikingly, a classifier trained to recognize BRCA basal cancers can predict type III bladder cancers with nearly 90% accuracy (AUC 0.89; $p < 10^{-5}$ label permutation test). This strongly supports the notion of an intrinsic connection between these disease. We also find a smaller, but still significant association between the luminal-A and the type II bladder cancers (AUC 0.78; $p < 10^{-5}$), which could suggest an estrogen-

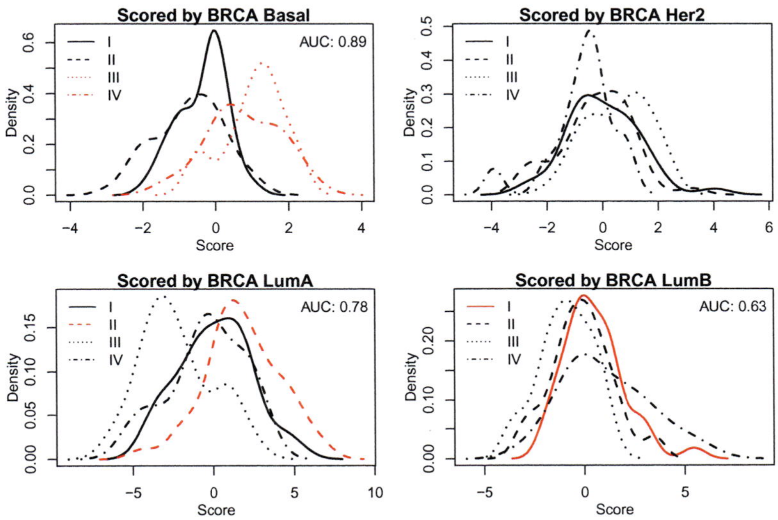

Fig. 4. One-class models trained on TCGA BRCA applied to TCGA BLCA. For each BRCA subtype, we present the distribution of scores from the corresponding one-class model across the four bladder subtypes. Bladder subtypes known to have molecular similarity to the given breast cancer subtype are highlighted in red.

or other hormone-driven component to the type II bladder cancers. Interestingly, the Her2-Amplified breast signature matched best with class III. Some bladder cancers have been found with amplification of the ERBB2 locus,[21] so it would be interesting to check if the type III are indeed enriched for this copy number event.

3.4. *One-class models identify a stemness signal in Basal breast cancer*

We applied a one-class logistic regression model trained on PCBC embryonic stem cell samples to score TCGA breast cancer (BRCA) samples. The scores are presented in Figure 5. Note the enrichment of Basal samples on the positive side and Luminal samples on the negative side. To measure the significance of this enrichment, we applied a Kolmogorov-Smirnov (KS) test similar to the one used by the Gene Set Enrichment Analysis method.[22] If there were no association with stemness we would expect the Basal, Luminal and Her2 samples would be equally likely to be encountered anywhere in the distribution of scores. We used `Bioconductor` package `piano` (http://www.sysbio.se/piano/) to compute the deviation between the expected probability of encountering a sample from the subtype of interest and the observed frequency as one "sweeps" across the score values. The largest deviation was reported as the *enrichment score*. The enrichment scores were positive for Basal (p-value $< 1e\text{-}5$), Her2 (p-value < 0.0069), and Luminal B (p-value $< 5e\text{-}5$) and negative for Luminal A (p-value $< 1e\text{-}5$).

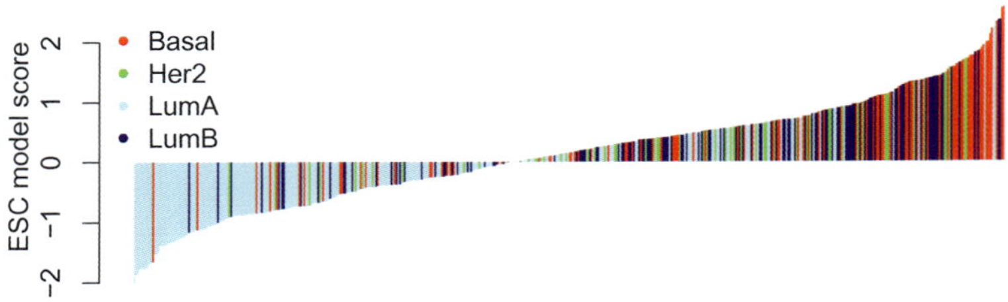

Fig. 5. TCGA BRCA samples scored by a one-class logistic regression model trained on Embryonic Stem Cells. The samples are ordered by score from highest to lowest and colored by the breast subtype.

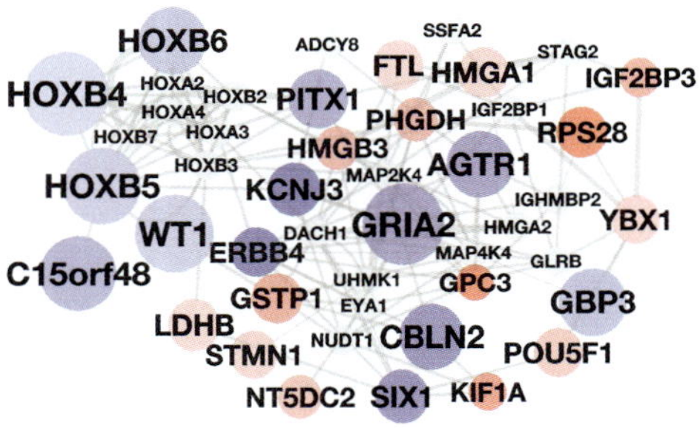

Fig. 6. Top 30 most-concordant genes between the stemness and the basal one-class models. Genes with positive weights in both signatures are shown in red, while those with negative weights are in blue. The size of the node represents the level of concordance (absolute value of the weight product). GeneMANIA[23] was applied to the 30 selected nodes to identify protein-level interactions. "Linker" genes identified by GeneMANIA are shown in gray.

Given the significant association between basal tumors and the stemness signature, we used the result to probe what processes might be shared in basals with undifferentiated cells. To reveal possible mechanisms of reprogramming at work in basal tumors, we took the component-wise product of the weights from the stemness predictor with the basal predictor to help uncover genes predictive of both cell types. This resulted in a list of genes reflecting specific involved pathways underlying the transformation processes in basal breast tumors. We then identified a connected mechanism using the GeneMANIA tool.[23] A few themes emerged from this analysis. First, and not surprising, several genes were identified that reflect the proliferative potential of the basal tumors including KIF1A and STMN1 (Figure 6).

The well-known OCT4 transcription factor (POU5F1), one of the Yamanaka factors with the ability to reprogram cells into a pluripotent state, and PROM1 have been shown to be correlated with aggressive cancers. In the case of PROM1, expression of this surface glycoprotein actively suppresses differentiation pathways and is associated with poor survival in colorectal cancers and recently in malignant papillary breast cancers.[24] Lower expression of differentiation genes such as FOXA1 and some of the HOX-family genes are also documented to play roles in aggressive forms of the disease.

Several genes reflect the metabolic state shared between stem cells and basal cancers. Stem cells often occupy low-oxygen niches and use anaerobic means to break down sugars. Intriguingly, the PHGDH gene identified as a common predictor by the one-class method, which catalyzes the first step in metabolizing serine downstream of glycolysis, was also recently implicated in breast cancers as a survival mechanism in hypoxic conditions.[25] Its expression has also been shown to be associated with ER-negative tumors, consistent with our results[26] even though its role is non-essential, suggesting tumors have a way to bypass this mechanism.

4. Conclusions

The collection and summarization of cell type signatures for precision medicine applications, especially in cancer, promises to greatly enhance the treatment of patients. For example, some patients present "cancers of unknown primary" (CUP) where metastatic advanced disease has already manifested itself when a patient first reports to a hospital. Often these cases are treated with generic protocols, but evidence suggests their outcomes could be improved significantly by first identifying the tissue of the primary tumor. Similarly, the detailed characterization of cell-of-origin signatures in heterogeneous biopsy specimens would improve the resolution by which treatments or treatment combinations could be matched to tumor subtypes.

A clear data science-inspired direction for precision medicine is to amass "dictionaries" of disease and normal signatures to help characterize patient specimens. Previous approaches have shown the power of this idea. For example, signatures trained from published expression datasets can be used to suggest repurposing drugs for new diseases.[27] Error correcting belief propagation has been used to predict normal and disease cell states in a comprehensive compilation of gene expression signatures.[27–30] These approaches use standard machine-learning classifiers as inputs to the inference strategy. To our knowledge, little investigation of the optimal approaches has been done to determine the best base-level classifiers. Instead, most approaches choose either a custom (e.g. standardized differential vectors as in[27,28]) or a popular standard (e.g. SVMs[29,30]). Thus, an open question remains about how best to build signature dictionaries.

One-class models provide a scalable approach to contribute cell type signatures to such dictionaries. Because they have no need for a set of negative examples in training, they can be updated in an online fashion without the need for a representative background set. Indeed, not requiring a contrasting set makes the learned models less arbitrary to nuances in any particular database, so one can expect the models to remain robust as more samples are added to a training dataset.

We demonstrated the strength of the one-class approach for detecting latent cell types in cancer samples. One-class predictors clearly outperform the use of dichotomous classifiers in our study that simulated "contamination" of the negative set with an unknown amount of positive examples. As the proportion increases, the dichotomous classifiers' performance degrades due to a loss in the distinction between the classes during training. However, one-class methods are immune to this influence because they use only the positive class for training.

One-class signatures had a clear advantage for use in the cell type detection problem in our study here of the major breast cancer subtypes. These models confirmed the recently reported commonality between the breast and bladder cancer subtypes. Finally, one-class signatures could detect stemness signatures in breast cancer tumor samples, supporting the observation that basal breast cancers are more likely to exhibit stem cell-like properties. The association suggests the wiring of basal cells may be set up to respond to similar developmental queues as progenitor cells with increases pluripotency. The genes identified as common between the basal and stemness signatures could therefore suggest novel putative targets for therapy.

Acknowledgements: Research supported by a Stand Up to Cancer – Prostate Cancer Foundation – Prostate Dream Team Translational Cancer Research Grant. This research grant is made possible by the generous support of the Movember Foundation. Stand Up To Cancer is a program of the Entertainment Industry Foundation administered by the American Association for Cancer Research.

References

1. J. Ahn, Y. Yuan, G. Parmigiani *et al.*, *Bioinformatics* , p. btt301 (2013).
2. J.-P. Brunet, P. Tamayo, T. R. Golub and J. P. Mesirov, *PNAS* **101**, 4164 (2004).
3. B. Schölkopf, R. C. Williamson, A. J. Smola *et al.*, *NIPS* **12**, 582 (1999).
4. D. M. Tax and R. P. Duin, *Machine learning* **54**, 45 (2004).
5. M. Yuan and Y. Lin, *Journal of the Royal Statistical Society: Series B* **68**, 49 (2006).
6. H. Zou and T. Hastie, *Journal of the Royal Statistical Society: Series B* **67**, 301 (2005).
7. J. Friedman, T. Hastie and R. Tibshirani, *Journal of statistical software* **33**, p. 1 (2010).
8. Schölkopf, Tsuda and Vert, *Kernel methods in computational biology* (MIT press, 2004).
9. C.-C. Chang and C.-J. Lin, *ACM Trans. on Intelligent Systems and Technology* **2**, 27:1 (2011).
10. W.-C. Chang, C.-P. Lee and C.-J. Lin, *A revisit to SVDD*, tech. rep. (2013).
11. X. Li and B. Liu, *IJCAI* **3**, 587 (2003).
12. S. Mei and H. Zhu, *Scientific reports* **5** (2015).
13. C. J. Burges, *Data mining and knowledge discovery* **2**, 121 (1998).
14. A. Prat, E. Pineda, B. Adamo *et al.*, *Breast* **15**, S0960 (2015).
15. M. Kleppe and R. L. Levine, *Nature medicine* **20**, 342 (2014).
16. M. T. Shekhani, A.-S. Jayanthy, N. Maddodi and V. Setaluri, *A. J. of Stem Cells* **2**, p. 52 (2013).
17. K. A. Hoadley, C. Yau, D. M. Wolf *et al.*, *Cell* **158**, 929 (2014).
18. W. Choi, B. Czerniak, A. Ochoa *et al.*, *Nature Reviews Urology* **11**, 400 (2014).
19. M. A. Knowles and C. D. Hurst, *Nature Reviews Cancer* **15**, 25 (2015).
20. D. J. McConkey, W. Choi and C. P. Dinney, *European urology* **66**, 609 (2014).
21. Cancer Genome Atlas Research Network *et al.*, *Nature* **507**, 315 (2014).
22. A. Subramanian, P. Tamayo, V. K. Mootha *et al.*, *PNAS* **102**, 15545 (2005).
23. D. Warde-Farley, S. L. Donaldson, O. Comes *et al.*, *Nucleic acids research* **38**, W214 (2010).
24. C.-H. Lin, C.-H. Liu, C.-H. Wen, P.-L. Ko and C.-Y. Chai, *Virchows Archiv* **466**, 177 (2015).
25. R. Possemato, K. M. Marks, Y. D. Shaul *et al.*, *Nature* **476**, 346 (2011).
26. J. Chen, F. Chung, G. Yang *et al.*, *Oncotarget* **4**, p. 2502 (2013).
27. N. S. Jahchan, J. T. Dudley, P. K. Mazur *et al.*, *Cancer Discovery* **3**, 1364 (2013).
28. H. Huang, C.-C. Liu and X. J. Zhou, *PNAS* **107**, 6823 (2010).
29. Y.-s. Lee, A. Krishnan, Q. Zhu and O. G. Troyanskaya, *Bioinformatics* **29**, 3036 (2013).
30. D. Amar, T. Hait, S. Izraeli and R. Shamir, *Nucleic Acids Research* (2015).

RDF SKETCH MAPS - KNOWLEDGE COMPLEXITY REDUCTION FOR PRECISION MEDICINE ANALYTICS

NATTAPON THANINTORN[1], JUEXIN WANG[2], ILKER ERSOY[1], ZAINAB AL-TAIE[2], YUEXU JIANG[2], DUOLIN WANG[2], MEGHA VERMA[1], TRUPTI JOSHI[2,3,4], RICHARD HAMMER[1], DONG XU[2,3], DMITRIY SHIN[1,3,2]*

[1]Department of Pathology and Anatomical Sciences, [2]Department of Computer Science and Christopher S. Bond Life Sciences Center, [3]MU Informatics Institute, [4]Department of Molecular Microbiology and Immunology and School of Medicine – Office of Research, University of Missouri, Columbia, MO 65203, USA
**Email: shindm@health.missouri.edu*

Realization of precision medicine ideas requires significant research effort to be able to spot subtle differences in complex diseases at the molecular level to develop personalized therapies. It is especially important in many cases of highly heterogeneous cancers. Precision diagnostics and therapeutics of such diseases demands interrogation of vast amounts of biological knowledge coupled with novel analytic methodologies. For instance, pathway-based approaches can shed light on the way tumorigenesis takes place in individual patient cases and pinpoint to novel drug targets. However, comprehensive analysis of hundreds of pathways and thousands of genes creates a combinatorial explosion, that is challenging for medical practitioners to handle at the point of care. Here we extend our previous work on mapping clinical omics data to curated Resource Description Framework (RDF) knowledge bases to derive influence diagrams of interrelationships of biomarker proteins, diseases and signal transduction pathways for personalized theranostics. We present RDF Sketch Maps – a computational method to reduce knowledge complexity for precision medicine analytics. The method of RDF Sketch Maps is inspired by the way a sketch artist conveys only important visual information and discards other unnecessary details. In our case, we compute and retain only so-called RDF Edges – places with highly important diagnostic and therapeutic information. To do this we utilize 35 maps of human signal transduction pathways by transforming 300 KEGG maps into highly processable RDF knowledge base. We have demonstrated potential clinical utility of RDF Sketch Maps in hematopoietic cancers, including analysis of pathways associated with Hairy Cell Leukemia (HCL) and Chronic Myeloid Leukemia (CML) where we achieved up to 20-fold reduction in the number of biological entities to be analyzed, while retaining most likely important entities. In experiments with pathways associated with HCL a generated RDF Sketch Map of the top 30% paths retained important information about signaling cascades leading to activation of proto-oncogene BRAF, which is usually associated with a different cancer, melanoma. Recent reports of successful treatments of HCL patients by the BRAF-targeted drug vemurafenib support the validity of the RDF Sketch Maps findings. We therefore believe that RDF Sketch Maps will be invaluable for hypothesis generation for precision diagnostics and therapeutics as well as drug repurposing studies.

1. Introduction

Basic science discoveries coupled with tremendous advances in "omics" technologies have triggered a paradigm shift in today's biomedicine. The idea of precision and personalized medicine is viewed by many as a solution to improve patient care by addressing disease complexity and heterogeneity [1]. It is especially evident in the direction that modern medical diagnostics and therapeutics, jointly coined as *theranostics*, is progressing. Pathway-based diagnostics is promising to open up a view at internal biological mechanisms of complex interplay of clinical biomarkers, diseases, signal transduction and other processes to be able to more precisely describe differences in individual patient cases [2]-[10]. Generation of a mechanistic picture of such processes can help develop combinatorial therapies utilizing novel drugs, small molecules inhibitors, cytotoxic and differentiating agents and other interventional techniques. And, even though, precision theranostic approaches have not yielded significant advances yet due to limited drug options, the number of successful clinical cases using targeted therapies being reported is increasing [11]-[15]. For instance, in some cases deeper analysis of signal transduction pathways revealed an alternative activation of carcinogenic mechanisms, which mandated a use of novel combinatorial therapies. In other cases, unconventional drugs have been used to treat patient exhibiting no response to conventional regimens. For example, a successful unconventional therapy of Hairy Cell Leukemia (HCL) with vemuratenib, a drug usually associated with melanoma cancers and targeting BRAF proto-oncogene, has been reported in several clinical cases [16]-[19].

The great challenge here in our view is the difficulty of conducting a comprehensive precision theranostic study due to limitations of individual practitioners' knowledge of biological processes. An inter-expert collaboration, while being able to expand the knowledge space to a certain extent, is still not an effective solution. For instance, a number of reported cases indicate that current attempts to practice precision and personalized medicine reflect more *descriptive* rather than *predictive* approaches. Pathologists and oncologists are trying more to *describe* the successful application of unconventional drugs by analyzing biopsies and linking proteomic expression to signal transduction and known mutations rather to *predict* patient-specific disease mechanisms based on clinical omics data. We strongly believe that new methods for clinical hypothesis generation for precision theranostics are needed to increase the chance of having more successes similar to the use of vermuratenib in HCL.

To this end, we have been investigating advanced inference methods to map clinical biomarkers data to biological pathways to recreate interplay of signaling proteomic networks for individual patient cases [20]. Our new computational formalism called *Resource Description Framework (RDF)-induced Influgrams* (RIIG) has been shown in a recent proof-of-concept study to exhibit qualities sufficient to provide case-specific reasoning for theranostics [10]. RIIG takes advantage of vast amounts of publicly available *curated biological knowledge* represented as the RDF format. The importance and utility of use of RDF knowledge bases (KBs) in biomedicine have been demonstrated in a number of publications [21]-[24].

The application of RIIG on the set of all pathways involved can dramatically reduce RIIG performance and result in reduction of its practical utility in a medical setting. A number of studies related to biological pathway data processing have been focused on pathway curation [25], visualization [26] and analysis [3], [27], [28]. There have been also some studies to construct a skeleton from complex networks by pruning edges [29]. The general idea of maintaining most informative nodes by finding shortest path in a directed network has been explored in metabolic engineering [30]-[32]. By searching all possible reactions between compounds, these methods output several minimum cost paths by defining different penalties of reaction type, compound type and atom mapping. However, simplifying networks in the context of precision medicine has not yet been investigated.

Here, we present the RDF Sketch Maps – a new computational method to reduce complexity of RDF-formatted knowledge networks to improve theranostic analyses in precision medicine settings. The method of RDF Sketch Maps is inspired by the way a sketch artist conveys only important visual information, while leaving out other unnecessary details. In our case, we compute and retain only so-called RDF Edges – places with highly important diagnostic and therapeutic information. To do this the method traverses knowledge networks and scores paths according to an objective function that incorporates information about a set of known diagnostic and therapeutic biomarkers (e.g. disease-associated genes and drug targets). The paths are then ranked by decreasing values of the scores. A set of *exploratory* genes that could possibly be useful in explaining patient-specific disease heterogeneity is used to compute the enrichment score for ranked paths for each version of the objective function. The top paths with high enrichment score are selected to form an RDF Sketch Map. The resulting maps are used for further analysis by computational methods or visualized for human analysis.

2. Methods

2.1. *Construction of RDF knowledge base*

For preliminary experiments we have constructed a knowledge base (KB) consisting of 35 signaling pathway maps from Kyoto Encyclopedia Genes and Genomes (KEGG), including pathways associated with molecular interactions, genetic information processing, environmental information processing, cellular processes, organismal systems, human diseases, and drug development [33]. We preferred KEGG maps over other pathway databases such as Biocarta [34] and Reactome [35] because of KEGG's inclusion of a variety of different types of signal transduction interactions (e.g. phosphorylation, methylation, ubiquitination, and glycosylation) that are relevant to cancer theranostics (Table 1).

KEGG however was initially designed as a set of manually-drawn pathway maps for human consumption. The electronic version of KEGG maps introduced later in the form of XML-like KGML files merely represents serialization of graphical artifacts. The high rate of inaccuracies and omissions (up to 30% comparing to graphical maps) in some KGML files makes them unacceptable for use in precision medicine applications, which requires high levels of accuracy of underlying facts and reliable knowledge provenance. We, therefore, set a goal to transform KEGG KGML files into highly accurate machine processable KB with inference capabilities. To do that we (i) designed a KEGG RDF ontology that models the relationships among biological entities and allows description logic inference, (ii) converted KGML files into RDF data set using in-house developed graphical curation tool and a set of scripts, and (iii) added information about biological processes from Gene Ontology (GO)[36] and proteomic information from UniProt database [37].

KEGG RDF ontology specifies the type and constraints of interactions among biological entities as well as their class/sub-class hierarchical relationships. For instance, we decided to preserve specific paths of propagation of signal transduction in individual maps through the notion of a "gene instance". The underlying reason behind it is that certain reactions (e.g. phosphorylation) occur under specific circumstances (e.g. presence of specific enzymes or involving specific protein domains). However, since gene instances coming from different maps are modeled as sub-classes of an "abstract gene", we can combine individual maps into an integrated semantic "mash-up" KB. This allows one to potentially recreate a systems view of signal transduction in individual patient cases. A similar approach is utilized while modeling gene groups, which represents protein complexes at the proteomic level. The constructed KEGG RDF KB was loaded into AllegroGraph RDF store [38] for querying and processing. To optimize performance of running RDF Sketch Maps algorithm we use AllegroGraph's internal SNA RDF graph processing only to resolve aliases and run description logic inference. The RDF Sketch algorithm is run on "static" graph serialization derived from the AllegroGraph KEGG RDF KB.

Table 1. Modeled KEGG interactions.

RDF predicate name	Modeling purpose
activates	Molecular interaction
binds_associates	Molecular interaction
changes_state	Molecular interaction
dephosphorylates	Molecular interaction
dissociates	Molecular interaction
expresses	Molecular interaction
glycosylates	Molecular interaction
indirectly_affects	Molecular interaction
inhibits	Molecular interaction
methylates	Molecular interaction
misses_interaction	Molecular interaction
phosphorylates	Molecular interaction
represses	Molecular interaction
ubiquitinates	Molecular interaction
deubiquitinates	Molecular interaction
phosphorylates_activates	Molecular interaction
phosphorylates_inhibits	Molecular interaction
dephosphorylates_activates	Molecular interaction
dephosphorylates_inhibits	Molecular interaction
ubiquitinates_activates	Molecular interaction
ubiquitinates_inhibits	Molecular interaction
deubiquitinates_activates	Molecular interaction
deubiquitinates_inhibits	Molecular interaction
methylates_activates	Molecular interaction
methylates_inhibits	Molecular interaction
glycosylates_activates	Molecular interaction
glycosylates_inhibits	Molecular interaction
indirectly_affects_activates	Molecular interaction
involved_in	Inference
is_part_of	Inference
contains	Inference
crosstalks_with	Inference

2.2. *Computation of RDF Sketch Maps*

The essential goal of RDF Sketch Maps method is to reduce knowledge complexity for theranostic analysis. In the case with the integrated RDF KEGG KB we have a "hairball" of myriad of molecular interactions that needs to be simplified. To do that we first define a model of a particular biological phenomenon. For our experiments in personalized theranostics we define a *cancer model* that reflects propagation of biological signal transduction from intercellular space through surface proteomic receptors all the way into the nuclear space where specific activated protein complexes regulate gene expression. In our cancer model we identify Start and End genes, with *Start* genes being surface receptors, proto-oncogenes and tumor suppressor genes and *End* genes being genes associated with biological processes involved in carcinogenesis (Table 2). An example of such model is shown in Figure 1.

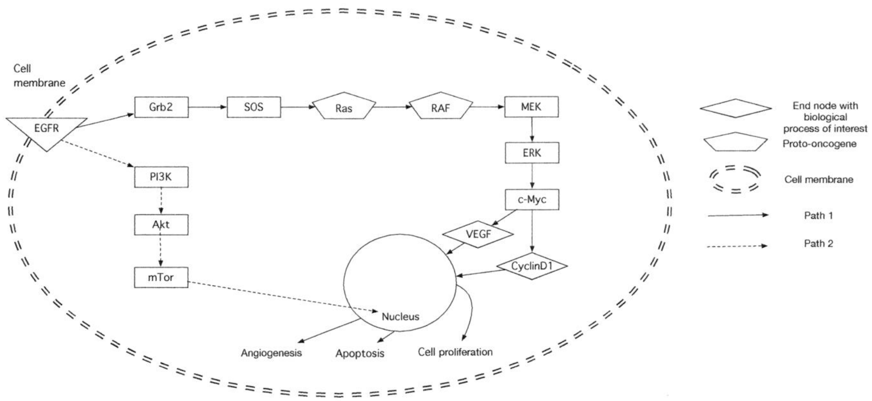

Figure 1. An example of cancer model.

We represent genes and their relationships in our KEGG RDF KB as a directed graph $G = (V, E)$, where V is a set of vertices representing KEGG genes and E as a set of edges representing gene-gene interactions. Adjacency matrix $A(i, j)$ describes whether there is a directed edge between vertices v_i and v_j. For the sake of simplicity, $A(i, j) \in \{0,1\}$. We define the *Start Gene* set as $SG = \{v_1, ..., v_m\}$, where m is the total number of start genes. The *End Gene* set is defined as $EG = \{v_1, ..., v_n\}$, where n is the total number of end genes. We also define a set of genes used as diagnostic, prognostic, and therapeutic biomarkers for specific cancer phenotypes. This gene set is called the *Confidence Gene* set *CG*. For each vertex $v \in V$, binary operator $confidence()=\{0,1\}$ indicates membership of v in *CG*, i.e. $v \in V$, if $confidence(v)=1$.

Table 2. Modeled biological processes.

GO ID	GO Definition
GO:0001525	Angiogenesis
GO:0006915	Apoptotic process
GO:0008150	Biological process
GO:0008283	Cell proliferation
GO:0008284	Positive regulation of cell proliferation
GO:0008285	Negative regulation of cell proliferation
GO:0016525	Negative regulation of angiogenesis
GO:0042127	Regulation of cell proliferation
GO:0042981	Regulation of apoptotic process
GO:0043065	Positive regulation of apoptotic process
GO:0043066	Negative regulation of apoptotic process
GO:0045765	Regulation of angiogenesis
GO:0045766	Positive regulation of angiogenesis
GO:0048518	Positive regulation of biological process
GO:0048519	Negative regulation of biological process
GO:0050789	Regulation of biological process

We then identify directed paths from the Start Genes to End Genes in the graph guided by the *Confidence Gene* set *CG*. In order to solve the problem, we formulate the problem as an *M-N* problem, i.e., finding the optimum paths from *M Start Genes* to *N End Genes*. To divide and conquer, we also define the sub-problem of the *M-N* problem as *1-1* problem, which aims to find *K* best paths from one source gene to one sink gene in the graph, $K>=1$. In contrast to finding only one optimal path, defining *K* optimal paths in the *1-1* problem could provide much more depth in each single path, and these alternative paths could illustrate much more information incorporating all these paths together in the *M-N* problem. In the *1-1* problem, comparing with the classical path finding problem in graph theory, which aims to find the shortest path defined by the adjacent matrix, the involvement of the *Confidence Genes* affects the path finding. The optimum paths we intend to find should be shortest and involve as many of *Confidence Gene* as possible. We could define a path with length l from a *Start Gene path(1)* to an *End Gene path(l)*, $path(i) \in V(i=1...l)$. In the classical shortest path finding problem, the objective function of the optimum path in the graph can be written as:

$$f(path) = \sum_{i=1}^{l-1} A(path(i), path(i+1)), \tag{1}$$

which only considers the topological distance. So, the problem is to find a path having *min f(path)*. By including *Confidence Genes*, we could redefine the objective function of the optimum path as:

$$f(path) = \frac{\sum_{i=1}^{l} confidence(path(i))}{\sum_{i=1}^{l-1} A(path(i), path(i+1))}, \tag{2}$$

From Eq. (2), the path with the shortest distance and more *Confidence Genes* involved should be our optimum path, and the problem is redefined to find a path having *max f(path)*. Hence, the *1-1* problem is changed to the well-known *k*-shortest path problem with modified objective function. In our case, as the KEGG pathway graph contains many cycles, and since genes in such cycles might be important, we do not make the acyclic restriction. We implement Eppstein's algorithm [39] with a replaced objective function to solve the *1-1* problem in polynomial time, which requires only computational complexity of $O(|E| + |V|log|V|+K)$. The *M-N* problem could be treated as an exhaustive combination of all possible *1-1* problems with defined K in the graph. For each gene in the *Start Genes* and each gene in the *End Genes*, we obtain the K optimum paths on each pair of *Start* and *End Genes*. In total, we have M times N of *1-1* combinations, which are $M \times N \times K$ paths. We then map these paths to the graph, and merge them together. We use Procedure 1 to solve the *M-N* problem, as described below. In all of these paths, the importance of each path is evaluated using the objective function *f(path)*. Hence, the importance of each node in the graph is calculated by sum of the paths going through the node. The total computational complexity of the *M-N* problem is $O(MN(|E| + |V|log|V|+K))$.

M-N problem:
Input: Start Genes set *SG*, End Gene set *EG*, Confidence Genes set *CG*, directed adjacency matrix *A(i, j)*
Output: List of ranked paths by their decreasing objective function *f(path)* values

Procedure 1:
1: For all $v_i \in SG$ do
2:   For all $v_j \in EG$ do
3:     Compute *path(i, j)* by solving *1-1* problem for (v_i, v_j, K)
4:   End for
5: End for
6: Rank paths by *f(path)* values
7: Output of top specified percentage of paths as RDF Sketch Map

According to the definition of the problem, the choice of objective function plays critical role in finding optimum paths. In practical usage, Eq. (2) may have limitations in favoring path with shorter length and having larger values of objective function. For instance, in a hypothetical case with a path of length 1 having a *Start Gene* to be a *Confidence Gene*, the value of objective function *f(path)* will be maximal (i.e. 1). Such a path will be given preference over other, perhaps larger but more biologically important paths that can have many more *Confidence Genes*. To overcome this bias, we also define several other objective functions *f(path)* as Eqs. (3-5).

$$f(path)=\frac{\sum_{i=1}^{l} confidence(path(i))}{\sum_{i=1}^{l-1} A(path(i),path(i+1))}-\frac{\sum_{i=1}^{l-1} A(path(i),path(i+1))}{\gamma}+\delta, \quad (3)$$

Comparing with Eq. (2), Eq. (3) adds a penalty term of the current path length divided by the maximum path length. γ is the estimated maximum path length, a predefined non-negative value. δ is a non-negative predefined value to guarantee *f(path)*>0. In our case δ=1.

$$f(path)=\frac{\sum_{i=1}^{l} confidence(path(i))}{\sum_{i=1}^{|V|} confidence(i)}-\frac{\sum_{i=1}^{l-1} A(path(i),path(i+1))}{\gamma}+\delta, \quad (4)$$

To enforce the impact of confidence genes and reduce the redundancy of the multiple usage of path length in Eq. (3), Eq. (4) introduces the fraction of confidence genes included in the path in the left term at the right side of the equation. The path length information occurs only in the right term as in Eq. (3).

$$f(path)=\left(\frac{\alpha+\sum_{i=1}^{l} confidence(path(i))}{\alpha+\sum_{i=1}^{|V|} confidence(i)}\right)\left(\frac{\log\left(\frac{\sum_{i=1}^{l-1} A(path(i),path(i+1))}{\sum_{i}^{|V|}\sum_{j}^{|V|} A(i,j)}\right)}{\log\left(\frac{1}{\sum_{i}^{|V|}\sum_{j}^{|V|} A(i,j)}\right)}\right), \quad (5)$$

Like Eq.(4), the left term at the right side of Eq.(5) also describes how many confidence genes are included in the path. α is a small non-negative predefined value, such that $0<\alpha<1/|V|$. α is defined to make sure the left term's values are within the interval [0,1]. The right term at the right side of Eq. (5) defines the influences of path length. Logarithm is used to favor large changes in short path length. The beneficial property of the objective function defined by Eq. (5) is normalization of the objective function values to the interval [0,1]. Even in the absence of any *Confidence Genes*, the algorithm can still be operational and compute the shortest paths.

3. Results and Discussion

Our preliminary experiments with RDF Sketch Maps method were performed on two sets of KEGG maps associated with signal transduction pathways related to leukemic cancers such as HCL and CML.

Table 3. Reduction of 7 KEGG maps of 1,597 nodes.

	Obj. function Eq. (3)		Obj. function Eq. (4)		Obj. function Eq. (5)	
	Uncollapsed	Collapsed	Uncollapsed	Collapsed	Uncollapsed	Collapsed
Top 10%	74	17	167	72	182	79
Top 30%	119	39	177	75	227	105

For each set we ran RDF Sketch Maps algorithm for three versions of the objective function *f(path)*, described by the Eqs. (3), (4), and (5). We then counted the number of nodes in the resulting graphs involving gene instances as well as in the transformed graphs where gene instances were collapsed and represented by their respective abstract genes. The number of nodes of RDF Sketch Maps representing the 7 KEGG maps' experiment consisting of 1,597 nodes is shown in Table 3 and the resulting graphs of top 10% of paths are shown in Figure 2. The number of nodes of RDF Sketch Maps representing the extended set of 18 KEGG maps consisting of 2,873 nodes is shown in Table 4 and the resulting graphs of top 30% paths are shown in Figure 3.

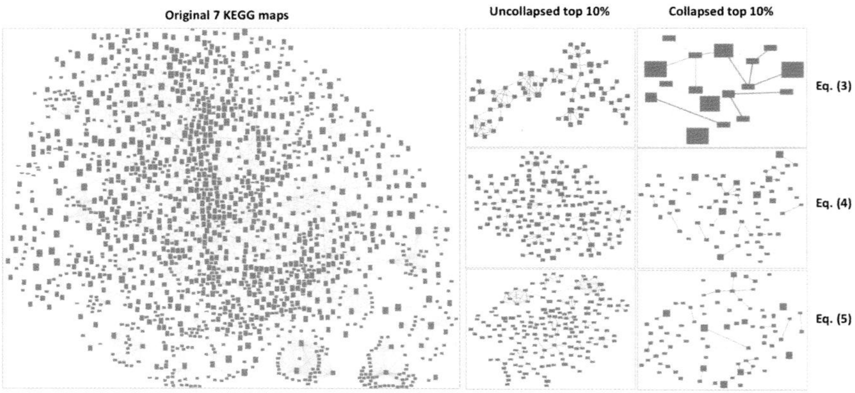

Figure 2. Reducing complexity of 7 integrated KEGG maps of 1,597 nodes. Top 10%.

Table 4. Reduction of 18 KEGG maps of 2,873 nodes.

	Obj. function Eq. (3)		Obj. function Eq. (4)		Obj. function Eq. (5)	
	Uncollapsed	Collapsed	Uncollapsed	Collapsed	Uncollapsed	Collapsed
Top 10%	666	220	596	188	669	224
Top 30%	778	298	661	220	791	283

It might be readily seen from the results that the overall reduction of nodes can reach 20 folds as in case with 7 KEGG maps' experiment and top 10% of collapsed gene instances using objective function of Eq. (3) (Table 3 and Figure 2). However, the practical utility of the RDF Sketch Maps is not defined by a mere reduction of the number of biological entities to be analyzed but by its retention of important entities that can explain subtle variations in patient-specific disease mechanisms. The objective function *f(path)* is biased toward inclusion of *Confidence Genes* in the resulting graphs. However, the *Confidence Genes,* as we noted before, are disease-associated biomarker genes that *are already known to be related to specific disease* for which analysis is performed. To uncover new, possibly unknown mechanisms, specific to individual patient cases, the resulting graphs should retain other important biological entities *not previously associated with the disease in question.*

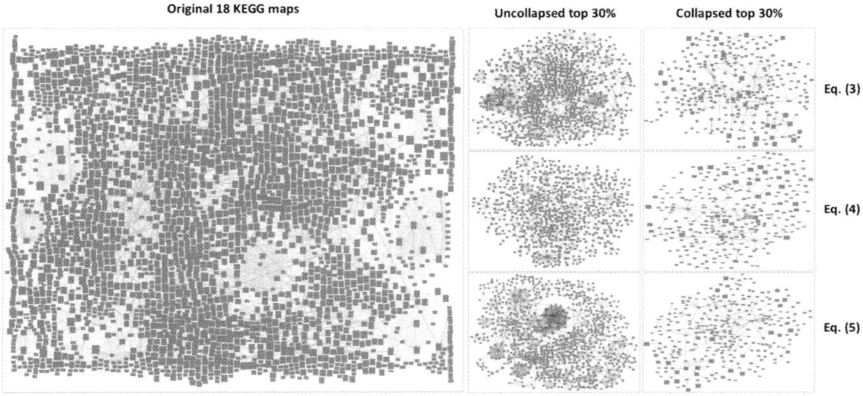

Figure 3. Reducing complexity of 18 integrated KEGG maps of 2,873 nodes. Top 30%.

To assess this quality of the RDF Sketch Maps method we define a set of *Exploratory Genes* – genes that are not directly implicated with disease in question but could possibly be useful in explaining its patient-specific disease heterogeneity (e.g. melanoma associated BRAF biomarker chosen as exploratory gene in an HCL case). We then estimate the inclusion of *Exploratory Genes* in the resulting maps.

In our preliminary experiments with leukemias the assessment procedure is done in the following way. A set of *Exploratory Genes EG* is defined as a set of genes implicated in other types of cancers. We then compute an enrichment score of *EG* genes in the resulting RDF Sketch Maps.

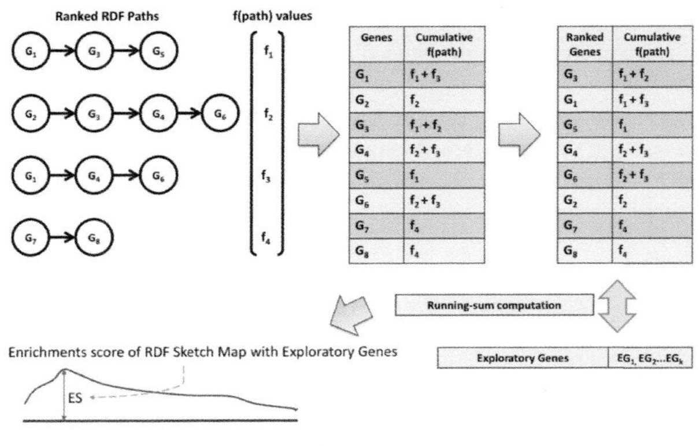

Figure 4. Extension of GSEA for RDF Sketch Maps.

To do that we extend Gene Set Enrichment Analysis (GSEA) [40]. First, we transform a ranked paths' list (see *Procedure 1*) into a ranked gene list according to their cumulative objective function *f(path)* values (Figure 4). Then we compute a running-sum while walking down the ranked gene list in the fashion similar to the original GSEA, adding *f(path)* value when current gene G is present in the *Exploratory Gene* set *EG*. The maximum deviation from zero is exported as an Enrichment Score for a specific RDF Sketch Map and *Exploratory Gene* set EG. The computed GSEA plot for objective function defined in Eq. (5) is shown in Figure 5.The non-normalized enrichment score is 0.728. The resulting RDF Sketch Map retained significant number of *Exploratory Genes* in top 10% of RDF Sketch Map paths.

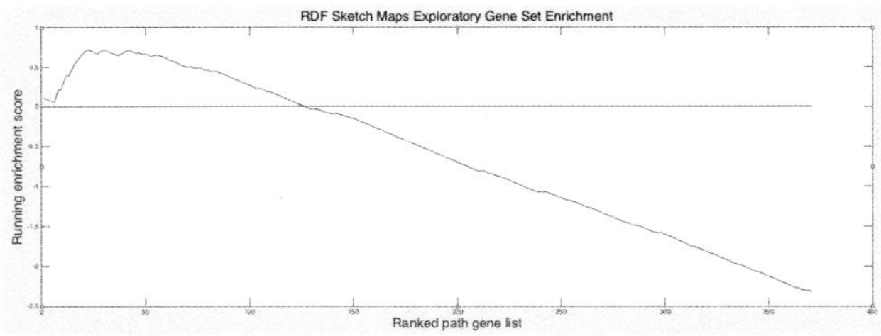

Figure 5. RDF Sketch Maps Exploratory Gene Set Enrichment Plot.

Another example of the exploratory power of RDF Sketch Maps is the fact that RDF Sketch Map of top 30% paths retained important information about signaling cascades leading to activation of proto-oncogene BRAF, which is usually associated with a different cancer – melanoma being the prototype. An increased number of successful treatments of HCL patients by BRAF-targeted drug *vemurafenib* were recently reported. The mechanisms of the involvement of BRAF in leukemias and other tumors are now being studied. We argue here that similar hypotheses to the BRAF drug-repurposing case could be generated by using our method.

4. Conclusions and Future Directions

Our preliminary experiments have demonstrated that RDF Sketch Maps can be invaluable for hypothesis generation in precision diagnostics and therapeutics as well as for drug repurposing studies. However, we identified several directions for RDF Sketch Maps improvement. Other disease models need to be explored. Initial pruning of RDF KB networks might help to increase performance of the algorithm. Many diverse types of *Exploratory Genes* need to be investigated, such as potential drug targets. And finally, new variations of objective function need to be studied.

5. References

[1] "FACT SHEET: President Obama's Precision Medicine Initiative."

[2] H. Wang, H. Cai, L. Ao, H. Yan, W. Zhao, L. Qi, Y. Gu, and Z. Guo, "Individualized identification of disease-associated pathways with disrupted coordination of gene expression.," *Briefings in bioinformatics*, p. bbv030, May 2015.

[3] C. J. Vaske, S. C. Benz, J. Z. Sanborn, D. Earl, C. Szeto, J. Zhu, D. Haussler, and J. M. Stuart, "Inference of patient-specific pathway activities from multi-dimensional cancer genomics data using PARADIGM.," *Bioinformatics*, vol. 26, no. 12, pp. i237–45, Jun. 2010.

[4] Y. Drier, M. Sheffer, and E. Domany, "Pathway-based personalized analysis of cancer.," *Proc Natl Acad Sci U S A*, vol. 110, no. 16, pp. 6388–6393, Apr. 2013.

[5] T. Ahn, E. Lee, N. Huh, and T. Park, "Personalized identification of altered pathways in cancer using accumulated normal tissue data.," *Bioinformatics*, vol. 30, no. 17, pp. i422–9, Sep. 2014.

[6] S. Ng, E. A. Collisson, A. Sokolov, T. Goldstein, A. Gonzalez-Perez, N. López-Bigas, C. Benz, D. Haussler, and J. M. Stuart, "PARADIGM-SHIFT predicts the function of mutations in multiple cancers using pathway impact analysis.," *Bioinformatics*, vol. 28, no. 18, pp. i640–i646, Sep. 2012.

[7] D. Bertrand, K. R. Chng, F. G. Sherbaf, A. Kiesel, B. K. H. Chia, Y. Y. Sia, S. K. Huang, D. S. B. Hoon, E. T. Liu, A. Hillmer, and N. Nagarajan, "Patient-specific driver gene prediction and risk assessment through integrated network analysis of cancer omics profiles," *Nucleic Acids Res*, vol. 43, no. 7, pp. gku1393–e44, Jan. 2015.

[8] T. Kessler, H. Hache, and C. Wierling, "Integrative Analysis of Cancer-Related Signaling Pathways," *Front. Physiol.*, vol. 4, Jun. 2013.

[9] P. Khatri, M. Sirota, and A. J. Butte, "Ten years of pathway analysis: current approaches and outstanding challenges.," *PLoS Comp Biol*, vol. 8, no. 2, p. e1002375, Feb. 2012.

[10] D. Shin, G. Arthur, M. Popescu, D. Korkin, and C.-R. Shyu, "Uncovering influence links in molecular knowledge networks to streamline personalized medicine.," *J Biomed Inform*, vol. 52, pp. 394–405, Dec. 2014.

[11] R. E. Brown, "Morphogenomics and morphoproteomics: a role for anatomic pathology in personalized medicine," *Arch Pathol Lab Med*, vol. 133, no. 4, pp. 568–579, Apr. 2009.

[12] J. Liu and R. E. Brown, "Morphoproteomics demonstrates activation of mTOR pathway in anaplastic thyroid carcinoma: a preliminary observation," *Ann Clin Lab Sci*, vol. 40, no. 3, pp. 211–217, 2010.

[13] C. F. Streckfus, R. E. Brown, and J. M. Bull, "Proteomics, morphoproteomics, saliva and breast cancer: an emerging approach to guide the delivery of individualised thermal therapy, thermochemotherapy and monitor therapy response," *Int J Hyperthermia*, vol. 26, no. 7, pp. 649–661, 2010.

[14] J. Liu and R. E. Brown, "Morphoproteomics demonstrates activation of mammalian target of rapamycin pathway in papillary thyroid carcinomas with nuclear translocation of MTOR in aggressive histological variants," Aug. 2011.

[15] V. Subbiah, A. Naing, R. E. Brown, H. Chen, L. Doyle, P. LoRusso, R. Benjamin, P. Anderson, and R. Kurzrock, "Targeted morphoproteomic profiling of Ewing's sarcoma treated with insulin-like growth factor 1 receptor (IGF1R) inhibitors: response/resistance signatures," *PloS one*, vol. 6, no. 4, p. e18424, 2011.

[16] S. Dietrich, J. Hüllein, M. Hundemer, N. Lehners, A. Jethwa, D. Capper, T. Acker, B. K. Garvalov, M. Andrulis, C. Blume, C. Schulte, T. Mandel, J. Meissner, S. Fröhling, C. von Kalle, H. Glimm, A. D. Ho, and T. Zenz, "Continued Response Off Treatment After BRAF Inhibition in Refractory Hairy Cell Leukemia," *JCO*, vol. 31, no. 19, pp. e300–e303, Jul. 2013.

[17] G. A. Follows, H. Sims, D. M. Bloxham, T. Zenz, M. A. Hopper, H. Liu, A. Bench, P. Wright, M. B. van't Veer, and M. A. Scott, "Rapid response of biallelic BRAF V600E mutated hairy cell leukaemia to low dose vemurafenib," *British Journal of Haematology*, vol. 161, no. 1, pp. 150–153, Apr. 2013.

[18] F. Peyrade, D. Re, C. Ginet, L. Gastaud, and M. Allegra, "Low-dose vemurafenib induces complete remission in a case of hairy-cell leukemia with a V600E mutation," ..., 2013.

[19] L. Arcaini, S. Zibellini, E. Boveri, R. Riboni, S. Rattotti, M. Varettoni, M. L. Guerrera, M. Lucioni, A. Tenore, M. Merli, S. Rizzi, L. Morello, C. Cavalloni, M. C. Da Vià, M. Paulli, and M. Cazzola, "The BRAF V600E mutation in hairy cell leukemia and other mature B-cell neoplasms," *Blood*, vol. 119, no. 1, pp. 188–191, Jan. 2012.

[20] D. Shin, G. Arthur, C. Caldwell, M. Popescu, M. Petruc, A. Diaz-Arias, and C.-R. Shyu, "A pathologist-in-

the-loop IHC antibody test selection using the entropy-based probabilistic method.," *Journal of pathology informatics*, vol. 3, no. 1, p. 1, 2012.
[21] S. S. Sahoo, O. Bodenreider, J. L. Rutter, K. J. Skinner, and A. P. Sheth, "An ontology-driven semantic mashup of gene and biological pathway information: application to the domain of nicotine dependence.," *J Biomed Inform*, vol. 41, no. 5, pp. 752–765, Oct. 2008.
[22] S. S. Sahoo, K. Zeng, O. Bodenreider, and A. Sheth, "From "glycosyltransferase" to 'congenital muscular dystrophy': integrating knowledge from NCBI Entrez Gene and the Gene Ontology," *Stud Health Technol Inform*, vol. 129, no. 2, pp. 1260–1264, 2007.
[23] M. E. Holford, H. Rajeevan, H. Zhao, K. K. Kidd, and K.-H. Cheung, "Semantic Web-based integration of cancer pathways and allele frequency data.," *Cancer Informatics*, vol. 8, pp. 19–30, 2009.
[24] K. M. Livingston, M. Bada, W. A. Baumgartner, and L. E. Hunter, "KaBOB: ontology-based semantic integration of biomedical databases.," *BMC Bioinformatics*, vol. 16, no. 1, p. 126, 2015.
[25] A. R. Pico, T. Kelder, M. P. van Iersel, K. Hanspers, B. R. Conklin, and C. Evelo, "WikiPathways: pathway editing for the people.," *PLoS Biol.*, vol. 6, no. 7, p. e184, Jul. 2008.
[26] M. Streit, A. Lex, M. Kalkusch, K. Zatloukal, and D. Schmalstieg, "Caleydo: connecting pathways and gene expression.," *Bioinformatics*, vol. 25, no. 20, pp. 2760–2761, Oct. 2009.
[27] S. Ekins, Y. Nikolsky, A. Bugrim, E. Kirillov, and T. Nikolskaya, "Pathway mapping tools for analysis of high content data.," *Methods Mol. Biol.*, vol. 356, pp. 319–350, 2007.
[28] "Home - Ingenuity."
[29] F. Zhou, S. Mahler, and H. Toivonen, "Simplification of Networks by Edge Pruning," in *Bisociative Knowledge Discovery*, vol. 7250, no. 13, Berlin, Heidelberg: Springer Berlin Heidelberg, 2012, pp. 179–198.
[30] M. Latendresse, M. Krummenacker, and P. D. Karp, "Optimal metabolic route search based on atom mappings," *Bioinformatics*, vol. 30, no. 14, pp. btu150–2050, Mar. 2014.
[31] E. Pitkänen, P. Jouhten, and J. Rousu, "Inferring branching pathways in genome-scale metabolic networks," *BMC Systems Biology*, vol. 3, no. 1, p. 103, Oct. 2009.
[32] T. Blum and O. Kohlbacher, "MetaRoute: fast search for relevant metabolic routes for interactive network navigation and visualization.," *Bioinformatics*, vol. 24, no. 18, pp. 2108–2109, Sep. 2008.
[33] M. Kanehisa and S. Goto, "KEGG: Kyoto Encyclopedia of Genes and Genomes," *Nucleic Acids Res*, vol. 28, no. 1, pp. 27–30, 2000.
[34] "BioCarta - Charting Pathways of Life," *biocarta.com*. [Online]. Available: http://www.biocarta.com/genes/index.asp.
[35] D. Croft, G. O'Kelly, G. Wu, R. Haw, M. Gillespie, L. Matthews, M. Caudy, P. Garapati, G. Gopinath, B. Jassal, S. Jupe, I. Kalatskaya, S. Mahajan, B. May, N. Ndegwa, E. Schmidt, V. Shamovsky, C. Yung, E. Birney, H. Hermjakob, P. D'Eustachio, and L. Stein, "Reactome: a database of reactions, pathways and biological processes," *Nucleic Acids Res*, vol. 39, no. Database issue, pp. D691–D697, 2011.
[36] M. Ashburner, C. A. Ball, J. A. Blake, D. Botstein, H. Butler, J. M. Cherry, A. P. Davis, K. Dolinski, S. S. Dwight, J. T. Eppig, M. A. Harris, D. P. Hill, L. Issel-Tarver, A. Kasarskis, S. Lewis, J. C. Matese, J. E. Richardson, M. Ringwald, G. M. Rubin, and G. Sherlock, "Gene ontology: tool for the unification of biology. {T}he {G}ene {O}ntology {C}onsortium," *Nature Genetics*, vol. 25, no. 1, pp. 25–29, 2000.
[37] T. U. Consortium, "The Universal Protein Resource (UniProt) in 2010," *Nucleic Acids Res*, vol. 38, no. 1, pp. D142–D148, Jan. 2010.
[38] J. Aasman, "Allegro graph: RDF triple database," 2006.
[39] D. Eppstein, *Finding the k shortest paths*. IEEE, 1994, pp. 154–165.
[40] A. Subramanian, P. Tamayo, V. K. Mootha, S. Mukherjee, B. L. Ebert, M. A. Gillette, A. Paulovich, S. L. Pomeroy, T. R. Golub, E. S. Lander, and J. P. Mesirov, "Gene set enrichment analysis: A knowledge-based approach for interpreting genome-wide expression profiles," *Proc Natl Acad Sci U S A*, vol. 102, no. 43, pp. 15545–15550, 2005.

REGULATORY RNA

DRENA DOBBS
Department of Genetics, Development & Cell Biology, Iowa State University
Ames, IA 50011 USA
Email: ddobbs@iastate.edu

STEVEN E. BRENNER
Department of Plant & Microbial Biology, University of California
Berkeley, CA 94720 USA
Email: brenner@compbio.berkeley.edu

VASANT G. HONAVAR
Departments of Genomics & Bioinformatics and Neuroscience, Pennsylvania State University
University Park, PA 16802 USA
Email: vhonavar@ist.psu.edu

ROBERT L. JERNIGAN
L. H. Baker Center for Bioinformatics & Biological Statistics,
Department of Biochemistry, Biophysics and Molecular Biology, Iowa State University
Ames, Iowa, 50011, USA
Email: jernigan@iastate.edu

ALAIN LAEDERACH
Department of Biology, University of North Carolina
Chapel Hill, North Carolina 27514, USA
Email: alain@unc.edu

QUAID MORRIS
Departments of Molecular Genetics and Computer Science, University of Toronto
Toronto, ON M5S 3E1, Canada
Email: quaid.morris@utoronto.ca

Advances in both experimental and computational approaches to genome-wide analysis of RNA transcripts have dramatically expanded our understanding of the ubiquitous and diverse roles of regulatory non-coding RNAs. This conference session includes presentations exploring computational approaches for detecting regulatory RNAs in RNA-Seq data, for analyzing *in vivo* CLIP data on RNA-protein interactions, and for predicting interfacial residues involved in RNA-protein recognition in RNA–protein complexes and interaction networks.

Discoveries over the past decade have revealed the previously unsuspected diversity of non-coding RNAs, which are ubiquitous in living organisms and play key roles in regulating gene expression and in organizing genomes [1-2]. The last time PSB had an RNA-focused session was in 2010. Since then the field of RNA regulation has been transformed by new data on RNA-based regulation (e.g., ENCODE and ENCORE); new experimental methods for genome-wide probing of RNA structure [e.g., 3] and translation [e.g., 4]; new knowledge about RNA-protein interactions [e.g., 5-8]; and new computational techniques to predict RNA regulation [e.g., 9].

Even though the human genome encodes nearly as many "non-coding" RNAs as it does protein-encoding mRNAs, the cellular functions of most ncRNAs remain unknown [10]. Also, despite impressive recent progress in annotating RNA-binding proteins [6, 11-12] and characterizing their binding sites [7, 13], our understanding of the roles of RNA-binding proteins and the determinants of RNA-protein recognition lag far behind our understanding of the mechanisms and regulatory roles of transcription factors. What we know at present is that ncRNAs have important functions in both *pre-* (e.g., epigenetic) and *post*-transcriptional regulation of gene expression in eukaryotes, as well as in bacteria [14] and viruses [15]. The expanding roles of ncRNAs in normal development and disease [16-17], the discovery that the oldest known alternative RNA splicing is for gene regulation, rather than producing alternate protein isoforms [18], the unanticipated abundance of circular RNAs in the brain [19], and their emerging roles in human disease [20], and novel potential applications of small RNAs in personalized medicine [21] all present extraordinary opportunities for productive interactions and collaborations among biologists and computer scientists.

The goal of this session is to bring together scientists who investigate structures, functions and dynamics of RNA, RNA-protein complexes, and RNA-protein interaction networks, with a focus on identifying and functionally annotating regulatory RNAs and RNPs. This is a data-rich field within molecular biology, where many terabytes of transcriptomics data have already been collected but remain largely unanalyzed. The technical advances mentioned above have resulted in an enormous expansion of available RNA sequences, structures and expression data – and highlighted an increasing gap in our functional understanding. This *Regulatory RNA* discussion session will provide a timely opportunity to discuss recent successes in computational, biochemical, biophysical and genetic approaches for studying the roles of non-coding RNAs, RNPs and RNA-protein interaction networks. This is a first step toward addressing both an urgent need and outstanding prospects for integrating computational and experimental methods in the quest for a better understanding of the fascinating structures and functions of regulatory RNAs and RNA-protein complexes.

Session Contributions

Jennifer Doudna, a Howard Hughes Medical Institute Investigator at the University of California, Berkeley, will provide the invited lecture. Doudna's laboratory has provided key insights into RNA-mediated gene regulation, including landmark discoveries on the molecular mechanisms of RNA interference and translational control in eukaryotes and on CRISPR-Cas immunity in bacteria.

One of the major challenges in studying regulatory RNAs is accurately detecting and quantifying them on a genome-wide scale. Many individual groups have grappled with both the biological (RNA decay and fragmentation) and quantitative (substring detection) aspects of the problem. The first paper in our session, from **Pena-Castillo** et al., focuses on the computational detection of small non-coding RNAs (sRNAs). sRNAs are bacterial regulatory RNAs that play important roles in stress response, virulence and a variety of other cellular processes. The authors present a systematic comparative assessment of available computational methods for detecting sRNAs in RNA-Seq data. They provide a valuable and balanced summary of the relative strengths and weaknesses of several approaches and propose a novel approach, ***sRNA-Detect***, which provides a higher retrieval rate than other methods.

Many regulatory RNAs carry out their functions by interacting with proteins, other RNAs or DNA. Recent technical advances in the detection of RNA-protein interactions *in vivo* include the development of protocols such as PAR-CLIP, HITS-CLIP and i-CLIP. The second paper in our session, from **Kassuhn** et al., provides a comparative evaluation of data processing tools for experimental CLIP-data. The paper describes a novel tool, ***Cseq-Simulator***, which can generate simulated datasets that can serve as valuable surrogates for an experimental gold standard CLIP dataset. Their comparisons point out which software tools are most useful for different purposes and identify some of the common pitfalls in genome-wide analysis of RNA-protein interactions.

Despite the tremendous advances in experimental approaches for detecting RNA-protein interactions, the sequence and structural determinants of specific recognition in RNA-protein complexes are still not well understood. In the final paper in our session, **Muppirala** et al. describe a computational method for predicting interfacial residues in RNA-protein complexes. They propose a machine learning approach that exploits sequence motifs found in the binding interfaces of known RNA-protein complexes. They propose a "partner-specific" method, ***PS-PRIP***, which simultaneously predicts interfacial residues for both the RNA and the protein components of complexes and is capable of predicting different interfaces for different binding partners. This method should be useful for identifying potential interfacial residues in RNA-protein complexes where structural information is not available.

The papers in this session provide an introduction to critical challenges facing researchers in the rapidly expanding field of regulatory RNA, and emphasize that this subject will undoubtedly continue to increase in importance for both basic and translational biology.

Acknowledgements

We thank all of the authors who submitted papers for this session and all of the reviewers who contributed their time and expertise. We acknowledge the Center for RNA Systems Biology P50 GM102706 to SEB and NIH R01 GM066387 to VGH, RLJ and DD for support. We are grateful to the PSB organizers for their support and especially Tiffany Murray for meeting coordination.

References

1. K. V. Morris and J. S. Mattick, *Nat Rev Genet.* **15**, 423 (2014).
2. J. W. Pek and K. Okamura, *Wiley Interdiscip Rev RNA*. **Oct 1.** doi: 10.1002/wrna.1309 (2015).

3. Y. Wan, K. Qu, Q. C. Zhang, R. A. Flynn, O. Manor, Z. Ouyang, J. Zhang, R. C. Spitale, M. P. Snyder, E. Segal and H. Y. Chang., *Nature.* **505**, 706 (2014).
4. N. T. Ingolia, L. F, Lareau and J. S. Weissman, *Cell* **147**, 789 (2011).
5. A. G. Baltz, M. Munschauer, B. Schwanhäusser, A. Vasile, Y. Murakawa, M. Schueler, N. Youngs, D. Penfold-Brown, K. Drew, M. Milek, E. Wyler, R. Bonneau, M. Selbach, C. Dieterich, and M. Landthaler, *Mol Cell.* **46**, 674 (2012).
6. A. Castello, B. Fischer, K. Eichelbaum, R. Horos, B. M. Beckmann, C. Strein, N. E. Davey, D. T. Humphreys, T. Preiss, L. M. Steinmetz, J. Krijgsveld and M. W. Hentze, *Cell* **149**, 1393 (2012).
7. D. Ray, H. Kazan, K. B. Cook, M. T. Weirauch, H. S. Najafabadi, X. Li, S. Gueroussov, M. Albu, H. Zheng, A. Yang, H. Na, M. Irimia, L. H. Matzat, R. K. Dale, S. A. Smith, C. A. Yarosh, S. M. Kelly, B. Nabet, D. Mecenas, W. Li, R. S. Laishram, M. Qiao, H. D. Lipshitz, F. Piano, A. H. Corbett, R. P. Carstens, B. J. Frey, R. A. Anderson, K. W. Lynch, L. O. Penalva, E. P. Lei, A. G. Fraser, B. J. Blencowe, Q.D. Morris and T. R. Hughes, *Nature* **499**, 172, (2013).
8. K. B. Cook, T. R. Hughes and Q. D. Morris, *Brief Funct Genomics* **14**, 74 (2015).
9. H. Y. Xiong, B. Alipanahi, L. J. Lee, H. Bretschneider, D. Merico, R. K. Yuen, Y. Hua, S. Gueroussov, H.S. Najafabadi, T. R. Hughes, Q. Morris, Y. Barash, A. R. Krainer, N Jojic, S. W. Scherer, B. J. Blencowe, B. J. Frey, *Science* **347,** 1254806 (2015).
10. J. S. Mattick and J. L. Rinn, *Nat Struct Mol Biol.* **22**, 5 (2015).
11. A. N. Brooks, M. O. Duff, G. May, L. Yang , M. Bolisetty, J. Landolin, K. Wan, J. Sandler, S. E. Celniker, B. R. Graveley and S. E. Brenner, *Genome Res.* **Aug 20.** pii: gr.192518.115. (2015).
12. A. Re, T. Joshi, E. Kulberkyte, Q Morris and C. T. Workman, *Methods Mol Biol.* **1097,** 491, (2014).
13. X. Li, H. Kazan, H. D. Lipshitz and Q. D. Morris, *Wiley Interdiscip Rev RNA.* **5,** 111 (2014).
14. A. K. Chaudhary, D. Na and E. Y. Lee, *Biotechnol Adv.* **33**, 914 (2015).
15. K. T. Tycowski, Y. E. Guo, N. Lee, W. N. Moss, T. K. Vallery, M. Xie and J. A. Steitz, *Genes Dev.* **29**, 567 (2015).
16. P. J. Batista and H. Y. Chang, *Cell,* **153**, 1298 (2013).
17. H. Ling, K. Vincent, M. Pichler, R. Fodde, I. Berindan-Neagoe, F. J. Slack and G. A. Calin, *Oncogene* **34,** 5003, (2015).
18. L. F. Lareau and S. E. Brenner, *Mol Biol Evol.* **32**, 1072 (2015).
19. A. Rybak-Wolf, C. Stottmeister, P. Glažar, M. Jens, N. Pino, S. Giusti, M. Hanan, M. Behm, O. Bartok, R. Ashwal-Fluss, M. Herzog, L. Schreyer, P. Papavasileiou, A. Ivanov, M. Öhman, D. Refojo, S. Kadener, N. Rajewsky. *Mol Cell.* **58**, 870 (2015).
20. S. Qu, X. Yang, X. Li, J. Wang, Y. Gao, R. Shang, W. Sun, K. Dou and H. Li, *Cancer Lett.* **365,** 141 (2015).
21. A.C. Solem, M. Halvorsen, S. B. Ramos and A. Laederach, *Wiley Interdiscip Rev RNA.* **6,** 517 (2015).

CSEQ-SIMULATOR: A DATA SIMULATOR FOR CLIP-SEQ EXPERIMENTS

WANJA KASSUHN

Max Delbrück Center for Molecular Medicine, Berlin Institute for Medical Systems Biology
13125 Berlin, Germany
Email: wanja.kassuhn@mdc-berlin.de

UWE OHLER*

Max Delbrück Center for Molecular Medicine, Berlin Institute for Medical Systems Biology
13125 Berlin, Germany
Email: uwe.ohler@mdc-berlin.de

PHILIPP DREWE*

Max Delbrück Center for Molecular Medicine, Berlin Institute for Medical Systems Biology
13125 Berlin, Germany
Email: philipp.drewe@mdc-berlin.de

CLIP-Seq protocols such as PAR-CLIP, HITS-CLIP or iCLIP allow a genome-wide analysis of protein-RNA interactions. For the processing of the resulting short read data, various tools are utilized. Some of these tools were specifically developed for CLIP-Seq data, whereas others were designed for the analysis of RNA-Seq data. To this date, however, it has not been assessed which of the available tools are most appropriate for the analysis of CLIP-Seq data. This is because an experimental gold standard dataset on which methods can be accessed and compared, is still not available. To address this lack of a gold-standard dataset, we here present Cseq-Simulator, a simulator for PAR-CLIP, HITS-CLIP and iCLIP-data. This simulator can be applied to generate realistic datasets that can serve as surrogates for experimental gold standard dataset. In this work, we also show how Cseq-Simulator can be used to perform a comparison of steps of typical CLIP-Seq analysis pipelines, such as the read alignment or the peak calling. These comparisons show which tools are useful in different settings and also allow identifying pitfalls in the data analysis.

1. Introduction

RNA-binding proteins (RBPs) play a central role in post-transcriptional gene regulation (e.g. in splicing, RNA-degradation or translation). However, the mechanisms by which RBPs regulate RNA-processing are still poorly understood. This is partially due to the challenges in quantifying protein-RNA interactions. Therefore, the recent introduction of cross-linking immunoprecipitation-high-throughput sequencing (CLIP-Seq) protocols that allow measuring protein-RNA interactions at a nucleotide level, such as PAR-CLIP [1], HITS-CLIP [2] or iCLIP [3], present a great advance as they allow getting an accurate picture of the RBP binding-landscape.

The approach of the CLIP-protocols is, to first UV-crosslinking RBPs to their bound RNA [4]. Subsequently, the RNAs are fragmented and the protein-RNA complexes are immunoprecipitated, in order to extract the complexes that involve the RBP of interest. Next,

*To whom correspondence should be addressed.

the RBP is digested using Proteinase K, but typically leaving cross-linked amino acids at the crosslinking site. Finally, the RNA-fragments are reverse transcribed to produce a cDNA library that can, at the end, be sequenced. The amino acids that are still linked to the cross-linking sites can introduce errors during the reverse transcription in the cDNA (diagnostic events) at the cross-linking site. For PAR-CLIP, these errors are predominately Thymine to Cytosine conversions (T-C conversion), whereas short deletions are introduced in HITS-CLIP and truncations in iCLIP experiments. As the diagnostic events occur at the crosslinking site, the events can be used to infer with single nucleotide-resolution the interaction site.

After sequencing the library, the resulting reads are aligned to the genome. A difference of CLIP-Seq reads and RNA-Seq reads is, however, that CLIP-Seq reads tend to be shorter than RNA-Seq reads (typically around 25 bp) and that they additionally can contain diagnostic events. Consequently, these two differences make alignment of CLIP-Seq reads more challenging than the alignment of RNA-Seq reads. To our knowledge there exists no spliced-alignment tool that is specifically design to align this data. Therefore, various aligners for gapped or ungapped alignments such as Bowtie2 [10], BWA [11], BWA-PSSM [12], STAR [18] or TopHat2 [19] are used to map the reads. The aligned reads can then be used in order to determine the sites of protein-RNA interactions. For this, sites where the CLIP-Seq reads are enriched are identified (peak calling). More sophisticated approaches, such as PARalyzer [5] or wavClusteR [7], make additionally use of the diagnostic events in order go get more accurate predictions.

However, for the read alignment and the subsequent peak calling, a systematics evaluation of the tools to perform the analyses has not been performed yet. This is partially due to the fact that there is no dataset available for which the ground truth is known and on the basis of which the tools can be benchmarked. A potential surrogate for such a dataset could be a realistic simulated dataset. However, there exist only simulators for RNA-Seq data (e.g. Flux Simulator [8]) but to our knowledge, there does not exist a realistic simulator for CLIP-Seq protocols.

In this work, we therefore present the CLIP-sequencing Simulator (Cseq-Simulator), a software to simulate data for various CLIP-protocols to address this lack of CLIP-Seq data simulators. We show that our simulation pipeline can be used to generate CLIP-Seq datasets that have the same characteristics as real datasets. Furthermore, we exemplify how this simulator can be used to assess the performance of various alignment tools for CLIP-Seq data. Finally, we study how the choice of the alignment algorithms influences the peak calling and identify potential pitfalls in the CLIP-Seq data analysis.

2. Material and Methods

2.1. *Read simulation approach*

Simulated data can provide a useful approximation to real dataset in cases where experimental determination of the ground truth on a large scale is infeasible. However, for the simulation to be useful, it is critical that it has the same characteristics as real datasets. Otherwise, the insights gained on the simulated data may not be transferable to real data. A challenge in the data simulation is, however, that the underlying processes are often only partially understood.

Thus, assumptions on the modelled processes have to be made, which may not be valid and can result in differences between simulated and real data.

In the Cseq-Simulator, we mimic key steps of the CLIP-Seq protocols in order to simulate CLIP-Seq data that is as realistic as possible. This is done in the following manner (see Fig. 1):

First, we determine the transcriptomic RNA-binding site of the RBP of interest. To this end, we use a position weight matrix (PWM) of the RBP of interest in order to predict its binding sites using FIMO [9]. The binding sites are called on the positive strand of annotated transcripts. As an alternative to the prediction of binding sites the user may also provide a list of transcriptomic binding sites. This can be useful when the RBP has an unspecific sequence motif, binding depends not only on the sequence or a high-quality set of experimentally determined binding sites is available.

After determining the binding sites, we simulated the raw reads (i.e. the reads without the diagnostic events). The PAR-CLIP, iCLIP and HITS-CLIP protocol share many steps with the standard RNA-Seq protocol. Therefore, we use components of the Flux Simulator [8], a simulator that has been shown to generate realistic RNA-Seq data, for simulation of steps of CLIP-Seq protocols that are similar to the RNA-Seq protocol. Specifically, we use the Flux-Simulator to first simulate the transcript abundances if the abundances are not provided by the user. As the transcripts that are not bound by the RBP are not of interest for the simulation, we set their expression to zero and readjust the other transcripts in order to speed up the simulation. This is done such that the overall number of transcripts remains constant.

Next, we use Flux Simulator to generate a library based on the transcript abundances. Then, we remove all the fragments in the library that do not contain a RBP binding site. This yields a library of RNA-fragments that have a RBP-binding site. Subsequently, we use the Flux Simulator to simulate the library amplification and sequencing of the library. This results in the raw reads. The advantage of using Flux simulator is that effects such as PCR-duplicates and sequencing errors can be simulated.

Finally, in order to generate the CLIP-Seq reads, we induce the diagnostic events (e.g. T-C conversions, deletions and truncations) in the raw reads. To this end, we sample the diagnostic events in the reads according to user specified distribution (diagnostic event profile) that is centred on the binding site. The resulting CLIP-Seq reads are returned in the FASTA-format.

2.2. *Dataset generation*

For the CLIP-Seq reads generation, we used our pipeline (see Sec. 2.1). We simulated reads for the GRCh38 *human* genome using the GENCODE release 21 gene annotation. To call Pumilio homolog 2 (PUM2) binding sites, we used the PWM that we obtained from [5]. For the read simulation we used the T-C conversion event profile of PUM2 from [5]. For the simulation of deletions, we assumed a uniform diagnostic event profile at all locations in the read that were a Thymine. To simulate truncations, we assumed that they occur at random at distance. However, we only introduced the truncation, if the location that was sampled had a distance of at least 4 bp the binding site.

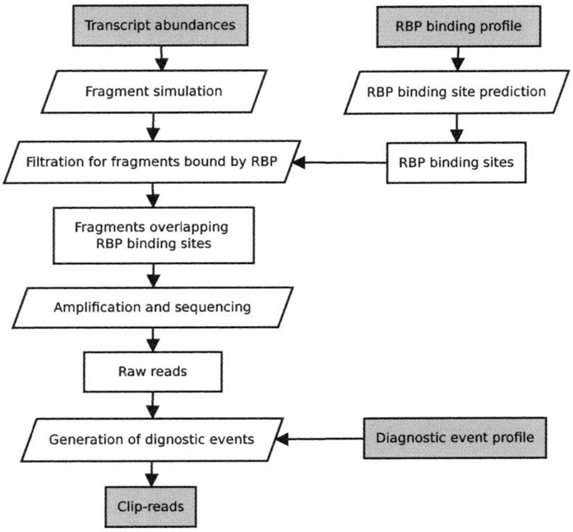

Fig. 1. Shown is a flow chart of the Cseq-Simulator read simulation. Shown in dark grey are the input and output data.

2.3. *Alignment algorithms*

In this study, we used the following aligners to align the CLIP-Seq reads to the hg38 *human* genome: Bowtie2 [10], BWA [11], BWA-PSSM [12], HISAT [15], PalMapper [16], Segemehl [17], STAR [18] and TopHat2 [19]. The tools have been selected to cover the commonly used short read alignment tools for CLIP-Seq data and aligners that are well suited for CLIP-Seq data alignment. For all the tools, we allowed in general for two mismatches and one indel during the alignment. To have a comparison that is less affected by the appropriateness of default parameters for CLIP-Seq data, we contacted the authors to obtain optimised parameter settings. For the tools where the authors did not reply, we used the parameters that were recommended by experts working with PAR-CLIP. For BWA-PSSM we used the error-profile that was provided for PAR-CLIP-Seq data. A list of the non-default parameters is given below:

Bowtie2: `-f -p 1 -L 15 -N 1 --very-sensitive --end-to-end`
BWA: `-k 1 -n 3 -t 1`
BWA-PSSM: `-l 15 -m 400`
HISAT: `-f -p 1 --mp 3,1 --pen-cansplice 0 --known-splicesites-infile splicesites.txt --max-intronlen 10000`
PalMapper: `-M 2 -n 3 -l 10 -E 3 -m 3 -S -min-spliced-segment-len 6 -include-unmapped-reads -report-gff-init annotation.gff3 -qpalma parameter.qpalma -I 10000 -no-ss-pred`
Segemehl: `-S -D 2 -M 3 -Z`

STAR: `--alignIntronMax 1 --sjdbGTFfile annotation.gtf --outSAMunmapped "Within" --outFilterMultimapNmax 3 --outFilterMismatchNmax 2 --seedSearchStartLmax 6 --winAnchorMultimapNmax 10000 --alignEndsType EndToEnd`

TopHat2: `--report-secondary-alignments --read-mismatches 2 --read-edit-dist 3 --min-anchor-length 10 --splice-mismatches 1 --max-intron-length 10000 --no-coverage-search --segment-mismatches 1 --max-multihits 3 --segment-length 10 --no-convert-bam`

2.4. *Alignment evaluation*

To evaluate the alignment of a set of reads, we determined for each read whether the read was mapping to multiple locations (multimapping) or to only one position. If the latter was the case we further determined whether the alignment was correct (i.e. it was mapped to the read's origin) or whether its mapping location was incorrect (mismapped).

2.5. *Alignment filtering*

For the filtering of multimappers, we only kept the best alignment for a read when the second best alignment had more than one mismatch more than the best alignment. Otherwise, all alignments for the respective read are discarded. In the later case, the read was treated as an unaligned read in the alignment evaluation. If read aligned only once, we kept it.

2.6. *Peak caller*

To call peaks from the CLIP-Seq reads, we used three tools: wavClusteR [7], PARalyzer [5] and BMix [?]. The tool Piranha [6] was not included in our evaluation as it could not be applied to all alignments. As the peak calling tools had different requirements to the input SAM-format, we standardized the SAM-files such that they were accepted by all peak callers. This was done by discarding all unmapped reads and alignments with "MD"-tags that included other operations than nucleotide substitutions. The peak calling tools were run with their default parameters. We defined the called peaks to be correct when they entirely overlapped the RBP binding sites.

3. Results

3.1. *Read generation*

We generated reads for PAR-CLIP, HITS-CLIP and iCLIP experiments of PUM2 using the Cseq-Simulator as described below. For the read simulation, we used all transcriptomic PUM2 binding-sites that were predicted by FIMO (FDR$\leq$ 0.1). Of the 23362 detected binding sites, 5233 were in transcripts that were simulated to be expressed. To simulate the reads, we first simulated the raw reads without diagnostic events for seven different read length: 14, 16, 18, 20, 24, 28 and 32. Overall, this resulted between 0.66×10^6 and 2.74×10^6 reads per library (see Tab. 1). We used these reads as templates to simulate three different groups of reads: Reads with T-C conversions, deletions or truncations. This resulted in seven sets of reads for each type of diagnostic event. From the reads for which we simulated T-C conversions, 75% had at least one diagnostic event. For the reads with simulated deletions and truncations between 85% and 91% resp. 15% and 69% had a diagnostic event. The high variation in the fraction of reads having a truncation was due to the fact that we set a boundary around the motif where truncation sites where there were no truncations. Therefore, many short reads were not truncated.

To determine whether the simulated PUM2-dataset had a realistic diagnostic event distribution, we analysed the diagnostic event distribution for the simulated data. For this, we compared the fraction of reads that had a T-C conversion at a given position relative to the predicted binding site with the diagnostic event profile from [5], which was used for the simulation (See Fig. 2). Overall, we found that the two profiles were very similar. We noticed, however, that there were subtle differences at the positions where the motif indicated a high preference for A. We believe, that these differences are due to the fact that the binding site prediction did not predict binding sites with a T at these positions. Consequently, a T-C conversion could not be simulated.

Table 1. Library sizes for the different read lengths.

Read length (bp)	*14*	*16*	*18*	*20*	*24*	*28*	*32*
Number of reads ($\times 10^6$)	0.66	1.13	1.16	1.35	1.92	2.22	2.74

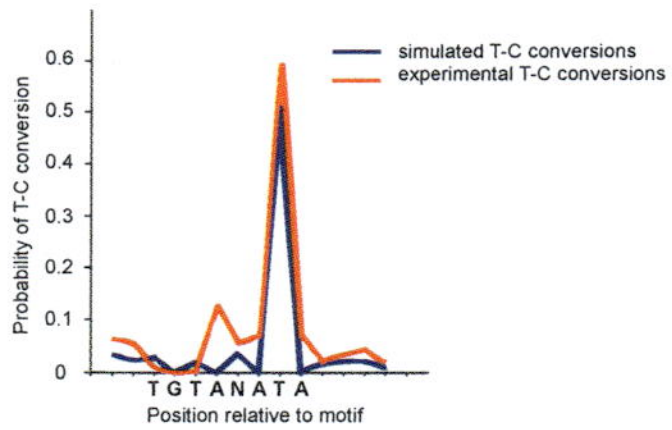

Fig. 2. Shown is T-C diagnostic event profile that was used for the simulation (red) and the fraction of reads that have a T-C conversion at a given position (blue) relative to the motif (bold letters).

3.2. *Assessment of alignment tools for CLIP-Seq data*

Short read alignment tools are used in most bioinformatics pipelines for the analysis of CLIP-Seq data. However, the influence of the choice of the aligner on the outcome is an aspect that has received little attention. Here, we exemplify how we can use the Cseq-Simulator to asses the performance of aligners for PAR-CLIP, HITS-CLIP and iCLIP data. For this, we aligned reads for a selection of commonly used short read aligners, namely Bowtie2, BWA, BWA-PSSM, HISAT, PalMapper, Segemehl, STAR and TopHat2. We then analysed different aspects of the alignment tools.

We first studied the sensitivity of the aligners, i.e. how many of the alignments are correct for reads with T-C conversions, deletions and truncations. To have a fair comparison between aligners that can produce split-alignments and the other methods, we only considered the reads that were unspliced in this analysis. We found, as it was expected, that the sensitivity of the aligners increased as the reads got longer (see Fig. 3). Moreover, we found that the sensitivity decreased as the number of diagnostic events increased.

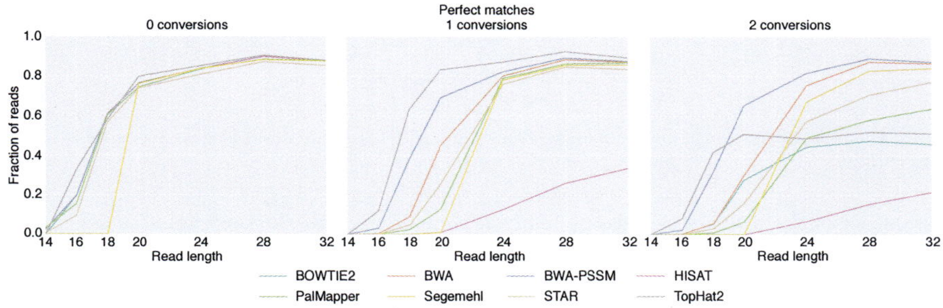

Fig. 3. Shown is the fraction of unspliced reads with 0 (left), 1 (middle) and 2 (right) conversions that map perfectly.

For reads with T-C conversions, we found that TopHat2 and BWA-PSMM had the highest sensitivity, although the sensitivity of TopHat2 for reads having two mismatches plateaued as the reads got longer. Furthermore, we found that the performance of HISAT was suboptimal for reads with mismatches. This was not surprising as it was developed to work with longer reads. We assume that the good performance of TopHat2 can be attributed to the strategy of TopHat2, to align reads to the transcriptome first and only perform alignment of reads to the whole genome when no good transcriptomic hit was found. This reduces the number of potential mapping locations significantly, thus also reducing the number of misalignments. To confirm that the high sensitivity of TopHat2 for short read lengths can indeed be attributed to the TopHat2 alignment strategy, we ran TopHat2 without providing a gene annotation (data not shown). This forced TopHat2 to align to the whole genome. We did this for the 16 bp long read-dataset. By doing this, the number of unspliced perfectly mapping reads dropped by 96%, showing that the transcriptome alignment was indeed responsible for the good performance on the short libraries. This suggests that the two-step alignment strategy might also be promising for CLIP-Seq data alignment pipelines that are using other aligners than TopHat2.

Next, we analysed the performance of the aligners for reads with simulated deletions (see Fig. 4). As we expected, the aligners achieved the same sensitivity on the reads without deletions as on the reads without T-C conversion. This was expected because we used the same reads as basis for the simulation of all three types of diagnostic events. For the reads with a deletion, we found that all algorithms could align less than 10% of the reads of length 20 and shorter to the correct location. For the reads that were 24 bp and longer, Segemehl had the highest sensitivity. The sensitivity of the other tools was considerably lower than in the T-C conversion setting. We assume that the high sensitivity of Segemehl may be the

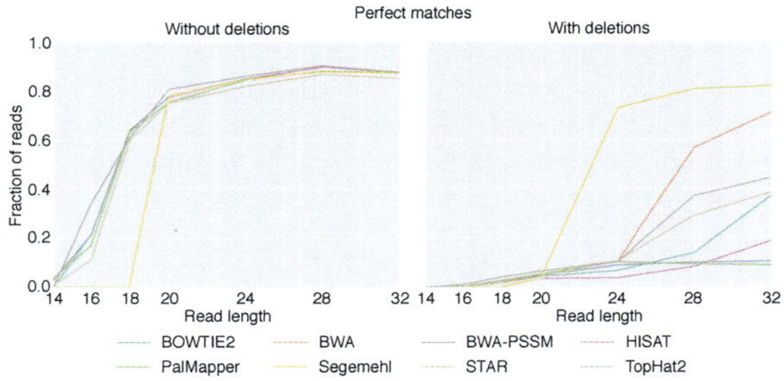

Fig. 4. Shown is the fraction of unspliced reads without and with deletions (left and right, respectively) that map perfectly.

consequence of the strategy to search, already at the seeding stage, for matches with deletions and insertions.

After this, we analysed the performance of the aligners for reads with truncations (see Fig. 5). For these, the performance of all the alignment tools on the reads with truncations reflected their performance on the reads without diagnostic events. This is because the libraries with truncations were basically a mixture of libraries for shorter read lengths without diagnostic event.

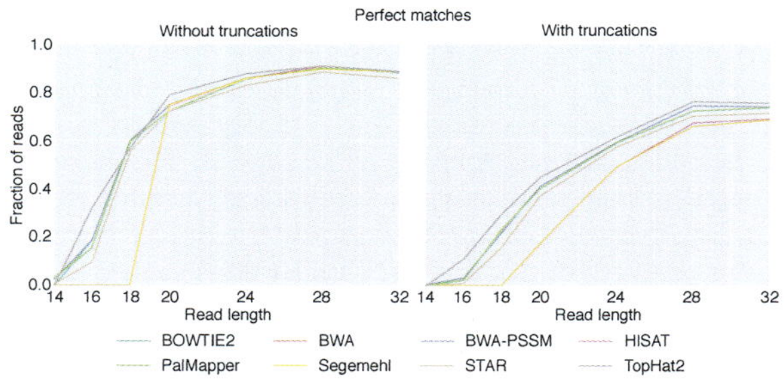

Fig. 5. Shown is the fraction of unspliced reads without and with truncations (left and right, respectively) that map perfectly.

In order to picture the overall performance of the selected alignment tools, we determined the fraction of reads that could not be mapped (unmapped), that mapped to multiple loci (multimapping), mapped to the wrong locus (errors) and both for spliced and unspliced reads, the reads that were correctly mapping (correct unspliced resp. spliced mapping). We performed this analysis for the reads with T-C conversions of length 32. Overall, we found that the fraction of reads in the different categories, varied substantially between the aligners (see Fig. 6). We found for example that the there were differences in the specificity of aligners.

Both BWA and BWA-PSSM could align a large fraction of the reads (95.6% resp. 97.4%). However, a substantial fraction of these alignments (9.0% resp. 10.3%) was mapping to the wrong location. In contrast, aligners such as STAR and TopHat2 were more conservative (i.e. did report more of the reads as unmapped): STAR and TopHat2 mapped 88.9% resp. 89.8% of the reads from which only 1.5% resp. 4.5% were mismapped.

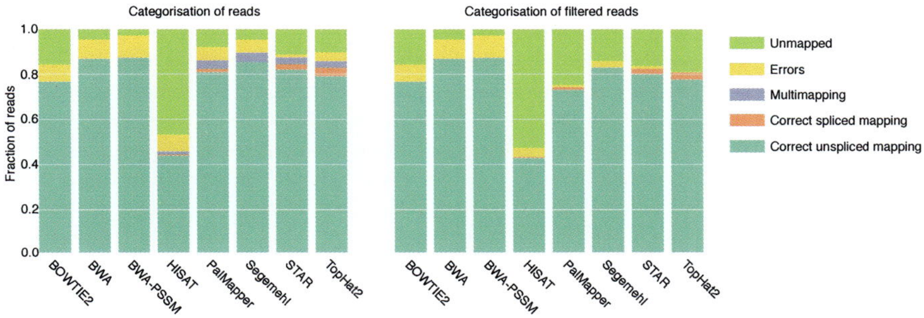

Fig. 6. Shown is the fraction of reads that were unmapped (light green), errors (yellow), multimapping (blue), correct spliced alignments (orange) and correct unspliced alignments (turquoise). Shown on right are the results for the T-C conversion dataset for reads of length 32. Shown on the left are the results for the dataset after filtering for multimappers.

3.3. *Alignment filtering*

A post-processing step that is commonly performed after the alignment, is removal of reads that map to multiple loci. The rational behind this is that for multimapping reads at least one alignment is wrong. This means that in the multimapping reads at least 50% map to a wrong location, thus their removal typically increases the quality of the alignment.

In order to understand how such filtering affects the CLIP-Seq data analysis, we first studied the influence of read-filtering on different categories of the alignments (as defined in the previous section). To this end, we filtered the reads of length 32 with T-C conversions for multimappers (see Sec. 2.5). We found, that the filtering affected the different alignment categories differently (see Fig. 6). Bowtie2, BWA and BWA-PSSM were not affected as with the parameters used, they only reported one alignment. For the other alignment tools, we observed that the filtering reduced the number of perfect matches and more strongly also the numbers of errors. This difference in reduction between the perfectly mapping reads and wrongly mapping reads was most striking for TopHat2, where only 5.8% of the correct mappings were removed but 93.8% of the errors. Overall, the filtering increased the specificity of the alignments. This suggests that filtering for multimappers is beneficial in settings where a high specificity is required.

3.4. *Peak calling*

The second step in the typical analysis of CLIP-Seq data analysis pipelines is the determination of binding sites by a peak calling. To analyse how the peak calling depends on the alignment tool that was used to align the reads, we applied PARalyzer, wavClusteR and BMix to all alignments of the 32 bp long reads with T-C conversions. We found that the number of peaks that were reported by the tree methods varied between the different alignments (see Fig. 7) and that PARalyzer had in general the lowest false positive rate. In our comparison wavClusteR detected between 4080 and 5867 peaks of which between 58% and 87% overlapped the true binding sites (n=5233). PARalyzer detected between 3336 and 4770 peaks of which between 74% and 95% overlapped the true binding sites and BMix detected between 3534 and 5948 peaks of which between 66% and 89% overlapped the true binding sites. Furthermore, we found that the fraction of true positives in the intersection of the peaks of all programs was in general higher than the fraction of true positives in the calls for each program (data not shown).

This shows that the choice of the alignments tool has a profound influence on the number of peaks that are called and suggest that it is important to use the same alignment strategy when comparing the performance of peak callers. We further investigated whether the difference in the number of peaks was due to the different numbers of reads that were aligned. Therefore, we evaluated the peak calling when the same number of reads was used from each aligner (the number of alignments in the smallest library). In order to exclude confounding of the result by the number of multimappers, we used the reads that have been filtered for multimappers.

We found that the number of clusters still showed a large variation (see Fig. 7). For wavClusteR, the number of peaks varied by 837 clusters, for PARalyzer by 973 and for BMix by 1093 clusters. This suggests that there are also systematic differences between results of the alignment tools. Furthermore, we found that filtering increased the fraction of true positives for all libraries for all tools.

4. Software

We have released the read simulation pipeline in a tool called Cseq-Simulator. This tools can be used under the GNU general public licence v.3. The tool can be obtained at: `https://ohlerlab.mdc-berlin.net/software/Cseq-Simulator_\%28Crosslinked-sequence_Simulator\%29_129/`

5. Discussion

In this work, we have presented Cseq-Simulator, a simulator for different types of CLIP-Seq data such as PAR-CLIP, HITS-CLIP or iCLIP. This simulator allows generation of datasets with known ground truth that exhibits several characteristics of real data, e.g. the read length or the diagnostic event profiles. In order to achieve a high resemblance of simulated and real data, we model different steps of the CLIP-Seq protocol and build on components of an existing RNA-Seq read simulator. For the binding site that are used for the simulation we provide two

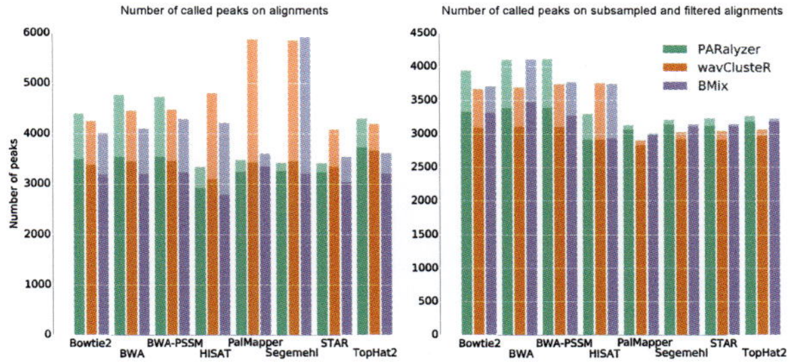

Fig. 7. Number of peaks that are called by PARalyzer, wavClusteR and BMiX. Light colors indicate the number of peaks that are called and dark colors the number true peaks that are called (n=5233). Shown on the right are the number of peaks that are called for reads with T-C conversions of length 32. Shown on the left are the number of peaks the multimapper-filtered and subsampled reads with T-C conversions of length 32.

options: (1) Prediction of the binding sites using a PWM. We expect that this provides a good approximation to RBP-binding in the case where the binding is mainly determined by the sequence. (2) A user provided list of binding site, which allows users to provide binding sites that are experimentally determined or derived using other models. We believe that the second option is particularly helpful when the RBP has a low sequence specificity or binding depends also on the secondary structure. Overall, Cseq-simulator allows modelling many aspects of CLIP-Seq datasets and can therefore be applied to simulate data for a broad range of RBPs and CLIP-Seq protocols.

Additionally, we have exemplified here, how simulated dataset can be used to assess the steps of a typical CLIP-Seq analysis pipelines. These analyses were performed for the read alignment, the peak calling and the interdependence of the two. In this assessment of the tools, we have made several interesting observations: For example, we have found that there was no best alignment tools for all CLIP-Seq data and that there was also a significant variation in the sensitivity and specificity of the alignments. When we compared PARalayzer, wavClusteR and BMix, we have observed that the number of peaks that were discovered, strongly depended on the choice of the alignment tool and that this was not only due to the different number of aligned reads. Overall, these observations show the potential of the Cseq-Simulator to inform decision on which tools to use for an CLIP-Seq data analysis.

A shortcoming of the benchmarking that we have carried out in this study is that we have mostly relied on default parameters for the tools. Therefore, the results might not reflect the optimal performance of the tools when tuned to a specific task. We would like to mention, however, that the simulated data is also valuable resource to improve the performance of the respective tools for CLIP-Seq data.

Another important point to mention is that, as the exact properties of CLIP-Seq data have not been characterised entirely, our simulations may not capture all aspects of this data. We did for example not simulate any biases. Therefore, the insights that have been gained on the basis of simulated data might not be entirely transferable to real data. However, we believe that independent of this shortcoming, important pitfalls in the data analysis can be identified, which could otherwise not be identified. In the future, we plan to extend Cseq-Simulator in order to also simulate stochastic binding and biases, e.g. the ones introduced by the choice of the restriction enzymes.

6. Acknowledgment

We would like to thank Neelanjan Mukherjee, Hans-Hermann Wessels and Alina-Cristina Munteanu and the reviewers of the manuscript for helpful discussions, comments and questions.

References

[1] M. Hafner, M. Landthaler, L. Burger, M. Khorshid, J. Hausser, P. Berninger, A. Rothballer, M. Ascano, A.-C. Jungkamp, M. Munschauer, A. Ulrich, G. S. Wardle, S. Dewell, M. Zavolan and T. Tuschl, *J Vis Exp* (2010).
[2] D. D. Licatalosi, A. Mele, J. J. Fak, J. Ule, M. Kayikci, S. W. Chi, T. A. Clark, A. C. Schweitzer, J. E. Blume, X. Wang, J. C. Darnell and R. B. Darnell, *Nature* **456**, 464 (Nov 2008).
[3] J. Konig, K. Zarnack, G. Rot, T. Curk, M. Kayikci, B. Zupan, D. J. Turner, N. M. Luscombe and J. Ule, *J Vis Exp* (2011).
[4] J. König, K. Zarnack, N. M. Luscombe and J. Ule, *Nat Rev Genet* **13**, 77 (Feb 2011).
[5] D. L. Corcoran, S. Georgiev, N. Mukherjee, E. Gottwein, R. L. Skalsky, J. D. Keene and U. Ohler, *Genome Biol* **12**, p. R79 (2011).
[6] P. J. Uren, E. Bahrami-Samani, S. C. Burns, M. Qiao, F. V. Karginov, E. Hodges, G. J. Hannon, J. R. Sanford, L. O. F. Penalva and A. D. Smith, *Bioinformatics* **28**, 3013 (Dec 2012).
[7] F. Comoglio, C. Sievers and R. Paro, *BMC Bioinformatics* **16**, p. 32 (2015).
[8] T. Griebel, B. Zacher, P. Ribeca, E. Raineri, V. Lacroix, R. Guigó and M. Sammeth, *Nucleic Acids Res* **40**, 10073 (Nov 2012).
[9] C. E. Grant, T. L. Bailey and W. S. Noble, *Bioinformatics* **27**, 1017 (Apr 2011).
[10] B. Langmead and S. L. Salzberg, *Nat Methods* **9**, 357 (Apr 2012).
[11] H. Li and R. Durbin, *Bioinformatics* **26**, 589 (Mar 2010).
[12] P. Kerpedjiev, J. Frellsen, S. Lindgreen and A. Krogh, *BMC Bioinformatics* **15**, p. 100 (2014).
[13] G. G. Faust and I. M. Hall, *Nat Methods* **9**, 1159 (Dec 2012).
[14] T. D. Wu and S. Nacu, *Bioinformatics* **26**, 873 (Apr 2010).
[15] D. Kim, B. Langmead and S. L. Salzberg, *Nat Methods* **12**, 357 (Apr 2015).
[16] G. Jean, A. Kahles, V. T. Sreedharan, F. De Bona and G. Rätsch, *Curr Protoc Bioinformatics* **Chapter 11**, p. Unit 11.6 (Dec 2010).
[17] S. Hoffmann, C. Otto, S. Kurtz, C. M. Sharma, P. Khaitovich, J. Vogel, P. F. Stadler and J. Hackermüller, *PLoS Comput Biol* **5**, p. e1000502 (Sep 2009).
[18] A. Dobin, C. A. Davis, F. Schlesinger, J. Drenkow, C. Zaleski, S. Jha, P. Batut, M. Chaisson and T. R. Gingeras, *Bioinformatics* **29**, 15 (Jan 2013).
[19] D. Kim, G. Pertea, C. Trapnell, H. Pimentel, R. Kelley and S. L. Salzberg, *Genome Biol* **14**, p. R36 (2013).

A MOTIF-BASED METHOD FOR PREDICTING INTERFACIAL RESIDUES IN BOTH THE RNA AND PROTEIN COMPONENTS OF PROTEIN-RNA COMPLEXES

USHA MUPPIRALA[†*]

Genome Informatics Facility, Iowa State University
Ames, Iowa, 50011, USA
Email: usha@iastate.edu

BENJAMIN A LEWIS[†]

Department of Computer Science, Truman State University
Kirksville, Missouri, 63501, USA
Email: benlewis@truman.edu

CARLA M. MANN

Bioinformatics and Computational Biology Program, Iowa State University
Ames, Iowa, 50011, USA
Email: cmmann@iastate.edu

DRENA DOBBS

Department of Genetics, Development and Cell Biology, Iowa State University
Ames, Iowa, 50011, USA
Email: ddobbs@iastate.edu

Efforts to predict interfacial residues in protein-RNA complexes have largely focused on predicting RNA-binding residues in proteins. Computational methods for predicting protein-binding residues in RNA sequences, however, are a problem that has received relatively little attention to date. Although the value of sequence motifs for classifying and annotating protein sequences is well established, sequence motifs have not been widely applied to predicting interfacial residues in macromolecular complexes. Here, we propose a novel sequence motif-based method for "partner-specific" interfacial residue prediction. Given a specific protein-RNA pair, the goal is to simultaneously predict RNA binding residues in the protein sequence and protein-binding residues in the RNA sequence. In 5-fold cross validation experiments, our method, PS-PRIP, achieved 92% Specificity and 61% Sensitivity, with a Matthews correlation coefficient (MCC) of 0.58 in predicting RNA-binding sites in proteins. The method achieved 69% Specificity and 75% Sensitivity, but with a low MCC of 0.13 in predicting protein binding sites in RNAs. Similar performance results were obtained when PS-PRIP was tested on two independent "blind" datasets of experimentally validated protein-RNA interactions, suggesting the method should be widely applicable and valuable for identifying potential interfacial residues in protein-RNA complexes for which structural information is not available. The PS-PRIP webserver and datasets are available at: http://pridb.gdcb.iastate.edu/PSPRIP/.

[†] Contributed equally to the work
* Corresponding author

1. Introduction

Despite the important roles of protein-RNA interactions in many biological processes, including transcription, translation, viral replication and pathogen resistance [1-2], the mechanisms and regulation of protein-RNA recognition are not yet fully understood. The Protein Data Bank (PDB) is a valuable resource for studying protein-RNA complexes, but protein-RNA complexes constitute less than 1% of the total structures in the database [3]. Recently, high-throughput (HTP) methods for identifying the *in vivo* targets of specific RNA binding proteins - and the RNA motifs they bind - have provided a wealth of information about the determinants of sequence recognition in protein-RNA complexes [4-6]. Data from both the PDB and HTP experiments have been exploited to develop several computational methods for predicting interfacial residues in protein-RNA complexes [reviewed in 7-10] as well as a few methods for predicting interaction partners in protein-RNA complexes and interaction networks [reviewed in 11-13].

Most computational approaches for predicting interfacial residues have focused on the protein side of the interface. Methods for predicting RNA-binding amino acid residues in proteins fall into two major classes: i) methods that use only sequence information, and ii) methods that take advantage of structural information, when available [8]. Only one published method [14-15] takes into account information regarding the RNA partner; the rest are "non-partner-specific" predictors of interfacial residues. Computational prediction of protein-binding ribonucleotides in RNA is a more difficult problem. The low per-character information content of the 4-ribonucleotide alphabet of unmodified RNA (i.e., ignoring modified ribonucleotides) makes this problem more challenging. One approach to overcoming this limitation is to expand the RNA alphabet by using known or predicted RNA secondary structure [16]. Another approach, taken in the current study, is to exploit short sequence motifs that occur in the interfaces of known protein-RNA complexes.

Here, we report a preliminary large scale analysis of contiguous RNA sequence motifs present in the interfaces of protein-RNA complexes and propose a new "partner-specific" motif-based method to simultaneously predict RNA-binding residues in the protein component and protein-binding ribonucleotides in the RNA component of a given protein-RNA pair.

2. Methods

2.1. *Generating interfacial sequence motifs*

To generate interfacial sequence motifs with which to scan target protein and RNA sequences, a dataset of 1,408 protein-RNA complex structures deposited in the Protein Data Bank (PDB) as of September 2012 was analyzed to find short strings of amino acids or ribonucleotides, contiguous in the primary sequence and composed entirely of interacting residues in either the protein or RNA chains. The sequences of these interfacial segments were extracted as 'n-mer motifs', where n can vary between 3 and 8. No requirement was made for motifs to be bounded by non-interacting residues; therefore, overlapping motifs were included. Thus, a 5-mer motif necessarily contains two 4-mer motifs and three 3-mer motifs.

2.2. *Datasets for interface prediction*

To generate datasets for evaluating the utility of motifs for interface prediction, interacting protein and RNA chains were extracted from protein-RNA complexes in the PDB with at least 3.5Å resolution. In one dataset, RPInt327, proteins of length < 25 amino acids and RNAs of length < 100 ribonucleotides were excluded. This dataset was used for training and cross-validation tests. The interaction information (i.e., interfacial residues) for these chains was downloaded from PRIDB [17]. Several additional fully independent datasets were generated to evaluate the performance of the classifier on RNAs of different lengths, e.g., RPInt79 (RNAs > 250 nts) and RPInt83 (RNAs 50-100 nts). The interfacial residues for these chains were computing using contact-chainID [18]. For both datasets, residues in protein and RNA chains were defined as interacting if any heavy atom in one chain lies within a 5Å distance cutoff of any heavy atom in the other chain. Based on BLASTClust results, redundant protein sequences (i.e., with ≥ 30% sequence identity) in complexes with similar RNA sequences (i.e., with ≥ 30% sequence identity) were discarded; RNA sequences in such redundant complexes were also discarded. For RPInt327, this resulted in a non-redundant dataset containing a total of 1,637 interacting protein-RNA pairs. 327 pairs were kept aside for independent evaluation and 5-fold cross-validation was performed on the remaining 1,310 pairs. Datasets RPInt79 and RPInt83 were reserved as a fully independent test datasets and were not used for training or cross-validation in this study.

2.3. *Generating a protein-RNA interface motif lookup table*

As illustrated in Figure 1, the protein-RNA interface motif lookup table consists of pairs of protein and RNA interfacial sequence motifs that are known to contact one another in a characterized protein-RNA complex. Entries in the lookup table were obtained as follows: First, the protein sequences in the non-redundant dataset of 1,637 protein-RNA pairs were scanned for interfacial sequence motifs (identified as described above) using a sliding window approach. Similarly, RNA sequences were scanned for interfacial sequence motifs (Fig. 1A). Second, each pair of protein-RNA sequences in the training dataset of known protein-RNA complexes was examined to identify cases in which there exists at least one physical contact (<5Å) between a heavy atom in any of the amino acids and any heavy atom in any of the ribonucleotides in a corresponding pair of sequence motifs (Fig. 1B & C). If a physical interaction is detected, that particular protein-RNA sequence motif pair is added to the lookup table (Fig. 1D).

2.4. *Motif-based prediction of interfacial residues in both RNA and protein*

After generating the protein-RNA interface motif lookup table, prediction of interfacial residues in a query protein-RNA pair is done in a single step. The protein and RNA sequences are scanned simultaneously for the presence of motif pairs in the lookup table. If any motif pair is present, those amino acids and ribonucleotides are marked as "interfacial" in the given query sequences. The remaining residues and ribonucleotides are marked as non-interfacial residues. For example, using the lookup table in Figure 1, if 'TRTYR' is found in the query protein and

'UUAAU' is found in the query RNA, the corresponding amino acids and ribonucleotides are predicted as interfacial residues.

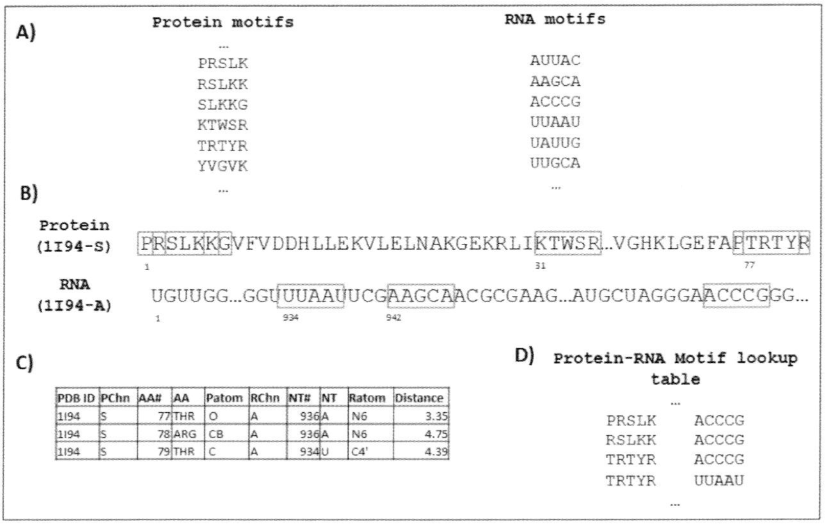

Fig. 1. Generation of the protein-RNA motif lookup table. **A)** A sample subset of the protein and RNA interfacial motifs used to scan target sequences. **B)** The protein and RNA sequences of each protein-RNA pair in the training dataset are scanned with the interfacial motifs. For the purpose of illustration, only a small portion of the example sequences and a subset of the interfacial motifs (indicated in boxes) are shown. **C)** Interacting residues within a distance threshold of 5Å are identified. Only a subset of interactions identified in this example is shown. **D)** Only protein and RNA motif pairs that contain at least one such interaction between them are added to the protein-RNA motif lookup table. Of the eighteen possible protein-RNA motif pairs illustrated in this example, only four satisfy this criterion and are added to the lookup table.

2.5. *Performance evaluation*

We used the following measures to evaluate the performance of motif-based prediction of interfacial residues on both proteins and RNAs. True Positive (TP) refers to the number of interfacial residues correctly identified as such by the method. False Positive (FP) refers to the number of non-interfacial residues misclassified as interfacial residues. False Negative (FN) refers to the number of interfacial residues misclassified as non-interfacial residues. True Negative (TN) refers to the number of non-interfacial residues correctly identified as such by the method. Note that here our definition of Sensitivity (true positive rate) is the same as Recall. We compute both Specificity (true negative rate), here as defined as in medical statistics literature, and Precision, which is referred to as Specificity in the machine learning literature [19].

$$Sensitivity(recall) = \frac{TP}{TP + FN} \quad (1)$$

$$Specificity = \frac{TN}{TN + FP} \quad (2)$$

$$Precision = \frac{TP}{TP + FP} \quad (3)$$

$$MCC = \frac{TP \times TN - FP \times FN}{\sqrt{(TP + FP)(TP + FN)(TN + FP)(TN + FN)}} \quad (4)$$

3. Results

3.1. *Motif-based partner-specific prediction of interfacial residues*

To evaluate whether an interface motif lookup table can be used to predict interfacial residues in specific protein-RNA pairs, we first performed preliminary experiments in which we tested the effect of varying the length of protein motifs from 4 to 6 amino acids, and the length of RNA motifs from 4 to 8 ribonucleotides (see *Methods*). As expected, using shorter motifs resulted in a larger number of false positive predictions, whereas using longer motifs resulted in larger number of false negative predictions. Based on these results, we determined that a protein motif of length 5 provides a good balance between prediction specificity and sensitivity. Although there are 20^5 different potential combinations of amino acid 5-mers, only 0.3% (11,269) of the theoretically possible 5-mer motifs were observed in interfaces extracted from known protein-RNA complexes (1,408 complexes, comprising 17,385 protein chains) in the PRIDB database [17].

To predict RNA-binding residues in the protein component of a given protein-RNA pair, we used a protein motif size of length 5 and varied the RNA motif lengths from 4 to 6. Table 1 summarizes results obtained using a 5-fold cross validation approach, in which 80% of the data was used to generate the protein-RNA motif lookup table and predictions were made on the remaining 20% of the data. There is little difference in the *Specificity* or *Matthews correlation coefficient* (MCC) using RNA motifs of length 4 and 5. Although using an RNA motif of length 6 resulted in higher *Specificity* (0.94), it resulted in lower *Sensitivity* and *MCC*. Using an RNA 4-mer resulted in higher *Sensitivity* (0.65) compared with using 5- and 6-mers.

Table 1. RNA-binding residue prediction performance using 5-fold cross-validation on a non-redundant dataset of 1,130 protein-RNA pairs

Protein motif length	RNA motif length	Specificity	Sensitivity	MCC
5	4	0.90	0.65	0.58
5	5	0.92	0.61	0.58
5	6	0.94	0.54	0.54

To predict which ribonucleotides in the RNA component of a given protein-RNA pair participate in protein binding, we again used a protein motif size of length 5 and varied the RNA motif lengths from 4 to 6. Table 2 summarizes the prediction results obtained in 5-fold cross-validation experiments. Again, as the RNA motif size is increased, the *Specificity* increased, but with the expected decrease in *Sensitivity*. A high *Specificity* of 0.91 is obtained using an RNA motif length of 6, but the corresponding *MCC* is much lower than that obtained for RNA binding site prediction (Table 1].

Table 2. Protein-binding residue prediction performance using 5-fold cross validation on a non-redundant dataset of 1,310 protein-RNA pairs

Protein Motif length	RNA motif length	Specificity	Sensitivity	MCC
5	4	0.35	0.89	0.07
5	5	0.69	0.75	0.13
5	6	0.91	0.55	0.21

3.2. *Performance evaluation on independent test sets*

To more rigorously test the performance of the method, we evaluated it on several independent datasets of known protein-RNA pairs (See *Methods*). As summarized in Table 3, on the RP327 dataset (which contains 327 protein-RNA pairs), using protein and RNA motifs of length 5, we obtained 92% *Specificity* and 64% *Sensitivity* in predicting RNA-binding residues. In predicting protein-binding ribonucleotides, the *Specificity* was 67% and *Sensitivity* was 79%. Thus, performance on the independent test set was comparable to that obtained in cross-validation experiments. This suggests that our proposed "partner-specific" method for predicting protein-RNA interfaces using sequence motifs, which we call PS-PRIP, should be generally applicable.

To investigate the influence of RNA length on performance, we also evaluated the classifier on several additional independent datasets of complexes containing RNAs of different lengths (Table 4 and data not shown). Although PS-PRIP performs very well on complexes containing RNAs longer than 250 nts, performance is poor on complexes containing shorter RNAs.

Table 3. Prediction performance on an independent test set of 327 protein-RNA pairs using protein and RNA motifs of length 5.

Prediction	Specificity	Sensitivity	MCC
RNA-binding amino acids in proteins	0.92	0.64	0.59
Protein-binding nucleotides in RNA	0.67	0.79	0.13

Table 4. Performance in predicting RNA binding residues in protein-RNA complexes containing RNAs of different lengths

Prediction	Specificity	Sensitivity	MCC
RNAs > 250 nts (83 total)	0.88	0.75	0.62
RNAs 50-100 nts (79 total)	0.99	0.03	0.03

3.3. *Comparison with other interface prediction methods*

Only one other published study has addressed the prediction of interfacial residues in protein and RNA components of protein-RNA complexes simultaneously. The catRAPID method proposed by Bellucci *et al.* [20] divides the protein and RNA sequences into a number of fragments and calculates interaction propensities between each pair of protein-RNA fragments. Because binding site prediction on a per residue basis was not reported, we could not directly compare our method with catRAPID.

A method for predicting protein-binding sites in RNAs was reported by Choi and Han [14-15]. We have not been able to make direct performance comparisons with this method because neither the test dataset nor a working webserver is available, and we did not attempt to re-implement it in order to provide a direct comparison with our method. In an earlier report, Choi and Han also proposed a partner-specific RNA binding site prediction method, in which the RNA sequence is encoded as the sum of the normalized positions of each nucleotide (A, C, G and U) in the sequence [14]. When we examined the dataset used in that study, we noticed that all except one RNA sequence was less than 100 nucleotides in length, and approximately half of the dataset consists of very short RNAs (< 15 nts). Because the minimum length of the RNA used in our training dataset is 100 nt, and, as discussed in the next section, our method is not suitable for small RNAs, we did not compare PS-PRIP with Choi and Han's method. Choi and Han reported prediction performance of 91% specificity and 60.7% sensitivity with a CC of 0.24 on a dataset of 267 interacting protein-RNA pairs [14].

We were able to compare the performance of our *partner-specific* PS-PRIP method with existing *non-partner* specific sequence-based methods for predicting RNA-binding residues in proteins. Walia *et al.* [8] performed a systematic comparison of existing methods for predicting RNA-binding residues and showed that PSSM-based methods had the best performance among published sequence-based approaches. Thus, we directly compared the performance of PS-PRIP with RNABindRPlus [21], which combines homology-based predictions with predictions from an optimized SVM classifier that uses a PSSM-based approach. Because homology-based methods exploit existing structures and interfaces, and our independent test set was extracted from the PDB, we expected the homology-based method to perform very well. Homology-based methods fail, however, when the query sequence has no homologs in the PDB. We also compared our method with the SVM component of RNABindRPlus and the results are also shown in Table 6. PS-PRIP has better performance in terms of *Specificity* (0.92), but lower *Sensitivity* (0.64) compared to RNABindRPlus. RNABindRPlus had the highest *MCC* (0.71); the MCCs for the other two methods were similar (0.59 vs 0.61). A larger difference is seen in the *Precision* (or positive prediction rate) of the two methods: PS-PRIP has higher *Precision* (0.80) than RNABindRPlus (0.76), when evaluated on this dataset.

Table 6. Performance comparison of PS-PRIP and RNABindRPlus in predicting of RNA binding residues.

Method	Specificity	Sensitivity	Precision	MCC
PS-PRIP	0.92	0.64	0.80	0.59
RNABindRPlus	0.85	0.88	0.76	0.71
RNABindRPlus (SVM-only)	0.74	0.90	0.65	0.61

4. Discussion

This study suggests that specific subsets of short contiguous interfacial motifs are over-represented relative to others within the sequences of both protein and RNA components of known protein-RNA complexes. A large number of interfacial amino acid motifs occur only once in the dataset analyzed here. This may be a consequence of the criteria for generating the short RNA-binding motifs in this study: all residues in an interfacial motif must be contiguous in sequence and must interact with at least one atom in a ribonucleotide within a 5 Å distance cutoff. It is striking that a simple lookup table of motif pairs, identified in a training set of protein-RNA complexes, can be used to accurately predict interfacial residues in an independent set of complexes. Although we have not yet directly calculated the interface propensities of these motifs (i.e., the over-representation of these motifs in interfacial versus non-interfacial regions of the protein and RNA sequences), it should be possible to improve prediction of interfacial residues by focusing on motifs with high interface propensity.

The interface prediction results reported here demonstrate that an ribonucleotide motif of length 5, while not informative on its own, can be highly informative when used in combination with an amino acid motif of length 5. From the non-redundant dataset of protein-RNA complexes used in this study, we generated a lookup table of 55,154 protein-RNA motif pairs, comprising 3,275 unique protein motifs and 835 unique RNA motifs. Using a non-redundant dataset is the appropriate way to evaluate and compare interface prediction methods, but doing so is expected to exclude some informative motif combinations. Thus, we created a motif lookup table *without* discarding redundant motifs. As expected, many additional protein-RNA motif pairs were identified: a total of 88,994 protein-RNA motif pairs, comprising 4,035 protein motifs and 893 RNA motifs.

Our results indicate that binding partner information, which has been largely ignored for predicting interfacial residues in protein-RNA complexes, can be valuable for making "partner-specific" interface predictions. Figures 2 and 3 illustrate this with an example. In the *E. coli* ribosome, the 16S rRNA in the small subunit interacts with various protein components of the 30S subunit, using different binding sites. Interaction of S4 and S11 proteins with a segment of the 16S

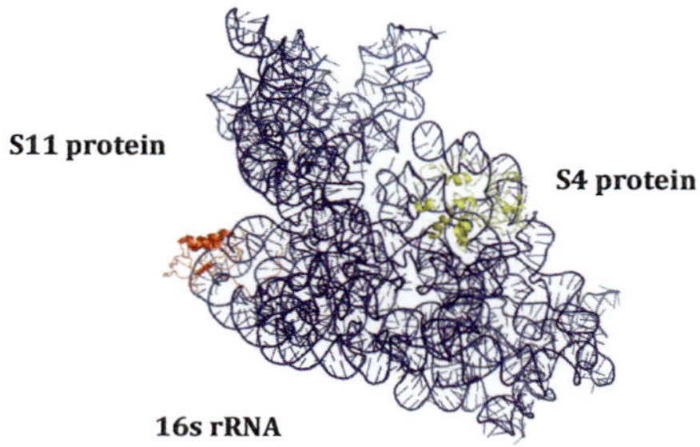

Fig. 2. *E. coli* 16S ribosomal RNA (blue) interaction with S4 ribosomal protein S4 (yellow) and ribosomal protein S11 (red). (PDB ID: 4GAS)

A. 16S rRNA interface with S4

UGAUGCAGCCAU**GCCGCGUGUAUGAAGAA**GGCCUUCGG**GUUGUA**AAGU**ACCU**

```
------------+++++++++++++++---------+++++----++++
```

B. 16S rRNA interface with S11

UGAUGCAGCCAUGCCGC**GUGUA**UGAAGAAGGCCUU**CGGGU**UGUAAAGUACCU

```
-----------------+++++--------------+++++------------
```

TP = True positives	**FP = False positives**
TN = True negatives	FN = False negatives

Fig. 3. Partner-specific interface prediction in the *E, coli* 16S ribosomal RNA (PDB ID: 4GAS). Different protein-binding residues are predicted for the same RNA sequences between nt 386-437 of 16S RNA when the segment is paired with two different protein partners. **A.** Ribosomal protein S4 protein. **B.** Ribosomal protein S11 protein. '+' indicates "positive/binding" and '-' indicates "negative/non-binding" predictions.

ribosomal RNA (PDB 4GAS) is shown in Figure 2. In this structure, the majority of 16S rRNA nucleotides that bind the S4 protein are located in the region between 400 – 440 nt. In contrast, the region between 670 – 720 nt of 16S rRNA contains most of the S11 protein-binding residues. Whereas a *non*-partner specific method would not be able to distinguish between these, Figure 3 shows that PS-PRIP makes distinct binding site predictions for the S4 and S11 proteins. In the 16S rRNA sequence between 386 – 437 nt, many S4 binding residues are correctly predicted. In the same region (where S11 does *not* bind), a few residues are incorrectly predicted as interacting with S11 in complexes that contain short RNAs (< 100 nts). Short RNAs, which often correspond to interface-containing fragments of the much longer RNAs present in native complexes, are common in structurally characterized protein-RNA complexes in the PDB. Thus, the likelihood that every ribonucleotide in such an RNA is an interfacial residue is very high compared to the situation for longer RNAs, in which only a small fraction of the ribonucleotides directly contact the bound protein(s). Because of this, we excluded RNAs <100 nts in length for generating motifs (see *Methods*), which results in a bias in our training set for RNA-protein pairs derived from ribosomes. In our experiments, PS-PRIP performed well on RNAs >100 nts in length (Table 3), but poorly when tested on RNAs < 100 nts (Table 4). Thus, PS-PRIP can be used to predict protein-binding sites in mRNAs, rRNAs, long non-coding RNAs and many short ncRNAs, but predictions on RNAs less than 100 nts are likely to be unreliable. Current work is directed at generating "custom" classifiers trained on datasets containing RNAs of variable length to obtain optimal performance on RNAs of different lengths and different functional classes (e.g., non-ribosomal ncRNAs, including sRNAs, sncRNAs, etc.)

In future work, we plan to evaluate the effect of incorporating predicted RNA secondary structure in the RNA sequence representation, which is expected to lead to better performance in predicting protein-binding residues in RNA [16]. In addition, we plan to test whether exploiting the extensively characterized resource of structural motifs in RNAs [22-23], can provide further improvement.

5. Conclusions

We have developed a new method for predicting partner-specific interfacial residues in protein-RNA complexes using short sequence motifs. PS-PRIP can simultaneously predict interfacial residues in both the protein and RNA components of a complex, albeit with much greater reliability for the protein component. An RNA motif of length 5, in combination with a protein motif of length 5, can be used to predict interfacial residues with high specificity (0.92 for RNA-binding residues in proteins; 0.67 for protein-binding residues in RNA), indicating that PS-PRIP can be a valuable tool for experimentalists who wish to target interfaces in specific protein-RNA complexes or to perturb specific interactions in protein-RNA interaction networks. A PS PRIP webserver and all training and test datasets used in this study are freely available online at: *http://pridb.gdcb.iastate.edu/PSPRIP/*.

Acknowledgements

Authors wish to thank Rasna Walia for valuable discussions and NIH R01 GM066387 for funding to support portions of this work.

References

1. D.D. Licatalosi and R.B. Darnell, *Nat. Rev. Genet.* **11,** 75 (2010).
2. A. Re, T. Joshi, E. Kulberkyte, Q. Morris and C. T. Workman, *Methods Mol. Biol.* **1097,** 491 (2014).
3. H. M. Berman, J. Westbrook, Z. Feng, G. Gilliland, T. N. Bhat, H. Weissig, I. N. Shindyalov and P. E. Bourne, *Nucleic Acids Res.* **8,** 235 (2000).
4. C. A. McHugh, P. Russell and M. Guttman, *Genome Biol.* **15,** 203 (2014).
5. D. Ray, H. Kazan, K. B. Cook, M. T. Weirauch, H. S. Najafabadi, X. Li, S. Gueroussov, M. Albu, H. Zheng, A. Yang, H. Na, M. Irimia, L. H. Matzat, R. K. Dale, S. A. Smith, C. A. Yarosh, S. M. Kelly, B. Nabet, D. Mecenas, W. Li, R. S. Laishram, M. Qiao, H. D. Lipshitz, F. Piano, A. H. Corbett, R. P. Carstens, B. J. Frey, R. A. Anderson, K. W. Lynch, L. O. F. Penalva, E. P. Lei, A. G. Fraser, B.J. Blencowe, Q. D. Morris and T.R. Hughes, *Nature.* **499,** 172 (2013).
6. K. B. Cook, H. Kazan, K. Zuberi, Q. Morris, T. R. Hughes, *Nucleic Acids Res.* **39,** D301 (2011).
7. T. Puton, L. Kozlowski, I. Tuszynska, K. Rother and J. M. Bujnicki, *J. Struct. Biol.* **179,** 261 (2012).
8. R. R. Walia, C. Caragea, B. A. Lewis, F. Towfic, M. Terribilini, Y. El-Manzalawy, D. Dobbs and V. Honavar, *BMC Bioinformatics.* **13,** 89 (2012).
9. H. Zhao, Y. Yang and Y. Zhou, *Mol. BioSyst.* **9** 2417 (2013).
10. J. Yan, S. Friedrich, and L. Kurgan, *Brief Bioinform.* **May 1**. doi: 10.1093/bib/bbv023 (2015).
11. U. K. Muppirala, B. A. Lewis and D. Dobbs, *J. Comput. Sci. Syst. Biol.* **6**, 182 (2013).
12. D. Cirillo D, C.M. Livi, F. Agostini, and G.G. Tartaglia, *Mol. BioSyst.* **10**, 1632 (2014).
13. U. K. Muppirala, C. M. Mann and D. Dobbs, *Methods Mol. Biol.* In press (2015).
14. S. Choi and K. Han, *BMC Bioinformatics.* **12,** S7 (2011).
15. S. Choi and K. Han, *Comp. Biol. Med.* **43(11),** 1687 (2013).
16. X. Li, H. Kazan, H. D. Lipshitz and Q. Morris, *Wiley Interdiscip Rev RNA.* **5,** 111 (2014).
17. B. A. Lewis, R. R. Walia, M. Terribilini, J Ferguson, C. Zheng, V. Honavar and D. Dobbs, *Nucleic Acids Res.* **39,** D277 (2011).
18. C. Dominguez, R. Boelens and A. M. J. J. Bonvin. *J. Am. Chem. Soc.* **125**, 1731 (2003)
19. P. Baldi, S. Brunak, Y. Chauvin, C. A. F. Andersen and H. Nielsen, *Bioinformatics,* **16,** 412 (2000).
20. M. Bellucci, F. Agostini, M. Masin and G.G. Tartaglia, *Nat. Methods.* **8,** 444 (2011).
21. R. R. Walia, L. C. Xue, K. Wilkins, Y. El-Manzalawy, D. Dobbs and V. Honavar, *PLoS ONE,* **9(5),** e97725 (2014).
22. A. I. Petrov, C. L. Zirbel and N. B. Leontis, *RNA.* **19**, 1327 (2013)
23. G. Chojnowski, T. Walen, and J. M. Bujnicki. *Nucleic Acids Res.* **42,** D123 (2014)

DETECTION OF BACTERIAL SMALL TRANSCRIPTS FROM RNA-SEQ DATA: A COMPARATIVE ASSESSMENT

LOURDES PEÑA-CASTILLO*[,1,2], MARC GRÜLL[2], MARTIN E MULLIGAN[3], and ANDREW S LANG[2]

[1] *Department of Computer Science, Memorial University of Newfoundland,*
[2] *Department of Biology, Memorial University of Newfoundland,*
[3] *Department of Biochemistry, Memorial University of Newfoundland,*
St. John's, NL, Canada
* *E-mail: lourdes@mun.ca*

Small non-coding RNAs (sRNAs) are regulatory RNA molecules that have been identified in a multitude of bacterial species and shown to control numerous cellular processes through various regulatory mechanisms. In the last decade, next generation RNA sequencing (RNA-seq) has been used for the genome-wide detection of bacterial sRNAs. Here we describe sRNA-Detect, a novel approach to identify expressed small transcripts from prokaryotic RNA-seq data. Using RNA-seq data from three bacterial species and two sequencing platforms, we performed a comparative assessment of five computational approaches for the detection of small transcripts. We demonstrate that sRNA-Detect improves upon current standalone computational approaches for identifying novel small transcripts in bacteria.

Keywords: RNA-seq, bacterial small transcripts detection, sRNA

1. Introduction

In the last decade, RNA sequencing (RNA-seq) methods have been used to identify small non-coding RNAs (sRNAs) on a genome-wide scale in numerous bacterial species. A key step in the detection of sRNAs from RNA-seq experiments is the analysis of RNA-seq data to assemble and identify expressed transcripts. Most studies identify sRNAs by manual inspection of the sequencing data (e.g., Refs. 1,2), developing their own in-house computational approach (e.g., Ref. 3) or a combination of both (e.g., Ref. 4). Manual identification is a strenuous task and hard to reproduce independently. The last issue also applies to small transcript identification by an in-house computational approach unless the program is made publicly available. In the past three years, several computational approaches to identify novel transcripts from prokaryotic RNA-seq data (e.g., Refs. 5,6) and from eukaryotic RNA-seq data (e.g., Ref. 7) have become available; however, a systematic side by side comparison of their performance has yet to be carried out. In this work, we compared the performance of four standalone computational approaches and our own approach (sRNA-Detect) to identify small transcripts from prokaryotic RNA-seq data.

2. Approaches for the Detection of Small Transcripts From RNA-Seq Data

Rockhopper[5] is a system that supports several phases of bacterial RNA-seq data analysis, including mapping sequencing reads to a reference genome, data normalization, assembling transcripts and identifying their boundaries, quantifying gene expression, testing for differential gene expression, characterizing operon structures, and visualizing results. This system specifically models bacterial transcriptome features such as operon structures, sRNAs and dense genomes with overlapping genes, and uses annotated genes as seeds to generate a transcriptome map. Rockhopper's capabilities to identify sRNAs were tested on *Neisseria gonorrhoeae* RNA-seq data.[8] In that study, Rockhopper detected 34 small transcripts of which 4 had previously been identified and 11 were experimentally confirmed by Northern blot.

TruHMM[6] is a Hidden Markov Model-based algorithm for assembling full-length transcripts in bacteria using directional RNA-seq short reads. TruHMM was designed to assemble transcripts with non-uniform read coverage and to address the problem of transcribed regions not covered by sequencing reads. TruHMM has been reported to achieve high sensitivity (recall) in assembling antisense RNAs (asRNAs) and non-coding RNAs (ncRNAs) in *Escherichia coli* K12 where it was able to recover 102 (91%) of 112 known asRNAs and ncRNAs.[6]

RNA-eXpress[7] is a system to perform feature annotation, comparison, sequence extraction and abundance quantification from eukaryotic RNA-seq data independently of current annotations. Several algorithms were implemented within RNA-eXpress to identify various types of features. Two of these algorithms are the Transcripts algorithm which performs comprehensive transcript identification, and the TLA algorithm which searches for features that have a certain minimum depth coverage and a minimum length. RNA-eXpress was found to achieve 96% accuracy identifying transcripts on region E2 of mouse chromosome 5.[7]

DETR'PROK pipeline[9] is a workflow to detect ncRNAs and untranslated regions (UTRs) in a reference genome from bacterial RNA-seq data using the Galaxy framework.[10] Starting from aligned reads and a genome annotation, the DETR'PROK workflow clusters overlapping reads, compares these clusters to the genome annotation, and classifies them into sRNAs, asRNAs, UTRs and operon spacers. DETR'PROK pipeline consists of more than 40 steps, requires a local installation of Galaxy, and detects ncRNAs based on distance between transcripts, their size and read coverage. DETR'PROK pipeline is reported to recover 56% of 63 known *E. coli* sRNAs.[9]

Finally, there are several web applications for the analysis of eukaryotic small RNA-seq data that include the detection of novel transcripts. Some of these web applications are Oasis,[11] MAGI[12] and CPSS.[13]

3. Our Approach: sRNA-Detect

We designed sRNA-Detect under the assumptions that small transcripts (< 250 nt) exhibit relatively uniform read coverage as their whole sequence may fit into a single read, and that RNA-seq data may contain debris from longer transcripts enclosing the small transcript. These assumptions might also be valid for eukaryotic microRNAs, as microRNAs are similar in size to sRNAs and should exhibit uniform coverage; however, unlike microRNAs, sRNAs are usually neither processed nor cleaved to a shorter form. In sum, sRNA-Detect searches for features that have a given minimum depth coverage, are within a given length range, and exhibit low depth coverage variation through their whole sequence. The input to sRNA-Detect is sequencing reads aligned to a reference genome in the sequence alignment/map (SAM) format and the output is a list of detected transcripts in the gene transfer format (GTF).

Fig. 1 depicts sRNA-Detect's algorithm. Basically, sRNA-Detect constructs a coverage vector from the aligned reads for each strand and reference sequence (i.e., chromosomes and

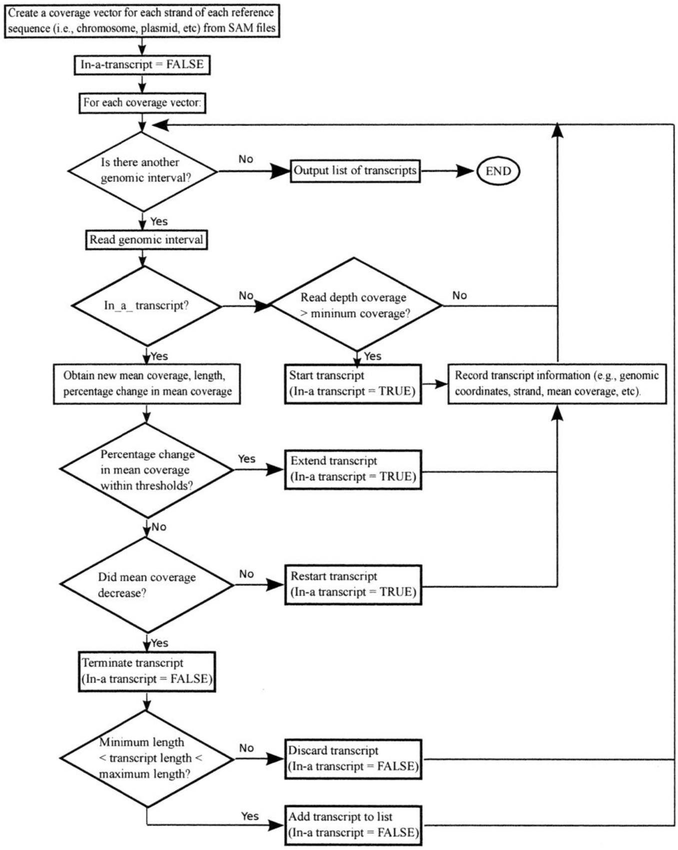

Fig. 1. Schematic flowchart of sRNA-Detect approach. The input to sRNA-Detect is a set of aligned reads in SAM format. sRNA-Detect's results are output as GTF files.

plasmids in the reference genome) using the HTSeq python library.[14] A coverage vector is a vector of genomic intervals where each interval represents a consecutive genomic stretch with constant coverage. Then sRNA-Detect goes through each coverage vector looking for genomic intervals with a given minimum number of reads aligned to them to start potential small transcripts. Small transcripts are extended into the following genomic intervals as long as their mean depth coverage does not decrease below a given percentage threshold. If the mean depth coverage of a potential transcript decreases, the transcript is terminated and added to the list of detected transcripts if its length is within the specified range. If the mean depth coverage of the potential transcript increases above a specified percentage threshold, then the transcript's start coordinate is reset to the start of the current genomic interval. Novel transcripts in the list of detected transcripts can be identified using the corresponding genome annotation and BEDtools.[15] As fragments of other longer RNA molecules may be detected, identified sRNAs should be further examined to distinguish authentic sRNAs based on their similarity to RNAs in public databases such as Rfam[16] and their predicted secondary structure. sRNA-Detect is available under the GNU GPL license at `www.cs.mun.ca/~lourdes`.

4. Comparative Assessment Methodology

We carried out a comparative assessment of the performance of standalone computer systems for detecting small transcripts from prokaryotic RNA-seq data. This comparative assessment included Rockhopper, TruHMM, the TLA and Transcripts algorithms implemented in RNA-eXpress and our own approach sRNA-Detect (see sections 2 and 3).

Rockhopper (version 2.02) was executed with the replicons provided for the corresponding bacteria and default values. To execute TruHMM, we followed the instructions provided at `http://bioinfolab.uncc.edu/TruHmm_package`. During TruHMM training, only genomic coordinates of protein-coding genes were provided as input to TruHMM (i.e., tRNAs and rRNAs were not included in the list of genes). To reconstruct operons and sRNAs, TruHMM's window size was left to its default value. TruHMM's predictions per sample were merged using the mergeBed tool available in BEDtools (version 2.16.2). Detection of small transcripts using RNA-express (version 1.4.4) was performed twice: once with the TLA algorithm and once with the Transcript algorithm. For the TLA algorithm the height (read coverage) was set to 10 for the *Rhodobacter capsulatus* case study and to 15 for the other two case studies, and the width (transcript length) was set to 20. For the Transcripts algorithm the height was set to 10, width to 20, penalty to -2 and tolerance to 500. sRNA-Detect was executed with default parameters; namely, minimum transcript length to 20 nucleotides, maximum transcript length to 210 nucleotides, allowed percentage change in mean coverage from -10% to 30%. Minimum coverage was set to 10 reads for the *R. capsulatus* case study and to 15 for the other two case studies. Algorithms' parameters were not optimized for the data sets used in the assessment.

To quantify the correctness of an inferred small transcript, we calculated two different measurements: minimum percentage sequence overlap and minimum percentage reciprocal sequence overlap. Minimum percentage sequence overlap indicates that at least that percentage of the sequence of the predicted transcript lies within the boundaries of an actual transcript. Minimum percentage reciprocal sequence overlap indicates that there is at least that percentage of agreement between the sequences of the predicted transcript and an actual transcript. Fig. 2 illustrates both correctness measurements.

To evaluate the performance of the various approaches, we calculated their recall, specificity, and accuracy at several levels of the two transcript correctness measurements. Recall indicates the proportion of actual expressed small transcripts that is detected by a given approach (i.e., true positives (TP) divided by the total number of positive instances (P)). Specificity is the proportion of small transcripts absent from the RNA-seq data that is undetected by a given approach (i.e., true negatives (TN) divided by the total number of negative

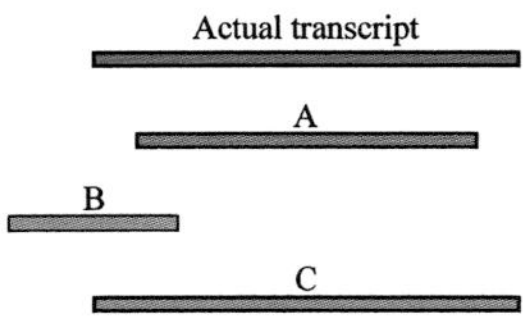

Detected transcript	Min % sequence overlap	Min % reciprocal sequence overlap
A	100%	80%
B	50%	20%
C	100%	100%

Fig. 2. Measurements to quantify the correctness of the detected transcripts. Minimum percentage sequence overlap quantifies what percentage of the sequence of the detected transcript lies within the boundaries of the actual transcript. Minimum percentage of reciprocal sequence overlap quantifies the congruence between the detected and the actual transcript. For example, detected transcripts A and C have both a minimum 100% sequence overlap but only detected transcript C has a minimum 100% reciprocal sequence overlap. Half of the sequence of detected transcript B is within the boundaries of the actual transcript having therefore a minimum 50% sequence overlap, but it only covers 20% of the actual transcript sequence having thus a minimum 20% reciprocal sequence overlap.

instances (N)). Accuracy is the number of correct results (TP + TN) divided by the total number of actual transcripts considered (P + N); i.e., proportion of results that are correct.

4.1. *Datasets*

We assessed the performance of the various approaches using sequence reads from RNA-seq experiments conducted in three different bacteria: *Rhodobacter capsulatus*, *Erwinia amylovora* and *Deinococcus radiodurans*, and generated with two sequencing platforms: Ion Torrent and Illumina.

4.1.1. Rhodobacter capsulatus *RNA-seq data*

RNA was isolated from *R. capsulatus* strain SB1003 during stationary phase and size-selected for RNAs shorter than 200 nt long to enrich for small transcripts. The isolated small RNAs were used for RNA library preparation for sequencing using an Ion Torrent Personal Genome Machine (PGM) system. In total, data from four RNA-seq experiments containing more than 6 million reads were used. After quality trimming the reads, reads were aligned to the *R. capsulatus* genome (assembly accession GCA_000021865.1) using the Torrent mapper tmap (version 3.0.1). Tmap was executed with the parameters: -B 18 -a 2 -v stage1 map1 map2 map3. These sequencing data were also used in another study for the genome-wide identification of candidate sRNAs in *R. capsulatus* (Peña-Castillo, Grüll, *et al.*, submitted for publication).

To create a test data set for our comparative assessment, we obtained the genome annotation for *R. capsulatus* (release 21.75) from EnsemblBacteria.[17] *R. capsulatus* gene models were generated by Ensembl using the Ensembl Bacteria pipeline.[18] Then we selected all annotated transcripts shorter than 200 nucleotides in length and we counted the number of reads aligned to these 176 small transcripts using the script htseq-count available in HTSeq (version 0.5.4p5). The htseq-count script was executed using mode intersection-nonempty, feature type gene, and all other parameters set to default values. We then classified these small transcripts as "expressed" or "not-expressed" based on the number of reads aligned to them across the four samples. Small transcripts (including tRNAs, rRNAs and mRNAs) with at least 10 reads aligned to them were classified as "expressed", while transcripts with fewer than 5 reads aligned to them were classified as "not-expressed". In our evaluation, the 90 expressed small transcripts were counted as positive instances, while the 74 not-expressed transcripts were counted as negative instances.

4.1.2. Erwinia amylovora *RNA-seq data*

We obtained *E. amylovora* RNA-seq data containing more than 53 million reads from NCBI-GEO[19] (accession numbers GSM1300251, GSM1300250, GSM1300248, and GSM1300247). Details of the *E. amylovora* RNA-seq experiments are described elsewhere.[20] In sum, sRNA libraries from *E. amylovora* Ea1189 were constructed from total RNAs using the Illumina TruSeq small RNA sample preparation kit and sequenced using an Illumina HiSeq 2000 system.[20] We aligned reads to the *E. amylovora* ATCC 49946 genome (assembly accession GCA_000027205.1) using Bowtie2[21] (version 2.1.0) with the preset parameters of the –sensitive option. Seventeen *E. amylovora* sRNAs have had their expression and size confirmed by Northern blot and one more sRNA that was not detected by Northern blot has had its transcription start site mapped by 5' RACE assay.[20,22] These 18 sRNAs are listed in Table 1 of Ref. 20. We considered these 18 sRNAs as true sRNAs and used them to estimate the recall of the computational approaches.

4.1.3. Deinococcus radiodurans *RNA-seq data*

We obtained *D. radiodurans* sequencing data containing more than 36 million reads from NCBI-GEO (accession number GSE64952). Details of the *D. radiodurans* RNA-seq experiments have been published previously.[23] In sum, cDNA libraries were prepared from total RNAs that were extracted from irradiated or non-irradiated *D. radiodurans* R1 cells and used to construct a cDNA library using a NEBNext Small RNA Library Prep Set for Illumina. The library was then sequenced with an Illumina HiSeq 2000 system.[23] We aligned reads to the *D. radiodurans* genome (assembly accession GCA_000008565.1) using Bowtie2 (version 2.1.0) with the preset parameters of the –sensitive-local option and allowing one mismatch in a seed alignment (-N 1). Tsai *et al.*[23] confirmed by Northern blot and/or RT-PCR the expression of 33 sRNAs detected from the RNA-seq data. We considered these 33 sRNAs as true sRNAs and used them to estimate the recall of the computational approaches.

5. Results and Discussion

In the following sections we present and discuss the performance of the five computational approaches to identify bacterial small transcripts in three case studies.

5.1. *Case study 1: Detection of annotated small transcripts in* R. capsulatus

Using the aligned *R. capsulatus* reads as input, we executed each of the five computational approaches and identified the transcripts in the test data set predicted as expressed by each approach at ten different levels of overlap (from 10% to 100%). To do this we used the intersectBed tool available in BEDtools. Transcripts reported by Rockhopper to have at least an expression measurement of 10 were considered predicted as expressed. Those transcripts detected by TruHMM in at least three samples were considered predicted as expressed. RNA-eXpress and sRNA-Detect both had a parameter specifying a minimum of 10 read coverage across all samples to report a transcript as expressed, thus no extra filtering was done to their output.

Fig. 3 shows the recall, specificity and accuracy of the five systems as a function of the correctness of the predicted transcript. Transcript correctness was measured as either minimum percentage sequence overlap or minimum percentage reciprocal sequence overlap (see Fig. 2). The recall and accuracy of all systems (except Rockhopper) decreased as the required correctness of the predicted transcript increased. As Rockhopper uses the genomic coordinates of the annotated transcripts provided in the corresponding replicons to guide its search for expressed transcripts, its performance measurements remained nearly constant across the various overlapping levels. Among the systems determining the transcript boundaries directly from the sequencing data, sRNA-Detect had the least pronounced drop in performance as the required transcript correctness increased.

At a minimum of 80% sequence overlap Rockhopper and sRNA-Detect are able to retrieve at least 63 (or 70%) of the 90 expressed transcripts, while all other approaches detected less than 23 (or 25%) of the expressed transcripts. In terms of specificity, most systems (except Rockhopper) were able to discriminate all negative instances as "not-expressed" at a minimum of 80% sequence overlap, while Rockhopper incorrectly identified as expressed 22 (or 29.7%) of the 74 "not-expressed" annotated transcripts bringing its specificity down to 70%. In terms of accuracy, sRNA-Detect outperforms all other systems up to a minimum 80% sequence overlap or a minimum 50% reciprocal sequence overlap. Above those transcript correctness levels, Rockhopper is the most accurate system for the identification of expressed known small transcripts.

We looked at the length of the transcripts detected as expressed by each computational approach with a minimum 10% sequence overlap. Fig. 4 shows the length distribution of the predicted transcripts per computational approach. As can be seen from this figure, Rockhopper, sRNA-Detect and the RNA-eXpress TLA algorithm predicted transcripts that were mostly in the correct length range, while TruHMM and the RNA-eXpress Transcript algorithm predicted transcripts that were well above the average actual length of the expressed transcripts. This is likely due to the fact that both of these approaches try to join fragments that may belong to the same transcript but are disconnected in the RNA-seq data because of gaps in read coverage. Based on this finding and on the low recall rate achieved by TruHMM and the RNA-eXpress Transcript algorithm, we concluded that these two approaches were not suitable for the detection of small transcripts from prokaryotic sequencing data and excluded them from subsequent evaluations.

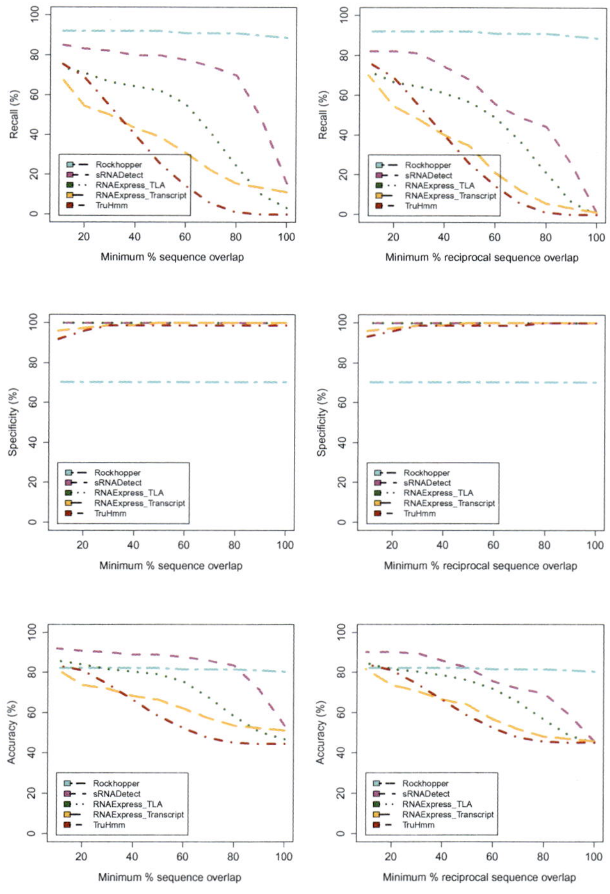

Fig. 3. Performance results for the identification of annotated small transcripts from *R. capsulatus* RNA-seq data. Left side plots show performance measurements as a function of the minimum percentage sequence overlap. Right side plots show performance measurements as a function of the minimum percentage reciprocal sequence overlap (see Fig. 2). Note that Rockhopper uses the genomic coordinates of the annotated small transcripts as input, while all other approaches estimate transcript boundaries directly from the RNA-seq data.

5.2. *Case study 2: Detection of experimentally confirmed novel sRNAs in* E. amylovora

In this case study we assessed how many experimentally confirmed sRNAs Rockhopper, sRNA-Detect and the RNA-eXpress TLA algorithm were able to retrieve from *E. amylovora* RNA-seq

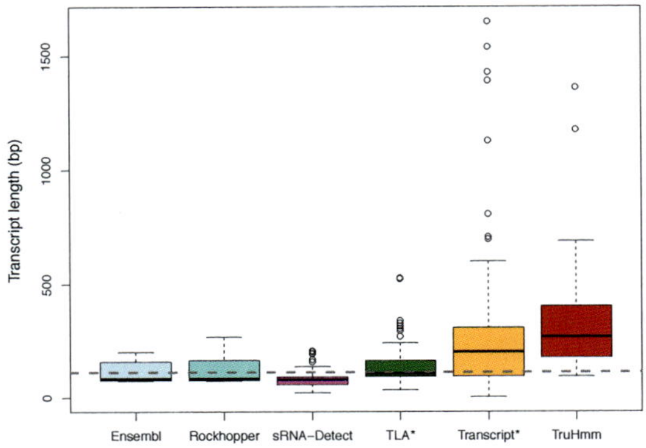

Fig. 4. Length (nt) distribution of *R. capsulatus* expressed and annotated small transcripts. The box labelled as Ensembl shows the length of the small transcripts as provided in *R. capsulatus* genome annotation (release 21.75). The dashed horizontal line indicates the average length of the small transcripts based on the genome annotation used. * Algorithms available in RNA-eXpress.

Table 1. Number of detected transcripts in *E. amylovora* RNA-seq data per method

Method	Total number of detected transcripts	Total number of detected novel transcripts
RNA-eXpress TLA	16,458	1,680
Rockhopper	2,646	167
sRNA-Detect	42,364	4,086

data. As the genomic coordinates of the confirmed sRNAs are not included in *E. amylovora* genome annotation, all systems (including Rockhopper) had to determine the sRNA boundaries directly from the RNA-seq data and thus their recall rate deteriorated as the required level of transcript correctness increased. Fig. 5 shows the systems' recall rate as a function of transcript correctness. The total number of detected transcripts and the number of detected novel transcripts per approach are provided in Table 1. Novel transcripts are those transcripts that do not overlap with known annotated features such as mRNAs and rRNAs. As the RNA-eXpress TLA algorithm and sRNA-Detect end transcripts when a gap in coverage is encountered, both of them might detect multiple fragments corresponding to a single longer transcript, and hence report a larger number of detected transcripts.

sRNA-Detect retrieved all 18 confirmed sRNAs up to a minimum percentage sequence overlap of 90%, while Rockhopper retrieved 6 (or 33.3%) of the 18 confirmed sRNAs at a minimum 90% sequence overlap and the RNA-eXpress TLA algorithm recovered one confirmed sRNA at the same minimum percentage sequence overlap. When transcript correctness is measured as minimum percentage reciprocal overlap, sRNA-Detect detected more confirmed sRNAs than those detected by the two other approaches up to a minimum reciprocal sequence

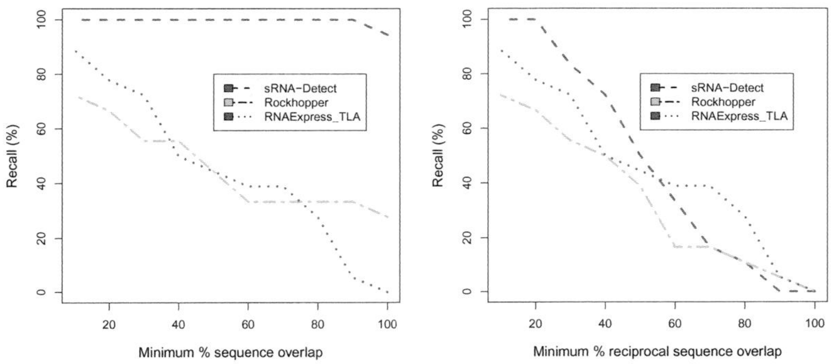

Fig. 5. Recall achieved by each computational approach during the identification of 18 experimentally confirmed *E. amylovora* sRNAs at various levels of transcript correctness.

overlap of 50%, above this point the RNA-eXpress TLA algorithm detected more confirmed sRNAs up to a minimum reciprocal sequence overlap of 90%. No computational approach was able to detect a single sRNA at a minimum 100% reciprocal overlap. These results indicate that the sRNAs predicted by sRNA-Detect are within the boundaries of the actual sRNA but fail to cover the whole sRNA actual sequence. This test also suggests that Rockhopper recall rates decrease when *a priori* knowledge of transcript boundaries is unavailable.

5.3. *Case study 3: Detection of experimentally confirmed novel sRNAs in* D. radiodurans

Here we assessed the systems' recall based on the number of experimentally confirmed sRNAs that the computational approaches were able to retrieve from *D. radiodurans* RNA-seq data. Fig. 6 shows the systems' recall rate as a function of transcript correctness. At a minimum sequence overlap of 90%, sRNA-Detect recovered 29 (or 87.9%) of the 33 confirmed sRNAs, while the RNA-eXpress TLA algorithm retrieved 3 (or 9.1%) of the confirmed sRNAs and Rockhopper recovered one of them. Following the same trend as that seen in the *E. amylovora* case study, when transcript correctness is measured as minimum percentage reciprocal overlap, the RNA-eXpress TLA algorithm recovered more confirmed sRNAs than those recovered by sRNA-Detect and Rockhopper at a minimum percentage reciprocal overlap of 40% and above. However, when transcript correctness is measured as minimum percentage overlap, sRNA-Detect has higher recall rates than those of the other two approaches. The total number of detected transcripts and the number of detected novel transcripts per approach are provided in Table 2.

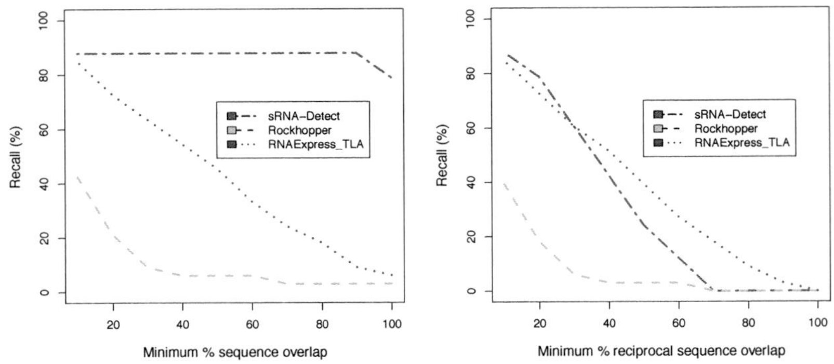

Fig. 6. Recall achieved by each computational approach during the identification of 33 experimentally confirmed *D. radiodurans* sRNAs at various levels of transcript correctness.

Table 2. Number of detected transcripts in *D. radiodurans* RNA-seq data per method

Method	Total number of detected transcripts	Total number of novel detected transcripts
RNA-eXpress TLA	17,687	1,168
Rockhopper	2,302	47
sRNA-Detect	6,740	828

6. Conclusions

Our sRNA-Detect approach demonstrated higher recall rates than those of other standalone systems. Small transcripts predicted by sRNA-Detect tend to lie within actual transcript boundaries and be smaller in length than the actual transcripts as indicated by the drop in recall rates when transcript correctness is measured as minimum percentage reciprocal sequence overlap. The results from our first case study indicate that most approaches show high specificity suggesting that false positives might not be a critical concern in the systems' performance; however, false positives may be detected if fragments from larger transcripts are present in the sequencing data. Thus, identified sRNAs should be further examined to distinguish authentic sRNAs. Criteria to consider in determining authentic sRNAs include secondary structure, phylogenetic conservation, and genomic context. If sRNAs are partially degraded, sRNA-Detect and the RNA-eXpress TLA algorithm may split transcripts into fragments with uniform read coverage. Based on our results, there is still room for improvement in the computational detection of bacterial small transcripts from RNA-seq data, especially in terms of achieving full recovery of transcript sequence.

Acknowledgements

LPC's and ASL's research programs are supported by NSERC Discovery grants. MG is supported by a fellowship of Memorial University's School of Graduate Studies.

References

1. O. A. Soutourina, M. Monot, P. Boudry, L. Saujet, C. Pichon, O. Sismeiro, E. Semenova, K. Severinov, C. Le Bouguenec, J.-Y. Coppée, B. Dupuy and I. Martin-Verstraete, *PLoS Genet* **9**, p. e1003493 (May 2013).
2. I. Wilms, A. Overlöper, M. Nowrousian, C. M. Sharma and F. Narberhaus, *RNA Biol* **9**, 446 (Apr 2012).
3. A. Mentz, A. Neshat, K. Pfeifer-Sancar, A. Pühler, C. Rückert and J. Kalinowski, *BMC Genomics* **14**, p. 714 (2013).
4. M. J. Moody, R. A. Young, S. E. Jones and M. A. Elliot, *BMC Genomics* **14**, p. 558 (2013).
5. R. McClure, D. Balasubramanian, Y. Sun, M. Bobrovskyy, P. Sumby, C. A. Genco, C. K. Vanderpool and B. Tjaden, *Nucleic Acids Res* **41**, p. e140 (Aug 2013).
6. S. Li, X. Dong and Z. Su, *BMC Genomics* **14**, p. 520 (2013).
7. S. C. Forster, A. M. Finkel, J. A. Gould and P. J. Hertzog, *Bioinformatics* **29**, 810 (Mar 2013).
8. R. McClure, B. Tjaden and C. Genco, *Front Microbiol* **5**, p. 456 (2014).
9. C. Toffano-Nioche, Y. Luo, C. Kuchly, C. Wallon, D. Steinbach, M. Zytnicki, A. Jacq and D. Gautheret, *Methods* **63**, 60 (Sep 2013).
10. J. Goecks, A. Nekrutenko, J. Taylor and Galaxy Team, *Genome Biol* **11**, p. R86 (2010).
11. V. Capece, J. C. Garcia Vizcaino, R. Vidal, R.-U. Rahman, T. Pena Centeno, O. Shomroni, I. Suberviola, A. Fischer and S. Bonn, *Bioinformatics* **31**, 2205 (Jul 2015).
12. J. Kim, E. Levy, A. Ferbrache, P. Stepanowsky, C. Farcas, S. Wang, S. Brunner, T. Bath, Y. Wu and L. Ohno-Machado, *Bioinformatics* **30**, 2826 (Oct 2014).
13. Y. Zhang, B. Xu, Y. Yang, R. Ban, H. Zhang, X. Jiang, H. J. Cooke, Y. Xue and Q. Shi, *Bioinformatics* **28**, 1925 (Jul 2012).
14. S. Anders, P. T. Pyl and W. Huber, *Bioinformatics* **31**, 166 (Jan 2015).
15. A. R. Quinlan and I. M. Hall, *Bioinformatics* **26**, 841 (Mar 2010).
16. S. W. Burge, J. Daub, R. Eberhardt, J. Tate, L. Barquist, E. P. Nawrocki, S. R. Eddy, P. P. Gardner and A. Bateman, *Nucleic Acids Res* **41**, D226 (Jan 2013).
17. P. J. Kersey, J. E. Allen, M. Christensen, P. Davis, L. J. Falin, C. Grabmueller, D. S. T. Hughes, J. Humphrey, A. Kerhornou, J. Khobova, N. Langridge, M. D. McDowall, U. Maheswari, G. Maslen, M. Nuhn, C. K. Ong, M. Paulini, H. Pedro, I. Toneva, M. A. Tuli, B. Walts, G. Williams, D. Wilson, K. Youens-Clark, M. K. Monaco, J. Stein, X. Wei, D. Ware, D. M. Bolser, K. L. Howe, E. Kulesha, D. Lawson and D. M. Staines, *Nucleic Acids Res* **42**, D546 (Jan 2014).
18. Ensembl Bacteria Pipeline, `http://ensemblgenomes.org/info/data/bacteria_pipeline`.
19. T. Barrett, S. E. Wilhite, P. Ledoux, C. Evangelista, I. F. Kim, M. Tomashevsky, K. A. Marshall, K. H. Phillippy, P. M. Sherman, M. Holko, A. Yefanov, H. Lee, N. Zhang, C. L. Robertson, N. Serova, S. Davis and A. Soboleva, *Nucleic Acids Res* **41**, D991 (Jan 2013).
20. Q. Zeng and G. W. Sundin, *BMC Genomics* **15**, p. 414 (2014).
21. B. Langmead and S. L. Salzberg, *Nat Methods* **9**, 357 (Apr 2012).
22. Q. Zeng, R. R. McNally and G. W. Sundin, *J Bacteriol* **195**, 1706 (Apr 2013).
23. C.-H. Tsai, R. Liao, B. Chou and L. M. Contreras, *Appl Environ Microbiol* **81**, 1754 (Mar 2015).

SOCIAL MEDIA MINING FOR PUBLIC HEALTH MONITORING AND SURVEILLANCE

MICHAEL J. PAUL,[1*] ABEED SARKER,[2] JOHN S. BROWNSTEIN,[3] AZADEH NIKFARJAM,[2] MATTHEW SCOTCH,[2] KAREN L. SMITH,[4] GRACIELA GONZALEZ[2]

[1] *Department of Information Science, University of Colorado, Boulder, CO 80304, USA*
[2] *Department of Biomedical Informatics, Arizona State University, Tempe, AZ 85287, USA*
[3] *Department of Pediatrics, Harvard Medical School, Boston MA 02115, USA*
[4] *School of Pharmacy, Regis University, Denver, CO 80221, USA*
**E-mail: michael.j.paul@colorado.edu*

This paper describes topics pertaining to the session, "Social Media Mining for Public Health Monitoring and Surveillance," at the Pacific Symposium on Biocomputing (PSB) 2016. In addition to summarizing the content of the session, this paper also surveys recent research on using social media data to study public health. The survey is organized into sections describing recent progress in public health problems, computational methods, and social implications.

Keywords: Social media; data mining; natural language processing; public health.

1. Background

Social media platforms have seen unprecedented worldwide growth. For example, as of June 30, 2015, Twitter has over 300 million active monthly users, 77% of whom are outside of the US.[1] Social networks form a platform for people to share and discuss their views and opinions, and many share their health-related information both in general-purpose social media (such as Twitter, Facebook or Instagram) and in health-related social networks (communities focusing specifically on health issues, such as DailyStrength or MedHelp). Advances in automated data processing, machine learning and natural language processing (NLP) present the possibility of utilizing these massive data sources for public health monitoring and surveillance, as long as researchers are able to address the methodological challenges unique to this media.

Numerous studies have been published recently in this realm, including studies on pharmacovigilance,[2] identifying smoking cessation patterns,[3] identifying user social circles with common experiences (like drug abuse),[4] monitoring malpractice,[5] and tracking infectious disease spread.[6–8] A systematic review[9] conducted in 2014 found numerous attempts to use this user-generated data, but none yet integrated in national surveillance programs, noting the promise and challenges of the field quite succinctly:

"*More direct access to such [social media] data could enable surveillance epidemiologists to detect potential public health threats such as rare, new diseases or early-level warnings for epidemics. But how useful are data from social media and the Internet, and what is the potential to enhance surveillance? The challenges of using these emerging surveillance systems for infectious disease epidemiology, including the specific resources needed, technical requirements, and acceptability to public health practitioners and policymakers, have wide-reaching implications for public health surveillance in the 21st century.*"[9]

The use of social media for health monitoring and surveillance indeed has many drawbacks and difficulties, particularly if done automatically. For example, traditional NLP methods that are applied to longer texts have proven to be inadequate when applied to short texts, such as those found in Twitter.[2] Something seemingly simple, such as searching and collecting relevant postings, has also proven to be quite challenging, given the amount of data and the diverse styles and wording used by people to refer to the topic of interest in colloquial terms (semantic heterogeneity) inherent to this type of media.

The goal of this session was to attract researchers that have explored automatic methods for the collection, extraction, representation, analysis, and validation of social media data for public health surveillance and monitoring, including epidemiological and behavioral studies. It serves as a unique forum to discuss novel approaches to text and data mining methods that respond to the specific requirements of social media and that can prove invaluable for public health surveillance. Research topics presented at this session include:

- Early detection of disease outbreaks[10]
- Medication safety, including drug interactions[11] and dietary supplement safety[12]
- Health behaviors, including diet success[13] and smoking cessation[14]
- Individual well-being,[15] which affects mental and physical health

This paper first summarizes the current state of, and recent advances in, social media mining for health monitoring, focusing on examples of promising research areas (Section 2), technical challenges (Section 3), and societal implications and considerations (Section 4). We then provide an overview of the research presented at this session in Section 5, with concluding remarks in Section 6.

2. Expanding the Frontiers of Public Health

We begin by summarizing recent research in some key areas of public health for which social media mining has been especially popular and fruitful, with an emphasis on how these focus areas are evolving to increase public health impact.

2.1. *Disease Surveillance: Beyond Influenza*

Disease surveillance is one of the longest-running use cases for social media mining. Some of the earliest work using web data for public health surveillance was to estimate influenza prevalence from search query volumes.[16] This idea was made famous with Google's widely-used Flu Trends service.[17,18] Google Flu Trends recently ended their service (as of August 2015), but Google will continue to share their data with academic research labs.[19] While search queries were the original data sources for web-based disease surveillance, social media has since become a popular data source for influenza monitoring, including weblogs[20] and microblogs, especially Twitter.[21–25]

Influenza has been by far the most commonly surveilled disease, in part due to its widespread prevalence—it affects millions of people each year (causing 3,000–50,000 yearly deaths in the US[26]), making it both an important disease to monitor and a disease that is widely discussed in social media. The original motivation for using web data to estimate influenza prevalence is that it can be estimated in real-time, in contrast to traditional government systems—the national surveillance coordinated by the Centers for Disease Control and Prevention in the US, for example, is one to two weeks out of date. However, it has been argued that social media-based influenza surveillance has limited utility in many scenarios, as many agencies and institutions already conduct timely influenza surveillance.[27]

More recently, web-based disease surveillance research has moved in new directions with potentially higher impact:

Other infectious diseases More recent social media research has considered disease surveillance for infectious diseases other than influenza. For example, a number of researchers have used search and tweet data to track dengue fever.[28–31] Others have used Twitter to monitor cholera,[7] *E. coli*,[32] and ebola.[33,34]

Forecasting While most early work on web-based disease surveillance focused on estimating the current week disease prevalence (referred to as "nowcasting"), more recent work has attempted to *forecast* disease prevalence, using web data to predict prevalence weeks into the future.[35–38] The ability to accurately predict future levels of disease prevalence will greatly help with planning and preparedness.

High-impact locations Much work with influenza has focused on surveillance at the national level in countries such as the US, but more recent work has focused on locations that would benefit more from real-time surveillance: countries with fewer existing surveillance resources[39] and fine-grained locations, such as hospitals[40,41] and mass gatherings.[42]

2.2. *Pharmacovigilance*

Pharmacovigilance, which primarily involves the monitoring of adverse reactions caused by medications, is another established use case of social media.[43] Users discuss their health-related experiences, including the use of prescription drugs, side effects and treatments on social media, which makes social networks unique and robust sources of information about health, drugs and treatments. Research has focused on the detection of user posts mentioning adverse reactions and the extraction of drug-adverse reaction association signals, utilizing data from specialized health communities and forums,[2,44–46] online reviews of drugs[47] and generic networks such as Twitter.[48–51]

Adverse reaction detection A number of studies focus on the automatic classification of user posts to determine if adverse reactions are mentioned. Common approaches involve utilizing annotated data sets to perform supervised classification to identify adverse reaction assertive posts and/or personal experiences of adverse reactions.[46,48,52,53] Supervised classification approaches require manually annotated data and recent advances in research have seen the creation of such data sets.[52–54] One important challenge that has been frequently discussed in supervised learning tasks is the data imbalance in social media data.[52,53]

Discovering drug-adverse reaction associations Some research has concentrated on extracting specific adverse reaction mentions (and their lexical variants) and identifying associations between specific drugs and adverse reactions. Most past approaches are lexicon-based[2,45,55] and recent approaches have applied supervised learning techniques for extraction.[51] Following the extraction of concepts, co-occurrence metrics have been applied for quantifying drug-adverse reaction associations.[44]

2.3. *Behavioral Medicine*

Another rapidly expanding area of social media surveillance is understanding behaviors that affect health, such as smoking and diet. It has been argued that behavioral medicine will play a prominent role in the digital surveillance revolution, because there is a large knowledge gap in many areas of behavioral medicine.[56] We summarize recent research in a few key areas.

Smoking and substance abuse One of the major uses of social media to study behavioral medicine has been to understand smoking and tobacco use.[57] Social media can be used to understand availability of and interest in various nicotine and tobacco products,[58] including electronic cigarettes, which are a rapidly evolving market for which social media has provided much faster intelligence than traditional sources.[59–61] Social networks have also been analyzed to understand smoking cessation and online social support for cessation.[3,62–64]

Other substance abuse issues have been studied as well, including trends in alcohol use[65,66] and problem drinking.[67,68] Some researchers have focused on using social media data for monitoring prescription drug abuse.[4,69–73] Specialized social networks have been used for analyzing the effects of drug reformulation[74] or the phases of drug abuse recovery.[75] Among generic social networks, Twitter is becoming increasingly popular for monitoring patterns of specific prescription medication abuse.[4,69]

Diet and fitness A number of researchers have analyzed food consumption patterns in Instagram[76,77] and Twitter,[78,79] including seasonal patterns in weight loss.[80] Researchers have also studied physical activities in Twitter,[81,82] including measuring outcomes of fitness goals.[83]

3. Technical Challenges of Social Media Mining

There are a number of challenges with automated text analysis, particularly when working with data from social media. We describe some of the key analytic tasks needed for public health mining, along with recent advances in these technologies.

3.1. *Processing Informal Text*

A key challenge with automatic data mining of social media is that standard NLP tools, which are traditionally trained on formal text (e.g., newswire), do not adapt well to the informal, non-standard language used online. Some researchers have created NLP tools, such as part-of-speech taggers and named entity recognizers, specifically for Twitter.[84,85] This can help researchers apply NLP to tweets, although this is not a general-purpose solution: tools tailored to Twitter may not work well on other social media platforms.

A particular challenge in the domain of health is that laypeople on social media may not use accurate medical terminology. One solution to this issue is to analyze text that mentions common symptoms, rather than references to specific illnesses.[86] There has also been research to correct and normalize medical terminology,[87] and there is a large body of research on general language normalization for social media text.[88]

3.2. *Sentiment Analysis*

A particular branch of NLP that shows promise for social media monitoring is *sentiment analysis*.[89] Sentiment analysis involves automatically ascribing positive, negative, or neutral sentiment to portions of text that express opinions. Sentiment analysis has been applied to social media in interesting ways to understand important public health issues. We provide a few examples here.

Sentiment analysis has been used to understand public attitudes toward vaccination by analyzing Twitter messages.[90] For example, the study in Ref. 91 found that negative sentiment toward vaccines spreads through social networks more than positive sentiment. Sentiment has also been analyzed in the context of drug abuse, in order to understand public interest in drugs. For example, researchers have measured shifts in public attitudes toward marijuana.[92,93] Sentiment analysis is particularly applicable to online reviews, which have been analyzed for public health in the domain of online doctor and healthcare provider reviews, to understand patient perceptions of care quality.[94,95] The studies in Refs. 96,97 found that sentiment inferred from reviews is significantly correlated with existing provider quality metrics.

However, sentiment analysis does not work as well for short text, such as tweets.[98] Sentiment classification is an active area of NLP research, and improvements in this technology will lead to improvements in understanding public opinion and awareness.

3.3. *Richer Language Understanding*

Much of the research on social media mining for health monitoring has used relatively simple methods of text analysis, such as dictionary associations. While simple approaches can work reasonably well, there is an upper limit to their performance, and future improvements will require NLP tools that can extract richer meaning from text.

Richer NLP can even improve seemingly simple tasks. For example, while early research showed that tweets with keywords such as "flu" are well-correlated with influenza prevalence,[21] more recent research has shown that flu is discussed in different ways on Twitter, for example, whether a user is describing a personal experience or simply sharing news of the flu season, and whether a user is personally sick or whether they are describing a family member or co-worker.[25,99] These distinctions can affect the performance of influenza surveillance, and such distinctions require NLP systems that incorporate richer n-gram and linguistic features.[25]

Richer NLP techniques have also been applied to concept extraction tasks, such as adverse drug reaction mention extraction. Early techniques primarily focused on lexicon-based approaches, where the natural language mentions of the elements of interest are encoded in lexicons and these are utilized to detect their mentions in text.[2,45,46,55] These techniques have led to the development of health-related lexical resources from social media sources (e.g., the Consumer Health Vocabulary[100]). The use of colloquial language, however, limits the performance of such approaches. With the creation of annotated data in recent years, supervised machine learning approaches are becoming increasingly popular, and they have also shown promising performance in quantitative evaluations.[51,101]

4. Societal Implications and Considerations

There are a number of social and societal implications of using social media for public health. We briefly discuss some key considerations here.

4.1. *Impact of Social Media Monitoring*

There is currently a gap between what is possible with social media monitoring—which many studies have demonstrated successfully, as described in Section 2—with what is being done in practice. As noted in the review in Ref. 9, existing social media systems have not widely integrated with national surveillance. However, the landscape is beginning to change. The US government has expressed interest in using social media for health surveillance, with both the CDC and the Department of Health and Human Services (HSS) soliciting submissions of systems that monitor social media for health issues.[102,103] Private companies, such as Sickweather, make social media monitoring available to the general public.

One hurdle in bringing social media monitoring to practice is gaining trust of practitioners and the public. For example, trust in web-based disease surveillance was eroded after researchers showed significant failings of the popular Google Flu Trends system.[104] More time will be needed to understand how such systems perform in practice. In the meantime, researchers must validate their social media models carefully to ensure progress is being made.[105]

4.2. *Ethics of Social Media Research*

There are a number of ethical considerations to keep in mind when using social media data for health research. One of the key concerns hinges on the extent to which social media data should be treated as public versus private data.[106] Even though social media data are publicly available, social media users may not intend or wish for their data to be used for research.[107] Users may not be aware that their social media data is publicly available,[108] and may have expectations of privacy even in public settings.[109] The distinction between public and private data becomes additionally complicated by the fact that machine learning algorithms can make inferences about private attributes, even if not explicitly stated in public data.[110]

Addressing these issues involves an ongoing conversation among Internet researchers,[111] and a number of scholars have written about using big data for research.[112] For more discussion of social media ethics in public health research, see Refs. 113–115.

5. Session Overview

This session hosted cutting-edge research in many of the public health areas described in Section 2. We briefly summarize the contributions below.

5.1. *Disease Surveillance*

Ofoghi *et al.*[10] presented research on disease-related emotion detection in tweets, suggesting that emotion tweets can be utilized to detect and monitor disease outbreaks. This work introduced NLP classifiers to categorize tweets into various *emotions* (e.g., "anger", "surprise"). The distributions of emotions were then analyzed in datasets of tweets pertaining to the ebola epidemic in 2014–2015. The authors found that the distributions differed among tweets at the time and place of an outbreak compared to outside tweets. These results suggest that emotion classification could help distinguish outbreak-related tweets from other disease discussion.

This research is an example of using richer NLP models to categorize disease-related tweets in useful ways, as discussed in Section 3.3: it is not enough to know that a tweet discusses ebola, but rather *how* ebola is being discussed.

5.2. *Pharmacovigilance*

The session hosted two papers on pharmacovigilance.

Correia *et al.*[11] investigated the utility of Instagram—an increasingly active social media platform—as a source of information for adverse drug reactions (ADRs). Instagram constitutes a potentially novel data source, in contrast to most social media-based ADR research which has focused on platforms such as Twitter and Facebook. This study analyzed, and introduced visualization tools for, Instagram messages mentioning various drugs used for depression. The results show that health issues are commonly discussed on Instagram, and there is potential for identifying ADRs, including interactions with other drugs and products.

Sullivan *et al.*[12] focused on adverse reactions to dietary supplements, which are products that are not currently well-monitored. This study analyzed Amazon.com reviews of nutritional supplements, and used a topic modeling system to categorize products based on their potential danger, as suggested in reviews. In the study, the proposed automated system agreed with human annotators 69.4% of the time, suggesting that automated methods can potentially be used to flag dangerous products.

5.3. *Behavioral Medicine*

The session included multiple studies that fall broadly in the category of behavioral medicine.

Aphinyanaphongs *et al.*[14] analyzed tweets for mentions of e-cigarette use. Because e-cigarettes constitute a relatively new product and public health phenomenon, real-time surveillance is needed to better understand usage patterns in the population. This work developed classifiers to identify tweets which mention e-cigarettes, as well as tweets which mention using e-cigarettes to support smoking cessation. The study developed a baseline classification performance of up to .90 AUC for detecting use and .94 AUC for detecting smoking cessation intent. The results show potential for measuring e-cigarette use from Twitter.

Weber and Achananuparp[13] analyzed public food diaries from the application, MyFitnessPal, and constructed models to predict whether users will or will not meet their daily caloric goals. By analyzing the predictive features, this study provides insights into what features are predictive of diet success or failure. Some results are expected, such as oil and butter contributing to diet failure and fruits contributing to diet success, while some insights are non-trivial, such as differences between types of meat. Future work points to insights from more complex features, such as the interactions of dietary groups.

Schwartz *et al.*[15] developed models to predict the state of well-being of individuals from their Facebook data, where *well-being* reflects positive mood as well as additional constructs such as meaning in life and engagement in activities. Using n-gram and topic features, the authors built classifiers to estimate various metrics of well-being at the level of individual Facebook messages, as well as the aggregate level of a user's entire stream. The goal of such research is to improve our understanding of the determinants and consequences of well-being, which is correlated with outcomes of both mental and physical health.

6. Concluding Remarks

The goal of this session was to create a single venue for cross-disciplinary researchers to present research on social media mining for public health monitoring and surveillance. The session provided a forum to share new research in a variety of important public health areas, including the detection of disease outbreaks and awareness; pharmacovigilance, including interactions with natural products and dietary supplements; and various issues related to behavioral medicine, including weight loss, e-cigarette use, and well-being. Through these projects, researchers also advanced the technology needed to understand social media text, for example by developing new NLP classifiers, new topic model variations, and new visualization systems. Given the ever-increasing amount of social media data around the world, interest in such systems will only increase over time.

References

1. Twitter: Company Facts `https://about.twitter.com/company`.
2. R. Leaman and L. Wojtulewicz, Towards internet-age pharmacovigilance: extracting adverse drug reactions from user posts to health-related social networks, in *Proceedings of the 2010 Workshop on Biomedical Natural Language Processing*, 2010.
3. L. L. Struik and N. B. Baskerville, *J. Med. Internet Res.* **16**, p. e170 (2014).
4. L. C. Hanson, B. Cannon, S. Burton and C. Giraud-Carrier, *J Med Internet Res* **15**, p. e189 (September 2013).
5. A. Nakhasi, R. J. Passarella, S. G. Bell, M. J. Paul, M. Dredze and P. J. Pronovost, Malpractice and Malcontent: Analyzing Medical Complaints in Twitter, in *AAAI Fall Symposium on Information Retrieval and Knowledge Discovery in Biomedical Text*, 2012.
6. M. J. Paul and M. Dredze, You are what you Tweet: Analyzing Twitter for public health, in *International Conference on Weblogs and Social Media*, 2011.
7. R. Chunara, J. R. Andrews and J. S. Brownstein, *American Journal of Tropical Medicine and Hygiene* **86**, 39 (2012).
8. D. A. Broniatowski, M. J. Paul and M. Dredze, *PLoS ONE* **8**, p. e83672 (2013).
9. E. Velasco, T. Agheneza, K. Denecke, G. Kirchner and T. Eckmanns, *Milbank Q* **92**, 7 (Mar 2014).
10. B. Ofoghi, M. Mann and K. Verspoor, Towards early discovery of salient health threats: A social media emotion classification technique, in *PSB*, 2016.
11. R. B. Correia, L. Li and L. M. Rocha, Monitoring potential drug interactions via network analysis of Instagram user timelines, in *PSB*, 2016.
12. R. Sullivan, A. Sarkar, K. O'Connor, A. Goodin, M. Karlsrud and G. Gonzalez, Monitoring dietary supplements: Challenges and promises of mining user comments for adverse events, in *PSB*, 2016.

13. I. Weber and P. Achananuparp, Insights from machine-learned diet success prediction, in *PSB*, 2016.
14. Y. Aphinyanaphongs, A. Lulejian, D. P. Brown, R. Bonneau and P. Krebs, Classification for automatic detection of e-cigarette use and use for smoking cessation from twitter: a feasability pilot, in *PSB*, 2016.
15. H. A. Schwartz, M. Sap, M. L. Kern, J. C. Eichstaedt, A. Kapelner, M. Agrawal, E. Blanco, L. Dziurzynski, G. Park and L. H. Ungar, Predicting individual well-being through the language of social media, in *PSB*, 2016.
16. G. Eysenbach, Infodemiology: tracking flu-related searches on the web for syndromic surveillance (2006).
17. J. Ginsberg, M. H. Mohebbi, R. S. Patel, L. Brammer, M. S. Smolinski and L. Brilliant, *Nature* **457**, 1012 (2009).
18. S. Cook, C. Conrad, A. L. Fowlkes and M. H. Mohebbi, *PLoS ONE* **6**, p. e23610 (2011).
19. Google Flu Trends, The next chapter for Flu Trends `http://googleresearch.blogspot.com/2015/08/the-next-chapter-for-flu-trends.html`, (2015), Accessed 2015-08-29.
20. C. D. Corley, D. J. Cook, A. R. Mikler and K. P. Singh, *International Journal of Environmental Research and Public Health* **7**, 596 (2010).
21. A. Culotta, Towards detecting influenza epidemics by analyzing Twitter messages, in *ACM Workshop on Soc.Med. Analytics*, 2010.
22. A. Signorini, A. M. Segre and P. M. Polgreen, *PLoS ONE* **6**, p. e19467 (2011).
23. E. Aramaki, S. Maskawa and M. Morita, Twitter catches the flu : Detecting influenza epidemics using Twitter the university of tokyo, in *Proceedings of the 2011 Conference on Emperical Methods in Natural Language Processing*, 2011.
24. V. Lampos and N. Cristianini, *ACM Trans. Intell. Syst. Technol.* **3**, 1 (2012).
25. A. Lamb, M. J. Paul and M. Dredze, *Proceedings of NAACL-HLT 2013* , 789 (2013).
26. Estimating Seasonal Influenza-Associated Deaths in the United States: CDC Study Confirms Variability of Flu `http://www.cdc.gov/flu/about/disease/us_flu-related_deaths.htm`.
27. D. R. Olson, K. J. Konty, M. Paladini, C. Viboud and L. Simonsen, *PLoS Computational Biology* **9** (2013).
28. E. H. Chan, V. Sahai, C. Conrad and J. S. Brownstein, *PLoS Neglected Tropical Diseases* **5** (2011).
29. B. M. Althouse, Y. Y. Ng and D. A. T. Cummings, *PLoS Neglected Tropical Diseases* **5** (2011).
30. R. T. Gluskin, M. A. Johansson, M. Santillana and J. S. Brownstein, *PLoS Negl Trop Dis* **8**, p. e2713 (February 2014).
31. J. Gomide, A. Veloso, W. Meira, Jr., V. Almeida, F. Benevenuto, F. Ferraz and M. Teixeira, Dengue surveillance based on a computational model of spatio-temporal locality of Twitter, in *Proceedings of the 3rd International Web Science Conference*, 2011.
32. E. Diaz-Aviles and A. Stewart, Tracking Twitter for epidemic intelligence: Case study: Ehec/hus outbreak in germany, 2011, in *Proceedings of the 4th Annual ACM Web Science Conference*, 2012.
33. M. Odlum and S. Yoon, *Am J Infect Control* **43**, 563 (Jun 2015).
34. M. Odlum, How Twitter can support early warning systems in ebola outbreak surveillance, in *Annual Meeting of the American Public Health Association*, 2015.
35. A. F. Dugas, M. Jalalpour, Y. Gel, S. Levin, F. Torcaso, T. Igusa and R. E. Rothman, *PLoS ONE* **8**, p. e56176 (2013).
36. J. Shaman, A. Karspeck, W. Yang, J. Tamerius and M. Lipsitch, *Nat Commun* **4**, p. 2837 (2013).
37. E. Nsoesie, M. Mararthe and J. Brownstein, *PLoS Currents* **5**, 1 (2013).
38. M. J. Paul, M. Dredze and D. Broniatowski, *PLOS Currents Outbreaks* (2014).

39. M. Paul, M. Dredze, D. Broniatowski and N. Generous, Worldwide influenza surveillance through Twitter, in *AAAI Workshop on the World Wide Web and Public Health Intelligence*, 2015.
40. D. A. Broniatowski, M. Dredze, J. M. Paul and A. Dugas, *JMIR Public Health Surveill* **1**, p. e5 (2015).
41. O. M. Araz, D. Bentley and R. L. Muelleman, *Am J Emerg Med* **32**, 1016 (Sep 2014).
42. E. Yom-Tov, D. Borsa, I. J. Cox and R. A. McKendry, *J. Med. Internet Res.* **16**, p. e154 (2014).
43. A. Sarker, R. Ginn, A. Nikfarjam, K. O'Connor, K. Smith, S. Jayaraman, T. Upadhaya and G. Gonzalez, *Journal of Biomedical Informatics* **54**, 202 (2015).
44. A. Nikfarjam and G. H. Gonzalez, Pattern mining for extraction of mentions of adverse drug reactions from user comments, in *AMIA Annual Symposium*, 2011.
45. A. Benton, L. Ungar, S. Hill, S. Hennessy, J. Mao, A. Chung, C. E. Leonard and J. H. Holmes, *J Biomed Inform* **44**, 989 (Dec 2011).
46. A. Yates, N. Coharian and O. Frieder, Extracting adverse drug reactions from forum posts and linking them to drugs, in *SIGIR Workshop on Health and Discovery*, 2013.
47. A. Yates and N. Goharian, ADRTrace: Detecting expected and unexpected adverse drug reactions from user reviews on social media sites, in *Proceedings of the 35th European Conference on Advances in Information Retrieval*, (Berlin, Heidelberg, 2013).
48. J. Bian, U. Topaloglu and F. Yu, Towards large-scale Twitter mining for drug-related adverse events, in *International Workshop on Smart Health and Wellbeing*, 2012.
49. K. O'Connor, P. Pimpalkhute, A. Nikfarjam, R. Ginn, K. L. Smith and G. Gonzalez, *AMIA Annu Symp Proc* **2014**, 924 (2014).
50. C. C. Freifeld, J. S. Brownstein, C. M. Menone, W. Bao, R. Filice, T. Kass-Hout and N. Dasgupta, *Drug Saf* **37**, 343 (May 2014).
51. A. Nikfarjam, A. Sarker, K. O'Connor, R. Ginn and G. Gonzalez, *J Am Med Inform Assoc* (March 2015).
52. R. Ginn, P. Pimpalkhute, A. Nikfarjam, A. Patki, K. O'Connor, A. Sarker and G. Gonzalez, Mining Twitter for adverse drug reaction mentions: a corpus and classification benchmark, in *Proceedings of the Fourth Workshop on Building and Evaluating Resources for Health and Biomedical Text Processing (BIOTXTM)*, 2014.
53. A. Sarker and G. Gonzalez, *Journal of Biomedical Informatics* **53**, 196 (2014).
54. I. Segura-Bedmar, R. revert and P. Martinez, Detecting drugs and adverse events from spanish health social media streams, in *Proceedings of the 5th international workshop on health text mining and information analysis (LOUHI)*, 2014.
55. S. Yeleswarapu, A. Rao, T. Joseph, V. G. Sapradeep and R. Srinivasan, *BMC Medical Informatics and Decision Making* **14** (2014).
56. J. W. Ayers, B. M. Althouse and M. Dredze, *JAMA* **311**, 1399 (2014).
57. K. W. Prier, M. S. Smith, C. Giraud-Carrier and C. L. Hanson, Identifying health-related topics on Twitter: An exploration of tobacco-related tweets as a test topic, in *Proceedings of the 4th International Conference on Social Computing, Behavioral-cultural Modeling and Prediction* SBP'11 (Springer-Verlag, 2011).
58. M. Myslin, S. H. Zhu, W. Chapman and M. Conway, *J. Med. Internet Res.* **15**, p. e174 (2013)
59. J. W. Ayers, K. M. Ribisl and J. S. Brownstein, *American Journal of Preventive Medicine* **40** 448 (2011).
60. J. Huang, R. Kornfield, G. Szczypka and S. L. Emery, *Tob Control* **23 Suppl 3**, 26 (July 2014)
61. H. Cole-Lewis, A. Varghese, A. Sanders, M. Schwarz, J. Pugatch and E. Augustson, *J. Med Internet Res.* **17**, p. e208 (2015).
62. N. K. Cobb, A. L. Graham, M. J. Byron, D. B. Abrams and Workshop Participants, *Journal of Medical Internet Research* **13** (2011).

63. M. Rocheleau, R. S. Sadasivam, K. Baquis, H. Stahl, R. L. Kinney, S. L. Pagoto and T. K. Houston, *J. Med. Internet Res.* **17**, p. e18 (2015).
64. J. J. Prochaska, C. Pechmann, R. Kim and J. M. Leonhardt, *Tob Control* **21**, 447 (Jul 2012).
65. J. H. West, P. C. Hall, K. Prier, C. L. Hanson, C. Giraud-Carrier, E. S. Neeley and M. D. Barnes, *Open Journal of Preventative Medicine* **2** (2012).
66. Y. Aphinyanaphongs, B. Ray, A. Statnikov and P. Krebs, Text classification for automatic detection of alcohol use-related tweets, in *International Workshop on Issues and Challenges in Social Computing*, 2014.
67. A. K. Fournier and S. W. Clarke, *Journal of Psychosocial Research on Cyberspace* **5** (2011).
68. M. A. Moreno, D. A. Christakis, K. G. Egan, L. N. Brockman and T. Becker, Associations between displayed alcohol references on facebook and problem drinking among college students (2012).
69. C. L. Hanson, S. H. Burton, C. Giraud-Carrier, J. H. West, M. D. Barnes and B. Hansen, *J. Med. Internet Res.* **15**, p. e62 (2013).
70. N. Genes and M. Chary, Twitter discussions of nonmedical prescription drug use correlate with federal survey data, in *Medicine 2.0 Conference*, 2014.
71. B. Chan, A. Lopez and U. Sarkar, *PLoS ONE* **10**, p. e0135072 (08 2015).
72. T. K. Mackey, B. A. Liang and S. A. Strathdee, *J. Med. Internet Res.* **15**, p. e143 (2013).
73. P. M. Coloma, B. Becker, M. C. J. M. Sturkenboom, E. M. van Mulligen and J. A. Kors, *Drug Safety* **38**, 921 (2015).
74. E. C. McNaughton, P. M. Coplan, R. A. Black, S. E. Weber, H. D. Chilcoat and B. S. F., *Journal of Medical Internet Research* **16** (May 2014).
75. D. MacLean, S. Gupta, A. Lembke, C. Manning and J. Heer, Forum77: An analysis of an online health forum dedicated to addiction recovery, in *Proceedings of the 18th ACM Conference on Computer Supported Cooperative Work & Social Computing*, CSCW '15 (ACM, New York, NY, USA, 2015).
76. S. Sharma and M. De Choudhury, Detecting and Characterizing Nutritional Information of Food and Ingestion Content in Instagram, in *WWW*, 2015.
77. Y. Mejova, H. Haddadi, A. Noulas and I. Weber, #foodporn: Obesity patterns in culinary interactions, in *Proceedings of the 5th International Conference on Digital Health 2015*, 2015.
78. S. Abbar, Y. Mejova and I. Weber, *CoRR* **abs/1412.4361** (2014).
79. D. Fried, M. Surdeanu, S. Kobourov, M. Hingle and D. Bell, Analyzing the language of food on social media, in *IEEE International Conference on Big Data*, 2014.
80. G. M. Turner-McGrievy and M. W. Beets, *Transl Behav Med* **5**, 160 (Jun 2015).
81. N. Zhang, S. Campo, K. F. Janz, P. Eckler, J. Yang, L. G. Snetselaar and A. Signorini, *Journal of medical Internet research* **15** (2013).
82. V. L. D. Reis and A. Culotta, Using matched samples to estimate the effects of exercise on mental health from Twitter, in *AAAI*, 2015.
83. E. Kiciman and M. Richardson, Towards decision support and goal achievement: Identifying action-outcome relationships from social media, in *KDD*, 2015.
84. K. Gimpel, N. Schneider, B. O'Connor, D. Das, D. Mills, J. Eisenstein, M. Heilman, D. Yogatama, J. Flanigan and N. A. Smith, Part-of-speech tagging for Twitter: Annotation, features, and experiments, in *Association for Computational Linguistics (ACL)*, 2011.
85. A. Ritter, S. Clark, Mausam and O. Etzioni, Named entity recognition in tweets: An experimental study, in *Proceedings of the Conference on Empirical Methods in Natural Language Processing (EMNLP)*, 2011.
86. F. Gesualdo, G. Stilo, E. Agricola, M. V. Gonfiantini, E. Pandolfi, P. Velardi and A. E. Tozzi, *PLoS ONE* **8**, p. e82489 (2013).
87. L. Nie, M. Akbari, T. Li and T.-S. Chua, A joint local-global approach for medical terminology

assignment, in *SIGIR 2014 Workshop on Medical Information Retrieval*, 2014.
88. J. Eisenstein, What to do about bad language on the internet, in *Proceedings of the North American Chapter of the Association for Computational Linguistics (NAACL)*, 2013.
89. B. Pang and L. Lee, *Foundations and Trends in Information Retrieval* **2**, 1 (2008).
90. M. Salathe and S. Khandelwal, *PLoS Comput Biol* **7**, p. e1002199 (2011).
91. M. Salathé, D. Q. Vu, S. Khandelwal and D. R. Hunter, *EPJ Data Science* **2** (2013).
92. P. A. Cavazos-Rehg, M. Krauss, S. L. Fisher, P. Salyer, R. A. Grucza and L. J. Bierut, *J Adolesc Health* **56**, 139 (Feb 2015).
93. L. Thompson, F. P. Rivara and J. M. Whitehill, *Cyberpsychology, Behavior, and Social Networking* **18**, 311 (2015).
94. A. López, A. Detz, N. Ratanawongsa and U. Sarkar, *J Gen Intern Med* **27**, 685 (Jun 2012).
95. S. Brody and N. Elhadad, Detecting salient aspects in online reviews of health providers, in *AMIA Annual Symposium*, 2010.
96. J. Segal, M. Sacopulos, V. Sheets, I. Thurston, K. Brooks and R. Puccia, *J. Med. Internet Res.* **14**, p. e50 (2012).
97. B. C. Wallace, M. J. Paul, U. Sarkar, T. A. Trikalinos and M. Dredze, *Journal of the American Medical Informatics Association (JAMIA)* (2014).
98. A. Agarwal, B. Xie, I. Vovsha, O. Rambow and R. Passonneau, Sentiment analysis of Twitter data, in *Proceedings of the Workshop on Languages in Social Media*, 2011.
99. R. Nagar, Q. Yuan, C. C. Freifeld, M. Santillana, A. Nojima, R. Chunara and J. S. Brownstein, *J. Med. Internet Res.* **16**, p. e236 (2014).
100. Consumer Health Vocabulary `http://consumerhealthvocab.org/`.
101. H. Sampathkumar, X. wen Chen and B. Luo, *BMC Medical Informatics and Decision Making* **14** (October 2014).
102. Now trending `https://nowtrending.hhs.gov/`, Accessed: 2015-08-23.
103. C. for Disease Control and Prevention, Predict the influenza season challenge `https://www.federalregister.gov/articles/2013/11/25/2013-28198/announcement-of-requirements-and-registration-for-the-predict-the-influenza-season-challenge`, (2013).
104. D. Lazer, R. Kennedy, G. King and A. Vespignani, *Science* **343**, 1203 (2014).
105. T. Bodnar and M. Salathé, Validating models for disease detection using Twitter, in *Proceedings of the 22Nd International Conference on World Wide Web Companion*, 2013.
106. R. McKee, *Health policy (Amsterdam, Netherlands)* **110**, 298 (2013).
107. J. M. Hudson and A. Bruckman, *Inf. Soc.* **20**, 127 (2004).
108. Y. Liu, K. P. Gummadi, B. Krishnamurthy and A. Mislove, Analyzing facebook privacy settings: User expectations vs. reality, in *Proceedings of the 2011 ACM SIGCOMM Conference on Internet Measurement Conference*, 2011.
109. J. C. H. Bromseth, Public places – public activities? methodological approaches and ethical dilemmas in research on computer-mediated communication, in *Researching ICTs in Context*, ed. A. Morrison (Inter/Media Report, 2002) pp. 33–61.
110. E. Horvitz and D. Mulligan, *Science* **349**, 253 (Jul 2015).
111. C. Fiesler, A. Young, T. Peyton, A. S. Bruckman, M. Gray, J. Hancock and W. Lutters, Ethics for studying online sociotechnical systems in a big data world, in *Proceedings of the 18th ACM Conference Companion on Computer Supported Cooperative Work & Social Computing*, 2015.
112. D. boyd and K. Crawford, *Information, Communication & Society* **15** (2012).
113. C. M. Rivers and B. L. Lewis, *F1000Research* **3** (2014).
114. M. Conway, *Journal of medical Internet research* **16** (2014).
115. E. Vayena, M. Salathé, L. C. Madoff and J. S. Brownstein, *PLoS Comput Biol* **11**, p. e1003904 (2015).

TEXT CLASSIFICATION FOR AUTOMATIC DETECTION OF E-CIGARETTE USE AND USE FOR SMOKING CESSATION FROM TWITTER: A FEASIBILITY PILOT

YIN APHINYANAPHONGS, ARMINE LULEJIAN

NYU Langone Medical Center
New York, NY 10016, USA
Email: yin.a@nyumc.org, Armine.Lulejian@nyumc.org

DUNCAN PENFOLD BROWN

New York University Social Media and Political Participation lab
New York, NY 10012, USA
Email: dpenfoldbrown@gmail.com

RICHARD BONNEAU

Simons Center for Data Analysis,
New York, NY, 10010, USA
Email: rb133@nyu.edu

PAUL KREBS

NYU Langone Medical Center
New York, NY 10016, USA
Email:paul.krebs@nyumc.org

Rapid increases in e-cigarette use and potential exposure to harmful byproducts have shifted public health focus to e-cigarettes as a possible drug of abuse. Effective surveillance of use and prevalence would allow appropriate regulatory responses. An ideal surveillance system would collect usage data in real time, focus on populations of interest, include populations unable to take the survey, allow a breadth of questions to answer, and enable geo-location analysis. Social media streams may provide this ideal system. To realize this use case, a foundational question is whether we can detect ecigarette use at all. This work reports two pilot tasks using text classification to identify automatically Tweets that indicate e-cigarette use and/or e-cigarette use for smoking cessation. We build and define both datasets and compare performance of 4 state of the art classifiers and a keyword search for each task. Our results demonstrate excellent classifier performance of up to 0.90 and 0.94 area under the curve in each category. These promising initial results form the foundation for further studies to realize the ideal surveillance solution.

1. Introduction

1.1. *E-cigarettes*

The use of e-cigarettes has been rapidly increasing since their introduction onto the market a few years ago. Sales of e-cigs and refillable vaporizers more than doubled to $1.7 billion in 2013.[1] Indeed, the trend has become so popular that 'vape' was voted word of the year for 2014 by the Oxford Dictionaries.[2] A limited, yet growing body of literature suggests that e-cigarettes and vaporizers can create potentially harmful byproducts including heavy metals[3] and formaldehyde,[4] and product failure can result in severe injury and burns. Very little is known, however, regarding the use, prevalence, and characteristics of e-cigarettes. Two surveys among youth have indicated rapid increases in use since 2011,[5] and recent results from the 2014 Monitoring the Future survey indicated that 17% of 12th graders have used an e-cigarette in the past 30 days, surpassing the number who used combustible cigarettes.[6] Even less information on adult use exists, with the only national data being one consumer-research web survey,[7] indicating that 8.5% of adults have tried e-cigarettes with a rate of 36% among combustible cigarette users. No large-scale surveys have yet assessed more in-depth opinions about e-cigarette use, such as reasons for use or beliefs about harm.

1.2. *Surveillance*

Survey results are necessary to understand usage trends, establish national and regional health goals and inform regulations and prevention campaigns. These surveys – while excellent in many ways – have several limitations. First, there is a time lag before new products of abuse are incorporated into the surveys.[8] For example, neither the BRFSS,[9] the National Health Interview Survey,[10] nor the National Survey on Drug Use and Health (NSDUH)[11] ask about e-cigarette use yet. Second, the time lag in collection and analysis may delay timely policy interventions. Third, the surveys are sized to capture general trends across demographics and may lack focus for specific populations. Fourth, surveys have limitations in detecting usage by minors as most are not allowed to take the surveys. Fifth, surveys may contain limited content for any specific question as every additional question competes against other questions for time and space in the survey. Sixth, surveys capture high level geo-located information of use. Continuing use of high-quality national surveys to inform prevention and treatment services is critical, yet new technologies may address some of these limitations.

An ideal surveillance solution could capture new drugs of abuse, collect data in real time, focus on populations of interest, include populations unable to take the survey, allow a breadth of questions to answer, and enable geo-location analysis. We believe that social media streams may provide one solution. Social media, in this case, specifically Twitter, may include up to date vernacular for drugs of abuse, is inherently real time in how Tweets are broadcast, includes many potential populations of interest and their demographic characteristics, has populations such as minors who may not qualify for surveys, contains Tweets that indicate other potentially risky behaviors, and includes geo-locations. To realize using social media for surveillance, a foundational question is whether we can detect drug use at all. This work addresses this foundational concern and reports two pilot tasks for e-cigarettes. In the first, we identify automatically e-cigarette Tweets that indicate *e-cigarette use*. In the second we identify automatically Tweets that indicate *e-cigarette use for smoking cessation.*

1.3. *Our Contribution*

This feasibility paper explores state of the art machine learning based text classification methodologies for identifying e-cigarette use tweets. This paper makes several key contributions:

1. Defines a novel classification task for identifying e-cigarette use.
2. Defines a novel classification task for identifying e-cigarette use for smoking cessation.
3. Defines a process for labeling tweets to identify e-cigarette use and use for smoking cessation.
4. Establishes baseline classification results for these tasks.
5. Distributes these labeled datasets for general use by the community.

2. Background

2.1. *Twitter As a Data Source*

Among social media platforms, Twitter offers unique potential to serve as a tool for tracking substance use. Twitter is a micro-blogging service (with posts limited to 140 characters) through which users can send messages to a set of followers. It has over 600 million users worldwide with 46% of users logging on daily. In a recent Pew Research survey conducted August-September 2013, 18% of US adults use Twitter. [12] A higher percentage of Blacks/African-Americans (29%) use Twitter compared with Whites (16%) and Hispanics (16%). Of Twitter subscribers, 31% are 18-29 and 19% are 30-49 years old.[12] Interestingly, there are relatively no differences in use by education level, gender, or income suggesting that use cuts across socioeconomic differences.

2.2. *Twitter and E-cigarettes*

A few studies have specifically addressed e-cigarettes via Twitter. Clark et al.[13] used 700,000 tweets collected from January 2012 to July 2014 to survey the general popularity and sentiment of consumer opinions regarding e-cigarettes.[14] In a follow up publication, they focused on approximately 20,000 geo-located tweets to characterize density and sentiment surrounding tobacco and e-cigarette tweets and link prevalence of word choices to tobacco and e-cigarette use at various localities.[14] In another publication, Huang et al.[15] labeled 73,672 tweets related to e-cigarettes to characterize how e-cigarettes are marketed, and Harris et al.[16] conducted a manual content analysis of tweets related to Chicago's regulation of e-cigarettes. While these studies produced a one-time picture of e-cigarette sentiment, neither the methodology of identifying e-cigarettes with a simple term search nor using manual coding are useful for ongoing surveillance purposes. A system that harnesses social media posts could serve as a low-cost method of examining usage trends and attitudes toward particular products.

In these previous studies, a common theme for analysis is the manual labeling of tweets. Manual labeling requires (1) time, (2) expertise, and (3) consistency. In addition, the samples must be small enough to allow feasible manual labeling which inherently is limited to the snapshot in time when the tweets are collected. Our aim in this paper is to use text classification machine learning techniques to address these limitations and convert these manual classifications to automated classifications. With tweet volumes of nearly 500 million per day, automation is the only realistic and feasible solution.

3. Methods

3.1. *Corpus Construction*

A challenge for building a labeled training corpus from Twitter is the low prevalence of Tweets in a target category. To enrich the e-cigarette use target category, we filtered Tweets by e-cigarette brand followers and hashtags. Specifically, we downloaded a 28.6 million tweet collection in January 2015. The tweet collection represents the tweets of 29,410 followers of the largest e-cigarette brands @v2cigs, @VaporFl, @HaloCigs, @bluecigs, @NJOYVape, @KRAVEeCig, and @LogicECig. To further increase the probability of encountering tweets about e-cigarette use and e-cigarette use for smoking cessation, we filtered the 28.6 million tweets with the Boolean OR of #vape #mod_ #vapeing #vaping #flavhub #eliquid #ejuice #pureclass or #ecigs. This corpus had 5,435 Tweeters covering a time span from Jan 2010 to Jan 2015 representing 228,145 Tweets. From these Tweets, we build a final corpus consisting of 13,146 randomly selected Tweets to label for our classifiers as outlined in section 3.2. The remaining 214,999 Tweets were not used for this pilot work and limitations with labeler time constrained the number of labeled Tweets.

3.2. *Corpus Labels*

3.2.1. *Task 1 – E-cigarette Use*

We defined e-cigarette use according to a similar protocol we developed for alcohol use.[17] Specifically, we considered tweets as positive if they indicate intent to use, the act of using, or sequel from use. Table 1 (on the following page) outlines our labeling protocol with examples.

3.2.2. *Task 2 – E-cigarette Use for Smoking Cessation*

We defined e-cigarette use for smoking cessation as tweets that indicate use for smoking cessation. These tweets were by definition a subset of the e-cigarette use tweets. In other words, a tweet for e-cigarette use for smoking cessation is also a tweet for e-cigarette use. Some example tweets indicating smoking cessation are shown in Table 2.

Table 1. Defining Tweets that indicate e-cigarette use (@mentions and urls removed)

Definitions of E-cigarette Use	Tweet	Label
Current behavior of using	#vaping with my new #vamo...carto tank filled with Cups 'O Peanut Butter from So good.	Positive
Owning or discussing paraphernalia and products	Loving the Nautilus Mini. Definitely my work/driving setup. No more stopping to drip when I'm dry! #Vapelif...	Positive
Entering contests to get products	I just entered to win 120mL of #ejuice from #vapelife #vape #vapers #ecig #vaping #ecigs # Check it out!	Positive
Use to quit smoking	See Table 2	Positive
Liking a brand or product	So far I'm loving my new #iStick #Eleaf #vape #vapelife #malibu	Positive
Asking others to gather to use	Vaper meet this saturday in London, team smokium will see you all there! Message me for details if your interested in coming. #ukvape #vape	Negative
Advertising	Kanger pro tank 3 (duel coil) is now in stock at our Jarrow Shop. #kanger #vapourvapourjarrow #ecigs	Negative
Announcing news reports	Louisville Delays Vote to Ban E-cigarettes in Outdoor Public Places #Spinfuel #Vape #Vaping #ecig #ecigs	Negative
Promotion of other's use	Happy Friday JJuice Friends! Stay Rad and Enjoy Yourselves this weekend. #vape #ecig...	Negative

Table 2: Tweets that indicate e-cigarette use (@mentions and urls removed)

Tweet
170 days no real cigarettes #vaping
I haven't used an ashtray in about 1 1/2 years...................................because #VAPING!
I'm an Ex-Smoker now thanks to #Ecigs. Public Health > #H3639 #EcigsSaveLives #mapoli

3.3. *Corpus Label Protocol*

We implemented the same procedure for both categories, *(1) e-cigarette use, and (2) e-cigarette use for smoking cessation.* To label tweets initially, two authors (YA and PK) independently coded a random subsample of 1,000 tweets using a draft coding protocol. Both authors held a consensus meeting to discuss labels that did not agree, and to refine the coding protocol.

To confirm the quality of the coding protocol, both authors blindly labeled 1,000 additional tweets. Because of the widely varying prevalence in classes, we calculated Siegel & Castellan's bias adjusted kappa. The resulting kappa was calculated at 0.87. [18] This high kappa suggested that the protocol and task were sufficiently generalizable. Corpus statistics are listed in Table 3. Finally authors YA and PK split and independently labeled the remaining 11,146 tweets.

3.4. *Tweet Preprocessing*

We relied on several preprocessing steps used successfully in other Twitter classification studies. [19, 20] For each tweet, we removed screen names (e.g. @britney), and urls. We then produced 4 encodings of the tweets as shown in Table 4 using the libshorttext program.[21] Each tweet is represented as a feature vector of counts for each token.

Table 3. Tweet Corpus Labeled

Descriptor	Value
Number of tweets labeled	13,146
Number of tweeters in labeled set	2,147
Number of tweets for task 1 (e-cigarette use)	728
Number of tweets for task 2 (e-cigarette use for smoking cessation)	73

Table 4. Encoding of Datasets

Encoding Name	Word Tokens Stemmed?	Stopwords Removed?	Unigram or Bigram	Number of Features
unigram	No	No	Unigram	17,371
bigram	No	No	Bigram	109,213
stem_unigram	Yes	No	Unigram	14,509
stem_bigram	Yes	No	Bigram	101,617
stop_unigram	No	Yes	Unigram	17,021
stop_bigram	No	Yes	Bigram	90,037
stop_stem_unigram	Yes	Yes	Unigram	14,186
stop_stem_bigram	Yes	Yes	Bigram	82,464

3.5. *Algorithms*

We use one baseline text classification and three state of the art text classification algorithms. We chose the baseline algorithm to establish the general difficulty of the task and three of the most recent state of the art classification algorithms.

3.5.1. *Naïve Bayes*

This algorithm is the text classification algorithm that typically serves as a baseline measure of text classification performance. This algorithm directly applies Bayes theorem to the classification task and assumes that the probability distribution of a feature is independent of another feature, given the class labels. We used the Multinomial Naïve Bayes [22] implementation in the mallet package. [23]

3.5.2. *Liblinear*

We employed a linear Support Vector Machine (SVM) classification algorithm as implemented in the liblinear package. The linear SVM's calculate maximal margin hyperplane(s) separating the two classes of the data. For text data, the linear SVMs demonstrated superior text classification performance compared to other methods [24], and this motivated our use of them. The liblinear implementation is an optimized version of the support vector machine optimized for quickly finding a linear separating hyperplane. We used liblinear as implemented in libSVM v1.96. [25] We used the default solver of L2-regularized L2-loss and the default penalty parameter of 1.

3.5.3. *Bayesian Logistic Regression*

We employed Bayesian logistic regression. This algorithm demonstrated superior performance in text classification benchmarks and thus motivates our inclusion of them. This algorithm constrains the coefficients using a Laplace prior and thus allows an efficient solution to the convex optimization. We used the bbrtrain [26] implementation for this study. We used the autosearch option to optimize the regularization parameter. This option does a grid search using 10 fold cross validation across the lambda parameters of 0.01 to 316 in multiples of the square root of 10.

3.5.4. *Random Forests*

We employed the random forest implementation in the fest [27] program. Random forests [28] are an ensemble classification method. The method produces a classification tree at each iteration. This classification tree is built from a random subset of the data, and at each node in the tree, a random subset of predictor variables are selected. Multiple trees are constructed in this fashion until at test time, the classification of these individual trees are combined to form a final prediction. We use the default settings that produces 100 trees with a maximum depth of 1000.

3.5.5. *Keyword Comparisons*

We compared the machine learning models to a simple keyword based approach for identifying tweets. Based on our definition, we asked PK to look at our protocol and the portion of the dataset that he reviewed and generate a Boolean keyword set that would provide a relative non machine learning baseline for this classification task. We added this analysis to address whether this task is difficult and to counter the claim that a human could craft a wordset that performs as well as the machine learning models. To simplify the comparison, we compare the keywords to the "unigram" encoding in one 10% split of the data. We used the keyword "OR" searches shown in Table 5.

Table 5: Keyword Searches

Category	Keyword Searches
E-cigarette Use*$	vape OR ecig OR ecigarette OR vaping OR ejuice OR vapers OR (drip AND tip) OR dripping OR (eliquid AND flavor) OR (e AND juice) OR (e AND liquid)
E-cigarette Use for Smoking Cessation*	(smoke AND free) OR (off AND cigarettes) OR (ex AND smoker) OR (no AND analogs) OR (I AND quit)

* - note that we do not consider phrases in these keyword searches. We assume that a bigram phrase is equivalently represented as a Boolean AND. This assumption seems reasonable considering how short tweets are.

$ - the query (e AND juice) is the preprocessed version of the word token "e-juice" with punctuation removed.

4. Results/ Discussion

4.1. *Task 1 – Ecigarette Use*

4.1.1. *Learning Algorithms for Task 1*

Table 6 shows the results for identifying e-cigarette use. Bayesian Logistic Regression, Liblinear, and Random Forests perform with high area under the receiver operating curve. The performances of each classifier are in line with expected text classification performances as in prior studies. [29]

These results highlight performance differences in encoding the tweets. Using unigram or bigram representations demonstrate little performance differences within each classifier. Including stopwords increases performances as shown by comparison between the top 4 and bottom 4 rows . Stemming seems to have a marginal effect in this classification task. In addition, the ranges across the 10 folds are relatively stable likely reflecting the homogeneity of content for this labeled task.

Table 6: E-cigarette Use - 10 Fold Cross Validation Area Under the Receiver Operating Curve Performance (Range of Performances Across 10 folds)

Encoding Name	Naïve Bayes	Liblinear	Bayesian Logistic Regression	Random Forests
unigram	0.80 (0.77-0.86)	0.86 (0.83-0.90)	0.90 (0.86-0.92)	0.89 (0.86-0.92)
bigram	0.79 (0.75-0.83)	0.86 (0.83-0.90)	0.90 (0.87-0.93)	0.88 (0.85-0.91)
stem_unigram	0.82 (0.78-0.85)	0.87 (0.85-0.89)	0.90 (0.85-0.92)	0.88 (0.85-0.91)
stem_bigram	0.78 (0.73-0.82)	0.87 (0.84-0.90)	0.90 (0.87-0.93)	0.88 (0.85-0.91)
stop_unigram	0.79 (0.74-0.84)	0.83 (0.81-0.85)	0.86 (0.84-0.89)	0.83 (0.80-0.88)
stop_bigram	0.77 (0.72-0.80)	0.83 (0.78-0.88)	0.86 (0.84-0.89)	0.83 (0.80-0.86)
stop_stem_unigram	0.79 (0.75-0.85)	0.83 (0.79-0.86)	0.85 (0.82-0.87)	0.84 (0.77-0.88)
stop_stem_bigram	0.77 (0.71-0.81)	0.83 (0.80-0.86)	0.86 (0.83-0.89)	0.83 (0.82-0.87)

4.1.2. *Keyword Comparisons for Task 1*

Keyword based searches are inferior to the machine learning methods. The keyword search returns a sensitivity and specificity while the machine learning methods return a ranked result. To make the comparison, we take one split from the 10 fold cross validation and obtain the sensitivity and specificity for the keyword search. The keyword search has a sensitivity of 0.75 and a specificity of 0.36. At 0.75 sensitivity, the best learning algorithm performs at 0.87 specificity (compared to 0.36). At 0.36 specificity, the best algorithm performs at 0.99 sensitivity (compared to 0.75). Task 1 of defining e-cigarette use through a keyword search benefits from machine learned models.

4.2. *Task 2 – E-cigarette Use for Smoking Cessation*

4.2.1. *Learning Algorithms for Task 2*

Table 7 (on the following page) shows the results for identifying e-cigarette use for smoking cessation. Random Forests perform with high area under the receiver operating curve. The high performance of this classifier for this task is possibly attributable to ensembling [30] that captures non-linearities and interactions effectively.

These results highlight performance differences in encoding the tweets. Using unigram or bigram representations demonstrate little performance differences within each classifier. Including stopwords increases performances as shown by comparison between the top 4 and bottom 4 rows that reflect keeping and removing stopwords respectively. Stemming seems to have a marginal effect in this classification task.

In contrast to the previous task, the ranges across the 10 folds are wide. These results likely reflect the small positive sample size of 73 in this dataset (even less in each fold) and the suspected heterogeneity (e.g. the many ways of communicating e-cigarette use for smoking cessation) in this labeled task.

For both tasks, retaining stopwords improves performance. This observation runs contrary to most other text classification tasks where removing stopwords typically does not affect performance. Stopwords can make a difference and prior researchers have shown that these words can affect performance depending on the task. [31] Further study is needed to examine which stopwords are important for classification in these tasks.

4.2.2. *Keyword Comparisons for Task 2*

Keyword based searches are inferior to the machine learning methods. The keyword search returns a sensitivity and specificity while the machine learning methods return a ranked result. To make the comparison, we take one split from the 10 fold cross validation and obtain the sensitivity and specificity for the keyword search. The keyword search has a sensitivity of 0.29 and a specificity of 0.99. At 0.29 sensitivity, the best learning algorithm performs at 0.99 specificity (compared to 0.99). At 0.99 specificity, the best algorithm performs at 0.37 sensitivity (compared to 0.29). Task

Table 7: E-cigarette Use for Smoking Cessation - 10 Fold Cross Validation Area Under the Receiver Operating Curve (AUC) Performance (Range of Performances Across 10 folds)

Encoding Name	Naïve Bayes	Liblinear	Bayesian Logistic Regression	Random Forests
unigram	0.57 (0.45-0.71)	0.78 (0.68-0.92)	0.88 (0.74-1.0)	0.94 (0.81-0.97)
bigram	0.53 (0.38-0.68)	0.75 (0.65-0.87)	0.87 (0.73-0.95)	0.93 (0.86-0.98)
stem_unigram	0.59 (0.51-0.78)	0.80 (0.55-0.90)	0.89 (0.67-0.98)	0.93 (0.82-0.99)
stem_bigram	0.50 (0.38-0.71)	0.76 (0.65-0.88)	0.89 (0.83-0.95)	0.94 (0.86-0.97)
stop_unigram	0.59 (0.49-0.74)	0.71 (0.40-0.91)	0.87 (0.76-0.97)	0.90 (0.81-0.96)
stop_bigram	0.59 (0.42-0.70)	0.69 (0.44-0.82)	0.86 (0.71-0.98)	0.88 (0.81-0.98)
stop_stem_unigram	0.60 (0.51-0.72)	0.70 (0.41-0.86)	0.85 (0.72-0.95)	0.86 (0.79-0.96)
stop_stem_bigram	0.57 (0.46-0.66)	0.69 (0.41-0.90)	0.83 (0.37-0.94)	0.87 (0.80-0.97)

2 of defining e-cigarette use for smoking cessation through a keyword search is not trivial and benefits from machine learned models.

5. Limitations

5.1. *Generalizability*

The models we built focused on Tweets from followers of e-cigarette brands that contain the specific hashtags. The excellent classification performance in both tasks lay the groundwork for a much larger study that will sample from the larger Tweet population and consider Tweets that do not contain the hashtags or whom are not followers of the top e-cigarette brands.

5.2. *Validity*

This study does not establish directly the validity of the e-cigarette use behavior. Because someone Tweets about use does not mean they actually used. An additional study would survey the Tweeters and about their use habits. We could then compare the tweets about use to actual reported behavior by these Tweeters.

5.3. *Users versus tweets*

In this study, we focused on identifying Tweets automatically. We did not uncover the users associated with the e-cigarette use Tweets. A logical next step will identify the users whom Tweet the use. It is theoretically possible that many Tweets about use originate from a small number of users. Further analysis is necessary.

5.4. *Applications*

In this study, we did not explore specific applications. The eventual driver of performance will dictate the necessary performance. This performance depends on the specific application. For example, if we used the tweet classifications to identify the tweet locations (from the subset of geo-located tweets), we could be more liberal in choosing a threshold that allows false positives as the classifier over time will identify tweets and thus locations of e-cigarette use.

5.5. *Comprehensiveness*

In this pilot, we focused on testing the feasibility of automatic Tweet classification for this task. In later work, we would aim to produce the best classifiers or comprehensively compare classifier performance.

6. Conclusion

This pilot shows that we can successfully build models to identify tweets indicating e-cigarette use and e-cigarette use for smoking cessation. These promising initial results form the foundation to build an ideal surveillance system that can collect data in real time, focus on populations of interest by place and characteristic, include populations unable to take the survey, allow a breadth of questions to answer, and enable geo-location analysis

Supported in part by the NYU CTSA grant UL1TR000038 from NCATS, NIH.
Sharing: The labeled classifications for both e-cigarette use and e-cigarette use for smoking cessation are available at https://github.com/yina/2015-amia-ecig-twitter-labeled as per [32].

References

1. **E-Cig Sales Slide as Regular Smokers Return to Real Thing** [http://www.bloomberg.com/news/articles/2014-07-16/e-cig-sales-slide-as-regular-smokers-return-to-real-thing]
2. Chappell B: **Take It In: 'Vape' Is The Oxford Dictionaries Word Of The Year**. In: *The Two Way*. NPR; 2014.
3. Goniewicz ML, Knysak J, Gawron M, Kosmider L, Sobczak A, Kurek J, Prokopowicz A, Jablonska-Czapla M, Rosik-Dulewska C, Havel C *et al*: **Levels of selected carcinogens and toxicants in vapour from electronic cigarettes**. *Tobacco control* 2014, **23**(2):133.
4. Jensen RP, Luo W, Pankow JF, Strongin RM, Peyton DH: **Hidden Formaldehyde in E-Cigarette Aerosols**. *New England Journal of Medicine* 2015, **372**(4):392-394.
5. CDC: **Notes from the field: electronic cigarette use among middle and high school students - United States, 2011-2012**. *MMWR Morbidity and mortality weekly report* 2013, **62**(35):729-730.
6. **E-cigarettes surpass tobacco cigarettes among teens** [http://www.monitoringthefuture.org/data/14data.html - 2014data-cigs]
7. King BA, Patel R, Nguyen K, Dube SR: **Trends in Awareness and Use of Electronic Cigarettes among U.S. Adults, 2010-2013**. *Nicotine & tobacco research : official journal of the Society for Research on Nicotine and Tobacco* 2014.
8. Richtel M: **E-Cigarettes, by Other Names, Lure Young and Worry Experts**. In: *New York Times*. New York, NY: New York Times; 2014.
9. **Behavioral Risk Factor Surveillance System Questionnaire** [http://www.cdc.gov/brfss/questionnaires.htm]

10. **National Health Interview Survey** [http://www.cdc.gov/nchs/nhis/quest_doc.htm]
11. **Results from the 2010 National Survey on Drug Use and Health: Summary of National Findings** [http://www.samhsa.gov/DATA/NSDUH/2K10NSDUH/2K10RESULTS.HTM - APPB]
12. **Social Media Update 2013** [http://www.pewinternet.org/2013/12/30/social-media-update-2013/]
13. Clark EM, Jones C, Gaalema D, Redner R, White TJ, Kurti A, Schneider A, Couch M, Dodds P, Danforth C: **Electronic Cigarettes and Twitter: Sentiments, Categorization, and Hedonometrics**. In.; 2014.
14. Clark EM, Jones C, Gaalema D, White TJ, Redner R, R E, Dodds P, Couch M, Danforth C: **SoCial Media MeetS PoPulation health: a SentiMent and deMoGRaPhiC analySiS oF tobaCCo and e-CiGaRette uSe aCRoSS the "tWitteRSPheRe"**. *Value in Health* 2014:A603.
15. Huang J, Kornfield R, Szczypka G, Emery SL: **A cross-sectional examination of marketing of electronic cigarettes on Twitter**. *Tobacco control* 2014, **23**.
16. Harris J, Moreland-Russell S, Choucair B, Mansour R, Staub M, Simmons K: **Tweeting for and Against Public Health Policy: Response to the Chicago Department of Public Health's Electronic Cigarette Twitter Campaign**. *J Med Internet Res* 2014, **16**(10):e238.
17. Aphinyanaphongs Y, Ray B, Statnikov A, Krebs P: **Text Classification for Automatic Detection of Alcohol Use Related Tweets**. In: *WICSOC: 2014; Redmond City, CA.*
18. Siegel S CN: **Nonparametric statistics for the behavioral sciences**: NY; McGraw Hill; 1988.
19. Kouloumpis E, Wilson T, Moore J: **Twitter Sentiment Analysis: The Good the Bad and the OMG!** *Proceedings of the Fifth International AAAI Conference on Weblogs and Social Media* 2011.
20. Pak A, Paroubek P: **Twitter as a Corpus for Sentiment Analysis and Opinion Mining**. *Proceedings of the Seventh International Conference on Language Resources and Evaluation* 2010.
21. H.-F. Yu C-HH, Y.-C. Juan, C.-J. Lin: **LibShortText: A Library for Short-text Classification and Analysis**. 2013.
22. A.M. Kibriya EF, B. Pfahringer, and G. Holmes: **Multinomial naive bayes for text categorization revisited**. *Lecture notes in computer science* 2004.
23. **MALLET: A Machine Learning for Langauge Toolkit** [http://mallet.cs.umass.edu]
24. Joachims T: **Learning to Classify Text Using Support Vector Machines: Methods, Theory and Algorithms (The Springer International Series in Engineering and Computer Science)**. In., 1 edn: Springer; 2002: 228.
25. Fan R, Chang K, Hsieh C: **LIBLINEAR: A library for large linear classification**. In: *The Journal of Machine ....* 2008.
26. Genkin A, Lewis DD, Madigan D: **Large-Scale Bayesian Logistic Regression for Text Categorization**. In: *Technometrics.* vol. 49; 2007: 291-304.
27. **Fast Ensembles of Sparse Trees (FEST)** [http://lowrank.net/nikos/fest/]
28. Breiman L: **Random forests**. In: *Machine Learning.* vol. 45; 2001: 5-32.
29. Aphinyanaphongs Y, Fu LD, Li Z, Peskin ER, Efstathiadis E, Aliferis CF, Statnikov A: **A comprehensive empirical comparison of modern supervised classification and feature selection methods for text categorization**. In: *J Assn Inf Sci Tec.* 2014.
30. Seni G, Elder JF: **Ensemble Methods in Data Mining: Improving Accuracy Through Combining Predicitons**: Morgan and Claypool Publishers; 2010.
31. Riloff E: **Little words can make a big difference for text classification**. In: *Proceedings of the 18th annual international ACM ....* 1995.
32. McCreadie R, Soboroff I, Lin J, Macdonald C, Ounis I, McCullough D: **On building a reusable twitter corpus**. In: *Proceedings of the 35th international ACM SIGIR conference on Research and development in information retrieval.* ACM; 2012: 1113-1114.

MONITORING POTENTIAL DRUG INTERACTIONS AND REACTIONS VIA NETWORK ANALYSIS OF INSTAGRAM USER TIMELINES

RION BRATTIG CORREIA[1,2], LANG LI[3] and LUIS M. ROCHA[1,4,*]

[1] *School of Informatics & Computing, Indiana University, Bloomington, IN 47408 USA*
**rocha@indiana.edu*

[2] *CAPES Foundation, Ministry of Education of Brazil, Brasília, DF 70040-020, Brazil*

[3] *Department of Medical and Molecular Genetics, Indiana University School of Medicine, Indianapolis, IN 46202 USA*

[4] *Instituto Gulbenkian de Ciência, Oeiras 2780-156, Portugal*

Much recent research aims to identify evidence for Drug-Drug Interactions (DDI) and Adverse Drug reactions (ADR) from the biomedical scientific literature. In addition to this "Bibliome", the universe of social media provides a very promising source of large-scale data that can help identify DDI and ADR in ways that have not been hitherto possible. Given the large number of users, analysis of social media data may be useful to identify under-reported, population-level pathology associated with DDI, thus further contributing to improvements in population health. Moreover, tapping into this data allows us to infer drug interactions with natural products—including cannabis—which constitute an array of DDI very poorly explored by biomedical research thus far.

Our goal is to determine the potential of *Instagram* for public health monitoring and surveillance for DDI, ADR, and behavioral pathology at large. Most social media analysis focuses on *Twitter* and *Facebook*, but *Instagram* is an increasingly important platform, especially among teens, with unrestricted access of public posts, high availability of posts with geolocation coordinates, and images to supplement textual analysis.

Using drug, symptom, and natural product dictionaries for identification of the various types of DDI and ADR evidence, we have collected close to 7000 user timelines spanning from October 2010 to June 2015. We report on 1) the development of a monitoring tool to easily observe user-level timelines associated with drug and symptom terms of interest, and 2) population-level behavior via the analysis of co-occurrence networks computed from user timelines at three different scales: monthly, weekly, and daily occurrences. Analysis of these networks further reveals 3) drug and symptom direct and indirect associations with greater support in user timelines, as well as 4) clusters of symptoms and drugs revealed by the collective behavior of the observed population.

This demonstrates that *Instagram* contains much drug- and pathology specific data for public health monitoring of DDI and ADR, and that complex network analysis provides an important toolbox to extract health-related associations and their support from large-scale social media data.

Keywords: Complex Network Analysis; Social Media; Drug Interaction; Public Health; Instagram; relational inference

1. Introduction

The analysis of social media data has recently allowed unprecedented access to collective human behavior. The new field of Computational Social Science has brought together Informatics and Complex Systems methods to study society via social media and online data in a quantitative manner not previously possible. From studying social protest[1] to predicting the Stock Market,[2] most of the work has focused on *Twitter*—though *Facebook*[3] and *Instagram*[4] have also received some attention lately. This approach shows great promise in monitoring public health, given the ability to measure the behavior of a very large number of human subjects.[5] For instance, several studies have shown that social media analysis is useful to track and predict influenza spread,[5–7] as well as the measurement of depression.[8] In particular, the potential for adverse drug reaction (ADR) extraction from *Twitter* has been recently demonstrated.[9,10]

There is still, however, much work to be done in order to fulfill the potential of social media in the monitoring of public health. For instance, analysis of social media data may be useful to identify under-reported pathology, particularly in the case of conditions associated with a perceived social stigma, such as mental disorders.[11] Given access to an extremely large population, it is reasonable to expect that social media data may provide early warnings about potential drug-drug interactions (DDI) and ADR.[9] These unprecedented windows into collective human behavior may also be useful to study the use and potential interactions and effects of natural products—including cannabis. The pharmacology of such products constitute an array of DDI and ADR very poorly explored by biomedical research so far, and thus an arena where social media mining could provide important novel discoveries and insight.

Most work on social media pertaining to public health monitoring that we are aware of has relied on data from *Twitter* or *Facebook*. However, *Instagram* is an increasingly important platform, with unrestricted access of public posts, high availability of posts with geolocation coordinates, and images to supplement textual analysis. While Instagram has been used to qualitatively observe the type of content people post regarding health situations such as Ebola outbreaks,[12] its potential for large-scale quantitative analysis in public health has not been established. *Instagram* currently has more than 300 million users.[13] It surpasses *Twitter* and *Facebook* for preferred social network among teens (12-24) in the US. In 2014 there were approximately more than 64 million active users in the US and this number is to surpass 111 million in 2019.[14] Therefore, our goal here is to explore the potential of this very important social media platform for public health monitoring and surveillance of DDI, ADR, and behavioral pathology at large. Specifically, we use literature mining and network science methods to automatically characterize and extract temporal signals for DDI and ADR from a sub-population of Instagram users.

We focused on posts and users with mentions of drugs known to treat depression (e.g. `fluoxetine`). The methodology developed can be easily replicated for different clinical interests (e.g. epilepsy drugs). The goal is to show that Instagram is a very rich source of data to study drug interactions and reactions that may arise in a clinical context of choice, and not depression per se. Using four different multi-word dictionaries (drug and pharmacology, natural products, cannabis, and ADR terminology), we have collected close to 7000 user timelines spanning from October 2010 to June 2015. We analyzed co-mentions in three distinct time-windows: monthly, weekly and daily. This allows the potential extraction of ADR and DDI that manifest at different time scales. From this data, we demonstrate that *Instagram* user timelines contain substantial data of interest to characterize DDI, ADR, and natural product use. To explore this data we have developed a monitoring tool to easily observe user-level timelines associated with drug and symptom terms of interest, which we describe below. To explore population-level associations at the different temporal scales, we compute knowledge networks that our previous work has shown to be useful for automated fact-checking,[15] protein-protein interaction extraction,[16] and recommender systems.[17,18] To illustrate the potential of data-driven, population-level associations, we use spectral methods to reveal network modules of symptoms and drugs, for instance those involved in psoriasis pathology. Our *Instagram* analysis relies on the distance closure of complex networks[18] built at distinct time resolutions, which is a novel development from related approaches to uncover ADR in *Twitter*.[9]

2. Data and Methods

We harvested from *Instagram* all posts containing hashtags that matched 7 drugs known to be used in the treatment of depression (# posts): `fluoxetine` (8,143), `sertraline` (574), `paroxetine` (470), `citalopram` (426), `trazodone` (227), `escitalopram` (117), and `fluvoxamine` (22). Synonyms were resolved to the same drug name according to *DrugBank*;[19] for instance, `Prozac` is resolved to `fluoxetine`, see supporting information (SI) for table of synonyms used. This resulted in a total of 9,975 posts from 6,927 users, whose complete timelines, spanning the period from October 2010 to June 2015, were collected. In total, these timelines contain $5,329,720$ posts, which is the depression timeline dataset we analyze below.

A subset of a previously developed pharmacokinetics ontology[20] was used to obtain a drug dictionary. The full ontology contains more than 100k drugs, proteins and pharmacokinetic terms. Here we used only names of FDA-approved drugs, along with their generic name and synonyms, resulting in 17,335 drug terms. The natural product (NP) dictionary was built using terms from the list of herbal medicines and their synonyms provided by MedlinePlus.[21] It contains 179 terms (see SI). The Cannabis dictionary was assembled by searching the web for terms known to be used as synonyms for cannabis, resulting in 26 terms (see SI) optimized for precision and recall on a subset of posts (data not shown). The symptom dictionary was extracted from BICEPP[22] by collecting all entities defined as an Adverse Effect, with a few manual edits to include more synonyms; it is comprised of 250 terms.

Timeline posts were tagged with all dictionary terms (n-grams) for a total of 299,312 matches. Uppercase characters were converted to lowercase, and hashtag terms were treated like all other harvested text for the purpose of dictionary matches. We found matches for 414 drugs, 133 of which with more than 10 matches. These numbers are 148/99 and 74/46 for symptoms and NP, respectively, for a total of 636 terms. This is a substantial number of dictionary terms, given that only 7 drugs prescribed for depression were used to harvest the set of timelines. The top 25 matches for each dictionary are provided in SI. Notice that the term '`depression`' was removed because of its expected high appearance. Matches in the cannabis dictionary (e.g. 420, marijuana, hashish) were aggregated into the term cannabis to be treated as a NP. The top 10 mentions are (counts shown): `cannabis` (66,540), `anorexia` (26,872), `anxiety` (26,309), `pain` (15,677), `suicide` (11,616), `mood` (11,532), `fluoxetine` (9,961), `suicidal` (8,909), `ginger` (7,289), `insomnia` (5,917).

Given the set X of all matched terms ($|X| = 636$), we first compute a symmetric co-occurrence graph $R_w(X)$ for time-window resolutions $w = 1$ month, 1 week and 1 day. These graphs are easily represented by adjacency matrices R_w, where entries r_{ij} denote the number of time-windows where terms x_i and x_j co-occur, in all user timelines. A matrix R_w is computed for each time-window resolution independently. To obtain a normalized strength of association among the set of terms X, we computed *proximity graphs*,[18] $P_w(X)$ for each time-window resolution w. Thus, the entries of the adjacency matrix P_w of a proximity graph are given by:

$$p_{ij} = \frac{r_{ij}}{r_{ii} + r_{jj} - r_{ij}}, \quad \forall_{x_i, x_j \in X} \tag{1}$$

where $p_{ij} \in [0, 1]$ and $p_{ii} = 1$; $p_{ij} = 0$ for terms x_i and x_j that never co-occur in the same time-window in any timeline, and $p_{ij} = 1$ when they always co-occur. This measure is the probability that two terms are mentioned in the same time window, given that one of them was mentioned.[17,18] To ensure enough support exists in the data for proximity associations, we computed proximity weights only when $r_{ii} + r_{jj} - r_{ij} \geq 10$; if $r_{ii} + r_{jj} - r_{ij} < 10$, we set $p_{ij} = 0$.

Proximity graphs are *associative knowledge networks*. As in any other co-occurrence method, the assumption is that items that frequently co-occur are associated with a common phenomenon. These knowledge networks have been used successfully for automated fact-checking,[15] protein-protein interaction extraction,[16] and recommender systems.[17,18] Here we use them to reveal strong associations of DDI-related terms for public health monitoring. We also compute distance graphs $D_w(X)$ for the same time-window resolutions, using the map:

$$d_{ij} = \frac{1}{p_{ij}} - 1 \tag{2}$$

In some of our analysis below, we compute the metric closure $D_w^C(X)$ of the distance graphs, which is isomorphic to a specific transitive closure of the proximity graph.[18] The metric closure is equivalent to computing the shortest paths between every pair of nodes in the distance graph. Thus, d_{ij}^C is the length (sum of distance edge weights) of the shortest path between terms x_i and x_j in the original distance graph $D_w(X)$, and is known to scale well.[15]

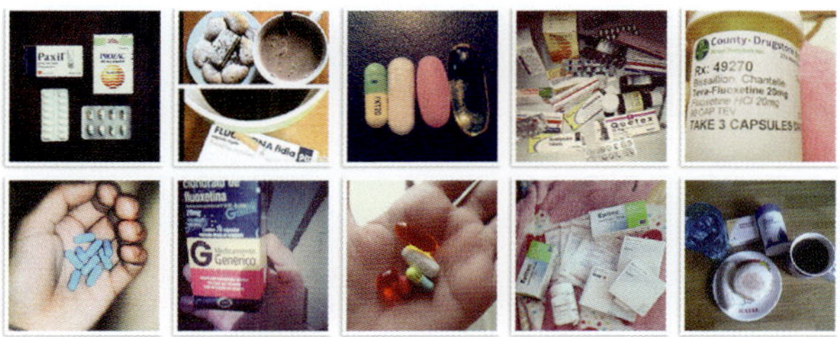

Fig. 1. Sample of images from collected posts related to `fluoxetine`.

3. A Monitoring tool for user-level behavior

From the analysis of user timelines, it is clear that *Instagram* is a social media platform with much data relevant for public-health monitoring. Users often discuss personal health-related information such as diagnoses and drugs prescribed. Photos posted (e.g. Figure 1) often depict pills and packaging, along with discussions of intake schedules, expectations and feelings.

- User A on May 25, 2014:

  "#notmypic .. Say hello to my new friend! Fluoxetina! Side effects by now are a bit of nausea and inquietude.. Better than zoloft! Yesterday night i started to cry while i was with my 2 friends because my ex, bulimia's stress.. I'm sure they thought i'm crazy so i felt like i had to explain my reasons with one of those friends.. Now i'm terrified of his reaction, he is even a friend of my ex.. Don't know what to expect.. It's so hard telling someone about ED and bulimia . I'm also thinking about a b/p session today after 2 days clean, maybe it's not the right solution. Idk. #bulimia #bulimic #mia #ed #edfamily #eatingdisorder #prorecovery #bingepurge #purge #binge #fat #prozac #fluoxetine #depression #meds"

- User B on May 13, 2015:

  "I start fluoxetine tomorrow, the doctor switched me from citalopram to this so let's hope it goes better this time #anxietymeds #depressionmeds #citalopram #fluoxetine #anxiety #depression"

- User B on May 14, 2015 (one day later):

  "ok so I don't know if it's the tablets that are doing this but I feel the lowest I've ever felt and I'm hoping it's not the tablets. Hopefully it's just a bad day, not that there are many good days I hope tomorrow is a better day for everyone, especially if you are feeling the same way I am. #fluoxetine #depression #anxiety #depressionmeds #anxietymeds"

- User C on Feb 05 2014:

  "i survived another trip to the clinic, saw a specialist, did a test that explained i'm an INFJ (introvert) which is apperently only 1% of the population. Added risperidone and upped ritalin as well as prozac. considering this keeps me 'sane' and able to assimilate into the chaos of everyday life i think this counts as my #100happydays today #findhappinessineachday #bipolar #borderlinepersonalitydisorder #INFJ #manicdepression #goinggovernment #prozac #lamotragine #ritalin #risperidone"

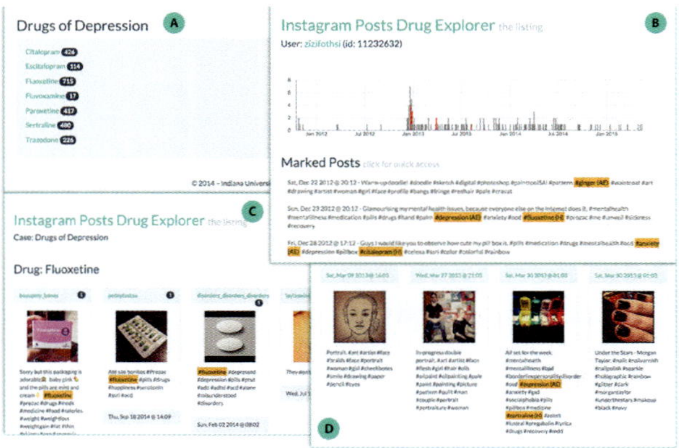

Fig. 2. Instagram Drug Explorer. See text for explanation.

Given the rich data users post on *Instagram*, from the perspective of public-health monitoring, it is useful to be able to quickly navigate and extract posts and user timelines associated with drugs and symptoms of interest. For that purpose, we developed the *Instagram Drug*

Explorer[a], a web application to explore, tag, and visualize the data. This tool also allows downstream improvement of our dictionaries by observing important discourse features not tagged. Figure 2 shows four screenshots with some of the current features: A) the possibility of defining multiple drugs of interest per project; B) a user timeline view that tags class-specific dictionary matches and displays post frequency in time and where individual posts can be quickly selected to be C) visualized separately; D) a summary of posts from user timelines of interest. Another feature (not shown) is the display of geo-located posts using overlay maps, which can be useful, for instance, to monitor users in places of interest, such as schools, clinics, and hospitals. Using this tool to inspect and select timelines with high number of matches, we were able to identify particularly relevant user timelines such as the one depicted in Figure 3, which contains matches from all four dictionaries, and varying post frequency.

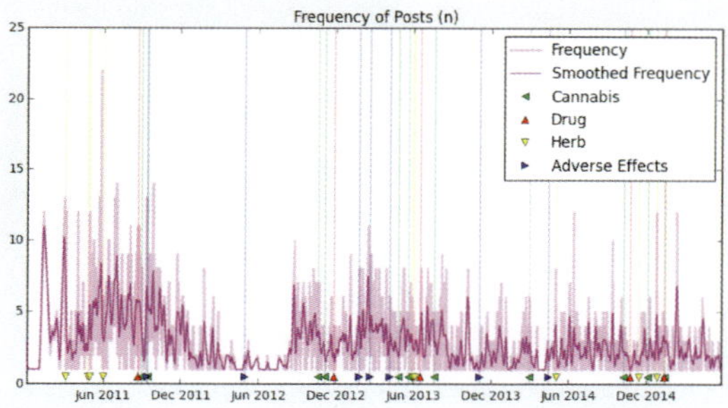

Fig. 3. User timeline showing daily frequency of posts in time; dictionary terms from are tagged in time.

4. Network analysis of associations in population-level behavior

Using the proximity or the isomorphic distance graphs (§2), we can explore strong pairwise term associations that arise from the collection of 5,329,720 posts from the population of 6,927 users in the study. The assumption is that dictionary terms that tend to co-occur in a substantial number of user timelines may reveal important interactions among drugs, symptoms, and natural products. Moreover, because we computed these knowledge networks at different time resolutions, we can explore term associations at different time scales: day, week, and month. Naturally, a statistical term correlation is not necessarily a causal interaction; also a drug-symptom association may reveal a condition treated by the drug, rather than an adverse reaction. But large-scale analysis of social media data for relational inference must start with the identification of multivariate correlations, which can be subsequently refined, namely with supervised classification and NLP methods. Here, as a first step in the analysis of *Instagram* data for public health monitoring, we use unsupervised network science methods to extract term associations of potential interest.

[a]`http://informatics.indiana.edu/rocha/IDE.`

Consider the proximity networks $P_w(X)$ for time resolution $w = 1$ week. The full network contains $|X| = 636$ terms (see Figure 5A for its largest connected component); Figure 4 (left) lists the top 25 drug/NP vs symptom associations, as well as the adjacency matrix of the distance subgraph $D_w(X)$ for these drug/NP and symptom pairs (right). The proximity and distance graphs are isomorphic (§2), but proximity edge weights (left) are directly interpretable as a co-occurrence probability (eq. 1), while the isomorphic nonlinear map to distance (eq. 2) provides greater discrimination in the visualization of the adjacency matrix (right).

Top 25 edges of `Drug/NP vs. Symptom`. Only co-mentions within 1W.

drug	symptom	proximity
naloxone	hypotension	0.4286
allopurinol	hypotension	0.4118
montelukast	hypotension	0.3077
belimumab	neuropathy	0.2989
lidocaine	hypotension	0.2917
hydroxychloroquine	neuropathy	0.2895
ribavirin	angina	0.2857
ribavirin	hypotension	0.2857
tiotropium	angina	0.2857
tiotropium	hypotension	0.2857
vasopressin	rhinitis	0.2727
allopurinol	angina	0.2632
norethindrone	aggression	0.2541
liraglutide	rhinitis	0.25
naloxone	angina	0.25
verapamil	hypotension	0.2333
hydroxyurea	blurred vision	0.2308
adalimumab	gastritis	0.2271
histamine phosphate	hypotension	0.2143
conjugated estrogens	dyspepsia	0.2105
donepezil	rhinitis	0.2
scopolamine	neurotoxicity	0.2
ribavirin	myopathy	0.2
tiotropium	myopathy	0.2
vardenafil	hypotension	0.1875

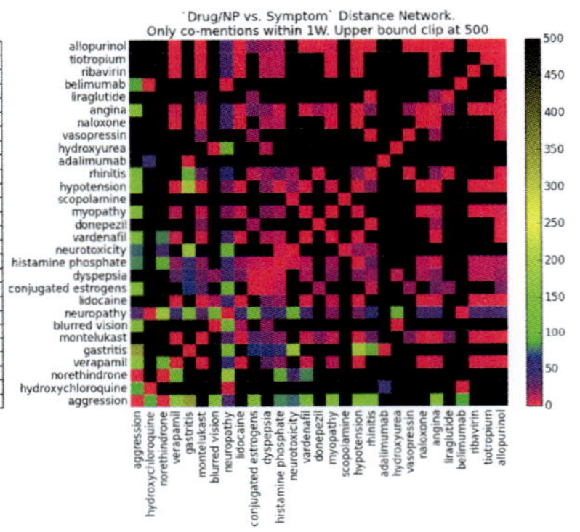

Fig. 4. drug/NP vs symptom subnetwork: (left) Top 25 pairs with largest proximity correlation. (right) adjacency matrix of distance subnetwork; nearest (furthest) term pairs in red (black).

Of the 25 to associations listed in Figure 4 (left), 12 are known or very likely ADR, 7 do not have conclusive studies but are deemed possible ADR from patient reports, 4 refer to associations between drugs/NP and symptoms they are indicated to treat, 1 has been shown to not be ADR, and 1 is unknown (evidence in SI). Thus, the strongest edges in the 1 week resolution network are relevant drug/NP-symptom associations. Furthermore, our methodology allows an analyst to collect (via the Drug Explorer tool §3) all the individual timelines and posts that support every association (edge) in the proximity networks, supporting a much more detailed study of the affected population—including for the purpose of fine-tuning dictionaries and mining techniques to better capture the semantics of specific populations.

The proximity networks $P_w(X)$ also allow us to visualize, explore and search the "conceptual space" of drugs, symptoms, and NP as they co-occur in the depression timeline dataset. The largest connected component of the proximity network for $w = 1$ week is shown in Figure 5A. The network representation allows us to find clusters of associations, beyond term pairs, which may be related via the same underlying phenomenon. Many multivariate and network analysis methods can be used to uncover modular organization.[23] To exemplify, here we use the Principal Component Analysis (PCA)[24] of the proximity network adjacency matrix, which reveals potential phenomena of interest.

For instance, Figure 5, depicts a set of terms correlated with principal component (PC)

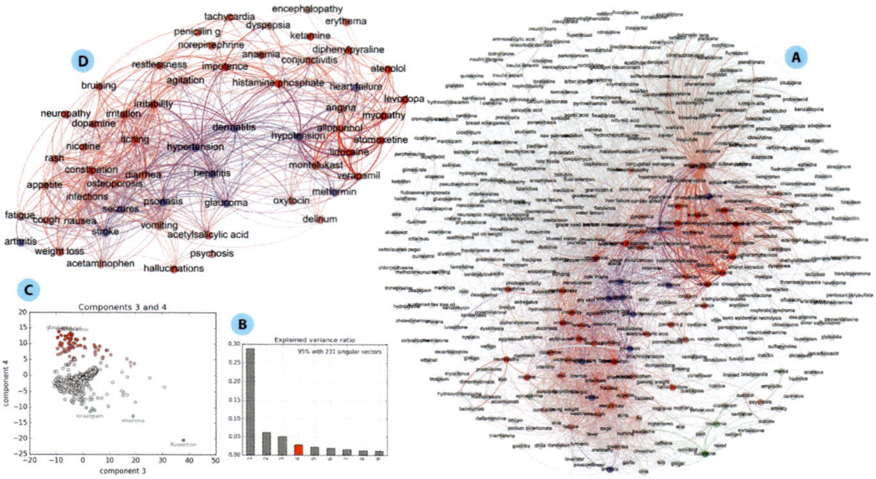

Fig. 5. A. Largest connected component of the proximity network for 1 week time resolution; weights shown only for $p_{ij} \geq 0.05$ with unconnected terms removed. Edges are colored according to correlation with PC 4. B. Spectrum of the PCA of the proximity network adjacency matrix. C. Biplot of correlation of terms with PC 3 and 4; red (green) terms are most (anti-) correlated with PC4. D. Subgraph depicting the network of terms most correlated with PC4, which is related to `Psoriasis`; blue nodes depict conditions linked to this complex disease (see text for details); weights shown only for $p_{ij} \geq 0.05$.

4 (red)—others could be chosen (see SI). The subnetwork of these terms is depicted in Figure 5D. and it reveals a set of terms denoting a complex interaction of conditions which are coherent with what is becoming known about `Psoriasis`. Several of the edges associate terms related to heart disease, stroke, hypertension, hypotension, and diabetes which are high risks for `Psoriasis` patients,[25] including potential drug interactions (`Metformin` for Diabetes, `Verapimil` for high blood pressure and Stroke). This subnetwork also reveals associations with `Psoriasis` which are currently receiving some attention, such as with viral hepatitis[26] and seizure disorder.[27] Naturally, the network also includes many terms associated with skin infections and immune reactions. The `Psoriasis` subnetwork is just an example of a multi-term phenomenon of interest that is represented in the whole network; other PCA components are shown in SI, including additional analysis of the Psoriasis subnetwork. Importantly, we can identify users who may be experiencing this cluster of symptoms by following the posts and timelines behind the weights in the subnetwork, which is useful for public health monitoring.

While the `Psoriasis` subnetwork was discovered purely by data-driven analysis, another way to use these networks is to to query them for specific terms most associated with a set of drugs or symptoms of interest. This problem of finding which other items $A \subseteq X$ are near a set of query items $Q \subseteq X$ (including a subnetwork of interest) is common in recommender systems and information retrieval.[17] The answer set A can be computed as:

$$A \equiv \left\{ x_j : \forall_{x_i \in Q} \underset{x_j \in X-Q}{\Phi} (p_{ij}) \geq \alpha \right\} \tag{3}$$

where Φ is an operator of choice, p_{ij} is the proximity weight between terms x_i and x_j (§2), and α is a desired threshold. If we are interested in a set of terms A which are strongly related to *every* term in query set Q, then we use $\Phi = \min$. If we are interested in terms strongly related to *at least one* term in Q, then $\Phi = \max$. For a compromise between the two, we can use $\Phi = \text{avg}$ (average). Consider the query $Q = \{$`fluoxetine`, `anorexia`$\}$ on the network of Figure 5A ($w = 1$ week). Using $\Phi = \min$, we obtain an answer set with terms strongly related to both query terms (ordered by relevance): $A = \{$`suicidal`, `suicide`, `anxiety`, `pain`, `mood`, `cinnamon`, `insomnia`, `soy`, `headache`, `mania`, `chia`, `cannabis` $\}$. For the query $Q = \{$`psoriasis`, `heart failure`, `stroke` $\}$ using $\Phi = \text{avg}$, we obtain (ordered by relevance): $A = \{$`infections`, `diarrhea`, `hypertension`, `seizures`, `hepatitis`, `constipation`, `dermatitis`, `glaucoma`, `vomiting` $\}$, which relates to the discussion above. Additional query examples and details of the network search interface are shown in SI.

Proximity $P_w(X)$ networks are useful to discover associations between terms which co-occur in time windows w of user timelines (§4). But they are also useful to infer *indirect associations* between terms. In other words, terms that do not co-occur much in user timelines, but which tend to co-occur with the same other terms. In network science indirect associations are typically obtained via the computation of shortest path algorithms on the isomorphic distance graphs $D_w(X)$.[18] Terms which are very strongly connected via indirect paths, but weakly connected via direct edges, break transitivity criteria.[18] We have previously shown that such indirect paths are useful to predict novel trends in recommender systems,[18] and are also instrumental to infer factual associations in knowledge networks.[15] In this context, the hypothesis is that strongly indirectly associated terms may reveal unknown DDI and ADR.

Top 25 semi-metric edges of 'Drug/NP vs. Symptom'.
Only co-mentions within 1W.

drug	symptom	proximity
cinnarizine	hypotension	0.21
bethanechol	hypotension	0.21
cabergoline	hypotension	0.21
primidone	hypotension	0.1963
desmopressin	hypotension	0.1963
sevelamer	hypotension	0.1963
atomoxetine	hypotension	0.1875
vasopressin	hypotension	0.1842
montelukast	angina	0.1802
lidocaine	angina	0.1768
liraglutide	hypotension	0.1736
naloxone	rhinitis	0.1667
cinnarizine	angina	0.163
bethanechol	angina	0.163
cabergoline	angina	0.163
allopurinol	myopathy	0.1579
cilostazol	hypotension	0.1556
follitropin beta	hypotension	0.1556
tetrabenazine	hypotension	0.1556
nevirapine	hypotension	0.1556
flumazenil	hypotension	0.1556
alendronate	hypotension	0.1556
betahistine	hypotension	0.1556
natamycin	hypotension	0.1556
prasugrel	hypotension	0.1556

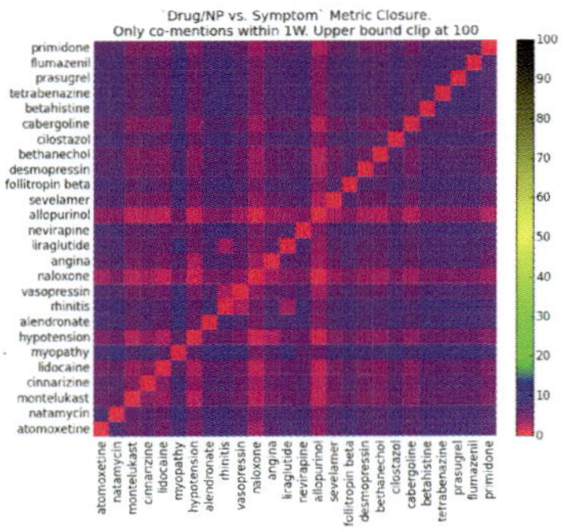

Fig. 6. drug/NP vs symptom subnetwork after shortest path calculation. (left) Top 25 non-transitive term pairs. (right) adjacency matrix of distance subnetwork after shortest path calculation.

To find the term pairs that most break transitivity we compute all shortest paths in the networks (via Dijkstra's algorithm): the metric closure $D_w^C(X)$. Figure 6 lists the top 25 drug/NP vs symptom associations which most break transitivity. In other words, these are term pairs which are very strongly associated via indirect paths, but very weakly associated directly. Of the extracted associations listed in the table of Figure 6, 6 are known or likely ADR, 3 are possible ADR from patient reports but no conclusive study, 2 refer to associations between drugs/NP and symptoms they are indicated to treat, and all other 14 are unknown (evidence provided in SI). Thus, unlike the case of direct associations (Figure 4), there is less evidence for the indirect associations in the literature. This could be because they are false associations, or because they have not been discovered yet. Validating these associations empirically is left for forthcoming work; here the goal is to show how network analysis methods can be used to select such latent associations which are highly implied by indirect paths (transitivity) but are not directly observed in user post co-mentions.

Similarly to what was done with direct associations above, we can also query the proximity network obtained after shortest path computation $P_w^C(X)$ (the isomorphic proximity graph to $D_w^C(X)$ via eq. 2). For instance, if we query the original $w = 1$ week proximity network $P_w^C(X)$ (the one depicted in Figure 5A) with $Q = \{$`psoriasis,metformin`$\}$ (a type 2 diabetes drug), using $\Phi = \min$, we obtain $A = \{$`montelukast` , `hypertension`, `dermatitis`, `hypotension`, `hepatitis`$\}$ as the top 5 terms—`montelukast` is a drug used to treat allergies. If we now use the same query Q on the metric closure network $P_w^C(X)$ instead, the top 5 answer set becomes $A^C = \{$`montelukast`, `hypotension`, `naloxone`, `allopurinol`, `hypertension`$\}$ (full query results in SI). In other words, after computing shortest paths,`naloxone` (a synthetic opiate antagonist used to reverse the effects, including addiction, caused by narcorics) and `allopurinol` (a drug used to treat gout, kidney stones, and decrease levels of uric acid in cancer patients), become more strongly associated with the query terms. These indirect associations to do not occur very strongly in the observed *Instagram* timeline data, but are strongly implied by indirect paths in the network of term proximity. In this case, the *latent* associations may provide additional evidence supporting recent observations that psoriasis (an autoimmune condition) is linked to heart disease, cancer, diabetes and depression.[25]

5. Discussion and Future Directions

Our preliminary analysis demonstrates that there exists a substantial health-related user community in *Instagram* who posts about their health conditions and medications. The drug, NP and symptom dictionaries we employed extracted a large number of posts with such data, enough to build knowledge networks of hundreds of terms representing the pharmacology and symptomatic "conceptual space" of *Instagram* users posting about depression. Our results and software further demonstrate that such space can be navigated for public health monitoring, whereby analysts can search and visualize user timelines of interest. Furthermore, the network representation of this space allows us to extract population-level term associations and subnetworks of terms arising from underlying (modular) phenomena of interest—such as the Psoriasis network involving various related conditions. Thus, *Instagram* data shows great potential for public health monitoring and surveillance for DDI and ADR.

Direct associations in the knowledge networks are substantiated by actual co-mentions in posts from user timelines, which can subsequently be retrieved by public health analysts using our drug explorer application. In our preliminary work, the top extracted direct associations are shown to be backed by the literature, but we intend to pursue the systematic validation of such associations in future work. Network methods also allow us to uncover indirect associations among terms. These may reveal latent, yet unknown, associations, and as such, very relevant for public health monitoring. Studying the network of indirect associations can be further used to understand community structure as well as redundancy in the data, which we intend to study next.

We have analyzed posts and user timelines related to depression only. Adding additional conditions of interest (e.g. epilepsy or psoriasis) to extract additional posts would monitor different communities, and would likely improve the overall extraction of associations, which we intend to test in the near future. While the drug dictionary is quite well developed already, the NP and symptoms dictionaries need to be further developed, especially towards increasing the terminology associated with symptoms as well as on catching particular linguistic expressions of symptoms in Instagram. The development of named entity recognition tailored to Instagram is another avenue we intend to pursue, starting from and expanding what has already been done for Twitter.[10]

The methodology we describe here allows us to discern drug, NP and symptom associations derived from user timeline co-mentions at different timescales. All the results displayed pertain to a one week window, however we also computed day and month windows. The comparison of results at different timescales would allow, in principle, the discovery of more immediate as well as more delayed interactions. Such a comparison is also something we intend to pursue in forthcoming work. Finally, the timeseries analysis of user timelines can be used to detect discernible changes in behavior for users and groups of users. One could track, for instance, critical changes in mood associated with the onset of depression,[28] which constitutes yet another exciting avenue to pursue with this line of research.

Our preliminary analysis demonstrates that *Instagram* is a very powerful source of data of potential benefit to monitor and uncover DDI and ADR. Moreover, our work shows that complex network analysis provides an important toolbox to extract health-related associations and their support from large-scale social media data.

Acknowledgments

This work was supported by a grant from the National Institutes of Health, National Library of Medicine Program, grant 01LM011945-01 "BLR: Evidence-based Drug-Interaction Discovery: In-Vivo, In-Vitro and Clinical," and a grant from Persistent Systems. RBC is supported by CAPES Foundation Grant No. 18668127. The funders had no role in study design, data collection and analysis, decision to publish, or preparation of the manuscript.

References

1. O. Varol, E. Ferrara, C. L. Ogan, F. Menczer and A. Flammini, Evolution of online user behavior during a social upheaval, in *Proc. 2014 ACM Conference on Web Science*, WebSci '142014.
2. J. Bollen, H. Mao and X. Zeng, *Journal of Computational Science* **2**, 1 (2011).

3. E. Bakshy, S. Messing and L. A. Adamic, *Science* **348**, 1130 (May 2015).
4. E. Ferrara, R. Interdonato and A. Tagarelli, Online popularity and topical interests through the lens of instagram, in *Proc. 25th ACM Conf. on Hypertext and Social Media*, HT '142014.
5. H. Kautz, Data mining social media for public health applications., in *23rd Int. Joint Conf. on Artificial Intelligence (IJCAI 2013)*, (AAAI Press, 2013).
6. A. Signorini, A. M. Segre and P. M. Polgreen, *PLoS ONE* **6**, p. e19467 (2011).
7. A. Sadilek, H. Kautz and V. Silenzio, Modeling spread of disease from social interactions, in *Sixth AAAI Int. Conf, on Weblogs and Social Media (ICWSM)*, (AAAI Press, 2012).
8. M. D. Choudhury, S. Counts and E. Horvitz, Social media as a measurement tool of depression in populations, in *Proc. 5th Annual ACM Web Science Conf.*, WebSci'13 (ACM, 2013).
9. A. A. Hamed, X. Wu, R. Erickson and T. Fandy, *J. of biomedical informatics* **56**, 157 (2015).
10. A. Sarker and G. Gonzalez, *Journal of biomedical informatics* **53**, 196 (2015).
11. B. A. Pescosolido, *Annual Review of Sociology* (2015).
12. E. Seltzer, N. Jean, E. Kramer-Golinkoff, D. Asch and R. Merchant, *Public Health* **129**, 1273 (September 2015).
13. Instagram Blog, 300 million. `http://blog.instagram.com/post/104847837897`.
14. Statista, Number of monthly active instagram users from january 2013 to december 2014 (in millions). `http://www.statista.com/statistics/253577/`.
15. G. L. Ciampaglia, P. Shiralkar, L. M. Rocha, J. Bollen, F. Menczer and A. Flammini, *PLoS ONE* **10**, p. e0128193 (2015).
16. A. Abi-Haidar, J. Kaur, A. Maguitman, P. Radivojac, A. Rechtsteiner, K. Verspoor, Z. Wang and L. M. Rocha, *Genome Biology* **9**, p. S:11 (September 2008).
17. L. M. Rocha, T. Simas, A. Rechtsteiner, M. D. Giacomo and R. Luce, Mylibrary@lanl: Proximity and semi-metric networks for a collaborative and recommender web service, in *2005 IEEE/WIC/ACM International Conference on Web Intelligente (WI'05)*, (IEEE Press, 2005).
18. T. Simas and L. M. Rocha, *Network Science* **3**, 227 (6 2015).
19. D. Wishart, C. Knox, A. Guo, D. Cheng, S. Shrivastava, D. Tzur, B. Gautam and M. Hassanali, *Nucleic Acids Res* **36**, D901 (January 2008).
20. H.-Y. Wu, S. Karnik, A. Subhadarshini, Z. Wang, S. Philips, X. Han, C. Chiang, L. Liu, M. Boustani, L. M. Rocha, S. K. Quinney, D. Flockhart and L. Li, *BMC Bioinformatics* **14**, 1 (2013).
21. MedlinePlus, Herbal medicine. `http://1.usa.gov/1IF33ng`.
22. F. P.-Y. Lin, S. Anthony, T. M. Polasek, G. Tsafnat and M. P. Doogue, *BMC Bioinformatics* **12**, p. 112 (April 2011).
23. S. Fortunato, *Physics Reports* **486**, 75 (2010).
24. M. E. Wall, A. Rechtsteiner and L. M. Rocha, Singular value decomposition and principal component analysis, in *A practical approach to microarray data analysis*, (Springer, 2003) pp. 91–109.
25. WebMD, Psoriasis linked to heart disease, cancer. studies also show link to increased risk of diabetes and depression. `http://wb.md/1IF3hL3`.
26. A. D. Cohen, D. Weitzman, S. Birkenfeld and J. Dreiher, *Dermatology* **220**, 218 (2010).
27. O. M, K. IS, C. T, G. MP and M. KD, *JAMA Neurology* **71**, 569 (2014).
28. I. A. van de Leemput, M. Wichers, A. O. Cramer, D. Borsboom, F. Tuerlinckx, P. Kuppens, E. H. van Nes, W. Viechtbauer, E. J. Giltay, S. H. Aggen *et al.*, *PNAS* **111**, 87 (2014).

TOWARDS EARLY DISCOVERY OF SALIENT HEALTH THREATS: A SOCIAL MEDIA EMOTION CLASSIFICATION TECHNIQUE

BAHADORREZA OFOGHI[1] and MEGHAN MANN[1] and KARIN VERSPOOR[1,2]

[1]*Department of Computing and Information Systems*
[2]*Health and Biomedical Informatics Centre*
The University of Melbourne
Parkville, Victoria 3010, Australia

Online social media microblogs may be a valuable resource for timely identification of critical ad hoc health-related incidents or serious epidemic outbreaks. In this paper, we explore emotion classification of Twitter microblogs related to localized public health threats, and study whether the public mood can be effectively utilized in early discovery or alarming of such events. We analyse user tweets around recent incidents of Ebola, finding differences in the expression of emotions in tweets posted prior to and after the incidents have emerged. We also analyse differences in the nature of the tweets in the immediately affected area as compared to areas remote to the events. The results of this analysis suggest that emotions in social media microblogging data (from Twitter in particular) may be utilized effectively as a source of evidence for disease outbreak detection and monitoring.

Keywords: Twitter, Ebola, Emotion classification, Shift detection

1. Introduction

Syndromic surveillance involves monitoring of public health information resources, to facilitate early detection of disease outbreaks, and to monitor the size, spread, and tempo of epidemic outbreaks.[1] Many jurisdictions have regulations for reporting on infectious diseases to public health officials, for instance requiring that laboratory-confirmed cases of influenza be notified to the government (see, e.g., the Australian National Notifiable Diseases Surveillance System[a]). However, it is important to have surveillance mechanisms in place that identify weaker signals of disease activity, in particular for diseases with potentially severe public health consequences, such as Botulism or Ebola, that public health officials want to be able to respond to quickly. Social media posts are a major source of uncurated user-generated feedback, that may have a positive impact on critical applications related to public health and safety.[2]

There have been a number of efforts to develop computational approaches that enable automated monitoring and *early warning* systems making use of online resources. In recent work,[3–6] prediction of near future Influenza events as well as the spread of N1H1 and Ebola cases were studied using descriptive statistics extracted from Twitter messages as well as utilizing data from Google Flu Trends.[7] This work supports the usefulness of Twitter data for pandemic event surveillance. However, it mostly focuses on descriptive statistics at the level of single tweets (or single sentiments) over time and does not consider the *combination* or *distribution* of sentiments across a *collection* of tweets as an early warning signal.

RSS feeds have also been classified as relating to certain pathogens without necessarily having explicit evidence or mention of the pathogen (i.e., from reported symptoms).[8] Such

[a]`www.health.gov.au/internet/main/publishing.nsf/Content/cda-surveil-nndss-nndssintro.htm`

systems have been demonstrated to produce similar predictions to that of government health organizations.[7,8] However, these methods may not be appropriate for identifying salient outbreaks where only a small number of people are infected, such as isolated Ebola cases, while an influx of web and social media messages regarding the disease is encountered.

We approach disease outbreak detection from the perspective of the emotional stance of a user towards a disease. The underlying hypothesis of our work is that a proximal disease incident will trigger the expression of concerns about the incident, and that these expressions will differ qualitatively (emotionally and linguistically) from the typical chatter around a distant or less immediate threat. We propose a model, building on this hypothesis, to detect a shift in the nature of the conversations around a specific disease on the basis of changes in the distribution of emotions expressed in tweets containing some response to a public health incident. Public mood has been demonstrated to relate to major socio-economic events,[9] and identifying shifts in emotions may also provide a useful early indicator that a new public health incident has occurred. This strategy removes the need for classification of textual documents into pre-defined syndromes or explicit prediction of future events; instead, it has a focus on the distribution of emotional expressions in the texts of microblogs in specific periods of time.

We therefore explore the relationship between public mood and salient public health threats in this paper. We believe that users express different emotions, thoughts and speculations and may post different types of informational links and resources at times prior to and following major epidemic incidents. This may be particularly true when a user feels directly impacted by an incident, e.g., due to geographical proximity to an event. We do not pre-suppose that there are specific emotions that will be consistently identifiable across distinct public health issues, but rather focus on whether there is a change in the distribution of emotions.

We examine the distribution of emotion classes in tweets to estimate the differences between emotional features before and after likely outbreaks with two component strategies:

- Emotion classification of tweets, using a trained classification model to assign each tweet to one of several emotion classes.
- Emotion shift detection through statistical analysis of tweet corpora, comparing the distribution of emotions expressed in tweets immediately prior to and after relevant incidents.

To explore our hypothesis, a case study of two recent events in London, United Kingdom where a health worker was found to have been exposed to the Ebola virus is provided. The emotions of all tweets in London around these events explicitly mentioning Ebola were analysed. We demonstrate that by monitoring Twitter microblogs, it is possible to capture likely outbreaks through detection of emotional shifts in user tweets.

2. Emotion Classification of Tweets

We begin by developing an emotion classifier for outbreak-related tweets, using a new annotated data set and an emotion inventory that adopts Ekman's six basic emotions[10] ("anger", "disgust", "happiness", "sadness", "surprise", and "fear"), and extends it with three additional "attitudinal' classes, "sarcasm", "news-related", and "criticism".

Emotion detection from textual data has been previously tackled using various unsupervised[11] and supervised approaches.[12] Aman and Szpakowicz[13] utilized corpus-based unigrams, emotion-related words extracted from Roget's thesaurus, and features derived from WordNet-Affect to train a supervised emotion classifier. They employed Ekman's inventory plus a no-emotion class in their work. Wang et al.[14] utilized an overlapping emotion inventory and similar features. They found improvements in tweet emotion classification through consideration of the sentiment of words (positive or negative) as features.

We approach emotion detection using similar textual features to these previous studies, testing both lexicon-based (unsupervised) and supervised methods. We developed a binary classifier for each emotion class, experimenting with several representations of tweet texts, as will be discussed in the following sections.

2.1. *Lexicon-Based Classification*

A simple unsupervised baseline emotion classifier was implemented for each emotion class, using a lexicon-based vector model. We constructed a *reference vocabulary* consisting of terms corresponding to each emotion class.[b] These terms include emotion-related terms from Emotion Vocabulary,[15] lexical units derived from the FrameNet[16] frame *Judgement-communication* (for class "criticism"), emotion terms from the Profile of Mood States,[9,17] and emoticons. We also include terms specifically for the "news-related" category, corresponding to popular international news agency names. The resultant reference vocabulary contains 499 terms. Each emotion class is represented as a binary vector with respect to this reference vocabulary; any term from the vocabulary relevant to the emotion was marked 1 and irrelevant terms 0.

For classification, tweets were also mapped into this lexical vector representation, with a 1 indicating that the tweet contains a given term from the vocabulary. The cosine similarity score between this tweet vector and each of the nine emotion class vectors was computed; the class with the highest similarity was returned as the classification of the tweet.

2.2. *Machine Learning-Based Classification*

A Naïve Bayes classifier, implemented in MALLET toolkit,[18] was used for our machine learning-based classification. The basic features used to represent tweets were bag-of-words. This set of features was augmented in a feature engineering step. The extra features included: i) the lexicon-based similarity score for each of the nine classes obtained from the baseline lexicon-based classifier, ii) emotion vocabulary from the same reference vocabulary that the lexicon-based classifier made use of, iii) emoticons, iv) punctuations including question and exclamation marks, and v) sentiment classification of the tweet text from the Stanford Sentiment Analyzer (i.e., negative, neutral, or positive).[19]

2.3. *Data*

To train the emotion classifier on tweets relevant to an active public health threat, we collected recent tweets regarding Ebola using the Twitter API. A total of 12,101 tweets that contained

[b]The vocabulary is available at: `https://bitbucket.org/readbiomed/socialsurveillance`

Table 1. The distribution of the nine emotion classes over the 4,405 *Ebola* tweet set labelled by Mechanical Turk workers.

Class	Sarcasm	News-rel.	Criticism	Fear	Surprise	Anger	Happiness	Disgust	Sadness
#Tweets	1,322	2,572	166	81	67	62	61	51	23

the word "ebola" were collected from all over the world in the second half of March 2015. Non-English tweets were filtered out, leaving 7,039 tweets. After initial pre-processing of the tweets and removal of redundant (identical) tweets, 4,405 tweets remained.

The tweet texts were normalized in the pre-processing step. All URLs, email addresses, mentions (i.e., @replies and @usernames), and hash tags were replaced by "url", "emailAddress", "atSign", and "hashTag", respectively. Only "#Ebola" tags were retained and converted to "ebola" in order to preserve mentions of the disease. The "RT" tags at the beginning of the re-tweets were also removed and any redundant tweets (e.g., re-tweets of the same text) were then filtered out. This resulted in preserving only those re-tweets for which the original tweets were missed in the time frame when data capture was in process.

Amazon's Mechanical Turk[20] was used to acquire human judgements of the emotion labels for each tweet in the set of 4,405 *Ebola* tweets. The qualification criteria for Mechanical Turk workers who labelled the data included: i) they were "categorization masters", ii) located in the US (as a proxy to ensure their English was of reasonable standard), and iii) achieved at least 90% accuracy on a test that involved labelling of 10 tweets in to one of the nine emotion classes. Table 1 summarizes the distribution of classes over the resultant tweet set.

A second round of pre-processing was carried out on the labelled dataset before training the binary Naïve Bayes classifiers for the nine classes. This included tokenization, lowercasing of tokens, removal of stop-words, and lemmatization.

2.4. *Experiments and Discussion*

The two classifiers were applied to the Ebola emotion dataset. The baseline classifier, as it is unsupervised, was tested on the full dataset. The ML classifier was trained and tested in a 10-fold cross-validation scenario. The macro average of precision, recall, and F1 measures were calculated over the nine classes for each classifier, with each feature set (see section 2.2). Table 2 summarizes the results (some results not shown for clarity).

The results in Table 2 demonstrate that the baseline lexicon-based classifier is strongly outperformed by the ML-based Naïve Bayes classifier, even with the basic bag-of-words features. Adding features beyond the bag-of-words features to the ML classifier had an incremental effect on the performance of the classifier. In general, bag-of-words features may result in higher classification performances as the number of input texts grow.[19] Since user tweets are mostly short pieces of text, the incremental effect of additional features is expected. On the other hand, in most cases, lemmatization of tweet tokens had only marginal impact on the results, so we have elided results without lemmatization from Table 2, except for the scenario with the overall best performance. When lemmatization was off, the highest classification performance was achieved by the ML-based classifier that utilizes all the possible feature sets. Among the different feature sets used in combination with bag-of-words features, the

Table 2. Binary emotion classification results on the set of 4,405 tweets with different feature sets. The Lexicon-Based (LB) measures were obtained on the entire data set as the test set while the Naïve Bayes (NB) measures were calculated using 10-fold cross validation. Note: bow=bag-of-words, LBsim=lexicon-based similarity measure (see section 2.1), eVoc=emotion vocabulary, emt=emoticon, punc=punctuation, sent=sentiment, rest=LBsim+sent+punc, p=precision, r=recall, f=F1 score, M.avg.=macro average, +[*] means NB/bow+[*], and lem=lemmatization. Except for +sent, all results are with +lem only.

Class	Metric	LB	NB/bow	+LBsim	+eVoc	+emt	+punc	+sent(iment)		+rest
								-lem	+lem	
Sarcasm	p	.517	.782	.793	.786	.787	.801	.798	.784	.794
	r	.910	.781	.791	.785	.787	.801	.798	.783	.791
	f	.659	.781	.790	.784	.786	.800	.797	.783	.791
News-rel.	p	.0	.827	.821	.824	.823	.828	.835	.829	.830
	r	.0	.821	.820	.822	.816	.825	.830	.825	.830
	f	.0	.823	.820	.823	.819	.826	.832	.827	.830
Anger	p	.466	.799	.726	.815	.807	.760	.776	.789	.769
	r	.774	.780	.726	.802	.805	.742	.768	.790	.755
	f	.582	.777	.720	.800	.798	.731	.747	.771	.733
Criticism	p	.0	.679	.688	.636	.636	.662	.675	.675	.661
	r	.0	.680	.683	.637	.636	.665	.670	.667	.660
	f	.0	.670	.679	.633	.633	.660	.663	.665	.652
Surprise	p	.473	.489	.566	.617	.609	.681	.658	.624	.707
	r	.791	.512	.546	.590	.603	.646	.618	.619	.677
	f	.592	.472	.486	.574	.560	.626	.594	.591	.659
Fear	p	.513	.711	.688	.697	.665	.695	.653	.717	.666
	r	.963	.701	.702	.681	.654	.687	.652	.709	.653
	f	.669	.678	.671	.664	.611	.676	.629	.673	.642
Happiness	p	.487	.717	.714	.760	.740	.668	.745	.842	.758
	r	.918	.702	.735	.750	.726	.658	.758	.820	.747
	f	.636	.681	.669	.735	.710	.646	.722	.824	.741
Disgust	p	.471	.742	.681	.695	.648	.686	.667	.743	.692
	r	.784	.716	.666	.681	.635	.684	.674	.716	.690
	f	.588	.661	.614	.654	.603	.661	.626	.696	.679
Sadness	p	.537	.829	.675	.785	.767	.729	.717	.821	.646
	r	.956	.829	.717	.771	.771	.729	.712	.754	.708
	f	.687	.799	.646	.720	.724	.686	.656	.739	.622
M.avg.p		.385	.731	.706	.735	.720	.723	.725	**.758**	.725
M.avg.r		.677	.725	.710	.724	.715	.715	.720	**.743**	.723
M.avg.f		.491	.705	.677	.710	.694	.701	.696	**.730**	.706

sentiment features (with lemmatization) resulted in the highest classification macro average values (precision=0.758, recall=0.743, and F1=0.730). This classification setting was therefore selected for the next experiments to detect shifts in emotions expressed in user tweets.

3. Unsupervised Emotion Shift Detection

We then explored the measurement of emotional shifts in user tweets around public health incidents using the best-performing emotion classification model. We propose a method for shift detection, and test it on a focused dataset collected for the study.

3.1. *Data*

3.1.1. *Ebola Incident Tweet Corpora*

We collected Twitter microblogs (tweets) around the time of two reported cases of possible Ebola infection in London. The first event involved a health worker named Pauline Cafferkey, who was diagnosed with Ebola in Glasgow on December 29, 2014 and transferred to London the following day. News reporting of the event began on December 30, 2014. The second event involved another healthcare worker who suffered a needle-stick injury in Sierra Leone and was flown to London for treatment. News reporting of the event began on January 31, 2015.

Specifically, we collected sets of tweets from London containing a mention of the word "ebola", for a period of 7 days prior and 7 days after each event. In this way, the time windows for tweets analysed for each of these events are disjoint. We refer to these datasets as the *ebola-event* datasets, and the subsets corresponding to the two time periods under study as the *pre-event corpus* (7 days prior) and the *post-event corpus* (7 days after), maintained separately for each event. We consider the day the event was reported as the split point. Tweets on that day are included in the *post-event corpus.*

To establish a reference dataset, we then downloaded a set of tweets in a similar way to the *ebola-event* datasets, but at a time period distinct from the events in London, and from a region remote to those events. We selected tweets mentioning Ebola from Australia in the time period December 09-22, 2014. As Australia has had no known cases of Ebola infection. and it is an issue that likely does not directly impact the Twitter users writing the collected tweets, this dataset should capture "normal", background dialogue about Ebola. We refer to this as the *ebola-background* dataset. We divide this dataset into two subsets, arbitrarily at the mid-point of the time period to obtain two subsets of Ebola-related tweets representing a comparable time frame to each of the *ebola-event* datasets. We refer to these subsets as the *pre-ebola-background* corpus and the *post-ebola-background* corpus. Figure 1 summarizes the tweet datasets that were collected and analyzed in this work.

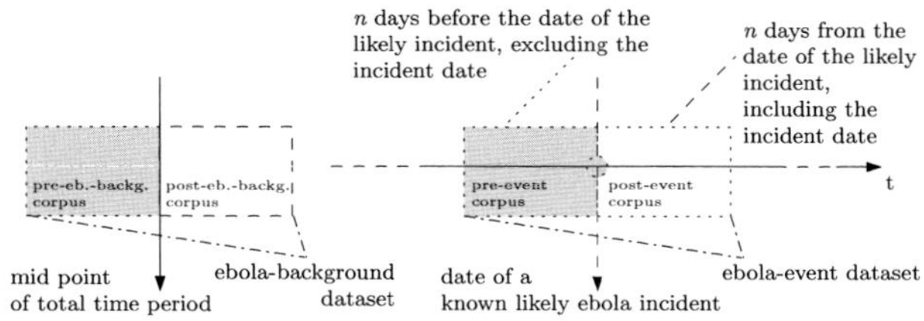

Fig. 1. The tweet collection schema used for analyzing and monitoring changes in user expressed emotions.

3.1.2. *Tweet Retrieval and Processing*

The tweets for the *ebola-event* datasets and the *ebola-background* dataset were collected using the Twitter API. The geo-codes of the tweets were used to retrieve only tweets posted by the

Table 3. Tweet and vocabulary statistics of the *ebola-event* datasets and *ebola-background* dataset with a window of 7 days before and after set dates, tied to likely Ebola incident in the region under study for the *event* data. The size of the vocabulary (|vocab.|) is equal to the number of distinct tokens.

Dataset	Date (±7)	pre-corpus		post-corpus	
		#tweets	|vocab.|	#tweets	|vocab.|
ebola-event-1	Dec/29/14	73	204	337	906
ebola-event-2	Jan/31/15	165	700	90	417
ebola-backg.	Dec/16/14	429	1453	340	1208

users from the specific regions under study. A radius of 200 kilometers was used around a specific geo-code, which for a city roughly corresponds to the geographic center of the city. Re-tweets and non-English tweets were excluded from the retrieved set of tweets. Only tweets containing a mention of the specific keyword "ebola" (either as hashtag or an individual word) were retained for both the *ebola-event* datasets and *ebola-background* dataset. No further analysis was performed for either finding the location of tweets that did not have explicit geo-code tags, or identifying any other tweets that may have been related to Ebola with no explicit mention of the disease.

The *ebola-event* datasets and *ebola-background* dataset were organized by retrieving tweets using the above query parameters, restricted to 7-day windows. This is the maximum number of days that one can move back in history of tweets and retrieve microblogs at any given time when using the current Twitter API. All of the tweets retrieved for each time window were put together in one text corpus, i.e., two tweet corpora were created for each dataset; one containing the tweets related to up to 7 days prior to the likely incident and another containing 7 days of tweets starting from the date of the likely incident in the region. Table 3 shows the statistics of the two datasets.

Textual modeling of retrieved tweets required some pre-processing of the tweet corpora, including tokenization, surface normalization, and removal of stop-words from the dictionary of terms for each tweet corpus. This processing was performed using MALLET.[18]

3.2. *Experiments and Discussion*

Our experiments assess the emotion class distributions in tweet corpora, in order to determine whether there are discernible differences in the emotions expressed in user tweets on the topic of an infectious disease that arise when the threat shifts from being abstract to being more immediate. To examine this, we considered the differences *within* the various datasets that we have collected — comparing the pre-event and post-event corpora.

3.2.1. *Corpus-Level Emotion Distribution Analysis*

Each tweet in each of the tweet corpora was first classified into one of the nine emotion and non-emotion classes introduced in section 2. For this, the highest-performing emotion classifier model trained on the distinct set of labelled Ebola tweets (see Section 2.4) was utilized to predict an emotion class label for each tweet. Table 4 summarizes the results of this step. Then, the differences in the distributions of classes between pairs of corpora were measured using

Table 4. Distribution of nine classes over the different tweet corpora obtained using the best-performing emotion classifier. For each data set we report two numbers, (X,Y): the number of instances classified as positive=X and the number of instances classified as negative=Y, Critic.=Criticism, Happ.=Happiness, pre/post-e-x=pre/post-ebola-event-x, pre/post-bkg.=pre/post-ebola-background.

Dataset	Sarcasm	News-rel.	Anger	Critic.	Surprise	Fear	Happ.	Disgust	Sadness
pre-e-1	71,2	73,0	73,0	73,0	73,0	73,0	73,0	73,0	63,10
post-e-1	298,39	322,15	300,37	325,12	265,72	288,49	308,29	298,39	277,60
pre-e-2	133,32	162,3	114,51	165,0	113,52	115,50	105,60	138,27	112,53
post-e-2	77,13	88,2	79,11	89,1	75,15	85,5	76,14	79,11	77,13
pre-bkg.	428,1	428,1	373,56	395,34	338,91	405,24	362,67	407,22	395,34
post-bkg.	255,85	332,8	255,85	295,45	283,57	230,110	308,32	309,31	294,46

Table 5. Statistical paired t-test analysis of class distributions in the different datasets in terms of positive and negative classified instances. A † shows a statistically significant p-value at the 5% level.

Classes	Dataset	p-value
6 emotions	ebola-event-1	0.004†
	ebola-event-2	0.002†
	ebola-backg.	0.259
6 emotions + 3 non-emotions	ebola-event-1	0.009†
	ebola-event-2	0.007†
	ebola-backg.	0.079

statistical paired t-tests. The statistical significance analysis of the differences between class distributions was performed for two groups of instances per class per pair of tweet corpora: group 1) all of the instances that were classified as positive (e.g., happy), and group 2) all of the instances that were classified as negative (e.g., not-happy). This analysis was done in terms of the pure-emotion classes (i.e., the six basic emotions) as well as all of the nine emotion and non-emotion classes.

The distribution differences *within* each tweet corpus (i.e., the two *ebola-event* datasets and the *ebola-background* dataset) were then calculated. Each time-delimited corpus of each dataset was compared against its neighboring counterpart; that is, the *pre-ebola-event-1* and the *post-ebola-event-1* corpora were compared with each other, the *pre-ebola-event-2* and the *post-ebola-event-2* corpora were compared with each other; and finally the two subsets of the *ebola-background* dataset were compared. Table 5 summarizes the results of this experiment.

In Table 5, all of the p-values obtained for comparing the tweet corpora before and after the incidents in London indicate statistically significant differences (at the 5% level) between class distributions. On the other hand, none of the comparisons between the tweet corpora in the *ebola-background* dataset shows a statistically significant difference. This suggests that the distribution of six basic emotions and/or the nine emotion and non-emotion classes in user tweets shift significantly as a result of salient health incidents such as Ebola.

Table 6. KL-divergence analysis of emotion classification distributions for the three datasets. Since KL-divergence is non-symmetric, X,Y values mean X=KL-divergence of tweet corpus before vs. after, and Y=KL-divergence of tweet corpus after vs. before.

Class	ebola-event-1	ebola-event-2	ebola-backg.
Sarcasm	0.077,0.119	0.013, 0.012	0.395, 1.377
News-rel.	0.066, ∞	0.001, 0.001	0.023, 0.048
Anger	0.168, ∞	0.175, 0.139	0.063, 0.074
Criticism	0.052, ∞	0.016, ∞	0.020, 0.024
Surprise	0.347, ∞	0.096, 0.083	0.010, 0.009
Fear	0.227, ∞	0.436, 0.278	0.312, 0.494
Happiness	0.130, ∞	0.186, 0.154	0.028, 0.024
Disgust	0.177, ∞	0.011, 0.010	0.016, 0.019
Sadness	0.009, 0.009	0.144, 0.119	0.022, 0.026

3.2.2. *Emotion-Level Distribution Analysis*

To understand how the emotions expressed in user tweets shift as a result of likely Ebola incidents, further analysis was carried out. Here, we measure the distribution of the tweets that were classified positive vs. negative with respect to each of the nine emotion classes before and after the likely incidents.

For this analysis, Kullback-Leibler divergence[21] was utilized. KL-divergence, also known as cross-entropy or information divergence, is a non-symmetric measure for the difference in two probability distributions P and Q over the same event space. On a finite set χ, the KL-divergence between the two probability distributions P and Q is calculated using Equation 1. In this case, P and Q represent probability distributions of positive and negative instances of a specific emotion class for tweet corpora prior to and after the likely incidents. The measure has been shown to be useful for comparing linguistic corpora in prior work.[22,23]

$$D_{kl}(P\|Q) = \sum_{x \in \chi} P(x) \log_n \frac{P(x)}{Q(x)} \tag{1}$$

Table 6 shows the results of KL-divergence analysis of the distribution of the results of emotion classifications (per emotion) between the two tweet corpora in each dataset. In addition to revealing the distributions of positive and negative classified instances per emotion class in each dataset, the results in Table 6 demonstrate how the differences in classifications distributions across the datasets vary. For instance, it can be observed that the KL-divergence values for class "surprise" are larger in the two event datasets (0.347,∞ and 0.096,0.083 for positive and negative classified instances) compared with those between the two corpora of the *ebola-background* dataset (0.010,0.009). In this particular case, the KL-divergence measures re-confirm our findings in previous sections; differences in class distributions are only significant between the tweet corpora in the two *ebola-event* datasets (note that smaller KL-divergence values indicate more similar probability distributions).

4. Limitations

This work has tackled the problem of understanding emotional shift as a result of likely disease outbreaks in particular regions of the world. However, it is important to note several limitations

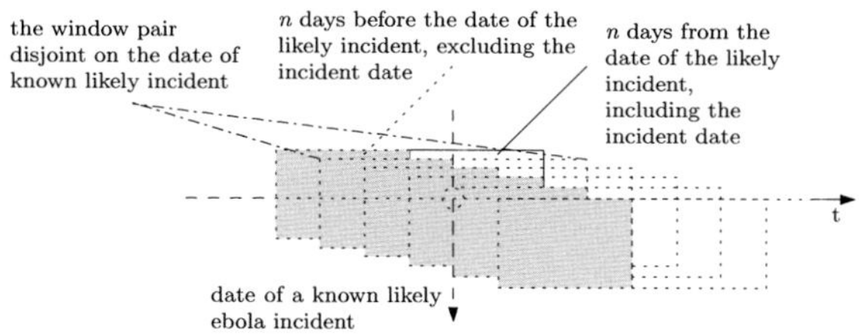

Fig. 2. The detection architecture for capturing microblogs and monitoring emotional changes in user posts

of the study, which we will address in future work.

First, the three tweet datasets that were collected may have been more directly comparable if they were collected from exactly the same region. While the two *ebola-event* datasets were from the same urban center in Europe (i.e., London, UK), *ebola-background* dataset was collected from Australia. While this allowed us to compare the event-related tweets with a neutral background set, a dataset from the same geographical region but separate in time from the events would provide a better assessment of the methods. An additional dataset from Australia collected over the same time period as the active events would also have been preferable as a background set. Together, these datasets would allow us to contrast geographical separation and temporal separation in terms of vocabulary. Due to the history restrictions on the Twitter API, it is not possible to re-create such datasets after the fact.

Second, due to the sparseness of geo-location meta-data in tweets, a number of related microblogs from the specific region were missed and not included in the three datasets we collected. Other researchers have investigated this problem in other contexts.[24] An extension of this study would be to utilize other tweet features for locating microblogs to improve the data collection procedure.

5. Towards a Detection Model

The long-term objective of our work is to provide practical evidence for early discovery and timely alarming of localized pandemic outbreaks and salient health threats. We propose that this aim can be achieved through continuous monitoring of user microblogs, specifically through identification of sudden emotion shifts. Figure 2 depicts the architecture of the proposal; in which microblog emotions are analyzed and monitored for changes in the distribution of emotions. The size and significance of the changes in emotion distributions can subsequently be utilized, either individually or in combination with other sources of evidence, to detect likely incidents or outbreaks that are of concern to the public. It is expected that the proportion of emotional shift reaches its highest value for the two time windows that are disjoint on the specific date when a putative disease-related incident or outbreak occurs. Although our focus has been on early identification of localized incidents, more generally, the proposed methodology can be utilised to detect any wave of panic in public related to other phenomena.

We have taken the initial steps towards reaching this goal by validating the underlying assumption that it is possible to observe emotion changes in neighboring sets of Twitter microblogs across a given time-point corresponding to the start of a reported health threat. We have established the viability of the approach, although further experiments are required to explore its application to real-time streaming data from Twitter, and to determine its effectiveness for early detection. Of particular interest is whether Twitter provides a meaningful information source for detecting concern about major diseases ahead of news reporting.

We also intend to capture a larger number of tweets and tweet corpora over time to further our understanding of the nature of *vocabulary or lexical shifts* around health threats, in addition to emotion shifts already studied in this work. We would like to implement an active monitoring and detection procedure over specific regions of the world for any outbreaks of Ebola or similar pandemic threats that may be both emotionally and lexically monitored and detected.

6. Conclusions

We have analyzed the variation in emotion in Twitter microblogs that are posted by users prior to and after an identified health threat, building on a text-based emotion classifier to produce a statistical assessment of emotion distributions. The combined classification and corpus analysis approach has promising application in online monitoring and detection of outbreaks in streaming textual data.

Different strategies for emotion classification in the context of serious public health events were studied in this work, including an unsupervised lexicon-based approach and a supervised machine learning-based classifier. Our experiments on a large set of Ebola tweets demonstrated that the ML-based classifier achieved the highest emotion classification performance when the tweets were represented using sentiments derived from the Stanford Sentiment Analyzer, combined with lemmatized bag-of-words features.

We considered differences in the distributions of emotion class labels assigned to microblogs across tweet corpora collected from two recent salient Ebola threats, examining variations in both corpus-level emotion and emotion-level changes. In our experiments, we found that there were statistically significant differences in the distribution of emotions in the tweet corpora that belong to the time periods before and after likely incidents of Ebola. There were no such differences for the two tweet corpora in a background dataset that was not aligned to any Ebola incident. This suggests that the distribution of predicted emotion class labels for tweets, based on Ekman's six basic emotions plus the three non-emotion classes "sarcasm", "news-related", and "criticism", can be used as an indication of the occurrence of pandemic health threats.

We will explore the broader capacity of this work to detect emergent health threats of concern to a localized community, prior to formal reporting, in future work. Currently, we are developing this study to consider lexical shifts in the tweet corpora alongside the emotion class distributions discussed in this work. We are working to identify the lexical items that distinguish tweets from before and after likely health threats.

7. Acknowledgments

We thank the Victoria State Government's Department of Health and Human Services in Melbourne, Victoria, Australia for funding support.

References

1. K. J. Henning, *Morbidity and Mortality Weekly Report (MMWR)* **53(suppl)**, 5 (2004).
2. N. Elhadad, L. Gravano, D. Hsu, S. Balter, V. Reddy and H. Waechter, Information extraction from social media for public health, in *KDD at Bloomberg: The Data Frameworks Track*, (New York City, US, 2014).
3. M. J. Paul, M. Dredze and D. Broniatowski, *PLOS Currents: Outbreaks* (2014).
4. C. Chew and G. Eysenbach, *PloS one* **5**, p. e14118 (2010).
5. J. Gomide, A. Veloso, W. Meira, V. Almeida, F. Benevenuto, F. Ferraz and M. Teixeira, *Proceedings of the ACM WebSci'11, June 14-17 2011, Koblenz, Germany.* , 1 (2011).
6. M. Odlum and S. Yoon, *American Journal of Infection Control* **43**, 563 (2015).
7. J. Ginsberg, M. H. Mohebbi, R. S. Patel, L. Brammer, M. S. Smolinski and L. Brilliant, *Nature* **457**, 1012 (2009), 10.1038/nature07634.
8. N. Collier, S. Doan, A. Kawazoeand, R. M. Goodwin, M. Conway, Y. Tateno, Q. Ngo, D. Dien, A. Kawtrakul, K. Takeuchiand, M. Shigematsu and K. Taniguchi, *Bioinformatics* **24**, 2940 (2008)
9. J. Bollen, A. Pepe and H. Mao, *CoRR* **abs/0911.1583** (2009).
10. P. Ekman, Universals and cultural differences in facial expression of emotion, in *Nebraska Symposium on Motivation*, (Lincoln, Nebraska, 1972).
11. A. Agrawal and A. An, Unsupervised emotion detection from text using semantic and syntactic relations, in *Proceedings of the The 2012 IEEE/WIC/ACM International Joint Conferences on Web Intelligence and Intelligent Agent Technology - Volume 01*, 2012.
12. S. Aman and S. Szpakowicz, Using roget's thesaurus for fine-grained emotion recognition, in *Proceedings of the Third International Joint Conference on Natural Language Processing*, 2008.
13. S. Aman and S. Szpakowicz, Using Rogets thesaurus for fine-grained emotion recognition, in *Proceedings of the Third International Joint Conference on Natural Language Processing* (Hyderabad, India, 2008).
14. W. Wang, L. Chen, K. Thirunarayan and A. P. Sheth, Harnessing Twitter "big data" for automatic emotion identification, in *Proceedings of the International Conference on Privacy, Security, Risk and Trust and the 2012 International Confernece on Social Computing (SocialCom)*, (Amsterdam, 2012).
15. T. Drummond, Emotion vocabulary http://www.sba.pdx.edu/ faculty/ mblake/ 448/ FeelingsList.pdf.
16. C. F. Baker, C. J. Fillmore and J. B. Lowe, The Berkeley FrameNet project, in *Proceedings of the 36th Annual Meeting of the Association for Computational Linguistics and 17th International Conference on Computational Linguistics - Volume 1*, ACL '98 (Association for Computational Linguistics, Stroudsburg, PA, USA, 1998).
17. D. McNair, M. Loor and L. Droppleman, *Profile of Mood States* 1971.
18. A. K. McCallum, MALLET: A Machine Learning for Language Toolkit http://mallet.cs.umass.edu (2002).
19. R. Socher, A. Perelygin, J. Y. Wu, J. Chuang, C. D. Manning, A. Y. Ng and C. Potts, Recursive deep models for semantic compositionality over a sentiment Treebank, in *Proceedings of the Conference on Empirical Methods in Natural Language Processing (EMNLP 2013)*, (Seattle, USA, 2013).
20. M. Buhrmester, T. Kwang and S. D. Gosling, *Perspectives on Psychological Science* **6**, 3 (2011).
21. S. Kullback and R. Leibler, *The Annals of Mathematical Statistics* **22**, 79 (1951).
22. P. Rayson and R. Garside, Comparing corpora using frequency profiling, in *Proceedings of the Workshop on Comparing Corpora, held in conjunction with ACL 2000*, 2000.
23. K. Verspoor, K. B. Cohen and L. Hunter, *BMC Bioinformatics* , p. 10:183 (2009).
24. J. Mahmud, J. Nichols and Drews, *ACM Trans. Intell. Syst. Technol.* **5**, 47:1 (July 2014).

PREDICTING INDIVIDUAL WELL-BEING THROUGH THE LANGUAGE OF SOCIAL MEDIA

H. ANDREW SCHWARTZ
Stony Brook University, Computer Science
Stony Brook, NY 11794

MAARTEN SAP, MARGARET L. KERN*, JOHANNES C. EICHSTAEDT, ADAM KAPELNER†, MEGHA AGRAWAL, EDUARDO BLANCO‡, LUKASZ DZIURZYNSKI, GREGORY PARK, DAVID STILLWELL§, MICHAL KOSINSKI¶, MARTIN E.P. SELIGMAN, AND LYLE H. UNGAR
University of Pennsylvania,
Positive Psychology Center & Computer and Information Sciences
Philadelphia, PA 19104

We present the task of predicting *individual* well-being, as measured by a life satisfaction scale, through the language people use on social media. Well-being, which encompasses much more than emotion and mood, is linked with good mental and physical health. The ability to quickly and accurately assess it can supplement multi-million dollar national surveys as well as promote whole body health. Through crowd-sourced ratings of tweets and Facebook status updates, we create message-level predictive models for multiple components of well-being. However, well-being is ultimately attributed to people, so we perform an additional evaluation at the user-level, finding that a multi-level cascaded model, using both message-level predictions and user-level features, performs best and outperforms popular lexicon-based happiness models. Finally, we suggest that analyses of language go beyond prediction by identifying the language that characterizes well-being.

1. Introduction

As human beings, we desire "the good life". When the British Broadcasting Cooperation (BBC) asked 1,001 Britons what the prime objective of their government should be – "greatest happiness" or "greatest wealth" – 81% answered with happiness.[1] In other studies, an average of 69% of people globally rate well-being as more important than any other life outcome.[2]

Beyond its popular appeal, another reason to consider well-being is that it is linked with positive life outcomes, including health and longevity.[3–6] Although it is not clear if well-being *causes* good health, it provides an indication of healthier or riskier individual trajectories, long before health problems develop.[7] Thus, the focus on well-being offers a preventative approach to public and personal health, with important economic consequences.

Well-being is more than simply positive emotion or mood. Psychologists, organizations, and governments measuring well-being are now using multi-dimensional measures that include a range of factors including meaning in life, engagement in activities, and the state of one's relationships, in addition to positive emotion.[8] While some language analyses have explored "happiness" based on emotion or mood,[9–14] modeling the broader construct of well-being is a relatively unexplored task.

In this paper, we present the task of predicting well-being based on natural language use. We develop a system that predicts the *satisfaction with life* (an overall evaluation of well-being) of Facebook *users* based on simple lexical and topical features. However, since individual *messages* themselves are the units of expression, we investigate the use of message-level models to improve

*University of Melbourne, Graduate School of Education, Australia
†Queens College, Department of Mathematics
‡University of North Texas, Computer Science
§Psychometrics Centre, University of Cambridge
¶Stanford Business School

user-level predictions. The message data, which was easy to come by through Amazon's Mechanical Turk, allowed us to supplement our *satisfaction with life* data, as well as explore other aspects of well-being, *PERMA* (discussed below). Lastly, for a human-level attribute like well-being, insights toward greater understanding is potentially just as important as prediction. Toward this end, we identify topics that most strongly correlate with well-being as clues to achieving "the good life".

Our unique contributions include: (a) the introduction of the task of predicting *individual* well-being, (b) the finding of a two-level, message-to-user model to perform better than models based on either independently, (c) the development of annotated well-being data across various constructs. Further, (e) we provide an analysis of the linguistic features that we find most significantly associated to individual *satisfaction with life*, and (e) we also release a well-being language model available for researchers (available to download at `wwbp.org/data.html`).

2. Background: Well-Being

Despite a desire for "the good life",[‖] well-being has traditionally been measured *indirectly* as a lack of problems (e.g., lack of depression and psychological disorder, low crime/disease/poverty rates) or by economic prosperity (i.e., gross domestic product). Reference 15 aptly notes: "if our interest is in the good life, we must look explicitly at indices of human thriving" (p. 144). Government agencies around the world are beginning to shift their attention toward *directly* measuring well-being.[16] In 2011, the U.K. Office of National Statistics piloted four well-being questions in their annual national survey, and similar efforts are now underway in Australia, Canada, France, Mexico, South Africa, and other places around the world. These initiatives follow a long history of academics who have attempted to rethink the notion of progress, deemphasizing the sole reliance on economic indicators and arguing instead that the welfare of a nation must be understood more holistically, with consideration of aspects such as social belonging, meaning, and optimism in the population.[8,17,18]

Satisfaction with Life (*SWL*) is a well-established representation of well-being, representing a person's cognitive evaluation of their own life. Measures of life satisfaction have been used reliably for several decades, and are increasingly being utilized by governments and organizations around the world as informative social indicators for policy decisions.[19] It is assessed by asking people to indicate the extent to which they agree with statements such as "In most ways my life is close to my ideal".[20] *SWL* draws on the subjects' evaluative judgments and seems to be highly comparative both within nations and between nations.[21] Overall life satisfaction strongly correlates with other well-being domains, such as meaning in life, relationships, and emotions.

PERMA. Beyond an overall evaluation of well-being, other psychologists break well-being into separate domains.[8,22] In such "dashboard approaches" — just like the "state" of an airplane is not given by a single indicator but instead by a variety of different indicators (altitude, speed, heading, fuel consumption) — well-being is best measured as separate, correlated dimensions.[8,23–25] In his well-being theory, Ref. 8 suggests five major pillars that together contribute to a person's sense of well-being: *Positive Emotions, Engagement, Relationships, Meaning*, and *Accomplish-*

‖Although debatable by well-being theorists and philosophers, for the purposes of this work, we consider the "good life", "well-being", and "satisfaction with life" synonymous. "Happiness" at times is equated with well-being, but often is used to denote positive emotion/ mood alone, so we only use this term when referencing other works that use this term.

ment (*PERMA*). Other "dashboard approaches" seek to capture subtle psychological notions such as "autonomy" or "self-acceptance".[22] Although greater specificity could be delineated, we chose the *PERMA* constructs since they capture fairly explicit and often foregrounded ends people pursue (i.e., things discussed in social media).

Positive emotion includes positively valenced emotions such as joy, contentment, and excitement. *Engagement* is a multi-dimensional construct that includes behavioral, cognitive, and affective components. It can refer to involvement and participation in groups or activities, enthusiasm and interest in activities, commitment and dedication to work, and focused attention to tasks at hand.[26,27] For our purposes here, we define it in terms of passion and involvement in life, as opposed to apathy and boredom. *Relationships* (or positive relationships) includes trusting others, perceiving others as being there if needed, receiving social support, and giving to others.[28] Considerable evidence identifies the importance of positive relationships for supporting health, longevity, and other important life outcomes.[29] *Meaning* in life captures having a sense of purpose, significance, and understanding in life.[30,31] It can also include transcending the self, feeling a sense of connection to a higher power or purpose, and provides goals or a course of direction to follow.[32] *Accomplishment* is often defined in terms of awards, honors, and other objective markers of achievement.[25,33] For our purposes here, we focus on the subjective side, in terms of a personal sense of accomplishment. It includes a sense of mastery, perceived competence, and goal attainment.[34]

Both PERMA and SWL are typically measured through Likert scales,[35] where people are presented a statement (e.g. "I am satisfied with my life") and asked the extent to which they agree or disagree with it.

3. Related Tasks

Among others, predicting emotion, mood, personality, and classic sentiment analyses are related to our task. In this section, we provide a cursory review of analyses with social media and related approaches. To our knowledge, well-being prediction as more than positive emotion is a novel task, and considering *message*-level predictions to improve a *user*-level model has not been explored.

Although sentiment analysis usually takes place at the single document or sentence level,[36–39] some have explored multi-level approaches. For example, Ref. 40 utilized a cascaded model where sentence level subjectivity classifiers are used to determine whether a sentence should be included for document-level analysis. Reference 41 expanded the idea by incorporating a joint-model of sentence and document level sentiment annotations. We find this sentence-to-document prediction analogous to our multi-level approach. However, such efforts typically aim at predicting the sentiment of text; they do not model any feature at the *user* level (human attributes). Our average user writes 123 messages, far more than the number of sentences in the typical sentiment-analyzed texts.

Some sentiment tasks do address human-level attributes. For example, signals used in distant supervision,[42,43] where heuristics replace manual annotations (e.g., ":(" = negative polarity), somewhat captures people's emotional states. Reference 44 attempts to differentiate between the emotions of the writer and the reader of content on a microblogging platform with social network characteristics. More directly, some have looked at predicting emotion of text[9] or more specifically learned the language of happy and sad blogposts based on self-annotated moods in an online live journal.[10] Reference 45 identified thesaurus-based topics related to the emotional expressions from an English blog corpus. Reference 46 constructed models from a large corpora of world knowledge to identify the affective tone across six basic emotions in texts.

While mood and sentiment analysis aim at text annotations, there have been a few tasks looking at modeling human-level conditions by the language one uses. Work on personality follows the widely accepted Big-Five personality traits.[47] Reference 48 analyzes correlations with basic Facebook features (number of friends, photos, tags, likes, etc.) and Ref. 49 with LIWC[50] word categories (pronouns, cognitive process, etc.). Personality prediction efforts use LIWC categories coupled with other shallow features[51,52] and n-grams.[53] Other tasks include finding indicators of psychological health,[54,55] or predicting political orientation.[56,57] In 2015, a shared task was organized to detect if a Twitter user suffered from depression or PTSD.[58] While these works predict attributes at the user level, they do not incorporate message-level features.

Recently, Ref. 11 used a premade lexicon of positive and negative emotion words to measure "gross national happiness" and both Refs. 12 and 39 used MTurk ratings of individual words for positive or negative valence. Dodds' MTurk application served as a model for our MTurk data collection, but rather than asking workers to annotate individual words, we had whole messages annotated, thus putting words into context. References 11,12 and 39 demonstrated face validity (i.e., people are happier on the weekends or sad after celebrities die), though they lacked an empirical evaluation of the degree to which they accurately measure happiness. By using these approaches as baselines for *SWL* prediction, we provide an empirical evaluation here. Furthermore, in this paper we model well-being as more than emotion and mood.

4. Method

We build predictive models of well-being, as measured through the *satisfaction with life* (*SWL*)[20] and *PERMA*[25] scales. We describe message-level models, a user-level model, and then a cascaded model whereby message predictions inform the user-level predictions.

As the first work attempting the prediction of user-level *SWL* using lexical features, we explore a moderately sized and consistent feature space for both user-level and message-level models:

ngrams. We used both unigrams and bigrams as features in this task, which we extracted using an informal text tokenizer ** that handles social media content and markup such as emoticons. Trigrams were not included in order to keep the number of features smaller.

topics. We used the 2000 topics released by Ref. 59, created by running *latent dirichlet allocation* (*LDA*) over a set of 18 million Facebook status updates from the MyPersonality application.[60] These topics were derived from the same domain, and thus add a more coarse-grained Facebook lexical feature to our models. A user, u's, usage of a topic, t, was calculated as: $p(t|u) = \sum_{w \in words_u} p(t|w) * p(w|u)$, where $p(t|w)$ is the probability of a topic given a word (a value provided by the generated topic model) and $p(w|u)$ is a user's probability of mentioning word w. Additionally, beyond simply prediction utility, topics provide insight into the latent categories of language that characterize well-being.

lexica. We also included the manually developed categories of words from Linguistic Inquiry and Word Count (LIWC)[61] as well as the weighted lexica from Dodd's Hedonometer.[12] While LDA

** `http://wwbp.org/data.html`

topics provide a data-driven set of categorical features, these lexica provide features grounded in psychological and linguistic theory and human judgment. LIWC, in particular, was developed over decades with many iterations[61] and its *positive emotion* and *negative emotion* categories are widely used, including being used by Ref. 11 in his measure of "gross national happiness" (*GNH*) and by Ref. 62 to track diurnal mood variation. *GNH* along with Dodd's *Hedonometer* – both lexica – also function as baselines for our predictive models.

Each feature is included as binary (1 if mentioned at least once, 0 otherwise) as well as in relative frequency over a its message or user ($\frac{freq(feature)}{\sum_{word \in doc} freq(w)}$). This results in hundreds of thousands of features. Thus, to reduce the change of overfitting, we filtered out infrequent features defined as those used by less than 10% of users or in less than 0.1% of messages.

4.1. *Message-level Models.*

We explore models for finding expressions of both *SWL* and positive and negative expressions of the five *PERMA* components. The features described above are aggregated at the message-level, where *n-grams* are encoded as booleans (i.e., whether they exist or not in the message) and the others are encoded as frequencies (probability over all words in message). We then use Randomized Principal Component Analysis[63] (*RPCA*) to transform the space to a more manageable size for ridge regression. Specifically, we reduce the feature matrix to $\frac{1}{4} * train\_size$ components for all models utilizing more than that many features. The projection matrix from *RPCA* is stored as part of the model such that prediction / test data is transformed based on a projection matrix fit to training data. Over the training data, we tested other prediction algorithms, such as Lasso (L1 penalized) regression,[64] which works well with sparse data, but ridge regression with RPCA performed better. Given that we have annotations for SWL and both negative and positive aspects of *PERMA*, we train a model for each outcome, resulting in 11 regression models total.

We have released a version of the PERMA language model without RPCA in the form of a weighted lexicon, extracted using the method described in Ref. 65 (due to Facebook policy restrictions, we are not able to release the annotated messages).

4.2. *User-Level Models.*

Our basic user-level model fits ridge regression[66] of the ngram, topic, and lexicon features to *SWL* scores. Just like with message level models, we use *RPCA* to reduce the dimension of the ngram feature space. We also tried Lasso (L1 penalized) regression,[64] which works well with sparse data (such as ngrams), but *RPCA* and combined with Ridge Regression yielded better results.

4.3. *Cascaded Message-to-User Level Well-Being Prediction*

Although well-being is attributed to people, we believe it might be possible to capture expressions of it at the message level and to pass this information along to the user level. Our cascaded model aggregates predictions of all message-level attributes across all of each user's messages and incorporates them as the mean prediction across a user's messages. For instance, $SWL_{user} = \frac{1}{\#msgs} \sum_{msg \in user} SWL_{msg}$. This in turn becomes a feature supplied to the user-level model. For example, if we train message-level predictors for both polarities for each of the five domains of

PERMA, this results in 10 features within the cascaded model: 5 domains $*$ 2 polarities. User-level attributes were not distributed to the messages because our annotated messages did not have any user-level attributes. This same cascading concept could be used in reverse, when the goal is message-level prediction, rather than user-level attributes. All algorithms were carried out using their SciKit-Learn implementation.[67]

5. Data Acquisition

Message-Level Data. We used Amazon's Mechanical Turk (MTurk) to acquire *PERMA* and *SWL* annotations for 5,100 public Facebook status updates. The status updates were randomly selected from among 230 million public Facebook messages that contained at least 50% English words according to the ASpell Official English dictionary.[††]

On MTurk, the largest online task-based labor market, tasks are completed by an on-demand labor force composed of "Turkers".[68] We set up a MTurk task where workers received $0.01 per annotation. Upon entry, turkers completed a research consent and were shown a video that explained the well-being category they were rating. Upon passing a quiz testing their understanding of the concepts, they were qualified to annotate the particular component of well-being for which they were trained.

For each of the 10 *PERMA* components, turkers indicated the "extent to which [a message] expresses" the particular component, by using a slider with rating scale that ranged from "none" (0) to "very strongly" (6). For *SWL*, workers indicated their agreement that the message indicates life satisfaction (0 = strongly disagree, 3 = neutral, 6 = strongly agree). Some examples statuses can be seen in Tables 1 and 2.

We decided to opt for getting more messages rather than getting more ratings per messages, utilizing two ratings for most messages. A third rating was brought in for disagreements (defined as those outside 1 standard deviation, 2 points, of each other). Between the two initial ratings, intra-class correlations[69] were in the "moderate" to "substantial"[‡‡] agreement range (i.e., .4 to .8) for all but two domains; positive engagement and negative accomplishment were in the "fair" agreement range (i.e., .2 to .4) suggesting these categories were more difficult to annotate and thus necessitating a third rating to improve accuracy (less accurate ratings only make our task of improving user-level predictions more difficult). In the end we used the mean rating for each message as the gold-standard.

We found the PERMA categories each contain different but related information. Considering all positive PERMA domains and then all negative domains, intercorrelations ranged from 0.36 to 0.68 (Pearson's r). Consistent with the psychological literature,[71] the highest correlations were between positive relationships ($R+$) and positive emotion ($P+$). There was an inverse but weak relationship between the positive and negative dimensions of PERMA ($r = -0.04$ to -0.39), supporting the idea of the two polarities being orthogonal. Given this, it is possible that each of the 10 PERMA categories will contribute independent information for predicting well being.

††http://misc.aspell.net/wiki/English_Dictionaries; we plan to make this data available upon ethics board approval for public sharing.
‡‡range labels provided by Ref. 70

Table 1. Examples of statuses with contrastive values for each PERMA category.

Status Update	P^+	P^-	E^+	E^-	R^+	R^-	M^+	M^-	A^+	A^-
Celebrating this amazing day.. lmao.. first of many	6.0	0.0	3.0	0.0	4.5	0.0	3.5	0.5	5.5	0.0
I wanna thank GOD for letting me see another BDAY...LOVE YA BIG MAMA I KNO U SMILING DOWN ON ME!!!!	3.0	1.7	4.0	0.0	5.0	0.0	5.5	0.0	3.0	0.0
Goin to laundry mat got hella laundry to do uuuhhhh.......just did a major clean up take him out...take him out of the game already..	0.0	5.5	3.0	0.0	4.5	1.7	0.0	0.0	0.0	0.0
I have such an amazing bf he took good care of me at the hospital.which he always takes good care of me.Im so blessed to have him.	5.0	1.0	1.7	0.0	6.0	1.0	2.0	0.5	0.0	0.0

Table 2. Example statuses expressing both polarities for categories engagement, relationships, and meaning.

Status Update	Cont. Category			
I hate wen you watch a movie and the ending is sooo predictable :/	E^+	3.0	E^-	2.7
Just when I thought my whole world had crumbled into a million pieces, you came along and brought me crazy glue and band-aids.	R^+	2.7	R^-	2.0
Another FUN bright photoshoot coming up in my future! Cant WAIT! :)	M^+	2.5	M^-	1.3

User-Level Data. Our user-level data was acquired through the MyPersonality Facebook App[60] from users who agreed to share their status updates for research purposes. We focused on users who took the Satisfaction with Life scale[20] (*SWL*), which previous research has shown has high internal consistency (reported alphas range from .79 to .89), and moderate temporal stability (reported test-retest correlations over two month intervals range from .50-.82).

Mean *SWL* in our sample was 4.3 (on a 7-point scale), consistent with the mean *SWL* reported in North American college and adult samples (typically 4.8). A subset of our sample ($N = 157$) retook the *SWL* scale six months later, and the resulting test-retest correlation ($r = .62$) was similar to those reported in past studies.[72] Test-retest correlation forms an upper-bound on the predictive accuracy we should expect to get from our models.

We further refined the data to only include users' status updates made in the 6 months prior to their taking the *SWL* questionnaire and only from those who wrote at least 500 words in that time. This resulted in a dataset of 2,198 individuals, having collectively written 260,840 messages. The messages in this dataset and those from the message-level dataset are completely disjoint. Thus, when creating the cascaded model, the message-level models will not have observed any of the users' messages before, avoiding overfitting issues. Though the message-level data set is smaller in number of messages (5,100 versus 260,000 for the user-level one), our hope is that since it has more labels (5,100 messages versus 2,200 users) can supplement to user-level data to improve accuracy. Furthermore, the benefit of a cascaded model might become even greater if fewer labeled users are available.

6. Evaluation

We evaluated our message-level, user-level, and cascaded message-to-user predictive models over a corpora of Facebook status updates. The second and third evaluation are over a user-level corpus in which volunteers from Facebook took the *SWL* questionnaire and shared their status updates. The

SWL questionnaire, an accepted metric of well-being, is used as a gold-standard for evaluating models at the user-level. Each corpus was divided randomly into 80% training/development instances and 20% test. We then test whether our models do better than a baseline of happy/hedonic lexica,[11,12] and whether a cascaded message-to-user level model improves upon the pure user-level model.

Message-level Results. Table 3 shows the results of each message-level regression model, reported using the Pearson correlation coefficient (r) between the predicted score and the annotated score. The annotated messages were randomly divided into 80% training and development (4080 messages) and a 20% test set (1020 messages). Interestingly, for some categories, *topics* were better than *ngrams*, while for others *ngrams* were better than *topics*. We also noticed that positive relationships seems to be the easiest component to predict, higher than both positive emotion and *SWL*.

Table 3. Message-level prediction scores as the Pearson correlation coefficient (r) across the PERMA and SWL categories.

features	$\mathbf{P^+}$	$\mathbf{P^-}$	$\mathbf{E^+}$	$\mathbf{E^-}$	$\mathbf{R^+}$	$\mathbf{R^-}$	$\mathbf{M^+}$	$\mathbf{M^-}$	$\mathbf{A^+}$	$\mathbf{A^-}$	**SWL**
ngrams	0.563	0.412	0.349	0.347	0.604	0.365	0.405	0.279	0.363	0.285	0.479
topics	0.500	0.430	0.252	0.405	0.500	0.417	0.400	0.286	0.260	0.322	0.495
lexica	0.413	0.442	0.205	0.292	0.445	0.376	0.217	0.205	0.147	0.292	0.430
ngrams+topics	0.598	0.492	0.369	0.421	0.641	0.401	**0.451**	0.314	0.380	**0.398**	0.550
ngrams+lexica	0.613	0.526	0.346	0.375	0.621	0.446	0.435	**0.331**	**0.414**	0.361	0.543
topics+lexica	0.509	0.464	0.247	0.376	0.525	0.423	0.372	0.260	0.307	0.268	0.505
ngrams+topics+lexica	**0.617**	**0.504**	**0.374**	**0.422**	**0.655**	**0.427**	0.441	0.311	0.402	0.352	**0.566**

User-level and Cascaded Results. Both the user-level and cascaded message-to-user-level models were evaluated over the same user-level corpus, divided such that 80% (1758 users) was used for training and development while 20% was held out for testing (440 users). The scores represent the correlation (Pearson r) between predictions over the test data and the users' scores from the *SWL* scale. Since we are studying well-being, Pearson correlation allows us to frame the results with a metric used widely in social sciences. To put our results in perspective, subjective (user-level) psychological variables typically have a "correlational upper-bound" in the range of $r = 0.3$ to $r = 0.4$ with human behaviors such as language use.[73]

Table 4 shows the user-level predictions for all combinations of features as well as the cascaded models. We see that an ngram model out-performs baselines of methods used previously for happiness prediction,[11,12,39] while our best user-level results come from a combination of *ngrams*, *topics*, and the *lexica* (though they are not significantly better than *ngrams* and *topics* alone).

Cascading models significantly boost performance, increasing it from .301 with the user-level language features alone to .333 with cascaded models. This is quite surprising, considering the message-level models were only based on 5,100 messages, but there were over 200,000 messages across all the users.

Analyses of individual message-level predictors for each component of PERMA showed all message-level predictors add to the prediction, with out-of-sample correlations ranging from $r = .15$ to $r = .247$. All domains of *PERMA*, as modeled through the language of annotated messages, had an impact on user-level *SWL*.

Table 4. User-level prediction scores as the Pearson correlation coefficient (r) . *message predictions*: message-level regression feeding cascaded model; *user features*: all user level language features (*ngrams + topics + lexica*. bold: significant ($p < .05$; p is Bonferroni corrected for multiple comparisons[74]) improvement over user-level features alone.

user-level models	*ngrams*	.262
	topics	.254
	lexica	.198
	ngrams + topics	.299
	ngrams + lexica	.269
	topics + lexica	.252
	ngrams + topics + lexica	.301
baselines	(mean)	.000
	lexica: GNH	.210
	lexica: Hedonometer	.108
cascaded models	*message predictions* alone	.236
	user features alone	.301
	message predictions + user features	**.333**

7. Discussion: Well-Being Insights

The LDA topics most highly correlated with user-level SWL shed light on the mechanisms which may contribute to a person's satisfaction with their life in a way that is in line with the psychological literature. Figure 1 shows four of the top ten positively correlated topics, and the two leading negative ones. Several of the topics tap various aspects of engagement. *excited, super, tomorrow* references

Fig. 1. The top 4 topics positively correlated (blue) and 2 topics negatively correlated (red) with *SWL*.

affective and psychological states which suggest that people are happily engrossed in activities of life.[75] *meeting, conference, staff* hints at communal engagement, which overlaps with involvement, dedication, and organizational citizenship behavior.[76] The *bored, bore, text* topic reflects the converse; disengagement merges as one of the strongest negative predictors of SWL. Engagement — a core component of $PERMA$[8] — is considered a key part of healthy aging.[77]

The *skills, management, business* topic corroborates theories that county level SWL is linked to employment in the "professional" occupation sector.[78] Theoretical psychology[79] suggests that people in high value-creation occupations, in which continuous learning, roles of responsibility and skill development are valued, would be more satisfied with their lives. The *family, friends, wonderful* topic supports the well-established idea that good relationships are a strong predictor of well-being.[29]The swearing topic emerges as the single strongest (negative) topic predictor of SWL.

8. Conclusions

We presented the task of predicting well-being, a multidimensional construct, based on natural language use. We developed predictive models of well-being, as measured through the *satisfaction with life(SWL)* scale, over Facebook volunteers. Our models significantly out-predict baselines of popular happy and hedonistic lexica.[11,12]

We created both message-level models as well as user-level models, and found a cascaded model, in which message-level predictions inform user-level predictions, gave the best performance. Additionally, we introduced corpora with annotated well-being data, and show that for such human-level information, language analysis can go beyond prediction and demonstrate insight into what leads to "the good life".

Well-being prediction is a worthwhile task for the social media mining community as the construct is gaining popularity and it is known to be linked with health, economics, and longevity. People and governments have started to recognize that economic measures alone do not capture the welfare of societies. Methodologically, there is much to explore, such as more sophisticated joint models of user and message-level information or the use of syntactic structure as features. Additionally, we suggest that language-based analyses of well-being need not end with *prediction*. Links between *SWL* and the everyday language in social-media enriches our *understanding* of well-being and its determinants, indicators and consequences.

Acknowledgment. We thank the Templeton Religion Trust for their generous support of this study.

References

1. M. Easton, *BBC News* **2** (2006).
2. E. Diener, *American psychologist* **55**, p. 34 (2000).
3. E. Diener and M. Y. Chan, *Applied Psychology: Health and Well-Being* **3**, 1 (2011).
4. R. T. Howell, M. L. Kern and S. Lyubomirsky, *Health Psychology Review* **1**, 83 (2007).
5. S. Lyubomirsky, L. King and E. Diener, *Psychological Bulletin* **131**, p. 803 (2005).
6. S. D. Pressman and S. Cohen, *Psychological bulletin* **131**, p. 925 (2005).
7. H. S. Friedman and M. L. Kern, *Psychology* **65**, p. 719 (2014).
8. M. E. P. Seligman, *Flourish: A Visionary New Understand of Happiness and Well-being* (Free Press, 2011).
9. C. Alm, D. Roth and R. Sproat, Emotions from text: machine learning for text-based emotion prediction, in *Proceedings of the conference on Empirical Methods in Natural Language Processing*, 2005.
10. R. Mihalcea and H. Liu, A corpus-based approach to finding happiness, in *Proceedings of the AAAI Spring Symposium on Computational Approaches to Weblogs*, 2006.
11. A. Kramer, An unobtrusive behavioral model of gross national happiness, in *Proc of the* 28th *int conf on Human factors in comp sys*, 2010.
12. P. S. Dodds, K. D. Harris, I. M. Kloumann, C. A. Bliss and C. M. Danforth, *Diversity* , p. 26 (2011).
13. J. Bollen, H. Mao and A. Pepe, Modeling public mood and emotion: Twitter sentiment and socio-economic phenomena., in *ICWSM*, 2011.
14. S. M. Mohammad, #emotional tweets, in *Proceedings of the First Joint Conference on Lexical and Computational Semantics-Volume 1: Proceedings of the main conference and the shared task, and Volume 2: Proceedings of the Sixth International Workshop on Semantic Evaluation*, 2012.
15. C. Peterson and N. Park, *Psychological Inquiry* **14**, 143 (2003).
16. J. E. Stiglitz, A. Sen and J.-P. Fitoussi, *The Commission on the Measurement of Economic Performance and Social Progress website. Available online at: http://www. stiglitz-sen-fitoussi* (2009).

17. M. Abramovitz, T. Scitovsky and A. Inkeles, *Bulletin of the American Academy of Arts and Sciences* , 11 (1973).
18. R. Layard, *Economics and Happiness* , 147 (2005).
19. E. Diener, R. Inglehart and L. Tay, *Social Indicators Research* , 1 (2012).
20. E. Diener, R. A. Emmons, R. J. Larsen and S. Griffin, *Journal of personality assessment* **49**, 71 (1985).
21. E. Diener, W. Ng, J. Harter and R. Arora, *Journal of Personality and Social Psychology* **99**, p. 52 (2010).
22. C. D. Ryff and C. L. M. Keyes, *Journal of personality and social psychology* **69**, p. 719 (1995).
23. J. E. Stiglitz, A. Sen and J.-P. Fitoussi, *The Commission on the Measurement of Economic Performance and Social Progress website. Available online at: http://www. stiglitz-sen-fitoussi* (2009).
24. B. S. Frey and A. Stutzer, *Happiness and economics: How the economy and institutions affect human well-being* (Princeton University Press, 2010).
25. M. J. C. Forgeard, E. Jayawickreme, M. Kern and M. E. P. Seligman, *International Journal of Wellbeing* **1**, 49 (2011).
26. J. J. Appleton, S. L. Christenson and M. J. Furlong, *Psychology in the Schools* **45**, 369 (2008).
27. M. Csikszentmihalyi, *Creativity: Flow and the psychology of discovery and invention* (Harper Perennial, 1997).
28. S. E. Taylor, *The Oxford Handbook of Health Psychology* , 189 (2011).
29. L. Tay, K. Tan, E. Diener and E. Gonzalez, *Applied Psychology: Health and Well-Being* (2012).
30. M. F. Steger, T. B. Kashdan, B. A. Sullivan and D. Lorentz, *Journal of Personality* **76**, 199 (2008).
31. V. E. Frankl, *Man's search for ultimate meaning* (Insight Books/Plenum Press, 1997).
32. S. M. Schueller and M. E. Seligman, *The Journal of Positive Psychology* **5**, 253 (2010).
33. K. A. Ericsson, *The pursuit of excellence through education* , 21 (2002).
34. L. I. Pearlin and C. Schooler, *Journal of health and social behavior* , 2 (1978).
35. R. Likert, *Archives of psychology* (1932).
36. P. Turney, Thumbs up or thumbs down? semantic orientation applied to unsupervised classification of reviews, in *Proceedings of 40th Annual Meeting of the Association for Computational Linguistics*, (Association for Computational Linguistics, Philadelphia, Pennsylvania, USA, July 2002).
37. B. Pang, L. Lee and S. Vaithyanathan, Thumbs up? Sentiment classification using machine learning techniques, in *Proceedings of the 2002 Conference on Empirical Methods in Natural Language Processing (EMNLP)*, 2002.
38. C. Strapparava and R. Mihalcea, Semeval-2007 task 14: Affective text, in *Proceedings of the 4th International Workshop on Semantic Evaluations*, 2007.
39. S. M. Mohammad, S. Kiritchenko and X. Zhu, *arXiv preprint arXiv:1308.6242* (2013).
40. B. Pang and L. Lee, A sentimental education: Sentiment analysis using subjectivity summarization based on minimum cuts, in *Proceedings of the 42nd Annual Meeting on Association for Computational Linguistics*, 2004.
41. R. McDonald, K. Hannan, T. Neylon, M. Wells and J. Reynar, Structured models for fine-to-coarse sentiment analysis, in *Annual Meeting-Association For Computational Linguistics*, (1)2007.
42. J. Read, Using emoticons to reduce dependency in machine learning techniques for sentiment classification, in *Proceedings of the ACL Student Research Workshop*, (Association for Computational Linguistics, Ann Arbor, Michigan, June 2005).
43. D. Davidov, O. Tsur and A. Rappoport, Enhanced sentiment learning using twitter hashtags and smileys, in *Coling 2010: Posters*, (Coling 2010 Organizing Committee, Beijing, China, August 2010).
44. Y.-j. Tang and H.-H. Chen, *Sentiment Analysis where AI meets Psychology (SAAIP)* , p. 11 (2011).
45. S. Aman and S. Szpakowicz, Identifying expressions of emotion in text, in *Text, Speech and Dialogue*, 2007.
46. C. Strapparava and R. Mihalcea, Learning to identify emotions in text, in *Proceedings of the 2008 ACM symposium on Applied computing*, 2008.
47. R. R. McCrae and O. P. John, *Journal of Personality* **60**, 175 (1992).

48. Y. Bachrach, M. Kosinski, T. Graepel, P. Kohli and D. Stillwell, *Web Science* (2012).
49. T. Yarkoni, *Journal of Research in Personality* **44**, 363 (June 2010).
50. J. W. Pennebaker, M. E. Francis and R. J. Booth, *Word J. Of The Int Ling Assoc* (2001).
51. F. Mairesse, M. Walker, M. Mehl and R. Moore, *Journal of Artificial Intelligence Research* **30**, 457 (2007).
52. A. Gill, S. Nowson and J. Oberlander, *Proc. of AAAI ICWSM* (2009).
53. S. Nowson, Identifying more bloggers: Towards large scale personality classification of personal weblogs, in *In Proceedings of the International Conference on Weblogs and Social*, 2007.
54. S. Saleem, R. Prasad, S. Vitaladevuni, M. Pacula, M. Crystal, B. Marx, D. Sloan, J. Vasterling and T. Speroff, Automatic detection of psychological distress indicators and severity assessment from online forum posts, in *Proceedings of COLING 2012*, (The COLING 2012 Organizing Committee, Mumbai, India, December 2012).
55. M. De Choudhury, M. Gamon, S. Counts and E. Horvitz, Predicting depression via social media, in *ICWSM*, 2013.
56. D. Rao, D. Yarowsky, A. Shreevats and M. Gupta, Classifying latent user attributes in twitter, in *Proceedings of the 2nd international workshop on Search and mining user-generated contents*, 2010.
57. M. Pennacchiotti and A.-M. Popescu, *ICWSM* **11**, 281 (2011).
58. G. Coppersmith, M. Dredze, C. Harman, K. Hollingshead and M. Mitchell, Clpsych 2015 shared task: Depression and ptsd on twitter, in *Proceedings of the Workshop on Computational Linguistics and Clinical Psychology: From Linguistic Signal to Clinical Reality, Denver, Colorado, USA, June. North American Chapter of the Association for Computational Linguistics*, 2015.
59. H. A. Schwartz, J. C. Eichstaedt, M. L. Kern, L. Dziurzynski, S. M. Ramones, M. Agrawal, A. Shah, M. Kosinski, D. Stillwell, M. E. Seligman *et al.*, *PloS one* **8**, p. e73791 (2013).
60. M. Kosinski and D. Stillwell, mypersonality project, in *http://www.mypersonality.org/wiki/*, 2012.
61. J. W. Pennebaker, C. Chung, M. Ireland, A. Gonzales and R. Booth, *Austin, TX, LIWC. Net* (2007).
62. S. Golder and M. Macy, *Science* **333**, 1878 (2011).
63. N. Halko, P.-G. Martinsson and J. A. Tropp, *SIAM review* **53**, 217 (2011).
64. R. Tibshirani, *Journal of the Royal Statistical Society. Series B (Methodological)* , 267 (1996).
65. M. Sap, G. Park, J. Eichstaedt, M. Kern, L. Ungar and H. A. Schwartz, Developing age and gender predictive lexica over social media, in *Proceedings of the 2014 Conference on Empirical Methods in Natural Language Processing, EMNLP*, 2014.
66. A. Hoerl and R. Kennard, *Technometrics* **12**, 55 (1970).
67. F. Pedregosa, G. Varoquaux, A. Gramfort, V. Michel, B. Thirion, O. Grisel, M. Blondel, P. Prettenhofer, R. Weiss, V. Dubourg *et al.*, *The Journal of Machine Learning Research* **12**, 2825 (2011).
68. D. Chandler and A. Kapelner, *Arxiv* (2010).
69. P. E. Shrout and J. L. Fleiss, *Psychological bulletin* **86**, p. 420 (1979).
70. J. R. Landis and G. G. Koch, *Biometrics* , 159 (1977).
71. E. Diener and M. E. Seligman, *Psychological Science* **13**, 81 (2002).
72. W. Pavot and E. Diener, *Psychological Assessment* **5**, p. 164 (1993).
73. G. Meyer, S. Finn, L. Eyde, G. Kay, K. Moreland, R. Dies, E. Eisman, T. Kubiszyn and G. Read, *American psychologist* **56**, p. 128 (2001).
74. O. J. Dunn, *Journal of the American Statistical Association* **56**, 52 (1961).
75. W. B. Schaufeli, A. B. Bakker and M. Salanova, *Educational and psychological Measurement* **66**, 701 (2006).
76. J. J. Appleton, S. L. Christenson, D. Kim and A. L. Reschly, *Journal of School Psychology* **44**, 427 (2006).
77. J. W. Rowe and R. L. Kahn, *Science* (1987).
78. N. M. Lawless and R. E. Lucas, *Social Indicators Research* **101**, 341 (2011).
79. R. M. Ryan and E. L. Deci, *American psychologist* **55**, p. 68 (2000).

FINDING POTENTIALLY UNSAFE NUTRITIONAL SUPPLEMENTS FROM USER REVIEWS WITH TOPIC MODELING

RYAN SULLIVAN*, ABEED SARKER, KAREN O'CONNOR, AMANDA GOODIN, MARK KARLSRUD and GRACIELA GONZALEZ

Department of Biomedical Informatics, Arizona State University,
Scottsdale, AZ 85259, USA
**E-mail: rpsulli@asu.edu, abeed.sarker@asu.edu, karen.oconnor@asu.edu, agoodin@asu.edu, mkarlsru@asu.edu, graciela.gonzalez@asu.edu*

Although dietary supplements are widely used and generally are considered safe, some supplements have been identified as causative agents for adverse reactions, some of which may even be fatal. The Food and Drug Administration (FDA) is responsible for monitoring supplements and ensuring that supplements are safe. However, current surveillance protocols are not always effective. Leveraging user-generated textual data, in the form of Amazon.com reviews for nutritional supplements, we use natural language processing techniques to develop a system for the monitoring of dietary supplements. We use topic modeling techniques, specifically a variation of Latent Dirichlet Allocation (LDA), and background knowledge in the form of an adverse reaction dictionary to score products based on their potential danger to the public. Our approach generates topics that semantically capture adverse reactions from a document set consisting of reviews posted by users of specific products, and based on these topics, we propose a scoring mechanism to categorize products as "high potential danger", "average potential danger" and "low potential danger." We evaluate our system by comparing the system categorization with human annotators, and we find that the our system agrees with the annotators 69.4% of the time. With these results, we demonstrate that our methods show promise and that our system represents a proof of concept as a viable low-cost, active approach for dietary supplement monitoring.

Keywords: Dietary Supplements, Pharmacovigilance, Natural Language Processing, Latent Dirichlet Allocation, Public Health Surveillance, Social Media Mining.

1. Introduction

According to the Dietary Supplement and Health Education Act (DSHEA),[1] dietary supplements (often referred to as nutritional products) are intended to supplement diet, intended for oral use, contain one or more dietary ingredients or their constituents, and are labeled on the packaging as dietary supplements. 50% to 70% of the general population in the United States uses a dietary supplement either for their purported benefits in maintaining good health or for the treatment of various diseases.[2–5] Evidence from multiple surveys suggests that dietary supplement users are more likely than non-users to adopt a number of positive health-related habits.[6] Thus, dietary supplements have become an integral part of health and wellness, and many health professionals and dietitians use and recommend their use.[4]

Despite the usefulness of dietary supplements, their widespread usage, and the perception that they are safe for use, they have been identified as causative agents for a variety of adverse

This work was funded (in part) by the University of Maryland, Joint Institute for Food Safety and Applied Nutrition through a cooperative agreement with the U.S. Food and Drug Administration (or FDA), #FDU001418.

reactions. For example, consumption of Chinese herbs that contain aristolochic acid (Mu Tong) has been reported to be associated with an increased risk of urinary tract cancer,[7] and more recently, the product OxyElite Pro® was recalled by the U.S. Food and Drug Administration (FDA) in November 2013[a] after possible links between the product and both liver failure and non-viral hepatitis were discovered.

Currently in the United States, the FDA regulates both finished dietary supplement products and dietary ingredients under a different set of regulations than those covering conventional food and drug products (prescription and over-the-counter).[8] Under the DSHEA[1] a manufacturer is responsible for ensuring that a dietary supplement or ingredient is safe before it is marketed. The FDA is responsible for taking action against any unsafe dietary supplement product after it reaches the market, and intervening if there is misleading product information. Generally, manufacturers do not need to register their products with the FDA nor do they need to get FDA approval before producing or selling dietary supplements. The responsibility of the manufacturer is to ensure that product label information is truthful and not misleading, that the product complies with the Dietary Supplement Current Good Manufacturing Practices (cGMPS) for quality control, and to submit to the FDA all serious adverse events[b] reports associated with use of the dietary supplement in the United States.

Under current adverse event monitoring protocols drug manufacturers and consumers can report adverse events caused or suspected to be caused by a dietary supplement using the Safety Reporting Portal.[c] Safety reports can be voluntarily submitted by manufacturers, packers, holders, researchers, or end users. However, numerous pharmacovigilance studies have revealed the ineffectiveness of self-reporting systems,[9] with some studies showing that only about 10% of adverse reactions generally reported.[10] There are many possible reasons for the low reporting numbers; a manufacturer may be reluctant to admit fault, or users may not report events (particularly for non-lethal events) to the manufacturer or even health care providers. Furthermore, even when a consumer has a serious event and goes to a poison center and a report is created, the FDA may not receive it. A 2013 Government Accountability Office report found that from 2008 through 2010 poison centers received over 1000 more reports than the FDA.[11] These facts clearly demonstrate that active surveillance is essential to the FDA's public health mandate with respect to dietary supplements. Although alternative sources (such as user comments from health forums or tweets) have been shown as potential sources for monitoring adverse reactions associated with prescription drugs,[12] there is still a research gap on active monitoring of dietary supplements.

---

[a] `http://www.fda.gov/ForConsumers/ConsumerUpdates/ucm374742.htm`,
`http://www.fda.gov/Food/RecallsOutbreaksEmergencies/Outbreaks/ucm370849.htm`.
Accessed: 7/10/2015.

[b] A serious adverse event is defined by the FDA as any adverse dietary supplement experience occurring at any dose that results in any of the following outcomes: death, a life-threatening adverse dietary supplement experience, inpatient hospitalization or prolongation of existing hospitalization, a persistent or significant disability/incapacity, or a congenital anomaly/birth defect.
(`https://www.federalregister.gov/articles/2007/06/25/07-3039/current-good-manufacturing-practice-in-manufacturing-packaging-labeling-or-holding-operations-for#h-493`. Accessed: 7/29/2015.)

[c] `https://www.safetyreporting.hhs.gov/fpsr/WorkflowLoginIO.aspx`. Accessed 7/10/2015.

Due to the strong motivation for active, low-cost monitoring systems for dietary supplements, we focused our study on extracting signals indicating the safety of dietary supplements from publicly available data on the Internet. In particular, we collected and automatically processed a large set of Amazon.com reviews, and used that information to predict the safety of the products. Our approach generates topics for each dietary supplement product based on its reviews, and uses these topics, with the assumption that the topics capture the semantic concepts associated with adverse effects, to rank the relative safety of individual products as compared to others in the same product class.

To generate the topics, we use a fully unsupervised variant of Latent Dirichlet Allocation (LDA).[13] Our approach biases the topic model by guaranteeing that tokens that match adverse reactions, based on ADRs listed in the SIDER database[d], will be limited to a sub-set of topics, and uses the topic distribution of a given product's reviews to score and rank that product. Essentially, the topic distributions are used as weights to score each product based on how much of the texts in its reviews appeared to be *generated* by the adverse reaction topics.

We consider three categories for each product: "high potential danger", "average potential danger" and "low potential danger," and compare the predictions of our system to a small set of 18 products categorized by human annotators. We find that our system agrees with the human rankings 69.4% of the time. Figure 1 visually illustrates our pipeline. We discuss the different components of the pipeline in the following sections, commencing with an overview of related literature.

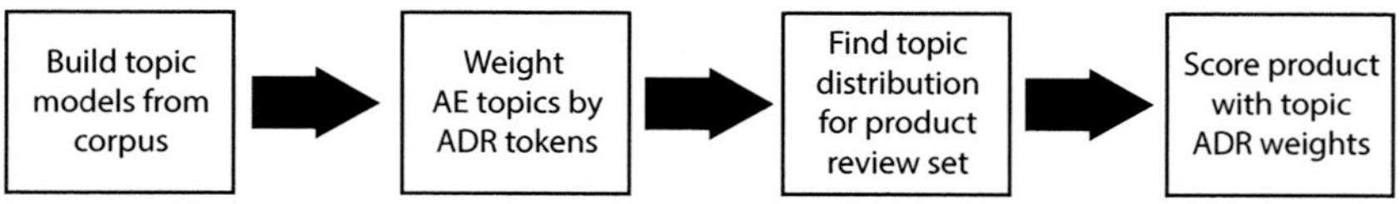

Fig. 1. System pipeline.

2. Related work

For public health issues, mining user-generated content has been shown to be a valuable resource of information, particularly because of the large volume and the possibility of real-time analysis.[14–16] Due to the underutilization of traditional reporting avenues,[17] detecting prescription drug ADR mentions in social media posts is an area that has seen a flurry of recent research. Leaman *et al.*[12] performed some of the earliest research in this area, using data from DailyStrength[e] to determine the feasibility of using lexicons for finding and extracting ADRs from user comments. Subsequent research was performed by Benton *et al.*[18] and Yates and Goarian,[19] and also used keyword based approaches, supplemented by synonym sets of lay vocabulary, to identify drug adverse events from social media sites.

[d] `http://sideeffects.embl.de/`
[e] `http://www.dailystrength.org/`

Current research in this space has also utilized NLP and ML techniques to overcome the shortcomings of lexicon-based approaches. For example, Nikfarjam and Gonzalez[20] and Yang *et al.*[21] both use association rule mining for ADR-related tasks using user-generated health text. Supervised text classification approaches have also been popular, particularly the use of Support Vector Machines (SVMs) (*e.g.*, Bian *et al.*,[22] Sarker and Gonzalez[23]).

Recent research has also seen the application of unsupervised approaches. For example, the study by Yang *et al.*[24] showed that LDA can be combined with a partially supervised classification approaches to create a classifier to locate consumer ADR messages in on-line discussion groups, and a study by Li *et al.*[25] showed that adding topics generated by LDA as a feature for an assertion classifier lead to a significant improvement in classification. Furthermore, Bisgin *et al.*[26] demonstrated that topics generated by LDA using drug labels as documents could be used to group drugs in a statistically significant way, which could be useful for discovering new relationships between drugs.

3. Methods

Our approach involves learning a probabilistic topic model that is partially based on background knowledge in the form of a dictionary of adverse reactions. We then build a weight map for our topic model where each topic is mapped to a value estimating how much each topic represents the ADRs. Finally, we use our topic model and our weight map to assign a single score to each product indicating the extent to which the reviews can be attributed to adverse reactions.

3.1. *Data*

Using a web crawler, we created a corpus of approximately 40,000 Amazon.com reviews from 2708 products[f]. The products chosen were those categorized by Amazon.com as "Herbal Supplements," "Sports Nutrition," "Supplements" and "Weight Loss." Our corpus consists of all products and from all their subcategories. Sample reviews for two products are shown in Table 1. These examples are representative of what is found across the review corpus and present examples of adverse reactions and indications. Furthermore, these examples show the varying seriousness of adverse reactions within the reviews and also give an example of a reviewer talking about a AE, as opposed to mentioning the event.

3.2. *LDA using background knowledge*

Our approach is driven by a variant of LDA.[13] LDA is an unsupervised technique, generally used for topic modeling, which builds a generative model for the data. Generative models are models which, given some parameters, could have randomly generated the observed data. In our specific case, we attempt to estimate the document-topic distributions and the topic-token distributions from which it would be possible to generate our text corpus. A document is generated by an LDA model one token at a time. The process begins by sampling the per-document topic distribution to choose a topic and then sampling the token distribution for

[f]Reviews were captured on 5th March 2014 from `http://www.amazon.com/b?node=37644410`

Table 1. Sample user reviews for two dietary supplement products.

Product: **batch5 Extreme Thermogenic Fat Burner**	Product: **NOW Foods Bromelain**
This pills dont work at all. Its just another pill with to much caffeine and makes you cranky, edgy and nervous.	This is just fine.....not sure what it was for. I do believe it is helping with my sinus problems, at least I haven't had any lately.
I take this product before i work out and i feel more energetic and i get a feeling of well being and it last long after im done working out. I definitely recommend B4.	the product caused adverse reactions for me and could not tolerate, had back pain and right right kidney pain and decreased urine output was not good for me.
I felt awful after I took it got a terrible niacin rush would never take it again side effects are scary	This product has helped me with the pain I have in my joints due to arthritis. My knees and hands were so bad before, but after just a couple of weeks I have gotten amazing relief.

the chosen topic to pick a word. The chosen word is added to document, and the process is repeated for the length of the document.

Our process is a variant of LDA which seeks to take advantage of background knowledge, which in our case is a dictionary of adverse reactions. Our intent is to generate topics that are semantically similar to the adverse reactions. We accomplish our goal by developing an LDA variant which uses a second per-document topic Dirichlet distribution ($Dirichlet(\alpha')$), which when sampled, will return a multinomial distribution over a sub-set of topics. This distribution over a subset of topics is then sampled to generate words that are known to be from our dictionary. This variant can be thought of as two parallel instances of topic modeling. One instance consisting of the tokens found in the dictionary and encompassing a subset of topics, and a second instance of the standard LDA for all topics and all non-ADR tokens.

Formally, our approach can be described as follows:

Let D be a collection of documents. For each $d \in D$ of length N, let $f_d : \{1, \ldots, N\} \rightarrow \{0, 1\}$ be an indicator function that maps an index in d to 1 when the word at the index is part of the background knowledge. To generate the collection D:

(1) For each topic k, draw a multinomial token distribution ϕ_k from $Dirchlet(\beta)$
(2) For each Document $d \in D$:
    (a) Draw a multinomial topic mixture θ from $Dirchlet(\alpha)$
    (b) Draw a multinomial topic mixture θ_{sub} from $Dirchlet(\alpha')$
    (c) Choose a document length N
    (d) For each token $0 \leq i < N$ in document d
        i. if $f_d(i) = 1$ choose topic z_i from θ_{sub}, else choose topic z_i from θ
        ii. Choose word w_i from ϕ_{z_i}

Figure 2 presents the plate notation for this variation of LDA.

This method is based on the general assumption that tokens which match the entries in

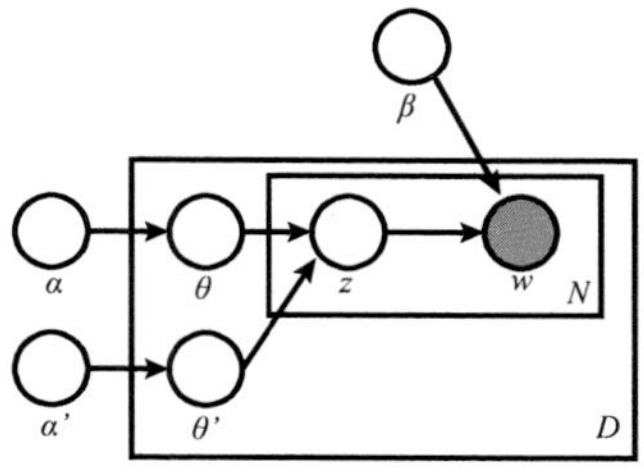

Fig. 2. Plate notation of our LDA model

the ADR dictionary could only have been chosen from a marked subset of topics. Though we do not label topics, we guarantee that tokens that match the tokens in the ADR dictionary are restricted to those subsets. The intent of this restriction is that the subset of topics containing ADR tokens will also contain tokens that are semantically similar to ADRs, but do not appear in the ADR dictionary.

Our approach was developed as an extension of the ParallelTopicModel class within the Mallet machine learning toolkit.[27] The ParallelTopicModel class is a implementation of the algorithm presented by Newman, et al.[28] and can be viewed as an approximation to Gibbs-sampled LDA. Our pre-processing consisted of removing stop words, and representing every instance of multi-token dictionary ADRs in the text as a single token. We chose to use 100 topics, and chose a subset size of 10. For our priors we chose standard values, $\alpha = 0.1$, $\alpha' = 0.1$ and $\beta = 0.01$. To learn our model we chose to use 10000 iterations of Gibbs Sampling and use a burn in of 1000 iterations. Table 2 provides examples of the top 15 tokens from selected topics from the category "SportsNutrition/Thermogenics/Fat Burners." The ADR topics are in bold.

3.3. *Scoring products based on topics*

Our system uses the Topic Models of the review set to generate a score for each product. Each topic is a distribution of tokens, so every token within a topic carries a weight as to how important that token is within its topic. We sum the weights of our known ADR tokens within each topic, and for each topic create a topic ADR weight. These topic ADR weights are the primary component of our scoring system.

To score each product, we first represent the product as a single document containing all the reviews. We then use the Mallet Topic Inferencer to estimate the distribution of topics for the product reviews. This provides us with information about how much of the review text was likely 'generated' by each topic, or what percent of the reviews can be explained by each given topic. We combine the topic percentages with the per-topic ADR weights to score each product and then normalize the product scores across all products within a category. An example of our scoring can be seen in table 3.

We choose to score products with respect to their Amazon category. That is, as opposed to building a topic model based on the full corpus, we build topic models for each category

Table 2. Tokens from Selected Topics of 'Fat Burners':

Topic 0	stomach, gas, doesn, problems, issues, give, product, upset, don, bloating, system, digestive, easy, bad, products
Topic 1	energy, boost, give, feel, extra, day, product, workout, focus, gave, jitters, work, workouts, felt, level, feeling, caffeine
Topic 3	blood, sugar, levels, body, cancer, health, diabetic, problems, insulin, liver, people, research, due, level, heart
Topic 47	oil, punch, meat, red, chicken, fish, fruit, eat, eggs, veggies, fruits, eating, vegetables, tropical, vegetable
Topic 75	lost, pounds, lbs, ve, weeks, months, weight, week, lose, taking, month, days, gained, started, pound
Topic 98	free, gluten, lactose, soy, intolerant, dairy, organic, milk, grass, fed, cows, wheat, product, gmo, products

we wish to evaluate. We also only score products in relation to other products in the same category. This was done because when the full corpus is used to generate topic models, we found that when one product has a strong co-occurrence with one type of ADR, the topics related to that ADR became more of a topic for the product class. In those cases, the ADR topics would represent the products with those adverse events, and not the adverse events within the product reviews. We also chose to only score the products that had at least 25 reviews because products with a low number of reviews do not have enough text for scoring to be accurate.

Table 3. ADR Score for product: Dexatrim Max Comple-7

Topic	Topic ADR Weight	ADR examples from topic	Topic Percent	Topic ADR weight
Topic 0	30	birth_defects(6.0), chest_pains(4.0)	0.01378	0.413
Topic 1	139	bloating(11.0), diarrhea(7.0)	0.00182	0.252
Topic 2	111	gas(14.0), headaches(12.0)	0.01138	1.263
Topic 3	41	liver_damage(6.0), loss_of_weight(4.0)	4.635 E-4	0.019
Topic 4	522	jittery(72.0), headache(67.0)	0.03975	20.749
Topic 5	202	gain_weight(18.0), feel_sick(16.0)	0.01276	2.577
Topic 6	131	jittery(46.0), heart_attack(12.0)	0.00283	0.370
Topic 7	53	hunger_pains(5.0), reduced_appetite(5.0)	0.01322	0.700
Topic 8	91	inflammation(10.0), joint_pain(7.0)	7.46 E-6	6.78 E-4
Topic 9	150	palpitations(21.0), high_blood_pressure(11.0)	0.016450	2.46
			ADR Score: 28.803	

4. Evaluation and results

Our primary goal is to develop a system to help identify potentially dangerous nutritional supplements. The majority of our evaluation is related to that primary goal. However, because such a large portion of our work is based on our variant on LDA, we feel it is necessary to provide an evaluation of that aspect of our methodology.

4.1. *Validation of background knowledge driven LDA*

To validate our methodology, we used the Twitter Adverse Drug Reaction corpus from Ginn *et al.*[29] and compared the ADR scores of the tweets annotated with adverse reactions to those tweets with no ADRs. We compared the ADR scores generated with topics from our variant to the scores generated with topics from standard LDA. We found that with our variant, tweets with an annotated adverse reaction on average had a ADR score 1.89 times bigger than the score of tweets without any adverse reactions. This can be compared to standard LDA, where the ADR tweets had a score on average of 1.56 times bigger than the non-ADR tweets. We also compared the tokens within the topics for both standard LDA and our variation. We examined the correlation between the weight of tokens from the SIDER database and tokens annotated as ADRs (not in the database). We found that the R-squared value for the correlation between known ADR tokens and annotated ADR tokens within topics was 0.356 for normal LDA and 0.445 for our variation. These results provide evidence that our variant on LDA does create topics which better capture adverse drug reactions.

4.2. *Evaluation of 'ADR Score'*

We evaluated our ADR Score results by having human annotators categorize products from within a category, and then comparing the categorization to our rankings. We chose to use the categories of "SportsNutrition/Thermogenics/Fat Burners" and "WeightLoss/AppetiteControl&Suppressants" for evaluation. From those categories we chose 9 random products, three from the top third, three from the middle, and three from the bottom third of the list of products within the category ranked by ADR score. Two human annotators then categorized each product, and we compared our automatically generated categorization to the annotator categorization.

4.2.1. *Human categorization of products*

The user comments for nine products from the "SportsNutrition/Thermogenics/Fat Burners" class and 'WeightLoss/AppetiteControl&Suppressants" class were manually reviewed by two expert annotators to assess the results of the classifier. For each product, the annotator classified the product as having either a high, average or low potential for ADRs. Each annotator assessed the ADR potential by a variety of indicators, including: comparing the number of comments with ADR mentions from the number of comments overall; the severity of the ADR mentioned; and the potential for adverse reactions from the ingredients in the supplement.

4.3. *Results*

Table 4 and Table 5 present the comparison of human annotated classification to the classification based on the 'ADR Score' for the class "SportsNutrition/Thermogenics/Fat Burners' and the class "WeightLoss/AppetiteControl&Suppressants." Over these two categories, The annotator agreement was 61.1 %. The system accuracy with respect to Annotator 1 is 66.6% and the accuracy with respect to Annotator 2 is 72.2%, and the average accuracy of our system is 69.4% over the two categories.

Table 4. Comparison of annotator categorizations with our systems categorizations for SportsNutrition/Thermogenics/Fat Burners.

Product	Human Annotator 1	Human Annotator 2	ADR Score	ADR Score Category
batch5 Extreme Thermogenic Fat Burner	High Potential	Average Potential	0.336	Average Potential
BPI Sports B4 Fat Burner	High Potential	High Potential	1.0	High Potential
Buy Garcinia Cambogia Extract With Confidence	Low Potential	Low Potential	0.129	Low Potential
Cellucor D4 Thermal Shock Thermogenic Fat Burner	High Potential	High Potential	0.614	High Potential
Garcinia Cambogia Drops	Low Potential	Low Potential	0.120	Low Potential
Liporidex MAX w Green Coffee Ultra	Average Potential	Average Potential	0.371	Average Potential
Raspberry Ketones The ONLY 250 mg PURE Raspberry Ketone Liquid	Low Potential	Low Potential	0.186	Low Potential
SafSlim Tangerine Cream Fusion	Low Potential	Average Potential	0.341	Average Potential
VPX Meltdown	Average Potential	High Potential	0.685	High Potential

5. Discussions and future work

The primary goal of this work is to use unsupervised NLP techniques for low-cost, active monitoring of dietary supplements, and with our results we have presented a promising proof-of-concept system. This system has shown to be reasonably accurate in identifying products with above-average potential for adverse reactions, especially when the results are considered with respect to the annotator agreement.

Through the process of error analysis, we found three important potential limitations of our system: The system treats all adverse reactions equally, it treats ADRs and indications equally, and it cannot differentiate real and fake reviews. In dietary supplement monitoring, a single serious adverse effect is given significantly more weight than multiple non-serious reactions. Currently, our system has no way to weigh the reactions, and thus numerous trivial reactions will generate a higher score than one serious adverse reaction. This particular case did lead to a disagreement between the annotators and the system, where one annotator found a product to have a higher potential than our system due to a small number of serious adverse reactions.

Table 5. Comparison of annotator categorizations with our systems categorizations for WeightLoss/AppetiteControl&Suppressants.

Product	Human Annotator 1	Human Annotator 2	ADR Score	ADR Score Category
Nature's Way Metabolic ReSet	Low Potential	Average Potential	0.385	High Potential
Burn + Control Weight-loss Gourmet Instant Coffee by Javita	Low Potential	Low Potential	0.058	Low Potential
Garcinia Cambogia Extract Pure (60% HCA)	Low Potential	Low Potential	0.0527	Low Potential
Garcinia Cambogia Extract Pure Premium Ultra	Low Potential	Low Potential	0.129	Average Potential
Life Extension Decaffeinated Mega Green Tea Extract	High Potential	Low Potential	0.366	High Potential
Garcinia Cambogia Liquid Weight Loss Diet Drops	Low Potential	Low Potential	0.0	Low Potential
LipoBlast Extreme Diet Pills/Energy Boosters/Appetite Suppressant	High Potential	Average Potential	0.128	Average Potential
MetaboLife Ultra, Stage 1	High Potential	Low Potential	0.335	High Potential
Saffron Extract - Appetite Suppressant	Average Potential	Low Potential	0.125	Average Potential

Indications can be defined as the reason why a consumer is taking a drug or supplement, and in many cases, indication tokens are adverse reaction tokens. The primary difference between indications and ADRs is how the reaction relates to the user with respect to the drug. For example, the ADR tokens in the phrase "*'This product has helped me with the pain I have in my joints due to arthritis*" are very similar to the ADR tokens in the phrase "*the product caused adverse reactions for me and could not tolerate, had back pain and right kidney pain and decreased urine output*," yet they are very different semantically. As the system currently works, a product that has many indications will be scored similarly to one with many adverse reactions, as the current system does not take into account the semantic relationship between a potential ADR term and the rest of the sentence.

Finally, the system is currently unable to identify fake reviews. Dietary supplement manufacturers are known to provide free products to those who write reviews, and for some products we found that there were a non-trivial amount of fake positive reviews. Because we are examining the percentage of the reviews that is generated by the ADR topics, these fake reviews affect our ranking.

The monitoring of dietary supplements is a challenging task due to both the sheer number of supplements on the market and the limited man-power of the FDA. Despite the current limitations, our system produces very promising results. In particular, this system shows the validity of an unsupervised NLP approach for this task, while also serving as a promising proof-of-concept system.

The current limitations also provide a roadmap for future work. We plan on exploring other variations of LDA such as the Topic Aspect Model[30] and multi-grain topic models,[31] to incorporate aspects of those approaches into our work. We feel these techniques are promising solutions that can help distinguish adverse reactions from indications.

We also plan on incorporating the work presented in Leaman et al.[12] to add ADR named entity recognition to our pipeline. This will allow our system to use more then just a dictionary of known adverse reaction tokens when learning the ADR topics. Furthermore, we plan on adding 'fake rule detection' to the pipeline, following the work of Lau et al.[32] In addition, we plan on expanding the system to include text from other sources, including Internet message boards. There are very active on-line communities which discuss nutritional supplements, and this textual data would add to our corpus and help increase the accuracy of our categorization. Finally, we plan to expand our evaluation and include a larger variety of product categories.

Our experiments show that large amounts of user-generated data, which is readily available, may be used to automatically identify high-risk dietary supplements. The identified supplements can then be marked for further investigation by the Center for Food Safety and Applied Nutrition (CFSAN). We hypothesize that this unsupervised NLP technique will provide valuable early signals of suspected associations between CFSAN-regulated products and adverse reactions. Based on our promising results, we envision that this technique will act as a crucial source for safety signals associated with dietary supplements and may eventually provide the ability to detect problematic supplements earlier and more cost effectively than current methods.

References

1. U. S. Food and Drug Administration, Dietary Supplement Health and Education Act of 1994 `http://www.fda.gov/RegulatoryInformation/Legislation/FederalFoodDrugandCosmeticActFDCAct/SignificantAmendmentstotheFDCAct/ucm148003.htm`, (1994).
2. K. Radimer, B. Bindewald, J. Hughes, B. Ervin, C. Swanson and M. F. Picciano, *American Journal of Epidemiology* **160**, 339 (2004).
3. B. B. Timbo, M. P. Ross, P. V. Mccarthy and C. T. J. Lin, *Journal of the American Dietetic Association* **106**, 1966 (2006).
4. A. Dickinson, L. Bonci, N. Boyon and J. C. Franco, *Nutrition Journal* **11** (2012).
5. K. Fisher, R. Vuppalanchi and R. Saxena, *Archives of Pathology and Laboratory Medicine* **139**, 876 (2015).
6. A. Dickinson and D. MacKay, *Nutrition Journal* **13** (2014).
7. M.-N. Lai, S.-M. Wang, P.-C. Chen, Y.-Y. Chen and J.-D. Wang, *Journal of the National Cancer Institute* **102**, 179 (2010).
8. U. S. Food and Drug Administration , Dietary Supplements `http://www.fda.gov/Food/DietarySupplements/` (April, 2015).
9. A. Sarker, A. Nikfarjam, K. OConnor, R. Ginn, G. Gonzalez, T. Upadhaya, S. Jayaraman and K. Smith, *Journal of Biomedical Informatics* **54**, 202 (February 2015).
10. R. Harpaz, W. DuMouchel, N. H. Shah, D. Madigan, P. Ryan and C. Friedman, *Clinical pharmacology and therapeutics* **91**, 1010 (2012).
11. U. S. G. A. Office, *DIETARY SUPPLEMENTS: FDA May Have Opportunities to Expand Its Use of Reported Health Problems to Oversee Products*, tech. rep., United States Government Accountability Office (03 2013).
12. R. Leaman, L. Wojtulewicz, R. Sullivan, A. Skariah, J. Yang and G. Gonzalez, Towards internet-

age pharmacovigilance: extracting adverse drug reactions from user posts to health-related social networks, in *Proceedings of the 2010 workshop on biomedical natural language processing*, 2010.
13. D. M. Blei, A. Y. Ng and M. I. Jordan, *the Journal of machine Learning research* **3**, 993 (2003).
14. M. J. Paul and M. Dredze, You are what you tweet: Analyzing twitter for public health., in *ICWSM*, 2011.
15. M. Szomszor, P. Kostkova and E. De Quincey, # swineflu: Twitter predicts swine flu outbreak in 2009, in *Electronic Healthcare*, (Springer, 2012) pp. 18–26.
16. Y. T. Yang, M. Horneffer and N. DiLisio, *Journal of public health research* **2**, p. 17 (2013).
17. L. Hazell and S. A. Shakir, *Drug Safety* **29**, 385 (2006).
18. A. Benton, L. Ungar, S. Hill, S. Hennessy, J. Mao, A. Chung, C. E. Leonard and J. H. Holmes, *Journal of biomedical informatics* **44**, 989 (2011).
19. A. Yates and N. Goharian, Adrtrace: detecting expected and unexpected adverse drug reactions from user reviews on social media sites, in *Advances in Information Retrieval*, (Springer, 2013) pp. 816–819.
20. A. Nikfarjam and G. H. Gonzalez, Pattern mining for extraction of mentions of adverse drug reactions from user comments, in *AMIA Annual Symposium Proceedings*, 2011.
21. C. C. Yang, L. Jiang, H. Yang and X. Tang, Detecting signals of adverse drug reactions from health consumer contributed content in social media, in *Proceedings of ACM SIGKDD Workshop on Health Informatics (August 12, 2012)*, 2012.
22. J. Bian, U. Topaloglu and F. Yu, Towards large-scale twitter mining for drug-related adverse events, in *Proceedings of the 2012 international workshop on Smart health and wellbeing*, 2012.
23. A. Sarker and G. Gonzalez, *Journal of Biomedical Informatics* **53**, 196 (2015).
24. M. Yang, M. Kiang and W. Shang, *Journal of biomedical informatics* **54**, 230 (2015).
25. D. Li, N. Xia, S. Sohn, K. B. Cohen, C. G. Chute and H. Liu, Incorporating topic modeling features for clinic concept assertion classification, in *Proceedings of the 5th International Symposium on Languages in Biology and Medicine*, 2013.
26. H. Bisgin, Z. Liu, H. Fang, X. Xu and W. Tong, *BMC bioinformatics* **12**, p. S11 (2011).
27. A. K. McCallum, Mallet: A machine learning for language toolkit, http://mallet.cs.umass.edu, (2002).
28. D. Newman, A. Asuncion, P. Smyth and M. Welling, *The Journal of Machine Learning Research* **10**, 1801 (2009).
29. R. Ginn, P. Pimpalkhute, A. Nikfarjam, A. Patki, K. O'Connor, A. Sarker and G. Gonzalez, Mining twitter for adverse drug reaction mentions: a corpus and classification benchmark, in *proceedings of the Fourth Workshop on Building and Evaluating Resources for Health and Biomedical Text Processing (BioTxtM). Reykjavik, Iceland*, 2014.
30. M. Paul and R. Girju, *Urbana* **51**, p. 61801 (2010).
31. I. Titov and R. McDonald, Modeling online reviews with multi-grain topic models, in *Proceedings of the 17th international conference on World Wide Web*, 2008.
32. R. Y. Lau, S. Liao, R. C. W. Kwok, K. Xu, Y. Xia and Y. Li, *ACM Transactions on Management Information Systems* **2**, 1 (2011).

INSIGHTS FROM MACHINE-LEARNED DIET SUCCESS PREDICTION

INGMAR WEBER
Qatar Computing Research Institute
Doha, Qatar
Email: iweber@qf.org.qa

PALAKORN ACHANANUPARP
Singapore Management University
Singapore
Email: palakorna@smu.edu.sg

To support people trying to lose weight and stay healthy, more and more fitness apps have sprung up including the ability to track both calories intake and expenditure. Users of such apps are part of a wider "quantified self" movement and many opt-in to publicly share their logged data. In this paper, we use public food diaries of more than 4,000 long-term active MyFitnessPal users to study the characteristics of a (un-)successful diet. Concretely, we train a machine learning model to predict repeatedly being over or under self-set daily calories goals and then look at which features contribute to the model's prediction. Our findings include both expected results, such as the token "mcdonalds" or the category "dessert" being indicative for being over the calories goal, but also less obvious ones such as the difference between pork and poultry concerning dieting success, or the use of the "quick added calories" functionality being indicative of over-shooting calorie-wise. This study also hints at the feasibility of using such data for more in-depth data mining, e.g., looking at the interaction between consumed foods such as mixing protein- and carbohydrate-rich foods. To the best of our knowledge, this is the first systematic study of public food diaries.

Keywords: MyFitnessPal; Calorie Counting; Weight Loss; Quantified Self

1. Introduction

In 2012, 30-50 million Americans were on a diet at any given point in time[a] for reasons ranging from lowering the risk of diseases to having a more positive self image. The annual revenue of the U.S. weight-loss industry is estimated at around $20 billion.[b] Clearly, dieting is not easy and new fashion diets come into existence every year.

In this paper we explore the practice of keeping an online food diary and its relation to dieting outcomes. Concretely, we turn to data from a large fitness and health application, MyFitnessPal (henceforth MFP), and look at the publicly logged food consumption of more than 4,000 users over several months. Figure 1 shows the interface through which users enter their consumed food. Users not only log their daily intake but they also specify a "daily calories goal" against which their consumption can be compared. Figure 2 shows this goal at the bottom of the screenshot. Though we cannot observe their actual weight progression, we are using the information on whether a user mostly consumes more or less calories than their self-declared goal as an indicator for dieting success.

[a]https://www.npd.com/wps/portal/npd/us/news/press-releases/the-npd-group-reports-dieting-is-at-an-all-time-low-dieting-season-has-begun-but-its-not-what-it-used-to-be/
[b]http://abcnews.go.com/Health/100-million-dieters-20-billion-weight-loss-industry/story?id=16297197

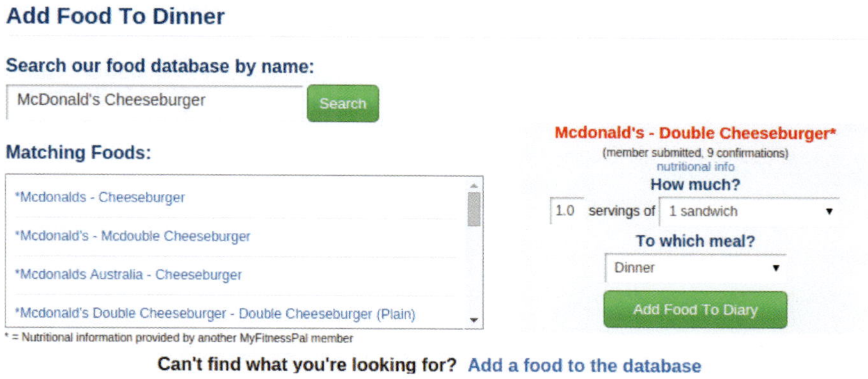

Fig. 1. Screenshot showing the food selection process in the MyFitnessPal web interface.

Using these user labels of being "above" or "below" we train a classifier to tell the two user groups apart using the types of food users have logged. Through a feature analysis of the classifier we gain insights into which foods are associated with diet success or failure.

Our findings are largely intuitive, e.g., logging food with "oil", "butter" or "mcdonalds" in the name is an indication of going above one's calorie goals. However, we also discover less obvious trends such as a distinction between pork (indicative for being "over") and poultry (indicative for being "under"). In addition, we describe general behavior related to food logging. E.g., users are least likely to log *any* meal on the weekend and, if they do, they are most likely to be above their weight goal.

To the best of our knowledge, this is the first systematic analysis of public food diaries. We believe that this paper helps to show the potential that this data holds for various health-related analyses.

2. Related Work

Our research is centered around "quantified self" data. The quantified self is a movement to voluntarily log personally relevant data for self knowledge and improvement. Users might decide to quantify or track their activity both for goal-driven or documentary motivation.[15] Striking the right balance between the amount of data that an application wants to collect and the amount of effort required on the user's end can be challenging.[11] Rusin *et al.*[16] offer a review of technologies used for logging food intake.

When it comes to weight loss and weight maintenance, certain practices related to the quantified self such as "think about how much progress you've made" (by having charts showing this), "weigh yourself" (through weight logging) and "read nutrition labels" (making use of the nutrition database that services such as MFP provide) have been among the few indicators of both successful weight loss and maintenance.[17] Other research has also found that "self monitoring" is important both for weight loss[2] and for successfully maintaining such loss.[7]

myfitnesspal

ABOUT | FOOD | EXERCISE | APPS | COMMUNITY | BLOG

[redacted]'s Food Diary For: Thursday, May 7, 2015

One	Calories	Protein	Fat	Carbs	Fiber	Sodium
Compliments - Mini Brownies, 40 grams (2 brownies)	179	2	9	21	1	100
Grimm's - Lean Turkey Pepperoni, 63 grams (1 stick)	105	13	6	2	0	609
Compliments - Paradise Trio Frozen Fruit, 154 grams	66	1	0	19	2	0
Fruit - Strawberries, 138 grams	44	1	0	11	3	1
Cheetos - Crunchy Cheetos, 16 grams (37 pieces)	90	1	6	10	0	141
Heritage - Hillbilly Kettle Corn, 9 grams	43	1	2	5	1	106
Quick Tools	527	19	23	68	7	957

Four						
Fruit - Grapes, 108 grams	72	1	0	18	1	2
Fruit - Oranges, General, 146 grams	72	1	0	18	3	0
Heritage - Hillbilly Kettle Corn, 9 grams	43	1	2	5	1	106
Quick Tools	187	3	2	41	5	108
Totals	**2,015**	**129**	**59**	**246**	**29**	**2,459**
[redacted] **Daily Goal**	**2,020**	**126**	**65**	**233**	**25**	**2,300**
Remaining	**5**	**-3**	**6**	**-13**	**-4**	**-159**
	Calories	Protein	Fat	Carbs	Fiber	Sodium

Fig. 2. Screenshot of a user's public food diary on MyFitnessPal. The user kept the default names for the meals ("One", "Two", ...) and only the first and last meal are included in the screenshot. The bottom shows the user's actual calories consumption of 2,015kcal and their daily goal of 2,020kcal.

These studies are, however, from the "pre-smartphone era" where calorie counting was done manually using pencil and paper. Still, they give credence to the potential benefits for both weight loss and weight management that apps such as MFP could offer. A currently ongoing study will also shed light on the effect of frequent weight control[10] on dieting outcomes. To date, few studies have, however, looked at the effectiveness of mobile apps for motivating health behavior changes.[13]

Concerning the analysis of food consumption through user-generated data, some studies have taken a public health approach and looked at regional or temporal patterns in food consumption. West *et al.*[21] used web search and click-through data to study seasonal changes in the recipes people search for. Server logs from a recipe site were used in[19,20] to study regional differences in food preferences. Their angle is less health-centric and more culture- or preference-centric with a focus on whether ingredients or regional influences dominate the recipe choice. Using *only* the ingredients of recipes, rather than regional or cultural knowledge, has been explored in[18] for the purpose of recipe recommendation.

Conceptually close to our study is the work by Abbar *et al.*[1] who look at food mentions on Twitter. Their analysis also includes bit on *individual* and not only public health. If users were to mention *everything* they ate on social media then the type of food diaries we are using would become redundant. However, we believe that this is unlikely to be the case for many users and our data is far cleaner and more structured and comes with user-defined daily calorie goals. Culotta[3] also used Twitter data to study obesity and other health issues but from a purely aggregate, public health perspective. Kuener *et al.*[8] use data from Yahoo Answers to look at issues of both mental and physical health of obese people. They find that "obese people residing in counties with higher levels of BMI may have better physical and mental health than obese people living in counties with lower levels of BMI". They do not, however, look at indicators related to the success or failure of weight loss.

As far as the idea of trying to predict dieting success is concerned, the work in[9] is related. The authors study data from an online weight loss community and offer insights into different phases of usage which, potentially, could help to predict who will engage in the long-term and hence have a higher chance of achieving and maintaining weight loss. Finally, Park *et al.*[12] study long-term sharing of MFP activity on Twitter. This also indirectly relates to dieting success as long-term engagement with a fitness application could be seen as beneficial for maintaining weight loss.

3. Data Acquisition and Preprocessing

3.1. *Obtaining Food Diaries and Constructing Food Taxonomy*

The construction of our dataset begins with the download of public food diary pages of MFP users. First, we extracted an initial list of 100K usernames from the 10 most popular MFP groups. Next, for each user, we retrieved up to the last 180 days of food diary pages (until March 2015 when the data collection took place). Ultimately, the food diary pages of 9,896 users are publicly accessible, resulting in 587,187 food diary pages retrieved. On average, each user has logged 59.3 days of diaries (S.D. = 54.6, median = 42) or 652.9 food entries in total (S.D. = 774, median = 366). According to the random samples (N = 200) of user profiles, the average age of users in the dataset is 36.6 years old (S.D. = 10.71). The vast majority of users are female (75%) and reside in the United States (67%). A small fraction of users (5.7%) do not provide any demographic information.

When adding an entry in a food diary, users can either search for existing foods in MFP database or enter a new food description and associated nutritional values. The lack of controlled vocabulary in the food data creation causes several data integrity issues, e.g., the same foods may be described slightly different by different users. To mitigate the problems, we needed ways to group related foods into semantic categories. Thus, we built a food taxonomy by compiling lists of food-related categories and page names from Wikipedia[c]. The taxonomy is manually organized into 18 main categories (e.g., staple food, meats, vegetables, etc.), 149 subcategories (e.g., wheat, rice, beef, etc.) and 4,233 entities, describing individual ingredients and meal types of food entries. For example, the entry "McDonald's - Premium Sweet Chili Chicken Wrap (Grilled)" will be annotated with the following set of {main category: subcategory: entity}: {Staple foods: Wheat: Wrap}, {Meats: Poultry: Chicken}, {Preparation Methods: Grill}, {Fast foods: McDonald's}.

3.2. *Data Preprocessing and Pruning*

To extract categories from a food entry, we first lemmatized all entity words in the taxonomy and the food entry's text. Then, we iterated through the main categories to find the maximal exact substring match between the taxonomic entities and the food entry's text. E.g., the entity "bean sprout" is a match in "Iga - bean sprouts" but not a match in "Sprouts - tiramisu espresso beans". After a match had been found, the corresponding main category, subcategory, and entity were added to the annotation. The taxonomy has a reasonable coverage with respect to our dataset. Out of 632,652 unique food entries, 88% were successfully annotated while 12% produced an empty annotation. The causes of failed annotation include misspelled (e.g., brocolli) and non-English names (e.g., huevos rancheros).

To describe a user's food intake, we used both the categories described above and single-word tokens extracted from the text of the diary entry. The tokenization steps are as follows: (i) splitting on non-word characters[d], (ii) lower casing everything, (iii) only considering tokens of length at least three, and (iv) requiring all characters to be alphabetical (a-z). Furthermore, for both tokens and categories, we required that more than 500 distinct users had to use it, leaving us with 1,720 distinct tokens and 392 distinct categories. Finally, only users who had at least 30 logged days with at least one non-zero feature were considered, ignoring days with less than 100 calories logged. This left us with 5,797 users.

3.3. *Labeling Calorie Goals "Success"*

The core of our analysis is centered around looking for differences between successful and unsuccessful users, where success is in relation to their self-declared daily calories goal. For each day, a user was assigned a label of "below", "on-target" or "above" depending on their calories goal and actual calories consumed as follows.

- below: (goal - actual) / goal > .2
- above: actual > goal
- on-target: otherwise

[c] http://en.wikipedia.org/wiki/Category:Foods
[d] The usual regular expression definition of non-word characters was used, i.e. [^a-zA-Z0-9_].

We chose to label a (user, day) pair where the user exceeds their goal (i.e., too much consumption) by even a single calorie as "above" to encode the inherent asymmetry in dieting to lose weight. Table 1 shows the trend of the (user, day) pairs when grouping by the day of the week and aggregating across users. As found in previous work looking at food consumption,[1] the weekend seems to be the worst period for dieting with (i) the highest fraction of "above" incidents, and (ii) the lowest number of logged days. Interestingly, web searches for recipes seem to follow the *opposite* trend.[21]

Table 1. This table shows the weekly logging trends for 9,896 users. The fraction of "above" increases slightly from its lowest on Mondays (19.0%) to its highest on Saturdays (24.9%). The number of total user-days logged also shows a drop on the weekend.

	Mon	Tue	Wed	Thu	Fri	Sat	Sun
% Above	19.1%	20.0%	20.6%	21.0%	23.3%	24.9%	23.7%
% Ontarget	33.5%	34.1%	33.6%	32.9%	29.9%	29.2%	31.3%
% Below	47.4%	46.0%	45.7%	46.0%	46.8%	46.0%	44.9%
# Total	93.0k	91.7k	88.7k	85.4k	80.7k	73.4k	73.6k

To have a single label for each user, the (user, day) label pairs were aggregated across days by taking the modal class, i.e., the biggest class. Note that this means a user could have as little as 34% of their days belonging to this class. Having a single label for each user, rather than modeling each (user, day) pair separately, had the advantages of (i) reducing noise due to the larger data set being considered, and (ii) interlinking behavior across days so that even "good" behavior on one day could be predictive of "bad" behavior on another day.

Using single-word tokens as features, with the above definition of user labels, 3,320 users were labeled "below", 1,546 were "on-target" and 931 "above". To have a clearer distinction between the classes, we chose not to consider users in the "on-target" group for further analysis. This left us with 4,251 users for the token-based analysis and * for the category-based analysis.

Table 2. Basic characteristics of our below-vs.-above data set containing 4,251 users after pruning (see text).

	min	10%	median	90%	max
Total days logged per user	30	36	77	168	186
% days "above" per user	0	1%	14%	53%	95%
% days "below" per user	0	18%	56%	86%	100%

4. Results

4.1. *Exploratory Cluster Analysis*

To gain a better understanding of the food consumption patterns, we performed a cluster analysis. Each of the 4,251 users for the token-based representation was mapped to a feature vector where each of the 1,720 dimension counted on how many distinct days a user used a specific token. These vectors were then normalized in 2-norm. We used X-Means[14] as implemented in Weka[5] as the clustering algorithm.[e] The algorithm automatically chose $k = 6$ as the optimal number of clusters (searching between $k = 2$ and $k = 10$).

Table 3. Summary of an XMeans clustering for the token-based normalized feature representation.

	Cluster 1 (n=545) below/above 375 (69%) / 170 (31%)	Cluster 2 (n=411) below/above 304 (74%) / 107 (26%)	Cluster 3 (n=686) below/above 575 (84%) / 111 (16%)	Cluster 4 (n=601) below/above 460 (76%) / 141 (24%)	Cluster 5 (n=878) below/above 716 (82%) / 162 (18%)	Cluster 6 (n=1,130) below/above 890 (79%) / 240 (21%)
Biggest Rank Gains	skimmed (+917)	grounds (+550)	creamer (+348)	coffee (+43)	great (+121)	almond (+84)
	sainsbury (+856)	brewed (+412)	packet (+83)	sandwich (+34)	value (+99)	organic (+75)
	semi (+832)	creamer (+361)	coffee (+74)	sausage (+34)	kraft (+59)	protein (+64)
	asda (+794)	from (+118)	protein (+58)	wheat (+28)	wheat (+40)	coffee (+49)
	tesco (+621)	packet (+76)	tsp (+50)	pizza (+28)	turkey (+30)	yogurt (+32)
	tea (+139)	coffee (+74)	sugar (+38)	turkey (+23)	cheddar (+26)	vanilla (+25)
	coffee (+46)	tsp (+52)	vanilla (+36)	chips (+23)	light (+25)	natural (+22)
	light (+20)	sugar (+32)	free (+32)	bacon (+17)	yogurt (+25)	banana (+15)
	banana (+17)	free (+30)	yogurt (+31)	peanut (+14)	free (+22)	eggs (+14)
	free (+16)	vanilla (+24)	natural (+22)	homemade (+12)	peanut (+17)	peanut (+13)

Table 3 shows a summary of the obtained clusters. The cluster sizes are reasonably balanced, ranging from 411 to 1,130. To understand the distinctive features for each cluster we looked at the tokens that went up the most in the ranking, compared to the global average. For example, a token that is globally ranked 1,000th in terms of its average user weight but is ranked 20th within a particular cluster has moved up 980 ranks. Furthermore, we required that the token had to end up in the top 40. This was done to ensure that the token is relatively frequent in the end. Note that the bigger clusters (Clusters 5 & 6) are closer to the "average" and so the relative change in ranking is smaller for them.

The fraction of "below" and "above" users shows moderate variations, ranging from 69% "below" to 84%. However, the discriminative tokens do not necessarily tell an intuitive story as, e.g., Cluster 4 with "pizza" and "bacon" has a *lower* fraction of "above" than Cluster 6 with "organic" and "natural". Clusters 2 and 3 also seem similar in that they both have many discriminative tokens related to coffee. However, Cluster 2 has an "above" percentage that is 10% above that of Cluster 3.

Overall, the unsupervised clustering did not yield practical insights as clusters seemed to be more influenced by things such as brand names or shopping at particular supermarket chains, than by healthy-vs.-unhealthy food categorization. Hence, we decided to look at *supervised* methods instead, where the above-vs.-below labels are central.

4.2. *Above-vs.-Below Machine Classification*

To understand the differences in food consumption between our above and below users, we trained a Support Vector Machine (SVM) classifier with a linear kernel and the default settings in the SVM-light[f] implementation.[6] We used the same general setup for both feature sets, tokens and categories, though the dimensionality of the feature space differed (1,720 vs. 392). In both cases, we trained the classifier on a *balanced* training set with a equal number of above and below instances, 931 for tokens and 919 for categories. The training set was then split into 10 folds, each with a 90%-10% train-test split. Table 4 summarizes the performance for the binary classifier.

[e] The exact parameters used were: Scheme:weka.clusterers.XMeans -I 2 -M 1000 -J 1000 -L 2 -H 10 -B 1.0 -C 0.5 -D "weka.core.EuclideanDistance -R first-las" -S 10

[f] http://www.cs.cornell.edu/people/tj/svm_light/

Table 4. Linear SVM classification results for the above-vs.-below calories target prediction in a 10-fold cross validation setting. The ± indicate standard deviations across 10 folds.

Features	Accuracy	Precision	Recall
tokens	67.3% ±2.9%	67.8% ±3.3%	66.1% ±5.1%
categories	64.7% ±3.9%	64.8% ±4.1	64.8% ±4.3

The classification performance was sufficient though the category-based model did not perform better than the token-based model. We also performed an analysis to look at when the classifier errs. For both feature sets, both the false positive and false negative instances furthest from the decision boundary had less than 50% in their modal class ("below" or "above"). In other words, their ground truth label had a low degree of confidence. Similarly, in all cases the individual instances furthest from the decision boundary had the correct labels assigned. This can also be seen in Figure 3 where the fraction of "above" days for users increases from left to right, i.e., from being classified most strongly as "below" to being classified most strongly as "above". Interestingly, the graph also shows that users with a higher fraction of "above" days also tend to have logged more days in the system.

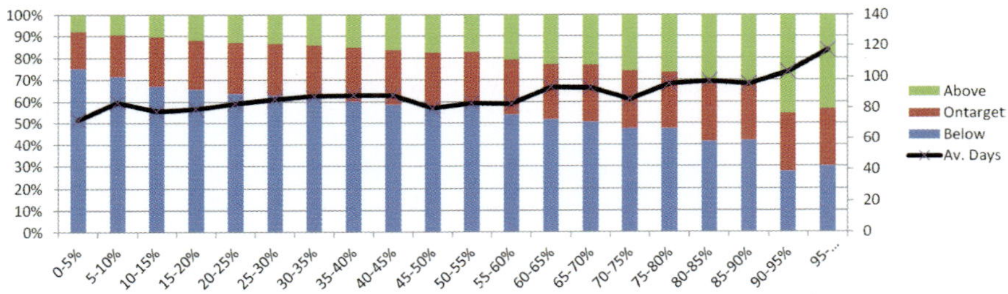

Fig. 3. 4,251 users are classified using the token-based SVM (see text) and then sorted based on the distance from the decision boundary. 0-5% refers to the 5% of users least likely to be labeled "above", similarly 95-100% refers to the 5% of users *most* likely to be labeled "above". The stacked plot shows the macro-averaged distribution across the logged days. The black line shows the average number of days logged for user in a given percentile group.

4.3. *Feature Analysis*

Our main motivation was *not* to predict if a user will be mostly above or below their weight goal, but rather, to understand the potential effect that different food choices might have on this outcome. To achieve this, we performed a feature analysis for the learned classification models. As we used a linear kernel, each feature is assigned a weight which can be directly interpreted with large and positive feature weights being indicate of "above" and large (in absolute value) and negative feature weights being indicative of "below".

Table 5 shows the features with the 10 most positive (negative) weights on the left (right).

Table 5. The 10 most discriminative features in the token-based linear SVM model. For each token, the foods logged by most users are listed.

Over Weight Goal		Under Weight Goal	
Token	Example Dish	Token	Example Dish
oil	oil - olive	cup	strawberries 1 cup*
wine	wine - table, red	kroger	sugar - kroger*
added	quick added calories	banana	turbana - banana
butter	salted butter*	grapes	grapes - raw
dairy	butter - dairy*	almond	almond milk*
original	original ranch*	value	value fries*
pieces	walnut pieces*	egg	whole egg*
container	mayonnaise container*	dole	dole banana*
lemon	lemon juice*	weight	weight control oatmeal*
mcdonalds	mcdo hash brown*	breast	turkey breast meat

Tokens such as "oil", "butter", or "mcdonalds" are all indicative for consuming more calories than planned. The "added" token in third position is mostly derived from using the "quick added calories" functionality. This functionality allows users to manually enter a summed caloric amount without having to enter each food item separately. Fruit tokens such as "banana", "grapes", or "lemon" are all indicative of staying below one's calorie goal.

Table 6. The 10 most discriminative categories in the category-based linear SVM model. For each category, the foods logged by most users are listed. A ".." indicates an omitted level for very long and multi-level categories

Over Weight Goal		Under Weight Goal	
Category	Example Dish	Category	Example Dish
beverage:alcohol	sabras - hummus	meat:..:turkey	sliced turkey*
dessert:cake	cheesecake	fruit	bananas - raw
preparation:fry	eggs - fried*	meat	turkey breast meat
staple:wheat:pizza	pepperoni pizza*	egg_dairy	eggs - fried*
meat:pork	ham - sliced*	meat:poultry	chicken breast*
dessert	cookies*	dessert:..:caramel	caramel - caramels
staple:other_cereal	fiber one bar*	fruit:..:banana	bananas - raw
..:..:coconut_oil	coconut oil*	..:milk_substitutes	almond milk*
staple:root_and_tuber	potatoes - baked*	preparation:bake	potatoes - baked*
staple:wheat:bread	bread - italian	snack:snack:donut	glazed donut*

Table 6 summarizes the category-based classification model. Overall, categories related to fruit, poultry, and baked foods are indicative of staying below one's calorie goals, whereas wheat, pork, and fried foods point towards going over.

It is interesting to see that desserts in general (denoted by the main category "dessert") are associated with logging too many calories, but caramels (denoted by the specific entity "dessert:confectionery:caramel") are associated with logging *less* categories than one's goal. However, the average usage of caramels corresponds to only 130kcal, compared to 173kcal for any logging entry under "dessert" and 195kcal for generic cakes ("dessert:cake"). The appearance of donuts ("snack:snack:donut"), with an average of 180 kcal, in the under-goal class is also unexpected. This may be potentially caused by the collinearity of certain features although the regularization term in the SVM usually deals with this.

Note that Table 6 shows that "sabras - hummus" is incorrectly categorized as beverage:alcohol in our taxonomy (see Section 3.1). A false positive is caused by the token "sabras" which was incorrectly matched to "Sabra", a liqueur produced in Israel, contained in the beverage category. Example foods in Table 5 & 6 that are marked with * were shortened to fit the table. The full list of names can be found in a footnote.[g]

5. Discussion

The analysis shown in this paper is preliminary in parts but still serves to show the value of using food diaries for studying dieting success in real-world settings. More complex machine-learning models could be used to, e.g., look at the *interaction* between food types. Conveniently, MFP provides this nutritional breakdown for the logged meals (see Figure 2). Such analysis could shed light on the success of dieting practices that advocate the separation of carbohydrates and protein or similar approaches.[h]

Our definition of whether a user is below or above their calories goal (see Section 3.3) is admittedly simple. For example, a user who is over their goal by 100% for one day, but then under for 10% for ten days would be labeled as "under" instead of "on-target". In fact, there are plausible alternative definitions but we do not expect them to change the results dramatically. For example, we had initially used a +/- 20% margin in *both* directions, not only towards below, and this gave similar list of discriminative tokens (Table 5).

More fundamentally, it is very difficult to determine if a shorter-than-usual log entry indicates a day of "food abstinence" or just an incomplete diary. Though we did not use it in this study, we could ideally obtain the weight loss goal, encoded in a picture posted in the profile page of some users. Having this information would also be helpful in distinguishing the small fraction of users on MFP who might be trying to *gain* or *maintain* weight rather than losing it. These users are not treated properly by our methodology though the analysis of "which food is linked to being above the weight goal" still holds.

Looking at the temporal patterns across a user's lifetime in the system, we did some preliminary analysis to see if users stopped logging food because of (i) achieving a set weight loss goal or (ii) getting frustrated by the failure to do so. For this, we looked at the fractions of users in a given "temporal percentile range", referring to a user's logging events buckets in 10% of their total time range. For each temporal bucket, we then assign it the modal

[g]The full names were "butter - salted butter", "butter - 1 pat - dairy", "hidden valley - original ranch" "walnuts - walnut pieces", "hellmann's - real mayonnaise 30fl oz container", "lemon juice - raw", "mcdonald's - hash brown from mcdonalds", "strawberries - raw, 1 cup", "light brown sugar - kroger", "almond milk - almond milk - vanilla - unsweetened", "wendy's - value french fries", "eggs - fried (whole egg)", "dole banana - bananas", "quaker oats - weight control instant oatmeal maple & brown sugar", "eggs - fried (whole egg)" "little caesars - pepperoni pizza", "ham - sliced, extra lean", "cookies - chocolate chip, soft-type", "fiber one - fiber one bar, oats & chocolate", "spectrum - coconut oil, unrefined", "potatoes - russet, flesh and skin, baked", "turkey, deli sliced - turkey", "eggs - fried (whole egg)", "protein - tyson chicken breast", "drinks - almond bilk (vanilla)", "potatoes - russet, flesh and skin, baked", "original glazed donut - krispy kreme".

[h]There are various types of "food combining" diets with the Hay Diet being one of the most prominent ones, despite the lack of success shown in randomized trials.[4]

label during that period. In aggregate, over a user's lifetime in the system their daily success ratios changed only slightly, from an initial below-vs.-on-target-vs.-above of 61%-23%-16%, to 63%-20%-17% for the penultimate 80-90% bucket. Only for the final 10% of logging events, the distribution changed to 69%-16%-15%. We are still unsure if this indicates (i) being most likely to "over-perform" by staying well below one's calories goal just before abandoning, or (ii) logging in a more and more incomplete manner at the end.

Weight control is related to controlling both calorie intake and energy expenditure. In this analysis, we only looked at the former of these. However, given that platforms such as MFP also provide a way to log the latter, we deem this worthwhile for future exploration.

Generally, having more access to user profile information could help predict "what type of diet will work for whom". What works could depend both on what type of foods a user has access to (e.g., due to income, geography, or working hours) but also relate to general lifestyles (e.g., with increased peer pressure when eating in a group). Automatically generated, *personalized* weight loss programs will definitely attract more attention in the future.

6. Conclusion

This paper presents a study that uses public food diaries of more than 4,000 long-term active MFP users. Our analysis is centered around a classifier that, given a list of foods consumed by a person, predicts if they will be below or above their self-defined calorie goal. While certain findings are expected ("oil" and "mcdonalds" being indicative of consuming too many calories) others are less obvious (poultry is linked to staying within one's goals, whereas pork indicates going above). Our results prove the feasibility of mining such data for health-related analysis. Especially with additional links to users' activity and general lifestyle patterns, automatically generated personalized and adaptive dieting seem a promising avenue to pursue. Health informatics is only starting to use the veritable gold mine that comes with public quantified-self data. This paper contributes to advances in this field by exploring how public food diaries can be mined to understand differences in unsuccessful and successful diets.

Acknowledgment

This work is supported by the National Research Foundation under its International Research Centre @ Singapore Funding Initiative and administered by the IDM Programme Office. In particular, we would like to thank Prof. Ee-Peng Lim for supporting a research visit by Dr. Weber in early 2015. This laid the foundation for this collaboration.

References

1. S. Abbar, Y. Mejova, and I. Weber. You tweet what you eat: Studying food consumption through twitter. In *Conference on Human Factors in Computing Systems (CHI)*, pages 3197–3206, 2015.
2. L. E. Burke, J. Wang, and M. A. Sevick. Self-monitoring in weight loss: A systematic review of the literature. *Journal of the American Dietetic Association*, 111:92—-102, 2011.
3. A. Culotta. Estimating county health statistics with twitter. In *Conference on Human Factors in Computing Systems (CHI)*, pages 1335–1344, 2014.
4. A. Golay, A.-F. Allaz, J. Ybarra, P. Bianchi, S. Saraiva, N. Mensi, R. Gomis, and N. de Tonnac.

Similar weight loss with low-energy food combining or balanced diets. *International Journal of Obesity*, 24(4):492–496, 2000.
5. M. A. Hall, E. Frank, G. Holmes, B. Pfahringer, P. Reutemann, and I. H. Witten. The WEKA data mining software: an update. *SIGKDD Explorations*, 11(1):10–18, 2009.
6. T. Joachims. Making large-scale SVM learning practical. In B. Schölkopf, C. Burges, and A. Smola, editors, *Advances in Kernel Methods - Support Vector Learning*, chapter 11, pages 169–184. MIT Press, Cambridge, MA, 1999.
7. J. L. Kraschnewski, J. Boan, J. Esposito, N. E. Sherwood, E. B. Lehman, D. K. Kephart, and C. N. Sciamanna. Long-term weight loss maintenance in the united states. *International Journal of Obesity*, 34:1644–1654, 2010.
8. M. Kuebler, E. Yom-Tov, D. Pelleg, R. M. Puhl, and P. Muennig. When overweight is the normal weight: An examination of obesity using a social media internet database. *PLOS ONE*, 8:e73479, 2013.
9. V. Li, D. W. McDonald, E. V. Eikey, J. Sweeney, J. Escajeda, G. Dubey, K. Riley, E. S. Poole, and E. B. Hekler. Losing it online: Characterizing participation in an online weight loss community. In *Conference on Supporting Group Work (GROUP)*, pages 35–45, 2014.
10. J. A. Linde, R. W. Jeffery, S. J. Crow, K. L. Brelje, C. R. Pacanowski, K. L. Gavin, and D. J. Smolenski. The tracking study: description of a randomized controlled trial of variations on weight tracking frequency in a behavioral weight loss program. *Contemporary Clinical Trials*, 40:199–211, 2015.
11. J. Meyer, S. Simske, K. A. Siek, C. G. Gurrin, and H. Hermens. Beyond quantified self: Data for wellbeing. In *Conference on Human Factors in Computing Systems (CHI)*, pages 95–98, 2014.
12. K. Park, I. Weber, M. Cha, and C. Lee. Persistent sharing of fitness app status on twitter. In *omputer-Supported Cooperative Work and Social Computing (CSCW)*, page to appear, 2016.
13. H. E. Payne, C. Lister, J. H. West, , and J. M. Bernhardt. Behavioral functionality of mobile apps in health interventions: A systematic review of the literature. *JMIR Mhealth Uhealth*, 3:e20, 2015.
14. D. Pelleg and A. W. Moore. X-means: Extending k-means with efficient estimation of the number of clusters. In *Conference on Machine Learning (ICML)*, pages 727–734, 2000.
15. J. Rooksby, M. Rost, A. Morrison, and M. C. Chalmers. Personal tracking as lived informatics. In *Conference on Human Factors in Computing Systems (CHI)*, pages 1163–1172, 2014.
16. M. Rusin, E. Arsand, and G. Hartvigsen. Functionalities and input methods for recording food intake: A systematic review. *International Journal of Medical Informatics*, 82:653–664, 2013.
17. C. N. Sciamanna, M. Kiernan, B. J. Rolls, J. Boan, H. Stuckey, D. Kephart, C. K. Miller, G. Jensen, T. J. Hartmann, E. Loken, K. O. Hwang, R. J. Williams, M. A. Clark, J. R. Schubart, A. M. Nezu, E. Lehman, and C. Dellasega. Practices associated with weight loss versus weight-loss maintenance. *American Journal of Preventive Medicine*, 41:159–166, 2011.
18. C.-Y. Teng, Y.-R. Lin, and L. A. Adamic. Recipe recommendation using ingredient networks. In *Web Science Conference (WebSci)*, pages 298–307, 2012.
19. C. Wagner, P. Singer, and M. Strohmaier. The nature and evolution of online food preferences. *EPJ Data Science*, 3(1), 2014.
20. C. Wagner, P. Singer, and M. Strohmaier. Spatial and temporal patterns of online food preferences. In *World Wide Web Conference (WWW)*, pages 553–554, 2014.
21. R. West, R. W. White, and E. Horvitz. From cookies to cooks: insights on dietary patterns via analysis of web usage logs. In *World Wide Web Conference (WWW)*, pages 1399–1410, 2013.

TRANSLATIONAL BIOINFORMATICS 101

JESSICA D. TENENBAUM
Department of Bioinformatics and Biostatistics, Duke University
Durham, NC 27715 USA
Jessie.Tenenbaum@duke.edu

SUBHA MADHAVAN
Innovation Center for Biomedical Informatics, Georgetown University
Washington, DC 20007 USA
Subha.Madhavan@georgetown.edu

ROBERT R. FREIMUTH
Department of Health Sciences Research, Mayo Clinic
Rochester, MN 55905 USA
Freimuth.Robert@mayo.edu

JOSHUA C. DENNY
Department of Biomedical Informatics, Vanderbilt University
Nashville, TN 37203 USA
josh.denny@Vanderbilt.Edu

LEWIS FREY
Public Health Sciences, Medical University of South Carolina
Charleston, SC 29425 USA
frey@musc.edu

1. Workshop Focus

This Workshop will give an overview of key topics in the young field of Translational Bioinformatics (TBI). TBI has been defined as the "development of storage, analytic, and interpretive methods to optimize the transformation of increasingly voluminous biomedical data, and genomic data, into proactive, predictive, preventive, and participatory health."[1] With PSB's stated focus on research in databases, algorithms, interfaces, natural language processing, and modeling, the bioinformatics aspect of TBI is a natural fit for this audience. Further, the US government's recently announced Precision Medicine Initiative makes it particularly timely for researchers to learn about and explore the translational side of our field. Specifically, this workshop provides context for how the various bioinformatics methods may be applied toward the enhancement of human health, enabling healthcare providers to deliver the right intervention for the right person at the right time.

This workshop covers major themes within the field of TBI, as put forth by a recent review in IMIA's (International Medical Informatics Association) Yearbook of Medical Informatics[2], and supplements those topics with a section on relevant data standards from the clinical and translational domain. Each presenter is a nationally or internationally recognized expert in their respective areas.

2. Workshop Agenda

Title	Speaker
Introduction	J Tenenbaum
Clinical "big data" I The use of EHR data for genomic discovery: the eMERGE network	J Denny
Clinical "big data" II Clinical3PO: Deep Phenotyping to Precision Medicine	L Frey
Omics for drug discovery and repurposing	J Tenenbaum
BREAK	
Intro to the clinic I Standards for Translational Bioinformatics	R Freimuth
Intro to the clinic II G-DOC *Plus*: A TBI platform for novel hypothesis generation in precision medicine research	S Madhavan
Personal genomic testing and related ethical, legal, and social issues	J Tenenbaum

3. Workshop Contributions

The use of EHR data for genomic discovery: the eMERGE network - J. Denny

Precision medicine offers the promise of improved diagnosis and more effective, patient-specific therapies. Typically, such studies have been pursued using research cohorts. Across the Electronic Medical Records and Genomics (eMERGE) Network, we have explored use of the electronic health records (EHRs) linked to DNA biobanks to do genomic and pharmacogenomic discovery. This combination allows study of the genomic basis of disease and drug response using real-world clinical data. Finding phenotypes in the EHR can be challenging, but the combination of billing data, laboratory data, medication exposures, and natural language processing has enabled efficient study of genomic and pharmacogenomic phenotypes. These studies have replicated many known associations as well as posited new genetic findings for diseases not yet studied via other methods. A particular advance may be for drug response traits, for which the EHR has proven cost efficient and effective. The EHR also enables the inverse experiment – starting with a genotype and discovering all the phenotypes with which it is associated – a phenome-wide association study (PheWAS). PheWAS requires a densely phenotyped population such as is found in the EHR. We have used PheWAS to replicate >300 genotype-phenotype associations, characterize pleiotropy, and discover new associations. We have also used PheWAS to identify characteristics with disease subtypes.

Collectively, EHR-linked biobanks across the U.S. alone are approaching 1 million people, portending a future in which these will play an increasingly important role. Indeed, the recently-announced presidential Precision Medicine Initiative highlights the role of EHR-based molecular study as an efficient and powerful longitudinal health data discovery platform. This national research cohort will enroll over 1 million individuals who are re-contactable and share biospecimens and health data. Many of these participants will likely share sensor and mobile technology data as well.

Clinical3PO: deep phenotyping to precision medicine - L. Frey

We will demonstrate components of the open source big data Clinical Personalized Pragmatic Predictions of Outcomes (Clinical3PO) platform, developed for the U.S. Department of Veterans Affairs (VA) along with its extension to typical healthcare environments. Its data representation is the Observational Medical Outcome Partnership (OMOP) common data model, which has been encoded to lower the barrier for cross-institutional data analysis. Using OMOP we will demonstrate the feature extraction module of Clinical3PO for deep phenotyping cohorts and feeding machine learning predictive analytic pipelines. The ability to support big data analytics for deep phenotyping using Clinical3PO applied to medical data will be described.[3] We will show how the pipeline can be used to create machine learning models focused on the care patterns of individuals. The path forward using the system to advance precision medicine will be discussed.

Omics for drug discovery and repurposing - J. Tenenbaum

Much has been written about the notoriously lengthy and expensive processes of drug discovery and FDA approval. From target identification to FDA approval, it is not uncommon for the process to take well over a decade and $1 billion. One major contribution of translational bioinformatics has been to apply a data-driven approach both to drug discovery, and drug repurposing: identifying existing FDA approved drugs that may help treat conditions for which they were not initially intended. High throughput omics technology can be used to identify promising pathways and molecular targets for a given disease, and also to identify molecular signatures of drug administration. These drug signatures can then be compared to signatures that characterize different diseases. Drugs that tend to have opposing effects on disease-related genes and pathways may be good candidates for treatment of those conditions. By effectively bypassing lead identification and Phase 1 trials, this approach can save both time and cost in FDA approval for new indications of existing drugs.

Standards for translational bioinformatics - R. Freimuth

The rapid growth of the biomedical domain has presented researchers and clinicians with more opportunities to share data and knowledge than ever before, but the diversity of data types, analysis methods, and contexts can pose significant challenges to the meaningful exchange, integration, and use of information. Standards can reduce barriers to semantic and syntactic interoperability. This presentation will review examples of existing and emerging standards within the translational bioinformatics community, including data, terminology, and message standards.

The generation, annotation, interpretation, and clinical reporting of genetic test results will serve as a use case throughout the presentation. Specific challenges in this process, such as variability in the representation of genetic data (e.g., nomenclature systems), and the impact that those challenges have on patient care and translational research will be discussed. Existing efforts by both international standards organizations and national consortia to develop normalized systems for the exchange of clinical genetic test results will be reviewed. Finally, methods for sharing knowledge, including that expressed in clinical genomic guidelines and decision support rules, will be summarized.

G-DOC *Plus*: A TBI platform for novel hypothesis generation in precision medicine research - S. Madhavan

G-DOC is a feature-rich shareable translational research infrastructure that allows physician scientists and translational researchers to mine and analyze a variety of "omics" data in the context of consistently defined clinical outcomes data for cancer patients.[4]

Scientists today are using not only a combination of clinical, NGS and omics data for analysis, but also medical and digital images for validation of analysis results. Currently, numerous tools and software exist that specialize in handling and processing of one or two "omics" data types, or only NGS data types, most of which need a bioinformatician to help with analysis. To drive hypothesis generation and validation of molecular markers for biologists and researchers, it would be convenient to have a "one–stop" system that can handle all these data types, including NGS and medical images, in one location without having to switch to other tools or resources for analysis.

With the goal of improving overall health outcomes through genomics research, we present G-DOC *Plus*, a web-based bioinformatics platform that enables the integrative analysis of multiple data types to understand mechanisms of cancer and non-cancer diseases at a systems level for systematic conduct of research in precision medicine. It currently holds data from over 10,000 patients selected from private and public resources including Gene Expression Omnibus (GEO), The Cancer Genome Atlas (TCGA) and the recently added datasets from REpository for Molecular BRAin Neoplasia DaTa (REMBRANDT), caArray studies of lung and colon cancer and the 1000 genomes data sets. G-DOC *Plus* allows researchers to explore clinical-omic data one sample at a time, as a cohort of samples; or at the level of population, providing the user with a comprehensive view of the data.

Three case studies in Pharmacogenomics, cancer variant search, and gene network analysis will be demonstrated to educate workshop attendees on features of G-DOC Plus for novel hypothesis generation to advance precision medicine research.

Direct to consumer genetic testing, and related ethical, legal, and social issues- J. Tenenbaum

Direct to consumer (DTC) genomic services enable individuals to obtain their own genetic data without a healthcare provider acting as intermediary either to order the test or interpret the results. For several years after these services were introduced, it was unclear whether or how they should be regulated by the government. In 2013, the FDA strongly asserted that these tests do indeed fall within their purview, and each health-related association must be separately validated. Their rationale for stepping in was that people might make healthcare decisions, drastic ones even, based on information obtained through these services.

In this portion of the workshop, we describe a pregnancy management case in which a treatment plan was modified based on a DTC result. A woman with no personal or family history of blood clotting-related complications learned through DTC testing about a heterozygous prothrombin (factor 2) gene mutation. Twice daily injections of enoxaparin were recommended throughout pregnancy for this patient based on this genetic information combined with other risk factors including advanced maternal age and pregnancy with twins. Genetically based medical guidelines are a moving target, however, and treatment of thrombophilic conditions in asymptomatic patients is controversial, with guidelines continuing to evolve.

We will also discuss ethical, legal, social, and economic issues raised by this case and its impact on the patient's subsequent efforts to obtain life insurance, which unlike health insurance, is not covered under the Genetic Information Nondiscrimination Act of 2008.

4. References

1. AMIA. *Translational Bioinformatics | AMIA*. 10/5/15]; Available from: http://www.amia.org/applications-informatics/translational-bioinformatics.
2. Denny, J.C., *Surveying Recent Themes in Translational Bioinformatics: Big Data in EHRs, Omics for Drugs, and Personal Genomics.* Yearb Med Inform, 2014. **9**: p. 199-205.
3. Frey, L.J., L. Lenert, and G. Lopez-Campos, *EHR Big Data Deep Phenotyping. Contribution of the IMIA Genomic Medicine Working Group.* Yearb Med Inform, 2014. **9**: p. 206-11.
4. Madhavan, S., et al., *G-DOC: a systems medicine platform for personalized oncology.* Neoplasia, 2011. **13**(9): p. 771-83.

COMPUTATIONAL APPROACHES TO STUDY MICROBES AND MICROBIOMES

CASEY S. GREENE[†]
Systems Pharmacology and Translational Therapeutics
Perelman School of Medicine
University of Pennsylvania
Philadelphia PA 19104, USA
Email: csgreene@upenn.edu

JAMES A. FOSTER
Institute of Bioinformatics and Evolutionary Studies
University of Idaho
Moscow, ID 83844 USA
Email: foster@uidaho.edu

BRUCE A. STANTON
Department of Microbiology and Immunology
The Geisel School of Medicine at Dartmouth
Hanover, NH 03755, USA
Email: Bruce.A.Stanton@dartmouth.edu

DEBORAH A. HOGAN
Department of Microbiology and Immunology
The Geisel School of Medicine at Dartmouth
Hanover, NH 03755, USA
Email: Deborah.A.Hogan@dartmouth.edu

YANA BROMBERG
Biochemistry and Microbiology, School of Environmental and Biological Sciences
Rutgers University
New Brunswick, NJ 08901, USA
Institute for Advanced Study, Technische Universität München
Garching, Germany
Email: yana@bromberglab.org

Technological advances are making large-scale measurements of microbial communities commonplace. These newly acquired datasets are allowing researchers to ask and answer questions about the composition of microbial communities, the roles of members in these communities, and how genes and molecular pathways are regulated in individual community members and communities as a whole to effectively respond to diverse and changing environments. In addition to providing a more comprehensive survey of the microbial world, this new information allows for the development of computational approaches to model the processes underlying microbial systems. We anticipate that the field of computational microbiology will continue to grow rapidly in the coming years. In this manuscript we highlight both areas of particular interest in microbiology as well as computational approaches that begin to address these challenges.

[†] To whom correspondence should be addressed.

1. Introduction

Microbes, including viruses, bacteria, and fungi are the most numerous organisms on earth. Bacteria alone are estimated to equal the biomass of plants on earth.[1]Moreover, they are the key drivers of life on earth by controlling the majority of Earth's biogeochemical fluxes.[2]

Microbial communities also play key roles in human health and disease.[3,4] While the role of microbes underlying certain illnesses has been widely recognized, we are also recognizing their role in normal physiology, and the role that they can play to restore normal physiology. For example, a diet of non-digestible but fermentable carbohydrates given to children affected by the Prader-Willi syndrome has been shown to lead to changes in the gut microbiome structure, contributing to reduction in weight, regardless of the continued presence of the primary driving forces.[5] In a more directed experiment, transplants of fecal microbiota has been used to alleviate chronic *Clostridium difficile* infections.[6,7]

Microbial communities were historically relatively difficult to survey and characterize. The development of fast and inexpensive sequencing methods has dramatically aided in this analysis.[8] We can now readily evaluate and describe communities that we could not easily catalog with other approaches.[9,10] These new experimental platforms are providing the basis of in depth surveys of the microbial components of our world. For example, the human microbiome project (HMP) was designed to catalog human-associated microbial communities,[11] producing an extensive bacterial catalog of over 200 adults.[12]

Many other studies are working towards identifying microbiome features that are important for health or disease. For example, a series of studies have characterized the microbiome in lungs of individuals with conditions such as cystic fibrosis (CF),[13–16] chronic obstructive pulmonary disease (COPD),[17] asthma,[3,18] and in the intestinal tract of individuals with CF[19], and diabetes.[4,20] In some cases it has been possible to identify pathogens and/or the expression of particular genes that are associated with positive or negative outcomes.[19,21] It is the hope that knowledge of the microbiome and gene expression can be leveraged to develop more targeted interventions and preventative treatments.

The wealth of microbial data is generating new challenges as well as new opportunities for computational microbiology. Some predict that genomic data will become the foremost example of big data, outpacing astronomy and other data-intensive fields within the next ten years.[22] Algorithms that address this challenge will transform microbiology, but to do so they will need to be accurate, scalable, and wrapped in software accessible to and usable by biologists.

2. Challenges in Microbiology and Computational Approaches

We discuss existing challenges in microbiology, and highlight computational approaches that address these challenges. We focus primarily on those areas that have been transformed by the wealth of sequencing data now available.

2.1. *Gene molecular function and process prediction*

While DNA and RNA sequencing has become substantially easier and less costly, the process of understanding the function of genes remains difficult. This process of functional determination has been facilitated by computational algorithms that aim to automatically annotate functions based on: the gene's nucleic acid sequence; the similarity of the gene's sequence to those with annotated functions;[23] how the gene is expressed;[24] the gene's interaction partners;[25,26] and other features.[27]

While there are many approaches for prediction, there are also many approaches for assessment, and the need for commonly accepted benchmarks has been highlighted as an area of need.[28] Recently, the Critical Assessment of Function Annotation (CAFA) was conducted to address this need.[29] While CAFA represents an important first step, the need for benchmark datasets, particularly those with comprehensive experimental validation and standardized assessment, remains high. This is particularly true in bacterial systems, which have not been well covered by CAFA challenges to date.[29] Ideally microbiologists will be able to both retrieve a best estimate for any gene of interest in an organism, and also receive a well-calibrated confidence score for that prediction.

2.2. *Microbes' molecular functionality and classification*

The overall sum of molecular functionality encoded in the genomes of microbes is representative of both their morphology and physiology – key features in bacterial taxonomic classification. Our interest in microbes is often focused on specific parts of their molecular abilities – their pathogenicity, toxicity and antibiotic resistance (to us and other species, *e.g.* for bio-pesticide purposes), as well as their ability to survive and thrive in extreme environments or with specific or limited nutrient sources (bioremediation and green energy). Thus, classification of microbes that implies similar treatment of similar organisms is important for industrial and clinical applications.

Current taxonomy is guided by evolutionary relationships,[30] which, however, ignores horizontal gene transfer (HGT) and, often, plasmid contributions and, therefore, does not guarantee functional similarity. Recent work[31] has shown the advantages of using microbial genome-guided predictions as proxies for functional comparisons. Microbial functional comparisons, informed by individual organisms' environmental preferences, highlight specific genes and functions responsible for particular environmental adaptations (*e.g.* functional studies of cyanobacteria clades identify sigma factors potentially responsible for salt tolerance).[31] However, despite significant recent efforts[32,33], only a third of the microbial genes (for which sequences are available) are explicitly functionally annotated,[31] and high-throughput experiments exploring temporal relationships between gene expressions are missing for the vast majority of (already fully sequenced) microorganisms, and annotations of molecular pathways are limited. Additionally, any available experimental tests only reflect a portion of overall bacterial functionality, with nearly three hundred tests only accessing 5–20% of the total functional potential.[30] Thus, significant further research is necessary to properly identify, describe, and use microbial functional abilities on a large scale. Within the confines of the current state and speed of the experimental art, computational approaches remain the sole, most significant means for producing new knowledge from existing data (*e.g.* computational studies on co-occurrence of specific functions encoded across genomes of organisms occupying similar environments could inform the necessary molecular pathways).

Microbial molecular functional abilities accurately reflect the environmental challenges faced by the individual subpopulations of microbes (ecotypes). In fact, the environment often has a more pronounced effect on the microbial genomes than does vertical descent. We further expect that a function-based approach at exploring environmental impact will be even more relevant to the study of entire microbial communities in light of their emergent functionalities (*i.e.* functions that are available to the diversity of microorganisms together occupying a single niche, but not to each individual organism within that niche).

2.3. *Microbes' responses to their environment*

Microbes must respond to their environment to adapt to changing conditions such as nutrient availability, changes in a host, new members of their microbial community and many other factors.[34] Sequencing-based methods allow the transcriptomes of organisms to be measured without the potentially time consuming and costly array-design process that was required in the past.[35] This has allowed for assays of a diverse array of organisms, including many microbes. Such assays readily allow for differential expression analyses, in which genes are ordered by the extent to which they differ between conditions, isolates, or environments. While differential expression analyses play an important role, being able to integrate newly performed experiments into the context of existing data provides a key opportunity.

There are now more than 1.8 million genome-wide assays freely available in repositories such as ArrayExpress[32] and NCBI's Gene Expression Omnibus[33] (GEO). In total, these repositories contain experiments for more than 2000 different organisms (Fig. 1A). More than 150 species had more than 500 assays publicly available as of July 1, 2015 (Fig 1B). We anticipate that the number of organisms with large amounts of transcriptomic data will continue to grow. The transcriptomes of nearly 45,000 single cells were recently sequenced in one experiment,[36,37] surpassing the number of transcriptomes available for many organisms. While such techniques cannot yet be readily applied to bacteria, we expect that new approaches will become available and rapidly expand the diversity and scale of available transcriptomic datasets for microbiological systems.

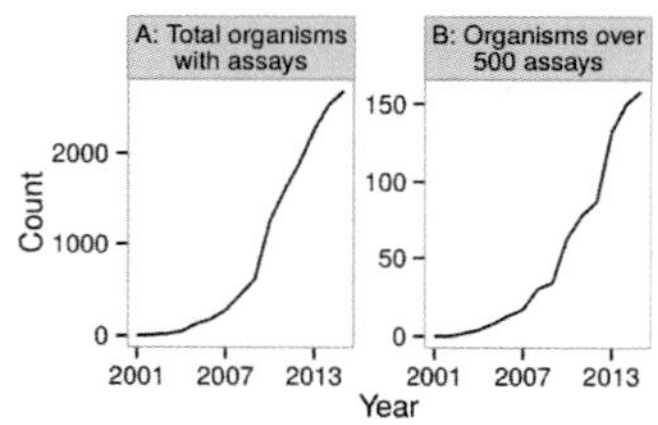

Fig. 1. The number of organisms with (A) genome wide data available and (B) those with more than 500 publicly available genome-wide expression assays. Counts for 2015 include assays from January through July.

We now have the opportunity to integrate and analyze these data to understand how the response to the environment in a specific or newly performed experiment relates to the response observed by others in past experiments. In well-characterized systems, data have been integrated using supervised methodologies that leverage extensively curated knowledgebases.[38,39] For many microbial systems, these knowledge-bases are limited or unavailable. To address this challenge we need to develop unsupervised algorithms capable of integrating data with diverse information, ideally across multiple platforms. Such algorithms are now being developed,[40] but significant work remains to be done to apply these to large-scale microbial data compendia.

2.4. *Host-microbe and microbe-microbe interactions*

While adaptive immune responses in the host and evasion strategies of the microbe have been extensively studied, we are still discovering new mechanisms of host-microbe interactions. For example, Lee et al. identified genetic variants in a specific bitter taste receptor that were associated with susceptibility to respiratory infections, and that these receptors could be activated by compounds produced by *Pseudomonas aeruginosa*.[41] Subsequent studies have continued to reveal roles for taste receptors in the innate immune response.[42,43] Similar sensing of microbe-produced compounds have been reported in plants.[44] Computational techniques that facilitate the analysis of host genetics in combination with the composition and metagenomic characteristics of microbial communities may continue to identify additional novel mechanisms of host-microbe interactions.

2.5. *Membership in microbial communities*

Microbial communities have now been extensively profiled.[10,45–47] For communities on human hosts, the HMP has provided a large-scale survey across many available surfaces.[12] In addition to this large-scale assessment, numerous surveys have been made of microbial communities in a multitude of specific sites.[48–51] Such analyses have been performed in both healthy individuals and those experiencing a variety of conditions.[52–57]

2.5.1. *Heterogeneity across microbial communities*

Analysis of datasets from the HMP and others has raised numerous questions. For example, the abundance of microbial taxa varies substantially between individuals and body sites, but the relative abundance of metabolic modules within the communities remains consistent.[12,58] Studies of twins have revealed differences in similarity between monozygotic and dizygotic twins, suggesting that an individual's genetics affect his or her microbial communities.[59,60] This observation raises the question: what are the drivers of these differences, and what are their implications for both the community and the host?

2.5.2. *Heterogeneity within microbial populations*

Prior to the advent of large-scale inexpensive sequencing, microbial communities were assessed through sequencing of portions of the 16S ribosomal subunit that provided a family, genus, or species level of resolution.[61] Metagenomic analysis of both mixed species communities and single species populations provides the opportunity to identify genetic heterogeneity at the sub-species level within microbial communities. Such genomic diversification is rapid and common in biofilms[62,63] and chronic disease[64] and needs to be incorporated into our models for microbial communities. For example, traditional microbiological analyses have found that *P. aeruginosa* mutants with increased alginate production or the loss of quorum sensing regulation are commonly selected for in the lungs of individuals with CF, and the appearance of these mutants is associated alterations in pathways known to be associated with host interactions[65] and have been associated with worse disease outcome[66]. The ability to associate genomic, transcriptomic, and phenotype or outcomes data will position us to understand which environmental factors drive the selection for certain variants and how these variants change the course of host-microbe and microbe-microbe interactions.

3. Conclusions

This is an exciting time in computational microbiology. Both the questions being asked and the experimental methodologies available to answer them are expanding in scope and diversity. We have highlighted a number of areas where we see particular opportunities for computational approaches. We anticipate that addressing these questions will require expertise in microbiology and the development, evaluation, and application of computational systems. The goal of our workshop is to provide a venue that brings these constituencies together.

4. Acknowledgments

C.S.G. is supported in part by the Gordon and Betty Moore Foundation's Data-Driven Discovery Initiative through Grant GBMF4552 and by the National Science Foundation under Grant Number 1458390 to C.S.G. B.A.S. is supported by the CF Foundation, NIH R01-HL074175, NIH P30-GM106394, NIEHS P42-ES007373 and R01-HL074175. D.A.H. is supported by NIH R01-AI091702. Y.B. is supported by the USDA-NIFA grant 1015:0228906 and the Technische Universität München – Institute for Advanced Study Hans Fischer Fellowship, funded by the German Excellence Initiative and the European Union Seventh Framework Programme, grant agreement 291763. Our workshop is supported by NIH P30-GM106394.

References

1. Whitman, W. B., Coleman, D. C. & Wiebe, W. J. Prokaryotes: the unseen majority. *Proc. Natl. Acad. Sci. U. S. A.* **95,** 6578–83 (1998).

2. Falkowski, P. G., Fenchel, T. & Delong, E. F. The microbial engines that drive Earth's biogeochemical cycles. *Science* **320,** 1034–9 (2008).

3. Dominguez-Bello, M. G. & Blaser, M. J. Asthma: Undoing millions of years of coevolution in early life? *Sci. Transl. Med.* **7,** 307fs39–307fs39 (2015).

4. Semenkovich, C. F., Danska, J., Darsow, T., Dunne, J. L., Huttenhower, C., Insel, R. A., McElvaine, A. T., Ratner, R. E., Shuldiner, A. R. & Blaser, M. J. American Diabetes Association and JDRF Research Symposium: Diabetes and the Microbiome. *Diabetes* (2015). doi:10.2337/db15-0597

5. Zhang, C., Yin, A., Li, H., Wang, R., Wu, G., Shen, J., Zhang, M., Wang, L., Hou, Y., Ouyang, H., Zhang, Y., Zheng, Y., Wang, J., Lv, X., Wang, Y., Zhang, F., Zeng, B., Li, W., Yan, F., *et al.* Dietary modulation of gut microbiota contributes to alleviation of both genetic and simple obesity in children. *EBioMedicine* **2,** 966–82 (2015).

6. Aas, J., Gessert, C. E. & Bakken, J. S. Recurrent Clostridium difficile colitis: case series involving 18 patients treated with donor stool administered via a nasogastric tube. *Clin. Infect. Dis.* **36,** 580–5 (2003).

7. Kassam, Z., Lee, C. H., Yuan, Y. & Hunt, R. H. Fecal microbiota transplantation for Clostridium difficile infection: systematic review and meta-analysis. *Am. J. Gastroenterol.* **108,** 500–8 (2013).

8. Mardis, E. R. The impact of next-generation sequencing technology on genetics. *Trends Genet.* **24,** 133–41 (2008).

9. Turnbaugh, P. J., Ley, R. E., Mahowald, M. A., Magrini, V., Mardis, E. R. & Gordon, J. I. An obesity-associated gut microbiome with increased capacity for energy harvest. *Nature* **444,** 1027–31 (2006).

10. Dinsdale, E. A., Edwards, R. A., Hall, D., Angly, F., Breitbart, M., Brulc, J. M., Furlan, M., Desnues, C., Haynes, M., Li, L., McDaniel, L., Moran, M. A., Nelson, K. E., Nilsson, C., Olson, R., Paul, J., Brito, B. R., Ruan, Y., Swan, B. K., *et al.* Functional metagenomic profiling of nine biomes. *Nature* **452,** 629–32 (2008).

11. Turnbaugh, P. J., Ley, R. E., Hamady, M., Fraser-Liggett, C. M., Knight, R. & Gordon, J. I. The human microbiome project. *Nature* **449,** 804–10 (2007).

12. The Human Microbiome Project Consortium. Structure, function and diversity of the healthy human microbiome. *Nature* **486,** 207–14 (2012).

13. Madan, J. C., Koestler, D. C., Stanton, B. A., Davidson, L., Moulton, L. A., Housman, M. L., Moore, J. H., Guill, M. F., Morrison, H. G., Sogin, M. L., Hampton, T. H., Karagas, M. R., Palumbo, P. E., Foster, J. A., Hibberd, P. L. & O'Toole, G. A. Serial analysis of the gut and respiratory microbiome in cystic fibrosis in infancy: interaction between intestinal and respiratory tracts and impact of nutritional exposures. *MBio* **3,** e00251–12– (2012).

14. Gifford, A. H., Alexandru, D. M., Li, Z., Dorman, D. B., Moulton, L. A., Price, K. E., Hampton, T. H., Sogin, M. L., Zuckerman, J. B., Parker, H. W., Stanton, B. A. & O'Toole, G. A. Iron supplementation does not worsen respiratory health or alter the sputum microbiome in cystic fibrosis. *J. Cyst. Fibros.* **13,** 311–8 (2014).

15. Carmody, L. A., Zhao, J., Kalikin, L. M., LeBar, W., Simon, R. H., Venkataraman, A., Schmidt, T. M., Abdo, Z., Schloss, P. D. & LiPuma, J. J. The daily dynamics of cystic fibrosis airway microbiota during clinical stability and at exacerbation. *Microbiome* **3,** 12 (2015).

16. LiPuma, J. J. Assessing Airway Microbiota in Cystic Fibrosis: What More Should Be Done? *J. Clin. Microbiol.* **53,** 2006–2007 (2015).

17. Garcia-Nuñez, M., Millares, L., Pomares, X., Ferrari, R., Pérez-Brocal, V., Gallego, M., Espasa, M., Moya, A. & Monsó, E. Severity-related changes of bronchial microbiome in chronic obstructive pulmonary disease. *J. Clin. Microbiol.* **52,** 4217–23 (2014).

18. Huang, Y. J., Nariya, S., Harris, J. M., Lynch, S. V, Choy, D. F., Arron, J. R. & Boushey, H. The airway microbiome in patients with severe asthma: Associations with disease features and severity. *J. Allergy Clin. Immunol.* (2015). doi:10.1016/j.jaci.2015.05.044

19. Hoen, A. G., Li, J., Moulton, L. A., O'Toole, G. A., Housman, M. L., Koestler, D. C., Guill, M. F., Moore, J. H., Hibberd, P. L., Morrison, H. G., Sogin, M. L., Karagas, M. R. & Madan, J. C. Associations between Gut Microbial Colonization in Early Life and Respiratory Outcomes in Cystic Fibrosis. *J. Pediatr.* **167,** 138–147.e3 (2015).

20. Delzenne, N. M., Cani, P. D., Everard, A., Neyrinck, A. M. & Bindels, L. B. Gut microorganisms as promising targets for the management of type 2 diabetes. *Diabetologia* **58,** 2206–17 (2015).

21. Filkins, L. M., Hampton, T. H., Gifford, A. H., Gross, M. J., Hogan, D. A., Sogin, M. L., Morrison, H. G., Paster, B. J. & O'Toole, G. A. Prevalence of Streptococci and Increased Polymicrobial Diversity Associated with Cystic Fibrosis Patient Stability. *J. Bacteriol.* **194,** 4709–4717 (2012).

22. Stephens, Z. D., Lee, S. Y., Faghri, F., Campbell, R. H., Zhai, C., Efron, M. J., Iyer, R., Schatz, M. C., Sinha, S. & Robinson, G. E. Big Data: Astronomical or Genomical? *PLOS Biol.* **13,** e1002195 (2015).

23. Wilson, C. A., Kreychman, J. & Gerstein, M. Assessing annotation transfer for genomics: quantifying the relations between protein sequence, structure and function through traditional and probabilistic scores. *J. Mol. Biol.* **297,** 233–49 (2000).

24. Greene, C. S. & Troyanskaya, O. G. PILGRM: an interactive data-driven discovery platform for expert biologists. *Nucleic Acids Res.* **39,** W368–W374 (2011).

25. Marcotte, E. M., Pellegrini, M., Ng, H.-L., Rice, D. W., Yeates, T. O. & Eisenberg, D. Detecting Protein Function and Protein-Protein Interactions from Genome Sequences. *Science (80-.).* **285,** 751–753 (1999).

26. Vazquez, A., Flammini, A., Maritan, A. & Vespignani, A. Global protein function prediction from protein-protein interaction networks. *Nat. Biotechnol.* **21,** 697–700 (2003).

27. Troyanskaya, O. G., Dolinski, K., Owen, A. B., Altman, R. B. & Botstein, D. A Bayesian framework for combining heterogeneous data sources for gene function prediction (in Saccharomyces cerevisiae). *Proc. Natl. Acad. Sci. U. S. A.* **100,** 8348–53 (2003).

28. Murali, T. M., Wu, C.-J. & Kasif, S. The art of gene function prediction. *Nat. Biotechnol.* **24,** 1474–5; author reply 1475–6 (2006).

29. Radivojac, P., Clark, W. T., Oron, T. R., Schnoes, A. M., Wittkop, T., Sokolov, A., Graim, K., Funk, C., Verspoor, K., Ben-Hur, A., Pandey, G., Yunes, J. M., Talwalkar, A. S., Repo, S., Souza, M. L., Piovesan, D., Casadio, R., Wang, Z., Cheng, J., *et al.* A large-scale evaluation of computational protein function prediction. *Nat. Methods* **10,** 221–7 (2013).

30. Garrity, G., Boone, D. R. & Castenholz, R. W. *Bergey's Manual of Systematic Bacteriology Volume 1: The Archaea and the Deeply Branching and Phototrophic Bacteria.* (Springer, 2001). doi:10.1007/0-387-29298-5

31. Zhu, C., Delmont, T. O., Vogel, T. M. & Bromberg, Y. Functional Basis of Microorganism Classification. *PLoS Comput. Biol.* **11,** e1004472 (2015).

32. Rustici, G., Kolesnikov, N., Brandizi, M., Burdett, T., Dylag, M., Emam, I., Farne, A., Hastings, E., Ison, J., Keays, M., Kurbatova, N., Malone, J., Mani, R., Mupo, A., Pedro Pereira, R., Pilicheva, E., Rung, J., Sharma, A., Tang, Y. A., *et al.* ArrayExpress update--trends in database growth and links to data analysis tools. *Nucleic Acids Res.* **41,** D987–90 (2013).

33. Barrett, T., Troup, D. B., Wilhite, S. E., Ledoux, P., Evangelista, C., Kim, I. F., Tomashevsky, M., Marshall, K. A., Phillippy, K. H., Sherman, P. M., Muertter, R. N., Holko, M., Ayanbule, O., Yefanov, A. & Soboleva, A. NCBI GEO: archive for functional genomics data sets--10 years on. *Nucleic Acids Res.* **39,** D1005–10 (2011).

34. Dey, N., Wagner, V. E., Blanton, L. V., Cheng, J., Fontana, L., Haque, R., Ahmed, T. & Gordon, J. I. Regulators of Gut Motility Revealed by a Gnotobiotic Model of Diet-Microbiome Interactions Related to Travel. *Cell* **163,** 95–107 (2015).

35. Ekblom, R. & Galindo, J. Applications of next generation sequencing in molecular ecology of non-model organisms. *Heredity (Edinb).* **107,** 1–15 (2011).

36. Macosko, E. Z., Basu, A., Satija, R., Nemesh, J., Shekhar, K., Goldman, M., Tirosh, I., Bialas, A. R., Kamitaki, N., Martersteck, E. M., Trombetta, J. J., Weitz, D. A., Sanes, J. R., Shalek, A. K., Regev, A. & McCarroll, S. A. Highly Parallel Genome-wide Expression Profiling of Individual Cells Using Nanoliter Droplets. *Cell* **161,** 1202–1214 (2015).

37. Klein, A. M., Mazutis, L., Akartuna, I., Tallapragada, N., Veres, A., Li, V., Peshkin, L., Weitz, D. A. & Kirschner, M. W. Droplet Barcoding for Single-Cell Transcriptomics Applied to Embryonic Stem Cells. *Cell* **161,** 1187–1201 (2015).

38. Wong, A. K., Park, C. Y., Greene, C. S., Bongo, L. A., Guan, Y. & Troyanskaya, O. G. IMP: a multi-species functional genomics portal for integration, visualization and prediction of protein functions and networks. *Nucleic Acids Res.* **40,** W484–90 (2012).

39. Greene, C. S., Krishnan, A., Wong, A. K., Ricciotti, E., Zelaya, R. A., Himmelstein, D. S., Zhang, R., Hartmann, B. M., Zaslavsky, E., Sealfon, S. C., Chasman, D. I., FitzGerald, G. A., Dolinski, K., Grosser, T. & Troyanskaya, O. G. Understanding multicellular function and disease with human tissue-specific networks. *Nat. Genet.* **47,** 569–576 (2015).

40. Tan, J., Ung, M., Cheng, C. & Greene, C. S. Unsupervised feature construction and knowledge extraction from genome-wide assays of breast cancer with denoising autoencoders. *Pacific Symp. Biocomput.* 132–43 (2015).

41. Lee, R. J., Xiong, G., Kofonow, J. M., Chen, B., Lysenko, A., Jiang, P., Abraham, V., Doghramji, L., Adappa, N. D., Palmer, J. N., Kennedy, D. W., Beauchamp, G. K., Doulias, P.-T., Ischiropoulos, H., Kreindler, J. L., Reed, D. R. & Cohen, N. A. T2R38 taste receptor polymorphisms underlie susceptibility to upper respiratory infection. *J. Clin. Invest.* **122,** 4145–59 (2012).

42. Lee, R. J., Kofonow, J. M., Rosen, P. L., Siebert, A. P., Chen, B., Doghramji, L., Xiong, G., Adappa, N. D., Palmer, J. N., Kennedy, D. W., Kreindler, J. L., Margolskee, R. F. & Cohen, N. A. Bitter and sweet taste receptors regulate human upper respiratory innate immunity. *J. Clin. Invest.* **124,** 1393–405 (2014).

43. Lee, R. J. & Cohen, N. A. Role of the bitter taste receptor T2R38 in upper respiratory infection and chronic rhinosinusitis. *Curr. Opin. Allergy Clin. Immunol.* **15,** 14–20 (2015).

44. Hartmann, A., Rothballer, M., Hense, B. A. & Schröder, P. Bacterial quorum sensing compounds are important modulators of microbe-plant interactions. *Front. Plant Sci.* **5,** 131 (2014).

45. Xie, W., Wang, F., Guo, L., Chen, Z., Sievert, S. M., Meng, J., Huang, G., Li, Y., Yan, Q., Wu, S., Wang, X., Chen, S., He, G., Xiao, X. & Xu, A. Comparative metagenomics of microbial communities inhabiting deep-sea hydrothermal vent chimneys with contrasting chemistries. *ISME J.* **5,** 414–26 (2011).

46. Mackelprang, R., Waldrop, M. P., DeAngelis, K. M., David, M. M., Chavarria, K. L., Blazewicz, S. J., Rubin, E. M. & Jansson, J. K. Metagenomic analysis of a permafrost microbial community reveals a rapid response to thaw. *Nature* **480,** 368–71 (2011).

47. Pope, P. B., Mackenzie, A. K., Gregor, I., Smith, W., Sundset, M. A., McHardy, A. C., Morrison, M. & Eijsink, V. G. H. Metagenomics of the Svalbard reindeer rumen microbiome reveals abundance of polysaccharide utilization loci. *PLoS One* **7,** e38571 (2012).

48. Segata, N., Haake, S. K., Mannon, P., Lemon, K. P., Waldron, L., Gevers, D., Huttenhower, C. & Izard, J. Composition of the adult digestive tract bacterial microbiome based on seven mouth surfaces, tonsils, throat and stool samples. *Genome Biol.* **13,** R42 (2012).

49. Willner, D., Haynes, M. R., Furlan, M., Hanson, N., Kirby, B., Lim, Y. W., Rainey, P. B., Schmieder, R., Youle, M., Conrad, D. & Rohwer, F. Case studies of the spatial heterogeneity of DNA viruses in the cystic fibrosis lung. *Am. J. Respir. Cell Mol. Biol.* **46,** 127–31 (2012).

50. Zhou, X., Brown, C. J., Abdo, Z., Davis, C. C., Hansmann, M. A., Joyce, P., Foster, J. A. & Forney, L. J. Differences in the composition of vaginal microbial communities found in healthy Caucasian and black women. *ISME J.* **1,** 121–33 (2007).

51. Hunt, K. M., Foster, J. A., Forney, L. J., Schütte, U. M. E., Beck, D. L., Abdo, Z., Fox, L. K., Williams, J. E., McGuire, M. K. & McGuire, M. A. Characterization of the Diversity and Temporal Stability of Bacterial Communities in Human Milk. *PLoS One* **6,** e21313 (2011).

52. Erb-Downward, J. R., Thompson, D. L., Han, M. K., Freeman, C. M., McCloskey, L., Schmidt, L. A., Young, V. B., Toews, G. B., Curtis, J. L., Sundaram, B., Martinez, F. J. & Huffnagle, G. B. Analysis of the lung microbiome in the 'healthy' smoker and in COPD. *PLoS One* **6,** e16384 (2011).

53. Spencer, M. D., Hamp, T. J., Reid, R. W., Fischer, L. M., Zeisel, S. H. & Fodor, A. A. Association between composition of the human gastrointestinal microbiome and development of fatty liver with choline deficiency. *Gastroenterology* **140,** 976–86 (2011).

54. Larsen, N., Vogensen, F. K., van den Berg, F. W. J., Nielsen, D. S., Andreasen, A. S., Pedersen, B. K., Al-Soud, W. A., Sørensen, S. J., Hansen, L. H. & Jakobsen, M. Gut microbiota in human adults with type 2 diabetes differs from non-diabetic adults. *PLoS One* **5,** e9085 (2010).

55. Qin, J., Li, Y., Cai, Z., Li, S., Zhu, J., Zhang, F., Liang, S., Zhang, W., Guan, Y., Shen, D., Peng, Y., Zhang, D., Jie, Z., Wu, W., Qin, Y., Xue, W., Li, J., Han, L., Lu, D., *et al.* A metagenome-wide association study of gut microbiota in type 2 diabetes. *Nature* **490,** 55–60 (2012).

56. Greenblum, S., Turnbaugh, P. J. & Borenstein, E. Metagenomic systems biology of the human gut microbiome reveals topological shifts associated with obesity and inflammatory bowel disease. *Proc. Natl. Acad. Sci. U. S. A.* **109,** 594–9 (2012).

57. Morgan, X. C., Tickle, T. L., Sokol, H., Gevers, D., Devaney, K. L., Ward, D. V, Reyes, J. A., Shah, S. A., LeLeiko, N., Snapper, S. B., Bousvaros, A., Korzenik, J., Sands, B. E., Xavier, R. J. & Huttenhower, C. Dysfunction of the intestinal microbiome in inflammatory bowel disease and treatment. *Genome Biol.* **13,** R79 (2012).

58. Abubucker, S., Segata, N., Goll, J., Schubert, A. M., Izard, J., Cantarel, B. L., Rodriguez-Mueller, B., Zucker, J., Thiagarajan, M., Henrissat, B., White, O., Kelley, S. T., Methé, B., Schloss, P. D., Gevers, D., Mitreva, M. & Huttenhower, C. Metabolic reconstruction for metagenomic data and its application to the human microbiome. *PLoS Comput. Biol.* **8,** e1002358 (2012).

59. Goodrich, J. K., Waters, J. L., Poole, A. C., Sutter, J. L., Koren, O., Blekhman, R., Beaumont, M., Van Treuren, W., Knight, R., Bell, J. T., Spector, T. D., Clark, A. G. & Ley, R. E. Human Genetics Shape the Gut Microbiome. *Cell* **159,** 789–799 (2014).

60. Hampton, T. H., Green, D. M., Cutting, G. R., Morrison, H. G., Sogin, M. L., Gifford, A. H., Stanton, B. A. & O'Toole, G. A. The microbiome in pediatric cystic fibrosis patients: the role of shared environment suggests a window of intervention. *Microbiome* **2,** 14 (2014).

61. Lane, D. J., Pace, B., Olsen, G. J., Stahl, D. A., Sogin, M. L. & Pace, N. R. Rapid determination of 16S ribosomal RNA sequences for phylogenetic analyses. *Proc. Natl. Acad. Sci.* **82,** 6955–6959 (1985).

62. Traverse, C. C., Mayo-Smith, L. M., Poltak, S. R. & Cooper, V. S. Tangled bank of experimentally evolved Burkholderia biofilms reflects selection during chronic infections. *Proc. Natl. Acad. Sci. U. S. A.* **110,** E250–9 (2013).

63. Penterman, J., Nguyen, D., Anderson, E., Staudinger, B. J., Greenberg, E. P., Lam, J. S. & Singh, P. K. Rapid Evolution of Culture-Impaired Bacteria during Adaptation to Biofilm Growth. *Cell Rep.* **6,** 293–300 (2014).

64. Sousa, A. M. & Pereira, M. O. Pseudomonas aeruginosa Diversification during Infection Development in Cystic Fibrosis Lungs-A Review. *Pathog. (Basel, Switzerland)* **3,** 680–703 (2014).

65. Hammond, J. H., Dolben, E. F., Smith, T. J., Bhuju, S. & Hogan, D. A. Links between Anr and quorum sensing in Pseudomonas aeruginosa biofilms. *J. Bacteriol.* JB.00182–15– (2015). doi:10.1128/JB.00182-15

66. Hoffman, L. R., Kulasekara, H. D., Emerson, J., Houston, L. S., Burns, J. L., Ramsey, B. W. & Miller, S. I. Pseudomonas aeruginosa lasR mutants are associated with cystic fibrosis lung disease progression. *J. Cyst. Fibros.* **8,** 66–70 (2009).

USE OF GENOME DATA IN NEWBORNS AS A STARTING POINT FOR LIFE-LONG PRECISION MEDICINE

STEVEN E. BRENNER

Department of Plant and Microbial Biology, University of California
Berkeley, CA 94720, USA
Email: brenner@compbio.berkeley.edu

STEPHEN KINGSMORE

Rady Pediatric Genomic and Systems Medicine Institute
San Diego, CA 92123, USA
Email: SKingsmoreMDDSc@rchsd.org

SEAN D. MOONEY

Department of Biomedical Informatics and Medical Education, University of Washington
Seattle, WA 98105, USA
Email: sdmooney@uw.edu

ROBERT NUSSBAUM

Invitae Corporation
San Francisco, CA 94107, USA
Email: robert.nussbaum@invitae.com

JENNIFER PUCK

Professor, Departments of Immunology and Pediatrics, UCSF
San Francisco, CA 94143, USA
Email: Jennifer.Puck@ucsf.edu

Rare genetic disorders affect millions of individuals worldwide. Many of these disorders can take decades to correctly diagnose. Because of this, genome sequencing of newborns raises a substantial opportunity to identify genetic disorders before they present symptoms, and to identify patient risks at the start of life. Many of these disorders can take decades to correctly diagnose. Because of this, genome sequencing of newborns raises a substantial opportunity to identify genetic disorders before they present symptoms, and to identify patient risks at the start of life. This workshop will report on efforts to screen newborns using genetic sequencing technologies, and attendant biomedical informatics and computational biology approaches.

1. Introduction

> *Over the course of the next few decades, DNA sequencing will lead to each baby's genome being sequenced, and used to shape a lifetime of personalized strategies for disease prevention, detection and treatment.*
>
> FS Collins, Wall Street Journal, July 8, 2014.

Within the past year, the director of the NIH updated the 1990s projection by Walter Gilbert, who postulated that by 2030-2040 the parents of every newborn child in the developed world would leave the hospital with their child's genome on a CD-ROM.

The analytic cost of whole human genome (WGS) and whole exome sequencing (WES) continues to out-pace Moore's law and is currently approaching $1,000 for WGS. Given the high burden of treatable genetic diseases in children, there is considerable interest in using WGS to diagnose and treat genetic diseases of infancy, both as an extension of current newborn screening programs (NBS) and as a pioneering implementation of precision medicine for affected infants. Early diagnosis of genetic diseases can dramatically decrease morbidity and mortality, as evidenced by current, federally mandated newborn screening (NBS) programs for ~60 treatable genetic diseases, which identifies ~5000 affected babies per year at ~10 days of life. With the emergence of WGS and whole exome sequencing, the National Institutes of Health and state NBS programs are considering the value of genome sequence information in addition to currently performed tests with the goal of improving accuracy, reducing false positives and false negatives, or identifying disorders not currently screened for.

For NBS programs, some diseases screened for currently have positive predictive values well below 5%, so improving specificity could have a substantial positive societal impact. In order to evaluate the use of deep sequencing within context of newborns has many technical, computational, informatics and ethical questions that have arisen. This workshop will focus on challenges of population based genetic sequence diagnosis for newborn disorders, the future possibilities of WGS as an enabling technology for lifelong precision medicine, and current experience with use of diagnostic WGS in the delivery of precision medicine in acutely ill infants. The challenges addressed include:

- Determination whether accurate genome or exome sequences can be obtained using existing blood spot material and testing pipelines.
- Improvement in the annotation of pathogenic genetic variants in genes associated with metabolic and other disorders.
- Evaluation of whether genetic information can reduce false positive and false negative rates by identifying cases of nonpathogenic variants and heterozygous carriers of who are not anticipated to have symptomatic disease.
- Collection of patient-level rich phenotypic data through ontologies (and other means) and integration into national repositories and the electronic medical record.

- Improved capture and diagnosis of undiagnosed and affected infants for whom early intervention is warranted to minimize morbidity and mortality.
- Selection of disorders for screening, including and beyond the 31 conditions recommend by the HRSA advisory committee
- Ethical challenges surrounding widespread sequencing and storing genetic data of children for clinical diagnosis.
- Public education and engagement.

2. Workshop Presenters

Robert Nussbaum, MD
Invitae Corporation

Title: Newborn screening and genome sequencing

Newborn screening for inborn errors of metabolism was first instituted in the 1960's for phenylketonuria using blood obtained from newborns that is soaked into filter paper. Screening expanded over the next 30 years by the introduction of individual tests designed to screen for individual disorders until the 1990s when tandem mass spectroscopy (MS/MS) was first applied. MS/MS allows the identification of many disorders in parallel using a single assay that measures a host of abnormal metabolites in the newborn blood spot. The NSIGHT project, a four-site project funded by NICHD and NHGRI, is now looking at a potential next step in newborn screening: identifying many more actionable, treatable diagnoses in the newborn period through exome or genome sequencing of newborn blood spot DNA. Outstanding questions include:

- the positive predictive value of abnormalities found in newborn screening,
- sequencing's utility and cost as compared to standard mass spectroscopic methods to see whether it could replace standard screening,
- if not replacing current screening, could sequencing complement standard screening methods and make them more specific, and
- finally to explore the ethical and legal aspects of generating sequence information from newborns under a nearly mandatory public health regimen when some of these data may have significant medical relevance but are not germane to the mission of newborn screening.

Biography: Robert L. Nussbaum, M.D. is Chief Medical Officer of Invitae Corporation and also on the clinical faculty of the Dept. of Medicine, UCSF. As a board-certified internist and medical geneticist, Dr. Nussbaum has dedicated his career to improving the care of individuals with hereditary disorders. He has played leadership roles in the Cancer Genetics and Cardiovascular Genetics Programs at UCSF as well as serving as inaugural PI of the UCSF U19 project to investigate the role of deep sequencing as an adjunct to newborn screening. Dr. Nussbaum is a member of the National Academy of Medicine and American Academy of Arts and Sciences and is past president of the American Society for Human Genetics. He was co-discoverer of the first inherited form of Parkinson's disease, and his basic research has focused on efforts to identify the pathogenesis and genetic contributions to the disease. Prior to joining UCSF, Dr. Nussbaum was chief of the Genetic Disease Research Branch of the National Human Genome Research Institute, National Institutes of Health.

Stephen Kingsmore, MB, ChB, BAO, DSc, FRCPath
President of the Rady Pediatric Genomic and Systems Medicine Institute.

Title: Integrating deep phenotyping and genome sequencing to enable precision medicine in neonatal intensive care units

Genetic diseases and congenital abnormalities are the leading cause of death both in neonatal intensive care units, and among infants in general (children aged <1 year). Until the advent of clinical WGS and whole exome sequencing (WES), timely molecular diagnosis of suspected genetic disorders had been largely precluded in acutely ill infants by virtue of profound clinical and genetic heterogeneity, and tardiness of results of standard genetic tests. However, it is now possible to decode an ill infant's genome in 24 hours (STATseq) and to use clinicopathologic correlation software to evaluate the likelihood that their symptoms are the result of any of the 4,500 known monogenic disorders. We and others have recently reported rates of molecular diagnosis of 25-73% in retrospective case series of infants and children with diseases of possible monogenic etiology by proband or trio WGS/WES. However, there are immense challenges in scaling STATseq for general use, and in timely implementation of STATseq as part of NICU workflows that have the potential to change outcomes in infants with genetic diseases.

Biography: Stephen Kingsmore grew up in Northern Ireland during the 'troubles'. He moved to the US after medical school and trained in internal medicine at Duke. After positionally cloning a few disease genes in academia in the '90s, he switched to the genomics industry for a decade. He was CEO of the National Center for Genome Resources in Santa Fe while it became a leader in the development of bioinformatic tools for next-gen sequencing for AgBiotech applications. From 2011-2015 he was the Director of Genomic Medicine and Executive Director of Medical Panomics at Children's Mercy Hospital in Kansas City. Recently he became President of the Rady Pediatric Genomic and Systems Medicine Institute. This is a new research institute which is affiliated with the Rady Children's Hospital and UCSD.

Jennifer Puck, MD
Professor, Departments of Immunology and Pediatrics, University of California San Francisco

Title: Using newborn dried blood spots to obtain deep sequence data for enhanced, early diagnosis

Our center has developed methodology for extracting DNA from dried blood spots obtained in newborn nurseries that is of sufficient integrity and quantity for whole exome or whole genome sequencing. Modifications to standard DNA processing have been made to optimize the generation of deep sequence data. Deep sequencing and analysis is possible using de-identified samples from infants with positive metabolic newborn screens. Moreover, additional archived newborn dried blood spot samples have been obtained with informed consent from individuals affected with clinically significant primary immune system disorders. The latter sample set addresses the question of whether newborn screening by deep sequencing could improve early detection of diverse immunodeficiency diseases, making possible optimal treatment and avoidance of infectious complications.

Biography: Dr. Puck is a Professor of Immunology in the Dept. of Pediatrics at UCSF and is also a member of the UCSF Inst. for Human Genetics and an associate of the Berkeley Innovative Genomics Initiative. Her basic and translational research program focuses on inherited human immune disorders. Noting the advantages in survival and outcome for infants with severe combined immunodeficiency (SCID) diagnosed early in life, Dr. Puck conceived and developed a newborn screening test using the universally collected dried blood spots to detect SCID. DNA extracted from the blood spots is assayed by PCR to quantitate T cell receptor excision circles (TRECs), a biomarker for the generation of a normal diverse repertoire of T cells. Absent or low TRECs suggest SCID. SCID screening, now adopted in over half of the states in the US, allows infants affected with SCID and other conditions with insufficient T cells to be detected early and treated. Dr. Puck directs the UCSF Jeffrey Modell Diagnostic Center for Primary Immunodeficiencies. She serves on the Medical Advisory Committee of the Immune Deficiency Foundation, the Committee on Primary Immunodeficiency Disease of the International Union of Immunological Societies, the Board of Scientific Councilors of NIAID, and the Steering Committees of the Primary Immune Deficiency Treatment Consortium (PIDTC) and the US Immunodeficiency Network (USIDNET). A member of the American Society of Clinical Investigation (ASCI), Association of American Physicians (AAP), American Pediatric Society (APS) and Institute of Medicine, she received the Abbot Award in Clinical and Diagnostic Immunology from the American Society of Microbiology in 2013 and the Colonel Harlan Saunders Award for Lifetime Achievement in Genetics from the March of Dimes in 2014.

Steven E. Brenner, PhD

Professor, Department of Plant and Microbial Biology, University of California, Berkeley and Adjunct Professor, Department of Bioengineering and Therapeutic Sciences, UCSF

Title: Analysis challenges of newborn genome sequences

A hallmark of current newborn screening programs is their outstanding sensitivity with false negatives on the order of one per million babies screened and impressive specificity. Both the sensitivity and the specificity are essential in order, respectively, to meet the public health goals of the program at a reasonable financial and human cost. This is achieved in large part through careful optimization of highly-sensitive mass spectrometry instruments which are detecting a molecular intermediate phenotype resulting from the genetic abnormalities. Identifying such diseases from genetic information is intrinsically more challenging. The analytic instrumentation for genome variant calling is less mature than the mass spec technology. More importantly, identifying disease phenotypes from genotypes can be vastly more challenging than identifying disease phenotypes from molecular phenotypes. Preparatory studies have revealed widespread limitations of even the most authoritative databases of mutations causing those diseases incorporated in newborn screening; they have numerous variants reported as pathogenic that are present in unaffected individuals. It seems likely that they also overlook variants that may cause disease. Computational predictive methods fare far worse. Health disparities may arise due to differential ability to reliably identify pathogenic mutations in individuals from different ethic backgrounds. This talk will include progress on methods to overcome these challenges, and approaches that may help present genome sequencing as a useful technology associated with newborn screening.

Biography: Steven E. Brenner is a Professor at the University of California, Berkeley, and also holds appointments at Lawrence Berkeley National Laboratory and at the University of California, San Francisco. As an undergraduate he studied in Walter Gilbert's laboratory at Harvard College. He received his M.Phil from the Department of Biochemistry at Cambridge University, and obtained a Ph.D. from the MRC Laboratory of Molecular Biology and Cambridge where he studied with Cyrus Chothia. After graduation Brenner had a brief fellowship at the Japan National Institute of Bioscience, followed by postdoctoral research supervised by Michael Levitt at Stanford University School of Medicine. Brenner's research is primarily in the area of computational genomics, covering topics in protein structure, RNA regulation, function prediction, metagenomics, and individual genome interpretation. He is founding chair of the Computational Biology graduate program at Berkeley. He is currently a director of the Human Genome Variation Society, and is a founding editor of PLoS Computational Biology. He has served two terms as a director of the ISCB and was a founding director of the Open Bioinformatics Foundation. His recognitions including being a Miller Professor, a Sloan Research Fellow, a Searle Scholar, an AAAS Fellow, and named the recipient of ISCB's Overton Prize.

Sean D. Mooney, PhD
Professor, Department of Biomedical Informatics and Medical Education, University of Washington

Title: Understanding the complex genetics of simple Mendelian traits

Most of the diseases included in newborn screening are considered classical Mendelian diseases. However, more careful study reveals that many if not all involve variable penetrance and expressivity, and several likely involve degrees of epistasis or other engagement of multiple genes. Our group has spent much effort development methods to identify and characterize pathogenic variants from sequencing projects. This talk will outline some of these challenges and successes in characterizing these more complex genetic relationships relevant to accurate newborn screening by genome sequencing. I will discuss our efforts to connect phenotype to disrupted molecular mechanisms to better predict clinically relevant mutations.

Biography: Prof. Sean Mooney is the Chief Research Information Officer (CRIO) of UW Medicine, a Professor in the Department of Biomedical Informatics and Medical Education, and an NIH funded researcher. Previous to his CRIO role, he was an Associate Professor and Director of Informatics at the Buck Institute for Research on Aging. He has a long history in managing the development of collaborative electronic systems supporting biomedical research. His interests focus on leading the next generation informatics tools for biomedical research and in understanding the underlying molecular causes of inherited genetic diseases and cancer. As an Assistant Professor, he was appointed in Medical and Molecular Genetics at Indiana University School of Medicine and was founder of the Indiana University School of Medicine Bioinformatics Core. In 1997, he received his B.S. with Distinction in Biochemistry and Molecular Biology from the University of Wisconsin at Madison. He received his Ph.D. in 2001 at the University of California in San Francisco under the mentorship of Dr. Teri Klein, and then was an American Cancer Society John Peter Hoffman Fellowship at Stanford University. He is funded by the National Library of Medicine and other NIH Institutes, mostly in the area of data science and translational medicine. He was part of the team that won the $150k 2000 Garage.com Student Business Plan Competition, where the proposed plan focused on web-based tools for drug discovery research.

3. Acknowledgments

We would like to thank the PSB 2016 chairs and Tiffany Murray of Stanford University for their support in organizing the meeting. Research was supported by NICHD and NHGRI under Award U19HD077627, NLM LM009722 (Mooney), the National Center For Advancing Translational Sciences of the National Institutes of Health under Award Number UL1TR000423 (Mooney) and by Tata Consultancy Services. The content is solely the responsibility of the authors and does not necessarily represent the official views of the National Institutes of Health.

WORKSHOP ON TOPOLOGY AND ABSTRACT ALGEBRA FOR BIOMEDICINE

ERIC K. NEUMANN
Foundation Medicine, Cambridge, MA 02139, USA
Email: eneumann@foundationmedicine.com

SVETLANA LOCKWOOD
School of Electrical Engineering and Computer Science,
Washington State University, Pullman, Washington, USA
Email: svetlana.lockwood@email.wsu.edu

BALA KRISHNAMOORTHY
School of Electrical Engineering and Computer Science,
Washington State University, Pullman, Washington, USA
Email: bkrishna@math.wsu.edu

DAVID SPIVAK
Department of Mathematics,
MIT, Cambridge, MA 02139, USA
Email: dspivak@math.mit.edu

The use of large-scale data analytics, aka Big Data, is becoming prevalent in most information technology discussions, especially for the life and health sciences. Frameworks such as MapReduce/Hadoop are offered as "Swiss-army knives" for extracting insights out of the terabyte-sized data. Beyond the sheer volume of the data, the complexity of the data structure associated with such data sets is another issue, and may not be so readily mined using only these technological solutions. Rather, the issues around data structure and data complexity suggest new representations and approaches may be required. The LinkedData standard (W3C, Semantic Web) has been promoted by some communities to address complex and aggregatable data, though it focuses primarily on querying the data and performing logical inferences on it, and its use in deep mining application is still in the early stages. In summary, there appears to be a gap between how we access structured data, and the deeper analyses we want to perform on it that preserve representation.

Over the last few years, an increasing number of examples from life science research have appeared that apply topological and algebraic forms to genomic and complex data problems (Isomap[†], PLEX[‡], Ayasdi, FQL[§], BioHaskell[**]). The relevance of finding structure in rich data has been underscored by the increasing efforts to combine

† http://isomap.stanford.edu
‡ http://www.math.colostate.edu/~adams/jplex/index.html
§ http://categoricaldata.net/fql.html
** http://biohaskell.org

clinical data with genomic analyses. Although much attention has been placed on Big Data and *batching* computational algorithms (e.g., MapReduce), understanding the structure of the data to better analyze, extract, and infer insights from it are also critical. These areas, however, are currently not supported sufficiently for the health and life sciences communities, and many possible applications are only recently being proposed[1,4,6].

Coming from a very different perspective, abstract algebra and algebraic topology (AAAT) may provide new powerful insights to biomedical data sciences. Historically, these algebra forms have been very successful in the study of many profound topics, yielding an understanding of rich mathematical and logic structures, as well as their relations to one another. Several key advances in the computer sciences over the last few decades (relational algebras (SQL), monadic structures (javascript), description logic (OWL), homotopy theory), have emerged from these fields of study. Yet, due to their mathematical generalities, other facets of abstract algebras have not easily been applied to domain-specific applications such as biomedical research. The potential is just beginning to emerge from limited cross-pollination as the landscape shifts to greater use of large, diverse data sets. What is yet lacking is a set of lucid, yet powerful examples from AAAT to biomedical applications that will help establish a bridge between these diverse disciplines.

Life science data is a mix of conceptual relations (aka knowledge, e.g., *proteins encoded by genes*) based on our current understanding of biology, and the data measurements gathered from applying large-scale chemical and genomic profiling technologies. The latter set is often assumed to "rest" on top of the conceptual entities (genes, proteins, mRNA, cellular structures), which have specific relations with each other (e.g., protein -> gene deterministic mappings). The logic associated with conceptual models that house data could be extended with additional AAAT theorems in order to enable a much deeper analysis of the data.

As an initial example, consider some concepts around topologies, which can be used to describe different "molecular spaces", including a sequence topology based on what makes one sequence similar or different from another. Here one element represents an entire genome for a given individual of a species, and the adjacent elements (neighborhood) are genomes from other individuals that differ in only a few bases. No scalar metric may exist in this space, but the overlap of subsets containing similar elements, and subsets that are only related by many different subset coverings provides a very discrete topology[††]. In addition to the elements, edges between the elements may be included that represent the incremental mutational transitions, unequal rearrangements, and reciprocal recombinations that may occur. Given a starting set of elements (genomes with only a few alleles), multiple applications of recombination to the elements will define a limited space of "accessible" genomes, known as a *closure*. A corollary from this is the *Founder Effect*

[††] It is enormous, since every 1000-base string has 3 * 1000 one-step neighbors and 9,000,000 (3000^2) two-step neighbors, and so on.

and H-W equilibrium for a limited starting population cut-off from larger allelic set. Only additional mutations can free genomes from these closures.

Topologies can obviously be applied to protein sequences as well, but proteins also offer additional relations including interactions to other proteins. One also realizes that such interactions depend (in complex ways) to their underlying sequence, so that the genome topology space captures the interaction graph of the proteins "fibered" above each coding region of the genome. One can continue to build upon these objects, to yield dynamic networks that can affect states and synthesis/degradation of all biomolecules. Eventually a topological mapping between genome space and phenotype spaces can begin to be formally represented[2], and to some extent, be possibly projected from the underlying genomic information.

Categorical Theory[3,4] (CT) is major device that originates from abstract algebras, and has several powerful features for organizing concepts and inherent system logic. Categories are composed of objects, morphisms (relations), and the ability to compose morphisms into transitive maps. Here *objects* are equivalent to what most if us call *classes*, and the morphisms define the relations between objects. The universal properties that come along with these entities allow combining objects, determining uniqueness, and establishing equivalencies between the objects and some fundamental morphisms, e.g., maps from an unique initial object to any other object defines exactly one relation per object called an *element*). These can be populated with a set of genes of interest and the relations they yield, including *commuting* paths. For example, for every protein p, the map (i) from p to its transcript, r, can be composed with the map (j) from r to the gene g it is expressed from, to yield the composition (k) = (j) ∘ (i)[‡‡]. Not only does (k) map a protein to a gene, but it is guaranteed to always have the same results as (j) ∘ (i), even though multiple proteins can map to the same transcript, and multiple transcripts can map to the same gene. We say that these relational structures *commute*.

One important feature of CT is the definition and use of *functors*[3], which not only transform objects to other objects (within or between categories), but also their morphisms to other morphisms. They become very useful when taking data structures of one model (e.g., a genome topology) to a more advantageous form for a different problem (e.g., a graph of molecular interactions). Since the relations (morphisms) come along as well, both the data and their semantics can be effectively transformed together. As will be described below, this applies to databases as well as analytical manipulations.

Another important area of topology is the representation of simplicial complexes, which are the compositions of ordered relations of entities for different dimensional objects: points, edges, faces, volumes, etc. Each n-dimensional object, or n-simplex, is composed of n+1 (n-1)-dimensional objects: a 3d tetrahedron has 4 2d-triangular

‡‡ The notation for composition is always read right-to-left, since they are operators.

faces, each with 3 1d edges, consisting of 2 0d points. If the edges between any 2 points are less than a distance ε, they can be chained into complexes. Furthermore, if any form a cycle of 3 edges, a face object is induced and identified with directionality (clockwise or counter-clockwise); the same method is applied to successfully generate and chain higher structures, such as tetrahedrons (3-simplex), and beyond. When applied to complex data that have some form of distance metric, they form clusters of multidimensional simplicial complexes chains. The analysis of such complexes yields understanding of the general structure, or homotopy, of the data field. The use of *barcode analysis*, or *persistent homology*, by current researchers[5,6,7] is one path of analysis that is helping identify complex relations with biomedical data.

Altogether, the above areas support data mining of complex "high-feature" data while also aligning it to established and hypothetical concepts/relations and the logic they induce on the data elements. Often data analytics is tied to data representations within a database or other kind of repositories. As stated before, AAAT has helped shape current tools and methodologies supporting schemas and ontologies. This can be further enhanced by recent work on Category Theory as applied to databases[8]. These can address important issues on data migration, schema changes, data integrity and normalization, and intelligent query strategies. Most database operations are some combination of three fundamental operations: project, join, union[8,9].

Combining the logical manipulations of biomolecular relations along with the data captured for these under select conditions, new data constructs can be produced or perhaps even automatically generated to address complex analytics. Time series and tissue-specific data (e.g., with gene expression) can be formally encoded as (Cartesian) products of simpler objects within a category, and inherit their logical relations directly from original set of morphisms by applying universal properties (e.g., limits) and *functors*. Analytic tools that understand such composite structures, as well as the multitude of properties linked to each object (gene, patient, tumor, study, etc.), can then perform deep analytics intelligently using decomposition rules on the data subsets (sigma algebras). The vision would be that biomedical data becomes less pre-structured by computer science, and more emergent in structure based on the rules for combining data, on analyzing data, and inferring hypotheses from data. What is most important now is to come together as a community and discuss what directions we should consider exploring, and to identify a few relevant examplar cases to work on in order to validate these ideas.

References

1. I. C. Baianu. "A Category Theory And Higher Dimensional Algebra Approach To Complex Systems Biology, Meta-Systems And Ontological Theory Of Levels: Emergence of Life, Society, Human Consciousness and Artificial Intelligence", Mathematics and Bioinformatics 30(12): Special Issue, 09/2012.

2. F. Mynard and G. J Seal, "Phenotype spaces", Journal of Mathematical Biology, 60(2):247-66, 2009.

3. Saunders Mac Lane, "Categories for the Working Mathematician (Graduate Texts in Mathematics)", Springer-Verlag, 1998.

4. David Spivak, "Category Theory for the Sciences", MIT Press, 2014.

5. G. Carlsson and A. Zomorodian, "Computing persistent homology", Journal of Discrete and Computational Geometry, 2004.

6. G. Carlsson, M. Nicolau and A. Levine, "Topology based data analysis identifies a subgroup of breast cancers with a unique mutational profile and excellent survival", Proceedings of the National Academy of Sciences, , April 26, 2011.

7. Vin de Silva, "Topological and Symmetrical Structures in Data Analysis", http://www.samsi.info/sites/default/files/deSilva_Lecture5_august2013.pdf

8. D. I. Spivak, R. Wisnesky Relational Foundations For Functorial Data Migration, CoRR vol. abs/1212.5303, 2012.

9. D. J. Abadi , A. Marcus , S. R. Madden , K. Hollenbach, Scalable semantic web data management using vertical partitioning, Proceedings of the 33rd international conference on Very large data bases, September 23-27, 2007, Vienna, Austria

SOCIAL MEDIA MINING SHARED TASK WORKSHOP

ABEED SARKER*
Department of Biomedical Informatics, Arizona State University,
Scottsdale, AZ 85259, United States of America
**E-mail: abeed.sarker@asu.edu*
diego.asu.edu

AZADEH NIKFARJAM
Department of Biomedical Informatics, Arizona State University,
Scottsdale, AZ 85259, United States of America
E-mail: anikfarj@asu.edu

GRACIELA GONZALEZ
Department of Biomedical Informatics, Arizona State University,
Scottsdale, AZ 85259, United States of America
E-mail:graciela.gonzalez@asu.edu

Social media has evolved into a crucial resource for obtaining large volumes of real-time information. The promise of social media has been realized by the public health domain, and recent research has addressed some important challenges in that domain by utilizing social media data. Tasks such as monitoring flu trends, viral disease outbreaks, medication abuse, and adverse drug reactions are some examples of studies where data from social media have been exploited. The focus of this workshop is to explore solutions to three important natural language processing challenges for domain-specific social media text: (i) text classification, (ii) information extraction, and (iii) concept normalization. To explore different approaches to solving these problems on social media data, we designed a shared task which was open to participants globally. We designed three tasks using our in-house annotated Twitter data on adverse drug reactions. Task 1 involved automatic classification of adverse drug reaction assertive user posts; Task 2 focused on extracting specific adverse drug reaction mentions from user posts; and Task 3, which was slightly ill-defined due to the complex nature of the problem, involved normalizing user mentions of adverse drug reactions to standardized concept IDs. A total of 11 teams participated, and a total of 24 (18 for Task 1, and 6 for Task 2) system runs were submitted. Following the evaluation of the systems, and an assessment of their innovation/novelty, we accepted 7 descriptive manuscripts for publication— 5 for Task 1 and 2 for Task 2. We provide descriptions of the tasks, data, and participating systems in this paper.

Keywords: Concept Extraction; Text Classification; Adverse Drug Reaction; Pharmacovigilance; Social Media Mining.

1. Background

Adverse drug reactions (ADRs), defined as accidental injuries resulting from correct medical drug use, present a serious and costly health problem contributing to 5.3% of all hospital admissions each year.[1] The process of detection, assessment, understanding, and prevention of these events is called pharmacovigilance.[2] To facilitate pharmacovigilance efforts, governments worldwide have diverse surveillance programs. One example, in the U.S., is MedWatch;[a]

[a]`http://www.fda.gov/Safety/MedWatch/default.htm` [Accessed Sep-28-2015]

it enables both patients and providers to manually submit ADR information. However, these programs are chronically underutilized. A systematic review encompassing 12 countries, estimated an 85-94% under-reporting rate of ADRs in local, regional, and national level reporting systems. To improve detection rates, researchers have begun turning to alternative sources of healthcare data, such as social media.[3,4] Recent studies suggest that 26% of adult Internet users discussed personal health issues online, with 42% of them discussing current conditions on social media and 30% reportedly changing their behavior as a result.[5] Studies have focused on automatic classification of ADR assertive user posts,[6–10] and the automatic extraction of ADR mentions from posts.[11–14] However, despite the proposal of various techniques for utilizing social media data in the past, public availability of data is scarce, making direct comparison of different approaches impossible. Our recent release of large annotated data sets prepared from Twitter data[8,10,15] has opened up the possibility to compare the performances of distinct approaches for social media based pharmacovigilance.

1.1. *Shared task and workshop design*

To facilitate research on social media based pharmacovigilance and social media text mining in general, we organized this workshop on social media text mining. Unlike traditional workshops, where manuscripts within the scope of the workshop are submitted, reviewed and chosen for acceptance, we ran this workshop as a shared task. The shared task consisted of three tasks: classification, extraction and normalization. We provided annotated data for the three tasks, and participants were required to develop and submit systems for evaluation on previously unreleased test data. Following the evaluation of the submitted systems, participants were required to submit short system descriptions. The system descriptions were reviewed by at least one peer and one member of the workshop organizing committee for selection.

In total, 11 teams registered for the shared task, and a total of 24 system runs were submitted. We received 18 submissions for Task 1, and 6 for Task 2. Unfortunately, perhaps due to the complex nature of the task, we did not receive any system submissions for Task 3. 9 teams were invited to submit system descriptions, of which 7 were eventually selected for publication.

In the following sections, we provide detailed descriptions of the task and submitted systems. For the rest of this section, we provide some background on social media, its use in the public health domain in general, and the challenges faced by text mining systems relying on this source of data.

1.2. *Pharmacovigilance from social media*

Over the last few years, social networks have seen massive growths (*e.g.*, as of 29th September 2015, Twitter has over 645,750,000 users and grows by an estimated 135,000 users every day, generating 58 million tweets per day[b]). Because of this heavy usage of social media to share information, users have begun to see it as a channel for obtaining and broadcasting information.[3] A large number of users post about health related information, particularly in

[b]`http://www.statisticbrain.com/twitter-statistic/` [Accessed Sep-29-2015]

online health communities.[16] A recent survey by the Pew Research Center[17] has elucidated the relevance of social media in modern day public health, explaining that 34% of caregivers and 20% of patients read or watch someone elses commentary or experience online. Additionally, 11% of caregivers and 6% of patients share experiences and post questions online. Health related social networks have been attracting many users, perhaps because it allows users of a particular health interest to exchange information. In such platforms (*e.g.*, DailyStrength,[18] MedHelp[19]), users discuss their health-related experiences, including use of prescription drugs, side effects and treatments. Due to the emergence of such platforms, and the abundance of data available through them, research on public health monitoring, including ADR monitoring, has focused on exploiting data from these sources in recent times.[11,20,21]

From the perspective of public health, social media has been utilized for studying smoking cessation patterns on Facebook,[22] identifying user social circles with common medical experiences (like drug abuse),[23] and monitoring malpractice,[24] to name a few. When different patients that suffer from a common disease, or use a specific medication, share information about their symptoms, treatments or drug outcomes, this information can provide valuable clinical insights for both patients and health-related industries that go beyond traditional communication methods.[25] Although specific information about a single user may not be available or usable for privacy reasons, various resources are currently available to perform some demographic analysis with social media data. Furthermore, over the last decade, a number of social media based surveillance systems have been developed, reviewed, and implemented locally, nationally, and globally.[26] The main value of social media, is not derived from individual posts, but from a large number of posts on a specific topic. Recent advances in the data processing capabilities of machines, and machine learning and NLP research present the possibility of utilizing this massive data source for a variety of purposes, including public health. The fact that it is a direct source of users personal experiences makes it a lucrative resource. According to Harpaz *et al.*,[27] social media offers new opportunities for public health monitoring due to the availability of large amounts of data that is internet-based, patient-generated, unsolicited, and up-to-date.

ADR monitoring research have seen significant strides towards the use of automatic NLP techniques for mining drugs and associated reactions from social media. User posts in social media contain information about treatment outcomes and provide early access to reported ADRs that could be beneficial for health and pharmaceutical industries. The type and volume of ADR information that social media makes available to the health industry may not be easily obtainable by other means. This includes the ADRs experienced by those with special conditions, such as patients with rare diseases, pregnant/nursing women, elderly people or patients with co-morbidities who are usually excluded from clinical trials.[28] It is now well established that social media data is rich in knowledge, which is drowned in large volumes of noise.

1.3. *Challenges of social media-based pharmacovigilance*

Various pros and cons of using social media for automatic ADR monitoring,[29,30] and more generally, for public health monitoring, have been mentioned in recent literature. We briefly outline the opportunities that social media presents, and the obstacles associated with its use for health-related research.

The drawbacks found when utilizing the user generated content of social media may include issues with the credibility, recency, uniqueness, frequency, and salience of the data.[3,31] Abbasi and Adjeroh[31] demonstrate the potential downside of each of these five points and the importance of selecting the right media channel for social media analytics. For example, the authors discuss the potential low salience of Twitter because of the short text limits. In addition to these general problems related to the data generated within social media, there are difficulties and challenges posed by the processing and extraction of relevant information using NLP techniques. A frequently encountered challenge is due to the fact that the data is generated by consumers, and they tend to use misspellings, non-medical, descriptive terms to discuss health issues. This reduces a systems ability to automatically extract mentions of relevant concepts and map them to suitable medical lexicons for further analysis.[11,12,15]

Traditional NLP methods that are used on longer texts have proven to be inadequate when applied to short texts, such as those found in Twitter.[32] Thus, recent research tasks have focused on developing NLP tools specifically for data from social media.[33] Some recent articles have reported the imbalance that exists in data coming from social media.[8,10,34] Only a small proportion of drug-associated data collected from social media tend to contain information associated with ADRs. This results in problems associated with annotations, since large volumes of data need to be annotated for the inclusion of sufficient numbers of posts containing ADRs. This data imbalance issue is a major problem for supervised machine learning approaches, particularly because it is the smaller class that is of primary interest for the research. While access to users personal experiences with prescription drugs is one of the key advantages of social media, automatic determination of true personal experiences is challenging. In addition to these, there are also technical, policy, and privacy challenges associated with the use of social media for pharmacovigilance, as pointed out by Edwards and Lindquist.[29]

2. Workshop Task Descriptions

The primary objective of the workshop is to promote the application of different techniques on a common social media based data set, so that useful approaches can be identified and utilized in the future. We divided the overall task of utilizing social media posts for identification of ADR signals into three subtasks:

(1) Automatic classification of ADR assertive user posts (tweets). The goal of this task is to efficiently separate the large amount of noise from posts presenting real ADR associated experiences.
(2) Automatic extraction of ADR mentions. The goal of this task is to apply information extraction techniques to extract text segments so that specific ADRs associated with a drug can be identified.
(3) Normalization of ADR mentions. The goal of this task is to normalize different lexical representations of the same ADR concepts into standard IDs.

To facilitate the shared task, we made available our large annotated Twitter data set. The overall shared task was designed to capitalize on the interest in social media mining and appeal to a diverse set of researchers working on distinct topics such as natural language processing, biomedical informatics, and machine learning. The different subtasks present a number of interesting challenges including the noisy nature of the data, the informal language of the user posts, misspellings, and data imbalance. The rest of this section details the nature of our data and annotations, and each task in detail.

2.1. *Data*

The data set made available for the shared task has been sourced from the social networking site Twitter. The corpus was created through two phases of annotations performed for a large study on ADR detection from social media that is currently in progress. Our finalized annotations are periodically made publicly available at: `http://diego.asu.edu/downloads`.

The tweets associated with the data were collected using the generic and brand names of the drugs, and also their possible phonetic misspellings,[35] since it is common for user posts on Twitter to contain spelling errors. Following the collection of the data, a randomly selected sample of the data was chosen for annotation. The data was annotated by two domain experts under the guidance of a pharmacology expert. Each tweet is annotated for the presence of ADRs (binary), spans of ADRs, indications, and beneficial effects. For each ADR, indication, and beneficial effect, the annotators also identified the most appropriate UMLS concept ID. Following the annotation of the full set, the disagreements were resolved by the pharmacology expert.

2.2. *Task 1: Adverse drug reaction classification*

The first task focuses on automatic classification of ADR assertive user posts. This task utilizes the binary annotations in the data. Participants were provided with a training/development set, containing a set of tweets with associated binary annotations indicating the presence or absence of ADRs. Evaluation was performed on a blind set not released prior to the evaluation deadline. Systems were evaluated on their ability to automatically classify ADR containing posts.

2.2.1. *Training and evaluation sets*

A total of 10,822 annotated tweets were made available [c]. The final data set made available for training is highly imbalanced, as one would expect, with 1,239 (11.4%) tweets containing ADR mentions and 9,583 (88.6%) containing no ADR mentions. Further details about the data set, at an intermediate stage of preparation, and annotations (in addition to the binary annotations) can be found in our past publications.[8,10]

The evaluation set consisted of 4,895 tweets with only 367 (7.4%) ADR instances.

[c]Because of Twitter's privacy policy, the actual tweets cannot be shared. Instead, we have made available a download script and Twitter userIDs and tweetIDs, which interested researchers can use to download the tweets, and associated meta-data.

2.2.2. *Inter annotator agreement*

A randomly chosen subset of the data (1082 tweets) was annotated by the pharmacology expert for the measurement of Inter Annotator Agreement (IAA). We used Cohens Kappa (κ)[36] to compute inter annotator agreement which is given by the following equation. We computed κ for all three pairs of agreements, and obtained an average of 0.71, which can be considered as significant agreement.[37] For the two annotators, $\kappa = 0.69$.

2.3. *Task 2: Adverse drug reaction extraction*

This sub-task is a Named Entity Recognition (NER) task, and the aim is to automatically extract the ADR mentions reported in user posts. This includes identifying the text span of the reported ADRs. Participants were encouraged to use advanced machine learning systems on the annotated training set to extract the mentions and correctly distinguish ADRs from similar non-ADR mentions.

2.3.1. *Training and evaluation sets*

The training data for this sub-task consisted of 2,131 tweets which are fully annotated for mentions of ADR and indications. This set contains a subset of the tweets from task 1 that were tagged as ADR assertive, plus a random set of non-ADR tweets. The non-ADR subset was annotated for mentions of indications, in order to allow participants to develop techniques to deal with this confusion class. To summarize, each instance may contain annotations of medical signs and symptoms with the following semantic types:

- adverse drug reaction– a drug reaction that the user considered negative;
- beneficial effect– an unexpected positive reaction to the drug;
- indication– the condition for which the patient is taking the drug; and
- other– any other mention of signs or symptoms.

Every annotation includes the span of the mention (start/end position offsets), the semantic type, the related drug name, and the corresponding UMLS (Unified Medical Language System) concept ID— assigned by manually selecting concepts in our in-house ADR lexicon.[14] The evaluation set consisted of 476 instances.

2.3.2. *Inter annotator agreement*

We measured inter annotator agreement on the whole training set. The calculated κ value for approximate matching of the concepts is 0.81 for Twitter, which can be considered high agreement.[37]

2.4. *Task 3: Normalization of adverse drug reactions*

This is a concept normalization task. Given an ADR mention in natural language (colloquial or other), participant systems were required to identify the UMLS concept ID for the mention. Unlike the other two tasks, there has not been prior work on normalization of concepts expressed in social media text. We expect immediate future research tasks to focus on this topic.

2.4.1. *Training and evaluation sets*

Training data consists of a set of ADR mentions and their corresponding, human-assigned UMLS concept IDs, as shown below:

```
schizophrenia          c0036341
tension in my nerves   c0027769
shaking                c0040822
```

2.5. *Evaluation metrics*

2.5.1. *Task 1 Evaluation*

For this task, the evaluation metric was the ADR F-score. The binary annotation consisted of two classes: ADR and non-ADR. The intent of this task was to devise automatic classification techniques for detecting ADR assertive user posts. As such, the evaluation was based on the harmonic mean of the recall and precision for the ADR class. The ADR F-score has been previously used for evaluation of systems performing this task.[10] The system with the highest ADR F-score on the test set was ranked first.

2.5.2. *Task 2 Evaluation*

F-score was also used as the metric for evaluation in this task. True positives, false positives and false negatives for a system were identified via approximate matching. The F-score was then computed from these values, as described in our past system evaluations.[14]

2.5.3. *Task 3 Evaluation*

For this task, the proposed evaluation metric was accuracy: $\frac{numberofcorrect}{total}$. In this evaluation scheme, a system prediction is considered correct if the predicted concept ID is identical, is a synonym, or has a is-a relationship to the gold standard concept.

3. Methods and Participating Systems

In this section, we summarize the methods used by a selected set of participating teams/systems. We discuss 5 teams' submissions for Task 1 and 2 teams' submissions for task 2. There were no submissions for task 3.

3.1. *Task 1 Systems*

All the submitted systems applied supervised classification approaches. The two best performing systems applied classifier ensembles. The following is a brief discussion of each system.

3.1.1. *Mayo-NLP*

The Mayo-NLP system[38] used an ensemble machine learning classifier to tackle the unbalanced distribution of the classes in the data provided for the task. A feature set containing unigrams, bigrams, and trigrams (a selected list, using mutual information), co-occurrence of drug and side effect, negation, and sentiment score were used to train Random Forest classifiers for identifying ADR assertive tweets. For training, the system obtained best results when the ratio of the training and test sets are balanced, via removal of a random set of negative instances. The system obtained a best F-score of 0.4195.

3.1.2. *TJZZF*

The TJZZF system[39] also uses an ensemble classification strategy. The system uses a weighted average ensemble of four classifiers: (1) a concept-matching classifier based on an ADR lexicon, (2) a maximum entropy (ME) classifier with n-gram features and a TF.IDF weighting scheme, (3) a ME classifier based on n-grams using Naïve Bayes (NB) log-count ratios as feature values, and (4) a ME classifier with word embedding features. This system showed the second best performance with an ADR F-score of 0.4182.

3.1.3. *ReadBioMed*

The READ-BioMed system[40] utilized a few lexical normalization processes and employed existing tools to enrich tweet texts before applying a machine learning-based classifier on the tweets. Unlike the Mayo-NLP system, the focus of this system is to reduce the number of errors caused by the lexical irregularities of tweets. The conceptual enrichment of tweets is based on the sentiment of the tweets, emotion classes, some UMLS Metathesaurus concepts, as well as drug, chemical substance, and disease mentions. The best performance of READ-BioMed on the official test set was achieved using Support Vector Machines (SVMs) trained on a bag-of-words representation for tweets which was enriched with sentiment analysis, emotion classes, and specific UMLS Metathesaurus concepts. The best ADR F-score obtained by this system is 0.358. Importantly, this system shows that enriching text segments via the incorporation of semantic information may be helpful for this task.

3.1.4. *NTTUMUNSW*

The NTTUMUNSW system[41] applied a linear SVM classifier. In addition to n-grams, the system uses a set of lexicon-based features, polarity cues, and topic models derived from the tweets. The best ADR F-score obtained by the system is 0.33. Importantly, the experiments performed using this system show that incorporating features based on topic models improve classification performance.

3.1.5. *SwissChocolate*

The SwissChocolate system[42] adapted a sentiment classification system to the ADR classification task by adding additional features and domain-specific resources. Features include word and character n-grams, POS tags, word clusters and embeddings, and a set of lexicon-based features. The system obtained a relatively low F-score compared to competing systems. However, this system produced very high recall scores.

3.2. *Task 2 Systems*

3.2.1. *DLIR*

The DLIR system[43] uses a very similar technique to the current state-of-the-art in social media based ADR extraction.[14] The system utilizes Conditional Random Fields (CRFs) trained on the annotated data. The system leverages word representations from large amount of unlabeled tweets, both drug related and generic. In addition to using vector representations of words, the system incorporates Part-of-speech tags, n-grams, lexicons, spell-checking, and negations. The best run of the system obtains F-score of 0.611.

3.2.2. *NTTUMUNSW*

The NTTUMUNSW extraction system[44] primarily focuses on token normalization and word representations, and their impacts on extraction. The system utilizes different word representation methods, including token normalization, and two state-of-the-art word embedding methods, namely word2vec and global vectors. The best system run achieved an F-score of 0.540.

4. Results and Discussions

Table 1 presents the results of the different runs of the systems discussed in this paper. The Mayo-NLP-2 system[38] and the TJZZF-1 system[39] achieved the best F-scores, 0.419 and 0.418 respectively. The two runs of the SwissChocolate system[42] performed significantly better than the other systems in terms of recall, but at the cost of precision.

Table 2 presents the results for Task 2. The DLIR system[43]runs significantly outperformed the NTTUMUNSW system runs.[44]

5. Conclusions

The primary aim of this workshop is to facilitate the development of state-of-the-art NLP and machine learning systems that can effectively utilize social media data. We received 11 registrations, of which we have discussed 7 selected system descriptions in this paper. The participating systems explored various interesting properties of social media text, and their impacts on pharmacovigilance oriented tasks.

This is the first time that a shared task is hosted at the Pacific Symposium on Biocomputing 2016. Considering the success of this style of workshop organization, we hope that we will host more of such shared task oriented workshops in the future.

Acknowledgments

Our work on social media mining for adverse drug reaction monitoring is supported by the National Institutes of Health (NIH) National Library of Medicine (NLM) grant number NIH

Table 1. Performances of selected system submissions for the Social Media Mining Shared Task 1. ADR F-scores were used to rank the systems. Best performing system shown in boldface.

System	Precision	Recall	ADR F-score	Accuracy
NTTUMUNSW-1	0.355	0.302	0.327	0.904
NTTUMUNSW-2	0.351	0.244	0.288	0.907
TJZZF-1	0.353	0.512	0.418	0.890
TJZZF-2	0.270	0.578	0.368	0.847
Read-BioMed-1	0.312	0.326	0.319	0.892
Read-BioMed-2	0.358	0.353	0.355	0.901
Read-BioMed-3	0.342	0.371	0.356	0.897
Read-BioMed-4	0.340	0.379	0.358	0.895
Read-BioMed-5	0.358	0.331	0.344	0.903
MayoNLP-1	0.380	0.430	0.403	0.902
MayoNLP-2	0.361	0.501	**0.419**	0.893
MayoNLP-3	0.392	0.408	0.400	0.906
MayoNLP-4	0.431	0.347	0.385	0.914
MayoNLP-5	0.459	0.270	0.338	0.919
SwissChocolate-1	0.202	0.741	0.317	0.754
SwissChocolate-2	0.202	0.743	0.317	0.754

Table 2. Performances of selected system submissions for the Social Media Mining Shared Task 2. ADR F-scores were used to rank the systems. Best performing system shown in boldface.

System	Precision	Recall	ADR F-score
NTTUMUNSW-1	0.782	0.412	0.540
NTTUMUNSW-2	0.718	0.416	0.526
NTTUMUNSW-3	0.778	0.414	0.540
DLIR-1	0.805	0.482	0.603
DLIR-2	0.806	0.485	0.606
DLIR-3	0.760	0.511	**0.611**

NLM 5R01LM011176. The content is solely the responsibility of the authors and does not necessarily represent the official views of the NLM or NIH.

References

1. C. Kongkaew, P. R. Noyce and D. M. Ashcroft, Hospital Admissions Associated with Adverse Drug Reactions: A Systematic Review of Prospective Observational Studies, in *Annals of Pharmacotherapy*, (7-8)2008.
2. *The Importance of Pharmacovigilance - Safety Monitoring of Medicinal Products* (World Health Organization, 2002).
3. A. Sarker, R. Ginn, A. Nikfarjam, K. O'Connor, K. Smith, S. Jayaraman, T. Upadhaya and G. Gonzalez, *Journal of Biomedical Informatics* **54**, 202 (2015).
4. S. Golder, G. Norman and T. K. Loke, *British Journal of Clinical Pharmacotherapy* **80**, 878

(October 2015).
5. J. Parker, Y. Wei, A. Yates, O. Frieder and N. Goharian, A framework for detecting public health trends with twitter, in *Proceedings of the 2013 IEEE/ACM International Conference on Advances in Social Networks Analysis and Mining*, ASONAM '13 (ACM, New York, NY, USA, 2013).
6. K. Jiang and Y. Zheng, *Advanced Data Mining and Applications* **8346**, 434 (2013).
7. J. Bian, U. Topaloglu and F. Yu, Towards large-scale twitter mining for drug-related adverse events, in *Proceedings of the 2012 international workshop on Smart health and wellbeing*, 2012.
8. R. Ginn, P. Pimpalkhute, A. Nikfarjam, A. Patki, K. O'Connor, A. Sarker, K. Smith and G. Gonzalez, Mining Twitter for Adverse Drug Reaction Mentions: A Corpus and Classification Benchmark, in *Proceedings of the Fourth Workshop on Building and Evaluating Resources for Health and Biomedical Text Processing*, 2014.
9. A. Patki, A. Sarker, P. Pimpalkhute, A. Nikfarjam, R. Ginn, K. O'Connor, K. Smith and G. Gonzalez, Mining Adverse Drug Reaction Signals from Social Media: Going Beyond Extraction, in *Proceedings of BioLinkSig 2014*, 2014.
10. A. Sarker and G. Gonzalez, *Journal of Biomedical Informatics* (2014).
11. R. Leaman, L. Wojtulewicz, R. Sullivan, A. Skariah, J. Yang and G. Gonzalez, Towards Internet-Age Pharmacovigilance: Extracting Adverse Drug Reactions from User Posts to Health-Related Social Networks, in *Proceedings of the 2010 Workshop on Biomedical Natural Language Processing*, 2010.
12. A. Nikfarjam and G. Gonzalez, Pattern Mining for Extraction of Mentions of Adverse Drug Reactions from User Comments, in *Proceedings of the American Medical Informatics Association (AMIA) Annual Symposium*, 2011.
13. A. Yates and N. Goharian, ADRTrace: detecting expected and unexpected adverse drug reactions from user reviews on social media sites, in *Proceedings of the 35th European conference on Advances in Information Retrieval*, 2013.
14. A. Nikfarjam, A. Sarker, K. O'Connor, R. Ginn and G. Gonzalez, *Journal of the American Medical Informatics Association (JAMIA)* (2014).
15. K. O'Connor, A. Nikfarjam, R. Ginn, P. Pimpalkhute, A. Sarker, K. Smith and G. Gonzalez, Pharmacovigilance on Twitter? Mining Tweets for Adverse Drug Reactions, in *Proceedings for the American Medical Informatics Association (AMIA) Annual Symposium*, 2014.
16. W. Chou, Y. M. Hunt, E. B. Beckjord, R. P. Moser and B. W. Hesse, *Journal of Medical Internet Research* **11**, p. e48 (2009).
17. The Pew Rsearch Center, The social life of health information `http://www.pewinternet.org/2011/05/12/the-social-life-of-health-information-2011/`, (2011).
18. Online Support Groups and Forums at DailyStrength `http://www.dailystrength.org`.
19. MedHelp Medical Support Communities `http://www.medhelp.org/forums/list`.
20. C. D. Corley, D. J. Cook, A. R. Mikler and K. P. Singh, *Advances in Computational Biology* (Springer New York, 2010), ch. using Web and Social Media for Influenza Surveillance, pp. 559–564.
21. T. Kass-Hout and H. Alhinnawi, *British Medical Bulletin* **108**, 5 (2013).
22. L. L. Struik and N. B. Baskerville, *J. Med. Internet Res.* **16**, p. e170 (2014).
23. L. C. Hanson, B. Cannon, S. Burton and C. Giraud-Carrier, *J Med Internet Res* **15**, p. e189 (September 2013).
24. A. Nakhasi, R. J. Passarella, S. G. Bell, M. J. Paul, M. Dredze and P. J. Pronovost, Malpractice and Malcontent: Analyzing Medical Complaints in Twitter, in *AAAI Fall Symposium on Information Retrieval and Knowledge Discovery in Biomedical Text*, 2012.
25. J. Sarasohn-Kahn, The Wisdom of Patients: Health Care Meets Online Social Media `http://www.chcf.org/publications/2008/`

`04/the-wisdom-of-patients-health-care-meets-online-social-media`, Accessed 29-Sep-2015.
26. D. M. Hartley, *the Milbank Quarterly* **92**, 34 (March 2014).
27. R. Harpaz, A. Callahan, S. Tamang, Y. Low, D. Odgers, S. Finlayson, K. Jung, P. LePendu and N. H. Shah, *Drug Safety* **37**, 777 (August 2014).
28. B. H. Stricker and B. M. Psaty, *BMJ* **329** (2004).
29. I. R. Edwards and M. Lindquist, *Drug Safety* **34**, 267 (2011).
30. W. Franzen, *Drug Safety* **34**, p. 793 (2012).
31. A. Abbasi and D. Adjeroh, *Intelligent Systems, IEEE* **29**, 60 (March-April 2014).
32. S. Tuarob, C. S. Tucker, M. Salathe and N. Ram, *Journal of Biomedical Informatics* **49**, 255 (March 2014).
33. O. Owoputi, B. O'Connor, C. Dyer, K. Gimpel, N. Schneider and N. A. Smith, Improved Part-of-Speech Tagging for Online Conversational Text with Word Clusters, in *Proceedings of the NAACL-HLT*, 2-13.
34. B. W. Chee, R. Berlin and B. Schatz, Predicting Adverse Drug Events from Personal Health Messages, in *Proceedings of the American Medical Informatics Association (AMIA) Annual Symposium*, 2011.
35. P. Pimpalkhute, A. Patki and G. Gonzalez, Phonetic Spelling Filter for Keyword Selection in Drug Mention Mining from Social Media, in *Proceedings of the American Medical Informatics Association (AMIA) Annual Symposium*, 2013.
36. J. Carletta, *Computational Linguistics* **22** (1996).
37. A. Viera and J. Garrett, *Family Medicine* **37**, 36 (2005).
38. M. Rastegar-Mojarad, Detecting signals in noisy data - can ensemble classifiers help identify adverse drug reaction in Tweets?, in *Proceedings of the Social Media Mining Shared Task Workshop at the Pacific Symposium on Biocomputing*, 2016.
39. Z. Zhang, J.-Y. Nie and X. Zhang, An ensemble method for binary classificaiton of adverse drug reactions from social media, in *Proceedings of the Social Media Mining Shared Task Workshop at the Pacific Symposium on Biocomputing*, 2016.
40. B. Ofoghi, S. Siddiqui and K. Verspoor, READ-BioMed-SS: Adverse drug reaction classification of microblogs using emotional and conceptual enrichment, in *Proceedings of the Social Media Mining Shared Task Workshop at the Pacific Symposium on Biocomputing*, 2016.
41. J. Jonnagaddala, T. R. Jue and H.-J. Dai, Binary classification of Twitter posts for adverse drug reactions, in *Proceedings of the Social Media Mining Shared Task Workshop at the Pacific Symposium on Biocomputing*, 2016.
42. D. Egger, F. Uzdilli, M. Cieliebak and L. Derczynski, Adverse Drug Reaction Detection using an adapted Sentiment Classifier, in *Proceedings of the Social Media Mining Shared Task Workshop at the Pacific Symposium on Biocomputing*, 2016.
43. W. Wang, Mining adverse drug reaction mentions in twitter with word embeddings, in *Proceedings of the Social Media Mining Shared Task Workshop at the Pacific Symposium on Biocomputing*, 2016.
44. C.-K. Wang, H.-J. Dai, J. Jonnagaddala, T. R. Jue, O. Singh, U. Iqbal and J. Y.-C. Li, NTTUMUNSW system for adverse drug reactions extraction in Twitter data, in *Proceedings of the Social Media Mining Shared Task Workshop at the Pacific Symposium on Biocomputing*, 2016.